Numerical Analysis and Scientific Computation

Textbooks in Mathematics

Series editors:

Al Boggess, Kenneth H. Rosen

Linear Algebra
An Inquiry-based Approach
Jeff Suzuki

The Geometry of Special Relativity
Tevian Dray

Mathematical Modeling in the Age of the Pandemic
William P. Fox

Games, Gambling, and Probability
An Introduction to Mathematics
David G. Taylor

Linear Algebra and Its Applications with R
Ruriko Yoshida

Maple™ Projects of Differential Equations
Robert P. Gilbert, George C. Hsiao, Robert J. Ronkese

Practical Linear Algebra
A Geometry Toolbox, Fourth Edition
Gerald Farin, Dianne Hansford

An Introduction to Analysis, Third Edition
James R. Kirkwood

Student Solutions Manual for Gallian's Contemporary Abstract Algebra, Tenth Edition
Joseph A. Gallian

Elementary Number Theory
Gove Effinger, Gary L. Mullen

Philosophy of Mathematics
Classic and Contemporary Studies
Ahmet Cevik

An Introduction to Complex Analysis and the Laplace Transform
Vladimir Eiderman

An Invitation to Abstract Algebra
Steven J. Rosenberg

Numerical Analysis and Scientific Computation, Second Edition
Jeffery J. Leader

https://www.routledge.com/Textbooks-in-Mathematics/book-series/CANDHTEXBOOMTH

Numerical Analysis and Scientific Computation

Second Edition

Jeffery J. Leader

CRC Press
Taylor & Francis Group
Boca Raton London New York

CRC Press is an imprint of the
Taylor & Francis Group, an **informa** business

A CHAPMAN & HALL BOOK

Fifth edition published 2022
by CRC Press
6000 Broken Sound Parkway NW, Suite 300, Boca Raton, FL 33487-2742

and by CRC Press
4 Park Square, Milton Park, Abingdon, Oxon, OX14 4RN

© 2022 Jeffery J. Leader

First edition published by Pearson 2004

CRC Press is an imprint of Taylor & Francis Group, LLC

ISBN: 978-0-367-48686-0 (hbk)
ISBN: 978-1-032-20483-3 (pbk)
ISBN: 978-1-003-04227-3 (ebk)

DOI: 10.1201/9781003042273

Publisher's note: This book has been prepared from camera-ready copy provided by the authors

To my wife,
Meg.

Contents

Preface

Preface to the 2nd Edition, Instructor

The fundamental law of computer science: As machines become more powerful, the efficiency of algorithms grows more important, not less. Nick Trefethen

Surely [one imagines], we compute only when everything else fails, when mathematical theory cannot deliver an answer in a comprehensive, pristine form and thus we are compelled to throw a problem onto a number-crunching computer and produce boring numbers by boring calculations. This, I believe, is nonsense. Arieh Iserles

This book represents the numerical analysis course I regularly teach at Rose-Hulman Institute of Technology, which is taken as an elective course by juniors and seniors majoring in computational science, computer science, mathematics, physics, and a variety of engineering disciplines. The students will have had three terms of calculus as well as differential equations with matrix algebra, but may not have had any real programming experience. The key features of my approach have been an immediate immersion in numerical methods, delaying a discussion of computer arithmetic and round-off error until students have gained some experience using numerical algorithms; a presentation of numerical linear algebra that more closely mirrors how it is actually implemented and used in modern practice; and an emphasis on analysis while still bringing out the practical hardware and software issues involved in modern scientific computing. In addition, optional MATLAB® subsections allow for students to teach themselves basic programming while experimenting with the methods of the section. I want students to come away with a strong understanding of the mathematical underpinning of modern scientific computing–and also be prepared to apply that knowledge in a practical way, whether they are doing the coding themselves or, more likely, using what they've learned to get better performance out of the software packages that are common in their field.

Chapter 1 covers root-finding in one-dimension, which should be a comfortingly familiar sort of problem for students. Next is numerical linear algebra (direct methods in Chapter 2 and iterative methods and eigenvalues in Chapter 3) because of its crucial importance both on its own and in other algorithms. Chapter 4 is devoted to polynomial interpolation and splines. Chapter 5 covers numerical quadrature, including Gaussian quadrature and adaptivity; the numerical solution of ODEs is covered in an intentionally similar manner in Chapter 6. Chapter 7 surveys nonlinear optimization methods.

The MATLAB subsections provide a tutorial resource that presents a slow introduction to MATLAB that, by the end of the text, will have resulted in a very broad knowledge of the software as well as substantive programming skills. Students step through computations involving the algorithms from the section, engage in some numerical experimentation, and sometimes see a new method or technique that isn't considered in the main text. Programming techniques are taught early on. Regardless, these subsections are optional and the text can absolutely be used without them.

The overall structure and approach of the book remain the same in this edition apart from merging of the two problem sets in each section into one. Some sections have been added and others removed or shifted. But there were a number of changes that were already underway when I wrote the first edition of this book that loom even larger now. Parallel methods and concerns are given more attention throughout; emerging hardware and software capabilities, such as half-precision arithmetic, are new to this edition; and applications to Computational Science and Data Science are pointed out where appropriate. (It's now common to see reference to Computational and Data-Enabled Science and Engineering, which serves as a reminder that numerical analysts and computational physicists were arguably the first individuals to work with "Big Data".) However, the text remains first and foremost an introduction to the mathematical discipline of numerical analysis at the upper-division level, emphasizing serial algorithms. At my institution we offer not only a course in Numerical Analysis at the senior level (taught from this text), but also an Introduction to Computational Science course and a lab-like Computational Modeling course, both at the junior level and both using a computational physics point-of-view. The subject matter overlaps, but the approaches are very much distinct, and this text is primarily directed at a numerical analysis audience.

In a single term I typically cover the following sections: 1.1-1.7; 2.1-2.9; 3.1-3.4 and 3.6; 4.1-4.4; 5.1-5.5; 6.1-6.6; and 7.1-7.3. The course I teach is oriented somewhat toward numerical linear algebra because I believe so many other problems, such as the numerical solution of PDEs, rely on a knowledge of this material, and that many of the problems in those other areas involve getting a linear system solver to work well. More material from Chapters 5 and 6 could be included at the expense of material from the latter half of Chapter 2. When I've used the text for an introductory numerical methods/computational science course I've covered 1.1-1.3 and 1.5-1.7; 2.1-2.5; 3.1-3.4; 4.1-4.4; 5.1-5.3; 6.1-6.3 and 6.5-6.6; and 7.1-7.6, with significantly greater use of MATLAB in class in lieu of the additional rigor. Because of dependencies it will probably be necessary to cover sections 1.1-1.4 and 1.7, 2.1-2.4, 4.1, and 5.1 in any case. A section corresponds to one class period in most cases.

Prerequisites are a year-long course in the calculus, the basics of matrix algebra, and for Chapter 6 the basics of first-order ODEs. Programming experience is not required, and the instructor may simply not assign the homework problems that involve programming if giving a more theoretical course.

I received helpful suggestions from a number of anonymous reviewers for the first edition. Since the publication of the text I have received corrections and suggestions from a large number of Rose-Hulman students and from instructors there and at other institutions, as well as the occasional hardy individual engaging in self-study. All errors that remain are of course my own (and deeply regretted).

I was very glad to learn from editor Bob Ross that CRC/Taylor and Francis was interested in publishing this book. It was absolutely in need of updating and I am so happy to have an up-to-date edition now available. Completing it during a pandemic was more challenging than expected! My thanks to Bob, Vaishali Singh, Michele Dimont, the proofreader, and everyone else at CRC/Taylor and Francis who were involved in the production of this text.

Jeffery J. Leader
Terre Haute, IN

Preface to the 2nd Edition, Student

The object of the engine is in fact to give the utmost practical efficiency *to the resources of* numerical interpretations *of the higher science of analysis, while it uses the processes and combinations of this latter.* Ada Lovelace

Since the programming is likely to be the main bottleneck in the use of an electronic computer we have given a good deal of thought to the preparation of standard routines of considerable generality for the more important processes involved in computation. By this means we hope to reduce the time taken to code up large-scale computing problems, by building them up, as it were, from prefabricated units. Jim Wilkinson

This book is meant to help you learn about the algorithms that underlie major scientific computing environments. Whether designing an aircraft, analyzing the structure of a crystal, modeling the climate, simulating the collision of two galaxies, recreating electrical pulses around the surface of the heart, and so on, these packages–a great many of which are numerically solving partial differential equations–rely on dependable, stable, well-understood methods to do all the basic tasks of calculus and matrix algebra: Finding roots of equations, solving linear systems, approximately solving really big linear systems, interpolating or approximating data or a function, integrating numerically, differentiating numerically and computationally solving ordinary differential equations, and finding the maximum or minimum of a function–precisely the tasks one learns to do symbolically in calculus and differential equations courses. Well, those are exactly the topics of the 7 chapters of this book, and they are precisely the things you need to understand to branch out into the methods of computational fluid dynamics, thermodynamics, electromagnetism, etc. Whatever you think your code is doing, the bulk of the work typically involves manipulating a medium-to-large matrix A, usually by needing to solve $Ax = b$. If you're going to do computational science and/or engineering, you'll constantly be running code that implements these techniques. By learning how the underlying algorithms work, you'll be better positioned to get the most out of your design/modeling/simulation software packages, even if you never write a line of code yourself. Of course, if you do want or need to write that code...then you *really* need to understand these ideas.

The same is true of the growing field of data science. Computational scientists think of themselves as the original "big data" people, manipulating the truly massive amounts of data generated and used in particle physics, astronomical observation, genetics, weather prediction, census data, and so on, as well as the data generated by computational simulations themselves. But most current data science techniques are using some variation of the data structures in Sec. 3.2. Neural networks themselves are essentially matrices, and regression analyses are based on the computational linear algebra techniques in Sec. 2.7-10, as are many rank-reduction techniques for clustering. (For example, latent semantic indexing is a conceptual data mining tool used in specialized searches and is a straight-forward implementation of the ideas in Sec. 2.10.) A Google search relies, at its core, on the method presented in Sec. 3.6. Nonlinear regression models use material from Sec. 7.2-6. While the primary audience of this text is meant to be scientists and engineers seeking to model physical phenomena, if you need to write data mining code then you need to know this material.

To be clear, this is not a text focused on any particular one of these areas of applications. This book presents a calculus-based approach to deriving, analyzing, and quantifying the accuracy of the fundamental methods used in *all* of these areas. Numerical analysis is the mathematical material at the heart of scientific computing. You'll come to understand these methods well enough to trust them, implement them, modify them, and verify them. We'll be studying the tools that virtually all scientific code uses.

Since it's likely you're studying this material in hopes of making practical use of it, there is also a lot of material on practical implementation issues that you're likely to encounter. 'Numerical analysis' refers to the calculus-based side of things, but 'scientific computation', the other part of book's title, is the computer science aspect of actually getting efficient, accurate, verified and valid code written. While this book emphasizes the former, it goes further than most in helping you with the latter. The goal is that you will be an extremely well educated user of scientific software–able to get better performance out of it and to write small but accurate and efficient programs where need be. For those who might go on to write significant amounts of scientific code–this course is indispensable.

Part of how the text achieves this goal is via MATLAB subsections in every section. Your instructor may or may not use them, but if you have access to MATLAB (or if you install the freeware alternative GNU Octave) then you're encouraged to work through these. You'll learn MATLAB programming, commonly-encountered difficulties with the methods and ways around them, and scientific computing tricks and techniques for better software.

Read the sections you're assigned. Before attempting to write any code for homework, work a small-sized example by hand. (The Problems provide plenty of these.) In many cases you'll have to choose your own starting point or make other small decisions that will make your final answer close to, but not precisely the same as, those of your fellow students.

The key to understanding most of these methods rests on the following observations: *i.*) Functions encountered in practice are almost always very expensive to compute, so we try to find algorithms that evaluate $f(x)$ as infrequently as possible. *ii.*) Related to this, what is described in the book as a "function $f(x)$" is frequently a large simulation program, and so you can't take its derivative, etc.; and usually x is a vector so even plotting f is apt to be unrealistic. *iii.*) The matrices that come up in science and engineering tend to be very big, but fortunately they usually have relatively few nonzero entries so it isn't quite as bad as it seems; still, just manipulating them with respect to storage and access is already hassle enough.

This is a subject that really brings together calculus and differential equations, matrix algebra, programming, and hardware. Few courses bring together so many ideas to such a practical result. Enjoy!

Jeffery J. Leader
Terre Haute, IN

Preface to the 1st Edition

Numerical analysis is the study of algorithms for the problems of continuous mathematics. -Lloyd Trefethen (1992)

Elementary numerical analysis is a marvelously exciting subject. With a little energy and algebra you can derive significant algorithms, code them up, and watch them perform. -G. W. Stewart (1998)

This book represents the numerical analysis course I have been teaching for the past nine years, first as a required course for sophomore engineering majors and currently as an elective course for juniors and seniors in mathematics, computer science, physics, and engineering. It's been strongly influenced by my summertime computational consulting work at the Army Research Lab. (Adelphi, MD), Naval Surface Warfare Center (Dahlgren, VA), and Air Force Research Laboratory (Dayton, OH), and other consulting and research experience; it has also been influenced by conversations with my colleagues and of course by other texts I have turned to over the years.

I hope that what sets my text apart from other introductory numerical analysis texts will be the following features:

- Quick introduction to numerical methods, with roundoff error and computer arithmetic deferred until students have gained some experience with real algorithms
- Modern approach to numerical linear algebra
- Explanation of the numerical techniques used by the major computational programs students are likely to use in practice (especially MATLAB, but also Maple and the Netlib library)
- Appropriate mix of numerical analysis theory and practical scientific computation principles
- Greater than usual emphasis on optimization
- Numerical experiments so students can gain experience
- Efficient and unobtrusive introduction to MATLAB

The MATLAB material is introduced in optional subsections at the end of each section. If you'd like your students to learn MATLAB as part of this course, these subsections provide a way to slowly introduce them to it and a chance for them to engage in actual numerical experimentation to build their intuition about scientific computing. If you're not using MATLAB in your course, you may encourage your students to skim these subsections for the extra material and foreshadowing they sometimes contain but you may use the text without using MATLAB. I have used this approach to teach MATLAB in my classes and had great success with it.

My experience in scientific computing has been that almost every problem reduces, at some level, to a problem in root-finding, a problem in nonlinear optimization, or a problem in numerical linear algebra, and that the first two cases frequently involve the solution of a linear system. The text starts with root-finding (chapter 1), principally in one-dimension, because students will be familiar with it from the calculus. It then moves straight into numerical linear algebra (chapters 2 and 3) because of its crucial importance and so that this material will be available as needed in later chapters. Chapter 4 introduces polynomial interpolation is introduced as a tool for deriving other methods and splines as a tool for representing curves given in terms of data. Chapter 5 introduces the basic techniques of numerical quadrature, emphasizing those algorithms employed by MATLAB and Maple. Quadrature of ODEs is given a similar treatment in chapter 6. Nonlinear optimization is covered in chapter 7, and some basic ideas and methods of approximation theory are discussed in chapter 8.

I try to avoid falling into the trap of first introducing a completely new mathematical idea, then teaching computational techniques for it. If students don't already know the mathematical problems–root-finding, finding maxima and minima, differentiation, and integration from the calculus, and first-order ODEs from the calculus or an ODEs course,–it will be hard for them to appreciate why they need to know numerical methods for them. Hence I have avoided teaching the Fourier transform, *then* its discrete version the DFT, then the computational method known as the FFT, for example. I've had to vary from this principle to introduce the QR and singular value decompositions as these are not yet standard in a matrix or linear algebra course.

In class, I ask students for ideas about approaching a problem like root-finding or about improving a method we have already seen. To mimic that approach in the text, in several sections I briefly discuss a second method in addition to the main method I introduce and develop in that section. These brief discussions offer students alternatives I want them to know about, even if we do not discuss them in detail in the class or in the text. By the end of the course, these are the kind of ideas I want them to start suggesting, such as replacing linear approximations with quadratic ones, for example. If you prefer to skip these brief discussions, of course you may.

In emphasizing the ideas behind methods and the mixing of methods I hope to encourage students to feel that they "own" these methods. Too often students feel they cannot tweak existing algorithms or software, when in fact, adding heuristics is often necessary. In my experience students need to be told that it's OK to fiddle with the methods. That's actually one of the big reasons that people who are only going to *use*, not *write*, numerical software need a course like this; they need to know enough of the ideas to be able to pick good initial guesses, to approximate needed derivatives, to select the right method or the right parameters.

Over the past decade or so the practice and teaching of numerical analysis have undergone a number of changes. Fewer people write significant amounts of code; there's more reliance on packages like MATLAB, the similar Octave, GAUSS, and more specialized packages. Also, people are solving bigger and bigger problems, thanks to advances in computer hardware and the availability of these powerful software from Netlib, NAG, and so on. This leads to a need for greater understanding of the ideas behind algorithms and how to tweak them to get them to converge in a timely fashion–choosing proper initial guesses, mixing methods intelligently, adjusting algorithmic parameters–but less need for understanding the algorithms in sufficient detail to write professional level code. (In fact, I believe that writing code should be left to *teams* of experts.) There is a greater emphasis on *scientific computation* as opposed to *numerical analysis* these days.

On the pedagogical side, at many institutions the theoretical content of calculus courses has been reduced and computer algebra systems have been introduced.

This creates a two-fold challenge: First, instructors must work around the student's lesser analytical background, and second, students used to seeing a computer algebra system spit out the answers to all their problems must be convinced of the need for *numerical*, as opposed to *symbolic*, techniques. Students today come to a numerical analysis course with less numerical computing experience than before, thanks to computer algebra systems. This book aims to respond to these pedagogical changes by emphasizing the ideas in a way that appeals to geometric thinking and by providing examples that cannot be done by computer algebra systems and hence the need for numerical methods.

I mention the possibility of parallelization for many methods but do not develop parallel algorithms in detail. I do not believe a first course in numerical analysis is the place for emphasizing parallel computing, although students should be aware of its existence. Even if students have access to such a machine, programming it is likely beyond their ability.

This text is *not* intended to be a reference. It is meant to be worked through by students so they learn the subject. I have avoided in many cases dry statements of theorems in favor of rigorous derivations that show the ideas behind the methods. I do not believe that students can appreciate many of the theoretical aspects of the material until they have some experience with numerical computing–all the more because of their experience with computer algebra systems. This text aims to produce students who can competently use standard computational software, including choosing the right method for the job and advising others on the major benefits and pitfalls of various methods. I have had success with this in my current position and hope it will work for you in yours.

Each section is meant to be one lecture, although you may find that you need more or less time. The Problems sections contain basic problems that can be used to ensure understanding and do not require new computational resources beyond those required in the preceding sections. The MATLAB subsections introduce MATLAB commands and explore the material through numerical experimentation. Students can use these subsections as self-teaching MATLAB tutorials. I ask students to hand in the output from the `diary` command and to annotate it with brief comments on the results. Some of the Additional Problems may require new computing tools equivalent to those covered in the MATLAB subsections. Additional Problems numbered 10 and above may be more challenging; the last two or three Additional Problems usually are relatively more difficult.

The text contains more material than can be covered in a single quarter or semester. In a ten-week quarter I typically cover the following sections: 1.1-1.9 and 1.10; 2.1-2.9; 3.1-3.6; 4.1-4.4; 5.1-5.3, plus ideas from 5.6 and 5.7; 6.1-6.3, plus ideas from 6.5 and 6.6; and 7.1-7.3 (plus 7.4 and 7.5 if time permits). The course I teach is oriented somewhat toward numerical linear algebra because I believe so many other problems, such as the numerical solution of PDEs, rely on a knowledge of this material, and that many of the problems in those other areas involve getting a linear system solver to work well. More material from chapters 5 and 6 could be included at the expense of material from the latter half of chapter 2. Because of dependencies it will probably be necessary to cover 1.1-1.4, 1.7, 2.1-2.3, and 4.1 in any event.

Prerequisites are a year-long course in the calculus, the basics of matrix algebra, and for chapter 6 the basics of first-order ODEs. Some of the material in chapter

8 is at a more advanced level and requires additional mathematical maturity. Programming experience is not required (if MATLAB is to be used).

My father-in-law wrote in the preface to one of his books that 'The making of a book is the work of many hands.' I have found this to be true. I want to thank Addison-Wesley for publishing this book and I want to acknowledge in particular Cindy Cody, Joe Vetere, and RoseAnne Johnson at Addison-Wesley; Chris Miller and her group at TechBooks; Louise Gache, the copyeditor; Michael Brown, who wrote the solutions, and the several accuracy checkers; and the many other individuals who have helped make this book.

I also want to acknowledge the helpful comments I have received from a number of anonymous reviewers who suggested changes to the order of presentation. commented on the clarity of various passages, and pointed out errors and omissions. Four years' worth of numerical analysis students at Rose-Hulman Institute of Technology have used the manuscript and were of great help in refining the material, and I gratefully acknowledge their assistance. Any errors that remain are of course my own.

My colleagues at Rose-Hulman Institute of Technology have been very supportive during the time I wrote this book; I want to thank in particular S. Allen Broughton, Ralph P. Grimaldi, and Robert Lopez (now at Maplesoft) for their encouragement and advice.

Over the years I've benefitted from the chance to study and work with a number of numerical analysts who have helped me grow in my understanding of the field. I especially want to acknowledge my advisor, Philip J. Davis, of Brown University; David Gottlieb, also of Brown University; and Bill Gragg of the Naval Postgraduate School.

I also gratefully thank two people who have been very helpful to me in very many ways over the course of my career: Robert L. Borrelli and Courtney S. Coleman, both of Harvey Mudd College.

Last but not least I thank my wife, Meg, for her help on this project, and I also thank her and our children Derek and Corrinne for their patience while their father worked long hours on this project.

Jeffery J. Leader
Terre Haute, IN

CHAPTER 1

Nonlinear Equations

1. Bisection and Inverse Linear Interpolation

The numerical solution or simulation of a scientific or engineering problem very frequently requires that we solve a root-finding problem as a part of the process, even if that's not the ultimate goal of the computation[1]. In this chapter we will look at the problem of finding a root of a single nonlinear equation; we will look at linear and then nonlinear systems of equations in Ch. 2, and then extend that material in Ch. 3.

The root-finding problem is to find a solution x^* of $f(x) = 0$. (There may be many solutions, but we are only seeking any one of them.) The solution is called a **root** of the equation or a **zero** of the function f. For nearly all methods it's an implicit assumption that the zeroes of the function f are **isolated**, that is, that there is an open interval that contains that zero but no other zeroes of the function. See Problem 14 for a case of a non-isolated zero.

For special cases there are often special approaches: The quadratic formula applies if f is quadratic, and the zeroes of $\sin(x)$ are well known. Sooner or later, however, it becomes necessary to solve a problem that does not fit into a known special case and hence cannot be done by hand or by a Computer Algebra System (CAS). The simplest examples are equations like

$$\cos(x) - x = 0$$

and polynomials of degree 5 or higher. (There is a cubic formula for degree 3 polynomials and a quartic formula for degree 4 polynomials, but it can be shown that there can be no similar formula for polynomials of degree 5 or higher. Unless a factorization is obvious, numerical methods must be employed.) A more complicated but quite common case is that of finding where the solution of a differential equation passes through zero, when the differential equation itself must be solved numerically.

Using trial and error is one possibility. There is nothing inherently wrong with this approach, but it lacks a theory that predicts how rapidly it will find a solution (to within a given tolerance), and it is difficult to automate. In practice, different root-finding problems may need to be solved many times within a single run of a program. Because of this we will need to find methods that are *fast*, *reliable*, and *easily implemented*, ideally without the need for a human to intervene and make a judgement at any stage of the process.

---

[1]"The purpose of computing is insight, not numbers." -Richard Hamming

DOI: 10.1201/9781003042273-1

The first method we will consider is the method of **exhaustive search** (also called **direct**, **graphical**, or **incremental search**). Suppose that f is continuous on some (not necessarily finite) interval. By the Intermediate Value Theorem, if we can find two points a,b such that $f(a)$ and $f(b)$ have opposite signs, then a zero of f must lie in (a, b). One way to find such a pair of points is to pick some number x_0 as an initial guess as to the location of a root and successively evaluate the function at

$$x_0, \; x_1 = x_0 + h, \; x_2 = x_0 + 2h, \; x_3 = x_0 + 3h, \; ...$$

where $h > 0$ is called the **step size** (or **grid size**). When a change of sign is detected, we say that we have **bracketed** a root in this interval of width h. (The interval $[x_i, x_{i+1}]$ over which the change of sign occurs is called a **bracket**; see Fig. 1, where the endpoints form a bracket.) At this point we may repeat the process over the smaller interval $[x_i, x_{i+1}]$ with a smaller h in order to bracket the root more precisely.

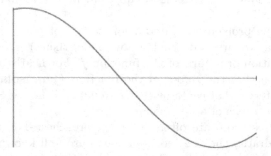

FIGURE 1. Bracketing a zero.

The exhaustive search method is equivalent to plotting the function and looking for an interval in which it crosses the x-axis. This is very inefficient, so let's look for a better approach. Suppose that we have found a bracket $[a, b]$ for a zero of a continuous function (by any means). Rather than searching the entire interval with a finer step size, we might reason that the midpoint

$$m = \frac{a + b}{2}$$

is a better estimate of the location of the true zero x^* of f than either a or b; after all, $|a - x^*|$ could be as large as the width $w = (b - a)$ of the interval if x^* is near b, and similarly for $|b - x^*|$, but

$$|m - x^*| \leq \frac{1}{2}w$$

since x^* lies either in the interval to the right or to the left of m (or both, in the extremely unlikely case that $x^* = m$). This is the idea behind the **bisection method** (or **binary search**): If f is continuous on some interval and $[x_0, x_1]$ is a bracket, that is, if $f(x_0)f(x_1) < 0$, then we set

$$x_2 = \frac{x_0 + x_1}{2}$$

and compute $f(x_2)$. If $f(x_2) = 0$, we are done; otherwise either $f(x_2)$ and $f(x_0)$ have opposite signs, in which case $[x_0, x_2]$ is a new bracket half the size of the previous one, or $f(x_2)$ and $f(x_1)$ have opposite signs, in which case $[x_2, x_1]$ is a new bracket half the size of the previous one. In either case we have reduced our uncertainty as to the location of the true zero x^* by 50% at the cost of a single new function evaluation (namely the computation of $f(x_2)$). We may now repeat this process on the new interval, finding its midpoint x_3 and then a smaller bracket with x_3 as an endpoint, and so on, until a sufficiently narrow bracket is obtained.

EXAMPLE 1: Consider the function $f(x) = \cos(x)$, which has a zero at $\pi/2 \doteq 1.5708$. (The symbol $\doteq$ means that the indicated value is correctly rounded to the number of significant figures shown.) Since $f(1) \doteq 0.5403$, $f(2) \doteq -0.4161$, and f is continuous, the interval $[1, 2]$ is a bracket. Its midpoint is $x_2 = 1.5$. Since the function value at the midpoint is

$$f(1.5) \doteq 0.0707$$

we have $f(1.5) > 0$, $f(2) < 0$ and so we replace $x_0 = 1$ by $x_2 = 1.5$, meaning that $[1.5, 2]$ is our improved bracket. The next midpoint is $x_3 = 1.75$, where the function value is

$$f(1.75) \doteq -0.1782$$

so that $[1.5, 1.75]$ is the new bracket. If we had to stop now, our estimate of the location of the true zero could be any value in $[1.5, 1.75]$; choosing its midpoint $x_4 = 1.625$ minimizes the worst case error. $\square$

Unless we have the misfortune to choose the wrong interval at some step because rounding error makes a positive value negative (or vice versa), this method must converge to some zero of f in the initial bracket, that is, as the iteration count k increases we must have

$$\lim_{k \to \infty} x_k = x^*$$

for some x^* that lies in the initial bracket and for which $f(x^*) = 0$. Furthermore, since the width of the bracket is halved at each step, we can predict how long it will take to achieve any desired precision. If the width of the initial interval is $w_0 = x_1 - x_0$, then after the first bisection step the width of the new bracket is $w_1 = w_0/2$, and after the second step it is $w_2 = w_0/4$, and, in general, after n steps the width of the resulting interval is equal to $w_n = w_0/2^n$. We might return to the user this interval, or we might find its midpoint and return that as a point estimate that is good to $\pm w_0/2^{n+1}$. In most scientific computing applications it's the new function evaluation to determine the next bracket that is expensive, while the cost of an addition and division to find the next midpoint is comparatively trivial.

Continuing Example 1, to reduce the initial interval of width 1 to an interval of width 10^{-4} we would need n to be large enough that

$$
\begin{aligned}
1//2^n &\leq 10^{-4} \\
10^4 &\leq 2^n \\
n &\geq \log_2(10^4) \\
&\doteq 13.2877
\end{aligned}
$$

so $n = 14$ iterations would suffice. Of course, although this guarantees that the width of the final interval is 2^{-14} (note that $2^{-14} \leq 10^{-4} \leq 2^{-13}$) and that its midpoint will be an estimate good to within 2^{-15}, the actual error may be much smaller. In fact, performing 14 iterations of the method on $f(x) = \cos(x)$ with the initial bracket $[1, 2]$ and using the midpoint $x_{15} \doteq 1.57077$ of the final interval as our estimate of the true zero $\pi/2 = 1.57079...$ gives an actual error of $2.6E10^{-5}$ as compared to the bound of $3.1E10^{-5}$. Error bounds represent worst-case scenarios, and often our results will be considerably better.

The sequence generated by the bisection method is guaranteed to converge to a root; exhaustive search is not (after all, it could step right over a pair of closely spaced roots if h is not sufficiently small; see Fig. 2). Bisection will also be much faster in general. However, we have paid a price for these advantages: Bisection requires us to find an initial bracket, whereas exhaustive search requires only a single initial guess lying to the left of the presumed root. Finding that initial bracket may well require an initial graphical search (or trial-and-error); see Sec. 1.6 for a method for searching for a bracket.

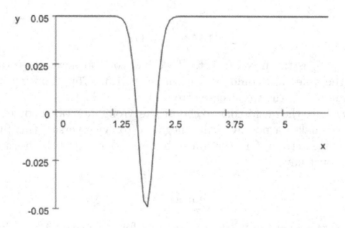

FIGURE 2. Exhaustive search misses a blip.

Consider again the function $f(x) = \cos(x)$, but now suppose that we had found the initial bracket $[0, 1.6]$ instead, for which we have $f(0) = 1.0000$ and $f(1.6) \doteq -0.0292$. Bisection will work, but noting that $f(1.6)$ is much smaller in size than $f(0)$ might incline us to choose a new point nearer $x_1 = 1.6$ than $x_0 = 0$. If we choose, say, $x_2 = 1.4$, then $f(x_2) \doteq 0.1700$ and so $[1.4, 1.6]$ is a new, smaller bracket; the bisection bracket would have been $[.8, 1.6]$ which is four times as wide. This suggests another method, known as **inverse linear interpolation** (also called **false position**): Given a continuous function f and a bracket $[x_0, x_1]$ for a zero of f, fit a straight line to the points $(x_0, f(x_0))$ and $(x_1, f(x_1))$. (See Fig. 3.) This line is said to interpolate f at these points, and if f is approximately linear over the interval then the zero of the line should be a good estimate of the zero of the function. (The method is referred to as *inverse* interpolation since we are using the interpolated line to find an x value, not a y value as usual. For this reason we write x as a linear function of y.) Since the y-values are of opposite sign, $f(x_0)$ is not equal to $f(x_1)$ and so the equation for the line may be written in slope-intercept form as

$$(1.1) \qquad x = \frac{x_1 - x_0}{f(x_1) - f(x_0)}y + \left(x_1 - \frac{x_1 - x_0}{f(x_1) - f(x_0)}f(x_1) \right)$$

and this may be solved for the x-intercept x_2 simply by setting $y = 0$:

$$(1.2) \qquad x_2 = x_1 - f(x_1)\frac{x_1 - x_0}{f(x_1) - f(x_0)}.$$

As before, either $[x_0, x_2]$ or $[x_2, x_1]$ is a new, smaller bracket. We now iterate until some **convergence criterion** is achieved, that is, until we meet some specified criterion for deciding that we are sufficiently close to the true answer. The width of the interval may not go to zero (if the approach to the root of the equation is one-sided; see Problem 1), so we can not use that as the only convergence criterion. For the same reason, unless the width of the bracket is very small the last computed endpoint is generally the best estimate of the location of the root, not the midpoint as before.

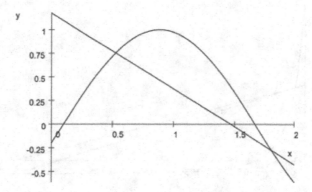

FIGURE 3. Inverse linear interpolation.

EXAMPLE 2: Consider again the function $f(x) = \cos(x)$ with bracket $[1, 2]$. Using Eq. (1.2) with $x_0 = 1$ and $x_1 = 2$ gives

$$\begin{aligned} x_2 &= x_1 - f(x_1)\frac{x_1 - x_0}{f(x_1) - f(x_0)} \\ &= 2 - \cos(2)\frac{2 - 1}{\cos(2) - \cos(1)} \\ &\doteq 1.5649 \end{aligned}$$

which is a pretty good point estimate of $\pi/2 = 1.57079...$, and as $\cos{(1.5649)} = 5.8963 \times 10^{-3}$ is positive it leads to the new bracket $[1.5649, 2]$ which is slightly smaller than the bracket $[1.5, 2]$ found by bisection after one step. The next step gives

$$
\begin{aligned}
x_3 &= x_2 - f(x_2)\frac{x_2 - x_1}{f(x_2) - f(x_1)} \\
&= 1.5649 - \cos{(1.5649)}\,\frac{1.5649 - 2}{\cos{(1.5649)} - \cos{(2)}} \\
&\doteq 1.5710
\end{aligned}
$$

(where we would be re-using the previously computed cosine values rather than re-computing them.) Since $\cos{(1.5710)} < 0$, the new bracket is $[1.5649, 1.5710]$. Its width is $1.5710 - 1.5649 = 0.0061$, a vast improvement over the corresponding width of .25 for bisection, and the midpoint of this interval is an excellent estimate of the root. □

Inverse linear interpolation will converge to some zero of the function in the bracket; however, the rate at which it converges will depend on how nearly linear $f(x)$ is near its zero. If $f(x)$ is well approximated by a straight line over the bracket, the method will usually be faster than bisection. (A function and bracket as in Fig. 4 will result in excruciatingly slow convergence until the bracket is very small.) We have given up the guaranteed slow-but-steady-wins-the-race speed of bisection for a likely but neither guaranteed nor easily predictable improvement.

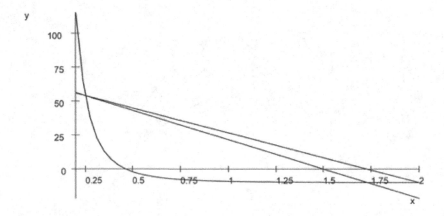

FIGURE 4. Tough function for inverse linear interpolation.

In the next section we will consider a much faster algorithm known as Newton's method. Once again a price will be paid: We will require that the function be differentiable as well. Faster methods will require more assumptions and will offer fewer guarantees. But first: MATLAB®.

MATLAB
Numerical analysis is the study and design of methods that may be used to solve the mathematical problems of engineering and the sciences by iterative algorithms

in a reasonable time and with desirable error properties[2]. This can be done without access to a computer, as will be clear from the names attached to some of these methods–such as Newton (1642-1727), Euler (1707-1783), and Gauss (1777-1855). For all the obvious reasons, however, it is best to gain experience with these methods by running them on a calculator or computer. These subsections will focus on the MATLAB environment; however, you may use C++, Fortran, Maple, Mathematica, or any other computational tool. Even if you are not using MATLAB in your course, it is important that you read these sections and attempt the computations in the environment you are using.

The MATLAB package is an interactive computing environment and also a programming language (from The MathWorks, Inc.). You should read this section at the computer and enter the commands as you read them; afterwards you should experiment, varying the functions used, etc. Run MATLAB by double-clicking on the MATLAB icon or typing `matlab`[3] at the prompt. You should see a shield with an image of a three-dimensional surface and then a prompt:
```
>>
```
The >> symbol (called the guillemotright) is the MATLAB prompt. We are now using MATLAB in what might be called immediate mode–that is, we are essentially using it as a calculator. Let's try a few simple calculations. At the >> prompt, type 3+4 followed by the enter key; that is, enter:
```
>> 3+4
ans =
     7
```
Let's keep playing; enter:
```
>> 3-4
ans =
     -1
>> 3*4
ans =
     12
>> 3/4
ans =
     0.7500
>> ans^2
ans =
     0.5625
>>
```
The variable `ans` is automatically assigned a value by MATLAB. Other variables may be assigned values in the obvious manner; enter:
```
>> x=5,y=2
x =
     5
y =
     2
```

---

[2]Lloyd Trefethen writes: "Numerical analysis is the study of algorithms for the problems of continuous mathematics."

[3]The precise command may vary at your location. For a simpler interface, try entering `matlab -nojvm` instead.

```
>> x+y,x^y
ans =
     7
ans =
    25
>>
```

(Notice that more than one command may be entered on a single line.) Let's try bisection on the function $f(x) = \sin(e^x)$. First we'll get a plot of the function. Enter:

```
>> x=0:.1:3
```

to make a row vector x (that is, a list of values) with entries ranging from 0 to 3 in steps of .1. Enter:

```
>> y=sin(exp(x));
```

to compute $\sin(e^x)$ for every x in the vector and assign those values to y. The semi-colon at the end suppresses printing of the results. The arguments of sin are assumed to be in radians. Enter:

```
>> y
```

(without a semi-colon) to see the y-values. Now enter:

```
>> plot(x,y)
```

to see a plot of the function. (The plot command automatically connects the dots to draw a continuous piecewise linear curve.) Enter:

```
>> grid
```

to overlay a grid on your plot. Look at the figure again; there are six zeroes evident in this interval. We have essentially performed graphical search.

This graph is too choppy. Let's try again with a finer grid. You will likely see the command that defined x in the "Command History" window and can click on it to retrieve it. You may also press the up arrow (cursor up ↑) key[4]. You should see the last command you entered reappear. Keep pressing the key until you find the line that defined x. Once you have it, use the left cursor key ⟵ to edit it to read x=0:.01:3 and then press enter; that is, enter:

```
>> x=0:.01:3;
```

(note the semi-colon at the end). This makes x a vector of values ranging from 0 to 3 in steps of .01. Now find the line defining y and re-enter it unchanged; it automatically uses the new x. Similarly, re-enter the plot and grid commands. The graph appears much smoother now, though if we zoomed in we'd see that it is as choppy as before.

Let's use bisection to try to find the first zero of $f(x)$ in $[0, 3]$. It appears to be near 1.2. We'll take $[1.1, 1.4]$ as our initial bracket. Enter:

```
>> x0=1.1,x1=1.4
x0 =
    1.1000
x1 =
    1.4000
>> m=(x0+x1)/2
m =
    1.2500
```

---

[4]There will be slight variations in how MATLAB responds on various systems (Unix vs. Windows, etc.).

```
>> sin(exp(m))
ans =
    -0.3417
```
From the graph, it's clear that $f(x_0) > 0$ and $f(x_1) < 0$, so our new bracket should be $[x_0, m] = [1.1, 1.25]$. Enter:
```
>> x1=m;
```
and re-enter the command for computing m and $f(m)$ using the up cursor key, that is, re-enter:
```
>> m=(x0+x1)/2
m =
    1.1750
>> sin(exp(m))
ans =
    -0.0964
```
The new bracket is $[1.1, 1.1750]$. On the next iteration you should find that $m = 1.1375$ and $f(m) \doteq 0.0226$ so that $[1.1375, 1.1750]$ is the current bracket; its midpoint is our best current estimate of the location of the zero.

You should repeat this using inverse linear interpolation. Enter:
```
>> clear
```
to clear the values of all variables. (Ask for the value of, say, x1 after entering the clear command. It should be undefined.) Then define x0 and x1 and define their corresponding function values; that is, enter:
```
>> x0=1.1,x1=1.4
>> y0=sin(exp(x0));y1=sin(exp(x1));
```
Now use x2=x1-y1*(x1-x0)/(y1-y0 to get the x-intercept, find the function value there, and then form your new bracket. Remember to update your y0 or y1 value when you change x0 or x1 to equal x2. You can check the current width of your interval; enter:
```
>> x1-x0
```
(If this is about 10^{-k} then we know the answer to about k decimal places; e.g., if $a = .1234$ and $b = .1233$ then $|a - b| = 1E - 4$ and so we conclude that these two values agree to about 4 decimal places.) Perform several iterations of inverse linear interpolation. When you are done, use the command quit or exit to exit MATLAB.

If you are used to using Maple or Mathematica, you may wonder if there is a command to save a worksheet. First, note that your previous commands will be available using the Command History window or the cursor up key. More generally, you may use the drop-down menu under *File* to find the *Save Workspace As* option and save all your work, including defined variables, in a file that can be re-loaded later using *File* and *Import Data* or the "Current Directory" window. You don't need to do this now but in the future you may wish to do so. Additionally, the diary command stores a text record of your \ work as the text file diary.txt. You enter diary to begin the record and read the file in a text reader later.

There is much more utility in MATLAB than we have even hinted at yet, including a programming language, as we'll see in later sections. By the end of the course you should feel very comfortable working in MATLAB, which is widely used in industry and research because of the ease and rapidity with which programs for scientific computing may be written, tested, and refined in it.

Problems

1.) Use four iterations of bisection with the initial bracket $[.5, 1]$ to approximate the sole positive root of $\cos(x) - x = 0$. Repeat with inverse linear interpolation, using the same initial bracket; note the one-sided approach to the solution.

2.) Show that there is a unique real solution of $5x^7 = 1 - 2x$. Use bisection to approximate it to four decimal places. Repeat with inverse linear interpolation.

3.) The function $f(x) = \cos(5x)$ has 7 zeroes in the interval $[0, 4.5]$. Use bisection and then inverse linear interpolation with that initial bracket (verify that it is in fact a bracket) to find a zero to at least three decimal places. Find the actual error in your estimates using the known locations of the zeroes of the cosine function. Comment on your results.

4.) Use bisection on $f(x) = x^2 - 5$ to approximate $\sqrt{5}$ to at least four decimal places. Compare your error estimate to the actual error.

5.) Use inverse linear interpolation with the initial bracket $[.25, 2]$ to approximate a zero of $f(x) = 1/x^3 - 10$ to three decimal places. Repeat with bisection. Comment.

6.) Use the bisection method on $\sin(x) = 0$ to approximate π to four decimal places.

7.) Estimate the number of iterations that would be required to find a root of $e^x - 4x = 0$ to four decimal place accuracy using bisection starting from the initial bracket $[0, 2]$. How many iterations are needed until the midpoint of the interval is actually correct to four decimal places?

8.) Estimate the number of iterations that would be required to find a solution of $\cos(\sin(x)) = .75$ to four decimal place accuracy using bisection starting from the initial bracket $[0, 2]$. How many iterations are actually needed until the midpoint of the interval is actually correct to four decimal places? How many iterations are needed for inverse linear interpolation to produce a value that is correct to four decimal places?

9.) Justify Eq. (1.1) and Eq. (1.2) by finding the equation for the line $x = my + b$ between the two points $(x_0, f(x_0))$ and $(x_1, f(x_1))$ and then finding its x-intercept.

10.) Use MATLAB to find the zeroes of $\sin(\ln(x))$ that lie in $(0, 30]$ to three decimal place accuracy by the method of exhaustive search. Include plots of the function near the zeroes.

11.) a.) A version of the cubic formula for finding the real roots of the cubic equation $x^3 + ax^2 + bx + c = 0$ is as follows: Find $A = (a^2 - 3b)/9$ and $B = (2a^3 - 9ab + 27c)/54$. If $B^2 - A^3 > 0$, there is a single real root given by $x_1 = -\operatorname{sgn}(B)(\alpha + A/\alpha) - a/3$, where $\alpha = \left(\sqrt{B^2 - A^3} + |B|\right)^{1/3}$ and $\operatorname{sgn}(\mu) = \mu/|\mu|$ if $\mu \neq 0$ and $\operatorname{sgn}(0) = 0$ is the signum (i.e.. sign) function. If instead $B^2 - A^3 \leq 0$, then there are three real roots given by $x_1 = -2\sqrt{A}\cos(\theta/3) - a/3$, $x_2 = -2\sqrt{A}\cos((\theta + 2\pi)/3) - a/3$, $x_3 = -2\sqrt{A}\cos((\theta + 4\pi)/3) - a/3$, where $\theta = \arctan(B/Q^{3/2})$. (There is a version of the cubic formula that can find complex roots also.) Use the cubic formula to find the real roots of the equation $x^3 - 2x^2 - 5x + 1 = 0$.

b.) Use the method of inverse linear interpolation to find any root of the equation $x^3 - 2x^2 - 5x + 1 = 0$.

c.) Use the cubic formula to find the real roots of the equations $x^3 - 8x^2 + 20.75x - 17.5 = 0$, $x^3 - 3x^2 + 3x - 1 = 0$.

d.) In the case $B^2 - A^3 \leq 0$ (three real roots), verify that the product of the roots is equal to $-c$.

12.) Use the MATLAB commands x=0:.1:1 and y=sin(x) to get a vector y of values of the sine function. Use standard linear interpolation of these values to estimate sin(.15), sin(.55), and sin(.95); that is, fit a straight line between adjacent points and use the values from the line to approximate the sine function. Compare your approximate values to the true values.

13.) What happens if inverse linear interpolation is applied to a linear function $f(x) = ax + b$? Is the same true for bisection?

14.) a.) How many zeroes does the function $f(x) = 0$ have in the interval $[0, 1]$? Are they isolated?

b.) How many zeroes does the function $\sin(1/x^2)$ have in the interval $(0, 1)$? Is each one isolated?

c.) If we define $g(x) = \sin(1/x^2)$ if $x \neq 0$ and $g(0) = 0$ then $g(x)$ is continuous everywhere. Is every zero of g isolated?

15.) Find a simultaneous solution of the linear system of equations $x^2 + y^2 = 1$, $1/x + 2y = 1$ by solving for y, substituting into the equation for x, and using bisection to find a solution to this equation in one unknown. Interpret your solution geometrically.

16.) Consider a projectile that moves horizontally through the atmosphere at a high speed. In this case the air resistance is typically modeled by $F_R = -kmv^2$ where m is the mass of the body and k is a positive constant that depends on the geometry of the body as well as other factors, and that must be determined experimentally.

a.) Show that Newton's Law $F = ma$ leads to the ODE IVP $dv/dt = -kv^2$, $v(0) = v_0$ and solve it for $v(t)$. Assume $v_0 > 0$.

b.) Solve the ODE IVP $x(t) = dv/dt$, $x(0) = x_0$ for $x(t)$.

c.) Suppose that the projectile is fired from $x_0 = 10$ m with an initial velocity of $v_0 = 10$ m/s. At $t = 3$ s the projectile has reached $x = 37$ m. Use a root-finding technique from this section to estimate k. Be sure to give the correct units for k.

2. Newton's Method

The bisection and inverse linear interpolation methods are too slow for many applications; typically these methods gain one additional correct significant figure for every several iterations. For example, if the error at step n of the bisection method is ε then to reduce it to $\varepsilon/10$, which corresponds to gaining one more correct significant figure in the answer, requires m additional bisections, where

$$
\begin{aligned}
\varepsilon/2^m &= \varepsilon/10 \\
2^m &= 10 \\
m &= \log_2(10) \\
&\doteq 3.3219.
\end{aligned}
$$

Every correct significant figure costs us about 3 iterations. The inverse linear interpolation method exhibits generally similar behavior.

These algorithms are sometimes used where program length is a consideration. For example, bisection was used in early calculators because the code for it takes so little memory. However, the slow convergence is a problem when function evaluations are expensive, and in many problems of interest the function evaluations are very expensive. (In fact, function evaluations taking a lot of time is virtually a defining feature of scientific computing.) Another difficulty is that these methods require an initial bracket. In Sec. 1.6 we'll look at one way an initial bracket might be found algorithmically. But we must address the question: What do we do if we cannot find one?

Suppose that f is a continuous function. If we pick two initial points x_0, x_1 which do not necessarily form a bracket, we can still interpolate a linear function to f,

$$y = \frac{f(x_1) - f(x_0)}{x_1 - x_0} x + \left(f(x_1) - \frac{f(x_1) - f(x_0)}{x_1 - x_0} x_1 \right)$$

and use its x-intercept

$$x_2 = x_1 - f(x_1) \frac{x_1 - x_0}{f(x_1) - f(x_0)}$$

as an improved estimate of the location of a zero of f. (Note that this is the same formula as for inverse linear interpolation.) If x_0 and x_1 do in fact form a bracket, this will be precisely an inverse linear interpolation step; if they do not form a bracket, this is an extrapolation to a point outside of the interval $[x_0, x_1]$ (or $[x_1, x_0]$ if $x_1 < x_0$). In either case we take x_1 and x_2 as our new pair of bracketing points. This is the **secant method**: Pick any two distinct points, draw the secant through them, and use the x-intercept of that secant line as the new estimate of the zero of the function, discarding the oldest point (see Fig. 1). It is immaterial to the algorithm whether or not the two current points form a bracket, although a programmer implementing it might decide to store the location of a bracket if the method did find one, allowing the program to return to the user some useful information even if it fails.

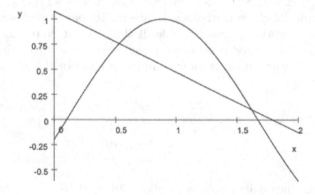

FIGURE 1. Secant method.

EXAMPLE 1: consider again the very simple function $f(x) = \cos(x)$ and the initial points $x_0 = 0, x_1 = 1$. Since $f(0) = 1.0000, f(1) \doteq 0.5403$, this is not a bracket, and so the bisection and inverse linear interpolation methods are not applicable. The secant method gives

$$\begin{aligned}
x_2 &= x_1 - f(x_1)\frac{x_1 - x_0}{f(x_1) - f(x_0)} \\
&= 1 - 0.5403(1 - 0)/(0.5403 - 1.0000) \\
&= 1 - (-1.1753) \\
&\doteq 2.1753
\end{aligned}$$

and so the new pair of points is $x_1 = 1.0000, x_2 \doteq 2.1753$. Since $\cos(x_2) \doteq -0.5684$, $[x_1, x_2]$ is actually a bracket, and the next step will be an inverse linear interpolation step, giving $x_3 \doteq 1.5728$. The true solution is $x^* \doteq 1.5708$. We have obtained a fairly good estimate fairly rapidly. $\square$

If provided with only a single initial guess x_0, we may easily generate a second point x_1 by choosing a point 10% greater, say, than x_0. If this fails, we might try a point 10% smaller. There is, of course, nothing special about the choice of 10% as an increment here; it's simply a heuristic choice intended to help a program implementing this method recover and continue without additional input or guidance from us.

As is evident from Table 1, which compares the bisection, inverse linear interpolation, and secant methods for the function $f(x) = x^2 - 2$, the secant method is considerably faster than the previous methods. (Notice from the table that a new correct decimal place is gained every one to three iterations for the bracketing methods, but that the secant method can pick up several correct decimal places each iteration. In the table, decimal places that agree with the answer are underlined.) We will soon quantify how much faster the secant method is.

Table 1
Comparison of methods for $f(x) = x^2 - 2$; $x_0 = 1, x_1 = 2$
Solution: $\sqrt{2} = 1.41421356237310$

Bisection Method	Inverse Linear Interpolation	Secant Method
1.5	1.33333333333333	1.33333333333333
1.25	1.40000000000000	1.40000000000000
1.375	1.41176470588235	1.41463414634146
1.4375	1.41379310344828	1.41421143847487
1.40625	1.41414141414141	1.41421356205732
1.421875	1.41420118343195	1.41421356237310
1.4140625	1.41421143847487	1.41421356237310

In addition to its speed advantage, the secant method does not require an initial bracket. The price paid is that it may fail to converge: If $f(x_0) = f(x_1)$, for example, then the interpolated line is horizontal and there is no x-intercept. For a function that is asymptotic to the x-axis, the sequence of iterates may diverge (see Fig. 2).

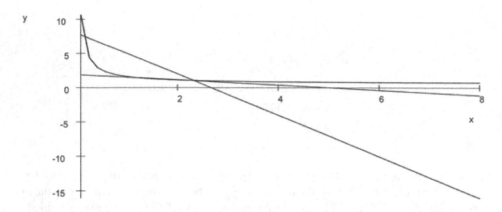

FIGURE 2. Secant method diverging.

We would really like to have a method that relies on only a single initial guess. What happens if we fix one of the two initial points in the secant method and slide the other one towards it? As $x_1 \to x_0$ with x_0 fixed, the slope of the interpolated line will tend to $f'(x_0)$ (assuming that f is in fact differentiable at x_0). It appears that, if f is differentiable, we might be able to choose a single initial point x_0, form the tangent line to f at that point

$$y = f'(x_0)x + (f(x_0) - f'(x_0)x_0)$$

(in slope-intercept form). If we now decide to set $y = 0$ in this equation so that we can find the x-intercept of the tangent line, namely

$$
\begin{aligned}
0 &= f'(x_0)x + (f(x_0) - f'(x_0)x_0) \\
x &= x_0 - f(x_0)/f'(x_0)
\end{aligned}
$$

(2.1)

(assuming $f'(x_0) \neq 0$), we obtain a new value of x that should be an improved estimate of the location of a zero of f, if the tangent line approximation is a good approximation near x_0. (See Fig. 3.) This is **Newton's method**: Pick a point x_0 on a differentiable function f. Compute

$$
\begin{aligned}
x_1 &= x_0 - f(x_0)/f'(x_0) \\
x_2 &= x_1 - f(x_1)/f'(x_1) \\
x_3 &= x_2 - f(x_2)/f'(x_2) \\
x_4 &= x_3 - f(x_3)/f'(x_3)
\end{aligned}
$$

$$\vdots$$

until some stopping criterion is met. The final value x_k is an estimate of the location of a zero of f.

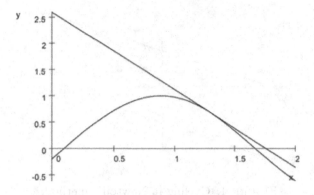

FIGURE 3. Newton's method.

EXAMPLE 2: Let $f(x) = \cos(x)$. We'll apply Newton's method with $x_0 = 1$ to attempt to find a zero of f. We have

$$
\begin{aligned}
x_1 &= x_0 - f(x_0)/f'(x_0) \\
&= 1 - \cos(1)/(-\sin(1)) \\
&= 1 + \cot(1) \\
&\doteq 1.6421 \\
x_2 &= x_1 - f(x_1)/f'(x_1) \\
&= 1.6421 + \cot(1.6421) \\
&\doteq 1.5707 \\
x_3 &= x_2 - f(x_2)/f'(x_2) \\
&= 1.5707 + \cot(1.5707) \\
&\doteq 1.5708
\end{aligned}
$$

which is already correct to the number of digits displayed. We would find that x_4 would be the same as x_3 (and so on for $x_5, x_6, x_7, ...$). This is much faster than the other methods we have considered so far. Of course, had we chosen $x_0 = 0$ instead then x_1 would have been undefined since the derivative of f is zero at the origin. In practice this uncommonly-occurring problem is easily addressed by perturbing the given x_0 slightly. $\square$

Newton's method can diverge (similar to the case for the secant method indicated in Fig. 2) and in principle it can cycle (see Fig. 4) though the latter is unlikely to occur in practice. (If we detect it happening we can simply average the two recurring points to get a new starting point.) Newton's method is a robust and powerful tool that is used in many areas of scientific computation; its importance can hardly be overstated.

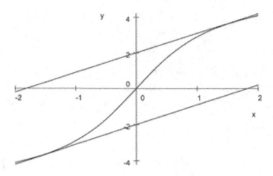

FIGURE 4. Cycling in Newton's method.

If we assume that f is twice continuously differentiable in the region of interest then we can derive Newton's method (Eq. (2.1)) in a different manner. (This derivation will be useful later.) For such an f we have the Taylor polynomial with remainder representation

$$f(x) = f(x_0) + (x - x_0)f'(x_0) + \frac{1}{2}(x - x_0)^2 f''(\delta_x)$$

(for some δ_x between x_0 and x)). We are looking for a point where $f(x)$ is zero; setting $f(x) = 0$ in the above equation gives while assuming that $f'(x_0)$ is nonzero, we find

$$
\begin{aligned}
0 &= f(x_0) + (x - x_0)f'(x_0) + \frac{1}{2}(x - x_0)^2 f''(\delta_x) \\
xf'(x_0) &= x_0 f'(x_0) - f(x_0) - \frac{1}{2}(x - x_0)^2 f''(\delta_x) \\
x &= x_0 - f(x_0)/f'(x_0) - \frac{1}{2}(x - x_0)^2 f''(\delta_x)/f'(x_0) \\
x &\approx x_0 - f(x_0)/f'(x_0)
\end{aligned}
$$

upon neglecting the quadratic term $(x - x_0)^2$, which should certainly be small when the initial guess x_0 is near the true solution x. This is once again the Newton's method formula.

Notice that Newton's method will have difficulty if $f'(x^*) = 0$. We say that a zero x^* of a differentiable function $f(x)$ is **simple** if $f'(x^*) \neq 0$. If x^* is not a simple zero of f, then the graph of f flattens out near the root because the derivative of the function is zero there. For a function like $f(x) = x^8$, that flattening effect will be quite severe.

In fact, a great many of the most common root-finding algorithms will have problems when applied to such a function. Consider $f(x) = x^2$, which has a zero that is not simple and hence can not be bracketed. Newton's method does converge for this function, but very slowly (see Fig. 5).

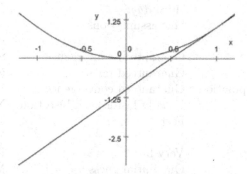

FIGURE 5. Newton's method for $f(x) = x^2$.

You may have noticed that we have contented ourselves with finding any one zero of the given function, not all zeroes of it on the whole real line. The latter problem is simply too difficult to solve for a general function. For one thing, we rarely know how many zeroes the function has so we do not know when to stop looking for them. Also, experience shows that methods like the secant method and Newton's method tend to have zeroes that they "like" and that they tend to return to again and again for a wide range of initial guesses, some quite far from the root; we may have to get very close to the other zeroes if we hope to have a reasonable chance of converging to them. (This phenomenon can be quite exasperating in practice.) If we have some way of bracketing each zero, or if the function is of a very special form such as a polynomial, then we will generally have a reasonable hope of being able to find all zeroes of the function; but for a general function, which may have zeroes that are nonsimple, or that are closely spaced and hence hard to resolve, or that may just barely cross the x-axis and then come up again so the function appears to be of one sign near the zeroes, this is an extremely challenging problem.

For these reasons, most root-finding programs only seek to return a single root of the equation. We would need to call the routine repeatedly, with different initial guesses, to find multiple roots.

We are going to need to develop new mathematical language to describe just how much faster the secant method and Newton's method are than the other methods that we have considered thus far. In the next two sections we will develop such language, and then go on to use it to prove a convergence theorem for Newton's method that will guide us regarding when we can expect this sort of extremely rapid convergence behavior to occur.

We end this section with a summary of the advantages and disadvantages of the methods that we have been discussing (see Table 2). For Newton's method, the most noticeable disadvantages are the need to find the formula for $f'(x)$ and then the extra computational effort required to evaluate $f'(x)$ at each iteration in addition to the evaluation of $f(x)$. Experience shows that–as might have been suspected from the product rule–if $f(x)$ is has a complicated expression and hence is expensive to compute, then the formula for $f'(x)$ is likely to be much more complicated and expensive.

Table 2

Method	Advantages	Disadvantages
Exhaustive Search	Few assumptions	Slow
		Inefficient
Bisection Method	Guaranteed convergence	Slow
	Guaranteed rate	Needs bracket
Inverse Linear Interpolation	Guaranteed convergence	Variable rate
	Usually faster than bisection	Needs bracket
Secant Method	Fast	Two initial guesses
		May diverge
Newton's Method	Very fast	Requires derivative
	One initial guess	May diverge

MATLAB

You may have experience with Computer Algebra Systems such as Maple, Mathematica, etc., that are capable of performing symbolic manipulation of algebraic quantities. There are ways to do this in MATLAB also, but for the types of large problems commonly encountered in practice this is of less practical use than it may seem at first. Symbolic differentiation is rarely a practical tool for real-world problems except possibly as we are deriving formulas to code up into a program in MATLAB, C++, Fortran, Python, etc.

Let's try Newton's method on the function $f(x) = \frac{1}{2}xe^x - 2x^2$. As before, let's start by plotting it to get a feel for the function. Enter:

```
>> x=0:.05:3;
```

and then enter:

```
>> y=.5*x.*exp(x)-2*x.^2;
```

Note the extra periods in the above line. The periods before the asterisk (multiplication) and caret (exponentiation) in the line defining y instruct MATLAB to perform the indicated operations entry-by-entry; otherwise, multiplication of the vector x by the vector exp(x) of values of e^{x_i} would be an undefined vector-vector product. The default operations of MATLAB mimic matrix algebra operations in a very natural way, so a special symbol is needed for this elementwise procedure. Now, enter:

```
>> plot(x,y),grid
```

The rapid growth of the exponential function makes it hard to see what's happening. Change x to go from 0 to 2.5 and recompute y, then plot the graph again. There is a zero at the origin, another near .4, and another near 2.2. Let's try to find the latter two zeroes to better precision using Newton's method.

To quickly define a function at the command line without writing a program we can create a MATLAB function handle, as follows:

```
>> f=@(x) .5*x.*exp(x)-2*x.^2
```

and you should see that MATLAB tells you it is indeed a function_handle. (A previous way of doing this, called an *inline function*, is being phased out by The MathWorks.) We may not need the ability to evaluate f at a vector x but leaving in the periods allows that as an option. The derivative of f is $\frac{d}{dx}(\frac{1}{2}xe^x - 2x^2) = \frac{1}{2}e^x + \frac{1}{2}xe^x - 4x$. Enter:

```
>> fp=@(x) .5*exp(x)+.5*x.*exp(x)-4*x;
```

The symbols f and fp now represent functions. We may evaluate them in the usual way, or indirectly via the command feval; enter:

```
>> f(3)
>> feval(f,3)
```

to find the value of $f(x) = \frac{1}{2}xe^x - 2x^2$ evaluated at $x = 3$, that is, $f(3)$. Now we're ready to use Newton's method. Let's start with the root near $x_0 = .4$; enter:

```
>> x0=.4;
>> x0=x0-f(x0)/fp(x0)
```

This gives $x_1 = 0.3611$. Cursor up and re-enter the last line to find $x_2 = 0.3574$. Cursor up and re-enter it again; the value does not change, no matter how many times we perform the calculation. To the precision displayed, the root is at 0.3574. To have more digits displayed, enter:

```
>> format long
>> x0=.4;
>> x0=x0-f(x0)/fp(x0)
```

This gives $x_1 = 0.36106860486687$. Iterating the last line by repeatedly entering it gives, successively, $x_2 = 0.35743611821915$, $x_3 = 0.35740295895202$, $x_4 = 0.35740295618139$, $x_5 = 0.35740295618139$, and from this point on the values no longer change. The correct digits were (when rounded as indicated) 0.36, 0.357, 0.35740296, 0.35740295618139 (if we assume that the last value provided to us by MATLAB was correct!). As a check, enter:

```
>> f(x0)            %x0 actually stores the most recent value now.
```

This is roughly $5E - 17$, which is quite small. As another check that we have indeed found a root, enter:

```
>> xleft=x0-1E-4;xright=x0+1E-4;
```

and evaluate f(xleft) and f(xright) to see that a zero is bracketed between xleft and xright. Conceivably round-off error is deceiving us by making one of these values appear to have a sign opposite to its actual sign, but this is very unlikely. You may now enter format short to return to the default short display of numeric values if desired. All calculations are still being done with about 16 digits of significance, but only 4 places after the decimal point will be displayed.

As mentioned above, the feval command can also be used with MATLAB-supplied functions and m-files (that is, MATLAB programs) that implement appropriate functions $f(x)$. For example, enter:

```
>> feval('sin',x0)
>> feval('sin',pi/2)
```

to evaluate the sine function at $x = x_0$ and then at $x = \pi/2$. The command feval('myfun',x0) or name='myfun';feval(name,x0) runs the MATLAB program myfun.m with the input argument x0 and returns its value at that argument. (See Sec. 1.4 for more about MATLAB programs.) In most situations in which we use a function handle, an m-file would work just as well, and in fact in practical applications it is highly likely that we would be calling an m-file instead. However, function handles are convenient for our demonstrations and we will see further uses of them later in the course.

Now use a different initial guess to find the root of $f(x) = \frac{1}{2}xe^x - 2x^2$ that lies near 2.2. Check your answer by computing the value of $f(x)$ at that point and by showing that it lies in a bracket. Of course, since we are working in one dimension we can also check our results by plotting the function near the root.

Problems

In each case compare your answer to the corresponding problem in the previous section and comment.

1.) Use Newton's method to approximate the sole positive root of $\cos(x) - x = 0$. Repeat with the secant method.

2.) Use Newton's method to approximate the sole real root of $5x^7 + 2x - 1 = 0$.

3.) Use Newton's method to find a root of $f(x) = \cos(5x)$ in the interval $[0, 5]$ to at least three decimal places. Find the actual error in your estimate using the known locations of the zeroes of the cosine function. Then use a different initial guess and find a different root.

4.) Use Newton's method on $f(x) = x^2 - 5$ to approximate $\sqrt{5}$ to at least four decimal places. Compare your error estimate to the actual error. What happens if you apply Newton's method to $f(x) = x^2$?

5.) Use Newton's method to approximate a zero of $f(x) = 1/x^3 - 10$ to three decimal places.

6.) Solve for the third smallest positive root of $x - \tan(x) = 0$ by bisection, inverse linear interpolation, the secant method, and Newton's method. Create a table with successive iterates of each method in separate columns. Comment on the speed of the methods.

7.) Use Newton's method to approximate all real zeroes of $p(x) = 2x^5 - 7x^4 - 3x^3 + 25x^2 - 23x + 6$. (Hint: Graph $p(x)$ first.) Does Newton's method perform equally well for all zeroes? Repeat with the secant method.

8.) Use Newton's method to approximate all real zeroes of $p(x) = x^7 - 6.65x^6 - 3.475x^5 + 112.3x^4 - 194.88x^3 - 361.02x^2 + 1306.4x - 999.02$. (Hint: Graph $p(x)$ first.) Does Newton's method perform equally well for all zeroes?

9.) Give an explicit formula and initial points for a function for that the secant method and Newton's method will diverge in the manner indicated in Figure 2.

10.) Apply Newton's method to $f(x) = (x-1)^2 + .1$ with initial condition $x_0 = 1.1$. Explain the results in terms of the graph of f.

11.) What happens if the secant method is applied to a linear function? What happens if Newton's method is applied to a linear function?

12.) a.) Bracket the roots of $x^4 - 6.7302x^3 - 15.608x^2 + 44.46x + 74.487 = 0$.

b.) Use the secant method to find all real roots of this equation.

c.) Use Newton's method to find all real roots of this equation.

13.) Use the secant method to approximate a zero of $f(x) = 1/x^3 - 10$ to six decimal places.

14.) Apply Newton's method to $f(x) = x^2 + 1$ using the complex initial condition $x_0 = 1 + i$. (This may be entered in MATLAB as x0=1+i; MATLAB will handle the complex arithmetic automatically.) Use another choice of x_0 to find the other root.

15.) Sketch the graph of a function for which you could choose x_0 very close to one root of an equation but Newton's method would converge to a root far from it.

16.) The **logistic model** for the size of a population is the solution of the **logistic equation** $dP/dt = kP(M - P)$. Here $M > 0$ is the maximum number of creatures that the region can support, and $k > 0$ is a constant such that, when the population is small, $dP/dt \approx (kM)P$ (the population size initially grows exponentially).

a.) Show that the logistic model is $P(t) = MP_0/(P_0 + (M - P_0)\exp(-kMt))$ in the range $dP/dt = kP(M - P)$ by separating variables and using partial fractions.

b.) Suppose a lake has 100 trout in it. A year later it has 347 trout in it. If the net birth rate $\beta = kM$ of the trout is believed to be 2.1 per year, what is the maximum population M of trout the lake can support under the logistic model? Use Newton's method to determine M.

3. The Fixed Point Theorem

It has been said that "a good proof is one that makes us wiser." We are going to study Newton's method in some detail and establish sufficient conditions for it to converge to a zero of the function in which we are interested. In doing so we will gain a better understanding of how and why the method works and just how rapidly it converges.

In using Newton's method we choose an initial guess x_0 and then iterate it in what is in essence a feedback loop: We always set

$$\begin{aligned} x_k &= x_{k-1} - f(x_{k-1})/f'(x_{k-1}) \\ &= N_f(x_{k-1}) \end{aligned}$$

$(k = 1, 2, ...)$, where the function

$$N_f(x) = x - f(x)/f'(x)$$

is sometimes called the **Newton transform** of f. An iteration of this form, where for some function g we perform the operation

(3.1) $$x_k = g(x_{k-1})$$

repeatedly, is called a **fixed point iteration** (or **functional iteration**), and any x^* such that

$$x^* = g(x^*)$$

is called a **fixed point** of g. A fixed point is therefore a point where the graph of $y = g(x)$ crosses the line $y = x$. Newton's method is of this form: Every simple zero of f corresponds to a fixed point of N_f. When Newton's method converges, it means that the fixed point iteration

$$x_k = N_f(x_{k-1})$$

is converging to a fixed point of N_f. That fixed point is a root of the underlying equation $f(x) = 0$ that we are actually trying to solve.

We will first consider fixed point iterations in more generality and then specialize them to Newton's method. This is useful because iterations in the form of Eq. (3.1) appear frequently in numerical analysis.

If g is continuous then a fixed point iteration based on g, if it converges, must converge to a fixed point of g. For, if $x_k = g(x_{k-1})$ is used to generate a sequence $\{x_n\}_{n=0}^{\infty}$ and

$$\lim_{n \to \infty} x_n = x^*$$

then

$$
\begin{aligned}
\lim_{n \to \infty} x_n &= \lim_{n \to \infty} g\left(x_{n-1}\right) \\
x^* &= g\left(\lim_{n \to \infty} x_{n-1}\right) \\
&= g\left(x^*\right)
\end{aligned}
$$

by continuity. However, a fixed point iteration will not always generate a convergent sequence for a given x_0.

There are many versions of the Fixed Point Theorem, which establishes conditions under which a function has a fixed point that may be found by fixed point iteration. The following version is appropriate for our work:

THEOREM 1 (FIXED POINT THEOREM): If g is continuous on $[a, b]$ and $g : [a, b] \to [a, b]$ then g has a fixed point in $[a, b]$. If in addition g is differentiable on (a, b) and there is a λ, $0 < \lambda < 1$, such that $|g'(x)| \leq \lambda$ on (a, b), then this fixed point is unique and the fixed point iteration $x_k = g(x_{k-1})$ converges to it for any $x_0 \in [a, b]$.

PROOF: Suppose g is continuous on $[a, b]$ and $g : [a, b] \to [a, b]$. If $g((a)) a = a$ or $g(b) = b$ we are done; hence let us assume that this is not so. Define $h(x) = g(x) - x$; $h(x)$ is continuous, and $h(x)$ has a zero in $[a, b]$ if and only if $g(x)$ has a fixed point in $[a, b]$. But

$$
\begin{aligned}
h(a) &= g(a) - a \\
&> 0
\end{aligned}
$$

since $g(a)$ is in $[a, b]$ and hence cannot be smaller than a, and we have assumed that $g(a)$ is not equal to a. Similarly,

$$
\begin{aligned}
h(b) &= g(b) - b \\
&< 0
\end{aligned}
$$

and so by the Intermediate Value Theorem there is a $c \in (a, b)$ such that $h(c) = 0$, that is, such that $g(c) = c$, as claimed.

Now suppose in addition that g is differentiable on (a, b) and that there is a $0 < \lambda < 1$ such that $|g'(x)| \leq \lambda$ on (a, b). If p and q are two distinct fixed points of g, then by the Mean Value Theorem

$$\frac{g(p) - g(q)}{p - q} = g'(\theta)$$

for some θ between p and q. But $g(p) = p$ and $g(q) = q$ since p and q are fixed points, so

$$g'(\theta) = \frac{g(p) - g(q)}{p - q}$$

$$= \frac{p - q}{p - q}$$

$$= 1$$

which is a contradiction. Hence there are not two distinct fixed points of g on $[a, b]$; the fixed point is unique.

It remains to show that we can find the fixed point x^* by fixed point iteration. Pick an x_0 in $[a, b]$. The sequence $x_k = g(x_{k-1})$ is well-defined since $g : [a, b] \to [a, b]$. Now

$$|x_n - x^*| = |g(x_{n-1}) - g(x^*)|$$

$$= |g'(\theta_n)||x_{n-1} - x^*|$$

$$\leq \lambda|x_{n-1} - x^*|$$

for some θ_n between x_{n-1} and x^*, by the Mean Value Theorem and the bound on $|g'(x)|$. Inductively, $|x_n - x^*| \leq \lambda^n|x_0 - x^*|$ and since $\lambda < 1$, $|x_n - x^*| \to 0$ as $n \to \infty$ and so $x_n \to x^*$. $\square$

EXAMPLE 1: On early computers, fixed point iteration based on $g(x) = \frac{1}{2}\left(x + \frac{2}{x}\right)$ was sometimes used to find $\sqrt{2}$ (and more generally, $g_\alpha(x) = \frac{1}{2}\left(x + \frac{\alpha}{x}\right)$ to find $\sqrt{\alpha}$, $\alpha > 0$). Does $g(x)$ have a fixed point? Note that $g(x)$ is continuous for $x > 0$ and $g(1) = 1.5, g(2) = 1.5$, and $g'(x) = \frac{1}{2}\left(1 - 2/x^2\right)$. Since $g'(x) = 0$ at $x = \pm\sqrt{2}$, g takes on its extreme values at these points, and in particular $g(\sqrt{2}) = \sqrt{2}/2 \doteq 1.4142$; hence $g : [1, 2] \to [\sqrt{2}, 1.5]$ and so, trivially, $g : [1, 2] \to [1, 2]$. Hence g has a fixed point on $[1, 2]$. Since $|g'(x)| \leq 1/2$ over $[1, 2]$ (verify this), the fixed point in $[1, 2]$ is unique. To find it we must either solve $x = g(x)$ in the equivalent form found by setting $f(x) = g(x) - x$ and finding a zero of f,

$$x = g(x)$$

$$x = \frac{1}{2}\left(x + \frac{2}{x}\right)$$

$$0 = \frac{1}{2}\left(x + \frac{2}{x}\right) - x$$

$$0 = x^2 - 2$$

to see that $\pm\sqrt{2}$ are indeed the fixed points of g, or use fixed point iteration by choosing, say, $x_0 = 1$ and forming the sequence

$$x_1 = \frac{1}{2}(x_0 + 2/x_0) = 1.5$$

$$x_2 = \frac{1}{2}(x_1 + 2/x_1) \doteq 1.41666666666667$$

$$x_3 = \frac{1}{2}(x_2 + 2/x_2) \doteq 1.41421568627451$$

$$x_4 = \frac{1}{2}(x_3 + 2/x_3) \doteq 1.41421356237469$$

($\sqrt{2} \doteq 1.41421356237310$). Notice the rapid convergence:

$$1$$
$$1.41$$
$$1.41421$$
$$1.41421356237$$

(where only those digits that agree with $\sqrt{2}$ are listed). This method for computing square roots was used on computers when only the four arithmetic operations were easily available, as it quickly finds the root using only those operations. □

From the Fixed Point Theorem, we know that under appropriate conditions a numerical method of the form $x_k = g(x_{k-1})$ will converge to a common value (the solution) for all sufficiently close initial guesses; the rate of convergence is controlled by $|g'(x)|$ near the fixed point. (We say that the fixed point is **attracting** when $|g'(x)| < 1$.) The smaller the derivative (in absolute value), the more rapid the convergence. For Newton's method,

$$
\begin{aligned}
g'(x) &= N_f'(x) \\
&= (x - f(x)/f'(x))' \\
&= 1 - [f(x)/f'(x)]' \\
&= \frac{f(x)f''(x)}{[f'(x)]^2}
\end{aligned}
$$

(if f is twice differentiable). If x^* is a simple zero of f (so that $f'(x^*)$ is nonzero) then

$$
\begin{aligned}
g'(x^*) &= \frac{f(x^*)f''(x^*)}{[f'(x^*)]^2} \\
&= \frac{0 \cdot f''(x^*)}{[f'(x^*)]^2} \\
&= 0
\end{aligned}
$$

and by continuity $|g'(x)| < 1$ in some neighborhood of the zero. This is the best possible case; sufficiently near x^*, the value of λ in the proof of the Fixed Point Theorem may be taken as small as desired. In this sense Newton's method may be viewed as a way of getting a best possible fixed point iteration for a given problem. Fixed point iterations that are not generated by Newton's method may also be used for root-finding, but if they don't also have a zero derivative at the root then we can't expect them to converge as rapidly as this method.

Let's finish applying the Fixed Point Theorem to Newton's Method. By continuity of $g'(x)$ $(= \frac{d}{dx}N_f(x))$, which requires that $f''(x)$ be continuous, that is, that f be twice continuously differentiable, there is a $\delta > 0$ such that $|g'(x)| \le \lambda < 1$ for $x \in [x^* - \delta, x^* + \delta]$. (That is, if $g'(x)$ is zero at x^* then $|g'(x)|$ is small near x^*.) But since the derivative of g is less than one in absolute value, it is growing less rapidly than the line $y = x$, and so if $x \in [x^* - \delta, x^* + \delta]$ then $gv \in [x^* - \delta, x^* + \delta]$. Taking $[a, b] = [x^* - \delta, x^* + \delta]$, $g(x) = N_f(x)$ meets all requirements of the Fixed Point Theorem. Hence:

THEOREM 2 (NEWTON'S METHOD CONVERGENCE THEOREM):
Suppose $f(x)$ has a simple zero x^* in $[a, b]$, and that f is twice continuously differentiable on $[a, b]$. Then there is an interval about x^* such that Newton's method converges to x^* for any x_0 in that interval.

There are many variations of a more general theorem, the Newton–Kantorovich theorem, regarding when Newton's method might converge under less stringent conditions. However, we will use this version for our purposes. Note carefully that our theorem only guarantees convergence to a *simple* zero if we are *sufficiently* close to it, and that it requires that f be *twice* continuously differentiable near the zero.

MATLAB

In working with fixed points it's convenient to be able to overlay plots. For example, let's look for the fixed points of $3\cos(x)$. Enter:
```
>> x=-5:.1:5;
>> plot(x,3*cos(x))
>> hold on
>> plot(x,x)
```
(Use the `grid` command in addition if you like.) The `hold on` command tells MATLAB to hold the current plot so that subsequent plots will be overlaid on the held plot. The command `hold off` releases the hold.

The command `help hold` will give more information on the `hold` command. In general, typing `help <command>` will give some basic information about that command. The built-in help is excellent and you should refer to it frequently as you learn MATLAB; it makes learning the language relatively easy. Use `help` frequently.

You may also use the help from the drop-down *Help* menu. Selecting *MATLAB Help* gives you access to demos, an introduction to MATLAB, and other "getting started" material, as well as search capabilities.

Type `help plot` and read about the `plot` command, including the color options for your plots. For example, enter:
```
>> plot(x,x,'c')
```
to overlay a new $y = x$ line on your previous one, but now in cyan. Close the previous plot. Enter:
```
>> plot(x,3*cos(x),'b+'),hold,plot(x,x,'yo')
```
You should get the cosine curve as blue pluses and the line as yellow circles. These options do not connect the dots. This is sometimes useful (though not here).

Let's do a fixed point iteration on $g(x) = 3\cos(x)$. The plot showed a fixed point near 1 and either one or two fixed points near -3 (we need better resolution to decide if it is one point of intersection or two). Enter:
```
>> x0=1;
```
We could simply enter x0=3*cos(x0) repeatedly (using the cursor up key), but MATLAB has looping structures. Enter:
```
>> for k=1:10,x0=3*cos(x0),end
```
(If the numbers go by too quickly, type `help more` for a useful option.) The method seems to have converged to a fixed point near -2.94. Cursor up and re-enter the line with the `for-end` loop; it will start over with the most recent x_0, x0=-2.9388[5], and do another ten iterations.

---

[5]In fact, MATLAB has stored this number to about 15 decimal places and will use *that* value in its calculations.

It appears that -2.9381 is a fixed point. Although we started with an initial guess near 1, we found another fixed point entirely. Enter:
```
>> -3*sin(1)
```
The reason this happened is that near $x = 1$, the derivative of $g(x) = 3\cos(x)$ is about -2.5244. The derivative is too large (in absolute value). Enter:
```
>> -3*sin(-3)
```
Near $x = -3$ the derivative: is 0.4234, which is small enough that if x_0 (or some later iterate) is sufficiently near it we will see convergence.

Close the open figures, if you haven't already (see **help close** for a way to do this using a MATLAB command). Enter:
```
>> x=-4:.01:-2;
>> plot(x,3*cos(x)),hold on,plot(x,x),grid
>> title('Fixed point of 3*cos(x)')
```
There's another fixed point near $x = -2.6$. But, enter:
```
>> -3*sin(-2.6)
```
The derivative is 1.5465. We won't be able to find this fixed point by simple fixed point iteration on $g(x)$. However, we can find any of the fixed points by Newton's method. Use Newton's method on $g(x) = 3\cos(x)$ to find the two remaining fixed points.

Problems

1.) a.) Show that the fixed point iteration of Example 1 is the result of applying Newton's method to $f(x) = x^2 - 2$.

b.) Use the method to approximate $\sqrt{2}$ to at least eight decimal places using the initial guesses $x_0 = 1, 1.4, 2, 10$.

2.) Solve the equation $x = 2^{-x}$ for x by performing fixed point iteration on it as given. (You may wish to plot $y = x$ and $y = 2^{-x}$ first to generate a good x_0.) Compare your results to using Newton's method on $f(x) = x - 2^{-x}$. Comment.

3.) Attempt to use fixed point iteration to solve $x^2 = 3$ by rewriting it as $x = x^2 + x - 3$. What happens? Why? Now rewrite it as $x = \frac{1}{2}(x + \frac{3}{x})$ (show that this is equivalent to $x^2 = 3$) and try again. Why does this version work better?

4.) a.) An early method for performing division b/a on computers was to apply Newton's method to $f(x) = 1/x - a$ to find $1/a$ and then multiply it by b. Show that Newton's method for $f(x)$ requires only multiplication and subtraction, and that it converges for all $x_0 \in (0, 1/a]$. (Hint: Plot $N_f(x)$ and sketch a few iterates.) What happens if $x_0 \gg 1/a$?

b.) Use this method with $a = 3$ to approximate $1/3$. Use the initial guesses $x_0 = .5, 1, 2, 4$. In each case make a table showing only those leading digits in your answer that agree with the solution. What can you say about the rapidity of convergence of Newton's method in this case? (This method is still commonly used on high-performance machines to implement quadruple precision division.)

5.) Let $g(x)$ be a continuously differentiable function with a fixed point x^* for which $|g'(x^*)| < 1$. Show that fixed point iteration of g will converge to x^* for any x_0 in some interval $[x^* - \delta, x^* + \delta]$ (with $\delta > 0$). What limits how large δ can be?

6.) a.) Estimate how many iterations it would take to find the sole fixed point of $\cos(x)$ to six decimal places by fixed point iteration $x_k = \cos(x_{k-1})$ starting from $x_0 = .8$. Do this by estimating λ from the proof of the Fixed Point Theorem.

b.) Use fixed point iteration on $g(x) = \cos(x)$ to find the fixed point to six decimal places. Compare your actual number of iterations to your estimate from part a.

c.) Use Newton's method on $f(x) = x - \cos(x)$ to find the fixed point to six decimal places. Compare the number of iterations for Newton's method to the number of iterations from part b.

7.) a.) Show that the functions $g_1(x) = x^3 - 9x^2 + 27x - 24$, $g_2(x) = 9 - 26/x + 24/x^2$, $g_3(x) = \sqrt{(9x^2 - 26x + 24)/x}$, $g_4(x) = (-x^3 + 9x^2 + 24)/26$, and $g_5(x) = \frac{2x^3 - 9x^2 + 24}{3x^2 - 18x + 26}$ have the same fixed points, and that these fixed points are exactly the zeroes of $f(x) = x^3 - 9x^2 + 26x - 24$.

b.) The zeroes of $f(x)$ are $2, 3, 4$. For each zero of f, determine which functions $g_1, ..., g_5$ will converge to that zero under fixed point iteration.

c.) Do any of $g_1, ..., g_5$ correspond to Newton's method for $f(x) = 0$?

8.) Consider again the function $f(x) = 1/x - a$ from problem 4. Using $x_0 = 10^{-10}$, how many iterations are needed to get six decimal place accuracy if $a = .5$? Does this contradict the claimed rapid convergence?

9.) a.) Show that if $\mu \in (1, 3)$ then there is an interval on which fixed point iteration must converge to a unique fixed point for $g(x) = \mu x(1-x)$. (What happens if $\mu = 1$ or $\mu = 3$?)

b.) Find a formula for the fixed point as a function of μ.

c.) Show that if $\mu > 3$ then this fixed point is not attracting. (If $|g'(x^*)| > 1$ we say that the fixed point is **repelling**.)

10.) a.) If $g(x) = \mu x(1 - x)$ with $\mu = 3.5$, find all 4 fixed points of $g(g(x))$. (If you have access to a computer algebra system such as Maple, assume $\mu > 3$ and find a formula for these points.) Show that two of them are fixed points of g and that the other pair form a cycle $x_b = g(x_a)$, $x_a = g(x_b)$.

b.) For $\mu = 3.5$, iterate $g(x) = \mu x(1 - x)$ starting with $x_0 = .5$. Is the sequence converging? (This is called a period two cycle; it is interesting in the study of mathematical chaos but represents a form of failure to converge in numerical analysis.)

11.) a.) The Fixed Point Theorem is a special case of a more general result known as the (Banach) Contraction Mapping Theorem. If D is a closed subset of $\mathbb{R}^\times$ and $T : D \to D$ then T is said to be a contraction (or to be contractive) on D if there is an s, $0 < s < 1$, such that $|T(x) - T(y)| \le s|x - y|$ for all $x, y \in D$. (We interpret $|\cdot|$ as a vector norm if necessary.) The constant s is called the contractivity factor. Show that if a function meets the hypotheses of the Fixed Point Theorem then it is a contraction on some $D \subseteq \mathbb{R}$. What is the contractivity factor s?

b.) Prove that a contraction T must be a continuous function.

c.) The **Contraction Mapping Theorem** states that every contraction mapping has a unique fixed point on its domain D, often called an **invariant set** or **attractor** in this context. Prove the Contraction Mapping Theorem.

d.) Show that the function on $\mathbb{R}^2$ defined by $T(x, y) \to (\frac{1}{2}x + 1, \frac{1}{2}y)$ is a contraction. What is its contractivity factor? Find its invariant set analytically, then verify that fixed point iteration converges to that fixed point.

12.) Solve $x^3 - 7.8x^2 + 16.4x - 9.6 = 0$ for the largest root using Newton's method; use MATLAB to plot the function in order to generate an initial guess. Would fixed point iteration in the form $x_{k+1} = x_k^3 - 7.8x_k^2 + 16.4x_k - 9.6$ be successful for any of these roots?

13.) a.) Create a convergent fixed point iteration for finding the positive root of $x^3 + 3.85x^2 - 4.9x - 17.2 = 0$ which is not Newton's method by algebraically rearranging the equation $x = x^3 + 3.85x^2 - 3.9x - 17.2$.

b.) Plot your $g(x)$ from part a. with the line $y = x$ on the same graph; indicate the location of the fixed point and the slope of g at the point of intercept.

14.) a.) Use the method of Example 1 to find an approximation to $\sqrt{10}$ that is correct to as many decimal places as you can display on your machine (use `format long` if working in MATLAB).

b.) Suppose the method of Example 1 is to be implemented as an automated square-root-finder, that is, as a program that accepts $\alpha > 0$ and returns $\sqrt{\alpha}$. Suggest a good choice of initial condition x_0; x_0 must be a simple arithmetic (addition, subtraction, multiplication, division) function of α, and the closer it is to $\sqrt{\alpha}$ the quicker the program will run.

15.) a.) The proof of the Fixed Point Theorem requires that $|g'(x)| \leq \lambda < 1$, that is, that the absolute value of the derivative is not just less than one but is bounded below one. What might fail if this condition was replaced by $|g'(x)| < 1$? What if it was replaced by $|g'(x)| \leq 1$?

b.) In using the Fixed Point Theorem to establish the Newton's Method Convergence Theorem we made the claim that since the derivative of g is less than one in absolute value, it is growing less rapidly than the line $y = x$, and so if $x \in [x^* - \delta, x^* + \delta]$ then $g(x) \in [x^* - \delta, x^* + \delta]$. Give a rigorous proof of this claim.

16.) In an article in the journal *Chemical Engineering Science*, Farr and Aris consider the number of steady states that can be achieved for two sequential reactions in a well-mixed (tank) reactor. One condition under which a change in the number of such states can occur is when v and γ are such that $v^2 = [(\gamma - 3)^2 + 6]/[3(\gamma - 2)^2]$ where $\gamma = E_1/R\bar{T}$ and $v = E_2/E_1$ are dimensionless parameters depending on the respective activation energies E_1 and E_2, the gas constant R, and the mean temperature $\bar{T}$.

a.) What is the form of Newton's method for finding a γ for a given value of v?

b.) Use Newton's method to find γ if $v^2 = 1, 2, 3$, if possible.

c.) Can γ be found algebraically?

4. Quadratic Convergence of Newton's Method

Newton's method generates, under appropriate conditions, a sequence $\{x_n\}_{n=0}^{\infty}$ with the property that x_n tends to a zero x^* of the function f as $n \to \infty$. But this is a theoretical result; will it still hold true on a computer, with finite precision arithmetic and a finite amount of time? The finite precision arithmetic is not so much of an issue with Newton's method, since if we are in the interval in which convergence is guaranteed, a small round-off error will likely leave us in that interval and convergence is still assured from that point. What we need is a **convergence criterion** (or **stopping criterion**) for the method, so that we know how to choose an n sufficiently large that x_n is approximately equal to the limit x^* of the sequence.

For bisection this was easily handled, since the width of the current bracket was a measure of the uncertainty in our knowledge of the root. More generally, we define the **absolute error** in an approximation x_n to a (typically unknown) true value x^* to be the quantity

$$\alpha = |x^* - x_n|$$

and we define the **relative error** in x_n as an approximation to x^* to be the quantity

$$\rho = |x^* - x_n|/|x^*|$$

if x^* is nonzero. This may be a more useful measure of the error when $|x^*|$ is far from unity, since it's generally unreasonable to ask for an absolute error of 10^{-8} when the answer is on the order of millions, for example. If ρ is about 10^{-k} then the computed value x_n is correct to about k decimal places.

For the secant method and Newton's method we do not have a bracket and so we cannot estimate α or ρ with certainty. Ideally we would like to terminate the algorithm when α (or ρ) is less than some tolerance τ (say, relative error of no more than 1%, that is, $\rho = .01$). It is tempting to use the **residual error** $|f(x_n)|$ as that is already computed by the algorithm,

$$|f(x_n)| < \tau$$

but it is possible for this to be small even though x is far from x^* (see Fig. 1, where $x^* = 6$). Because of this the criterion $|f(x)| < \tau$ should be used only in conjunction with at least one other criterion, or when there is sufficient knowledge of the f in question to be sure that this is a safe convergence criterion.

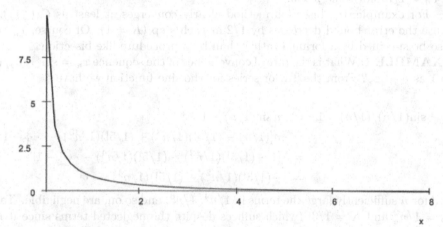

FIGURE 1. Small function values far from root.

If the method is working, x_{n+1} should be a better estimate of x^* than x_n was, so it might be reasonable to use

$$(4.1) \qquad \begin{aligned} \alpha_n &\approx |x_{n+1} - x_n| \\ \rho_n &\approx |x_{n+1} - x_n|/|x_{n+1}| \end{aligned}$$

to estimate the absolute and relative errors, respectively, in x_n. This is especially so if x_{n+1} is generally a *much* better estimate of x^* than x_n is. Then we could choose to terminate the iteration if α_n (or ρ_n) is less than the tolerance. To have confidence in such a criterion, we would like to know just how much better x_{n+1} is than x_n as an estimate of the true solution x^*.

Let's try to quantify how rapidly Newton's method converges to its limit. When studying the calculus you focused mostly on whether or not a limit existed; in

numerical analysis we also consider how rapidly that limit is approached. There are
two key ideas used in quantifying that speed. The first is the **rate of convergence**.
When considering the rate of convergence, we take a sequence $\{x_n\}_{n=0}^{\infty}$ generated
by a numerical method (or any other means) and a reference sequence $\{c_n\}_{n=0}^{\infty}$
to which we compare it. Suppose $x_n \to x^*$ and the reference sequence $\{c_n\}_{n=0}^{\infty}$
satisfies $c_n \to 0$. We say that $\{x_n\}_{n=0}^{\infty}$ converges to x^* with rate of convergence
$\{c_n\}_{n=0}^{\infty}$ if there is a constant $K > 0$ such that

$$|x_n - x^*| \leq K|c_n|$$

for all n sufficiently large. (Compare the Sandwich Theorem for limits. We want
a sandwiching sequence whose speed of approach to the limit we understand.) We
write

$$x_n = O(c_n)$$

(or say that $\{x_n\}_{n=0}^{\infty}$ converges to x^* as $O(c_n)$), called the "big oh" notation. The
reference sequence $\{c_n\}_{n=0}^{\infty}$ is commonly taken to be $c_n = a^n$ for some $0 < a < 1$,
or $c_n = 1/n^p$ for some $p > 0$.

For example, the bisection method clearly converges at least as $O((\frac{1}{2})^n)$, because the error bound decreases by $1/2$ at each step ($K = 1$). Of course, x_n could
also be specified by a formula rather than by a procedure like bisection:

EXAMPLE 1: What is the rate of convergence of the sequence $x_n = \sin(1/n)/(1/n)$
to 1 as $n \to \infty$? From the Taylor series for the sine function we have

$$
\begin{aligned}
\sin(1/n)/(1/n) - 1 &= n\sin(1/n) - 1 \\
&= n[(1/n) - (1/3!)(1/n^3) + (1/5!)(1/n^5) - \cdots] - 1 \\
&= [1 - (1/3!)(1/n^2) + (1/5!)(1/n^4) - \cdots] - 1 \\
&= -(1/3!)(1/n^2) + (1/5!)(1/n^4) - \cdots
\end{aligned}
$$

and for n sufficiently large the terms in $1/n^4$, $1/n^6$, and so on, are negligible. Taking
$c_n = 1/n^2$ and $K = 1/3!$ (which suffices despite the neglected terms since it is an
alternating series) shows that the rate of convergence is $O(1/n^2)$. □
Note that we needed to subtract the limit L from the value of x_n so as to isolate
the part of x_n that is actually diminishing to zero.

From the proof of the Fixed Point Theorem, we see that for a fixed point
iteration $x_{n+1} = g(x_n)$ satisfying its conditions,

$$|x_n - x^*| \leq K\lambda^n$$

where λ is the bound on the derivative of g, and K may be taken to be the distance
$|x_0 - x^*|$ from the initial guess to the actual root. Hence the rate of convergence
is $O(\lambda^n)$ for a convergent fixed point iteration that meets the conditions of the
theorem. In principle any $\lambda \in [|g'(x^*)|, 1]$ would work here, but taking $\lambda = |g'(x^*)|$
gives the best result.

The rate of convergence is a useful way to compare two methods when the two
reference series are well known and are themselves easily comparable; for example,
a $O(1/n^2)$ method certainly converges more rapidly for large n than a $O(1/n)$

method. But Newton's method is $O(\lambda^n)$ where $\lambda > 0$ may be taken arbitrarily small if we are sufficiently near the zero. While this is obviously a desirable quality, we will need another notion of convergence speed to better quantify this and to compare it to other methods with this property.

We say that a sequence $\{x_n\}_{n=0}^{\infty}$ that converges to a limit x^* does so with **order of convergence** 1 if there is a constant $0 < M < 1$ such that the limit

$$(4.2) \qquad \lim_{n \to \infty} \frac{|x_{n+1} - x^*|}{|x_n - x^*|} = M$$

exists. The constant M is called the **asymptotic error constant**. We say that the sequence, or the method that generates it, exhibits **linear convergence**. Evidently a general fixed point iteration has order of convergence 1 if it has a positive asymptotic error constant $M = \lambda$, since the absolute error $e_n = |x_n - x^*|$ satisfies

$$\frac{e_{n+1}}{e_n} \to \lambda$$

with $\lambda < 1$. This means that

$$e_{n+1} \approx \lambda e_n$$

for large n. For Newton's method, however, λ may be taken arbitrarily small in the limit, and so we may take M as small as desired. Indeed,

$$\lim_{n \to \infty} \frac{|x_{n+1} - x^*|}{|x_n - x^*|} = 0$$

for Newton's method (corresponding to $M = 0$ in Eq. (4.2), where this was disallowed). When we have $M = 0$ in Eq. (4.2) we say that the method or sequence is **superlinearly convergent** (or simply **superlinear**). Newton's method is superlinear for a simple zero.

In some cases that's where the story ends. But for other superlinear methods it can happen that there is some $p > 1$ such that

$$(4.3) \qquad \lim_{n \to \infty} \frac{|x_{n+1} - x^*|}{|x_n - x^*|^p} = M$$

for some $M > 0$. If e_n is the absolute error after step n then this is the same as

$$\lim_{n \to \infty} \frac{e_{n+1}}{e_n^p} = M$$

which means that

$$e_{n+1} \approx M e_n^p$$

which suggests why when $p > 1$ we do not need to require $M < 1$: Because, by assumption, x_n converges to x^*, it follows that the nonnegative sequence e_n tends to zero and so eventually we must have $e_n < 1$ for all remaining n, which insures that $e_n^p < e_n$.

When Eq. (4.3) is satisfied for some $p > 1$ and $M > 0$ we say that the sequence has order of convergence p and that M is the asymptotic error constant (or sometimes **rate** or **ratio**). In the special case $p = 2$ we say that the sequence, or the method that generates it, exhibits **quadratic convergence**. Note that p need not be an integer.

Might Newton's method be convergent of some order $p > 1$? Yes. Newton's method exhibits quadratic convergence, as we will now show. Consider a fixed point iteration $x_{n+1} = g(x_n)$ with the assumptions of the Fixed Point Theorem. Then

$$
\begin{aligned}
x_{n+1} - x^* &= g(x_n) - g(x^*) \\
&= g'(\theta_n)(x_n - x^*)
\end{aligned}
$$

(for some θ_n between x_n and x^*) by the Mean Value Theorem. Hence

$$
\frac{x_{n+1} - x^*}{x_n - x^*} = g'(\theta_n)
$$

and so

$$
\begin{aligned}
\lim_{n\to\infty} \frac{|x_{n+1} - x^*|}{|x_n - x^*|} &= \lim_{n\to\infty} |g'(\theta_n)| \\
&= |g'(x^*)|
\end{aligned}
$$

justifying the previous claim that a fixed point iteration has order of convergence $p = 1$ if $0 < |g'(x^*)| \le \lambda < 1$. What if $g'(x^*) = 0$ (as is the case for Newton's method)? The above argument shows that the fixed point iteration is then superlinear. Suppose that (in addition to the other hypotheses of the Fixed Point Theorem) g is twice continuously differentiable on some interval containing x^*. By Taylor's series with remainder,

$$
\begin{aligned}
g(x_n) &= g(x^*) + g'(x^*)(x_n - x^*) + g''(\xi_n)(x_n - x^*)^2/2 \\
&= g(x^*) + g''(\xi_n)(x_n - x^*)^2/2 \\
&= x^* + g''(\xi_n)(x_n - x^*)^2/2
\end{aligned}
$$

if $g'(x^*)$ is zero, for some ξ_n between x_n and x^*. Hence

$$
\begin{aligned}
g(x_n) - x^* &= g''(\xi_n)(x_n - x^*)^2/2 \\
x_{n+1} - x^* &= \frac{1}{2}g''(\xi_n)(x_n - x^*)^2 \\
\frac{x_{n+1} - x^*}{(x_n - x^*)^2} &= \frac{1}{2}g''(\xi_n)
\end{aligned}
$$

and so

$$
\begin{aligned}
\lim_{n\to\infty} \frac{|x_{n+1} - x^*|}{|x_n - x^*|^2} &= \lim_{n\to\infty} |\frac{1}{2}g''(\xi_n)| \\
&= \frac{1}{2}g''(x^*)
\end{aligned}
$$

by continuity of g and convergence of the method. So if $g''(x^*) \neq 0$ then the method is quadratically convergent, with asymptotic error constant $\frac{1}{2}g''(x^*)$. If $g''(x^*) = 0$ the order of convergence can be even higher; in general, if $g'(x^*)$ through $g^{(p-1)}(x^*)$ are all zero then the order is at least p, and the order is exactly p if $g^{(p)}(x^*) \neq 0$.

Let's apply this Newton's method for a root-finding problem $f(x) = 0$ in which we are seeking an isolated root x^*. We know already that $g'(x^*) = N_f'(x^*)$ is zero, so we consider

$$
\begin{aligned}
g''(x) &= N_f''(x) \\
&= (x - f(x)/f'(x))'' \\
&= (f(x)f''(x)/[f'(x)]^2)' \\
&= \frac{f'(x)f''(x) + f(x)f'''(x)}{[f'(x)]^2} - 2\frac{f(x)[f''(x)]^2}{[f'(x)]^3}
\end{aligned}
$$

if f is thrice continuously differentiable. Evaluating this at $x = x^*$ gives

$$
\begin{aligned}
g''(x^*) &= \frac{f'(x^*)f''(x^*) + f(x^*)f'''(x^*)}{[f'(x^*)]^2} - 2\frac{f(x^*)[f''(x^*)]^2}{[f'(x^*)]^3} \\
&= f''(x^*)/f'(x^*)
\end{aligned}
$$

after using the fact that $f(x^*) = 0$ (and also assuming that the root is simple so that $f'(x^*) \neq 0$).

Summarizing, if f is thrice continuously differentiable in some neighborhood N of a simple zero x^* of f, then Newton's method converges quadratically to x^* for all x_0 sufficiently close to x^*, that is, for some possibly smaller neighborhood of x^* lying within N. In particular, $f'(x^*) \neq 0$ combined with the continuity of $f'(x)$ in the neighborhood means we'll never have a division-by-zero error while attempting to calculate x_{n+1} once we are sufficiently near the root.

It can be shown that if the zero is nonsimple then Newton's method still converges, but only linearly; on the other hand, if $f''(x^*) = 0$ we expect even faster convergence. This explains why Newton's method works poorly to find the zero of the quadratic function $f(x) = x^2$ but is faster-than-expected for finding the solution $x^* = 0$ of $\sin(x) = 0$.

In Problems 8 and 15, methods that converge with order $p = 3$ (**cubic convergence**) generally, not just in special cases, are introduced. But quadratic convergence is usually the kind of convergence that we would usually like to have in a method. How good is quadratic convergence? Typically for Newton's Method, the asymptotic error constant is not greatly larger than one. Take $M = 1$. If the error at step n is 10^{-4}, then

$$
\begin{aligned}
e_{n+1} &\approx e_n^2 \\
&= (10^{-4})^2 \\
&= 10^{-8} \\
e_{n+2} &\approx e_{n+1}^2 \\
&= (10^{-8})^2 \\
&= 10^{-16}
\end{aligned}
$$

and so on. Looked at another way, if x_n is accurate to four decimal places, then x_{n+1} will be accurate to eight decimal places, and x_{n+2} will be accurate to sixteen decimal places; Newton's method roughly doubles the number of correct significant figures in the approximate solution at every iteration (once this limiting behavior begins). If we can get just a little bit of accuracy in our solution, the degree of accuracy will grow quite rapidly. This also means that any small round-off error in x_{n+1} will have negligible effect on the speed of the method as the doubling effect is so much greater than the round-off error could possibly be. Cubically convergent methods are a little bit faster, but are usually sufficiently more complicated to make them not competitive with a quadratically convergent method.

It's worth repeating the key fact: Once the region where the quadratic convergence behavior begins is reached, *Newton's method doubles the number of correct significant figures in the approximation at every iteration*. This applies to the relative error also even though we've been reasoning mostly in terms of the absolute error, and is one of the single most important take-aways from this text in terms of day-to-day scientific computing.

This certainly justifies using $\alpha_n = |x_{n+1} - x_n|$ (from Eq. (4.1)) to approximate the absolute error in x_n, and using $\alpha_n \leq \tau$ or, preferably, $\rho_n \leq \tau$ (for some small tolerance $\tau > 0$) as a convergence criterion for Newton's method. After all, if x_n is good to about 4 decimal places then x_{n+1} should be good to about 8 decimal places, making α_n or ρ_n about a 4 decimal place estimate of the error in x_n. If we are being careful, we should note that these are estimates of the error in the previous iterate x_n and not the most recent iterate x_{n+1}; but of course they're only good estimates if x_{n+1} is a much better estimate of x^* than is x_n. It would be correct in principle, but would usually be considered pedantic in practice, to insist on returning x_n with an error estimate of ρ_n rather than returning the better root x_{n+1} with the presumably overlarge error estimate ρ_n.

Unlike the case of bisection, a small α_n or ρ_n is not a guarantee that we are near a root, since Newton's method can remain near some point for several iterations despite the fact that it is not converging to it. But α_n and ρ_n are certainly good heuristics. We may wish to check that $|x_{n+1} - x_n|$ is less than the tolerance for at least two successive iterations to help guard against the possibility of being misled.
EXAMPLE 2: Consider applying Newton's method to find a root of $x^3 - 1.4x^2 + x - 1.4 = 0$, which must have at least one real root. (A sketch makes it look like there's only one.) The iteration is

$$x_{k+1} = x_k - (x_k^3 - 1.4x_k^2 + x_k - 1.4)/(3x_k^2 - 2.8x_k + 1)$$

and as plotting it suggests there is a root near $x_0 = 1.5$ we will use that as our initial guess. We have:

$$
\begin{aligned}
x_1 &= x_0 - (x_0^3 - 1.4x_0 + x_0 - 1.4)/(3x_0^2 - 2.8x_0 + 1) \\
&\doteq 1.40845070422535 \\
x_2 &\doteq 1.40006688770576 \\
x_3 &\doteq 1.40000000423180 \\
x_4 &\doteq 1.40000000000000
\end{aligned}
$$

and in fact 1.4 is the sole real root. The estimated and actual absolute errors are:

$$\begin{aligned}
\alpha_1 &= |x_2 - x_1| \\
&\doteq 0.00838381651959 \text{ (Actual: 0.00845070422535)} \\
\alpha_2 &= |x_3 - x_2| \\
&\doteq 6.688347396521799E - 5 \text{ (Actual: 6.688770576102065E - 5)} \\
\alpha_3 &= |x_4 - x_3| \\
&\doteq 4.231795580622588E - 9 \text{ (Actual: 4.231795802667193E - 9)}
\end{aligned}$$

and as expected Eq. (4.1) gives an excellent estimate of the absolute error; had we set a tolerance of $\tau = 10^{-4}$ then we could have stopped with x_3 since α_2 is already less than this τ. Note that

$$\begin{aligned}
\alpha_2/\alpha_1^2 &\doteq 0.9366 \\
\alpha_3/\alpha_2^2 &\doteq 0.9459
\end{aligned}$$

which suggests that the asymptotic error constant is roughly $M = .94$. For a method with order of convergence one this would have meant very slow convergence, but since Newton's method is quadratic we still have excellent convergence; $\alpha_1 \doteq 8E-3$, $\alpha_2 \doteq 7E - 5$, $\alpha_3 \doteq 4E - 9$, showing the roughly doubling number of correct significant figures at each iteration. $\square$

Although convergence of Newton's method is only guaranteed if we are sufficiently near a zero, in practice it is not uncommon for it to recover from a poor initial guess and converge anyway (by finding its way into the neighborhood of the zero where convergence is assured, sometimes in an apparently haphazard way). The method is notoriously robust in this way. Quadratic convergence is only achieved near the root in any event.

Newton's method can be viewed as an algorithm designed to achieve sure quadratic convergence for a root-finding problem rather than the linear, assuming the derivative is small enough, convergence properties of naive fixed point iteration. But sometimes we have no choice but to use a certain method, perhaps because it is already implemented in software and we haven't the time or ability to change it. In that case we may take the output of that program and apply an acceleration technique to improve the rapidity of convergence of the sequence. There are a great many such techniques, but we will only mention one: **Aitken's Δ^2 process** (or Aitken acceleration or extrapolation). If $x_k \to x^*$ *linearly* then it is a fact that the associated sequence

$$a_k = x_k - \frac{(x_{k+1} - x_k)^2}{x_{k+2} - 2x_{k+1} + x_k}$$

converges more rapidly to x^*, in the sense that $a_k \to x^*$ and

$$(4.4) \qquad \lim_{k \to \infty} \frac{a_k - x^*}{x_k - x^*} = 0$$

in this case (and in certain other cases). If the cost to produce x_k is large compared to the handful of arithmetic operations involved in computing a_k, or if a second processor is available to perform the a_k calculations in parallel (albeit delayed by two steps, as computing a_k requires x_{k+2}), this may be useful. In fact, this technique is often built directly into a method or program that is known to use a primary algorithm that produces linearly convergent results.

The name comes from the use of the **forward difference operator** Δ of the difference calculus, which performs the forward differencing operation

$$\Delta c_n = c_{n+1} - c_n$$

on a sequence. The action of Δ^2 is defined in the usual way in terms of Δ acting on Δc_n, i.e.,

$$\begin{aligned} \Delta^2 c_n &= \Delta(\Delta c_n) \\ &= c_{n+2} - 2c_{n+1} - c_n \end{aligned}$$

so that the accelerated sequence can be written compactly using difference calculus notation as $a_k = x_k - (\Delta x_k)^2 / \Delta^2 x_k$.

EXAMPLE 3: The sequence $x_k = 1/k$ converges to zero but it doesn't technically converge linearly because we would have $M = 1$ in Eq. (4.2). Let's try Aitken's Δ^2 process anyway. We have

$$\begin{aligned} a_k &= \frac{1}{k} - \frac{\left(\frac{1}{k+1} - \frac{1}{k}\right)^2}{\frac{1}{k+2} - \frac{2}{k+1} + \frac{1}{k}} \\ &= \frac{1}{2(k+1)} \end{aligned}$$

($k \geq 1$) and indeed

$$\frac{1}{2(k+1)} < \frac{1}{k}$$

so that a_k is less than x_k for all $k \geq 1$. In fact

$$\frac{1}{2(k+1)} < \frac{1}{k+2}$$

so that a_k is less than x_{k+2} for all $k \geq 1$, which–though not, strictly speaking, what Aitken acceleration guarantees (in the limit, a_k is better than x_k per Eq. (4.4))–is a more fair comparison, since a_k and x_{k+2} are the estimates that would be available at the same step. Of course it will rarely be possible to achieve such simplification of the formula for a_k. □

MATLAB

The time has come to learn how to write programs in MATLAB. You may use any text editor, or MATLAB's `edit` command (which launches the MATLAB editor). Using the MATLAB editor is certainly easiest unless you have a strong preference for some other editor. If you are using Windows, be sure that the editor you use saves your program as a plain text file.

Let's start with a very simple program that performs one step from the inverse linear interpolation method: Finding the x-intercept. Our program will input two pairs of points (x_0, y_0), (x_1, y_1) and return the x-intercept of the interpolated line. Type the following program in your editor:

```
function x=intercept(x0,y0,x1,y1)
```

```
%INTERCEPT   Return x-intercept of line from (x0,y0) to (x1,y1).
x=x1-y1*(x1-x0)/(y1-y0);
```

Save this as `intercept.m`. Now enter:
```
>> intercept(-1,2,3,1)
```
The x-intercept of the line connecting $(-1, 2)$ and $(3, 1)$ is produced. If you receive the error message:
```
???  Undefined function or variable intercept.
```
type `path` and make sure that you stored your file in one of the listed directories and that the file extension is .m.

Let's look at this simple program. The use of `function` in the first line lets MATLAB know that this is a function. The syntax is `function <output argument(s)>=<function_name>(<input argument(s)>)`. User-defined MATLAB functions are standard text files with the .m extension.

Anything following the percentage sign % is considered to be a comment. Type `help intercept`; MATLAB displays the first comment block of this program. Your program has been assimilated by the `help` command. Following the comment line are the MATLAB commands that compute the output argument. Notice that no end-of-program command is needed.

Let's add an error check. We need x0 and x1 to be distinct, y0 and y1 to be distinct, and y0*y1 negative. Edit your program to read as follows:

```
function x=intercept(x0,y0,x1,y1)
%INTERCEPT   Return x-intercept of line from (x0,y0) to (x1,y1).
if (x1==x0), error('Invalid x data'),end
if (y1==y0), error('No intercept'),end
if (y1*y0>=0), error('Not a bracket'),end
x=x1-y1*(x1-x0)/(y1-y0);
```

The first `if-end` pair (see `help if` for more details) prints an error message if the x-values are equal; MATLAB uses a double equals sign == to test for equality. The second `if-end` pair prints an error message if the y-values are equal. (In both of these cases it would be better to test for near equality, not just exact equality, of these values, as we'll discuss in a later section.) The third `if-end` pair prints an error message if the y-values are not of opposite sign; the combination of symbols >= tests if $y_1 y_0 \geq 0$. See `help ops` for more information on the relational operators. The `error` command prints the indicated error message and terminates execution of the program. (It's usually better to end the program more gracefully than this when a problem is detected, so use the `error` command sparingly. To simply print a message or variable use the `disp` command, e.g., `disp('hello world')` or `disp(x)`.) Choose data to make each message appear and check that your program behaves as intended.

This program may now be used like any other MATLAB function, which is an important aspect of MATLAB programming. For example,
```
>> z=cos(intercept(-1,log(1/2),3,log(5)))^2
```
is perfectly valid. This is a big deal: MATLAB does not distinguish between your programs and its own. In fact, many MATLAB-provided functions are just MATLAB programs themselves. Enter:
```
>> type fzero
```

to see the code for the MATLAB root-finding command. (Some functions are built-in and can't be typed.) Looking at the code of such functions is a good way to learn MATLAB programming techniques and tricks.

If you edit a program (m-file) after you have run it, you may need to enter the command `clear functions` in order to see your changes take effect. Although MATLAB programs are interpreted, not compiled, when you first run a program it pre-compiles that program for future use in that session. The command `clear functions` clears all pre-compiled functions so that the next time a program is run it will be checked (and pre-compiled) again. If you are editing a program and your changes don't seem to be having any effect, try clearing all functions.

The `if`-`end` pairs can be spread out over several lines, as can a `for`-`end` loop. (There is also a `while` command for use in a `while`-`end` loop.) That is, we may rewrite our function as:

```
function x=intercept(x0,y0,x1,y1)
%INTERCEPT   Return x-intercept of line from (x0,y0) to (x1,y1).
if (x1==x0)
      error('Invalid x data')
end
if (y1==y0)
      error('No intercept')
end
if (y1*y0>=0)
      error('Not a bracket')
end
x=x1-y1*(x1-x0)/(y1-y0);
```

Make the changes above and re-test your program. In many cases formatting the constructs like this makes the program more readable. The MATLAB editor will suggest such formatting automatically. Here's another example MATLAB program:

```
function x=secant(f,x0,x1,N)
%SECANT   Perform N iterations of the secant method on
%             the function f, starting from x0, x1.
if (N<1), error('N must be at least 1'), end
   for i=1:N,
      y0=feval(f,x0);y1=feval(f,x1);
      xnew=intercept(x0,y0,x1,y1);
      x0=x1;x1=xnew;
   end
   x=xnew;
```

Most of the error checking is performed in `intercept.m` and in fact the bracket check must be removed from it for the secant method code to work properly. You can comment out those three lines by simply placing a percent sign at the start of each line. See `help colon` for information on what will happen if N is not an integer; see `help punct` for information on other punctuation symbols. Enter the program, save it, and test it using `@sin` for the function, $x_0 = .1$, $x_1 = .2$, and $N = 2$, then $N = 3$, and so on. You should see very rapid convergence. This is unsurprising since the secant method uses a linear approximation and the sine function is well-approximated by a line near the origin as the quadratic term in its Taylor series is zero there.

Somewhere around $N = 7$ you should get the `'Invalid x data'` error from `intercept.m`. The current x-values in `secant.m` will have gotten so close together that, in fact, they are now numerically equal. This wouldn't happen in real arithmetic, but does happen on computers.

But again, we shouldn't be testing for exact equality on finite precision machines in the first place. For a cautionary note in this regard, enter:
```
>> 1-.8
>> 1-.8==.2
>> .2-(1-.8)
```
The test for equality fails due to the effects of finite precision arithmetic. Part of the reason is that, although the rational number .2 has a terminating expansion in base 10, it has a non-terminating expansion in base 2, and numbers are stored and manipulated in base 2 by the machine. (As a comparison, ask whether $1 - .75$ is equal to .25.) We'll discuss these issues in more depth throughout the text, but for now we note that a test for equality of quantities x_1, x_2 which are not known to be integers should usually be implemented in the form $|x_1 - x_2| < \tau$ for some small positive tolerance τ (possibly dependent on the size of x_1 and/or x_2).

The MATLAB package has some very sophisticated capabilities that can be augmented by optional specialized toolboxes for certain applications (e.g., signal or image processing). We won't get to all of the features of MATLAB in this text; learning MATLAB is important and useful for scientific computing but it is still secondary to the goal of learning numerical analysis methods and scientific computing principles that could be implemented in any programming language and will allow you to be a more knowledgeable user of a wide variety of numerical packages. Enter:
```
>> demo 'matlab'
```
to explore some of MATLAB's basic operations, and just `demo` to see a tour of some of its more sophisticated capabilities. The `help` command may be used to learn more about them. These open the MATLAB browser, but you can copy-and-paste the commands in to duplicate the demonstrations for yourself.

Problems

1.) Show that any sequence that satisfies Eq. (4.3) with $p > 1$ for some $M > 0$ is superlinearly convergent.

2.) Create a table that compares the errors for methods that are convergent of orders $p = 1, 2, 3$. Assume $M = .5$ and $e_0 = .5$ in each case and use $e_{n+1} \approx M e_n^p$. How much better is an order 3 method than an order 2 method?

3.) Show that if $x^* = g(x^*)$ and $g'(x^*) = g''(x^*) = 0$ then the corresponding fixed point iteration is convergent of order at least 3.

4.) Verify experimentally that the secant method has order of convergence $(1 + \sqrt{5})/2 \doteq 1.618$ (that is, that roughly 60% more accurate significant figures are gained at each iteration in the limit).

5.) A function $f(x)$ is said to have a zero of multiplicity m at x^* if $f(x^*) = f'(x^*) = \cdots = f^{(m-1)}(x^*) = 0$ but $f^{(m)}(x^*) \neq 0$.

a.) Show that if $f(x)$ has a zero x^* of multiplicity m then $f(x) = (x - x^*)^m h(x)$ where $h(x^*)$ is nonzero.

b.) Show that if $m > 1$ then Newton's method converges linearly, and give the asymptotic error constant.

c.) Show that if $m > 1$ then the method $x_{k+1} = x_k - mf(x_k)/f'(x_k)$ converges quadratically. (This is one of many methods known as **Schröder's method**.)

d.) Show that $f(x)/f'(x)$ has only simple roots. Write the Newton's method iteration for $f(x)/f'(x)$; what additional assumptions on f are needed?

6.) Show by example that a sequence can be superlinear but still not converge of order p for any $p > 1$.

7.) Show that the sequence $x_k = 2^{-2^k}$ converges to zero with order of convergence 2.

8.) Householder's method $x_{k+1} = x_k - \frac{f(x_k)}{f'(x_k)}\left(1 + \frac{f(x_k)f''(x_k)}{2[f'(x_k)]^2}\right)$ for $f(x) = 0$ has order of convergence 3 (cubic convergence) under appropriate assumptions on $f(x)$. Apply it to the functions $f(x) = 5x^7 + 2x - 1$ and $g(x) = 1/x^3 - 10$; compare it to Newton's method for these functions.

9.) a.) Show that $c_k = 1/k^2$ converges linearly to zero.

b.) Show that Aitken's Δ^2 process accelerates the convergence of this sequence to zero.

10.) If $x_k \to x^*$ and $\lim_{n\to\infty} \frac{|x_{k+1}-x^*|}{|x_k-x^*|} = 1$ then we say that the convergence is **sublinear**. Show that $x_k = 1/k$ converges sublinearly to zero.

11.) Justify the statement that if the relative error ρ in an approximation is about 10^{-k} then the approximation is correct to about k decimal places.

12.) a.) What is the rate (not order) of convergence of $(\cos(1/n) - 1)/(1/n)$ to zero as $n \to \infty$?

b.) What is the rate of convergence of $2/n^3$ to zero as $n \to \infty$? What is its order of convergence?

13.) a.) Assume that the secant method has a well-defined order of convergence, and that there is an α such that $|x_{k+1} - x^*| \approx \alpha|x_k - x^*||x_{k-1} - x^*|$ for large k. Reason that the secant method must have order of convergence $(1 + \sqrt{5})/2$. (The secant method is an example of a two-point iteration, dependent on two previous values, and so is more difficult to analyze than the one-point iterations we have been considering.) This is in fact its convergence order.

b.) Prove that if $f(x)$ is quadratic then the secant method has order of convergence $(1 + \sqrt{5})/2$.

14.) a.) Use Newton's method to find all real roots of $x^3 - 5.06x^2 - 10.392x + 46.2 = 0$ to within an absolute error of no more than 10^{-6}, as estimated by Eq. (4.1).

b.) Use Newton's method to find all real roots of $x^3 - 5.06x^2 - 10.392x + 46.2 = 0$ to within a relative error of no more than 10^{-6}, as estimated by Eq. (4.1).

c.) Use Newton's method to find all real roots of $x^3 - 1100x^2 + 100220x - 20000 = 0$ to within an absolute error of no more than 10^{-6}, as estimated by Eq. (4.1).

d.) Use Newton's method to find all real roots of $x^3 - 1100x^2 + 100220x - 20000 = 0$ to within a relative error of no more than 10^{-6}, as estimated by Eq. (4.1).

15.) a.) **Halley's method** is $x_{k+1} = H_f(x_k) = x_k - \frac{2f(x_k)f'(x_k)}{2(f'(x_k))^2 - f(x_k)f''(x_k)}$. Show that Halley's method has order of convergence 3 for simple zeroes by showing that $H'_f(x)$ and $H''_f(x)$ vanish at the zero.

b.) Apply Halley's method to $f(x) = 5x^7 + 2x - 1$ and $g(x) = 1/x^3 - 10$.

16.) The **Lanchester (combat) model** for two opposing military forces is, in its simplest form, given by $x' = -ay$, $y' = -bx$ where $x(t)$ and $y(t)$ are the sizes of the forces involved (numbers of combatants, or of tanks, or of planes, etc.). The constants a and b measure the fighting effectiveness of the forces. The Lanchester model and variations of it are for wargaming simulations.

a.) Use the chain rule to find a formula for dy/dx and then solve it to show that $a(y^2 - y_0^2) = b(x^2 - x_0^2)$ for all $t \geq 0$. This is the **Lanchester square-law model**.

b.) Show that $d^2y/dt^2 - aby = 0$ and use it to solve for $y(t) = y_0 \cosh(\sqrt{abt}) - x_0\sqrt{b/a}\sinh(\sqrt{abt})$. Do the same for $x(t)$.

c.) Take $a = .15$, $b = .1$. For the initial conditions $(x_0, y_0) = (10000, 9000)$, $(x_0, y_0) = (10000, 8166)$, and $(x_0, y_0) = (10000, 5000)$, use Newton's method to determine the time, if any, at which the opposing forces are of equal strengths. What are those strengths? How long does each battle last and who wins?

d.) Take $a = .144$, $b = .1$. For the initial conditions $(x_0, y_0) = (12000, 10000)$, use Newton's method to determine the time, if any, at which the opposing forces are of equal strengths. How long does the battle last and who wins?

5. Modifications of Newton's Method

Although we could construct a method with order of convergence higher than that of Newton's method, it's usually pointless to do so. Once Newton's method gets a few decimal places of accuracy it will only take a small number of additional iterations to get all the accuracy one might need. Even if the asymptotic error constant M is large, Newton's method is still as fast as one is likely to need a method to be to solve nonlinear equations in one variable in a practical situation. Of course, it could take a long time for the method to get this close to the solution, but Newton's method is an extraordinarily robust root-finder.

However, note that bisection, inverse linear interpolation, and the secant method each require us to compute a single new value of $f(x)$ each iteration, whereas Newton's method requires both a new value of $f(x)$ and a new value of $f'(x)$. Typically the computation of $f(x)$ (and $f'(x)$) is the most time-consuming step, by far, of any of these methods, and the time required for the other arithmetic operations performed in combining those values is negligible. If this is not the case–if $f(x)$ is cheap (in terms of computation time) to evaluate–virtually any method will be acceptable. But is not uncommon for values of $f(x)$ to be computed by *another* program, that could easily take as much as several minutes to solve an ODE or PDE numerically or to perform a simulation in order to produce a single value of $f(x)$. When $f(x)$ is expensive to compute, Newton's method might be less attractive than a method that uses more iterations but for which the total number of function evaluations required to achieve a desired tolerance is smaller overall. Focusing only on number of iterations, as we have thus far, can be misleading! A method that converges of order 3 or higher will typically require higher-order derivatives and will use more function evaluations per iteration. This means that such a method often is not actually an improvement over Newton's method in terms of run-time, even though it would use fewer iterations. Additionally such methods are usually more difficult to code.

Newton's method has other advantages. One which we will *not* pursue here that if x_0 is complex then Newton's method can converge to complex roots of $f(x) = 0$, which can be useful when such roots are desired. One of the biggest advantages of Newton's method is that it generalizes in a natural way to nonlinear *systems* of equations. If we must solve a nonlinear system such as

$$f(x,y) = 0$$
$$g(x,y) = 0$$

for a simultaneous root (x^*, y^*) of f and g, then attempting to bracket it is going to be quite difficult, but Newton's method (using partial derivatives with respect to x and y) works fine. While there are many choices for a one-dimensional (that is, single variable) zero-finding algorithm, in several dimensions a nonlinear root-finding problem is typically solved by some variant of Newton's method. For this reason we will now discuss several such variants. You should bear in mind that the real need for, and advantages of, these variants are most apparent in the several variables case (discussed later in the text).

We will address three types of difficulties with Newton's method in this section:

- It may not be possible to find a formula for $f'(x)$;
- We may be able to find a formula for $f'(x)$ but it may be too expensive to compute;
- Newton's method may fail to converge.

Let's start with the case where finding a formula for $f'(x)$ is not feasible. This may be due to the complicated nature of the formula for $f(x)$; it may be due to the formula being supplied by the user as input in a computing environment where symbolic differentiation is not available; it may be because $f(x)$ is measured data or is computed by another program that does not supply derivatives. The last case is very common; perhaps we have been asked where the solution of a certain differential equation passes through the origin, for example. Then every value of $f(x)$ required by our root-finding algorithm involves a call to a numerical differential equations solver that produces only values of $f(x)$, not a formula for it. We do not have access to the derivatives unless the ODE solver can optionally return them to us[6].

The obvious recourse in such a case is to use the secant method. The convergence order of the secant method is $p \doteq 1.6$ so convergence will still be quite rapid. We can request only a single initial guess x_0 from the user and use $x_1 = (1 + \delta)x_0$ (with some small δ) for the second guess in order to simulate the single-initial-guess nature of Newton's method.

An alternative to the secant method is to approximate the derivative $f'(x_k)$ by some means. One possibility is to use the approximation

$$(5.1) \qquad f'(x_k) \approx \frac{f(x_k + h) - f(x_k)}{h}$$

for some small $h > 0$. This is said to be a **finite difference** approximation to $f'(x_k)$ and if we use it in Newton's method the resulting scheme is sometimes called the **finite difference Newton's method** (or **Newton's method with finite differences**). Note that instead of computing $f(x)$ and $f'(x)$ at each iteration we

---

[6]When writing code that might be used as an objective function $f(x)$ by some other program, it's good practice to have your code optionally return $f'(x)$ and even $f''(x)$ if requested by the user (and if reasonable to do so). An ODE solver likely generates derivative estimates internally, for example, and can easily return those.

compute $f(x)$ for two different values of x, that will presumably be comparable in computational cost to computing $f(x)$ and $f'(x)$.

More generally, we can replace the term $f'(x_k)$ in Newton's method by any approximate slope s_k, giving the method

$$x_{k+1} = x_k - f(x_k)/s_k.$$

which is called a **quasi-Newton method**. The secant method approximates the slope using

$$s_k = \frac{f(x_k) - f(x_{k-1})}{x_k - x_{k-1}}$$

and the finite difference Newton's method uses

$$s_k = \frac{f(x_k + h) - f(x_k)}{h}.$$

(More generally we could let $h = h_k$ in the finite difference Newton's method, ideally forcing $h_k \to 0$ so that $s_k \to f'(x^*)$ as $k \to \infty$. Note how the secant method achieves this automatically when it is convergent.) Another possibility is the **constant-slope Newton's method**, which uses

$$s_k = c$$

for some constant c. In the problems you will be asked to show that, for appropriate choices of c, this gives linear convergence; the closer c is to $f'(x^*)$, the more rapid the convergence near the zero. There are much more sophisticated quasi-Newton methods for the case of root-finding in several dimensions.

What if it is possible to find a formula for (or otherwise compute values of) $f'(x)$ but the computations are so time-consuming that it isn't practical to do so every iteration? We know from the calculus that when the product rule must be used to take a derivative, the formula for it can quickly grow larger and hence more expensive to compute ("ugly functions tend to have uglier derivatives"). Once again we can use a quasi-Newton method such as the secant method, the finite difference Newton's method, or the constant-slope Newton's method as though $f'(x)$ were not available at all. The difference in this case, though, is that it may be reasonable to compute values of $f'(x)$ every several iterations. A common approach is to use the constant-slope Newton's method with c initially taken to be $f'(x_0)$, then after two to five iterations recompute $c = f'(x_k)$ and use this value for the next two to five iterations, and so on. This approach attempts to strike a compromise between the benefits of having an accurate slope and the cost of computing it.

Let's us look at the final difficulty mentioned previously, namely, failure of Newton's method to converge. One reason this can happen is because Newton's method is taking steps that are too large and so it overshoots the root. For example, in using Newton's method on

$$f(x) = 1/x - 7$$

to approximate 1/7, we find

$$
\begin{aligned}
N_f(x) &= x - (1/x - 7)/(-1/x^2) \\
&= x + x - 7x^2 \\
&= 2x - 7x^2
\end{aligned}
$$

and choosing $x_0 = 1$ gives

$$
\begin{aligned}
x_1 &= -5 \\
x_2 &= -185 \\
x_3 &= -239945
\end{aligned}
$$

and the method is clearly failing. We know that if we start sufficiently close to the answer we'll get the answer, but as we generally don't know the answer this fact may not be helpful. (The difficulty of generating a good initial guess so as to get convergence is a common complaint in practice; most strategies for addressing it are heuristic, often on a case-by-case basis, but a more soundly based approach will be discussed in Sec. 7.2.) The **damped Newton's method** tries to "slow down" Newton's method by using a damping factor μ_k to reduce the size of the step taken from the current point. That is, we use

$$(5.2) \qquad\qquad x_{k+1} = x_k - \mu_k f(x_k)/f'(x_k)$$

where $0 < \mu_k \leq 1$ is a constant (for each k), and $\mu_k \to 1$ as $k \to \infty$ (so that the method is simply Newton's method near the root). In the current case we have $x_0 = 1$ and

$$
\begin{aligned}
N_f(x_k) &= x_k - \mu_k \left(1/x_k - 7\right) / \left(-1/x_k^2\right) \\
&= x_k - \mu_k \left(7x_k^2 - x_k\right)
\end{aligned}
$$

for some choice of damping sequence μ_k. Less commonly we may wish to take $\mu_k > 1$ to take stretched, rather than damped, steps in hopes of finding a zero far away. We would still need $\mu_k \to 1$ in the limit in order to achieve convergence.

How do we choose the damping sequence? We generally want $|f(x_{k+1})| < |f(x_k)|$. If the method is converging then eventually the usual Newton's method behavior should kick in and we will be able to take $\mu_k = 1$ for k sufficiently large. In practice we often must experiment until a suitable choice of damping sequence is found. (In rare circumstances knowledge of the shape of $y = f(x)$, or at least the scale on which it varies, may guide us.) If we only have to solve this problem once that is probably wasteful; but if this is to be used to implement division on a high-performance computer, this is a design problem and time spent on it may be time well-invested. (Actually, for the division problem there is a rule that always generates a suitable initial guess.) It is common in practice to know that we will frequently be solving problems from a narrowly defined class of very similar problems. A heuristic such as a damping rule that is found for one such problem often works well for the others in that class; in such a case the time spent experimenting is well invested. For example, this might be a small part of a numerical wind tunnel package for aircraft design, and every small proposed change in the design of the aircraft's wing shape results in a different but similar problem. In such a case a heuristically determined damping rule will often be good for a wide range of cases. This is a situation where we may have to solve the problem to figure out how to solve the problem...but then also other problems very close to it.

EXAMPLE 1: Let's reconsider the problem of using Newton's method on $f(x) = 1/x - 7$ to approximate $1/7$. We'll use the same $x_0 = 1$ but we'll use the damping sequence (found by trial-and-error) $\mu_0 = .1, \mu_1 = .2, \mu_2 = .3,...$ with $\mu_k = 1$ for $k \geq 9$. This gives:

$$
\begin{aligned}
x_1 &= x_0 - \mu_0(7x_0^2 - x_0) \\
&= 1 - .1(7(1^2) - 1) \\
&= 0.4000 \\
x_2 &= 0.2560 \\
x_3 &= 0.1952 \\
x_4 &= 0.1666 \\
x_5 &= 0.1528
\end{aligned}
$$

and the method certainly appears to be converging to $1/7 \doteq 0.1429$; in fact, after the first few iterations the damping is no longer helping us but rather is slowing us down. Nonetheless damping the size of the step taken by Newton's method has had a beneficial effect: It has turned a divergent sequence into a convergent sequence. □

Damping is an important strategy in a number of numerical methods, though the damped Newton's method is of much greater interest when dealing with problems in several variables. In the multivariable case a parameter such as μ_k in Eq. (5.2) is very commonly used in Newton's method, either to provide damped steps or to ensure that early steps are not too small.

Another way to address the possibility that Newton's method may diverge is to use a *mixed method* that combines Newton's method with a bracketing method. Suppose we were using the bisection method to solve a problem because of its guaranteed convergence; after we found the zero to the desired tolerance, it would be sensible to do several Newton's method iterations on it. If the bracket is small enough then we are almost certainly in the range where Newton's method must converge and will double the number of correct figures at each iteration; since any number in the bracket is an estimate good to within the tolerance, we could use Newton's method and take its result as our best estimate if it stays within the bracket, and simply use the midpoint of the bracket if Newton's method should fail.

EXAMPLE 2: Consider finding the positive root of $x^6 - 10 = 0$. A plot suggests that the root is at roughly 1.5, and that $[1.4, 1.5]$ is a bracket. Indeed, if $f(x) = x^6 - 10$ then $f(1.4) = 1.4^6 - 10 = -2.4705$, $f(1.5) = 1.5^6 - 10 = 1.3906$. Two iterations of bisection gives $f(1.45) = 1.45^6 - 10 = -.7059$ (new bracket is $[1.45, 1.5]$), $f(1.475) = 1.475^6 - 10 = .2980$ (new bracket is $[1.45, 1.475]$). Now, Newton's method for this function is $x_{k+1} = x_k - (x_k^6 - 10))/(6x_k^5) = \frac{5}{6}x_k + \frac{5}{3}x_k^{-5}$. Taking $x_0 = 1.4625$ gives

$$x_1 = \frac{5}{6}x_0 + \frac{5}{3}x_0^{-5}$$
$$= 1.4678$$
$$x_2 = 1.4678$$

and further iterates do not change this value. (This is the fixed point property.) Since it is in the bracket, we take it as the best estimate of the location of the root; the midpoint 1.4625 has a smaller worst-case error, but the value from Newton's method is almost certainly a better estimate. □

We can do better than simply improving an estimate made by the slow-but-steady bisection method: We can actually combine Newton's method with bisection from the start. There are many ways to do this, sometimes called **modified**, **safeguarded**, or **globalized Newton's methods**. We will describe only one such approach. Suppose we have an initial bracket $[x_0, x_1]$ of a zero of a differentiable function f. Rather than use bisection or inverse linear interpolation, we might compute the Newton's method approximation of the zero

$$x_{new} = x_1 - f(x_1)/f'(x_1)$$

using either x_0 or x_1 as the initial point. If x_{new} is in the bracket, we use either $[x_0, x_{new}]$ or $[x_{new}, x_1]$, whichever is a bracket, as our new bracket; if x_{new} does not lie in $[x_0, x_1]$, we discard it and take a bisection step. Hence at each iteration we either take a Newton step–which should give excellent results in the limit–or a bisection step, which gives at least a 50% reduction in the bracket size. Typically most steps are Newton steps after the first few iterations. Notice that if the approach to the zero is one-sided the bracket width may not go to zero, so we must use additional convergence criteria (say, that successive values of x_{new} are very close together).

EXAMPLE 3: Consider again the problem of finding the positive root of $x^6 - 10 = 0$, for which $[1.4, 1.5]$ is a bracket. We'll use the globalized Newton's method. From Example 2, Newton's method for this function is $x_{k+1} = \frac{5}{6}x_k + \frac{5}{3}x_k^{-5}$. Taking $x_0 = 1.5$, the first Newton's step is $x_1 = \frac{5}{6}x_0 + \frac{5}{3}x_0^{-5} = 1.4695$ which is in the bracket; since $f(1.4695) \doteq 0.0697$, the new bracket is $[1.4, 1.4695]$. The next step is $x_2 = \frac{5}{6}x_1 + \frac{5}{3}x_1^{-5} = 1.4678$. (We could have used 1.4 as the starting point if we preferred, but 1.4695 is almost certainly a better estimate as it was found by Newton's method.) This is in the bracket, and $f(1.4678) \doteq 2.0084E - 4$ so the new bracket is $[1.4, 1.4678]$. We are converging rapidly to the root, but our bracket is not shrinking swiftly because our approach is one-sided. This is not a problem as long as width of the bracket is not our only convergence criterion. □

In the worst possible case, this globalized method repeatedly wastes time computing x_{new} by Newton's method only to take a bisection step, but even in that case we will be slowly converging to a zero of the function. In the best possible case it is essentially the same as Newton's method with the added guarantee of a bracket. This is certainly superior to damping but unlike damping it requires an initial bracket and, as brackets are only viable for one-dimensional problems, it cannot generalize to several variables.

We've looked at a number of variants of Newton's method in this section: The quasi-Newton methods (finite differences, secant, constant-slope) for handling the derivative; the damped Newton's method for preventing divergence; and the globalized Newton's method for insuring convergence. They highlight the difficulties that are encountered in working with the function f and the difficulty of finding a sufficiently good initial guess. Each has its place in practice, and each represents an idea that can be applied to other numerical methods as well.

MATLAB

All of our results assume infinite-precision arithmetic, but the methods will be implemented on finite precision machines. There are analytical methods for studying the effects of this which we must and will discuss[7] but for now we'll only investigate a danger lurking in the secant method and the finite difference Newton's method. Consider the latter algorithm, written as a quasi-Newton method $x_{k+1} = x_k - f(x_k)/s_k$ where $s_k = \frac{f(x_k+h)-f(x_k)}{h}$ for some small $h > 0$ (from Eq. (5.1)). Let's try approximating the derivative of $f(x) = \ln(x)$ at $x = 1$ in this way; of course, $f'(x) = 1/x$ so $f'(1) = 1$. Enter:
```
>> x=1;h=.1;
>> (log(x+h)-log(x))/h
```
We get 0.9531, which is not unreasonable for this size h. Enter:
```
>> h=10.^(-(1:20))
```
to create a vector of the values $10^{-1}, 10^{-2}, ..., 10^{-20}$. (Note that in forming the vector 1:1:20 the step size may be omitted if it is 1.) Enter:
```
>> (log(x+h)-log(x))./h
```
where the period ensures that the division is done entry-by-entry. The MATLAB response is:
```
ans =
 Columns 1 through 7
 0.9531 0.9950 0.9995 1.0000 1.0000 1.0000 1.0000
 Columns 8 through 14
 1.0000 1.0000 1.0000 1.0000 1.0001 0.9992 0.9992
 Columns 15 through 20
 1.1102 0 0 0 0 0
```
These are the estimates of the derivative for the corresponding values of h. We see that the error decreases for a while, as expected, but starting from about $h = 10^{-12}$ it begins growing again.

We'll discuss finite differences further in Sec. 6.1. For now we point out that there are two things going wrong in the above calculation when h is small: Subtraction of nearly equal numbers in the numerator, and division by a small number. These cause problems due to the fact that a computer represents numbers with only finitely many bits[8]. Enter:
```
>> format long
>> pi
ans =
 3.141592653589793
```

---

[7]Some authors *define* numerical analysis as the study of round-off errors; it's certainly a crucial part of the subject.

[8]In the widely used IEEE standard arithmetic, 32 bits are used in single precision (1 sign bit, 8 exponent, 23 mantissa), and 64 bits are used in double precision (1 sign bit, 11 exponent, 52 mantissa).

By default, MATLAB works in double precision, meaning about 16 digits accuracy is available. (The exponent is handled separately so the same number of digits is available over many orders of magnitude; see Sec. 1.7.) In particular, there is a largest and a smallest positive real number on the machine. Enter:

```
>> realmax
>> realmin
```

to see them[9]. Anything larger than the result of realmax is set to Inf, a special variable for "machine infinity" (we say the operation **overflows**) and anything too small **underflows**. The Inf variable behaves as expected; enter:

```
>> 1/0
>> exp(1000)
>> inf*2
>> inf-inf
```

The result of the last operation is NaN ("not a number") and is another special variable produced under certain circumstances. See help NaN for further information.

There are necessarily only finitely many machine numbers, which must we must use to model the entire real line. Enter:

```
>> eps
>> 1+eps
>> ans==1               %Does this equal 1?
>> 1+eps/2
>> ans==1
>> (1+eps/2)+eps/2
>> ans==1
>> 1+(eps/2+eps/2)      %Computer addition is not associative.
>> ans==1
```

The value eps is the distance from 1 to the next largest machine number and may vary from machine to machine, and in different precision settings. The distance between machine numbers is not constant; the gap between numbers is larger when the numbers are larger.

The value of eps is important as it determines what a small error is on the particular machine. The effects of eps can be seen; enter:

```
» x=1;for i=1:100;x=x+eps;end;x
» x=1;for i=1:100;x=x+eps/2;end;x
```

Since the spacing between adjacent machine numbers is smaller when the numbers are smaller and larger when the numbers are larger you will get different results if you try the above with a different initial x; try it! You can enter eps(x) to find the distance from x to the next larger floating point number.

Even evaluating a function can be subject to error. Given a function $f(x)$, we may wish to evaluate it at some point x_0. On the machine it may well be that x_0 cannot be represented exactly, however, and so we actually will evaluate $f(x_0 + \epsilon)$ for some small ϵ (not necessarily the machine epsilon ϵ_M). Assuming that f is a differentiable function which can be evaluated exactly, we expect to make an error $|f(x_0 + \epsilon) - f(x_0)| = \epsilon |f'(\zeta)|$ for some ζ lying between x_0 and $x_0 + \epsilon$ (by the Mean Value Theorem). Note that the larger the derivative (rate of change) of f, the larger the expected error. For example, we expect that evaluating $\sin(x)$ gives

---

[9]Actually realmin is the smallest *normalized* floating point number; it is not the smallest number representable. See Sec. 1.7.

more accurate results for x near $\pi/2$ (where the derivative is zero) than for x near 0, where the derivative is 1. Enter:

```
>> err=1E-10;
>> x1=.1;x2=pi/2;
>> abs(sin(x1+err)-sin(x1))
>> abs(sin(x2+err)-sin(x2))
```

As expected, the error is smaller near the point where the derivative of the function is smaller. Since the function's values are changing less rapidly at such a point, an error in the argument has less effect at such a point. (We sidestep for now the question of how the computer finds values of the sine function, and whether that affects our logic here.) Enter:

```
>> err=1E-6;
>> abs(sin(x1+err)-sin(x1))
>> abs(sin(x2+err)-sin(x2))
```

to see this effect more clearly. (Recall that $\sin(x) \approx x$ for x near zero and $\sin(x) \approx 1 - x^2/2 \approx 1$ for x near $\pi/2$.) Compare the relative errors in this case also.

We'll discuss the effects of finite precision arithmetic and round-off error in greater detail in Sec. 1.7. For now we note once again that computer arithmetic differs from standard arithmetic in a number of important respects.

Problems

1.) Perform three iterations of the finite difference Newton's method with $h = .1$ to approximate the sole real root of $5x^7 + 2x - 1 = 0$. Repeat with $h = .01$ and the same initial guess.

2.) The secant method has order of convergence $p \doteq 1.6$. If it is working well; the finite difference Newton's method has order of convergence $p \approx 2$. If function evaluations are expensive, which is preferable for one-dimensional root-finding? What about the globalized Newton's method?

3.) a.) Use the globalized Newton's method to approximate a zero of $f(x) = 1/x^3 - 10$ to three decimal places.

b.) Draw the graphs of several functions for which Newton's method would fail to converge. Does the globalized Newton's method handle all of these cases adequately?

4.) Prove that the constant-slope Newton's method converges linearly by treating it as a fixed-point iteration. What are the restrictions, if any, on the slope c? What value of c gives superlinear convergence?

5.) Write a MATLAB program which performs the constant-slope Newton's method. Input the function and the slope and initial guess; return the estimate of the root. Test it on several functions.

6.) a.) Show that Newton's method for $f(x) = 1/x - 10$ fails to converge if $x_0 = 10$. Give a sketch which demonstrates why this happens.

b.) Find a damping sequence which forces the damped Newton's method to converge for this function and initial guess.

c.) Does your damping sequence work if $x_0 = 100$?

7.) Write a complete algorithm (pseudo-code) for a safeguarded secant method.

8.) Approximate the derivatives of $\sin(x)$, $\cos(x)$, and $\exp(x)$ at $x = 1.5$ by finite differences with $h = 10^{-k}$, $k = 1, 2, ..., 20$. What value of h gives the smallest error?

9.) Approximate the derivatives of $\ln(x)$, $\cosh(x)$, and x^3 at $x = 1.5$ by finite differences with $h = 10^{-k}$, $k = 1, 2, ..., 20$. What value of h gives the smallest error?

10.) Consider the convergence theorem for Newton's method, which guarantees convergence in some interval about a simple root (in effect, a bracket), and the safeguarded Newton's method, which requires a bracket. We have stated the Newton's method and the secant method may be used when a bracket cannot be found, but the convergence theorem and the safeguarded methods seem to indicate that a bracket is still needed in some sense. Comment on the importance of brackets in root-finding.

11.) Write a MATLAB program that inputs a function and its derivative (both as function handles), a bracket of a zero of the function, and a positive integer N, and performs N iterations of the globalized Newton's method. Demonstrate your program on at least three root-finding problems.

12.) Use the constant-slope Newton's method to approximate a solution of $x^3 - 10 = 0$. Justify your choice of slope.

13.) a.) Use the constant-slope Newton's method to approximate a zero of $f(x) = e^{2x} - 2x^3 - 12$ to six decimal places. Use $f'(x_0)$ as the constant slope.

b.) Use the constant-slope Newton's method to approximate a zero of $f(x) = 1/x^3 - 10$ to six decimal places, but update your slope every five iterations; that is, use $f'(x_0)$ as the initial constant slope, then $f'(x_5)$ when it becomes available, then $f'(x_{10})$ (if needed), and so on.

c.) Repeat part c., but update every three iterations.

d.) Compare your results to the standard Newton's method for this problem.

e.) (Refer to Problem 4.) One way to improve the efficiency of this mixed Newton's method/constant-slope Newton's method strategy is to switch to a pure constant-slope Newton's method when the method is sufficiently near the root. The idea is that at that point the constant slope is very close to the true slope at the root x^*. One way to implement this idea is to consider the method as a fixed point iteration $x_{k+1} = g(x_k) = x_k - f(x_k)/c$ and monitor the size of $g'(x_k)$; when it is (and stays) sufficiently small the convergence is linear with ratio $|g'(x_k)|$. If $|g'(x_k)|$ is very small then it may not be worthwhile to evaluate $f'(x)$ to gain quadratic convergence. Discuss this method; suggest a specific strategy for switching to the pure constant-slope Newton's method.

14.) Why is it preferable to use bisection in the globalized Newton's method rather than inverse linear interpolation?

15.) a.) Let $f(x) = x^8$. If $x_k = 1/k$, how large must k be to insure that $f(x_k) < 10^{-6}$? How large must k be to insure that $x_k < 10^{-6}$? What does this imply for a test for convergence based on $f(x_k) \approx 0$?

b.) What happens if you apply Newton's method to $f(x) = e^{-2x}$ and test for convergence based on $|f(x_k)| < \tau$ for some tolerance $\tau > 0$?

c.) Show by example that $|f(x_k)|$ small does not imply that $|x_k - x^*|$ is small. Give an explicit function and values of x_k, $|f(x_k)|$, and x^*.

16.) The Schrödinger wave equation, as used in the Kronig-Penney model of the movement of electrons in the band theory of solids, has solutions only when $f(\alpha a) = P \operatorname{sinc}(\alpha a) + \cos(\alpha a)$ lies in $[-1, 1]$, where $\alpha = \sqrt{2mE/\hbar^2}$, $P = mV_0 ba/\hbar^2$, $b > 0$ is the width of potential wells in the solid (assumed to be in the form of a square wave), $V_0 > 0$ is their height, and $a > 0$ is the period of the potential of the crystal lattice of the solid. (The sinc function is defined by $\mathrm{sinc}(x) = \sin(x)/x$, with $\mathrm{sinc}(0) = 1$.) Here m is the mass of an electron and $E > 0$ is the energy of the electron; $\hbar = h/2\pi$ where h is Planck's constant.

a.) Plot $f(x)$ vs. x for $P = 100$.

b.) What value of αa gives a solution if $P = 100$ and we require that $f(\alpha a) = .5$? Use the finite difference Newton's method.

c.) If $a = 1\,\text{Å} = 10^{-10}$ m, what is the energy of the electron in this case?

6. Brent's Method

All of the Newton's method variants require that the derivative of f either be known or that it be approximated. This is not always convenient. Let's return to the idea of (inverse) linear interpolation, where we started with a bracket $[a, b]$ of a zero of f and estimated the location of that zero as the x-intercept of the line determined by those two points. If we choose a third point c somewhere in the bracket, we can interpolate a quadratic to the function and use the zero of the quadratic as an estimate of the zero of the function. This idea is known as **quadratic interpolation**.

More specifically, suppose that f is continuous on the interval $[x_0, x_2]$ and that $[x_0, x_2]$ brackets a zero of f. Pick some point x_1 in (x_0, x_2), giving a triple of points $[x_0, x_1, x_2]$. We seek a quadratic polynomial that interpolates $f(x)$ at these three points. Write the quadratic as $Q(x) = \alpha(x - x_2)^2 + \beta(x - x_2) + \gamma$; then we want

$$
\begin{aligned}
f(x_0) &= \alpha(x_0 - x_2)^2 + \beta(x_0 - x_2) + \gamma \\
f(x_1) &= \alpha(x_1 - x_2)^2 + \beta(x_1 - x_2) + \gamma \\
f(x_2) &= \alpha(x_2 - x_2)^2 + \beta(x_2 - x_2) + \gamma \\
&= \gamma
\end{aligned}
$$

and we see immediately that $\gamma = f(x_2)$. (This is one advantage of writing the quadratic in the form we chose; others are discussed in Sec. 4.1.) Since the remaining unknowns are α and β, we may write this as the linear system

$$
\begin{pmatrix} (x_0 - x_2)^2 & (x_0 - x_2) \\ (x_1 - x_2)^2 & (x_1 - x_2) \end{pmatrix} \begin{pmatrix} \alpha \\ \beta \end{pmatrix} = \begin{pmatrix} f(x_0) - f(x_2) \\ f(x_1) - f(x_2) \end{pmatrix}.
$$

It is a fact that there is a unique solution of this system, given by

$$
\begin{aligned}
(6.1) \qquad \alpha &= \frac{(x_1 - x_2)(f(x_0) - f(x_2)) - (x_0 - x_2)(f(x_1) - f(x_2))}{(x_0 - x_1)(x_1 - x_2)(x_0 - x_2)} \\
\beta &= \frac{(x_0 - x_2)^2(f(x_1) - f(x_2)) - (x_1 - x_2)^2(f(x_0) - f(x_2))}{(x_0 - x_1)(x_1 - x_2)(x_0 - x_2)}
\end{aligned}
$$

and so the quadratic must have a root in $[x_0, x_2]$, since $Q(x_0)Q(x_2) = f(x_0)f(x_2)$ is negative. The roots r_1, r_2 of $Q(x)$ may be found by the quadratic formula. Let x_3 be the root which lies in the bracket. Unless $f(x_3)$ is actually zero–which is extremely unlikely in practice–we now have a new, smaller bracket of the zero of

f, $[x_0, x_3]$ or $[x_3, x_2]$, and a new triple of points for the next iteration. This is the method of quadratic interpolation.

EXAMPLE 1: Suppose that $f(x) = \sin(x)$. Take $x_0 = -1$, $x_2 = 2$; then $f(x_0) = -0.8415$ and $f(x_2) = 0.9093$. Hence $[x_0, x_2]$ is a bracket. Choose an x_1, say, $x_1 = .5$. The quadratic is

$$\begin{aligned} Q(x) &= \alpha(x - x_2)^2 + \beta(x - x_2) + \gamma \\ &= \alpha(x - 2)^2 + \beta(x - 2) + 0.9093 \end{aligned}$$

and

$$\begin{aligned} \alpha &= \frac{(x_1 - x_2)(f(x_0) - f(x_2)) - (x_0 - x_2)(f(x_1) - f(x_2))}{(x_0 - x_1)(x_1 - x_2)(x_0 - x_2)} \\ &\doteq -.1980 \\ \beta &= \frac{(x_0 - x_2)^2(f(x_1) - f(x_2)) - (x_1 - x_2)^2(f(x_0) - f(x_2))}{(x_0 - x_1)(x_1 - x_2)(x_0 - x_2)} \\ &\doteq -0.0104 \end{aligned}$$

so that $Q(x) = -.1980(x - 2)^2 + -0.0104(x - 2) + 0.9093$. Write this as $Q(z) = -.1980z^2 + -0.0104z + 0.9093$; by the quadratic formula, the roots are $z_1, z_2 \doteq -2.1694, 2.1169$ so (from $x = 2 + z$) $r_1, r_2 \doteq -0.1694, 4.1169$. The root $r_1 = -0.1694$ is in the bracket so $x_3 = -0.1694$. Since $f(-0.1694) = \sin(-0.1694) = -0.1686$ the new bracket is $[x_3, x_2]$ and the current triple of points is $[x_3, x_1, x_2] = [-0.1694, .5, 2]$. This reduces the bracket width to

$$2 - (-0.1694) = 2.1694$$

which is not as good as what we would have had from bisection (which would have reduced the width to 1.5), but is comparable.

Let's do another step. Since x_2 hasn't changed, the form of the quadratic will once again be $Q(x) = \alpha(x - 2)^2 + \beta(x - 2) + 0.9093$, but now x_3 replaces x_0, giving

$$\begin{aligned} \alpha &= \frac{(x_1 - x_2)(f(x_3) - f(x_2)) - (x_3 - x_2)(f(x_1) - f(x_2))}{(x_3 - x_1)(x_1 - x_2)(x_3 - x_2)} \\ &\doteq -.3141 \\ \beta &= \frac{(x_3 - x_2)^2(f(x_1) - f(x_2)) - (x_1 - x_2)^2(f(x_3) - f(x_2))}{(x_3 - x_1)(x_1 - x_2)(x_3 - x_2)} \\ &\doteq -.1846 \end{aligned}$$

so that $Q(x) = -.3141(x - 2)^2 + -.1846(x - 2) + 0.9093$. Once again, write this as $Q(z) = -.3141z^2 + -.1846z + 0.9093$; by the quadratic formula, the roots are $z_1, z_2 \doteq -2.0205, 1.4328$ so (from $x = 2 + z$) $r_1, r_2 \doteq -0.0205, 3.4328$. The root $r_1 = -0.0205$ is in the bracket and so $x_4 = -0.0205$. Since $f(-0.0205) = \sin(-0.0205) = -0.0205$ (recall that $\sin(x) \approx x$ for x near 0), the new bracket is $[x_4, x_2]$ and the current triple of points is $[x_4, x_1, x_2] = [-0.0205, .5, 2]$. The new width is

$$2 - (-0.0205) = 2.021$$

which is very little improvement over the previous bracket. The method is working, but not rapidly. $\square$

If we do not necessarily start with and maintain a bracket and allow extrapolation (as was done in the secant method), using the root of the quadratic which lies closest to x_2, we have a method known as Müller's method. Müller's method is convergent of order approximately 1.84 and can find complex roots from real initial guesses (if the interpolated quadratic has no real roots). This can be advantageous—we can start with a real initial guess and if the method needs to use complex values to find a solution, it will simply generate them.

EXAMPLE 2: Consider again $f(x) = \sin(x)$ with $x_0 = -1$, $x_1 = .5$, and $x_2 = 2$. From Example 1, the roots of the interpolated quadratic are $r_1, r_2 \doteq -0.1694, 4.1169$. The one closer to x_2 is $r_2 \doteq 4.1169$ so we continue with $x_1 = .5$, $x_2 = 2$, and $x_3 = 4.1169$. Since $\sin(x_1) < 0$, $\sin(x_2) > 0$, and $\sin(x_3) < 0$, the outermost points $[x_1, x_3]$ no longer form a bracket. The quadratic through these points is

$$Q(x) = -.3061x^2 + 1.0519x + .0300$$

(in a program we would expand this polynomial around x_2). The roots are $r_1, r_2 \doteq -.0283, 3.4645$ and the root nearest the most recent point x_3 is $x_4 = 3.4645$. We now repeat with x_2, x_3, and x_4. In the limit the most recent point x_n should tend to a root, even if the width of the bracket does not go to zero. In this case we seem to be approaching the root π. Indeed, the next quadratic is $Q(x) = .0260x^2 + -.9795x + 2.7645$ (note, the concavity has switched) with roots $r_1, r_2 \doteq 3.0726, 34.6294$ giving $x_5 = 3.0726$, followed by $r_1, r_2 \doteq 3.1383, 8.4680$ giving $x_6 = 3.1383$, followed by $r_1, r_2 \doteq 3.14160, 27.12094$ giving $x_7 \doteq 3.14160$ in good agreement with $\pi \doteq 3.14159$ ($\alpha \doteq 1.2E - 5$). The next iteration gives $x_8 \doteq 3.14159265386786$ with absolute error $2.8E - 10$. The order of convergence of 1.84 is nearly as good as the quadratic convergence of Newton's method. $\square$

Quadratic interpolation converges superlinearly, as does inverse linear interpolation, though quadratic interpolation is generally faster. For quadratic interpolation vs. Müller's method and for inverse linear interpolation vs. the secant method we see that the bracketing method provides guaranteed convergence, but that giving up that guarantee gives greater speed and makes for easier analysis. For example, we can say exactly what the order of convergence is for Müller's method and the secant method. Quadratic interpolation will generally give rapid convergence to the root nonetheless.

Quadratic interpolation is generally quite fast. After all, if a function is sufficiently differentiable then a quadratic should be a good fit for it even before we reach a scale so small that the function is approximately linear (by Taylor series reasoning). However, quadratic interpolation can provide poor estimates of the location of the zero if the function is not nearly quadratic on the current bracketing interval. Once again we look to a mixed method. Suppose that f is continuous on the interval $[x_0, x_2]$ and that $[x_0, x_2]$ brackets a zero of f. We might try a quadratic interpolation step and see if the new bracket is much smaller than the original one. If so, we believe the quadratic interpolation method is working as expected. If it is not, we might reason that we are in a range where quadratic interpolation is not working well because we are in a region where a quadratic model is a poor approximation; hence we might on the next step try, say, an inverse linear interpolation step instead. If this gives little reduction in the width of the bracket we might drop down once more and try bisection, which gives a guaranteed 50% reduction in the width of the bracket–which may be enough to force the method into a range where the quadratic *will* be a good fit. We continue to try to use the quadratic interpolant when it seems to be working well, and use various heuristics to decide when to drop down to a simpler method.

In practice there is usually little gain in using the inverse linear interpolation step, so a method of this sort generally mixes just quadratic interpolation and bisection. If quadratic interpolation gives poor results then we switch to bisection for a guaranteed improvement rather than just hoping for meaningful improvement from inverse linear interpolation before, quite possibly, dropping down to bisection anyway.

This mix of the fast quadratic interpolation approach with the sureness of bisection is the key idea in **Brent's method** (also called the **Brent-Dekker method**), and there are many ways to implement it. Experience shows that it generally does not pay to switch from quadratic interpolation simply because a single iteration gives poor results, as the method may recover; hence we would usually switch only if at least two consecutive iterations gave poor reductions in the interval width (or if the average reduction over two or more iterations was poor). Similarly, one might reason that if things are bad enough that the method switched to bisection then it makes sense to do at least two bisection steps in order to move the method away from the uncooperative region. Many of these decisions are a matter of programmer's choice; the programmer tests various heuristics in order to see which ones seem to work well on the types of problems that will be solved by the program. The key points are that dropping the bracketing requirements of inverse linear interpolation and quadratic interpolation, which gives us the secant and Müller's methods, respectively, leads to greatly improved speed at the cost of possible failure to converge; and that the superlinear methods may be made more robust by adding in a safeguarding strategy based on the (linear) bracketing method of bisection, as in the globalized Newton's method and Brent's method. Such safeguarding approaches are widely employed in numerical algorithms.

We are actually using the term *Brent's method* in a generic sense to indicate a class of mixing strategies that maintain a bracket and do not use derivative information. In fact, what might be called the "classical" Brent's method, specifically designed to be derivative-free, uses **inverse quadratic interpolation**; that is, a quadratic $x = Q(y)$ is used as the interpolant (not $y = Q(x)$ as in quadratic interpolation). The desired x-intercept is then easily found as $x = Q(0)$. The computational benefit of using inverse interpolation in the form $x = P_n(y)$ with some polynomial P_n of degree n is that no matter what degree polynomial is used, finding the x-intercept $x = P_n(0)$ does not require that we solve a polynomial equation for x, as we would have to do if we fit $y = P_n(x)$ and then needed to solve $P_n(x) = 0$ for x. The resulting methods are generally referred to as **inverse interpolation methods**. (The benefits of this approach are best appreciated when interpolating with a polynomial of degree greater than 4–not that this is generally recommended. But cubic interpolation is a reasonable thing to try, and using inverse cubic interpolation is much easier than applying the cubic formula to find the three roots–two of which might be complex!) A slight complication is that the x-intercept need not lie in the bracket (consider Fig. 1 with a bracket of $[3,5]$), but that is easily handled by simply taking a bisection step. A well-behaved function should be increasingly well approximated by a quadratic as the interval shrinks.

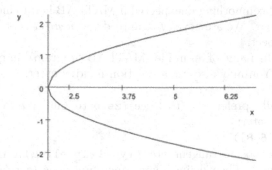

FIGURE 1. Inverse quadratic interpolation.

Brent's method with an appropriate switching strategy provides guaranteed convergence (given an initial bracket, and assuming no pernicious round-off error destroying the bracket) and good speed, with bisection slowly but steadily moving the method to a region where the quadratic fit will be effective. It is this author's preferred method for one-dimensional root-finding problems, and is the basis for the MATLAB univariate root-finding routine `fzero`; in fact, T.J. Dekker is credited in the comments in `fzero`, and a book by R.P. Brent is cited. For problems in several variables, a variant of Newton's method is generally the best approach.

There are heuristics for finding an initial bracket from a single initial guess. The guess itself establishes the scale on which the program should look for a solution. Here's one possibility, based on the idea of a single-variable optimization method due to Howard Rosenbrock (see Sec. 7.5): Given x_0, pick a step size δ that may be positive or negative and that is the initial scale for your search. (In `fzero`, the step is $\delta = x_0/50$.) If $f(x_0 + \delta)$ differs in sign from $f(x_0)$ then we have succeeded; if not, we replace x_0 by $x_0 + \alpha\delta$ for some $\alpha > 1$ and repeat. When $f(x_0 + \alpha^{k-1}\delta)$ differs in sign from $f(x_0 + \alpha^k\delta)$, we have found a bracket. If we do not find one in this way after searching for a predetermined number of function evaluations, we change the sign of δ and begin again, now searching in the opposite direction. If this fails, we can try a different δ and/or α, or simply return an indication that the search has failed to find a bracket. (Remember, the user may have provided a function that never changes sign so that no bracket actually exists!) We might take the $f(x_0 + \alpha^{k-1}\delta)$ that has the smallest absolute value and do a search from that point with a smaller δ and α, for example.

MATLAB

The MATLAB zero-finding program `fzero` is based on Brent's method (using inverse quadratic interpolation); enter `type fzero` to see the program listing. Look at it and find where the program is looking for a change of sign to find the initial bracket, and where it uses a Rosenbrock-style search with `dx` set to `x0/50`. You may wish to enter the `more on` command and re-enter the `type fzero` command to page through the program listing. By entering `help fzero` you will see that the calling sequence is `fzero(f,x0)` for some function `f` and initial guess `x0`. For example, enter:

```
>> fzero('sin',3)
```

There would have been nothing unexpected if MATLAB had found the root at, say, zero rather than at π. We are never guaranteed the *nearest* root, though certainly that's what we expect.

You may use the name of an m-file (MATLAB program), in quotes, in place of the sine function. You may also use a function handle; enter:

`>> fzero(@sin,3)`

and this is generally preferred. To force `fzero` to find a zero within a certain bracket, say $[6, 8]$, enter:

`>> fzero(@sin,[6 8])`

Try finding zeroes of cosine, tangent, etc. (Type `help elfun` for a list of elementary MATLAB functions.) Then define some other functions (e.g., polynomials) and find zeroes of them; for example, use `f=@(x)x.^3-x.^2-x-1` and plot it, then try to find its real root(s). In most applications of interest, the function would actually be defined by an m-file rather. Finding zeroes of polynomials is difficult enough that they are still used as test problems for root-finding programs.

There is also a MATLAB function which uses a specialized algorithm that attempts to find all roots of a polynomial equation, even if complex or repeated. The polynomial is entered as a vector. In MATLAB a row vector may be entered directly; enter:

`>> p=[1 2 1]`

The roots of the polynomial $a_n x^n + a_{n-1} x^{n-1} + \cdots + a_1 x + a_0$ can be found by entering the coefficients in descending order a vector such as p and using the `roots` command. Enter:

`>> roots(p)`

This returns a double root at -1 as expected. Enter:

`>> roots([1 2 2])`

This returns $-1 \pm i$ (the roots of $x^2 + 2x + 2 = 0$) as expected. The inverse of `roots` is `poly`; if `r` is a vector of roots of a polynomial, `poly(r)` is a vector of the coefficients of that polynomial. Hence `poly(roots(x))` and `roots(poly(x))` should both equal x...in infinite precision arithmetic.

As explained in `help poly`, the special case `poly(1:20)` is known as **Wilkinson's polynomial**. This is the polynomial of degree 20 with zeroes at $1, 2, 3, ..., 20$. Enter:

`>> roots(poly(1:20))`

This should return the vector `1:20`. As you can see, it does not–even though these roots are all simple and well-separated.

Looking ahead to Problem 8, note that we can plot $f(x) = (x-1)^8$ by defining an x vector (see `help linspace`) and using `plot`. Enter:

`>> x=linspace(.8,1.2,200);`
`>> plot(x,(x-1).^8),grid`

To plot its expanded form, use `c=poly([1 1 1 1 1 1 1 1])` to find the coefficients of the polynomial with roots at $x = 1$ with multiplicity 8, then enter:

`>> y=polyval(c,x);`

to evaluate this polynomial at all points in the x vector. Enter:

`>> plot(x,y),grid`

There is no visible difference at this scale, but as Problem 8 will show, there is a significant difference near the origin at smaller scales. Try finding it.

To see the effects of round-off error and poor scaling in another form, enter:
```
>> roots(poly([1 1 1 1 1 1 1]))
```
The numerical method, when applied to this entirely real problem with entirely real roots, generates answers with complex-valued round-off error. As distressing as this may seem at first, round-off error is round-off error and the fact that it happens to be complex-valued shouldn't concern us overly much. (The fact that the relative error is so large should concern us however; type `abs(ans-1)` to see the size of the errors.) This is not a failing of MATLAB. These problems are simply difficult.

Problems

1.) Perform three iterations of quadratic interpolation to approximate the sole positive root of $\cos(x) - x = 0$.

2.) Perform three iterations of quadratic interpolation to approximate the sole real root of $5x^7 + 2x - 1 = 0$.

3.) a.) Write a detailed algorithm (pseudo-code) for performing Müller's method. (Note that you must select a convergence criterion.) Terminate the algorithm if there is no real root of the quadratic.
b.) The order of convergence of Müller's method is (at least) the real root of $x^3 - x^2 - x - 1 = 0$. Find this value using Müller's method.

4.) Write a detailed algorithm (pseudo-code) for performing a simple version of Brent's method using quadratic interpolation (as in Eq. (6.1)). Assume that the method will switch to bisection if the last two quadratic interpolation steps have not reduced the function by at least as much as two bisection steps would have reduced it, and that when the method switches to bisection it always does two bisection steps before returning to quadratic interpolation.

5.) Use quadratic interpolation to approximate a zero of $f(x) = 1/x^3 - 10$ to three decimal places.

6.) Use your version of Brent's method from Problem 4 to approximate a root of $\cos(12x) - x/2 = 0$ with the initial bracket $[0, 4]$.

7.) Use your version of Brent's method from Problem 4 to approximate a root of $\cos(e^x) = 0$ with the initial bracket $[0, 4]$.

8.) Plot $f(x) = (x - 1)^8$ for $x \in [.9, 1.1]$ using a fine spacing. Repeat for $x \in [.995, 1.005]$. Then plot $g(x) = x^8 - 8x^7 + 28x^6 - 56x^5 + 70x^4 - 56x^3 + 28x^2 - 8x + 1$ over the same two ranges. (Type `help axis` for help on adjusting the display of the plot.) Given that $f(x) = g(x)$ for all x, can you explain what's going on? How many roots does $g(x)$ have in principle, and in practice?

9.) a.) Verify that if all three data points lie on a straight line then the quadratic interpolant (Eq. (6.1)) is indeed a line ($\alpha = 0$).
b.) Verify that Eq. (6.1) is correct (including that the solution is unique).

10.) Use Müller's method to find the zeroes of $f(x) = 8x^3 - 12x^2 - 50x + 75$. Check your answers with the MATLAB command `roots([8 -12 -50 75])`.

11.) Write a detailed algorithm (pseudo-code) for performing a simple version of Brent's method using inverse quadratic interpolation. You will need to derive a formula for the quadratic interpolant $x = Q(y)$. Recall that the x-intercept $x = Q(0)$ may fall outside of your bracket.

12.) a.) What are the relative advantages and disadvantages of the following root-finding methods: Bisection, inverse linear interpolation, secant method, Newton's method, quadratic interpolation, Müller's method.
b.) What are the relative advantages and disadvantages of Brent's method and the globalized Newton's method?
13.) a.) Use the MATLAB command `roots([1 -6.04 13.68 -13.76 5.196])` to find all roots of $x^4 - 6.04x^3 + 13.68x^2 - 13.76x + 5.196 \doteq (x - 1.52)(x - 1.48)(x - 1.5)(x - 1.54)$. Comment on your results.
b.) Find all real roots of $x^7 - 7x^6 + 21x^5 - 35x^4 + 35x^3 - 21x^2 + 7x - 1$.
14.) Write a MATLAB program that implements your version of Brent's method from Problem 3. Assume that $f(x)$ will be given as a function handle. Type `help while` if you would like to use a while loop in your program. Demonstrate your program on at least three root-finding problems.
15.) Use Müller's method to find all roots of $x^3 - x^2 - x - 1 = 0$. (Refer to Problem 3.) Note that you do not need to begin with complex values; they are generated by the method.
16.) The electric field $E_a(\theta)$ of a transmitting linear phased-array radar system can be approximated for small angles by $E_a(\theta) = \sqrt{N}\,\text{sinc}(\pi(D/\lambda)\sin(\theta))$ where $\text{sinc}(X) = \sin(x)/x$ and $\text{sinc}(0) = 1$. Here N is the number of elements in the array, D is the length of the array, and λ is the wavelength at which the system is operating. (Recall that $f\lambda = c$, where f is frequency and c is the speed of light.) If $N = 10$, $D = N \cdot 2\lambda$, and the frequency is 46 MHz, use Brent's method to determine what angle θ corresponds to a field strength of .5 V/m.

7. Effects of Finite Precision Arithmetic

In developing a numerical method for a given problem we begin with mathematical or physical intuition about the particular problem and use that intuition to devise a scheme that, we hope, will converge to a solution of the problem. Since most methods do not converge in a finite number of iterations we include convergence criteria that indicate when we have achieved practical convergence, that is, when we have found sufficiently accurate values for our purposes (or at least, as accurate as we can reasonably hope to find).

Whether a method does or does not converge in the usual mathematical sense, as a limit, and how rapidly it does so, is the first issue. The second issue is *stability*, by which we mean the sensitivity of the method to small errors such as round-off errors. If a method converges only when infinite precision arithmetic is used but not on a computer then it will not be of much interest to us.

Here's an example of a method that is not convergent: Consider the problem of finding a fixed point of $f(x) = 3\cos(x)$ near $x = 1$ (see the MATLAB subsection of Sec. 1.3). The direct application of fixed point iteration gives

$$x_{k+1} = 3\cos(x_k)$$

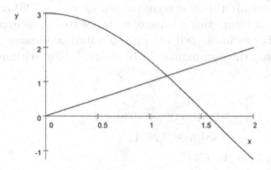

FIGURE 1. Fixed point of $y = 3\cos(x)$.

(see Fig. 1). There is a solution at $x^* \doteq 1.17$; however, since $f'(x^*) = -3\sin(x^*) \approx -2.76$ the fixed point is not attracting (i.e., it is **repelling**) and the method cannot converge for any x_0 in a deleted neighborhood of x^*, *even using infinite precision arithmetic*. The method is not convergent in the usual sense of the calculus; Newton's method, of course, would be convergent for the corresponding root-finding problem $f(x) - x = 0$.

Lack of stability is a different issue, however. To explore it, let's consider the two-step recurrence relation

$$(7.1) \qquad x_{k+1} = x_k + x_{k-1}$$

with the initial conditions $x_0 = 1, x_1 = (1 - \sqrt{5})/2 \doteq -0.6180$. Then

$$\begin{aligned} x_2 &= x_1 + x_0 \\ &= (1 - \sqrt{5})/2 + 1 \\ &= (3 - \sqrt{5})/2 \\ &= \left[\frac{(1 - \sqrt{5})}{2}\right]^2 \doteq 0.3820 \end{aligned}$$

(the last step is not obvious but it is easily verified to be correct). Similarly

$$\begin{aligned} x_3 &= x_2 + x_1 \\ &= (3 - \sqrt{5})/2 + (1 - \sqrt{5})/2 \\ &= (4 - 2\sqrt{5})/2 \\ &= \left[\frac{(1 - \sqrt{5})}{2}\right]^3 \doteq -0.2361 \end{aligned}$$

and in general

$$(7.2) \qquad x_k = \left[\frac{(1 - \sqrt{5})}{2}\right]^k$$

$(k = 0, 1, 2, ...)$. Certainly this converges to zero as $k \to \infty$. Of course, the obvious numerical method for computing a value of x_k is not to do this analysis first but just to use Eq. (7.1). For example, with the given initial conditions $x_0 = 1, x_1 = (1 - \sqrt{5})/2$ we have, using the approximately 16 decimal digit arithmetic of MATLAB,

$$
\begin{aligned}
x_2 &= x_1 + x_0 \\
&\doteq -0.61803398874989 + 1 \\
&\doteq 0.38196601125011 \\
x_3 &= x_2 + x_1 \\
&\doteq 0.38196601125011 + -0.61803398874989 \\
&\doteq -0.23606797749979 \\
x_4 &= x_3 + x_2 \\
&\doteq -0.23606797749979 + 0.38196601125011 \\
&\doteq 0.14589803375032
\end{aligned}
$$

and this sequence should decrease to zero in agreement with Eq. (7.2). But this is not what happens! As k increases, x_k initially decreases, as expected, but then begins increasing rapidly. By $k = 50$, x_k is on the order of 10^9; by $k = 100$, x_k is on the order of 10^{20}. Why? Look again at Eq. (7.1) which is called a difference equation, in analogy with differential equations; this particular difference equation is the **Fibonacci recurrence relation**. The general solution is given by the Binet formula

$$
(7.3) \qquad x_k = c_1 \left[\frac{(1 - \sqrt{5})}{2} \right]^k + c_2 \left[\frac{(1 + \sqrt{5})}{2} \right]^k
$$

where the values of c_1 and c_2 are determined by the initial conditions x_0 and x_1. In particular, the initial conditions $x_0 = 1, x_1 = (1 - \sqrt{5})/2$ that we have been using give $c_1 = 1$ and $c_2 = 0$. But if the values of x_0 and x_1 are not *precisely* equal to 1 and $(1 - \sqrt{5})/2$, respectively, then we will get different values of c_1 and c_2 and c_2 will almost certainly be nonzero. Choosing values so that $c_1 \approx 1$ and $c_2 \approx 0$, but with $c_2 \neq 0$, causes this behavior because of the small error in the approximation

$$
x_1 = \frac{1 - \sqrt{5}}{2} \doteq -0.61803398874989
$$

(an absolute error of about $5E - 15$, and a relative error of the same magnitude). What we have seen here is analogous to the problem of stiffness in systems of differential equations

This is the phenomenon of instability: A scheme that would give the correct answer in exact arithmetic gives very wrong answers when implemented on a computer because small errors are exaggerated. (Look at the huge effect of initial errors on the order of 10^{-15} in the Fibonacci example.) Throughout this text we shall have to modify many methods to change an unstable scheme, which would converge in exact arithmetic but not on a computer, to a stable scheme that converges even on a computer. To do so it will be helpful to understand a bit more about computer arithmetic and how it differs from real arithmetic.

Computers store and manipulate numbers in base two[10] using the binary digits 0 and 1, called **bits** in this context. Eight bits make a **byte**. In most cases, a number is stored in some integer number of bytes.

There are special ways of encoding numbers that are known in advance to be integers, but we will focus on the main method used to represent, as well as we can, arbitrary real numbers. In this approach, one of the bits is used to represent the sign of the number and the remainder are split between the mantissa (fractional part) and the exponent. The number has the form $N = \pm m \cdot b^e$, where N has mantissa m, b is a fixed base (typically 2 or 16), and e is the exponent of the number.

In the IEEE Standard for Floating-Point Arithmetic (IEEE 754), used in any computing environment you are reasonably likely to encounter, a **single precision** number is stored using 32 bits (four bytes, called a **word**). In single precision, 1 bit is devoted to the sign of the number, 8 bits are used for the exponent, and the remaining 23 bits are used to represent the significand (i.e., the mantissa), which contains the actual significant figures of the number. This is indicated in Fig. 2.

sign   exponent   significand

FIGURE 2

The number represented by these bits has the form $(mantissa) \times (base)^{exponent}$ with appropriately chosen sign, and can have magnitude as small as about 10^{-38} and as large as about 10^{38}. These are the **machine numbers**.

The situation is actually rather more complicated than this! There is a normalization criterion and a hidden bit for the mantissa, which means that a single precision number actually has 24 significant bits rather than just 23, and an offset is subtracted from the exponent in order to get a range of machine numbers from about $\pm 10^{-38}$ to $\pm 10^{38}$ (and of course 0). These are called the **normal** machine numbers; by relaxing the normalization criterion some numbers smaller than 10^{-38} in absolute value can be represented, though with lessened precision. These are called **subnormal** machine numbers.

In **double precision**, the number is stored using 64 bits (a double word). One bit is devoted to the sign of the number, 11 bits are used for the exponent, and the remaining 52 bits are used to represent the mantissa. Note that double precision actually devotes *more* than double the number of significant bits as compared to single precision–52 as opposed to 23–but does not double the number of bits in the exponent (8 in single precision as opposed to 11 in double precision, giving numbers that have magnitude as small as about 10^{-308} and as large as about 10^{308}). With the hidden bit, double precision has 53 significant bits. Double precision arithmetic operations may take up to slightly more than twice as long as corresponding single precision operations, depending on the machine architecture, but on most current machines the principal advantage of using single precision is in storage space, not speed. Single precision gives about 7 significant (decimal) figures and double precision gives about 16 significant (decimal) figures. Once again there are subnormal double precision numbers with magnitudes less than 10^{-308}.

---

[10]Calculators and some special-purpose computers may use other bases.

Note that single precision and double precision are defined by the machine, not the programming language. This means, for example, that the same computation performed on two different machines could give two different answers due to differences in how numbers are represented and in how arithmetic operations are implemented. This issue of portability of software has been a continual source of frustration for those who write and test numerical programs. Fortunately most computers use the same IEEE 754 compliant arithmetic systems. While there can still be slight variations even between implementations of IEEE standard arithmetic as the standard does not cover *all* possibilities, you are much less likely to encounter such issues nowadays.

The representation system that we have been discussing is said to use **floating point numbers** because the variable exponent allows the decimal (or binary) point to move within the significant figures. The machine arithmetic is then called **floating point arithmetic**. In controls engineering it is not uncommon to work with **fixed point numbers** and **fixed point arithmetic**, in which the decimal point has a fixed location, as was done on early computers.

For authors of high-quality numerical software, knowledge of the details of how computer numbers are stored and represented, as well as related hardware issues, is very important. For our purposes, however, what we need to know is the following: Computers represent numbers with a fixed number of significant bits (precision) and a limited exponent; programs using double precision arithmetic will take longer to run than ones using single precision arithmetic and will also require more memory; and floating point arithmetic differs from real arithmetic because the results must be computed and stored in finite precision.

Half precision numbers, long used in graphics applications, are increasingly being used in scientific computing because they reduce storage and energy needs, and because many applications involving simulations or data science can tolerate the lower precision. (Energy costs for operating a supercomputer can be substantial! It's common to see the energy cost per floating point operation used as a metric of performance.) In addition, depending on the underlying hardware, half precision computations can run considerably faster than, say, double precision computations.

There is more than one 16 bit, half-word format in use. The half-precision floating point format **binary16** specified by IEEE 754 uses 5 exponent bits and 10 significand bits (11 with the hidden bit); the **bfloat16** format uses 8 exponent bits in order to better approximate the range of single-precision values, leaving it only 7 bits (8 with the hidden bit) for storing the significand. Some computer manufacturers offer variations on these formats (e.g., giving up the ability to represent infinities or not-a-number in order to gain a more normalized numbers). We sometimes speak of half precision formats, and specialized precisions even lower than this, as **minifloats**.

Since MATLAB defaults to double precision floating point, we'll focus on that. (In fact, it defaults to double precision complex floating point numbers $\alpha + i\beta$, with the real numbers α and β each stored in double precision.) It's easy to convert between single and double precision in MATLAB (see the `single` and `double` commands, and more generally `help datatypes` for available data types), and in any computing environment it's wasteful to use more precision than is appropriate for your computation.

Let's look at some differences between floating point arithmetic and real arithmetic. For a given implementation of floating point numbers, define $\mathrm{fl}(x)$ to be the floating point number nearest the real number x (breaking ties by rounding in the usual way). Then it is usual to model the result of adding x and y on the computer as

$$x \oplus y = \mathrm{fl}(\mathrm{fl}(x) + \mathrm{fl}(y))$$

where $\oplus$ represents computer addition. In other words, our model for floating point addition is that the floating point sum of x and y is found by converting x to floating point form, converting y to floating point form, adding these floating point numbers correctly, then converting the result to floating point form. Similarly for subtraction, multiplication, and division. We assume that there are no limitations on the exponent for simplicity. This is a reasonable model for what happens inside most machines. The IEEE 754 standard specifies that the four basic arithmetic operations return the nearest floating point value to the actual operation performed on the stored operands, as though real arithmetic had been used and then rounded.

For the purpose of our examples here, we will reason as though the numbers are being stored in base 10 rather than in binary form. On such a machine the result of adding $x = \pi$ to $y = \pi$ would be $\mathrm{fl}(\mathrm{fl}(\pi) + \mathrm{fl}(\pi))$. If we had 6 decimal digits precision available then $\mathrm{fl}(\pi) = 3.14159$ on the machine in question then $x \oplus y = \mathrm{fl}(6.28318) = 6.28318$. Note that this is the correct value of $x \oplus y$ and an approximate value of $x + y$. If instead $x = \pi$ and $y = \pi/100$ then

$$x \oplus y = \mathrm{fl}(\mathrm{fl}(\pi) + \mathrm{fl}(z))$$

where $z = \pi \oslash 100 = \mathrm{fl}(\mathrm{fl}(\pi)/\mathrm{fl}(100)) = \mathrm{fl}(3.14159/100) = 0.0314159$. Then

$$
\begin{aligned}
x \oplus y &= \mathrm{fl}(\mathrm{fl}(\pi) + \mathrm{fl}(z)) \\
&= \mathrm{fl}(3.14159 + 0.0314159) \\
&= \mathrm{fl}(3.1730059)
\end{aligned}
$$

and so $\mathrm{fl}\,(3.1730059) = 3.17301$ is the result. For comparison,

$$\pi + \pi/100 \doteq 3.173008580125691$$

in double precision arithmetic. Hence, 3.17301 is the correctly rounded (to 6 decimal places) value, as expected.

Our hypothetical machine is working in base 10 and storing numbers in the form $.d_1 d_2 \cdots d_n \times 10^p$ where each d_i is a digit, n is a positive integer ($n = 6$ in the example above), and p is any integer. In this idealized system of n decimal digit floating point arithmetic, all numbers are rounded to their n-digit form after any operation. (In Maple the **digits** command may be used to simulate this system. See also **help digits** in MATLAB.) Calculators have traditionally manipulated numbers in base 10, though they may be stored in base 2 using a technique called **BCD (binary coded decimal)**.

To compute $\pi - 3.14158$ in 6 digit floating point arithmetic we imagine entering π, which will be rounded to 3.14159 (and represented as $.314159 \times 10^1$ since the mantissa must be between 0 and 1), the nearest floating point number to it, then subtracting these adjacent machine numbers to get $3.14159 - 3.14158 = .00001$. This must then be cast to the nearest floating point number to it, but it can be represented exactly as $.100000 \times 10^{-4}$. But if we compare $.100000 \times 10^{-4}$ to $\pi - 3.14158 = 1.26536 \times 10^{-05}$ (found using double precision arithmetic and then storing to 6 digit accuracy) we find that the relative error in our answer is nearly 21%! This is a really big error, and is an example of a common problem in numerical analysis: Loss of significant figures caused by the subtraction of nearly equal numbers. When we subtract two numbers that are nearly equal, we effectively cancel out the leading significant figures, leaving a much less precise number. On the hypothetical machine, the result $.314159 \times 10^1 - .314158 \times 10^1 = .100000 \times 10^{-4}$ indicates that we have lost five significant figures from our data; any further calculations can have no more than one significant figure of accuracy, and any additional digits in the result, as in the MATLAB result

```
>> 3.14159-3.14158
ans =
    1.000000000006551e-05
```

are just noise. (The result is different because the numbers are stored and manipulated in base 2.) The analogous 5-digit arithmetic example, $\pi - 3.1415$, gives $.0001$ in 5-digit decimal arithmetic and

```
>> 3.1416-3.1415
ans =
    9.999999999976694e-05
```

in MATLAB. The relative error in $.0001$ as an approximation to $\pi - 3.1415$ is nearly 8%. As an aside, this is a key reason why calculators may still use decimal operations: For many people, seeing this result for $3.1416 - 3.1415$ rather than the nice, round $.0001$ would be unsettling. However, round-off error is round-off error, and while the choice of base makes a difference in individual cases like this, overall it's immaterial which base we use.

Subtracting nearly equal numbers and the ensuing loss of significant figures is one of the most common problems in scientific computing. Unfortunately, as with the finite difference formula used with Newton's method in Sec. 1.5,

$$\frac{f(x_k + h) - f(x_k)}{h}$$

(for some small $h > 0$), doing so can be hard to avoid. Whenever possible, however, we try to re-write any computation to avoid or minimize subtracting nearly equal numbers. It's difficult to overestimate the importance of this: Subtraction of floating point values sharing several leading digits (technically, bits) is perhaps the largest source of loss of accuracy in numerical computations.

Let's look at an example of using our floating point arithmetic model for multiplication. Since e and π are both irrational, we can't hope to store the result exactly in finite precision arithmetic. Multiplication, however, is less problematic than subtraction.

EXAMPLE 1: To multiply e by π on a machine that uses 5 digit floating point arithmetic, the result will be $w = \mathrm{fl}(\mathrm{fl}(e) \cdot \mathrm{fl}(\pi))$, where

$$\begin{aligned}
\mathrm{fl}(e) &= \mathrm{fl}(2.7182818284590...) \\
&= 2.7183 \\
\mathrm{fl}(\pi) &= \mathrm{fl}(3.1415926535897...) \\
&= 3.1416
\end{aligned}$$

(note the rounding in each case). Hence

$$w = \mathrm{fl}(2.7183 \cdot 3.1416)$$
$$= \mathrm{fl}(8.53981128)$$
$$= 8.5398$$

with absolute error $|e\pi - 8.5398| \doteq 6.6E{-}05$ and relative error $|e\pi - 8.5398| \,/\, |e\pi| \doteq 7.7E{-}06$. The relative error is about what we'd expect given the available precision. □

It's very common in practice that we devise mathematically equivalent ways of re-writing a formula in order to achieve greater accuracy in floating-point arithmetic. The quadratic formula is a commonly occurring case.

EXAMPLE 2: If $y = ax^2 + bx + c$ with $a \neq 0$ then the roots of $y(x) = 0$ are given by $r_1, r_2 = \left(-b \pm \sqrt{b^2 - 4ac}\right)/2a$. If a is small then we may have a division by a small number problem. However, let's look instead at the case in which $b > 0$ and $b^2 \gg 4ac$. In this situation

$$r_1 = \left(-b + \sqrt{b^2 - 4ac}\right)/2a$$

will involve the subtraction of the nearly equal numbers b and $\sqrt{b^2 - 4ac} \approx b$ in the numerator. (This is not the case for r_2, which involves the addition of nearly equal numbers and so is not a concern; if $b < 0$, then r_2 is the tricky case and r_1 is fine.) The standard trick for this difficulty is called rationalizing the numerator, using the conjugate:

$$r_1 = \left(-b + \sqrt{b^2 - 4ac}\right)/2a \cdot \frac{\left(-b - \sqrt{b^2 - 4ac}\right)}{\left(-b - \sqrt{b^2 - 4ac}\right)}$$

$$= -\frac{2c}{b + \sqrt{b^2 - 4ac}}$$

(after some simplification). Like the formula for r_2, this involves only the addition of nearly equal numbers; in addition, we're dividing by the larger value $b + \sqrt{b^2 - 4ac} \approx 2b$ rather than the smaller value $2a$. Hence we find the roots using

$$r_1 = -\frac{2c}{b + \sqrt{b^2 - 4ac}}$$
$$r_2 = \frac{-\left(b + \sqrt{b^2 - 4ac}\right)}{2a}$$

where we can compute the quantity $b + \sqrt{b^2 - 4ac}$ just once and use it for both roots. Of course a similar trick works for r_2 when b is negative. Any code for computing the roots of a quadratic polynomial via the quadratic formula should be performing checks for issues like this and using appropriate formulas. □

Common sources of numerical error include the subtraction of nearly equal numbers and division by small numbers. (The finite difference formula above combines both of these worrisome issues.) We will need to verify that algorithms that must converge in infinite precision arithmetic exhibit stability with respect to the effects of round-off errors, and make appropriate changes to the algorithms when we find them to be unstable. In fact, round-off error is usually indistinguishable from errors in the values of the given data (say, measurement error in the initial conditions of an ODE initial value problem), so stability of a method with respect to small round-off errors also means stability with respect to small measurement errors, or the error associated with representing an initial condition like $y(0) = \pi$ using finite precision.

The MATLAB subsection for this section is relatively long but makes several important points. As you work through it, consider the differences between what you see, what you expected to see, and what would be true of arithmetic on the real numbers $\mathbb{R}$.

MATLAB

Let's plot the values found from Eq. (7.1) with initial conditions $x_0 = 2, x_1 = 1$ and the values found from Eq.(7.3) with $c_1 = 1, c_2 = 1$ (Problem 1). Enter:

```
>> x0=2,x1=1
>> x=[x0,x1]
```

(x is a row vector of two values, 2 and 1 in that order). Then enter:

```
>> for i=1:10;xnew=x1+x0;x=[x,xnew];x0=x1;x1=xnew;end
```

to compute 10 more values of x_k and append them to x. (This is not necessarily the most efficient way to do this.) Enter:

```
>> plot(x,'b'),grid
>> hold on
```

to plot them in blue and hold the plot. Now enter:

```
>> s1=(1-sqrt(5))/2;s2=(1+sqrt(5))/2;
>> z=s1.^(0:11)+s2.^(0:11)
>> plot(z,'r')
```

to evaluate $x_0, ..., x_{11}$ using the Binet formula from Eq. (7.3) and plot the resulting values in red. They overlay the blue values almost perfectly. Enter:

```
>> hold off
>> plot(z-x),grid
```

to plot the difference between these two vectors. The error is barely noticeable (note the vertical scale). The Binet formula does match the recurrence relation–in this case, with integer initial conditions $x_0 = 2, x_1 = 1$.

Now let's look at some idiosyncrasies of floating point arithmetic. Recall that MATLAB uses double precision by default. There is a largest machine number and a smallest (positive) normal machine number. Enter:

```
>> realmax,realmin
```

to see them. The value of **realmax** corresponds to the extreme case in Fig. 2, that is, **realmax** corresponds to all 1s in the mantissa and exponent. The value of **realmin** corresponds to the smallest normal number but is not the smallest number. Values larger than **realmax** are said to overflow and values smaller than realmin are said to underflow (and are set to subnormal numbers or zero). To ask for the corresponding values in single precision, enter:

```
>> realmax('single'),realmin('single')
```

Since there are only finitely many choices of bits in Fig. 2, there can be only finitely many machine numbers. Enter:

```
>> eps
>> 2^-52
```

to see the distance between 1 and the next nearest floating point number (called the **machine epsilon**). The distance between two normal floating point numbers varies with scale; the distance between 2 and the next nearest floating point number is twice the machine epsilon (enter `eps(2)` to see the value there), the distance between 4 and the next nearest floating point number is four times the machine epsilon (enter `eps(4)` to see the value there), and so on. Importantly, this gives the same *relative* accuracy at all scales but growing *absolute* inaccuracy as the scale increases. We'll see the significance of the machine epsilon recur in the text.

Suppose we wanted to compute $2436.7 - 2.3751 - 2438.2$ using our 5 digit floating point arithmetic. (Note that all of these numbers are perfectly represented in this base 10 model system.) Enter:

```
>> format long
>> 2436.7-2.3751
ans =
      2.434324900000000e+003
```

which becomes 2.4343×10^3 or 2434.3 in our system; to find the error, enter:

```
>> abs(2434.3-ans)
```

or about 0.025. Now let's finish the computation. We need to compute $2434.3 - 2438.2$; enter:

```
>> 2434.3-2438.2
ans =
      -3.89999999999964
```

or -3.9000 to 5 digits. Let's get the correct value; enter:

```
>> 2436.7-2.3751-2438.2
```

to see that it is -3.8751. Our answer matches to only two significant figures (-3.8751 is -3.9 to two significant figures), despite the fact that each of the original three numbers is represented perfectly on our hypothetical machine. The subtraction $2434.3 - 2438.2$ has canceled the other three significant figures, though the real problem is the initial subtraction $2436.7 - 2.3751$ which results in an error of .02. This is a small error on the scale of some of the numbers in the calculation but is not an insignificant error on the scale of the desired answer -3.8751. Any future calculations with this result can be no more accurate than this weakest link, which has only two significant figures. Try this again on $2436.7 - .23751 - 2438.2$; attempt to predict what will happen before you start.

We'll see other cases where subtraction of nearly equal numbers causes unacceptably large errors. For now note that performing the computation as $(2436.7 - 2438.2) - 2.3751$ gives the exact result. (Computer arithmetic is commutative but is not associative.) Although the problem is poorly scaled, re-writing it in this form addresses that problem.

Now suppose that we were to divide the result -3.9000 of our calculation by a small number ϵ. Since the true answer is $x = -3.8751$, we have $-3.9000/\epsilon = x/\epsilon + E/\epsilon$ where:

```
>> E=-3.9000-(-3.8751)
```

is the error -0.0249 in our computed answer. Hence any attempt to divide the result by ϵ will result in an additive error of $-0.0249/\epsilon$ in our solution. If ϵ is large this may not be a problem, but if ϵ is small this error could swamp the calculation, giving meaningless results. (Look again at the attempts to approximate $f'(1) = 1$ for $f(x) = \ln(x)$ with a very small h in the MATLAB subsection of Sec. 1.5.) For example, enter:

```
>> -3.9000/1E-10-(-3.8751/1E-10)
```

giving -249000000. This is a fairly large error on the scale of the true answer. This is the problem seen previously in the computation of the derivative of $\ln(x)$, and it can occur any time we divide by a small number (or multiply by a large number), especially when we are dividing the result of a previous calculation, which has its own error, by a small number. This is important: *Subtraction of nearly equal numbers and division by small numbers are to be avoided whenever possible.*

Since we cannot expect to evaluate functions with more relative accuracy than the machine epsilon, it would not be sensible to choose a convergence criterion of the form $\rho_n \leq \tau$ for τ about the size of $\in$ (or smaller); but even if a tolerance on the order of $\in$ was viable it would usually be wasteful to ask for 16 digit accuracy in our answers, given the amount of error likely to be included in the initial data and/or modeling assumptions in a physically motivated problem. Still, we can use $\in$ as part of the consideration of what tolerance to choose.

Let's try using the alternative version of the quadratic formula from Example 2. We want a case where $b > 0$ and b^2 is much greater than $4ac$. Enter:

```
>> a=1,b=10000,c=1;
>> x=(-b+sqrt(b^2-4*a*c))/(2*a)
```

giving $-1.0000e-004$ for the root using the traditional version of the formula. But, enter:

```
>> a*x^2+b*x+c
```

This gives a residual error of $-1.1177e-009$, which seems large for something that should be zero. The equivalent form $r_1 = -2c/(b + \sqrt{b^2 - 4ac})$ from using the conjugate gives:

```
>> xc=-2*c/(b+sqrt(b^2-4*a*c))
```

and again we see $-1.0000e-004$, which is the same as before *to the precision displayed*, but :

```
>> a*xc^2+b*xc+c
ans =
      0
```

so that this is a better estimate of the root, at least in terms of the residual error. (Using **format long** shows that this truly is zero.) A small amount of inspection of the formula for potential problems and application of algebra to re-work it has certainly paid off.

Problems

1. a.) Show that Eq. (7.2) is also valid for $k = 5$ and $k = 6$.

b.) Show that Eq. (7.1) with initial conditions $x_0 = 2, x_1 = 1$ gives the same values as Eq.(7.3) with $c_1 = 1, c_2 = 1$ for $k = 0, 1, 2, 3, 4, 5$.

2.) a.) Using 6 digit floating point arithmetic, compute e/π and π/e. Give the relative errors as well.

b.) Using 6 digit floating point arithmetic, compute $\sin(1) + \cos(1)$. Give the relative error as well.

3.) Using 5 digit floating point arithmetic, compute $6/9 - 2*(1/3)$ using the usual order of operations. Comment.

4.) Using 3 digit floating point arithmetic, perform the following sum from left to right: $801 + 150 + 72 + 14 + 4 + 3 + 2 + 1$. Repeat but perform the sum from right to left. Comment.

5.) a.) Floating point numbers in the form $(mantissa) \times (base)^{exponent}$ incorporate trade-offs made in deciding that, say, double precision numbers have 52 bit mantissas with 11 bit exponents, as opposed to the 46 bit mantissas and 17 bit exponents that would have resulted from a straight-forward doubling of the significand from single precision. Discuss the trade-offs between these two possible versions of double precision.

b.) How many bits would be used for a quadruple precision value? How many words does that correspond to? How would you suggest the bits be distributed (sign, exponent, mantissa)? Look up the IEEE 754 binary128 standard for quadruple precision and compare it to your suggested format.

c.) Discuss and compare the different possibilities suggested for half precision.

6.) In the notation of the MATLAB subsection, what formula would you use for r_2 in order to avoid the subtraction of nearly equal numbers when $b < 0$?

7.) Using the Mean Value Theorem, where would you expect evaluations of $\sin(\ln(x))$ to be relatively inaccurate for $x \in (0, 6)$, and where would they be relatively accurate?

8.) a.) Enter the line: `x=0;d=.1;for i=1:1000,x=x+d;end;x` in MATLAB. (Use `format long`.) Comment on your results, which are due to the fact that .1 does not have a terminating base 2 expansion. (This phenomenon caused the failure of a number of Patriot air defense missiles, which counted time in tenths of a second, to destroy incoming Scud missiles during the 1991 Gulf War.) Repeat for $d = .24, .25, .26$. Why is the error smallest for $d = .25$?

b.) Enter `realmax+1` and `realmax*2`. Explain why the former gives `realmax` again yet the latter gives `Inf`. Experiment to find how large x0 must be to make `realmax+x0` overflow to `Inf`. Does this match the claim that MATLAB gives about 16 digits precision?

9.) a.) Compute `Inf-Inf`, `Inf/Inf`, `3*Inf`, `1/Inf`, and several similar quantities. Recalling that `Inf` represents machine infinity and `NaN` represents not a number, do the results match your expectations?

b.) Does your computer conform to the IEEE standard? (See **help isieee** in MATLAB. The answer is almost certainly yes.)

10.) Create an example (in MATLAB or on a calculator) where the quadratic formula in the standard form gives a noticeably less accurate answer than the quadratic formula with rationalized numerator.

11.) Floating point arithmetic rounds numbers in the usual way; e.g., in 3 digit arithmetic, 1046.1 would become 1050 (i.e., $.105E4$). Rather than rounding one could also use chopping, which simply chops off extra digits or bits: In this case 1046.1 would become 1040. Perform several numerical experiments to determine the effects of chopping versus rounding, and report your results. Is rounding clearly superior to chopping?

12.) a.) In MATLAB, let $x = 3E200$ and $y = 4E200$ and compute $z = \sqrt{x^2 + y^2}$. Since x, y and z are each less than `realmax`, what went wrong?

b.) The problem of computing z is poorly scaled. Let $\rho = 1E200$ and compute z in the mathematically equivalent form $z = \rho\sqrt{(x/\rho)^2 + (y/\rho)^2}$.

c.) Repeat parts a. and b. with $x = 3E200$ and $y = 4$. Note that $(y/\rho)^2$ is less than `realmin`; is this a problem?

d.) How would you use this idea in the case $x = 3E - 200$ and $y = 4E - 200$?

e.) What is a reasonable rule for selecting a value of ρ?

13.) a.) If w, x, y, and z are numbers greater than one and on the same scale, suggest an appropriate formula for computing $(x - y) + (w + z)$.

b.) Repeat for $(x - y)^2 + (w - z)^2$.

14.) For almost every $x_0 \in (0, 1)$, the iteration $x_{k+1} = 2x_k \mod 1$ never reaches the fixed point at $x = 0$ for any k. What happens if you iterate this map on the computer? Experiment first, then give an explanation for your results.

15.) a.) Consider a 32 bit word and an integer x. As it is an integer we might store it as its bitstring, preceded by a 1 if negative or a 0 if positive. (This is called the **sign-and-modulus** form. No exponent is used so that the representation is exact.) What are the largest and smallest integers that may be represented in this form? How many different integers may be represented in this way? In how many ways may 0 be represented?

b.) If x is a negative number then it may be stored in the **twos complement** form in which $2^{32} - x$ is stored rather than x; if x is positive then it is stored as its bitstring in this scheme. Show that if $-2^{31} \le x \le 2^{31} - 1$ then the leftmost bit is 1 if x is negative and 0 if x is positive when x is stored in twos complement form. What are the largest and smallest integers that may be represented in this form? How many different integers may be represented in this way? In how many ways may 0 be represented?

c.) Show that the representation for x, when added to the representation for $-x$, gives 0 in the twos complement form if the overflow bit is neglected.

d.) Discuss the relative advantages and disadvantages of these two representation schemes.

16.) Calculators often work in base 10, using the binary coded decimal format. The calculations are done in base 10 although the digits are coded using bits via BCD. Which of the effects that we have discussed would be changed by working in base 10, and how?

8. Polynomial Roots

In general it is not reasonable to ask that a numerical method find all roots of even a continuous function of a single variable. There may be infinitely many such roots; there may be closely spaced roots (consider $\sin(1/x)$ near the origin); there may be a root at such a large value of x that the method simply never ends up looking for it there; the root may be a blip between two adjacent machine numbers with the function positive at all machine numbers; and so on. Even for well-behaved functions, a multiple root is hard to handle. If the multiplicity is even, as for $f(x) = x^2$, then the root is hard to see numerically, and whether it is even or odd, the fact that $f'(x) \approx 0$ in a neighborhood of the root means that the function is flat there, so points near it will seem like roots too.

Further, experience shows that methods like Newton's method sometimes have "favorite" zeroes, that is, if we find an approximate root x_a of $f(x) = 0$ starting from some initial guess x_0, then choose a new initial guess z_0, it is not uncommon for the method to find the same approximate root x_a from this new initial guess even if there are other roots to be found. One way to defeat this is to eliminate the newly found zero by dividing it out of the function, giving a new function

$$g(x) = \frac{f(x)}{x - x^*}$$

which has zeroes exactly where f does, save for x^* (assuming x^* is an exact root of $f(x) = 0$). This process is known as **deflation**. In practice $f(x_a)$ is not exactly zero and so

$$g(x) = \frac{f(x)}{x - x_a}$$

has a pole, that is, an infinitely large spike, next to a zero. Often the use of deflation is enough to drive the method away from the known zero near x^*, but not always. The pole-near-a-zero phenomenon can itself cause difficulties, and so the cure can be as bad as the disease in some cases. Still, repeated deflation of the form

$$g(x) = \frac{f(x)}{(x - x_a)(x - x_b)(x - x_c)}$$

(after three different approximate roots x_a, x_b, and x_c have been found) can be a useful tool. When $f(x)$ is a polynomial of known degree n, we know we can stop after we have found n zeroes of it.

It's worth repeating for emphasis: For a general function–especially a function of multiple variables–it's unrealistic to hope to find all zeros of a function that has several zeroes, even in the uncommon case that we know how many zeroes there are. Only when we know that there is a unique zero can we reasonably hope for full success.

In the special case of polynomials in a single real variable it may be reasonable to ask that we find *all* zeroes, or at least all distinct zeroes, of the function. One approach for finding all real zeroes of a polynomial is to use Newton's method to find a root, apply deflation, and repeat. (Because the function is known to be a polynomial we can use synthetic division in the deflation stage.) If complex-valued zeroes are also wanted we can use Newton's method with a complex initial guess. A similar approach is to use Müller's method; this method is extremely robust for polynomials and will find a zero for most initial guesses. Once a first zero is found, the polynomial may be deflated and we can use the resulting lower degree polynomial to find another zero, and so on. Even if, due to the accumulating effects of errors due to deflation, these estimates of the zeroes are not as precise as we might like, we may return to the original equation $f(x) = 0$ at any time and use Newton's method with these estimates as initial guesses to "clean them up"" as needed. This cleaning-up process will generally converge rapidly and hence improve the accuracy of the approximate zeroes.

To deflate a polynomial $p(x)$ with a root at x_0 we must be able to perform the operation $p(x)/(x - x_0)$ so as to find a lower-degree polynomial $q(x)$ plus a remainder (if x_0 is not a precise zero of p); that is, we must be able to find $q(x)$ such that $p(x) = (x - x_0)q(x) + q_n$. **Horner's method** computes the coefficients in $q(x) = q_0 x^{n-1} + q_1 x^{n-2} + \cdots + q_{n-2}x + q_{n-1}$ from the coefficients in $p(x) = a_0 x^n + a_1 x^{n-1} + \cdots + a_{n-1}x + a_n$ and the value of x_0 by recursion. We set $q_0 = a_0$, then

$$q_i = x_0 q_{i-1} + a_i$$

$i = 1, ..., n$. (This follows from synthetic division. A slight extension of it allows us to evaluate q' also.) Hence, we can actually perform the deflation in a floating point computing environment, without a CAS. We ignore the small error q_n that reflects the fact that x_0 is only an *approximate* zero.

If the roots are known to be real and well-separated, another possibility is to attempt to bracket them by sampling the function and looking for sign changes. Newton's method, possibly in the globalized form, may then be applied in each interval. This would likely require user intervention to obtain the brackets in non-trivial cases.

There are numerous specialized methods for finding all zeroes of a polynomial. Many of them work by finding a quadratic factor and then using synthetic division to reduce the degree of the polynomial by two. This is based on the **Fundamental Theorem of Algebra**: Every real polynomial can be written as a constant times a product of real factors of the form $(x - x_i)^{m_i}$ for each real root of multiplicity m_i and $(x^2 + c_j x + d_j)^{p_j}$ for each complex conjugate pair of roots of multiplicity p_j. We use quadratic factors at each stage so that we can use real arithmetic for polynomials with real coefficients (at least until we must actually factor a quadratic that has complex conjugate roots).

One such method is **Bairstow's method**, which is based on the observation that if we knew any two roots γ_1, γ_2 of the degree n polynomial

$$P(x) = a_0 x^n + a_1 x^{n-1} + \cdots + a_{n-1} x + a_n$$

($a_0 \neq 0$) then we could form $(x - \gamma_1)(x - \gamma_2) = (x^2 - rx - s)$ and factor $P(x)$ in the form

$$
\begin{aligned}
P(x) &= a_0 x^n + a_1 x^{n-1} + \cdots + a_{n-1} x + a_n \\
&= Q(x)(x^2 - rx - s)
\end{aligned}
$$

where $Q(x)$ is a polynomial of degree $n - 2$. But what if we chose some $x^2 - rx - s$ that is not necessarily a factor of $P(x)$? Then division of $P(x)$ by $x^2 - rx - s$ would leave a polynomial remainder of degree (at most) 1, which we could write as

$$(8.1) \qquad P(x) = Q(x)(x^2 - rx - s) + R(x)$$

where the remainder term has the form $R(x) = \alpha x + \beta$. The key observation is this: The remainder is zero exactly when $\alpha = \beta = 0$, but we can choose r and s, and those values of r and s affect the values of the coefficients $\alpha(r, s)$ and $\beta(r, s)$ in the remainder. What if we try to choose r and s so as to make both α and β zero simultaneously? That is, we will try to find a solution (r, s) of the 2×2 nonlinear algebraic system

$$
\begin{aligned}
\alpha(r, s) &= 0 \\
\beta(r, s) &= 0
\end{aligned}
$$

in order to force the remainder $R(x)$ to be zero. If we succeed, we can then use synthetic division to divide the quadratic factor $x^2 - rx - s$ out of $P(x)$ to get the lower degree polynomial $Q(x)$; the two (real or complex) roots of $x^2 - rx - s$, found by the quadratic formula, are roots of $P(x)$ (at least approximately). In the end we should have all n roots of the polynomial–including repetitions!

This scheme leaves us with one unresolved sub-problem, however: How shall we solve the 2×2 nonlinear algebraic system in r and s? There's more than one way this can be done, but most implementations of Bairstow's method use Newton's method for systems, which we will discuss in the next chapter. It is an iterative method in which an initial guess (r_0, s_0) is chosen and we continue until successive iterates are changing very little from iteration to iteration. In the MATLAB subsection we suggest a work-around for now: The system

$$\alpha(r, s) = 0$$
$$\beta(r, s) = 0$$

is solved by the same values of (r, s) that make minimize

$$F(r, s) = \alpha^2(r, s) + \beta^2(r, s)$$

as the global minimum of F is 0, assuming that both $\alpha(r, s) = 0$ and $\beta(r, s) = 0$ do actually occur simultaneously. Hence, if we have a routine–like the MATLAB command `fminsearch`–that minimizes functions, we can avoid explicitly performing the root-finding. In practice, we sometimes use this minimization trick on particularly difficult root-finding problems. When working the Problems, you could also use a CAS to find the remainder explicitly–an option unlikely to be feasible in most programming environments.

We will need to be able to compute values of $\alpha(r, s)$, $\beta(r, s)$, and, for Newton's method, their partial derivatives $\alpha_r(r, s)$, $\alpha_s(r, s)$, $\beta_r(r, s)$, and $\beta_s(r, s)$. The derivation of the required formulas is a tedious exercise in synthetic division–the analogue of Horner's method for division by quadratic factors–and we present only the results. Pick an (r, s). Let the polynomial $q(x)$ of Eq. (8.1) be written as

$$q(x) = q_0 x^{n-2} + q_1 x^{n-3} + \cdots + q_{n-3} x + q_{n-2}.$$

After some algebra, it follows from Eq. (8.1) that the recursion:

(8.2)
$$q_0 = a_0$$
$$q_1 = q_0 r + a_1$$
$$\vdots$$
$$q_k = q_{k-2} s + q_{k-1} r + a_k$$

$(k = 2, 3, ..., n-2)$ computes the coefficients of $q(x)$, and then

(8.3)
$$\alpha(r, s) = q_{n-3} s + q_{n-2} r + a_{n-1}$$
$$\beta(r, s) = q_{n-2} s + a_n$$

are the values of α and β for this (r, s). (Note that we are obtaining values for α and β, not formulae for $\alpha(r, s)$ and $\beta(r, s)$ in general.) Differentiating Eq. (8.1) and performing some algebra shows that:

$$\alpha_r(r,s) = r\alpha_s(r,s) + \beta_s(r,s)$$
$$\beta_r(r,s) = s\alpha_s(r,s)$$

in terms of $\alpha_s(r,s)$ and $\beta_s(r,s)$. Then the recursion (with $m = n - 2$)

$$t_0 = q_0$$
$$t_1 = t_0 r + q_1$$
$$\vdots$$
$$t_k = t_{k-2}s + t_{k-1}r + q_k$$

($k = 2, 3, ..., m - 2$) gives the values needed to compute

$$\alpha_s(r,s) = t_{m-3}s + t_{m-2}r + q_{m-1}$$
$$\beta_s(r,s) = t_{m-2}s + q_m$$

which provides the derivative values that will be needed by Newton's method. See one of the texts listed in the bibliographic references, such as the classic text by Stoer and Bulirsch, for the details.

Summarizing, each time we find a quadratic factor its roots are roots of $P(x)$, and we divide out this quadratic to deflate $P(x)$ and then start again. If $P(x)$ has real coefficients then the method will use only real arithmetic save when we apply the quadratic formula to a quadratic factor that corresponds to a complex conjugate pair. When (after successive deflations) we finally reach a polynomial of degree two or one, we simply find its roots by the quadratic formula or by finding the x-intercept of the line.

How does the MATLAB `roots` command work? In a matrix algebra course you may have been taught how to find the characteristic polynomial $p(\lambda)$ of an $n \times n$ matrix and use it to find the eigenvalues of the matrix by finding the roots of the characteristic equation $p(\lambda) = 0$. The `roots` command does exactly the opposite: It treats the polynomial $P(x)$ as if it were the characteristic polynomial of a certain matrix that has $P(\lambda)$ as its characteristic polynomial and finds all n of its eigenvalues by an iterative method acting on the matrix (see Ch. 3). It then reports the eigenvalues of the matrix as the desired roots.

As we have seen in computer experiments, finding all roots of a polynomial can be a challenging problem. One way to understand this is as follows: Let $p(x)$ be a polynomial of degree n and let x^* be a simple root of $p(x) = 0$. (Things only get worse at multiple roots.) We can imagine the effect of a slight change to the coefficients of $p(x)$–possibly just the error in representing those coefficients on the machine–as being due to the addition of another polynomial that has small coefficients

$$\widetilde{p}(x) = p(x) + \epsilon h(x)$$

called a **perturbation** of $p(x)$. How much of an effect can this change have on the location of the root? Certainly we expect that the root has most likely moved. If the parameter ϵ is small it can be shown that the location of the root depends smoothly on it. Let $\zeta(\epsilon)$ be the location of the root of $\widetilde{p}(x)$, so that $\zeta(0) = x^*$.

(Remember that $\zeta(\epsilon)$ may be complex-valued even if x^* and ϵ are real.) We seek the solution $\zeta(\epsilon)$ of

$$\tilde{p}(\zeta) = p(\zeta) + \epsilon h(\zeta) = 0$$

as parameterized by ϵ. Differentiating with respect to the parameter ϵ gives

$$\frac{d}{d\epsilon}\left(p(\zeta(\epsilon)) + \epsilon h(\zeta(\epsilon))\right) = \frac{d}{d\epsilon}(0)$$
$$p'(\zeta(\epsilon))\zeta'(\epsilon) + \epsilon h'(\zeta(\epsilon))\zeta'(\epsilon) + h(\zeta(\epsilon)) = 0$$

and at $\epsilon = 0$ we have

$$p'(\zeta(0))\zeta'(0) + h(\zeta(0)) = 0$$
$$p'(x^*)\zeta'(0) + h(x^*) = 0$$
$$\zeta'(0) = -h(x^*)/p'(x^*)$$

(recall that x^* is a simple zero of $p(x)$ so the denominator is nonzero). The rate of change of the location of the root of $p(x) = 0$ (corresponding to $\epsilon = 0$) may be large if the ratio $|h(x^*)/p'(x^*)|$ is large. The tangent line approximation at $\epsilon = 0$ is

$$\zeta(\epsilon) \approx x^* - \epsilon h(x^*)/p'(x^*)$$

and so even for a small value of ϵ we may see a large change $-\epsilon h(x^*)/p'(x^*)$ in the location of the root if $|h(x^*)/p'(x^*)|$ is sufficiently large compared to ϵ. In particular, the smaller the denominator $|p'(x^*)|$ is, the larger we expect this ratio to be. When it is the case that $|h(x^*)/p'(x^*)|$ is large we say that the root is **ill-conditioned**, or that this root-finding problem is ill-conditioned[11]. This means that small changes in the coefficients of $p(x)$ can cause disproportionately large changes in the locations of the zeroes of $p(x)$. Wilkinson's polynomial `poly(1:20)` (in MATLAB notation) is one example, but unfortunately it is not atypical in this respect. The degree of ill-conditioning is even worse for multiple roots.

MATLAB

Does MATLAB have a Bairstow's method program already implemented for us? If so, how would we find it? To be perfectly frank, it's common to simply search the web for something like "MATLAB Bairstow" and see if you can find some guidance on this.

However, the MATLAB environment itself provides a number of useful utilities for locating MATLAB-provided functions. To see if MATLAB has a Bairstow's method program, enter:
```
>> lookfor bairstow
```
(Presumably you found no results.) The `lookfor` command looks for its argument in the first comment line (called the H1 line) of all m-file on the path. Does MATLAB have a Newton's method program? Enter:
```
>> lookfor newton
```

---

[11]Of course $h(x)$ won't actually be known; for this and other reasons we use $|p'(x^*)|$ small as our criterion for ill-conditioning. Note that since $p(x)$ is a polynomial, $p'(x)$ is easy to evaluate and we can check its size near a root.

You will probably find several results but none that are Newton's method for root-finding. Enter:

```
>> lookfor newton -all
```

to look for the word Newton in the entire first block of consecutive comment lines. The search is not case sensitive. Both of these searches could have been done by typing **help** and the name, but it is not always the case that the name is obvious. You can investigate the purpose of these commands using **help**, **type**, or **doc** (for html documentation).

It's common at this point in the process to go to the website of The MathWorks, the makers of MATLAB, and search the large library there of user-provided m-files for a variety of tasks. If you use MATLAB a lot, you should get used to doing this. It's also possible to ask questions there and get answers from the community of MATLAB users and occasionally from a MathWorks employee, and of course you can search the archive of previous questions-and-responses. Again, as many programmers joke: Programming is googling. Don't ever be reluctant to search the web for help.

Given the lack of a built-in Newton's method for systems routine, we will solve the nonlinear system occurring in Bairstow's method using a transformation: To solve $\alpha(r,s) = 0$, $\beta(r,s) = 0$ simultaneously we will instead attempt to minimize $F(r,s) = \alpha^2(r,s) + \beta^2(r,s)$. If there is a solution to the root-finding problem, then it makes $F(r,s)$ zero, which is surely as small as it can be. The **fminsearch** command in MATLAB can be used when working the Problems if a system can't be solved by hand[12]. (This command uses an algorithm discussed in Sec. 7.6.) Like so many MATLAB commands, this minimization routine expects that its argument will be a vector so that it can work with maximum efficiency, so we will need to write our system as a function of $v = (r,s)^T$.

As an example, if $\alpha(r,s) = r^2 + rs$ and $\beta(r,s) = (r+1)^2 + (s-1)^2$ (for which $r = -1$, $s = 1$ is a solution of the system) then

```
>> F=@(v) (v(1)^2 +v(1)*v(2))^2+((v(1)+1)^2+(v(2)-1)^2)^2
>> F([-1 1])
```

which is indeed zero. (If you had entered F(-1,1) instead you would have gotten an error as the function expects a single, vector argument.) Enter:

```
>> fminsearch(F,[-2 3])
```

which finds the solution to about 5 decimal places. For now, use this method to solve the nonlinear system. By the end of Ch.2 we'll be able to bring Newton's method to bear on this problem.

It's common to approach some of the hardest root-finding problems for systems by transforming the problem from $F(x) = 0$ to minimizing $\|F(x)\|$. This is because minimization algorithms tend to be more robust that root-finding methods. Since we are minimizing $f_1^2(x) + ... + f_n^2(x)$, this is termed a nonlinear sum-of-squares (NLS) approach. When other methods fail, changing your root-finding problem to a minimization problem may succeed.

Let's end with a few more notes on MATLAB functionality. In addition to the help available from drop-down menus there is also tab completion. Type h and then strike the tab key twice; that is, type:

```
>> h
```

---

[12]The MATLAB command **fsolve** directly aattempts to solve for the roots of a nonlinear system in essentially this way.

followed by *tab*, *tab* rather than return. All commands beginning with the letter h are listed. Striking the tab key once (not twice) will complete a command if there is only one possible choice.

Additionally, MATLAB has a number of commands that relate to the operating system of your computer. Type `help which`, `help what`, and `help dir` to see how to locate the full path for a function, list the MATLAB files in the current directory, and list all files in the current directory. The `which` command is useful if two m-files happen to share the same name. (Of course, you should try to avoid this, but it sometimes happens when you add someone else's MATLAB programs to your machine, e.g., from a toolbox.) Type `help path`, then enter:

```
>> path
```

to see your current path. If two m-files share a name then the first one on the path is the one that will be executed. The help for `path` suggests ways in which the path may be manipulated.

Every time you start MATLAB it will look for an m-file `startup.m` and if it finds it it will execute it (see `help startup`). This script file is meant to enable you to set your own personal preferences for your session. It can be used to set your path, to specify format choices like `format long` and/or `format compact`, to initialize frequently used constants (e.g., Planck's constant $h = 6.62607015 \times 10^{-34}$ as `Planck = 6.62607015E − 34`; defining a long list of these in the startup file is quite common) and so on. Of course, you may want to load different constants depending on what type of work you're doing that day. You can achieve similar functionality via workspaces, which this text does not address (but you should be able to find help for it online easily).

The help command works on directories as well. When you enter `help` the entries in the left-hand column are directory names. Enter:

```
>> help matlab\polyfun
```

to see the functions available in the `polyfun` subdirectory, which are interpolation or polynomial routines. You may find `polyval` particularly useful for this section.

It's advisable to have your own directory or directories for your m-files. You might like to add m-file for common abbreviations like `ln.m` to compute the natural logarithm (which is the `log` function in MATLAB) for mnemonic convenience, say. You might want a list of electromagnetic constants defined by a script in one directory with material strengths in another.

The command `pwd` lists the current directory, and the command `cd` may be used to change it. These work on Windows machines as well as on Unix machines. If you enter the exclamation point `!` then any text that follows it is interpreted as a command to the underlying operating system; see `help punct`. For example,

```
>> !ls -la
```

lists files in a Unix environment. (Try `!dir` in a PC environment, and see also `help dos` and `help unix`.) There are many MATLAB commands that duplicate operating system functions, such as the `dir` command previously mentioned, `delete` for deleting files, and others. Of course, placing a deletion command in an m-file is potentially risky if there's any chance that you might delete another user's files.

Problems

1.) a.) Use `roots(poly(1:20))` to generate the approximate zeroes of Wilkinson's polynomial $(x - 1)(x - 2) \cdots (x - 20)$. Which numerically computed zero is most accurate? Which is least accurate? What happens for `roots(poly(1:40))`?

b.) Use a single step of Newton's method to clean up the approximations for the roots at 10, 11, and 12. How much improvement do you see?

2.) Use Horner's method to compute $q(x) = q_0 x^2 + q_1 x + q_2$ and q_3 for $(x^3 - 2x^2 + 3x - 1)/(x - 1)$. Check your result by computing $(x - 1)q(x) + q_n$.

3.) a.) Use Newton's method with deflation to find all real roots of $x^3 - 4.2x^2 + 5.6x - 2.4 = 0$. Deflate once and then apply the quadratic formula.

b.) This time find the first root, deflate, then use Newton's method again to find the next root. Then deflate the quadratic, and solve the resulting linear equation directly.

4.) Use the recursion in Eq. (8.2) and Eq. (8.3) to divide $5x^6 - 3x^5 - 4x^3 + 2x - 2$ by $2x^2 - 3x + 1$, giving a quartic polynomial $Q(x)$ and a linear remainder $R(x) = \alpha x + \beta$.

5.) Perturb $p(x) = (x - 1)^5$ by a "small" polynomial so as to demonstrate the ill-conditioning of the zero.

6.) Write a detailed algorithm (pseudo-code) to perform one iteration of Bairstow's method (that is, your algorithm should try to find one quadratic factor and returns its roots).

7.) a.) Use Bairstow's method to find one pair of roots of $x^4 - 2x^3 + 2x^2 - 2x + 1$. Then use the quadratic formula to find the other pair.

b.) Solve this problem again using Newton's method with deflation. Which is the better approach in this particular case?

8.) Use Bairstow's method to find one pair of roots of $x^4 + 2x^3 + 6x^2 + 8x + 8 = 0$ (verify that they are all complex-valued by plotting it). Then use the quadratic formula to find the other pair.

9.) Use Bairstow's method to find one pair of roots of $x^9 - 5x^8 - x^5 + 3x^4 + 2x^2 + 6x + 2 = 0$.

10.) Use Bairstow's method to find one pair of roots of $2x^8 + 3x^7 - 4x^6 + 3x^4 + 4x^3 + 6x + 2 = 0$.

11.) Write a MATLAB program that implements Bairstow's method.

12.) a.) Derive Horner's method by equating coefficients in $p(x) = (x - x_0)q(x) + q_n$.

b.) How could Horner's method be used to also compute derivative values $q'(x)$?

13.) Use Newton's method with deflation (implemented via Horner's method) to find all roots of $x^7 - 5.8x^6 + 55.43x^5 - 256.79x^4 + 188.76x^3 + 1711.6x^2 - 6188.1x + 7474.8 = 0$; when you can no longer find real roots, begin using complex initial guesses.

14.) Use Müller's method with deflation (by Horner's method) to find all roots of $x^7 - 5.8x^6 + 55.43x^5 - 256.79x^4 + 188.76x^3 + 1711.6x^2 - 6188.1x + 7474.8 = 0$; improve the quality of the estimates by using Newton's method with the results of Müller's method as initial guesses.

15.) The following polynomials have a simple zero at $x = 1$: $x^3 - 5x^2 + 8x - 4$, $x^3 - 1.75x + .75$, $3x^3 - 13.5x^2 + 19.5x - 9$. For which of these is $x^* = 1$ the most ill-conditioned?

16.) Devise an algorithm of your own to find polynomial roots. Implement it and test it.

CHAPTER 1 BIBLIOGRAPHY:

Numerical Analysis, 6th ed., Richard L. Burden and J. Douglas Faires, Brooks/Cole Publishing Company 1997
A traditional introductory undergraduate level textbook.
Iterative Methods for Linear and Nonlinear Equations, C. T. Kelley, SIAM 1995
A modern treatment of Newton methods.
Iterative Methods for the Solution of Equations, J. F. Traub, Prentice-Hall 1964
An older text on nonlinear equations and nonlinear systems.
Algorithms for Minimization Without Derivatives, Richard P. Brent, Dover 2002
Includes Brent's method for root-finding.
Methods for Solving Systems of Nonlinear Equations, Werner C. Rheinboldt, SIAM 1974
An older monograph on nonlinear equations and nonlinear systems.
Numerical Computing with IEEE Floating Point Arithmetic, Michael L. Overton, SIAM 2001
An introduction to floating point arithmetic for scientific computation.
MATLAB Guide, Desmond J. Higham and Nicholas J. Higham, SIAM 2000
An informal guide to the capabilities of MATLAB, with examples.
Numerical Recipes: The Art of Scientific Computing, William H. Press, Brian P. Flannery, Saul A. Teukolsky, and William T. Vetterling, Cambridge University Press 1986
A reference with many algorithms, emphasizing applications.
Numerical Computation 1: Methods, Software, and Analysis, Christoph W. Ueberhuber, Springer-Verlag 1997
A graduate level treatment of the fundamentals of scientific computing, including hardware issues.
Numerical Computation 2: Methods, Software, and Analysis, Christoph W. Ueberhuber, Springer-Verlag 1997
A graduate level treatment of numerical methods.
A First Course in Numerical Analysis, 2nd ed., Anthony Ralston and Philip Rabinowitz, Dover 2001
An older, comprehensive advanced text on the subject republished by Dover.
Numerical Methods for Scientists and Engineers, R.W. Hamming, Dover Publications 1986
An older, comprehensive text on the subject, republished by Dover.
Introduction to Numerical Analysis, J. Stoer and R. Bulirsch, Springer-Verlag New York 1980
The classic reference work.
Accuracy and Stability of Numerical Algorithms, Nicholas J. Higham, SIAM 1996
A modern treatment of error analysis.
Rounding Errors in Algebraic Processes, J. H. Wilkinson, Dover 1994
The classic text on error analysis.
A Brief Introduction to Numerical Analysis, Eugene E. Tyrtyshnikov, Birkhäuser 1997
A concise, modern, abstract introduction to the theory of numerical analysis.

Afternotes on Numerical Analysis, G.W. Stewart, SIAM 1996
An informal introduction to basic numerical analysis.

Afternotes Goes to Graduate School: Lecture Notes on Advanced Numerical Analysis, G.W. Stewart, SIAM 1998
An informal introduction to graduate-level numerical analysis.

An Introduction to Computational Science, Allen Holder and Joseph Eichholz, Springer 2019
An introductory textbook for Computational Science, including numerical methods and modeling.

CHAPTER 2

Linear Systems

1. Gaussian Elimination with Partial Pivoting

As stated previously, three of the most commonly occurring subproblems in scientific computing are the solution of a linear algebraic system, the solution of a nonlinear algebraic system, and the location of minima and maxima. One reason they occur so often is that many more sophisticated computations such as solving a system of differential equations often require that, say, a linear system must be solved for each point. We have discussed nonlinear scalar equations in Ch. 1. We now turn, in this chapter and the next, to the solution of linear systems of equations

$$(1.1) \qquad\qquad Ax = b$$

where, in general, A is an $m \times n$ matrix (we will assume real entries) and $b \in \mathbb{R}^n$. We will also return to nonlinear equations in this chapter so that we can generalize Newton's method to systems.

How shall we solve a linear system? In principle we could rewrite Eq. (1.1) in the equivalent root-finding form

$$(1.2) \qquad\qquad Ax - b = 0$$

and apply a method for solving nonlinear systems to $F(x) = Ax - b$. Why don't we? There are a number of reasons. One is that Newton's method for systems requires the solution of an equation like Eq. (1.1) at every iteration, leading to a circularity problem; we need a method for solving Eq. (1.1) before we can use Newton's method on a nonlinear system of equations. Another reason is that, especially if A is not square, there may be no solution or there may be infinitely many solutions, and it is important to be able to determine when this is the case. There are also numerical issues, as we will see.

But the over-riding reason that we consider the solution of linear systems separately from the solution of nonlinear systems is size. Linear systems have been studied extensively, and much is known about them; the very use of the term "nonlinear" implies "everything else" (all we know about such problems is that they are not linear). Because of what we know about linear systems from linear algebra, we can solve much larger *linear* systems, using techniques appropriate for them, than *nonlinear* systems. Large linear systems, with n in the thousands, occur constantly; handling them conveniently is why MATLAB® was developed. They may arise directly, as in a controls problem, but very often arise as a subproblem in a numerical method for a different type of problem, e.g., Newton's method for systems or the

DOI: 10.1201/9781003042273-2

discretization of a problem involving a partial differential equation. Large linear systems are ubiquitous in scientific computing.

In this chapter we will focus on two approaches to solving Eq. (1.1) (equivalently, Eq. (1.2)): The LU decomposition, based on Gaussian elimination, and the QR decomposition. In Ch. 3 we'll discuss other approaches to solving Eq. (1.1) as well as other aspects of computational matrix algebra.

EXAMPLE 1: Let's review the method of Gaussian elimination. Given an $m \times n$ matrix A and an n-vector b, we write the linear system $Ax = b$ in the form of an augmented matrix $[A \vdots b]$ and attempt to reduce it to the form $[U \vdots \widetilde{b}]$ using only the three **elementary row operations**, namely: *I.*) Subtracting a multiple of row i from row j and making this the new row j, *II.*) Interchanging rows i and j, and *III.*) Multiplying a row by a nonzero scalar. These operations leave the solution set unchanged, and (augmented) matrices that differ only by some number of these operations having been performed are said to be **(row) equivalent**. (We write $M \widetilde{\ } N$ to indicate that matrices M and N are row equivalent.) The final matrix $[U \vdots \widetilde{b}]$ must have U **upper triangular**, that is, all entries of U below the main diagonal must be zero. If, say, $A = [4 \ 2; 2 \ 4]$ and $b = (6, 6)^T$ then we have

$$\begin{bmatrix} 4 & 2 & \vdots & 6 \\ 2 & 4 & \vdots & 6 \end{bmatrix} \sim \begin{bmatrix} 4 & 2 & \vdots & 6 \\ 0 & 3 & \vdots & 3 \end{bmatrix}$$

upon performing $R_2 \leftarrow R_2 - \frac{1}{2}R_1$ (where R_i denotes row i); the new row 2 is the old row 2 minus 1/2 times row 1. The solution of $Ax = b$ is now easily found by **back substitution** (or **backwards substitution**), that is, since the matrix to the left of the partition is upper triangular, we start from the bottom and solve $3x_2 = 3$ to find $x_2 = 1$, then proceed to $4x_1 + 2x_2 = 6$ to find $x_1 = (6 - 2 \cdot 1)/4 = 1$. The solution is $x = (1, 1)^T$. □

From now on we will assume for the sake of simplicity that, unless otherwise stated, the coefficient matrix A in $Ax = b$ is **square** (meaning that $m = n$). Gaussian elimination as performed above is sometimes called "naive" Gaussian elimination because we do not include any checks for possible numerical problems. Such issues arise far too frequently to ignore, however. Consider the linear system $Ax = b$ given by

(1.3) $$\begin{bmatrix} .007 & -.8 \\ -.1 & 10 \end{bmatrix} \begin{bmatrix} x_1 \\ x_2 \end{bmatrix} = \begin{bmatrix} .7 \\ 10 \end{bmatrix}.$$

Naive Gaussian elimination requires that we perform $R_2 \leftarrow R_2 - (-.1/.007)R_1$, giving the equivalent (to the precision indicated) system

(1.4) $$\begin{bmatrix} .007 & -.8 \\ 0 & -1.4286 \end{bmatrix} \begin{bmatrix} x_1 \\ x_2 \end{bmatrix} = \begin{bmatrix} .7 \\ 20 \end{bmatrix}$$

and the solution is precisely $x_1 = -1500, x_2 = -14$ (as you may check by computing Ax). This is fine in principle, but we will be working in finite precision. If either A or b is the result of another calculation and hence has been affected by round-off errors, we have just magnified those errors by a factor with absolute value $.1/.007 \approx 14$ and carried them into Eq. (1.4). To see the effect of this in action, suppose that we were working in three decimal digit arithmetic; note that the original system Eq. (1.3) can be perfectly represented (i.e., without representation error) in such a system. We know that the $(2, 1)$ entry of the matrix in Eq. (1.4) must be zero, but the $(2, 2)$ entry is

$$
\begin{aligned}
10 - (-.1/.007)(-.8) &= 10 - (-14.2857...)(-.8) \\
&\doteq 10 - (-14.3)(-.8) \\
&= 10 - (11.44) \\
&\doteq 10 - 11.4 \\
&= -1.40
\end{aligned}
$$

which seems a reasonable approximation to the correct value -1.43 (to three digits); note however how the subtraction has canceled a digit (we get -1.40 but from $10 - (11.44)$ above it we should have gotten -1.44, which only requires three digits and would be a better approximation). Similarly, the second entry of the vector $\tilde{b}$ in Eq. (1.4) is

$$
\begin{aligned}
10 - (-.1/.007)(.7) &= 10 - (-14.2857...)(.7) \\
&\doteq 10 - (-14.3)(.7) \\
&= 10 + (10.01) \\
&\doteq 10 + 10.0 \\
&= 20.0
\end{aligned}
$$

which, despite an intermediate round-off error, is equal to the correct value 20; two wrongs have made a right. Finally, this gives $x_2 = 20/ - 1.4 \doteq -14.2857 \doteq -14.3$ (to three digits), a so-so approximation of the true value $x_2 = -14$ (the relative error is over 2%), and

$$
\begin{aligned}
x_1 &= (.7 - -.8 x_2)/.007 \\
&\doteq (.7 - -.8 \cdot -14.3)/.007 \\
&= (.7 - (11.44))/.007 \\
&\doteq (.7 - 11.4)/.007 \\
&= -10.7/.007 \\
&= -1528.57... \\
&\doteq -1530
\end{aligned}
$$

which is again at best a so-so approximation of the true value $x_1 = -1500$ (the relative error is 2%). Is it because we used three digit arithmetic? No, though using more digits would certainly give better results. But we can do better in three digit arithmetic already. Look again at Eq. (1.3):

$$
\begin{bmatrix} .007 & -.8 \\ -.1 & 10 \end{bmatrix} \begin{bmatrix} x_1 \\ x_2 \end{bmatrix} = \begin{bmatrix} .7 \\ 10 \end{bmatrix}.
$$

Although every value can be represented exactly in three digit arithmetic–in fact, every value can be represented exactly in *one* digit arithmetic–the entries of A

range over 5 orders of magnitude. Suppose we rewrite Eq. (1.3) by interchanging
the rows ($R_1 \longleftrightarrow R_2$):

$$\begin{bmatrix} -.1 & 10 \\ .007 & -.8 \end{bmatrix} \begin{bmatrix} x_2 \\ x_1 \end{bmatrix} = \begin{bmatrix} 10 \\ .7 \end{bmatrix}.$$

Now performing the row reduction step $R_2 \leftarrow R_2 - (.007/-.1)R_1$ involves multiplying any errors in row 1 by a quantity of magnitude $.007/.1 = .07$ rather than
of approximate magnitude 14; we won't be magnifying any errors. Let's see if this
helps: The new $(2,2)$ entry is

$$\begin{aligned} -.8 - (.007/-.1)(10) &= -.8 - (-.070)(10) \\ &= -.8 - (-.700) \\ &= -.1 \end{aligned}$$

and the new second entry of the vector $\widetilde{b}$ is

$$\begin{aligned} .7 - (.007/-.1)(10) &= .7 - (-.070)(10) \\ &= .7 - (-.700) \\ &= 1.40 \end{aligned}$$

(the fact that we didn't need to round anything off to three digits is a coincidence;
if we had, note that the lost values would have been on the order of hundredths or
smaller). The reduced system is

$$\begin{bmatrix} -.1 & 10 \\ 0 & -.1 \end{bmatrix} \begin{bmatrix} x_2 \\ x_1 \end{bmatrix} = \begin{bmatrix} 10 \\ 1.4 \end{bmatrix}.$$

giving $x_1 = 1.40/-.1 = -14.0$ (exact), and

$$\begin{aligned} x_2 &= (10 - 10x_2)/-.1 \\ &\doteq (10 - 10 \cdot -14.0)/-.1 \\ &= (10 + (140))/-.1 \\ &= (150)/-.1 \\ &= -1500 \end{aligned}$$

which is again exact. *The use of three digit arithmetic was not the fundamental
source of the errors*; it was instead the approach to the problem (namely, naive
Gaussian elimination) that was the source. A smarter approach gets the correct
answer even when limited to only three digits.

   In general we can't expect to get the answer correct to the last digit (or more
precisely, the **least significant bit**, LSB). However, we also can't always save a
calculation by simply doubling the precision in which we're working–and even if we
can, this means our calculation will take longer. We can work smarter instead of
harder by generalizing the trick above. The key idea is that at each stage of Gaussian
elimination we should interchange rows, if necessary, to insure that the diagonal
entry is the largest of all entries in that column that lie below the diagonal, so that
when we perform the $R_j \leftarrow R_j - (a_{j,i}/a_{i,i})R_i$ steps the quantity $m_{j,i} = a_{j,i}/a_{i,i}$,
called the (j,i) **multiplier**, is as small in magnitude as possible (at most 1 in

absolute value). This means that errors are muted, not magnified. Of course, since new errors are being introduced, we must expect some error to remain.

When we are processing column i in Gaussian elimination, the (i, i) position is called the **pivot position** and the entry in it is called the **pivot entry** (or simply the **pivot**). Algorithms for selecting pivots so as to decrease the effects of round-off error are called **pivoting strategies**. The idea described above is called (Gaussian elimination with) **partial pivoting** (GEPP). If A is an $n \times n$ matrix then the

method (as applied to the $n \times (n + 1)$ augmented matrix $[A\dot{:}b]$) may be described as follows:

Gaussian Elimination with Partial Pivoting:

1. Begin loop ($i = 1$ to $n - 1$):
2. Find the largest entry (in absolute value) in column i from row i to row n. If the largest value is zero, signal that a unique solution does not exist and stop.
3. If necessary, perform a row interchange to bring the value from step 2 into the pivot position (i, i).
4. For $j = i + 1$ to $j = n$, perform $R_j \leftarrow R_j - m_{j,i} R_i$ where $m_{j,i} = a_{j,i}/a_{i,i}$.
5. End loop.
6. If the (n, n) entry is zero, signal that a unique solution does not exist and stop. Otherwise, solve for x by back substitution.

Step 6 is of course the back substitution phase. In practice the row interchanges are not performed but are instead simulated by careful book-keeping via a permutation vector that tracks where the rows would have moved. This is because for large matrices, the time spent moving the rows in memory can be considerable. (Moving data in memory and moving data from memory to the CPU are relatively slow operations, and memory management issues are crucial to the intelligent implementation of methods for handling large matrices. The details tend to be hardware-specific, but data movement is usually the bottleneck for large matrix computations, not processor speed.) Typically in step 2 a tolerance is used and values below the tolerance (in magnitude) are zeroed out.

There are a number of other practical programming issues, but for now let us continue to concentrate on the numerics.

EXAMPLE 2: Let's use Gaussian elimination with partial pivoting to solve $Ax = b$ with the matrix $A = [1 \ 2 \ 2; 4 \ 4 \ 12; 4 \ 8 \ 12]$ and with $b = (1, 12, 8)^T$. We have:

$$
\begin{bmatrix}
1 & 2 & 2 & \vdots & 1 \\
4 & 4 & 12 & \vdots & 12 \\
4 & 8 & 12 & \vdots & 8
\end{bmatrix}
\sim
\begin{bmatrix}
4 & 4 & 12 & \vdots & 12 \\
1 & 2 & 2 & \vdots & 1 \\
4 & 8 & 12 & \vdots & 8
\end{bmatrix}
$$

$$
\sim
\begin{bmatrix}
4 & 4 & 12 & \vdots & 12 \\
0 & 1 & -1 & \vdots & -2 \\
0 & 4 & 0 & \vdots & -4
\end{bmatrix}
$$

$$
\sim
\begin{bmatrix}
4 & 4 & 12 & \vdots & 12 \\
0 & 4 & 0 & \vdots & -4 \\
0 & 1 & -1 & \vdots & -2
\end{bmatrix}
$$

$$
\sim
\begin{bmatrix}
4 & 4 & 12 & \vdots & 12 \\
0 & 4 & 0 & \vdots & -4 \\
0 & 0 & -1 & \vdots & -1
\end{bmatrix}
$$

$(m_{2,1} = 1/4,\ m_{3,1} = 1,\ m_{3,2} = 1/4)$. Back substitution gives $x_3 = -1/-1 = 1$, $x_2 = -4/4 = -1$, and $x_1 = (12 - 4x_2 - 12x_3)/4 = (12 - 4(-1) - 12(1))/4 = 1$. The solution is $x = (1, -1, 1)^T$. $\square$

We've kept the multipliers for a reason, as we'll see in the next section. As with interchanging rows, we wouldn't actually form a new augmented matrix in a program realizing Gaussian elimination with partial pivoting, but would perform the appropriate operations on b at each step.

There are several other pivoting strategies. **Complete pivoting** (or **maximal pivoting**) searches not just down the column but also to the right through the remaining columns and uses a combination of row interchanges (corresponding to renumbering equations) and column interchanges (corresponding to renaming variables) to bring to the pivot position the largest entry in the southeast part of the matrix (that part that remains to be processed). This is rarely necessary, and good results are usually obtained with partial pivoting, which is fortunate since complete pivoting is expensive due to the large number of searches required. It is possible to construct matrices for which partial pivoting gives arbitrarily bad results, using arbitrarily large matrices, but experience shows that these matrices almost never occur in practice. Partial pivoting or a more sophisticated strategy should *always* be employed when performing Gaussian elimination.

It's common to visualize these types of algorithms by writing generic matrices with an X representing an arbitrary nonzero value and a 0 (or blank) representing a value that must be zero. In this scheme the stages of Gaussian elimination (on, say, a 4×4 coefficient matrix A that is to be reduced to a triangular matrix U) are represented as follows

$$
\begin{bmatrix} X & X & X & X \\ X & X & X & X \\ X & X & X & X \\ X & X & X & X \end{bmatrix} \sim \begin{bmatrix} X & X & X & X \\ 0 & X & X & X \\ 0 & X & X & X \\ 0 & X & X & X \end{bmatrix}
$$

$$
\sim \begin{bmatrix} X & X & X & X \\ 0 & X & X & X \\ 0 & 0 & X & X \\ 0 & 0 & X & X \end{bmatrix}
$$

$$
\sim \begin{bmatrix} X & X & X & X \\ 0 & X & X & X \\ 0 & 0 & X & X \\ 0 & 0 & 0 & X \end{bmatrix}
$$

where the individual entries represented by Xs may change value from stage to stage. Such a visual representation is called a **Wilkinson diagram**, and is said to display the **zero structure** of the matrices, that is, the places where entries are necessarily zero.

MATLAB

The MATLAB package was designed for problems of this sort. It was originally intended to be an interactive interface to certain Fortran computational matrix algebra subroutines from the LINPACK and EISPACK subroutine libraries, later re-written in C. Starting with version 6, the newer LAPACK routines have replaced their older LINPACK and EISPACK counterparts. (See *www.netlib.org/lapack* for details on LAPACK.) The LAPACK routines use newer numerical algorithms in some cases (though generally they are based on the same ideas as before). Just as importantly, they utilize the Basic Linear Algebra Subroutines (BLAS). These machine-specific subroutines take advantage of the particulars of the computer's architecture, especially its memory structure, to greatly increase the speed of various matrix manipulations. The BLAS are divided into three classes: Level 1 (vector-vector operations), Level 2 (matrix-vector operations), and Level 3 (matrix-matrix operations). Higher levels generally give better speed, and are superior to the traditional element-by-element manipulations. (Of course, the BLAS themselves must use element-by-element operations on a standard computer; speedups in this case are achieved by arranging the computations for most efficient processing, including the details of retrieval of data from memory and the attempt to maximize the number of computations performed per retrieval of data from memory. On a vector machine the effects will be even more noticeable.) The LAPACK routines, originally designed for use on supercomputers, use (sub-)blocks of matrices to perform much faster. The key change in computer architecture that makes these changes worthwhile has been the widespread use of caches, that is, fast memory near the CPU. Cache management is an important practical issue that we will not be able to address in any detail because of its hardware-specific nature; by carefully using the BLAS and relying on the computer vendor to supply appropriate implementations optimized for their particular machine, the methods can be (and have been) made portable.

Let's start working with some MATLAB matrix commands. The basic way to enter a matrix in MATLAB is the same as for a vector, but with rows separated by semi-colons:

» M=[1 2;3 4]

The usual arithmetic operations are implemented naturally:

» N=[2 2;5 5]

» M+N,M-N,M*N,M^2

(Note that, in agreement with standard matrix algebra notation, M^2 is M*M and not an entry-by-entry squaring.) You can use the `size` command to check the size of a matrix:

» size([1 2 3;4 5 6])

The basic MATLAB matrix solver is implemented in the backslash; see `help slash`. If A is a nonsingular matrix and b is a column vector of the appropriate length then A\b is the solution of $Ax = b$ as computed by Gaussian elimination (with pivoting). For example, enter:

» format long

» A=[.00111 .00112;111 113];b=[.00223 224]';

» A\b

It's important to remember that this operation uses the *back*slash, not the forward slash of scalar division. In principle we could have used $A^{-1}b$; enter:

» inv(A)*b

but this will be less accurate, in general. The matrix inverse approach is also less efficient, and use of the numerical inverse of a matrix is always strongly discouraged. Note that A^{-1} is also badly scaled; enter:

» inv(A)

Using A^{-1} gives no advantage over the use of A in this case (nor in general). We can check our results by computing the norm of the **residual** $r = b - Ay$ where y is the computed approximation to x. Enter:

» y=A\b;r=b-A*y;norm(r)

Of course, it isn't perfectly clear that the fact that $\|r\|$ is small implies that the actual error $\|y - x\|$ is small as well; in fact, as we'll discuss in a later section, this need not be so. For now note that $\|r\|$ is quite small (it likely underflows to zero on your machine, that is, it's so small that it's set to 0), but enter:

» norm(y-[1 1]')

to see that $\|y - x\|$ is not zero; the absolute error is noticeably larger than the norm of the residual. Recall that sometimes a value in MATLAB may display as though it matches another number exactly, but will fail a test for equality due to the effect of extra bits retained internally by MATLAB but not displayed. Typically we do not expect equality between quantities computed using floating point numbers but only that they differ by less than some tolerance. That is to say, y is actually a pretty good estimate of x.

Let's try re-scaling the matrix in hopes of better balancing it (see Problem 4 part c.). Enter:

» S=[100000 0;0 1]

» M=S*A,b1=S*b

» M\b1

Notice that M is near to being a singular matrix in the sense that changing, say, the $(2, 2)$ entry by 1 to 112 gives a singular matrix; enter:

```
» Q=M;Q(2,2)=112
» det(Q)
```
(Recall that a matrix is singular if and only if its determinant is zero.) Matrices that are nearly singular tend to give poor numerical results. The determinant is not a good way to characterize this property of being nearly singular, though; we'll discuss an appropriate measure of near-singularity in Sec. 2.6.

As indicated above with the Q matrix, it is possible to select a single element of a matrix. Enter:
```
» y=[2 4 6 8]
» y(3)=-5
» W=[1 2;3 4]
» W(2,1)=-W(2,2)
```
and note the effects. Indexing single elements out of a vector or matrix is a relatively slow process and it is best to avoid it when possible. Rather than writing a loop that manipulates matrix elements successively, we try to find a corresponding vector command that has the same effect (called "vectorizing" the code).

The MATLAB command rref computes the reduced row echelon form of a matrix. It uses a tolerance to declare certain elements of the matrix zero if they are "negligible". This is usual in practice.

In MATLAB the transpose of a vector or matrix is obtained by using the apostrophe; for complex vectors this gives the conjugate transpose (that is, it performs the Hermitian operation). Enter:
```
» clear all
» x=[1 2 3]
» x'
» x=[1+i 1-i;2+i 2-i]
» x'
```
(see help punct and see also help transpose, help ctranspose, and help conj). The clear command is used here to clear the values of all variables. If i had been used as, say, an integer earlier, it might not have its default value of $\sqrt{-1}$. In accordance with electrical engineering notation, j also defaults to i. See also help real and help imag. Most MATLAB commands work automatically with matrices that have complex entries.

We'll briefly list a few more matrix-oriented MATLAB commands that may be of use in this chapter: ones (a matrix of all ones), zeros (a matrix of all zeroes), and diag (create a diagonal matrix). See help elmat for more matrix commands. Let's try a few of these. Enter:
```
» ones([3 3])
» size(ans)
» zeros(4,5)
» diag([1 2 3])
» rand([4 4])
» diag(ans)
```
There is also a dot product command dot; dot(x,y) is the same as x'*y if x and y are column vectors of the same length. (See help length.) For most purposes, MATLAB treats a vector as just another matrix.

We mention briefly that MATLAB allows arrays with more than two dimensions; that is, you may enter a $2 \times 3 \times 4$ matrix. Enter:

```
>> A=reshape(1:24,2,3,4)
```
This data structure represents a **multilinear array** (or **discrete tensor**). See `help shiftdim` and `help squeeze` for some related commands.

In MATLAB, empty arrays occur often as initial placeholders or markers of some state. They have other uses also. Enter:
```
>> v=[]
>> size(v)
```
The variable v is a 0×0 matrix with no elements. Enter:
```
>> v=[v,1]
>> v=[v,2]
>> c="    %Two single quotes, not one double quote.
>> size(c)
```
In a number of MATLAB commands, entering an empty matrix instructs MATLAB to use default values for the corresponding parameters. This is explained in the help for the command.

Early versions of MATLAB included a `flops` command that counted the total number of floating point operations (flops) used. This has been removed, as the BLAS do not usually maintain flop counts. Execution time may then be measured using `tic` and `toc`. Typing `tic` before a sequence of commands and `toc` afterwards gives the time elapsed since the `tic` command. Enter:
```
» tic;for i=1:10000,cos(i);end;t1=toc
```
The elapsed time is displayed. Repeat this command several times; especially on a shared machine, you will get different results. It's common to average the results of several runs. See also `help cputime`.

Counting the number of floating point operations used in matrix manipulations is an important technique for comparing their relative efficiency, but there are many caveats. Although addition and multiplication are both single floating point operations, multiplications (and especially divisions) take longer than additions and subtractions, and often we focus on the number of multiplications/divisions. The square root is usually counted as a single flop also, despite being noticeably slower to compute. Numerical analysts have traditionally used flop counts (the total number of floating point operations required) to compare algorithms but their importance is much less nowadays, when floating point operations are faster, computer architectures are more varied, and memory management is much more of an issue in run-time than it used to be. Consider comparing flop counts for standard machines versus parallel machines, or even for standard machines versus machines with a **fused add/multiply** operation (which computes $\alpha x + \beta$ as a single flop, in essence), or standard machines with and without use of the BLAS; flop counts and their relevance depend on the machine to be used. For many, perhaps most, large problems, the greatest amount of time is spent waiting for data to be moved in from slow memory. Despite all this, flop counts are still a common proxy for total run-time and overall efficiency and we will present them but will not emphasize them.

Problems

1.) a.) Use Gaussian elimination with partial pivoting and backward substitution to solve (by hand–you may use a calculator for the arithmetic if desired but step through the algorithm) the linear system $Ax = b$ where $A = [1\ 2\ 3; 4\ 5\ 6; 7\ 8\ 8]$

and $b = (1, 1, 1)^T$. Check your answer by computing Ax. List the three multipliers explicitly.

b.) Repeat with the same A but with $b = (2, -1, 2)^T$.

c.) Repeat with $A = [1\ 2\ 3; 4\ 5\ 6; 7\ 8\ 9]$ and $b = (1, 1, 1)^T$.

2.) a.) Write a detailed algorithm (pseudo-code) for solving a square system by Gaussian elimination with partial pivoting and backward substitution. (You will need to expand considerably on what is in the text.) Assume that the row interchanges are actually performed (not simulated).

b.) Draw the Wilkinson diagram for solving a linear system by Gaussian elimination with partial pivoting (assuming a 4x4 coefficient matrix, so that the augmented matrix is 4x5).

3.) a.) Repeat Example 2 with maximal pivoting. You will need to keep track of which column corresponds to which variable.

b.) Solve the linear system of Eq. (1.3) with maximal pivoting and three digit arithmetic.

4.) a.) Solve the linear system $Ax = b$ with $A = A = [.0002\ .2; 2\ 2]$ and $b = (.2, 4)^T$ using naive Gaussian elimination and three digit arithmetic. Compare your result to the true answer $(1.00\overline{1001}, .99\overline{8998})^T$.

b.) Solve the same system using Gaussian elimination with partial pivoting and three digit arithmetic.

c.) Repeat part a. with $b = (.2, 5.6)^T$ (for which $x^* \doteq (1.8018, .9982)^T$). Then multiply both sides of $Ax = b$ by the matrix $S = [10\ 0; 0\ 1]$, giving the equivalent system $SAx = Sb$ and then repeat part a. This corresponds to performing the elementary row operations of multiplying row 1 by a constant and row 2 by a constant before beginning the naive Gaussian elimination. Does it help?

d.) We say that a matrix exhibits **artificial ill-conditioning** if its poor behavior with respect to Gaussian elimination is due in large measure to it being poorly scaled, that is, the entries in the matrix vary widely, over several orders of magnitude, which is always a warning sign that we should be on the lookout for poor numerical behavior. Multiply both sides of Eq. (1.3) by $S = [10\ 0; 0\ 1]$ and solve the resulting system by naive Gaussian elimination in three digit arithmetic. Does this help? (If not, the matrix may be truly, not artificially, ill-conditioned.) This technique is sometimes referred to as **balancing** the matrix, that is, making all entries, or the sums along all rows (in absolute value), more nearly equal.

5.) a.) How many additions/subtractions and how many multiplications/divisions, in general, are required to perform back substitution after Gaussian elimination has been applied to an $n \times n$ matrix?

b.) How many additions/subtractions and how many multiplications/divisions, in general, are required to perform Gaussian elimination with partial pivoting on an $n \times n$ matrix?

6.) a.) Let $A = [.007\ \ -.8; -.1\ \ 10]$ be the matrix of Eq. (1.3). Show that if $P = [0\ 1; 1\ 0]$ then PA is the matrix obtained by switching row 1 and row 2 of A.

b.) Demonstrate that the MATLAB command A([2 1],:) produces PA.

c.) Choose a 2×4 matrix B and form PB. What is the effect of premultiplication by P?

c.) Choose a 4×2 matrix C and form CP. What is the effect of postmultiplication by P?

7.) a.) Using $A = [.007 \ -.8; -.1 \ 10]$ and $b = (.7, 10)^T$ from Eq. (1.3), find the solution of $Ax = b$ in MATLAB (use y=A\b).

b.) Find the residual vector r, the norm of the residual, and the absolute error (recall that the true solution is $x_1 = -1500$, $x_2 = -14$).

c.) Compute the ratio of the absolute error to the norm of the residual $\|y - x\|/\|r\|$.

8.) a.) Solve $Ax = b$ with $A = [.0005 \ 1; 9 \ 2]$ and $b = (1, 1)^T$ using three-digit arithmetic and naive Gaussian elimination. Compare this to the solution found using exact arithmetic.

b.) Repeat part a. using Gaussian elimination with partial pivoting.

9.) Solve $Ax = b$ (by hand) using Gaussian elimination with partial pivoting, where $A = [1 \ 2 \ 3 \ 3; 0 \ 2 \ -1 \ -2; 3 \ 4 \ -4 \ 3; -2 \ -3 \ 3 \ 6]$ and $b = (1, 1, 1, 1)^T$.

10.) Solve $Ax = b$ (by hand) using Gaussian elimination with partial pivoting, where $A = [-4 \ 5 \ 2 \ 1; 1 \ 3 \ -3 \ 0; 5 \ 1 \ 0 \ 0; 1 \ 0 \ 0 \ -3]$ and $b = (-1, 0, 0, 0)^T$.

11.) How many flops are required to compute the square $M^2 = MM$ of an $n \times n$ matrix? How many for element-by-element squaring of the entries of the matrix (M.^2)?

12.) In Gauss-Jordan elimination, zeroes are introduced both above and below the main diagonal so as to make the matrix U diagonal if possible. (This is the reduced row echelon form.) In Eq. (1.4), for example, this would mean using the $(2, 2)$ entry -1.4286 to eliminate the $(1, 2)$ entry $-.8$. Repeat Problem 1 part a. using Gauss-Jordan elimination (by hand). Note that you still may only use values below the pivot position for new pivots without destroying the diagonal structure of the northwest part of the matrix.

13.) Write a MATLAB program that implements Gaussian elimination with maximal pivoting. You may assume that the input matrix will be nonsingular and you may actually switch the rows and columns (this is easier but typically slower than simulating the switches). You may wish to use the fact that the MATLAB command C([k1,k2],:)=C([k2,k1],:) interchanges rows k1 and k2 of a matrix C, and C(:,[k1,k2])=C(:,[k2,k1]) interchanges columns k1 and k2 of this matrix. (The colon : signifies all columns or all rows, respectively.) The elementary row operation $R_j \leftarrow R_j - mR_i$ may be implemented by a command such as C(i,:)=C(i,:)-m*C(j,:). Your program will need to note which column interchanges have been performed so that the elements of your answer can be put back into the correct order. Demonstrate your program.

14.) The **Hilbert matrices of order N** ($N = 1, 2, ...$) form a class of well-known ill-conditioned matrices. The MATLAB command hilb(N) generates the Nth such matrix. Perform a numerical experiment that demonstrates this ill-conditioning for $N = 10, 11, 12$. You may wish to use the invhilb command as well.

15.) a.) How many floating point operations would be required to compute the determinant $|A|$ of a general $n \times n$ matrix by Laplace expansion (repeatedly expanding along a row or column in order to break the determinant down to many 2×2 matrices)?

b.) Estimate the time required to find the determinant of a 100×100 matrix on a typical machine using this method.

c.) Roughly how much time does MATLAB use to compute the determinant det(A) of a 100×100 randomly-generated matrix? Does it seem likely that it is using Laplace expansion?

16.) A standard model for the unforced motion of a mass in a mass-spring system leads to the ODE $mx'' + cx' + kx = 0$, where m is the spring's mass, c is the coefficient of friction (damping) of the medium through which the mass moves, and k is the Hooke's Law constant for the spring. If m, c, and k are positive, this leads to the model $x(t) = \exp(\alpha t)(A\cos(\beta t) + B\sin(\beta t))$ for the motion of the mass, where A and B are determined from the initial position x_0 and velocity v_0 of the mass, and $\alpha \pm i\beta$ are the roots of the characteristic equation $mr^2 + cr + k = 0$.
a.) Compute A and B using Gaussian elimination with partial pivoting if $m = 10$ kg, $k = 400$ N/m, and the force due to friction is proportional to the velocity with proportionality constant $c = .5$. Take $x_0 = .2$ m and $v_0 = .3$ m/s. Repeat using the same values of m, c, and k but use the initial data $x(1) = -.1$ m and $x'(1) = .2$ m (with time in seconds).
b.) Compute A and B using Gaussian elimination with partial pivoting if $m = 10$ kg, $k = 40$ N/m, and the force due to friction is proportional to the velocity with proportionality constant $c = .5$. Take $x_0 = .2$ m and $v_0 = .3$ m/s. Repeat using the same values of m, c, and k but use the initial data $x(1) = .1$ m and $x'(1) = -.3$ m.

2. The LU Decomposition

Gaussian elimination with partial pivoting is a good algorithm for small-to-medium sized problems; as usual, it's difficult to set a value for "small"" or "medium" across a variety of platforms and needs. In addition, most matrices that occur in practice possess some sort of **structure**. Structure is a generic term that refers either to the pattern of the locations of the zero elements versus nonzero elements within the matrix, referred to as the **sparsity structure**, or simply the **sparsity**, of the matrix, or to special properties that the matrix possesses, such as symmetry being triangular. A matrix with mostly zero entries is said to be **sparse** and one with mostly nonzero entries is said to be **dense** (or **full**). There are many special numerical methods for sparse matrices–a typical rule of thumb is 95% zero entries, which is much more common than you might think–and for matrices with special properties like symmetry.

As a trivial case, if the matrix is already upper triangular then we need only perform the back substitution (and if it is lower triangular, that is, if its transpose is upper triangular, we need only perform forward substitution). Back substitution requires $O(n^2)$ flops, which is reasonable even for large n. (As a rule we can't expect to do much better than that; after all, a general $n \times n$ matrix has n^2 distinct entries, each of which should figure into the solution.) For a general matrix, that is, an unstructured matrix, Gaussian elimination with partial pivoting requires about $2n^3/3$ flops (for large n): About $n^3/3$ additions/subtractions and about $n^3/3$ multiplications/divisions. This doesn't count the cost of the roughly $n^2/2$ down-the-column searches for the pivots or any associated bookkeeping operations.

Because it is common to consider matrices so large that $2n^3/3$ flops is unrealistic or at least very inconvenient, and because Gaussian elimination does not take advantage of (nor preserve) any structure the matrix might possess, we'll need other methods for large matrices. But for small to medium matrices, Gaussian elimination with a pivoting strategy remains the method of choice.

Despite the fact that it does not take advantage of any special structure of the coefficient matrix A, Gaussian elimination does have an advantage in the frequently occurring case where we must solve $Ax = b$ repeatedly for different choices of b. In some cases the b vector plays the role of a forcing vector, analogous to a forcing function for a differential equation, and we load the system represented by A with various choices of b; here A represents some system that remains unchanged, and b represents an input that will be changed often. Note that if we were to solve the linear system $Ax = b$ once by Gaussian elimination then we could in principle write down every row operation $R_j \leftarrow R_j - m_{j,i} R_i$ and every pivoting operation $R_k \leftrightarrow R_p$ which was performed to obtain the reduced form

$$\left[A \vdots b \right] \sim \left[U \vdots \widetilde{b} \right].$$

What if we now wish to solve the related system $Ax = d$ by performing Gaussian elimination with partial pivoting? Every decision made about which rows to interchange and what multipliers $m_{j,i}$ to use was based only on the entries of A, and *not* on the entries of b. This means that row reducing $\left[A \vdots d \right]$ must give

$$\left[A \vdots d \right] \sim \left[U \vdots \widetilde{d} \right]$$

for the same matrix U, arrived at by exactly the same steps. That's great! That means that we need only store U and the operations required to produce it from A the first time, and then perform the corresponding steps on d the next time. We do not need to reduce A to U again. If n is large, this represents a considerable savings of floating point operations.

EXAMPLE 1: Consider the linear system

$$\begin{bmatrix} 1 & 2 \\ 2 & 6 \end{bmatrix} \begin{bmatrix} x_1 \\ x_2 \end{bmatrix} = \begin{bmatrix} 1 \\ -2 \end{bmatrix}.$$

Gaussian elimination with partial pivoting gives

$$\left[\begin{array}{cc:c} 1 & 2 & 1 \\ 2 & 6 & -2 \end{array} \right] \sim \left[\begin{array}{cc:c} 2 & 6 & -2 \\ 1 & 2 & 1 \end{array} \right]$$

$$\sim \left[\begin{array}{cc:c} 2 & 6 & -2 \\ 0 & -1 & 2 \end{array} \right]$$

where we have performed, in order, $R_1 \leftrightarrow R_2$, $R_2 \leftarrow R_2 - \frac{1}{2} R_1$. This gives $x_2 = 2/-1 = -2$ and $x_1 = (-2 - 6x_2)/2 = 5$. If we now wish to solve

$$\begin{bmatrix} 1 & 2 \\ 2 & 6 \end{bmatrix} \begin{bmatrix} x_1 \\ x_2 \end{bmatrix} = \begin{bmatrix} 6 \\ 16 \end{bmatrix}$$

we need only operate on the new forcing vector $(6, 16)^T$, performing $R_1 \leftrightarrow R_2$ to get $(16, 6)^T$ followed by $R_2 \leftarrow R_2 - \frac{1}{2} R_1$ to get $(16, 6 - \frac{1}{2}16)^T = (16, -2)^T$. The reduced system is therefore

$$\left[\begin{array}{cc:c} 2 & 6 & 16 \\ 0 & -1 & -2 \end{array} \right]$$

which gives $x_2 = -2/-1 = 2$ and $x_1 = (16 - 6x_2)/2 = 2$. We did not need to reduce A to U again; for a 2x2 matrix this is not a big savings but for a large matrix it would be. $\square$

Once we've processed the coefficient matrix A once we need not process it again; we need only store the multipliers $m_{j,i}$, the row interchanges, and the final upper triangular matrix U. There's an efficient way to store this data that, additionally, relates the method of Gaussian elimination with partial pivoting back to matrix algebra by representing the result as a product of two matrices, namely, a certain lower triangular matrix L and the upper triangular matrix U. There are many ways to do this, and any means of writing a square matrix A as the product

$$(2.1) \qquad\qquad A = LU$$

of a lower triangular matrix L and an upper triangular matrix U is called an **LU decomposition** of A. This is no more than Gaussian elimination in matrix algebra form, and we largely use the terms Gaussian elimination and LU decomposition interchangeably.

It is a fact (to be established in Sec. 2.3) that if A is reduced to the upper triangular matrix U without any row interchanges, then we may take L in Eq. (2.1) to be the lower triangular matrix with ones on the main diagonal and the multipliers $l_{j,i} = m_{j,i}$ below the main diagonal. This gives a means of computing the decomposition in such a case.

EXAMPLE 2: If $A = [2\ \ 2; 1\ \ 2]$ then no pivoting is actually performed, and we simply row reduce A to U using $R_2 \leftarrow R_2 - m_{2,1}R_1$ (with $m_{2,1} = 1/2$) to find

$$U = \begin{bmatrix} 2 & 2 \\ 0 & 1 \end{bmatrix}$$

and we may take

$$L = \begin{bmatrix} 1 & 0 \\ m_{2,1} & 1 \end{bmatrix}$$
$$= \begin{bmatrix} 1 & 0 \\ 1/2 & 1 \end{bmatrix}$$

as the lower triangular matrix. (Note carefully that although we used $-m_{2,1}R_1$, the multiplier $+m_{2,1}$ is used in L.) Let's check this:

$$LU = \begin{bmatrix} 1 & 0 \\ 1/2 & 1 \end{bmatrix} \begin{bmatrix} 2 & 2 \\ 0 & 1 \end{bmatrix}$$
$$= \begin{bmatrix} 2 & 2 \\ 1 & 2 \end{bmatrix}$$
$$= A.$$

Yes, this checks. We have found an LU decomposition of this matrix. $\square$

If we wish to solve $Ax = b$ for only one value of b, there is no need to perform an LU decomposition. If we expect to work with A repeatedly, and it's not too large, an LU decomposition may be advisable.

Suppose we have found an LU decomposition of a matrix A. How exactly do we use it? To solve $Ax = b$ we proceed as follows:

$$
\begin{aligned}
Ax &= b \\
LUx &= b \\
L(Ux) &= b \\
Ly &= b
\end{aligned}
$$

because Ux (which we are now calling y) is simply some unknown vector. We can solve $Ly = b$ by forward substitution to find y, then solve $y = Ux$, that is,

$$Ux = y$$

for x by back substitution (as usual). This is a very computationally efficient process once the LU decomposition is known. Since finding a U is equivalent to performing Gaussian elimination, it's an $O(n^3)$ process, and this can't be avoided; but each subsequent solve involving A requires only two $O(n^2)$ back-solves, making them *much* more efficient, without the need for any additional storage (assuming compact form).

EXAMPLE 3: An LU decomposition of $A = [6 \ 4 \ 2; 3 \ -2 \ -1; 3 \ 4 \ 1]$ may be found as follows:

$$
\begin{bmatrix} 6 & 4 & 2 \\ 3 & -2 & -1 \\ 3 & 4 & 1 \end{bmatrix}
\sim
\begin{bmatrix} 6 & 4 & 2 \\ 0 & -4 & -2 \\ 0 & 2 & 0 \end{bmatrix}
\sim
\begin{bmatrix} 6 & 4 & 2 \\ 0 & -4 & -2 \\ 0 & 0 & -1 \end{bmatrix}
$$

$(m_{2,1} = 1/2, m_{3,1} = 1/2, m_{3,2} = -1/2)$, so

$$
U = \begin{bmatrix} 6 & 4 & 2 \\ 0 & -4 & -2 \\ 0 & 0 & -1 \end{bmatrix}
$$

$$
L = \begin{bmatrix} 1 & 0 & 0 \\ .5 & 1 & 0 \\ .5 & -.5 & 1 \end{bmatrix}
$$

(check that $LU = A$). To solve

$$
Ax = \begin{pmatrix} 12 \\ 0 \\ 8 \end{pmatrix}
$$

we first solve $Ly = b$, that is,

$$
\begin{bmatrix} 1 & 0 & 0 \\ .5 & 1 & 0 \\ .5 & -.5 & 1 \end{bmatrix}
\begin{bmatrix} y_1 \\ y_2 \\ y_3 \end{bmatrix}
=
\begin{pmatrix} 12 \\ 0 \\ 8 \end{pmatrix}
$$

by forward substitution, giving $y_1 = 12$, $y_2 = 0 - .5y_1 = -6$, $y_3 = 8 - .5y_1 + .5y_2 = 8 - 6 - 3 = -1$. We then solve $Ux = y$, that is,

$$
\begin{bmatrix} 6 & 4 & 2 \\ 0 & -4 & -2 \\ 0 & 0 & -1 \end{bmatrix}
\begin{bmatrix} x_1 \\ x_2 \\ x_3 \end{bmatrix}
=
\begin{pmatrix} 12 \\ -6 \\ -1 \end{pmatrix}
$$

by back substitution, giving $x_3 = -1/-1 = 1$, $x_2 = (-6 - -2x_3)/-4 = (-6 + 2)/-4 = 1$, $x_1 = (12 - 4x_2 - 2x_3)/6 = (12 - 4 - 2)/6 = 1$. Hence $x = (1, 1, 1)^T$. $\square$

Requiring the diagonal entries of L to be ones and computing the entries of L and U in the order we have described is called **Doolittle's method** of obtaining an LU decomposition. Other choices are possible. Note that A has n^2 possibly nonzero entries, and U and L each have

$$1 + 2 + \ldots + n = n(n+1)/2$$

possibly nonzero entries (1 in column 1, 2 in column 2, and so on). Hence U and L together have

$$\frac{n(n+1)}{2} + \frac{n(n+1)}{2} = n^2 + n$$

entries. This means that there are n extra entries between L and U that may be assigned freely. Doolittle's method arbitrarily makes the n diagonal entries of L each be one; **Crout's method** arbitrarily makes the n diagonal entries of U each be one (and changes the order of the computations). Another method uses the n constraints that the diagonals of L and U be equal. We'll continue to focus on Doolittle's method. The numerous variants exist so as to be able to take advantage of any structure of the matrix and also the specifics of the programming language and machine being used.

One of the advantages of using an LU decomposition is that we may replace the matrix A in memory by L and U. At each stage we overwrite A with the portion of U that has been formed already, and store the multipliers in the lower triangle of the matrix. The Wilkinson diagram would be

$$\begin{bmatrix} X & X & X & X \\ X & X & X & X \\ X & X & X & X \\ X & X & X & X \end{bmatrix} \rightarrow \begin{bmatrix} X & X & X & X \\ x & X & X & X \\ x & X & X & X \\ x & X & X & X \end{bmatrix}$$

$$\rightarrow \begin{bmatrix} X & X & X & X \\ x & X & X & X \\ x & x & X & X \\ x & x & X & X \end{bmatrix}$$

$$\rightarrow \begin{bmatrix} X & X & X & X \\ x & X & X & X \\ x & x & X & X \\ x & x & x & X \end{bmatrix}$$

where we are using X for an entry of U and x for an entry of L. We can extract L and U explicitly at any time if desired, filling in zeroes and (for L) ones as needed, and use them to recreate $A = LU$; more often we simply write the program in such a way that it draws the needed values from this array. We will refer to this as the **compact form** of the LU decomposition; it is also called an **in place** version because it puts L and U in the same place (in memory) as A was. This destroys the matrix A but it can be reformed if needed from L and U (with some slight loss of accuracy in the reconstructed version, of course). However, since the memory for A has already been allocated, this reduces storage–storing A, L, and U is like storing two copies of A, or three if L and U aren't stored in compact form–and using previously-allocated memory allows for much more rapid writing of the components of L and U back to memory, which can provide noticeable speed-up even if there is sufficient storage available. Many programming environments make it convenient to allocate memory space for matrices in advance, and this is usually advisable.

For similar reasons, rather than passing an entire matrix A to a subroutine and passing its LU decomposition back, which would mean that there were temporarily two copies of A in memory (with the second one local to the subroutine), we could pass a pointer to the beginning of A in memory to the subroutine, which could operate on A there, putting L and U in its place using this storage scheme rather than explicitly returning them[1].

EXAMPLE 4: Consider again the matrix

$$A = \begin{bmatrix} 6 & 4 & 2 \\ 3 & -2 & -1 \\ 3 & 4 & 1 \end{bmatrix}$$
$$= \begin{bmatrix} 1 & 0 & 0 \\ .5 & 1 & 0 \\ .5 & -.5 & 1 \end{bmatrix} \begin{bmatrix} 6 & 4 & 2 \\ 0 & -4 & -2 \\ 0 & 0 & -1 \end{bmatrix}$$
$$= LU$$

of Example 3. We would replace A in memory sequentially, as follows:

$$\begin{bmatrix} 6 & 4 & 2 \\ 3 & -2 & -1 \\ 3 & 4 & 1 \end{bmatrix} \rightarrow \begin{bmatrix} 6 & 4 & 2 \\ .5 & -2 & -1 \\ .5 & 4 & 1 \end{bmatrix} \rightarrow \begin{bmatrix} 6 & 4 & 2 \\ .5 & -4 & -3 \\ .5 & -.5 & -1 \end{bmatrix}$$

to put it in compact form. The final matrix has the not necessarily null entries of U in the upper triangle (including the diagonal) and the not necessarily null or unit entries of L in the (strict) lower triangle. □

An LU decomposition is also called an **LU factorization** since it factors A into a product of matrices (its factors). There are many other matrix factorizations that are of interest, and it pays to be able to write numerical matrix algebra algorithms in terms of formal matrix algebra both for analysis and for design of such methods. In the next section we'll use matrix algebra to justify some of what we've done so far and also to explain why we can form L from the multipliers and what happens if we do in fact need to perform row interchanges (pivoting), as we usually must.

The LU decomposition involves the multiplication of two matrices, although in practice we would rarely do that. But there are plenty of times when we need to multiply two matrices A and B. This is an $O\left(n^3\right)$ operation and involves accessing a lot of memory. Sometimes we can save some computation. Suppose we need to compute

$$ABx$$

where x is a vector. Multiplying a matrix with a vector is an $O\left(n^2\right)$ operation. That means that although

---

[1]Memory issues are often not as immediately apparent in the highly structured MATLAB environment as when working in, say, C++ or a similar language.

$$(2.2) \qquad\qquad (AB)x = A(Bx)$$

by the associativity of matrix multiplication, the left-hand side requires $O(n^3)$ flops while the right-hand side requires only $O(n^2)$. As always, it pays to consider whether any formula we are using should be re-written into an equivalent form that gives faster and/or more accurate results.

Incidentally, the most important open theoretical problem in numerical analysis is probably the question of how many operations are required to compute the matrix product AB where A and B are both $n \times n$. The conventional method is $O(n^3)$ but in 1969 Strassen developed an algorithm, Strassen multiplication (sometimes referred to as "fast matrix multiplication"), that is about $O(n^{\log_2(7)})$ (i.e., $O(n^{2.8074})$). The technique is only useful for unrealistically large matrices, but it opened the question of what the optimal order might be. The question is of great interest because Strassen also showed that Gaussian elimination (and finding the inverse or the determinant) have the same order as matrix multiplication. The best known bound as of this writing has order about $O(n^{2.3729})$, but the conjecture is that there is an $O(n^{2+\epsilon})$ algorithm for arbitrarily small $\epsilon > 0$.

MATLAB

The MATLAB `lu` command may be used to compute the LU decomposition of a (not necessarily square) matrix. Enter:

```
» A=[8 8 4;4 2 -1;2 2 2]
» [L,U]=lu(A)
>> L*U     %matches A
```

Notice that calling `lu` with two output arguments causes it to return two output arguments; calling it with one (or no) output arguments gives a different result:

```
» M=lu(A)
```

This is the compact form as in Example 4. If you use `lu` with a matrix that requires pivoting you'll get an L which is not exactly lower triangular. We'll explain why in Sec. 2.3.

If you were using the single output form M=lu(A) you could extract the upper and lower triangles using the MATLAB commands `triu` and `tril`. Enter:

```
» U1=triu(M)
» L1=tril(M)
```

and note that U1 is the same as U but L1 contains the entire lower triangle of M, including the diagonal. We could extract the diagonal if desired as a vector or as a matrix as follows:

```
» diag(L1)
» diag(diag(L1))
```

(The `diag` command takes a matrix and returns its diagonal as a vector, or takes a vector and makes a diagonal matrix from it; why does this make `diag(diag(L1))` work as it does?) We need to fix the diagonal of L1. There are many ways, including a loop of the form:

```
» for i=1:size(L1,1);L1(i,i)=1;end; L1
```

but using a loop is rarely the best approach in MATLAB. (In other languages it might be.) We can do it without loops as follows:

```
» L1=tril(M)    %Reset L1.
» L1=L1-diag(diag(L1))+eye(size(L1,1))
```

which subtracts out the diagonal of L1 and adds in the identity matrix I_n. While this doesn't have an explicit loop, forming the sparse (that is, mostly full of zeroes) matrices `diag(diag(L1))` and `eye(size(L1,1))` is terribly wasteful if n is large. Another possibility is to initially extract L1 from M as follows:

```
» L1=tril(M)      %Reset L1.
» L1=tril(M,-1)
```

This starts extracting from 1 diagonal below the main diagonal (called the **first subdiagonal**). Hence we may use:

```
» L1=tril(M,-1)+eye(size(L1,1))
```

to form L1. If n is large we probably don't want to actually create I_n however and should consider another approach. There's a useful trick: In MATLAB, you can also singly-index a matrix as though it were a vector. Importantly, MATLAB stores matrices in memory column-by-column, called **column-major** format. Enter:

```
» M,M(7)
```

The 7th entry of M is found by counting the 3 entries down the first column, then the 3 entries down the second column, and finally hitting on entry 7 in the top of the first column, in the $(1,3)$ position in the matrix. Enter:

```
» L1=tril(M)      %Reset L1.
» n=size(L1,1)
» diagindex=1:(n+1):(n*n)
```

(the parentheses are actually unnecessary here). Check that `diagindex` includes precisely the indices of the diagonal of the matrix, singly-indexed. Enter:

```
» L1(diagindex)=ones(n,1)
```

and we now have the desired diagonal. In fact, `L1(diagindex)=1` would have achieved the same result as MATLAB would automatically replicate the 1 as needed.

How does the `lu` command know whether to return a single matrix containing both L and U or the two separate matrices? Two useful permanent variables, **nargin** and **nargout**, are automatically set inside (and are local to) any program. The value of **nargin** is the number of inputs to the function, and the value of **nargout** is the number of output arguments which have been requested. You can see its use in **fminsearch**; enter;

```
» more on
» type fminsearch
```

and note that in the very first executable statements after the initial comments the value of **nargin** is checked and the variable `options` is set to a default value (the empty vector [] that has no elements) if **nargin** is less than 3, indicating that a value of `options` was not supplied by the user.

To demonstrate the use of these variables, suppose that we could only access the compact form that is the result of using the `lu` command with a single output, that is, we could only call it in the form M=lu(A). (This might happen if we used MATLAB to call a C++ or Fortran program that computed the LU decomposition by a different algorithm, for example. MATLAB can call C++ and Fortran executable programs and this is frequently useful or even necessary.) We could mimic the behavior of the actual `lu` command with the following wrapper program built around it:

```
function [L,U,x]=ludecomp(A,b)
%LUDECOMP  LU decomposition of A and optional solution of Ax=b.
%          Assumes that no pivoting need be performed.
```

```
[m n]=size(A);if m~=n | n<2, error('A must be square.'),end
L=lu(A);x=[];
if nargout>1
    U=triu(L);L=-tril(L,-1)+eye(n);
else
    L=triu(L)-tril(L,-1);   %Changes signs in strict lower triangle.
end
if nargin>1 & nargout>1
    if any(size(b)~=[n 1]), error('A and b must conform.'),end
    y=L\b;
    x=U\y;
    if nargout==2   %Prepare outputs.
        U=x;              %Assume L,x are desired.
    end
end
```

Let's test this program. Create a file ludecomp.m on your path and type in the code. (It's probably easiest to use the **edit** command to do this.) Then enter:

```
» clear all
» clc
» A=[8 8 4;4 2 -1;2 2 2]
» ludecomp(A)   %Should just return the compact form.
» [L,U]=ludecomp(A)   %Should return L and U.
» [L,x]=ludecomp(A,[1 1 1]')   %Should return L and x.
» [L,U,x]=ludecomp(A,[1 1 1]')   %Should return L, U, and x.
» [L,U,x]=ludecomp(A)   %Should return L, U, and empty vector x.
```

Now let's discuss `ludecomp.m`. The statement if `m~=n | n<2` checks for $m \neq n$, indicating a nonsquare matrix, or if the matrix happens to be a scalar. (The symbol | is MATLAB's relational operator *or*; see **help ops**.) The LU decomposition of a scalar is exactly what one would expect:

```
» [L,U]=lu(7)
L =
    1
U =
    7
```

but we rule out this trivial case in our program for convenience. We then find the LU decomposition of the matrix A and call it L rather than by some temporary name to avoid maintaining more matrices than we need. (Overwriting A by `lu(A)` would be sensible too.) If more than one output argument has been requested, we either will need both L and U or we will need L and x. We will find x if requested by the LU decomposition, so whenever **nargout** exceeds 1 we will need L and U.

If more than one input argument has been provided then we must solve $Ax = b$, and we do so using the LU decomposition (which we have already computed and extracted from the output of `lu`). The statement if `any(size(b)~=[n 1])` checks the size of b against the expected size $(n \times 1)$. We are checking a 2-vector against a 2-vector here so the result is a 2-vector of zeroes and ones, as in:

```
» [2 2]~=[2 3]
```

We wish to signal an error if either dimension does not match, so we use the **any** command to check if any element of the result of the comparison is true (that is, is equal to 1). If not, we solve using the backslash command and the LU decomposition.

If `ludecomp` is called with exactly two outputs requested then we assume that the desired outputs are L and U if the number of input arguments is 1, and that the desired outputs are the result of L and the solution x of $Ax = b$ if the number of input arguments is 2 (that is, if b is provided). We might have returned the compact form rather than just L, but *something* has to go in its place. After preparing the outputs in this manner we are done. The program would need to be properly commented so that these conventions are clear when someone uses the `help` command on it.

An improved version of our program would not use the `error` command to force a halt to all computations but would rather return a status variable that the calling routine can check. This is absolutely standard practice in scientific computing. This means that the first line of ludecomp.m would be something like `function [L,U,x,err]=ludecomp(A,b)` and we would set the variable `err` to, say, -1 to indicate that an error occurred, and 1 to signal that the program believes it has completed successfully. We could even return -1 for the first type of error in the code and -2 for the second, with this explained in the help comments. The calling program would request 4 output arguments–we'd need to insure that all 3 of the first arguments have values, even if they're meaningless given the error–and would check the status variable before trusting the output of the program. It could then decide how it wanted to proceed, given the failure of the ludecomp.m program.

The `nargin` and `nargout` commands provide great flexibility in structuring MATLAB calling sequences, and should be better documented (within the program) than in our example above. See also `help vararrgin` and `help varargout`.

The program uses the backslash command to solve the triangular systems $Ly = b$ and $Ux = y$. Is this faster than using `A\b`? Yes, it is. The verification by MATLAB that no reduction is needed is much quicker than actually performing the reduction. (If we were writing our own code we'd know that they were triangular and so we wouldn't need to check.) Let's see what happens:

```
» clear all
» A=rand(1000);b=ones([1000 1]);c=ones([1000 1]);
» tic;[L,U]=ludecomp(A);toc  %Get the LU factors.
» tic;x=A\b;toc
» tic;y=L\c;xx=U\y;toc
```

You should repeat the last two lines a few times to get a better idea of the average performance of each approach. Each time `A\b` is computed, the cost of row-reduction must be paid again. But that cost has been built into L and U in advance.

This is a somewhat unfair test in that most matrices of interest are far from random. The redundant variable `c` is intended to overcome the objection that `b` may have been loaded into fast memory and hence some of the speed increase in the last line may have been due to a savings in time moving data from memory. This experiment should demonstrate that once L and U are known, the use of the LU decomposition represents a considerable advantage–in general an order of magnitude savings in computational time. Adding in the cost of performing the initial LU decomposition, it appears that this is worthwhile (for sufficiently large matrices) if we will be solving as few as 2 linear systems involving A.

We used a new variable xx rather than reusing x because there would have been some speed advantage in having the vector to which the result will be assigned pre-allocated space in memory. Pre-allocation for a matrix can be done using zeros([M N]) and is often advantageous. (In other languages you may wish to use a different approach.) Enter:

» clear all
» y=zeros([1 1000]);
» tic;for i=1:1000,y(i)=sin(10*i);end;toc
» clear all
» tic;for i=1:1000,z(i)=sin(10*i);end;toc

to see this effect. Of course, if you re-enter the last line you should see much better speed, because the space for z is now allocated. Using explicit indexing into the vector (i.e., using y(i)) is itself a relatively slow operation; we might have set t=1:1000 initially and used y=sin(10*t).

Many MATLAB commands can return more than one output if it is requested. A useful command in writing a program that uses pivoting is the max command with two output arguments. Enter:

» [M,ind]=max([4 3 5 2 1])

to see that max returns not only the maximum value but also its location. Enter:

» [M,ind]=max([4 3 5 2 1 5])

to see that all locations of the maximum are returned in the index vector. (In pivoting, one must have a rule for breaking ties resulting from more than one entry having maximum modulus.) Similarly for min, median, and sort.

Problems

1.) a.) Find the determinant of the matrix A of Example 3 by Laplace expansion.
b.) Find the determinant of the matrix A of Example 3 by using its LU decomposition and the fact that $|M_1 M_2| = |M_1||M_2|$ for any square matrices M_1 and M_2 and the fact that the determinant of a triangular matrix is the product of the entries on its main diagonal.

2.) a.) Find the LU decomposition of $A = [8\ 8\ 4; 4\ 2\ -1; 2\ 2\ 2]$.
b.) Write the LU decomposition of A in the compact form of Example 4.
c.) Solve $Ax = (16, 4, 6)^T$ using the LU decomposition of A.
d.) Solve $Ax = (0, 0, -2)^T$ using the LU decomposition of A.

3.) a.) Show that if $A = [a\ b; c\ d]$ and $L_1 = [1\ 0; -m\ 1]$ then $L_1 A$ is the matrix obtained from A by performing $R_2 \leftarrow R_2 - mR_1$.
b.) Find L_1^{-1}.
c.) If $|a| \geq |c|$ and a is nonzero, show that pivoting is not necessary in row-reducing A and that choosing $m = -c/a$ makes $L_1 A$ an upper triangular matrix U. What is the LU decomposition of A in terms of L_1 and U?

4.) a.) Additively decompose $A = [1\ 2\ 3; 4\ 5\ 6; 7\ 8\ 9]$ into the sum $A = L + U$ of a lower triangular matrix L and an upper triangular matrix U in such a way that L and U are both nonsingular. Is A nonsingular?
b.) If A is an $n \times n$ matrix, how much freedom is there in your choice of L and U in the decomposition $A = L + U$?

5.) a.) The matrix $A = [0\ 0\ 1; 0\ 2\ 3; 1\ 4\ -4]$ can't be row-reduced without row interchanges so our LU decomposition method does not apply. Show that if $P = [0\ 0\ 1; 0\ 1\ 0; 1\ 0\ 0]$ then PA has an LU decomposition. Identify L and U explicitly, then write them in the form of Example 4.

b.) Repeat with $A = [0\ 1\ 1; 0\ 2\ 3; 1\ 4\ -4]$ and the same P.

6.) a.) Find the LU decomposition of the matrix $A = [4\ 5\ 6; 2\ 3.5\ 1; -1\ -1\ 2]$.

b.) Write the LU decomposition of A in the compact form of Example 4.

c.) Use the LU decomposition of A to solve the linear system $Ax = (12, 6, 1)^T$.

7.) a.) Find the LU decomposition of $A = [18\ 3\ -6; 6\ 19\ 16; -9\ 3\ 13.5]$.

b.) Write the LU decomposition of A in the compact form.

c.) Use the LU decomposition of A to solve the linear system $Ax = (21, 25, 6)^T$.

8.) a.) The LU decomposition represented by $[1\ 2\ 1; -.2\ 1\ 0; .4\ -.5\ 1]$ in compact form corresponds to what matrix?

b.) What matrix has the LU decomposition represented by $[5\ -1\ 3; -.5\ 2\ 1; .3\ .5\ 4]$ in compact form?

9.) a.) How many flops are required to multiply an $n \times n$ matrix by a vector of length n? Count additions/subtractions and multiplications/divisions separately.

b.) How many flops are required to multiply two $n \times n$ matrices? Count additions/subtractions and multiplications/divisions separately.

c.) Compare the flops count for both sides of Eq. (2.2).

10.) Use the Doolittle LU decomposition of $A = [8\ 8\ 4; 4\ 2\ -1; 2\ 2\ 2]$ from Problem 2 part a. to find a LU decomposition of A for which the diagonal entries of U are all equal to 1 (but the diagonal entries of L need not be).

11.) a.) Show that A is nonsingular if and only if its LU factors L and U are both nonsingular.

b.) Show that L is always nonsingular.

c.) Show that if A is nonsingular then $A^{-1} = U^{-1}L^{-1}$.

12.) a.) If $A = [2\ 1\ 0\ 0; 1\ 2\ 1\ 0; 0\ 1\ 2\ 1; 0\ 0\ 1\ 2]$, find its LU decomposition.

b.) If $A = [2\ -1\ 0\ 0; -1\ 2\ -1\ 0; 0\ -1\ 2\ -1; 0\ 0\ -1\ 2]$, find its LU decomposition.

13.) Solve the linear system in Problem 9 using Gaussian elimination with maximal pivoting, which has round-off error properties that are superior to those of Gaussian elimination with partial pivoting.

14.) a.) Given the LU decomposition of A, what is the LU decomposition of A^T?

b.) Can L and/or U ever be symmetric?

c.) If $A = LU$ and $U = L^T$, show that A is symmetric and that $x^T A x$ is nonnegative for all x.

15.) Write a MATLAB program that accepts as arguments a nonsingular matrix A and one or two vectors b_1, b_2. If one output argument is requested, the program should return the compact form of the LU decomposition of A if A is the sole input and the solution of the linear system $Ax = b_1$ (found by LU decomposition) otherwise. If two output arguments are requested, the program should return the LU decomposition L and U of A if A is the sole input and the solution of $Ax = b_1$ otherwise. If three output arguments are requested, the program should return the compact form of the LU decomposition of A if A and both b_1, b_2 are given and the LU decomposition L and U of A if A and b_1 are given. If four output arguments are requested, the program should return the LU decomposition L and U of A if A and both b_1, b_2 are given. Signal an error and terminate execution in any case not covered.

16.) A discrete predator-prey model relates the populations of two interacting species, a predator of population size F_i and its sole prey of population size R_i. (We think of the predator and prey as foxes and rabbits, respectively.) These are used, for example, to predict the effect of spraying for Mediterranean fruit flies on the population of aphids in California. The basic form of such a model is $F_{i+1} = \alpha F_i + \beta R_i$, $R_{i+1} = -\gamma F_i + \delta R_i$ where all parameters are positive and $\alpha < 1$ (the predator dies off in the absence of the prey) and $\delta > 1$ (the prey thrives in the absence of the predator). Take i to be in months.

a.) If $\alpha = .3$, $\beta = .25$, $\gamma = .4$, and $\delta = 1.3$, compute the population after six months if $R_0 = 1000$, $F_0 = 20$.

b.) For the same parameter values, if $R_6 = 900$, $F_6 = 40$, use Gaussian elimination to estimate R_5 and F_5. Repeat until you have an estimate for R_0 and F_0.

3. The LU Decomposition with Pivoting

Gaussian elimination is referred to as a **direct method** for solving a linear system, meaning a method that terminates in finitely many iterations. These are very different from our methods of Ch. 1; in infinite precision arithmetic, Gaussian elimination would always give us the right answer in finitely many operations (like the quadratic formula does for quadratic equations), whereas Newton's method would always give an approximation (even in infinite precision arithmetic). There are also **iterative methods** for linear systems that, like Newton's method, do not reach their answer in finitely many steps, and that are useful for certain types of large systems (as we will discuss in Ch. 3). We'll continue to focus on direct methods for now.

At this point our method for finding an LU decomposition works only when row-reducing the matrix does not require pivoting. There are types of matrices for which it can be shown that Gaussian elimination with partial pivoting will in fact never actually pivot (that is, interchange rows), but this is still a significant restriction. We need to remove it.

A matrix that does require pivoting need not have an LU decomposition at all. For example, consider the 2×2 matrix

$$\begin{bmatrix} 0 & 1 \\ 1 & 1 \end{bmatrix}.$$

This is a square, nonsingular, well-behaved matrix. However, to have

$$\begin{bmatrix} 0 & 1 \\ 1 & 1 \end{bmatrix} = \begin{bmatrix} 1 & 0 \\ m & 1 \end{bmatrix} \begin{bmatrix} a & b \\ 0 & c \end{bmatrix}$$

would require that (upon multiplying the matrices on the RHS) $a = 0$, $b = 1$, $am = 1$, and $mb + c = 1$. But $am = 1$ is not possible, since $a = 0$. Therefore this matrix does not admit an LU decomposition. Of course if we simply interchange the rows then it is essentially its own LU decomposition:

$$\begin{bmatrix} 1 & 1 \\ 0 & 1 \end{bmatrix} = \begin{bmatrix} 1 & 0 \\ 0 & 1 \end{bmatrix} \begin{bmatrix} 1 & 1 \\ 0 & 1 \end{bmatrix}.$$

(The sole multiplier is zero, and the LU decomposition takes the form $A = IA$.) We certainly want to find a way to include such matrices in the general scheme of LU decomposition.

To find a way to get a (near) LU decomposition of matrices that require pivoting–which, because of stability concerns, is the typical case, unlike the specially prepared examples in the previous section–let's revisit Gaussian elimination. We perform Gaussian elimination using two of the three elementary row operations. Each elementary row operation corresponds to an elementary matrix that implements that operation by premultiplication; this matrix is always the matrix that arises by applying the elementary row operation to the identity matrix. That is, if A is an $n \times n$ matrix then performing $R_i \leftrightarrow R_j$ to A is equivalent to forming the matrix product PA where P is the matrix obtained from I_n by interchanging rows i and j of it, and similarly for the other row operations.

EXAMPLE 1: Let $A = [1 \ 2 \ 3; 0 \ 5 \ 6; 0 \ 8 \ 9]$. Then to perform $R_2 \leftrightarrow R_3$ we may use the elementary matrix

$$P = \begin{bmatrix} 1 & 0 & 0 \\ 0 & 0 & 1 \\ 0 & 1 & 0 \end{bmatrix}$$

giving

$$PA = \begin{bmatrix} 1 & 0 & 0 \\ 0 & 0 & 1 \\ 0 & 1 & 0 \end{bmatrix} \begin{bmatrix} 1 & 2 & 3 \\ 0 & 5 & 6 \\ 0 & 8 & 9 \end{bmatrix}$$
$$= \begin{bmatrix} 1 & 2 & 3 \\ 0 & 8 & 9 \\ 0 & 5 & 6 \end{bmatrix}$$

(check this). The matrix P implements $R_2 \leftrightarrow R_3$ for 3×3 matrices. If we now want to use row 2 of PA to eliminate its $(3,2)$ entry we may use the elementary matrix corresponding to the operation $R_3 \leftarrow R_3 - (5/8)R_2$ which we find by performing $R_3 \leftarrow R_3 - (5/8)R_2$ on I_3, giving

$$E = \begin{bmatrix} 1 & 0 & 0 \\ 0 & 1 & 0 \\ 0 & -5/8 & 1 \end{bmatrix}.$$

(Notice that E is lower triangular.) Then

$$E(PA) = \begin{bmatrix} 1 & 0 & 0 \\ 0 & 1 & 0 \\ 0 & -5/8 & 1 \end{bmatrix} \begin{bmatrix} 1 & 2 & 3 \\ 0 & 8 & 9 \\ 0 & 5 & 6 \end{bmatrix}$$
$$= \begin{bmatrix} 1 & 2 & 3 \\ 0 & 8 & 9 \\ 0 & 0 & 3/8 \end{bmatrix}$$

(verify this by hand). The matrix EPA is in echelon (reduced) form; that is, the matrix

$$M = EP$$
$$= \begin{bmatrix} 1 & 0 & 0 \\ 0 & 1 & 0 \\ 0 & -5/8 & 1 \end{bmatrix} \begin{bmatrix} 1 & 0 & 0 \\ 0 & 0 & 1 \\ 0 & 1 & 0 \end{bmatrix}$$
$$= \begin{bmatrix} 1 & 0 & 0 \\ 0 & 0 & 1 \\ 0 & 1 & -5/8 \end{bmatrix}$$

has the effect of triangularizing A, meaning that it transforms A into a row equivalent (upper) triangular matrix U:

$$MA = \begin{bmatrix} 1 & 0 & 0 \\ 0 & 0 & 1 \\ 0 & 1 & -5/8 \end{bmatrix} \begin{bmatrix} 1 & 2 & 3 \\ 0 & 5 & 6 \\ 0 & 8 & 9 \end{bmatrix}$$
$$= \begin{bmatrix} 1 & 2 & 3 \\ 0 & 8 & 9 \\ 0 & 0 & 3/8 \end{bmatrix}$$

(check this). $\square$

The elementary matrices corresponding to the operations $R_j \leftarrow R_j - m_{j,i} R_i$ in Gaussian elimination are *always lower triangular* because j is always greater than i, so that its row lies below row i. For example, $R_4 \leftarrow R_4 - m_{4,2} R_2$ would correspond to premultiplication by

$$E_{4,2} = \begin{bmatrix} 1 & 0 & 0 & 0 & 0 \\ 0 & 1 & 0 & 0 & 0 \\ 0 & 0 & 1 & 0 & 0 \\ 0 & -m_{4,2} & 0 & 1 & 0 \\ 0 & 0 & 0 & 0 & 1 \end{bmatrix}$$

for $n = 5$. (Notice that even when n is fixed, $E_{4,2}$ is not a specific matrix; it depends on the particular $m_{4,2}$ needed to eliminate the $(4,2)$ entry of the matrix we are working on.) Now think again about the previous section, where we assumed that no row interchanges were needed to perform Gaussian elimination. That means that the process can be represented algebraically as

$$(3.1) \qquad\qquad E_{4,3} E_{4,2} E_{3,2} E_{4,1} E_{3,1} E_{2,1} A = U$$

($n = 4$). Look at Eq. (3.1): The first multiplication is by $E_{2,1}$ which eliminates the $(2,1)$ entry of A; then comes the multiplication by $E_{3,1}$ which eliminates the $(3,1)$ entry of A; then multiplication by $E_{4,1}$ to eliminate the $(4,1)$ entry of A. This completes the process of introducing zeroes in the first column, below a_{11}. We then move to column 2 and use $E_{3,2}$ and $E_{4,2}$ to make zero the entries beneath a_{22}, successively triangularizing the matrix as we move from left to right across it. Finally we use $E_{4,3}$ to introduce the final zero, in the $(4,3)$ position. The progression of the matrix A into U has the form:

$$\begin{bmatrix} X & X & X & X \\ X & X & X & X \\ X & X & X & X \\ X & X & X & X \end{bmatrix} \sim \begin{bmatrix} X & X & X & X \\ 0 & X & X & X \\ X & X & X & X \\ X & X & X & X \end{bmatrix} \sim \begin{bmatrix} X & X & X & X \\ 0 & X & X & X \\ 0 & X & X & X \\ X & X & X & X \end{bmatrix}$$

$$\sim \begin{bmatrix} X & X & X & X \\ 0 & X & X & X \\ 0 & X & X & X \\ 0 & X & X & X \end{bmatrix} \sim \begin{bmatrix} X & X & X & X \\ 0 & X & X & X \\ 0 & 0 & X & X \\ 0 & X & X & X \end{bmatrix}$$

$$\sim \begin{bmatrix} X & X & X & X \\ 0 & X & X & X \\ 0 & 0 & X & X \\ 0 & 0 & X & X \end{bmatrix} \sim \begin{bmatrix} X & X & X & X \\ 0 & X & X & X \\ 0 & 0 & X & X \\ 0 & 0 & 0 & X \end{bmatrix}.$$

Each step corresponds to a single matrix multiplication. Of course, note that applying $E_{3,1}$ to the matrix $E_{2,1}$,

$$E_{3,1}E_{2,1}$$

has the same effect as applying the row operation $R_3 \leftarrow R_3 - m_{3,1}R_{21}$ to $E_{2,1}$,

$$E_{3,1}E_{2,1} = \begin{bmatrix} 1 & 0 & 0 & 0 \\ 0 & 1 & 0 & 0 \\ -m_{3,1} & 0 & 1 & 0 \\ 0 & 0 & 0 & 1 \end{bmatrix} \begin{bmatrix} 1 & 0 & 0 & 0 \\ -m_{2,1} & 1 & 0 & 0 \\ 0 & 0 & 1 & 0 \\ 0 & 0 & 0 & 1 \end{bmatrix}$$

$$= \begin{bmatrix} 1 & 0 & 0 & 0 \\ -m_{2,1} & 1 & 0 & 0 \\ -m_{3,1} & 0 & 1 & 0 \\ 0 & 0 & 0 & 1 \end{bmatrix}$$

(check the multiplication). By extension,

$$E_{4,1}E_{3,1}E_{2,1} = \begin{bmatrix} 1 & 0 & 0 & 0 \\ -m_{2,1} & 1 & 0 & 0 \\ -m_{3,1} & 0 & 1 & 0 \\ -m_{4,1} & 0 & 0 & 1 \end{bmatrix}$$

$$= L_1$$

which performs all three eliminations in column 1 in a single matrix multiplication. Since the choice of the multiplier $m_{3,1}$ isn't affected by the action of $E_{2,1}$, and similarly for $m_{4,1}$, we can form this single matrix in place of $E_{4,1}E_{3,1}E_{2,1}$ in Eq. (3.1). But we cannot easily add on $E_{3,2}$ since its multiplier will depend on the effects of having introduced zeroes in column 1 (which affects the entries in column 2); we must wait until L_1 has been applied, before we can start deciding what the entries of $E_{3,2}$ and $E_{4,2}$ will be. But after we have transformed A into L_1A we can play the same game for column 2:

$$E_{4,2}E_{3,2} = \begin{bmatrix} 1 & 0 & 0 & 0 \\ 0 & 1 & 0 & 0 \\ 0 & -m_{3,2} & 1 & 0 \\ 0 & -m_{4,2} & 0 & 1 \end{bmatrix}$$

$$= L_2$$

and then

$$E_{4,3} = \begin{bmatrix} 1 & 0 & 0 & 0 \\ 0 & 1 & 0 & 0 \\ 0 & 0 & 1 & 0 \\ 0 & 0 & -m_{4,3} & 1 \end{bmatrix}$$

$$= L_3$$

allowing us to write Eq. (3.1) as

(3.2) $$L_3 L_2 L_1 A = U$$

or simply $MA = U$ with $M = L_3 L_2 L_1$. (Matrices such as L_1, L_2, and L_3 are called **Gauss transformations**.) Now comes a significant point: Every matrix L_i is a lower triangular matrix. We are using triangular matrices to triangularize A to U.

But beyond that, every matrix L_i is a lower triangular matrix with 1s on the main diagonal, called a matrix is **unit lower triangular**. It is a fact that a product of unit lower triangular matrices is itself unit lower triangular. Additionally, unit lower triangular matrices are nonsingular (after all, they have unit determinant). Hence we can solve Eq. (3.2) for A.

The inverse of a unit lower triangular matrix is in fact also a unit lower triangular matrix. Perhaps surprisingly, in the case of the Gauss transformations L_i^{-1} the inverse is easily given:

$$L_1^{-1} = \begin{bmatrix} 1 & 0 & 0 & 0 \\ m_{2,1} & 1 & 0 & 0 \\ m_{3,1} & 0 & 1 & 0 \\ m_{4,1} & 0 & 0 & 1 \end{bmatrix}$$

$$L_2^{-1} = \begin{bmatrix} 1 & 0 & 0 & 0 \\ 0 & 1 & 0 & 0 \\ 0 & m_{3,2} & 1 & 0 \\ 0 & m_{4,2} & 0 & 1 \end{bmatrix}$$

$$L_3^{-1} = \begin{bmatrix} 1 & 0 & 0 & 0 \\ 0 & 1 & 0 & 0 \\ 0 & 0 & 1 & 0 \\ 0 & 0 & m_{4,3} & 1 \end{bmatrix}$$

as is easily verified by computing, e.g., $L_3^{-1} L_3$, and seeing that it is the identity matrix. (Maybe this isn't so surprising since L_3^{-1} should reverse the action of L_3 and indeed, L_3 represents the row operation $R_4 \leftarrow R_4 - m_{4,3} R_3$ while L_3^{-1} represents the row operation $R_4 \leftarrow R_4 + m_{4,3} R_3$ which reverses it.) Hence we have

$$\begin{aligned}
MA &= U \\
L_3 L_2 L_1 A &= U \\
L_2 L_1 A &= L_3^{-1} U \\
L_1 A &= L_2^{-1} L_3^{-1} U \\
A &= L_1^{-1} L_2^{-1} L_3^{-1} U \\
&= LU
\end{aligned}$$

which is the LU decomposition, since L_i^{-1} is unit lower triangular and hence L is also. (We now see why the LU decomposition used $m_{j,i}$ rather than $-m_{j,i}$ in the last section.) The details do not change for a general n. It's crucial that the L_i^{-1} can be found easily, as seen in the previous section, and do not require us to form L_i and then invert it. We would continue to code the method for finding L and U according to the procedure in Sec. 2.2 and *not* by forming these large matrices at each step.

What happens if we must pivot? In the worst case we might need to pivot at every step. Since the identity matrix I_n represents a trivial row interchange $R_i \leftrightarrow R_i$, we can write Eq. (3.2) as

$$L_3 P_3 L_2 P_2 L_1 P_1 A = U$$

where each P_i is an elementary matrix representing a row interchange (as in Example 1). It may be that $P_i = I$ for some i (and possibly for all i, as in the examples of the previous section). Note that P_i must be its own inverse, since P_i interchanges two rows and hence $P_i P_i$ interchanges the rows and then switches them back again. But then $P_i P_i = I$ ($P_i P_i$ is the identity transformation), and so

$$P_i = P_i^{-1}.$$

Thus we could write

$$\begin{aligned}
L_3 P_3 L_2 P_2 L_1 P_1 A &= U \\
P_3 L_2 P_2 L_1 P_1 A &= L_3^{-1} U \\
L_2 P_2 L_1 P_1 A &= P_3 L_3^{-1} U \\
P_2 L_1 P_1 A &= L_2^{-1} P_3 L_3^{-1} U \\
L_1 P_1 A &= P_2 L_2^{-1} P_3 L_3^{-1} U \\
P_1 A &= L_1^{-1} P_2 L_2^{-1} P_3 L_3^{-1} U \\
A &= P_1 L_1^{-1} P_2 L_2^{-1} P_3 L_3^{-1} U
\end{aligned}$$

which gives a decomposition of the form

(3.3) $$A = \widetilde{L} U$$

where U is upper triangular but the matrix $\widetilde{L} = P_1 L_1^{-1} P_2 L_2^{-1} P_3 L_3^{-1}$ is not in general lower triangular. The MATLAB command [L,U]=lu(A) produces $\widetilde{L}$ and U for general A.

It is not hard to show that it is always the case that $\widetilde{L} = P^T L$ where L is a unit lower triangular matrix and $P = P_3 P_2 P_1$ is a **permutation matrix**, that is, P is the result of permuting (re-ordering) the rows of the identity matrix. For example,

$$P = \begin{bmatrix} 0 & 0 & 0 & 1 \\ 1 & 0 & 0 & 0 \\ 0 & 1 & 0 & 0 \\ 0 & 0 & 1 & 0 \end{bmatrix}$$

is a permutation matrix. It shuffles the position of all 4 rows of a 4×4 matrix but does not change the entries:

$$\begin{bmatrix} 0 & 0 & 0 & 1 \\ 1 & 0 & 0 & 0 \\ 0 & 1 & 0 & 0 \\ 0 & 0 & 1 & 0 \end{bmatrix} \begin{bmatrix} 1 & 1 & 1 & 1 \\ 2 & 2 & 2 & 2 \\ 3 & 3 & 3 & 3 \\ 4 & 4 & 4 & 4 \end{bmatrix} = \begin{bmatrix} 4 & 4 & 4 & 4 \\ 1 & 1 & 1 & 1 \\ 2 & 2 & 2 & 2 \\ 3 & 3 & 3 & 3 \end{bmatrix}$$

(PA is equal to A with its rows permuted in the same way that those of P are), though not all permutation matrices move every row. A square matrix is a permutation matrix iff P has exactly one 1 in each row and in each column. If P is a permutation matrix then P^{-1} is also a permutation matrix and $P^{-1} = P^T$ (permutation matrices are orthogonal). The matrices P_i used for pivoting are permutation matrices that only switch two rows.

We see from $\widetilde{L} = P^T L$ that $\widetilde{L}$ is just a permuted unit lower triangular matrix. We can now write Eq. (3.3) as $A = P^T LU$ or, equivalently, as

$$(3.4) \qquad\qquad\qquad PA = LU.$$

This is the form of the LU decomposition for matrices that require pivoting. Formally it is sometimes called the **PLU decomposition**, but since pivoting is nearly always applied when performing Gaussian elimination it's usual to just call it the LU decomposition.

We interpret Eq. (3.3) as follows: For any square matrix A, there is a permutation P of its rows such that PA admits an LU decomposition. Of course, we do not know P until we perform the actual reduction, so this observation is not directly useful in practice, but we are assured of the existence of a PLU decomposition of an arbitrary matrix, and we have an algorithm for finding it.

In practice we do not form the large and sparse matrix P, just as we don't form the L_i. In place of P we use a vector p that records the order in which the rows of I have been permuted and use it to simulate the effects of P. For example,

$$P = \begin{bmatrix} 0 & 0 & 0 & 1 \\ 1 & 0 & 0 & 0 \\ 0 & 1 & 0 & 0 \\ 0 & 0 & 1 & 0 \end{bmatrix}$$

might be encoded as the vector $p = [4\ 1\ 2\ 3]$ (with $p(1) = 4$ indicating that row 1 of the matrix P has its only nonzero entry in column 4, etc.).

EXAMPLE 2: Let's find the LU decomposition of $A = [1\ 2\ 3; 4\ 5\ 6; 7\ 8\ 9]$. We begin by performing the row interchange $R_1 \leftrightarrow R_3$:

$$P_1 A = \begin{bmatrix} 0 & 0 & 1 \\ 0 & 1 & 0 \\ 1 & 0 & 0 \end{bmatrix} \begin{bmatrix} 1 & 2 & 3 \\ 4 & 5 & 6 \\ 7 & 8 & 9 \end{bmatrix}$$

$$= \begin{bmatrix} 7 & 8 & 9 \\ 4 & 5 & 6 \\ 1 & 2 & 3 \end{bmatrix}$$

($p = [3\ 2\ 1]$). We now perform the reductions for column 1

$$L_1 P_1 A = \begin{bmatrix} 1 & 0 & 0 \\ -4/7 & 1 & 0 \\ -1/7 & 0 & 1 \end{bmatrix} \begin{bmatrix} 7 & 8 & 9 \\ 4 & 5 & 6 \\ 1 & 2 & 3 \end{bmatrix}$$

$$= \begin{bmatrix} 7 & 8 & 9 \\ 0 & 3/7 & 6/7 \\ 0 & 6/7 & 12/7 \end{bmatrix}$$

(check these multiplications). Next we pivot again to obtain

$$P_2 L_1 P_1 A = \begin{bmatrix} 1 & 0 & 0 \\ 0 & 0 & 1 \\ 0 & 1 & 0 \end{bmatrix} \begin{bmatrix} 7 & 8 & 9 \\ 0 & 3/7 & 6/7 \\ 0 & 6/7 & 12/7 \end{bmatrix}$$

$$= \begin{bmatrix} 7 & 8 & 9 \\ 0 & 6/7 & 12/7 \\ 0 & 3/7 & 6/7 \end{bmatrix}$$

(switch the corresponding entries in $p = [3\ 2\ 1]$ to get the updated permutation vector $p = [3\ 1\ 2]$). Then

$$L_2 P_2 L_1 P_1 A = \begin{bmatrix} 1 & 0 & 0 \\ 0 & 1 & 0 \\ 0 & -1/2 & 1 \end{bmatrix} \begin{bmatrix} 7 & 8 & 9 \\ 0 & 6/7 & 12/7 \\ 0 & 3/7 & 6/7 \end{bmatrix}$$

$$= \begin{bmatrix} 7 & 8 & 9 \\ 0 & 6/7 & 12/7 \\ 0 & 0 & 0 \end{bmatrix}$$

(note that A is singular since it is row equivalent to a matrix with zero determinant). So

$$U = \begin{bmatrix} 7 & 8 & 9 \\ 0 & 6/7 & 12/7 \\ 0 & 0 & 0 \end{bmatrix}$$

and

$$\begin{aligned}
\widetilde{L} &= (L_2 P_2 L_1 P_1)^{-1} \\
&= P_1^T L_1^{-1} P_2^T L_2^{-1} \\
&= P_1 L_1^{-1} P_2 L_2^{-1} \\
&= \begin{bmatrix} 0 & 0 & 1 \\ 0 & 1 & 0 \\ 1 & 0 & 0 \end{bmatrix} \begin{bmatrix} 1 & 0 & 0 \\ 4/7 & 1 & 0 \\ 1/7 & 0 & 1 \end{bmatrix} \begin{bmatrix} 1 & 0 & 0 \\ 0 & 0 & 1 \\ 0 & 1 & 0 \end{bmatrix} \begin{bmatrix} 1 & 0 & 0 \\ 0 & 1 & 0 \\ 0 & 1/2 & 1 \end{bmatrix} \\
&= \begin{bmatrix} 1/7 & 1 & 0 \\ 4/7 & 1/2 & 1 \\ 1 & 0 & 0 \end{bmatrix}
\end{aligned}$$

which is a permutation of a unit lower triangular matrix ($\widetilde{L} = P^T L$) as expected. We also have

$$\begin{aligned}
P &= P_2 P_1 \\
&= \begin{bmatrix} 1 & 0 & 0 \\ 0 & 0 & 1 \\ 0 & 1 & 0 \end{bmatrix} \begin{bmatrix} 0 & 0 & 1 \\ 0 & 1 & 0 \\ 1 & 0 & 0 \end{bmatrix} \\
&= \begin{bmatrix} 0 & 0 & 1 \\ 1 & 0 & 0 \\ 0 & 1 & 0 \end{bmatrix}
\end{aligned}$$

in agreement with the order specified by the permutation vector $p = [3 \ 1 \ 2]$. To find L we'll use $\widetilde{L} = P^T L$ to find

$$\begin{aligned}
L &= P\widetilde{L} \\
&= \begin{bmatrix} 0 & 0 & 1 \\ 1 & 0 & 0 \\ 0 & 1 & 0 \end{bmatrix} \begin{bmatrix} 1/7 & 1 & 0 \\ 4/7 & 1/2 & 1 \\ 1 & 0 & 0 \end{bmatrix} \\
&= \begin{bmatrix} 1 & 0 & 0 \\ 1/7 & 1 & 0 \\ 4/7 & 1/2 & 1 \end{bmatrix}
\end{aligned}$$

so that

$$PA = LU$$

$$\begin{bmatrix} 0 & 0 & 1 \\ 1 & 0 & 0 \\ 0 & 1 & 0 \end{bmatrix} \begin{bmatrix} 1 & 2 & 3 \\ 4 & 5 & 6 \\ 7 & 8 & 9 \end{bmatrix} = \begin{bmatrix} 1 & 0 & 0 \\ 1/7 & 1 & 0 \\ 4/7 & 1/2 & 1 \end{bmatrix} \begin{bmatrix} 7 & 8 & 9 \\ 0 & 6/7 & 12/7 \\ 0 & 0 & 0 \end{bmatrix}$$

$$\begin{bmatrix} 7 & 8 & 9 \\ 1 & 2 & 3 \\ 4 & 5 & 6 \end{bmatrix} = \begin{bmatrix} 7 & 8 & 9 \\ 1 & 2 & 3 \\ 4 & 5 & 6 \end{bmatrix}$$

which checks. We have found an LU decomposition of A. $\square$

There are many practical implementation issues that we will not address. One crucial issue, however, is this: Some programming languages store matrices by writing successive rows to memory (called storage in **row-major** form), and others store matrices by writing successive columns to memory (called **column-major** form). All matrix algorithms come in row-oriented and column-oriented versions, and using the appropriate version[2] will speed execution noticeably. (The reason is that it is faster to retrieve data near recently retrieved data in memory; if matrices are stored by rows then a_{11} and a_{12} are close but if matrices are stored by columns they need not be, though a_{11} and a_{21} will be. When a particular value is requested, nearby values are automatically loaded into the fast cache memory, and under a row-major storage scheme nearby means nearby when tracing by rows.) Another issue is the desire to use as many Level-3 BLAS as feasible (or Level-2 or Level-1 BLAS by designing appropriate vector-oriented versions of the method).

The MATLAB command `[L,U,P]=lu(A)` produces the matrices L, U, and P of the LU decomposition of a general square matrix. The relation $L = P\tilde{L}$ means that if `[L,U,P]=lu(A)` and `[L1,U1]=lu(A)` then `U` is equal to `U1` and `L` is equal to `P*L1`. Note that the LU decomposition is not unique and so this command gives *an* LU decomposition of the matrix.

Once we know L, U, and P we may solve the linear system $Ax = b$ in essentially the same manner as before. We premultiply both sides by P to obtain

$$
\begin{aligned}
PAx &= Pb \\
(PA)x &= d \\
LUx &= d
\end{aligned}
\tag{3.5}
$$

($d = Pb$) and proceed as before, solving $Ly = d$ and then $Ux = y$. Premultiplication by P simply reorders the equations that comprise the linear system. Once again, we would not form the matrix P but would rather shuffle the rows of A and the rows of b according to the permutation vector p.

For special matrices it may be possible to simplify or otherwise improve upon this scheme, and there are many variations of the LU decomposition. In some cases it is convenient to use an **LDU decomposition** in which A (or PA) is factored into the product $A = LDU$ where L is a (possibly permuted) lower triangular matrix, D is a diagonal matrix, and U is an upper triangular matrix.

We will close this section with a few words on the matrix inverse. This is almost never needed in computation, and whenever a matrix inverse is encountered in a formula it pays to study the equation to see if it is possible to rewrite it so that an equivalent linear system is solved rather than an inverse matrix computed[3]. In practice this is almost always possible–exceptions are exceedingly rare. Certainly in the formula

$$
x = A^{-1}b
$$

we would want to solve $Ax = b$ using the LU decomposition or some similar approach. To compute a matrix result $X = A^{-1}B$, where A and B are matrices, we can solve $AX = B$ column-by-column.

EXAMPLE 3: If we are asked to compute the product

---

[2]G.W. Stewart notes that "both [FORTRAN90 and C++] include powerful features that make it easy to write inefficient code."

[3]The same goes for matrices which may be the products of (simpler or better structured) matrices earlier on in a derivation. Bill Gragg stated that whenever one encounters a matrix, one would do well to ask from whence it came.

$$y = A^{-1}BCD^{-1}Ex$$

given the five $n \times n$ matrices A, B, C, D, and E and the conformable vector x, then we should proceed as follows: First, compute $u = Ex$. Next, solve

$$Dv = u$$

via Gaussian elimination (with pivoting) for $u = D^{-1}Ex$. Then, form $w = B\,(Cu)$, as parenthesized. Finally, solve

$$Ay = w$$

via Gaussian elimination (with pivoting) for y. □

Experience shows that many people feel that they need to compute a matrix inverse but few if any actually do[4]. However, on those occasions when it is desired to compute $x = A^{-1}b$ in a way that is both *less accurate* and *less efficient* than the use of the LU decomposition, the usual way to do so is to use the fact that column i of A^{-1} is the solution of

$$Ac_i = e_i$$

where e_i is column i of the identity matrix, that is, e_i is the column n-vector that is entirely zero except for entry i, which is equal to one. (This is equivalent to the traditional by-hands method of forming $\left[A \vdots I\right]$ and row-reducing it to $\left[I \vdots U\right]$.) We simply solve

$$P^T LUc_i = e_i$$

in the usual way for $i = 1, ..., n$ and then form $A^{-1} = [c_1 \cdots c_n]$ from the results. We will still need to do a matrix-vector multiplication afterwards to find $A^{-1}b$. There are algorithms for finding the inverse in place, and of course there are specialized methods for special matrices.

MATLAB

When dealing with the display of small matrices it may be useful to issue the command `format compact`, which suppresses extra linefeeds in the display of results. Enter:

```
» format loose    %This is the default.
» A=rand(13)
» format compact
» A
» format loose
```

to see the difference. The `format loose` command changes back to the default display with respect to linefeeds; other format settings (e.g., `format long`) are not affected, whereas simply entering `format` resets all formatting options to their default state.

The MATLAB command `prod` used in the Problems computes the product $\prod_{i=1}^{n} x_i$ of the elements of a vector, much like `sum` computes the sum $\sum_{i=1}^{n} x_i$ of

---

[4]G.W. Stewart suggests that one exception is when a researcher wishes to inspect an inverse matrix "to get a feel for its structure".

2. LINEAR SYSTEMS

the elements of a vector. For a triangular matrix the determinant is the product of its diagonal entries `prod(diag(U))`, that is, $\prod_{i=1}^{n} u_{i,i}$. This is a case where MATLAB notation is arguably simpler than standard mathematical notation[5].

As mentioned in the reading, the `lu` command with three outputs requested returns the L, U, and P of the PLU decomposition. Enter:

```
» A=[1 2 3;4 5 6;7 8 9]
» [L1,U1]=lu(A)
» [L,U,P]=lu(A)
```

to see the PLU decomposition factors of Example 2. The matrix L1 is clearly a permuted unit lower triangular matrix. To see that $L = P\widetilde{L}$ as claimed, enter:

```
» P*L1
» L==ans
```

and to see that $PA = LU$ as claimed, enter:

```
» P*A,L*U
```

and then enter:

```
» [L2,U2]=lu(P*A)
```

to see that PA can be row-reduced without pivoting, as claimed, because L2 is lower triangular (as opposed to permuted lower triangular matrix, as it would be if pivoting had been performed). In fact, if we request the values of L, U, and P by typing:

```
» [L2,U2,P2]=lu(P*A)
```

we find that P2 is the identity matrix (I_3), indicating that no pivoting has been performed in row-reducing PA to triangular form.

In experimenting with permutation matrices, the command `randperm` may be useful. This command generates a (pseudo-)random permutation of the integers $1, ..., n$. Try:

```
» v=randperm(6)
» v=randperm(6)
» M=eye(6); P=M(v,:)
```

The colon signifies all columns or all rows, respectively. To permute the entries of a vector we may index it in the desired order. For example, enter:

```
» v=2*(1:8),v(8:-1:1),v([2 4 6 8 1 3 5 7])
```

The `sort` command can make use of this feature (as discussed in the MATLAB subsection of Sec. 2.6). Enter:

```
» y=rand([1 8])
» [s,index]=sort(y)
» y(index)      %Sorted.
```

(Of course y(index) is equal to s.) We may work with matrices in a similar way. Enter:

```
» A=[1 2 3 4;5 6 7 8;9 10 11 12;13 14 15 16]
» A([1 3 2 4],:)
» A(:,[1 3 2 4])
```

The vector [1 3 2 4] is interpreted as a permutation vector, and the rows or columns, respectively, are permuted according to it.

---

[5]The programming language APL (now known as J in its modern form) was initially developed as a mathematical notation and later implemented as a computer language. It had many MATLAB-like features including simple notation for sums, products, etc., and a vector and array orientation.

Gaussian elimination with partial pivoting–that is, LU decomposition–stands with Newton's method as one of the most widely used numerical algorithms in existence. (It is somewhat atypical of numerical methods in that it always "converges" after finitely many operations.) While it can in principle exaggerate errors, in practice it does not. However, we will need to develop other methods for solving linear systems, especially large ones. For such matrices the $O(n^3)$ operation count of this method may be too expensive. In addition, in most cases of interest, large matrices have a structure and in particular are sparse: Most of their entries (greater than 95%, say) are zeroes. This *should* save us a lot of work, much as symmetry should in general cut our work in half. What happens when Gaussian elimination with partial pivoting is applied to a matrix with a structured pattern of null entries? Enter:

» A=eye(10);A(1,:)=ones([1 10]);A(:,1)=ones([10 1])

to create such a matrix. This is a symmetric matrix with nonnegative integer entries. Enter:

» sum(sum(A>0))/prod(size(A))

to compute the percentage of A that is nonzero (28%). Enter:

» A1=inv(A)

On the one hand, the inverse matrix has an obvious pattern; on the other hand, while A was 28% nonzero, its inverse A1 is 100% nonzero. The inverse of a sparse matrix is typically not sparse. (As usual, there are special classes of matrices that are exceptions to this general rule; the most obvious one is diagonal matrices.) This is another reason why we avoid inverting a matrix; there are special methods of storing and manipulating sparse matrices which give considerable speed advantages and these advantages are typically lost if the matrix is inverted, let alone the concern over whether we have pre-allocated sufficient memory. Enter:

» norm(A1*A-eye(10))

to see that the inverse is nonetheless quite accurate in this sense. Now enter:

» [L,U,P]=lu(A)

Note that L and U are each as full as they could be, save for one null entry. Their inverses have the same property. The act of performing Gaussian elimination with partial pivoting on the matrix A has destroyed its sparsity structure and with it any advantages in speed or storage that we might have had. For small matrices this may be acceptable, and for small-to-medium n we typically use Gaussian elimination with partial pivoting for both sparse and dense matrices, unless the matrix has some other structure that we can make use of to get a better algorithm. (Even then we usually use a variant of the LU decompositions.) For large sparse matrices we can do better, as will be discussed in Ch. 3 where we consider other types of methods that do not in general converge in finitely many steps. Large dense problems are problematic.

A matrix like the matrix A is sometimes called an **arrowhead matrix** (or **arrow matrix**). Enter:

» spy(A)

to see why. The spy command indicates graphically the location of the nonzero entries of a matrix (its sparsity structure or sparsity pattern). In fact the term arrowhead matrix is more commonly used to mean the matrix obtained as follows:

» B=flipud(fliplr((A)))

» spy(B)

The flip commands flip a matrix or vector up-down or left-right. For example, try:
» `fliplr(1:10)`
» `flipud(1:10)`
(Recall that `1:10` is a 1×10 matrix to MATLAB.) The command `rot90` rotates a matrix through a ninety degree angle. Enter:
» `spy(rot90(A))`
» `spy(rot90(A,2))`
(The command `rot90(A,2)` rotates `A` through 2 right angles, that is, it is the same as `rot90(rot90(A))`.) To try `spy` on a bigger matrix, we may use the `sprandn` command that generates a pseudo-random sparse matrix. (There are other commands to generate such matrices; see `help sprandn`.) Enter:
» `S=sprandn(100,100,.05);`
» `spy(S)`
to generate a 100×100 matrix that has about 5% nonzero entries. Enter
» `S`
and you will see that this matrix is displayed in a different form than usual; this is because MATLAB knows that matrices generated by `sprandn` should be treated as sparse matrices, which it stores and manipulates them in a more efficient way. The `sparse` command tells MATLAB to change the representation of a matrix to sparse form, and the `full` command changes back to the standard form. Enter:
» `F=full(S)`
The matrices `S` and `F` are equal but are stored and handled differently. Enter:
» `tic;F*F;toc`
» `tic;S*S;toc`
to see the difference. The matrices `S*S` and `F*F` are equal but using the sparse form is more efficient. (Repeat these commands and note that the time goes down considerably as MATLAB optimizes its access to them.) The difference is even greater when the sparsity pattern is not random as it is here.

Problems

1.) a.) Let $L_1 = [1 \ \ 0 \ \ 0; -1/2 \ \ 1 \ \ 0; 1/2 \ \ 0 \ \ 1]$, $L_2 = [1 \ \ 0 \ \ 0; 0 \ \ 1 \ \ 0; 0 \ \ 1/4 \ \ 1]$. Compute (by hand) the products $L_1 L_2$ and $L_2 L_1$.
b.) If A is a 3×3 matrix, describe the row operations that would be performed on A if it was premultiplied by L_1; if it was premultiplied by L_2; and if it was premultiplied by $L_2 L_1$.
c.) Find the inverses of L_1 and L_2.
d.) Let $A = [1 \ \ 1 \ \ 1; 1/2 \ \ 3/2 \ \ 3/2; -1/2 \ \ -3/4 \ \ 1/4]$. Compute $L_2 L_1 A$.
e.) What is the LU decomposition of A? Explain how you are using your answers from the previous parts of this problem to answer this question.
2.) a.) Find the LU decomposition of the matrix $A = [1 \ \ 2 \ \ 3; 4 \ \ 5 \ \ 6; 7 \ \ 8 \ \ 8]$ (by hand; follow Example 2).
b.) Use the LU decomposition of A to solve the linear system $Ax = b$ where $b = (1, 2, 3)^T$.
3.) Prove that a product of unit lower triangular matrices must be unit lower triangular. (Hint: Consider a typical dot product of a row from the first matrix and a column from the second matrix.)
4.) a.) Prove that the product of two elementary matrices that represent row interchanges is a permutation matrix.
b.) Prove that a permutation matrix is nonsingular and in fact orthogonal.

5.) Write a MATLAB program that accepts as arguments a nonsingular matrix A and a vector b (check that it is conformable) and solves $Ax = b$ by using the command [L,U,P]=lu(A) followed by the method of Eq. (3.5). (Use the backlash to solve the triangular systems $Ly = Pb$ and $Ux = y$.) Signal an error if U is found to be singular (use prod(diag(U)) rather than det(U) to check the determinant). Return x if only one output is requested and return x and the compact form of the LU decomposition (given by tril(L,-1)+triu(U)) if two output arguments are requested.

6.) a.) Find the LU decomposition of the matrix $A = [1\ 0\ -2; 4\ 3\ 8; 8\ 8\ -3]$ (by hand).

b.) Use the LU decomposition of A to solve the linear system $Ax = (-3, 4, -11)^T$.

7.) Give two LU decompositions of the zero matrix (with L unit lower triangular) and two LU decompositions of the identity matrix (with L unit lower triangular in at least one of the decompositions).

8.) Describe the set of all 3×3 unit lower triangular matrices L such that LU is a LU decomposition of $A = [1\ 0\ 0; 0\ 0\ 0; 0\ 0\ 0]$ (take $U = A$).

9.) Show that if L is unit lower triangular then so is L^{-1}.

10.) a.) Consider again the equation $L_3 P_3 L_2 P_2 L_1 P_1 A = U$ representing the general case of row-reducing a 4×4 matrix to upper triangular form. Define $\Lambda_3 = L_3$, $\Lambda_2 = P_3 L_2 P_3^{-1}$, $\Lambda_1 = P_3 P_2 L_1 P_2^{-1} P_3^{-1}$, and $P = P_3 P_2 P_1$. Show that Λ_1, Λ_2, and Λ_3 are unit lower triangular and hence $\Lambda = \Lambda_3 \Lambda_2 \Lambda_1$ is unit lower triangular.

b.) Show that $L_3 P_3 L_2 P_2 L_1 P_1 A = \Lambda_3 \Lambda_2 \Lambda_1 P \ (= \Lambda P)$.

c.) Justify the transition from Eq. (3.3) to Eq. (3.4) using the relation $\Lambda P A = U$ derived above.

d.) Let A be $n \times n$. Show that A has an LU decomposition $PA = LU$ with L unit lower triangular (that is, justify Eq. (3.3) and Eq. (3.4) starting from the first step of Gaussian elimination in this general case).

11.) a.) Find (by hand) the LU decomposition of $A = [2\ -1\ 1; -1\ 2\ -1; 1\ -1\ 2]$.

b.) Find (by hand) a decomposition of A of the form $A = LDL^T$ where L is unit lower triangular and D is diagonal with strictly positive diagonal entries.

c.) Show that if $A = LDL^T$ where L is unit lower triangular and D is diagonal with strictly positive diagonal entries then A is symmetric, and $x^T A x$ is always positive unless $x = 0$. (In fact this is an if-and-only-if scenario.) Matrices with these two properties are discussed in Sec. 2.5.

d.) Find (by hand) a decomposition of A of the form $A = LL^T$ where L is lower triangular with nonzero diagonal entries.

e.) Show that if $A = LL^T$ where L is lower triangular with nonzero diagonal entries then A is symmetric, and $x^T A x$ is always positive unless $x = 0$. (Again, this is an if-and-only-if.)

12.) For the matrix $A = [1\ 2\ 3\ 4; 5\ 6\ 7\ 8; 9\ 10\ 11\ 12; 13\ 14\ 15\ 16]$, write out all elementary matrices involved in row-reducing A to upper triangular form and indicate the order in which they would be applied to A (similar to Eq. (3.1)).

13.) a.) If A is square then it is possible to write it as $A = UTU^{-1}$ where U is unitary ($U^{-1} = U^T$) and T is upper triangular with the eigenvalues of A on its main diagonal. (This is called the **Schur decomposition**.) Note that U and T may be complex-valued even though A is real. Use the MATLAB commands [U1,T1]=schur(A);[U,T]=rsf2csf(U1,T1) to find the Schur decomposition of

$A = [0\ \ 1\ \ 0; -1\ \ 0\ \ 0; 0\ \ 0\ \ 1]$. (The `rsf2csf` command converts the output of `schur` to the form described above.)

b.) If you had the Schur decomposition of A pre-computed, how would you solve $Ax = b$? Assume that complex arithmetic is implemented automatically. (Note: The Schur decomposition is not in fact used to solve linear systems; it is used in numerical eigenvalue computations.)

14.) Use MATLAB to find the LU decomposition of `ones([10 10])`. Explain why L, U, and P have the form they do in terms of the Gaussian elimination with partial pivoting algorithm.

15.) a.) Use the MATLAB commands `lu` and `chol` to find the LU and Cholesky decompositions of $A = [2\ -1\ \ 1; -1\ \ 2\ -1; 1\ -1\ \ 2]$. The Cholesky decomposition is an LU decomposition that applies only to certain matrices (see Sec. 2.5); L is not necessarily *unit* lower triangular in this case, but $L^T = U$. Multiply out the results in each case ($L_1 U$ and $L_2 L_2^T$, respectively) to verify that they are correct.

b.) Use the Cholesky decomposition of A to solve $Ax = (1, -2, 1)^T$.

16.) In the analysis of linear systems in electrical engineering it is frequently necessary to find the roots of a polynomial that appears in the denominator of a rational function, the transfer function of the system. (In many cases it is only necessary to determine whether these roots, called **poles**, have positive or negative real part.) A common approach for high-degree monic polynomials $x^n + p_{n-1} x^{n-1} + ... + p_1 x + p_0$ is to form the **companion matrix** of the polynomial, which can be written as the matrix with first row $[-p_{n-1}\ -p_{n-2}\ \cdots\ -p_1\ -p_0]$ and all other rows are zero save that the subdiagonal entry is unity. For example, $x^3 + 4x^2 + 2x + 3$ would have companion matrix $C = [-4\ -2\ -3; 1\ 0\ 0; 0\ 1\ 0]$. (See `help compan` in MATLAB.) The eigenvalues of the companion matrix are the same as the roots of the polynomial. A scalar λ is an eigenvalue of C if and only if $Cx = \lambda x$ has a nonzero solution x, that is, if and only if $(C - \lambda I)x = 0$ admits a nontrivial solution.

a.) For the polynomial $x^4 - 10x^3 + 35x^2 - 50x + 24$, use the LU decomposition to determine which of the following are eigenvalues of C and hence roots of the polynomial: $-5, -4, -3, -2, -1, 0, 1, 2, 3, 4, 5$.

b.) Is there an efficient way to use the LU decomposition of $C - \lambda_1 I$ to find the LU decomposition of $C - \lambda_2 I$?

4. Newton's Method for Systems

We've been focusing on linear systems, but we are now prepared to discuss nonlinear systems. These occur very frequently in practice. An example from the calculus is finding the location of a minimizer of a differentiable function of two variables $w = F(x, y)$, requiring the solution of the system of equations defined by $\nabla F(x, y) = 0$, that is,

$$\frac{\partial}{\partial x} F(x, y) = 0$$

$$\frac{\partial}{\partial y} F(x, y) = 0$$

and we expect in general that both equations in this system will be nonlinear in both x and y. If F depended on three variables x, y, and z, we would have a system of three nonlinear equations in three unknowns.

Suppose $F : \mathbb{R}^n \to \mathbb{R}^n$ and we want to solve $F(x) = 0$. Based on our experience with Newton's method for scalar equations, we might start with a linearization of $F(x)$. The Taylor series for $F(x)$ has the form

$$F(x) = F\left(x^0\right) + J\left(x^0\right)\left(x - x^0\right) + R(x)$$

(assuming sufficient differentiability) for some remainder term $R(x)$. Here

$$J(x) = \begin{bmatrix} \dfrac{\partial f_1}{\partial x_1} & \dfrac{\partial f_1}{\partial x_2} & \cdots & \dfrac{\partial f_1}{\partial x_n} \\ \dfrac{\partial f_2}{\partial x_1} & \dfrac{\partial f_2}{\partial x_2} & \cdots & \dfrac{\partial f_2}{\partial x_n} \\ \vdots & \vdots & \ddots & \vdots \\ \dfrac{\partial f_n}{\partial x_1} & \dfrac{\partial f_n}{\partial x_2} & \cdots & \dfrac{\partial f_n}{\partial x_n} \end{bmatrix}$$

is called the **Jacobian matrix** (or just the **Jacobian**) of

$$F(x) = \begin{pmatrix} f_1(x_1, ..., x_n) \\ f_2(x_1, ..., x_n) \\ \vdots \\ f_n(x_1, ..., x_n) \end{pmatrix}$$

and it plays the role of the first derivative of F. We proceed just as for Newton's method in the one-dimensional case: Neglecting the remainder term, we have

$$F(x) \approx F\left(x^0\right) + J\left(x^0\right)\left(x - x^0\right)$$

and setting $F(x) = 0$ gives a (hopefully) improved estimate

$$\begin{aligned} 0 &\approx F\left(x^0\right) + J\left(x^0\right)\left(x - x^0\right) \\ J\left(x^0\right)x &\approx J\left(x^0\right)\left(x^0\right) - F\left(x^0\right) \\ x &\approx x^0 - \left[J\left(x^0\right)\right]^{-1} F\left(x^0\right) \end{aligned}$$

where $\left[J\left(x^0\right)\right]^{-1}$ is the inverse of $J(x)$. The iteration

$$x^{k+1} = x^k - \left[J\left(x^0\right)\right]^{-1} F\left(x^k\right)$$

is **Newton's method for systems**. Of course, by now we know that in practice we will want to avoid the use of an inverse. We re-write the equation as

(4.1)
$$\begin{aligned} J\left(x^k\right)\left(x^{k+1} - x^k\right) &= -F\left(x^k\right) \\ J\left(x^k\right)s^k &= -F\left(x^k\right) \end{aligned}$$

in which we solve for $s^k = (x^{k+1} - x^k)$, the step from x^k to x^{k+1}, by Gaussian elimination with pivoting, and then add it to x^k to obtain

$$
\begin{aligned}
x^{k+1} &= x^k + s^k \\
&= x^k + (x^{k+1} - x^k)
\end{aligned}
$$

rather than attempt to compute $\left[J\left(x^0\right)\right]^{-1}$. Clearly, if $J\left(x^k\right)$ is noninvertible for any k then the method fails, and if $J\left(x^*\right)$ is singular then poor performance of the method should be expected.

We can terminate by choosing any convenient vector norm and then using the approximate relative error

$$
\rho_{k+1} \approx \|x^{k+1} - x^k\| / \|x^{k+1}\|
$$

as a convergence criterion. In essence this says to terminate when the step s^k taken by Newton's method from x^k to x^{k+1} is small as compared to $\|x^{k+1}\|$; if iterations are not too expensive then it is preferable to stop when ρ_k and ρ_{k+1} are both small, that is, when two successive steps are small, in case the method has simply taken an unusually short step for some iteration.

EXAMPLE 1: To find the point of intersection of the circle $x^2 + y^2 = 1$ and the ellipse $\frac{1}{3}x^2 + \frac{1}{2}y^2 = 1$ we might write this as the nonlinear system

$$
\begin{aligned}
x^2 + y^2 - 1 &= 0 \\
\frac{1}{3}x^2 + \frac{1}{2}y^2 - 1 &= 0
\end{aligned}
$$

and seek a pair (x, y) that makes these two equations simultaneously vanish. We have

$$
\begin{aligned}
f_1(x, y) &= x^2 + y^2 - 1 \\
f_2(x, y) &= \frac{1}{3}x^2 + \frac{1}{2}y^2 - 1
\end{aligned}
$$

so

$$
J(x) = \begin{bmatrix} 2x & 2y \\ 2x/3 & y \end{bmatrix}
$$

is the Jacobian. If we use

$$
\begin{pmatrix} x_0 \\ y_0 \end{pmatrix} = \begin{pmatrix} 1 \\ 1 \end{pmatrix}
$$

for our initial guess then, using the known formula for a 2×2 inverse

$$
\begin{bmatrix} a & b \\ c & d \end{bmatrix}^{-1} = \frac{1}{ad - bc} \begin{bmatrix} d & -b \\ -c & a \end{bmatrix}
$$

in this trivially-sized case, we have

$$\begin{pmatrix} x_1 \\ y_1 \end{pmatrix} = \begin{pmatrix} x_0 \\ y_0 \end{pmatrix} - \begin{bmatrix} 2x_0 & 2y_0 \\ 2x_0/3 & y_0 \end{bmatrix}^{-1} F\begin{pmatrix} x_0 \\ y_0 \end{pmatrix}$$

$$= \begin{pmatrix} x_0 \\ y_0 \end{pmatrix} - \frac{1}{2x_0y_0/3} \begin{bmatrix} y_0 & -2y_0 \\ -2x_0/3 & 2x_0 \end{bmatrix} F\begin{pmatrix} x_0 \\ y_0 \end{pmatrix}$$

$$= \begin{pmatrix} 1 \\ 1 \end{pmatrix} - \frac{1}{2/3} \begin{bmatrix} 1 & -2 \\ -2/3 & 2 \end{bmatrix} \begin{pmatrix} 1 \\ -1/6 \end{pmatrix}$$

$$= \begin{pmatrix} -1 \\ 2.5 \end{pmatrix}$$

$$\begin{pmatrix} x_2 \\ y_2 \end{pmatrix} = \begin{pmatrix} x_1 \\ y_1 \end{pmatrix} - \begin{bmatrix} 2x_1 & 2y_1 \\ 2x_1/3 & y_1 \end{bmatrix}^{-1} F\begin{pmatrix} x_1 \\ y_1 \end{pmatrix}$$

$$\doteq \begin{pmatrix} -1 \\ 2.5 \end{pmatrix} - \begin{bmatrix} 3 & -6 \\ -2/3 & 2 \end{bmatrix} \begin{pmatrix} 6.25 \\ 2.4583 \end{pmatrix}$$

$$\doteq \begin{pmatrix} 1.0001 \\ 2.05 \end{pmatrix}$$

and so on. Further iterates show that the method is failing to converge; indeed, a sketch of the circle and the ellipse show that they have no points in common, so there can be no (real) solutions. □

A small relative error is a commonly used convergence criterion for Newton's method for systems. Another commonly used convergence criterion for well-behaved problems[6] is to stop the iteration when

$$\text{(4.2)} \qquad \|F\left(x^k\right)\| \le \tau_\rho \|F\left(x^0\right)\| + \tau_\alpha$$

where τ_ρ is a small tolerance that measures the relative change in the function norm and τ_α is a small tolerance that measures the absolute size of the function norm. The motivation is easier to see when we consider a function $f : \mathbb{R} \to \mathbb{R}$ and initially neglect τ_α. In that case the convergence criterion is $\|f\left(x^k\right)\| \le \tau_\rho \|f\left(x^0\right)\|$ and we have

$$\left|f\left(x^k\right) - f\left(x^{k-1}\right)\right| \le \left|f'(\theta_k)\right| \left|x^k - x^{k-1}\right|$$

by the Mean Value Theorem. If $|f'(\theta_k)|$ is very small then $|x_k - x_{k-1}|$ could be very large but $\left|f\left(x^k\right) - f\left(x^{k-1}\right)\right|$ could still be small; in fact, if $f'(x)$ is small near the root then the function looks nearly flat (i.e., constant) near the root so the knowledge that $f\left(x^k\right)$ is approximately zero conveys little information. For example, if the function is $y = x^3$ then at $x = .5$ the residual error is $.5^3 = .125$ but the absolute error is $.5$; if we had used $y = x^5$ instead then the residual error would have been $.5^5 = .03125$ but the absolute error would still have been $.5$. Using $y = x^m$ we could make the residual error arbitrarily small at $x = .5$ while keeping the absolute error at $\alpha = .5$.

But if instead $|f'(x)|$ is, say, about unity near the root x^* then near x^* we have a truly *linear* approximation

---

[6]Here well-behaved means that the Jacobian of F evaluated at the root is well-conditioned in the sense of Sec. 2.5.

$$f\left(x^k\right) \approx f(x^*) + f'(x^*)(x^k - x^*)$$

as opposed to an essentially constant approximation. Since $f(x^*)$ is zero, this becomes

$$f\left(x^k\right) \approx f'(x^*)(x^k - x^*)$$

which indicates that the size of $\left|f\left(x^k\right)\right|$ is indeed a good measure of the size of the absolute error $\left|x^k - x^*\right|$. Look at this equation again; it plainly relates the desired true absolute error α to the measurable quantity $f\left(x^k\right)$, the residual! This makes geometric sense; if $f'(x^*)$ is not too small then the function looks pretty much like a straight line crossing the x-axis at an angle that is not too small (so that we don't have extreme sensitivity of x-values to changes in y-values). Thus there is a roughly linear relationship between the size of x and the size of $f(x)$ in this region. This is stable, that is, it is not sensitive to small changes in the values of the quantities involved. The convergence criterion

$$\|f(x_k)\| \le \tau_\rho \|f(x^0)\| + \tau_\alpha$$

of Eq. (4.2) uses this fact to stop the iteration when $\|f\left(x^k\right)\|$ is sufficiently small, where small is defined in terms of the initial guess by $\tau_\rho \|f\left(x^0\right)\|$. The tolerance τ_α guards against cases where evaluation of f is inaccurate or where $\|f(x^k)\|$ is already very small and so significant further reduction is unlikely. As in any computer program, an upper limit on the number of iterations allowed is a good idea to prevent an infinite loop, commonly implemented as a limit on the maximum number of function evaluations allowed in numerical software.

Similar arguments in favor of Eq. (4.2) apply in the case of systems of nonlinear equations when the Jacobian $J(x)$ is well-behaved at the root (in a sense that will be made precise in Sec. 2.6). The same idea is sometimes implemented in the form $\|F\left(x^k\right)\| \le \max(\tau_\rho \|F\left(x^0\right))\|, \tau_\alpha)$. In this context we often refer to $\|F\left(x^k\right)\|$ as the **residual error** or simply the **residual** at iteration k and we say that our convergence criterion is based on the residuals rather than the errors (i.e., α_k or ρ_k).

Newton's method for systems shares most of the strengths and weaknesses of Newton's method for functions of a single variable, including the ability to find complex zeroes given complex initial guesses and the difficulty of finding and evaluating the Jacobian. It adds the additional expense of the solution of a linear system at each step.

It's hard to overestimate the importance of Newton's method for systems in scientific computing, but it can be a costly method to apply as derived here. The Jacobian can be very expensive to evaluate and/or approximate, and the solution of the linear system in Eq. (4.1) can be expensive; after all, we need to find a new LU decomposition at every step of the method, requiring $O\left(n^3\right)$ floating point operations. This makes the use of the natural generalizations of the secant method, finite difference Newton's method, and constant-slope Newton's method (from Sec. 1.5) very important, as well as other approximations of Newton's method. The constant-slope Newton's method (or **chord method** in this context)

$$x_{k+1} = x_k - [J\left(x^0\right)]^{-1} F\left(x^k\right)$$

is a particularly attractive alternative version because the Jacobian–a matrix of n^2 distinct partial derivatives–need only be computed once, and the solution of Eq. (4.1)

$$J\left(x^k\right)\left(x^{k+1} - x^k\right) = -F\left(x^k\right)$$

is made considerably more efficient since we know that we will be using the same matrix $J\left(x^0\right)$ repeatedly. We find the LU decomposition $LU = J\left(x^0\right)$ initially, and then solve

$$LU s_k = -F\left(x^k\right)$$

with a new $F\left(x^k\right)$ each time, at only $O\left(n^2\right)$ cost. This is why the LU decomposition is such an important reformulation of Gaussian elimination: Because situations in which we must solve $Ax = b$ for one A but many choices of b arise often. Even though the convergence is only linear this method may reach a desired tolerance in less time than Newton's method because each iteration is so much less computationally expensive, particularly when the tolerance is not especially small. For tighter tolerances it is often not fast enough.

In practice it is common to restart the chord method every several iterations with a freshly evaluated Jacobian matrix, that is, to use a hybrid strategy that mixes Newton's method and the constant-slope Newton's method. The decision to restart may be made adaptively by checking the progress of the algorithm; if it is taking increasingly small steps and achieving large reductions in the residuals, we might let it go. If we are seeing little change in the size of the residuals or if the steps are so large that the iteration appears to be failing to converge, however, we might choose to re-evaluate $J(x)$ at the current point. Remember, re-evaluating the Jacobian matrix involves both finding n^2 new derivative values, and then paying the $O\left(n^3\right)$ price of finding a new LU decomposition.

EXAMPLE 2: Let's apply the chord method to $(x^4 + y^4, xy + \sin(xy)) = (0, 0)$ with the initial guess $(x_0, y_0) = (1, 2)$. The Jacobian is

$$\begin{aligned}
J(x, y) &= \begin{pmatrix} 4x^3 & 4y^3 \\ y + y\cos(xy) & x + x\cos(xy) \end{pmatrix} \\
J(1, 2) &= \begin{pmatrix} 4 & 32 \\ 2 + 2\cos(2) & 1 + \cos(2) \end{pmatrix} \\
&\doteq \begin{pmatrix} 4 & 32 \\ 1.1677 & .5839 \end{pmatrix}
\end{aligned}$$

so we have (using $\overrightarrow{x} = (x, y)^T$)

$$\vec{x}_0 = \begin{pmatrix} 1 \\ 2 \end{pmatrix}$$

$$\vec{x}_1 = \vec{x}_0 - J^{-1}(\vec{x}_0)F(\vec{x}_0)$$

$$= \begin{pmatrix} 1 \\ 2 \end{pmatrix} - \begin{pmatrix} 4 & 32 \\ 1.1677 & .5839 \end{pmatrix}^{-1} F\begin{pmatrix} 1 \\ 2 \end{pmatrix}$$

$$\doteq \begin{pmatrix} -1.3742 \\ 1.7655 \end{pmatrix}$$

$$\vec{x}_2 = \begin{pmatrix} -1.3742 \\ 1.7655 \end{pmatrix} - \begin{pmatrix} 4 & 32 \\ 1.1677 & .5839 \end{pmatrix}^{-1} F\begin{pmatrix} -1.3742 \\ 1.7655 \end{pmatrix}$$

$$\doteq \begin{pmatrix} 1.6626 \\ 0.9708 \end{pmatrix}$$

$$\vec{x}_3 = \begin{pmatrix} 1.6626 \\ .9708 \end{pmatrix} - \begin{pmatrix} 4 & 32 \\ 1.1677 & .5839 \end{pmatrix}^{-1} F\begin{pmatrix} 1.6626 \\ .9708 \end{pmatrix}$$

$$\doteq \begin{pmatrix} -.5823 \\ .9849 \end{pmatrix}$$

and $x_{15} \doteq (.0036, .7337)^T$. The y component is changing very slowly. Let's recompute J at x_{15}:

$$\vec{x}_{16} = \vec{x}_{15} - J^{-1}(\vec{x}_{15})F(\vec{x}_{15})$$

$$\doteq \begin{pmatrix} .0036 \\ .7337 \end{pmatrix} - \begin{pmatrix} .0000 & 1.5798 \\ 1.4674 & .0072 \end{pmatrix}^{-1} F\begin{pmatrix} .0036 \\ .7337 \end{pmatrix}$$

$$\doteq \begin{pmatrix} .0009 \\ .5503 \end{pmatrix}$$

$$\vec{x}_{17} = \begin{pmatrix} .0009 \\ .5503 \end{pmatrix} - \begin{pmatrix} .0000 & 1.5798 \\ 1.4674 & .0072 \end{pmatrix}^{-1} F\begin{pmatrix} .0009 \\ .5503 \end{pmatrix}$$

$$\doteq \begin{pmatrix} .0005 \\ .4922 \end{pmatrix}$$

and $x_{20} \doteq (.0002, .4067)^T$. We could update J again to attempt to improve the speed of convergence of the method to the true solution $(0,0)^T$. $\square$

There is a convergence theorem for Newton's method for systems that generalizes our previous result:

THEOREM: If B is an open subset of $\mathbb{R}^n$ and all second-order partial derivatives of $F : B \to \mathbb{R}^n$ exist and are continuous, then for any zero x^* of F in B such that $J(x^*)$ is nonsingular there exists an open set $V \subseteq B$ containing x^* such that Newton's method with initial guess x^0 converges to x^* for all $x^0 \in V$. $\square$

The theorem states that if F is twice continuously differentiable in some region and if the derivative is nonsingular at the zero, then Newton's method converges for any initial guess sufficiently close to the zero. The convergence is quadratic under mild additional conditions, that is, typically we have

$$\|x^* - x^{k+1}\| \approx C\|x^* - x^k\|^2$$

(for some constant $C > 0$ depending only on the particular vector norm used) when using Newton's method for systems. Importantly, the convergence is quadratic *with respect to the number of iterations*; this doesn't account for how expensive each iteration is.

There are many different convergence theorems for Newton's method for systems, each requiring different hypotheses. A class of theorems called **Newton–Kantorovich theorems**[7] specify conditions under which the method can be guaranteed to converge even though the function is not twice continuously differentiable.

MATLAB

Let's write an m-file that implements Newton's method for systems. Use the `edit` command to invoke the editor. We will assume there will be user-provided m-files `newtfun.m` and `newtjac.m` or function handles `newtfun` and `newtjac` that compute the function and its Jacobian, respectively; that is, `newtfun` takes in a vector of length n and returns a row vector of length n, and `newtjac` takes in a vector of length n and returns an $n \times n$ matrix (the Jacobian). A simple version of Newton's method might read as follows:

```
function xnew=newtsys(fun,jac,x0,N)
%NEWTSYS Newton's method for systems applied to function
% fun with Jacobian jac and initial condition xo;
% N is the number of steps.
d=-feval(jac,x0)\feval(fun,x0);xnew=d+x0;
for i=1:N
d=-feval(jac,xnew)\feval(fun,xnew);
xnew=d+xnew;
end
```

When using this function, we would pass it `newtfun` and `newtjac` as function handle arguments, that is, the calling sequence is `newtsys(@newtfun,@newtjac,x0,tol)` if the function and Jacobian are m-files, or `newtsys(newtfun,newtjac,x0,tol` if they're function handles already.

The line that computes `d` and the line that follows it could also be expressed as:
`xnew=x0-inv(feval(jac,x0))*feval(fun,x0);`
using the `inv` for inverting a matrix, but as discussed previously this should absolutely be avoided. Let's apply this program to the problem in Example 2. Since the program expects a vector input, you can define $F(x)$ and $J(x)$ as, say
```
>> newtfun=@(x) [x(1)^4+x(2)^4;x(1)*x(2)+sin(x(1)*x(2))]
>> newtfun([1;2])      %Test
>> newtjac=@(x)[4*x(1)^3,4*x(2)^3;x(2)+x(2)*cos(x(1)*x(2)),...
   x(1)+x(1)*cos(x(1)*x(2))]
>> newtjac([1;2])      %Test
```
(you may be able to enter the definition of newtjac in a single line, without the ellipsis). Of course there are other ways; e.g., `prod(x)` is equivalent to `x(1)*x(2)`. Try this with $N = 0$ to duplicate the first step taken in Example 2. Then try it with $N = 1$; you should get a different result than in Example 2 as this is a true

---

[7]Also called simply a Kantorovich Theorem as the first of this type was published by Leonid Kantorovich in 1948, long after Newton's time.

Newton's method. As N increases, unfortunately, we begin to see divergence. Try again with $x^0 = (.1, .2)^T$ instead. This converges to the root at $(0, 0)^T$, but as the Jacobian is the zero matrix there and hence singular, the convergence is slower than expected.

Note how compactly an algorithm like this can be expressed in MATLAB; consider what would be involved in writing it in any other languages you may know. Of course, it would run faster in a compiled language, assuming the use of BLAS as appropriate. Rapid prototyping (or rapid development) is an advantage of MATLAB but while it is possible to create MATLAB executables (see `help mex`) it is often advisable to re-write programs that were developed in MATLAB, but will be used repeatedly, in a compiled language. The NETLIB mathematical subroutines may be called for specific computations.

There are many possible criticisms of our program. We should probably check that `newtfun.m` (`fun` in the program) and `newtjac.m` (`jac` in the program) return results that have the expected sizes and that N is nonnegative; we should attempt to detect failure to converge; we should have a convergence criterion rather than, or at least in addition to, a specified number of steps; we should return an optional status variable indicating success or failure; and we should check if the Jacobian is "nearly" singular and issue a warning if it is (see Sec. 2.6).

For most problems of practical interest symbolic differentiation is not a realistic option. It can be very time-consuming, and the evaluation of those derivatives may be too time-consuming, as the derivative formulas obtained are often inadequately simplified and may be numerically unstable. For large computing problems, symbolic computation is almost never feasible. Most often the function $F(x)$ is itself computed by another program (say, one that solves a PDE numerically) and hence is not symbolically differentiable anyway. Type `help numjac` for a function that computes the Jacobian numerically. It uses finite differences to approximate $J(x)$ and has a number of checks programmed in so as to yield an accurate Jacobian (hopefully).

If you are writing a program to compute a function $F(x)$ and it is not overly difficult to optionally return to the user $F'(x)$ (the Jacobian) and even $F''(x)$ (the Hessian matrix) then you should absolutely do so. This is foundational to writing numerical code. For many programs–like differential equation solvers–the first derivatives are apt to be computed by the program anyway; why not offer to share them with the user?

In some cases you may be able to use **automatic differentiation** or **algorithmic differentiation**, two closely related ideas. (The terminology is not consistent in this area so we will just sketch the general concepts.) There exist special programs that read in the code for a program that computes $F(x)$ and apply the rules of differentiation to the formula in the program to produce *another* program that computes $F'(x)$ (or to augment the original program to also return these). This is done in advance; the differentiating program simplifies formulas in the program, including loops, in what would be an exceedingly tedious operation for a human, and applies the product and chain rules as needed. Then the program that computes $F'(x)$ is compiled, making derivatives evaluations available. These software tools are available for major languages. A key observation in this area is that if you wish to compute a quantity like $f(x) + f(y)$ for some differentiable function f, and at the point x you know not only $f(x)$ but also $f'(x)$, and at the point y

you know not only $f(y)$ but also $f'(y)$, then you could define a new datatype for f (presumably an object) of the form $\Phi_1 = (f(x), f'(x))$, $\Phi_2 = (f(y), f'(y))$. Then instead of merely computing $f(x) \pm f(y)$ you could compute $\Phi_1 \pm \Phi_2$, where the $+$ or $-$ operator has been "overloaded" so that it automatically computes

$$\begin{aligned}\Phi_1 \pm \Phi_2 &= (f(x), f'(x)) \pm (f(y), f'(y)) \\ &= (f(x) \pm f(y), f'(x) \pm f'(y))\end{aligned}$$

instead. Similarly for $f(x) \cdot f(y)$ you could compute $\Phi_1 \cdot \Phi_2$:

$$(f(x), f'(x)) \cdot (f(y), f'(y)) = (f(x) \cdot f(y), f(x) f'(y) + f'(x) f(y))$$

where the product rule is built into the definition of multiplication for this new type of variable. Similarly for division; the derivative values automatically tag-along with the corresponding function values and are always available without any further calculation. Since they're simply numerical values, we are not evaluating $f(\cdot)$ nor $f'(\cdot)$ at any time after the initial evaluations of $f(x)$ and $f(y)$. Almost always we include second derivatives too. You can build in some functions too, say

$$\sin((f(x), f'(x))) = (\sin(f(x)), \cos(f(x)) f'(x))$$

and so on, as needed. These tools are not widely used as of this writing but it's always possible that you will find yourself in a situation in which it's worth going looking for them. Some MATLAB toolboxes do have automatic differentiation tools.

Problems

1.) Use Newton's method for systems to find all points where the hyperbola $y - 1/x = 0$ intersects the ellipse $(x/5)^2 + (y/8)^2 = 1$.
2.) a.) Use fixed point iteration to find a root of $F(x, y, z) = 0$ where $F(x, y, z) = (6x - y - z, x - 5y + z, 2x + 2y - 7z)$ by converting it to the form $v = v + F(v)$ (v is the vector with components $(x, y, z)^T$) and using the initial condition $(x_0, y_0, z_0)^T = (.5, .5, .5)^T$.
b.) What is Newton's method for solving $F(x, y, z) = 0$?
3.) What form does Newton's method take when applied to a linear system $F(x) = Ax - b = 0$ (with A an $n \times n$ matrix and b an n-vector)?
4.) If $F : \mathbb{R}^{20} \to \mathbb{R}^{20}$, how many separate derivatives must be found in order to be able to compute $J(x)$?
5.) Why can't the secant method be directly generalized to a method for functions of more than one variable?
6.) Use Newton's method to find the points of intersection of the circle of radius 10 centered at the origin and the circle of radius 6 centered at $(12, 1)$.
7.) Show that Newton's method for systems reduces to Newton's method if $n = 1$.
8.) a.) Do Problem 7 of Sec. 1.8 using Newton's method for systems as the root-finder.
b.) Do Problem 8 of Sec. 1.8 using Newton's method for systems as the root-finder.
9.) Augment the Newton's method for systems program to use a convergence criterion in addition to number of steps. (The user should input a tolerance.) Exit the loop early using the **break** command (see **help break**) when convergence is met.

Return an optional status variable signaling failure by a value of -1 if the loop completes and the convergence criterion having been successfully met by 1.

10.) a.) Use Newton's method on $(x^2 - 2xyz, x - y^2 - z, 2x - 2y - 2z^3)$ to find a zero of it. Use the LU decomposition (in MATLAB or in another computing environment).

b.) Repeat with the constant Jacobian (that is, constant-slope) version, using $J\left(x^0\right)$ at each iteration not $J\left(x^k\right)$. Explicitly show the LU decomposition.

11.) Verify experimentally the linear convergence of the constant Jacobian Newton's method for systems.

12.) Verify experimentally the quadratic convergence of Newton's method for systems. (Recall that $\|x^k - x^*\|$ is the error that should be converging quadratically to zero.)

13.) Use Newton's method to find a simultaneous zero of $f(x, y) = xy^2 - 2z + \sin(x) + 2$, $2x^4 + 3xy^2 + yz - 1$ and $g(x, y) = 2xy + \cos(xy) + 2z^3 - 3$.

14.) a.) What happens if you apply Newton's method for systems to a system of the form $F(x) = (f_1(x_1), f_2(x_2), ..., f_n(x_n))$?

b.) What happens if you apply fixed point iteration in the form $x = x + F(x)$ to a system of the form $F(x) = (f_1(x_1), f_2(x_2), ..., f_n(x_n))$?

15.) a.) Modify the Newton's method for systems program to only compute the Jacobian every $p + 1$ iterations, where p will be supplied by the user ($p = 0$ is Newton's method, $p = \infty$ is the constant-slope Newton's method); see `help inf` for one way to work with the $p = \infty$ case in MATLAB.

b.) We can define an iterative method in which each iteration consists of one Newton step followed by p constant-slope Newton's method steps; that is, given x^0, we set $z_{k+1}^{(0)} = x^k - [J\left(x^k\right)]^{-1} F x^k$ and then compute $z_{k+1}^{(\nu)} = z_{k+1}^{(\nu-1)} - [Jx^k]^{-1} F(z_{k+1}^{(\nu-1)})$ for $\nu = 1, ..., p$, and finally we set $x^{k+1} = z_{k+1}^{(p)}$. Each step of the iteration consists of one Newton's step and p constant-slope steps. This formulation of the constant-slope Newton's method with restarting is sometimes called **Shamanskii's method**. The convergence criterion is usually applied after the computation of either an x-value or a z-value (not just the z-values). What advantages, if any, are gained by this change of point-of-view?

16.) A material is considered to be an imperfect dielectric (that is, a poor conductor of electromagnetic energy) if $\sigma/\varpi\varepsilon \ll 1$ where $\sigma > 0$ is the material's conductivity, ϖ is the frequency of the electromagnetic radiation, and ε is the material's permittivity. For such a material the attenuation and phase constants α and β are given by $\alpha \approx \sigma\sqrt{\mu/\varepsilon}(1/2 - \sigma^2\varpi^{-2}\varepsilon^{-2}/16)$ and $\beta \approx \varpi\sqrt{\mu\varepsilon}(1 + \sigma^2\varpi^{-2}\varepsilon^{-2}/8)$. The values of α and β are directly measurable but those of μ and ε must be inferred.

a.) Write out Newton's method for solving for μ and ε given α, β, and ϖ.

b.) Can μ and ε be found algebraically (without using a numerical method)?

c.) A perfect dielectric has $\sigma = 0$ leading to $\alpha = 0$, $\beta = \varpi\sqrt{\mu\varepsilon}$. (Since α is zero, all of the electromagnetic energy is attenuated.) What would happen to Newton's method in part a. if the parameters were those of a perfect dielectric rather than an imperfect dielectric?

5. The Cholesky Decomposition

When we know something special about a matrix A involved in a computational linear algebra problem we can often exploit it for additional efficiency. The two most commonly occurring cases are those in which we know that the matrix is sparse, that is, most of its entries are zeroes, and those in which we know that the matrix is symmetric, that is, that $A = A^T$. In this section we'll focus on symmetric matrices that have an additional property: If x is a nonzero vector then

$$x^T A x > 0.$$

A symmetric matrix with this property is said to be **positive definite**[8]. An equivalent characterization is that a matrix is positive definite if and only if it is symmetric and all of its eigenvalues are strictly positive. (Hence it will be nonsingular, as no eigenvalue is null.) A symmetric matrix with the property that $x^T A x \geq 0$ for all x is said to be **positive semi-definite** and all its eigenvalues must be nonnegative. Similarly for **negative definite** ($x^T A x < 0$) and **negative semi-definite** ($x^T A x \leq 0$) matrices.

Positive definite matrices appear more frequently than one might expect in practice. The quantity $x^T A x$ is called a **quadratic form** in A and is the natural generalization of the scalar form ax^2; in this context it occurs in the generalization of Taylor series to functions of more than one variable where it is called the Hessian matrix. We'll see in Ch. 7 that having the Hessian be positive definite is the natural analog of asking that $f''(x)$ be positive in single-variable calculus. This means that positive definite matrices appear often in multivariable minimization problems. Positive definite matrices also appear in least squares curve-fitting problems (see Sec. 2.7) and in the numerical solution of partial differential equations.

Since positive definite matrices are symmetric, they contain only about half as many distinct entries as a general matrix of the same order, and so we ought to be able to save about half the work and half the storage when dealing with them. However, an arbitrarily selected method will usually not take advantage of this structure. For example, the LU decomposition $A = LU$ of a symmetric matrix cannot have L or U symmetric in general[9]. We'll focus on the more specific case of positive definite matrices. Here there is indeed a straight-forward procedure that produces an LU decomposition of the matrix in about half the time and using about half the storage, as we would expect.

The method is called the **Cholesky decomposition**, or **Cholesky factorization**, of the matrix A and it has the form $A = LL^T$ where L is a lower triangular matrix. (Either L or $U = L^T$ may be referred to as a **Cholesky factor** of the matrix.) The matrix L will not in general be *unit* lower triangular. Clearly, it suffices to store only L once we have the decomposition and so we will be able to cut our storage requirements about in half.

It'll be convenient to take a brief tangent first. Let A be a positive definite $n \times n$ matrix. Since A is symmetric, it's orthogonally diagonalizable; that is, we can write

---

[8]Some authors do not include symmetry in the definition of a positive definite matrix. The Cholesky decomposition requires that A be both symmetric and satisfy $x^T A x > 0$.

[9]Every symmetric matrix has an LDL decomposition, $A = LDL^T$, which is nearly as good as a Cholesky decomposition.

$$A = SDS^T$$

where S is the matrix whose columns are the orthonormalized eigenvectors of A, and D is a diagonal matrix with the eigenvalues of A along the main diagonal. This is sometimes called the **eigenvalue decomposition** of A. Since $D = \text{diag}([\lambda_1 \quad \lambda_2 \cdots \lambda_n])$ (mixing MATLAB and traditional notation) with each $\lambda_i > 0$, we may define $H = \text{diag}([\sqrt{\lambda_1} \quad \sqrt{\lambda_2} \cdots \sqrt{\lambda_n}])$ and write this as

$$
\begin{aligned}
A &= SHHS^T \\
&= SHH^T S^T \\
&= (SH)(SH)^T \\
&= MM^T
\end{aligned}
$$

($M = SH$). It's very unlikely that M will be lower triangular, so this is not the decomposition we seek. But if we choose to change this to

$$
\begin{aligned}
A &= SHHS^T \\
&= SH(S^T S)HS^T \\
&= (SHS^T)(SHS^T) \\
&= H_0 H_0 \\
&= H_0^2
\end{aligned}
$$

then we say that $H_0 = SHS^T$ is the **matrix square root** of A. Note that H_0 is itself positive definite; its eigenvalues are the positive numbers $\sqrt{\lambda_1}, \sqrt{\lambda_2}, \cdots \sqrt{\lambda_n}$ and it is symmetric, for

$$
\begin{aligned}
H_0^T &= (SHS^T)^T \\
&= (S^T)^T H^T S^T \\
&= SHS^T.
\end{aligned}
$$

The matrix square root appears in a number of physical applications but we will use it as a tool in our algorithm for computing the Cholesky decomposition.

Now back to the Cholesky decomposition. The key idea for finding it is this: When A is symmetric, any elementary matrix E used in row-reducing A to upper triangular form by premultiplication (EA) has the analogous effect on the columns of A if we postmultiply by the transpose, giving AE^T. (The transpose of an elementary row matrix, used on the left, is an elementary column matrix, used on the right, implementing an **elementary column operation**.) We can maintain symmetry while reducing A if we form the product EAE^T at each step.

EXAMPLE 1: Consider the positive definite matrix $A = [4 \ 2 \ 1; 2 \ 3 \ 1; 1 \ 1 \ 4]$. The first step in finding an LU decomposition of A would be to perform $A \to E_1 A$ where

$$
E_1 = \begin{bmatrix} 1 & 0 & 0 \\ -1/2 & 1 & 0 \\ -1/4 & 0 & 1 \end{bmatrix}
$$

is the product of the two elementary matrices that perform $R_2 \leftarrow R_2 - \frac{1}{2}R_1$ and then $R_3 \leftarrow R_3 - \frac{1}{4}R_1$. This gives

$$
\begin{aligned}
E_1 A &= \begin{bmatrix} 1 & 0 & 0 \\ -1/2 & 1 & 0 \\ -1/4 & 0 & 1 \end{bmatrix} \begin{bmatrix} 4 & 2 & 1 \\ 2 & 3 & 1 \\ 1 & 1 & 4 \end{bmatrix} \\
&= \begin{bmatrix} 4 & 2 & 1 \\ 0 & 2 & 1/2 \\ 0 & 1/2 & 15/4 \end{bmatrix}
\end{aligned}
$$

as expected. This is not symmetric. But if we then post-multiply by E_1^T, we have

$$
\begin{aligned}
E_1 A E_1^T &= \begin{bmatrix} 4 & 2 & 1 \\ 0 & 2 & 1/2 \\ 0 & 1/2 & 15/4 \end{bmatrix} \begin{bmatrix} 1 & -1/2 & -1/4 \\ 0 & 1 & 0 \\ 0 & 0 & 1 \end{bmatrix} \\
&= \begin{bmatrix} 4 & 0 & 0 \\ 0 & 2 & 1/2 \\ 0 & 1/2 & 15/4 \end{bmatrix}
\end{aligned}
$$

(check this). We started off with a symmetric matrix A, performed a step of Gaussian elimination and got a nonsymmetric matrix $E_1 A$, and then postmultiplied by E_1^T to get a symmetric matrix $E_1 A E_1^T$ once again. The next step gives

$$
\begin{aligned}
E_2 E_1 A E_1^T &= \begin{bmatrix} 1 & 0 & 0 \\ 0 & 1 & 0 \\ 0 & -1/4 & 1 \end{bmatrix} \begin{bmatrix} 4 & 0 & 0 \\ 0 & 2 & 1/2 \\ 0 & 1/2 & 15/4 \end{bmatrix} \\
&= \begin{bmatrix} 4 & 0 & 0 \\ 0 & 2 & 1/2 \\ 0 & 0 & 29/8 \end{bmatrix} \\
E_2 E_1 A E_1^T E_2^T &= \begin{bmatrix} 4 & 0 & 0 \\ 0 & 2 & 1/2 \\ 0 & 0 & 29/8 \end{bmatrix} \begin{bmatrix} 1 & 0 & 0 \\ 0 & 1 & -1/4 \\ 0 & 0 & 1 \end{bmatrix} \\
&= \begin{bmatrix} 4 & 0 & 0 \\ 0 & 2 & 0 \\ 0 & 0 & 29/8 \end{bmatrix}
\end{aligned}
$$

and so if we define

$$
\begin{aligned}
\tilde{L} &= E_2 E_1 \\
&= \begin{bmatrix} 1 & 0 & 0 \\ 0 & 1 & 0 \\ 0 & -1/4 & 1 \end{bmatrix} \begin{bmatrix} 1 & 0 & 0 \\ -1/2 & 1 & 0 \\ -1/4 & 0 & 1 \end{bmatrix} \\
&= \begin{bmatrix} 1 & 0 & 0 \\ -1/2 & 1 & 0 \\ -1/8 & -1/4 & 1 \end{bmatrix}
\end{aligned}
$$

then $\tilde{L}^T = (E_2 E_1)^T = E_1^T E_2^T$ and we have $\tilde{L}A\tilde{L}^T = D_0$ where D_0 is a diagonal matrix with strictly positive diagonal entries[10]. Since D_0 is (trivially) positive definite, it has a matrix square root H_0 such that $D_0 = H_0^2$, and H_0 is also a diagonal matrix with strictly positive diagonal entries. Hence we may write

$$\begin{aligned} \tilde{L}A\tilde{L}^T &= D_0 \\ A &= \tilde{L}^{-1}D_0(\tilde{L}^T)^{-1} \\ &= \tilde{L}^{-1}H_0^2(\tilde{L}^{-1})^T \\ &= L_0 H_0^2 L_0^T \end{aligned}$$

where $L_0 = \tilde{L}^{-1}$ is lower triangular since it is the inverse of a lower triangular matrix. (We have also used the fact that if a matrix B is nonsingular then $(B^T)^{-1} = (B^{-1})^T$). Since H_0 is symmetric,

$$\begin{aligned} A &= L_0 H_0 H_0 L_0^T \\ &= L_0 H_0 H_0^T L_0^T \\ &= L_0 H_0 (L_0 H_0)^T \\ &= LL^T \end{aligned}$$

where $L = L_0 H_0$. Postmultiplication by a diagonal matrix simply rescales columns and so L is lower triangular; this is a Cholesky decomposition $A = LL^T$ of A. In our example we have

$$\begin{aligned} \tilde{L} &= \begin{bmatrix} 1 & 0 & 0 \\ -1/2 & 1 & 0 \\ -1/8 & -1/4 & 1 \end{bmatrix} \\ L_0 &= \tilde{L}^{-1} \\ &= \begin{bmatrix} 1 & 0 & 0 \\ 1/2 & 1 & 0 \\ 1/4 & 1/4 & 1 \end{bmatrix} \\ H_0 &= \begin{bmatrix} \sqrt{4} & 0 & 0 \\ 0 & \sqrt{2} & 0 \\ 0 & 0 & \sqrt{29/8} \end{bmatrix} \\ &= \begin{bmatrix} 2 & 0 & 0 \\ 0 & \sqrt{2} & 0 \\ 0 & 0 & \frac{1}{2}\sqrt{29/2} \end{bmatrix} \end{aligned}$$

so

---

[10]These are not the eigenvalues of A.

$$L = L_0 H_0$$

$$= \begin{bmatrix} 1 & 0 & 0 \\ 1/2 & 1 & 0 \\ 1/4 & 1/4 & 1 \end{bmatrix} \begin{bmatrix} 2 & 0 & 0 \\ 0 & \sqrt{2} & 0 \\ 0 & 0 & \frac{1}{2}\sqrt{29/2} \end{bmatrix}$$

$$\doteq \begin{bmatrix} 2 & 0 & 0 \\ 1 & 1.4142 & 0 \\ .5 & .3536 & 1.9039 \end{bmatrix}.$$

Let's check:

$$LL^T = \begin{bmatrix} 2 & 0 & 0 \\ 1 & 1.4142 & 0 \\ .5 & .3536 & 1.9039 \end{bmatrix} \begin{bmatrix} 2 & 0 & 0 \\ 1 & 1.4142 & 0 \\ .5 & .3536 & 1.9039 \end{bmatrix}^T$$

$$= \begin{bmatrix} 2 & 0 & 0 \\ 1 & 1.4142 & 0 \\ .5 & .3536 & 1.9039 \end{bmatrix} \begin{bmatrix} 2 & 1 & .5 \\ 0 & 1.4142 & .3536 \\ 0 & 0 & 1.9039 \end{bmatrix}$$

$$= \begin{bmatrix} 4 & 2 & 1 \\ 2 & 3 & 1 \\ 1 & 1 & 4 \end{bmatrix}$$

which checks. This is the desired Cholesky factorization of A. $\square$

The method above is entirely general, *assuming* we don't need to pivot at any stage. We introduce the needed zeroes in a column of the working matrix, then do the same on the corresponding row. When we are done, the matrix

$$D_0 = E_{n-1} \cdots E_2 E_1 A E_1^T E_2^T \cdots E_{n-1}^T$$

will be diagonal with positive diagonal entries. (This is one reason why we need positive definiteness for A, not just symmetry: To guarantee that D_0 will have positive diagonal entries so that we can take its square root.) Then we set

$$\tilde{L} = E_{n-1} \cdots E_2 E_1$$

so that we have

$$D_0 = \tilde{L} A \tilde{L}^T$$
$$A = \tilde{L}^{-1} H_0^2 (\tilde{L}^{-1})^T$$
$$= LL^T$$

with $L = \tilde{L}^{-1} H_0$. This is the desired Cholesky factorization of A. There is exactly one lower triangular matrix L having positive diagonal entries for which $A = LL^T$, and we will use this convention when creating any Cholesky factor L.

What if we must pivot? It can be shown that naive (unpivoted) Gaussian elimination can always be performed on a positive definite matrix and a zero pivot will never be encountered. In fact, all the pivots will be positive. Of course, for a general matrix we use partial pivoting not just because of the possibility of a zero pivot but also because it controls the growth of round-off errors. However, when A is positive definite, partial pivoting is not necessary to control errors; naive Gaussian elimination may be safely employed and will not amplify round-off errors significantly. (This is by no means obvious.) Hence the method described above need not be modified to use pivoting, though sometimes it is for even greater accuracy.

As usual, the matrix formulation given above is useful for analytical work but would be inefficient if implemented in a literal manner. We want to compute L directly rather than finding its inverse first. There are several computationally efficient algorithms for doing this. We'll consider the inner product version. We start with the trivial case where A is a 1×1 matrix

$$A = (a_{11})$$

(with $a_{11} > 0$). The decomposition must have $L = (\sqrt{a_{11}})$ if we are requiring that the diagonal elements of L be positive. For a 2×2 matrix

$$A = \begin{pmatrix} a_{11} & a_{12} \\ a_{12} & a_{22} \end{pmatrix}$$

(where we have used the fact that $a_{21} = a_{12}$), we need a decomposition of the form

$$\begin{pmatrix} a_{11} & a_{12} \\ a_{12} & a_{22} \end{pmatrix} = \begin{pmatrix} l_{11} & 0 \\ l_{12} & l_{22} \end{pmatrix} \begin{pmatrix} l_{11} & l_{12} \\ 0 & l_{22} \end{pmatrix}$$

which means that we must have

$$
\begin{aligned}
a_{11} &= l_{11}^2 \\
a_{12} &= l_{11}l_{12} \\
a_{22} &= l_{12}^2 + l_{22}^2
\end{aligned}
$$

(5.1)

and it appears that we may take advantage of the decomposition from the 1×1 case here, by using the already known $l_{11} = \sqrt{a_{11}}$ when we solve $a_{12} = l_{11}l_{12}$ for l_{12}. We're then ready to solve $a_{22} = l_{12}^2 + l_{22}^2$ for l_{22}.

More generally, suppose that we wish to find the Cholesky decomposition of an $n \times n$ positive definite matrix. Maybe we can generalize the approach in Eq. (5.1). Write the decomposition in the form $A = R^T R$ where $R = L^T$ is upper triangular (also called **right triangular**; hence the use of the letter R). Define A_p to be the $p \times p$ **leading principal submatrix** of A, that is, the northwest $p \times p$ submatrix of A. It's a fact that a matrix is positive definite if and only if each leading principal submatrix of the matrix is also positive definite. Let's try to find a relationship between the Cholesky decompositions of A_1, A_2, A_3, ..., $A_n = A$. We know that the case

$$A_1 = (a_{11})$$

has the decomposition $R_1 = \sqrt{a_{11}}$. Hence suppose that for some $p \geq 2$ that we have a Cholesky decomposition $A_{p-1} = R_{p-1}^T R_{p-1}$ of the $(p-1) \times (p-1)$ leading principal submatrix of A. We need a decomposition of the $p \times p$ submatrix

$$A_p = \begin{pmatrix} A_{p-1} & c_{p-1} \\ c_{p-1}^T & a_{pp} \end{pmatrix}$$

where a_{pp} is a scalar (which must be positive) and c_{p-1} is a column vector with $(p-1)$ entries. That is, we need to find a column vector γ_{p-1} with $(p-1)$ entries and a positive scalar r_{pp} such that

$$\begin{pmatrix} A_{p-1} & c_{p-1} \\ c_{p-1}^T & a_{pp} \end{pmatrix} = \begin{pmatrix} R_{p-1}^T & 0 \\ \gamma_{p-1}^T & r_{pp} \end{pmatrix} \begin{pmatrix} R_{p-1} & \gamma_{p-1} \\ 0 & r_{pp} \end{pmatrix}$$

is the Cholesky decomposition of A. Analogous to Eq. (5.1), we have

(5.2)
$$\begin{aligned} A_{p-1} &= R_{p-1}^T R_{p-1} \\ c_{p-1} &= R_{p-1}^T \gamma_{p-1} \\ c_{p-1}^T &= \gamma_{p-1}^T R_{p-1} \\ a_{pp} &= \gamma_{p-1}^T \gamma_{p-1} + r_{pp}^2 \end{aligned}$$

which we would like to solve for γ_{p-1} and $r_{pp} > 0$. The first equation $A_{p-1} = R_{p-1}^T R_{p-1}$ is already known to be satisfied; we do not need to consider it. The next two equations are redundant, for if $c_{p-1} = R_{p-1}^T \gamma_{p-1}$ then $c_{p-1}^T = (R_{p-1}^T \gamma_{p-1})^T = \gamma_{p-1}^T R_{p-1}$. Thus we need to solve the equations

$$\begin{aligned} c_{p-1} &= R_{p-1}^T \gamma_{p-1} \\ a_{pp} &= \gamma_{p-1}^T \gamma_{p-1} + r_{pp}^2 \end{aligned}$$

for γ_{p-1} and r_{pp}. But this can be done, and it can be done very efficiently; for, the system

$$R_{p-1}^T \gamma_{p-1} = c_{p-1}$$

is a triangular nonsingular system that may be solved by forward substitution for γ_{p-1} (as R_{p-1} and c_{p-1} are known), and then we can form the inner product $\gamma_{p-1}^T \gamma_{p-1}$ in order to finally arrive at

$$r_{pp}^2 = a_{pp} - \gamma_{p-1}^T \gamma_{p-1}$$

which gives

$$r_{pp} = \sqrt{a_{pp} - \gamma_{p-1}^T \gamma_{p-1}}$$

and it is a property of positive definite matrices that $a_{pp} - \gamma_{p-1}^T \gamma_{p-1}$ is always positive, so that r_{pp} is real and positive. Summarizing, our algorithm for finding the Cholesky decomposition of A from Eq. (5.2) is as follows:

Cholesky Decomposition: Inner Product Algorithm:
1. Set $R_1 = \sqrt{a_{11}}$.
2. Begin loop ($p = 2$ to n):
3. Solve $R_{p-1}^T \gamma_{p-1} = c_{p-1}$ for γ_{p-1} by forward substitution.
4. Set $r_{pp} = \sqrt{a_{pp} - \gamma_{p-1}^T \gamma_{p-1}}$.
5. Set $R_p = \begin{pmatrix} R_{p-1} & \gamma_{p-1} \\ 0 & r_{pp} \end{pmatrix}$.
6. End loop.
7. Set $L = R_n^T$.

Again, it can be shown that in infinite precision arithmetic step 4 will always give a positive r_{pp} and that each A_p is positive definite and hence has its own Cholesky decomposition. Notice that R_p differs from R_{p-1} only by the addition of a last row and last column, so that in forming the factor R_p of A_p we do not change the entries of R_{p-1} but simply add more entries.

Our algorithm computes R column-by-column, which is equivalent to computing L row-by-row, and is actually the Cholesky-Crout version of the algorithm. The Cholesky-Banachiewicz version computes R row-by-row (equivalently, it computes L column-by-column).

Importantly, if A is not positive definite then this algorithm detects that fact by arriving at a nonpositive r_{pp} at some iteration. It is grossly inefficient to test the matrix for positive definiteness before running the algorithm in most cases; finding the eigenvalues to see if they're positive, say, is much more work than just running the algorithm and allowing it to signal an error if $r_{pp} \leq 0$ at some point. The algorithm is itself the test for positive definiteness.

The flops count for this method is about half that of Doolittle's LU decomposition, as would be expected since U is determined from L in this case. Square roots are typically counted as a single flop although they are significantly more expensive than, say, an addition, and the need to compute n square roots is a drawback of the method. As with any problem of interest, there are other algorithms for finding the Cholesky decomposition of a positive definite matrix.

A brief note on hardware. We've been assuming throughout most of this text that we will have only a single processor available with a single central processing unit (CPU) to execute floating point operations. It's increasingly common to have more than one CPU on even a laptop–a dual-core (two processors, and hence two CPUs) or quad-core (four processors and four CPUs) device, say. This is a very low-level version of a parallel computer, but it does open up a variety of possibilities with respect to matrix computing. On a quad-core machine, for example, we could use appropriate software commands to instruct that the matrix-vector product Ax be computed by having processor 1 multiply rows 1, 5, 9, ... by x, while processor 2 multiplies rows 2, 6, 10, ... by x, processor 3 multiplies rows 3, 7, 11, ... by x, and processor 4 multiplies rows 4, 8, 12, ... by x. The product $y = Ax$ can then be formed from these parts on processor 1 at the end, cutting the amount of computation time down by a factor of 4. Of course, the elapsed time won't be decreased by this much as, depending on the memory layout, there will likely be nonparallelizable costs of distributing the rows of A and a copy of x to the various processors initially and then collecting their contributions and forming the final result y at the end.

In the case of the Cholesky decomposition algorithm that we have developed here, an obvious candidate for parallelization is the inner product $\gamma_{p-1}^T \gamma_{p-1}$. If the matrix is sufficiently large and especially if all processors can access the same memory directly, it would be easy to use even a dual-core machine to compute $\gamma_{p-1}^T \gamma_{p-1}$ in about half the time.

In fact, many programming languages will allow you to run multiple processes on a single processor via **threading** (also called **multi-threading**). This is somewhat analogous to having multiple cores, but instead of sending different tasks to different processors a single processor is given multiple tasks, each of which is called a **thread**. For example, in computing Ax we might spawn two threads, one of which computes the dot products for odd-numbered rows of A and the other for even-numbered rows. When there is a lull in the computation of $R_1 \cdot x$, perhaps because only part of R_1 could be brought nearer the CPU, the second thread can run at least part of the computation of $R_2 \cdot x$ rather than simply standing idle while waiting. It's possible to outsmart oneself when using threads, but when done well this can really speed up a computation.

In working with large matrices we generally try to have the compute time "hidden" behind the memory-movement times; that is, since floating point operations run so quickly compared to memory accesses, we hope to not work like this: Access memory, compute, access memory, compute, access memory, compute. Rather, we hope to arrange to have: Access memory, access memory (while computing with data from the first access), memory (while computing with data from the second access), compute (with data from the third access). We have hidden two of the three compute stages behind the memory moves, thereby shortening the total time to complete the computation by those two compute stages.

MATLAB

The MATLAB command `chol` finds the Cholesky[11] factor R. The command may also be used to test for positive definiteness; enter:

» R=chol([4 2 1;2 3 1;1 1 4])
» R'*R
» [R,p]=chol([4 2 1;2 3 1;1 1 1])
» [R,p]=chol([4 2 1;2 3 1;1 1 0])

The first matrix is positive definite and its right triangular factor R is returned, along with $p = 0$. If the `chol` command is used with two output arguments R and p, then `chol` returns zero for p if the input is a positive definite matrix; if `chol` encounters a problem while performing the decomposition, indicating that A is not positive definite, then it sets p to a positive integer indicating when it encountered the problem and returns an R that is the factor of A_{p-1} (in the notation of the Inner Product Algorithm). In this way, p serves as a status variable. For example,

» [R,p]=chol(-2)

returns an empty matrix for R and sets $p = 1$ (error encountered on the very first submatrix). Enter

» [R,p]=chol([2 1;1 -1])

which returns $R_1 = \sqrt{2}$ and $p = 2$ (the algorithm failed while attempting to factor A_2; the fact that the $(2, 2)$ entry of the matrix is negative shows that A cannot be positive definite, but $A_1 = (2)$ is positive definite).

We've already looked at some of the numerical issues associated with the Cholesky decomposition, so let's briefly mention some other MATLAB matrix commands that may be useful. First, the ellipsis ... (entered as three periods) may be used as a continuation character when entering long vectors or matrices by hand. Within an M-file this may not be necessary; to see an example, enter:

» type rosser

Also recall that when dealing with a large matrix it is often advisable to preallocate memory for it using a command of the form A=zeros([m n]). For sparse matrices there is a special command `spalloc` that may be used to preallocate memory.

The transpose has appeared a lot in this section. Often it is possible to rearrange a formula algebraically so as to avoid actually transposing the matrix, and this can be valuable. Enter:

---

[11]When searching for information on this topic, be aware that this Russian name is spelled in a variety of ways in English; e.g., Choleskii.

```
» clear all
» A=rand(10000);
» tic;A=A';toc
```

to see that transposition is not cost-free. Suppose that A is stored but we must solve $A^T x = b$. Rather than using A'\b we might rewrite the equation as $x^T A = b^T$ and use the forward slash command b'/A to find x^T, then transpose it to find x=(b'/A)'. (See help slash for the forward slash.) This requires two vector transpositions, which is certainly preferable to a matrix transposition in the general case. Enter:

```
» clear all
» A=randn(10000);b=randn([10000 1]);  %Use normal dist.  for a change.
» tic;C1=A'\b;toc
» tic;C2=(b'/A)';toc    %Use forward slash /.
```

The difference may not be dramatic and may not be seen every time but if you repeat the experiment several times it should be noticeable; use a larger random matrix if necessary. The command whos (a more informative version of who) details memory usage information for your current variables.

The poly command, when applied to a matrix, returns the coefficients of its characteristic polynomial. Enter:

```
» A=rand(3);
» poly(A)
» eig(A)
» roots(poly(A))
» det(A)
» prod(roots(poly(A)))
```

(the eigenvalues of A are the roots of the characteristic polynomial and the determinant of A is the product of those roots). The determinant and characteristic polynomial occur infrequently in numerical work; the fact that the roots of a polynomial may be viewed as the eigenvalues of a related matrix is useful however. The matrix typically used for this is the **companion matrix** of the polynomial, which may be found using the MATLAB command compan. The eigs command may be used to find some but not all eigenvalues of a matrix; often we want only the largest one, two or, three eigenvalues (or smallest), and if the matrix is large then it is more efficient to use this command.

The reshape command may be used to change the shape of a matrix. For example, a 4×4 matrix could be reshaped to be 8×2. Enter:

```
» A=rand(4)        %4x4 matrix.
» reshape(A,8,2)  %8x2 matrix.
```

The repmat command may be used to create matrices that have a repetitive structure. The kron command forms a Kronecker product of matrices. There are numerous sparse matrix commands; type help sparfun for a list of these commands (and see Sec. 3.2).

It is also possible to form matrices from blocks of other matrices in MATLAB, if the blocks conform. Enter:

```
» A11=[1 2;4 5]
» A12=[3 6]'
» A21=[7]
» A22=[8 9]
```

» A=[A11 A12;A21 A22]

Note that A does not retain information about the subblocks from which it was formed, however. There are more complicated structures in MATLAB; see help cell, for example.

The trace command finds the trace of a matrix, that is, the sum of its diagonal elements. The pinv command finds the pseudo-inverse of a matrix (that is, a near-inverse for singular matrices). The hess command orthogonally reduces a matrix to **upper Hessenberg** form (where the lower triangle is zero except possibly for the first subdiagonal). We'll discuss this nearly-triangular decomposition in Chapter 3.

If you have certain MATLAB toolboxes installed, you may have many other matrix functions available. Type help for a list of topics; one useful toolbox for matrices is the Image Processing Toolbox (help images). The Signal Processing Toolbox (help signal) has many useful vector commands.

Problems

1.) a.) Use the Inner Product Algorithm (by hand) to find the Cholesky decomposition of the matrix $A = [4\ 2\ 1; 2\ 3\ 1; 1\ 1\ 4]$. Compare your results to Example 1.

b.) Use the Cholesky decomposition of A to solve the linear system $Ax = (3, 0, 4)^T$.

2.) Use the Inner Product Algorithm to find the Cholesky decomposition of the matrix $A = [4\ 1\ 0\ 1; 1\ 4\ 2\ -1; 0\ 2\ 4\ -2; 1\ -1\ -2\ 4]$.

3.) Draw a Wilkinson diagram for R when applying the Inner Product Algorithm to a 4×4 positive definite matrix.

4.) Write a detailed algorithm (pseudo-code) for performing the Inner Product Algorithm in place (that is, over-writing A or part of it with L). Explicitly include the back substitution steps.

5.) a.) Show that if A is symmetric then MAM^T is symmetric for any M of the same size.

b.) Show by example that it is possible for a nonsymmetric matrix to satisfy $x^T Ax > 0$ for all nonzero x.

6.) a.) Use the algorithm to find the Cholesky decomposition of the $A^T A$ where $A = [1\ 0\ 1; 1\ 0\ -1; 1\ 1\ 0; 1\ 1\ 1]$.

b.) Use your Cholesky decomposition from part a. to solve $A^T Ax = A^T b$ where $b = (1, -1, 1, -1)^T$.

c.) Show that for any matrix M, not necessarily square, MM^T and $M^T M$ are always positive semi-definite.

7.) a.) Prove that if A is positive definite then so is $E_2 E_1 A E_1^T E_2^T$ (with $E_{1,2}$ elementary).

b.) Prove that if A is positive definite then every leading principal submatrix of A is positive definite. (Hint: Recall that a symmetric matrix is positive definite if and only if $x^T Ax > 0$ for all nonzero x. Partition A and choose an appropriate x.)

c.) Prove that if A is positive definite then it has a factorization of the form $A = LDL^T$ where L is unit lower triangular and D is a diagonal matrix with positive diagonal entries. (You may use the fact that A has a Cholesky decomposition.)

8.) Write a detailed algorithm (pseudo-code) for decomposing a positive definite matrix into the form $A = LDL^T$ where L is unit lower triangular and D is a diagonal matrix with positive diagonal entries.

9.) Use the eigenvalue decomposition to show that a symmetric matrix is positive definite if and only if all its eigenvalues are positive, and is positive semi-definite if and only if all its eigenvalues are nonnegative.

10.) a.) Prove that if every leading principal submatrix of a symmetric matrix A is positive definite then A is positive definite.

b.) Prove that the diagonal elements of a positive definite matrix are positive.

c.) Prove that every leading principal submatrix of a positive definite matrix has a positive determinant. These determinants are called the **leading principal minors** of the matrix.

11.) a.) What is the Cholesky Decomposition of the identity matrix?

b.) Is it possible to write the zero matrix as $0 = LL^T$? Can it be done with the diagonal elements of L positive, as we have required in our Cholesky decomposition?

12.) a.) Corresponding to the inner product form of the algorithm there is also an outer product version. Derive it by writing $A = R^T R$ in the form

$$\begin{pmatrix} a & s \\ s^T & A_1 \end{pmatrix} = \begin{pmatrix} r & 0 \\ \rho^T & S_1^T \end{pmatrix} \begin{pmatrix} r & \rho \\ 0 & S_1 \end{pmatrix}$$

where a is a scalar, r is a scalar, and s and ρ are appropriately sized row vectors. By equating blocks on the LHS and RHS find an equation for r and ρ (the first row of R) and an equation relating $S_1^T S_1$, A_1 (which is now a southeast submatrix), and an outer product. (If v is a column vector, vv^T is an inner product resulting in a scalar value and $v^T v$ is an outer product, resulting in a matrix of rank one.) Explain how this technique may be applied iteratively to determine the Cholesky factor R.

b.) Write a MATLAB program that implements this method. Attempt to mimic the behavior of the chol command by checking for errors and returning a p value that indicates how far the method has proceeded if an error occurs.

13.) a.) Create a 100×100 positive definite matrix A by using the eigenvalue decomposition $A = SDS^T$ with D having positive diagonal entries and the MATLAB command S=orth(rand(100)). Use the chol command to factor A. Find the error in $R^T R$.

b.) Create a 100×100 positive definite matrix with known Cholesky decomposition by using the sequence of commands R=eye(100)+triu(rand(100));A=R'*R; and compare the computed Cholesky decomposition (using chol) to the known Cholesky factor R.

c.) Recall that a matrix is positive definite if and only if it can be written in the form $R^T R$ where R is a square upper triangular matrix with positive diagonal entries. Repeat the sequence of commands R=triu(rand(100));A=R'*R; [R,p]=chol(A);p at least 10 times. Are all these matrices found to be positive definite by the chol command? (Probably not.) Why not? (Hint: For one possibility, consider min(diag(R)) and see if the change R=1000*R;A=R'*R; improves matters when p is found to be nonzero.)

14.) Write a MATLAB program that implements the Inner Product Algorithm for the Cholesky decomposition. Attempt to mimic the behavior of the chol command by checking for errors and returning a p value that serves as a status variable by indicating how far the method has proceeded if an error occurs.

15.) Suppose you had a Cholesky decomposition that did not enforce the convention $r_{ii} > 0$ in $A = R^T R$. If M is the identity matrix save that wherever $r_{ii} < 0$ we set

$m_{ii} = -1$ then how can we use M to obtain a Cholesky decomposition $A = S^T S$
with S right triangular and $s_{ii} > 0$?

16.) a.) Prove that every symmetric matrix A has a factorization of the form LDL^T
where L is unit lower triangular. (This allows us to avoid taking square roots.)

b.) Show that if $A = LDL^T$ where L is unit lower triangular and D has positive
diagonal entries then A has a Cholesky decomposition and hence is positive definite.

c.) What kind of decomposition could we get for a negative definite matrix?

6. Conditioning

Although Gaussian elimination with partial pivoting usually destroys the sparsity structure of a matrix, it's still the method of choice for small-to-medium n.
The rapid growth of error that can in principle happen simply does not occur in
practice. One reason is that there is some cancellation of errors due to the fact
that some will be high (positive error) and some will be low (negative error); there
is a statistical theory of error analysis. This is not the only reason, however, and
numerical analysts do not fully understand why the method performs as well as it
does.

Gaussian elimination with partial pivoting does not solve the linear system
$Ax = b$ with the same degree of accuracy for every $n \times n$ matrix A, however.
Clearly $Ix = b$ will always be solved exactly, but as we have seen, this is not
typical. To understand this we must extend the concept of a norm to matrices.
This will involve a fair amount of new terminology but the concept is indispensable
for analyzing the algorithms of computational matrix algebra.

Recall that a vector norm on $\mathbb{R}^n$ is a real-valued function $\|x\|$ defined for all
$x \in \mathbb{R}^n$ that is positive definite ($\|x\| > 0$ unless $x = 0$ in which case $\|x\| =
0$), satisfies $\|\alpha x\| = |\alpha| \|x\|$ for all scalars α, and satisfies the triangle inequality
$\|x + y\| \le \|x\| + \|y\|$ ($x, y \in \mathbb{R}^n$). The most commonly used vector norms are the
Hölder p-norms

$$\|x\|_p = \left[\sum_{i=1}^{n} |x_i|^p\right]^{1/p}$$

($p \ge 1$). These are also called the l_p norms. The cases of greatest interest are the
l_1 norm

$$\|x\|_1 = \sum_{i=1}^{n} |x_i|$$

and the l_2 norm

$$\|x\|_2 = \left[\sum_{i=1}^{n} x_i^2\right]^{1/2}$$

which is of course simply the Euclidean root-mean-square (RMS) norm, for which
we also write $\|\cdot\|_E$, as well as the limiting case as $p \to \infty$ which can be shown to
give a valid vector norm

$$\|x\|_\infty = \max_{i=1}^n(|x_i|)$$

called the l_∞ or **max** or **sup** (for supremum) norm. Each of these may be used for measuring the size of an error vector such as the residual from solving a linear system

$$r = b - A\widehat{x}$$

(where $\widehat{x}$ is an approximate solution of $Ax = b$); r is a vector, so its norm $\|r\|$ gives a single scalar value that may be compared to some tolerance. (Note that the computation of r involves the subtraction of nearly equal quantities so cancellation of significant figures is an issue here.) The l_1 norm measures the total error, the l_2 norm measures the RMS error, and the max norm measures the worst case error over all components.

For a fixed n, all vector norms on $\mathbb{R}^n$ are **equivalent**, meaning that for any two norms $\|\cdot\|_\alpha$, $\|\cdot\|_\beta$ there exist positive scalars r_l and r_u, independent of x, such that

$$r_l\|x\|_\beta \le \|x\|_\alpha \le r_u\|x\|_\beta$$

for every $x \in \mathbb{R}^n$. (For emphasis, the constants r_l and r_u depend only on the norms $\|\cdot\|_\alpha$ and $\|\cdot\|_\beta$ but do *not* depend on the specific vector x.) In particular, if a sequence converges (or diverges) in one norm, it does so in all norms.

The norm of a vector is a measure of its size; if $\|x_k\| \to 0$ then $x_k \to 0$, for example. We need a similar measure of size for (square) matrices. We could treat an $n \times n$ matrix as a vector in $\mathbb{R}^{n^2}$ by unraveling it row by row into a vector of length n^2 but this neglects an important difference between vectors and matrices: Matrices can be multiplied, while vectors cannot. In addition to the triangle inequality for addition, we want a similar result for products of matrices as well.

A real-valued function $\|\cdot\|$ defined for all $n \times n$ matrices is said to be a **matrix norm** if it has the following properties:

(1) $\|A\| > 0$ unless A is the zero matrix, in which case $\|A\| = 0$ (positive definiteness)
(2) $\|\alpha A\| = |\alpha|\|A\|$ for all scalars α (homogeneity)
(3) $\|A + B\| \le \|A\| + \|B\|$ (the triangle inequality)
(4) $\|AB\| \le \|A\|\|B\|$ (the submultiplicativity property)

(A, B $n \times n$ matrices). Unraveling the matrix to an n^2 long vector and using a vector norm on $\mathbb{R}^{n^2}$ gives a function satisfying the first three properties but which will not in general be submultiplicative. (Any function satisfying the first three properties is said to be a **matrix subnorm**.) We interpret $\|A\|$ as some measure of the size of the matrix A, though this lacks the clear geometric interpretation of a vector norm as the size of the vector. If $\|\cdot\|$ is a matrix norm then $d(A, B) = \|A - B\|$ defines a distance between matrices.

Submultiplicativity is an important and useful property. It immediately gives us the following inequality:

$$\|A^2\| = \|AA\| \le \|A\|\|A\| = \|A\|^2$$

and, inductively, we also have $\|A^k\| \le \|A\|^k$ ($k = 1, 2, ...$). This allows us to bound the size of powers of A by powers of the real number $\|A\|$ representing its size.

We'll also need to work with matrix-vector products. If $\|\cdot\|_v$ is a vector norm on $\mathbb{R}^n$ and $\|\cdot\|_M$ is a matrix norm on $n \times n$ matrices then we say that the vector norm and the matrix norm are **consistent** (or **compatible**) if

$$\|Ax\|_v \leq \|A\|_M \|x\|_v$$

(similar to the submultiplicativity property). If we have consistent vector and matrix norms then we may estimate the size of Ax from the sizes of A and x. This will be useful.

How do we find matrix norms? If A is an $n \times n$ matrix with entries a_{ij} then the function

$$\|A\|_F = \sqrt{\sum_{i=1}^{n} a_{ij}^2}$$

defines a matrix norm, called the **Frobenius norm**. It can be shown that

$$\|A\|_F = \sqrt{\operatorname{tr}(A^T A)}$$

where $\operatorname{tr}(M)$ is the **trace** of M, that is, the sum of the diagonal entries of M. The most important matrix norm, however, is the **spectral norm**

$$\|A\|_s = \sqrt{\rho(A^T A)}$$

where $\rho(M)$ is the **spectral radius** of M, that is, the largest eigenvalue of M in absolute value.

The eigenvalues of $A^T A$ must be nonnegative as it is positive semi-definite. The nonzero eigenvalues of $A^T A$ are the same as the nonzero eigenvalues of $A A^T$, and the square roots of these are called the **singular values** of A. Hence we can also say that $\|A\|_s$ is equal to the largest singular value of A.

The motivation for the Frobenius norm is clear, but where does the spectral norm come from? In fact, every vector norm induces a matrix norm. If $\|\cdot\|_v$ is a vector norm on $\mathbb{R}^n$ then the function

$$\begin{aligned} \|A\|_M &= \sup_{x \neq 0}(\|Ax\|_v / \|x\|_v) \\ &= \max_{\|x\|_v = 1}(\|Ax\|_v) \end{aligned}$$

is a matrix norm, called the **induced norm** (or **natural norm**) associated with, or induced by, the vector norm $\|\cdot\|_v$. An induced matrix norm is always consistent with the vector norm that induces it, that is,

$$\|Ax\|_v \leq \|A\|_M \|x\|_v$$

if $\|\cdot\|_v$ induces $\|\cdot\|_M$. Furthermore, if $\|\cdot\|_\mu$ is any other matrix norm consistent with $\|\cdot\|_v$ then for every $n \times n$ matrix A,

(6.1) $$\rho(A) \leq \|A\|_M \leq \|A\|_\mu$$

so that $\rho(A)$ is a lower bound on $\|A\|_M$ for any natural norm $\|\cdot\|_M$ and furthermore the natural norm gives the smallest value of all norms consistent with a given vector norm. This is a desirable property.

The spectral norm arises as the natural norm induced by the Euclidean norm, and this is why it is of interest. The Frobenius norm is also consistent with the Euclidean vector norm but by Eq. (6.1) it must always be the case that

$$(6.2) \qquad\qquad \|A\|_s \le \|A\|_F.$$

(The Frobenius norm is not induced by any vector norm.) Because of this when we use the Frobenius norm it is only because it is easier to compute than the spectral norm, which is relatively expensive to compute.

EXAMPLE 1: Let $A = [2\ \ 1\ \ 0; 1\ \ 1\ \ 0; 0\ \ 0\ \ -1]$. The Frobenius norm of A is $\|A\|_F = [2^2+1^2+0^2+1^2+1^2+0^2+0^2+0^2+(-1)^2]^{1/2} = \sqrt{8} \doteq 2.8284$ and the spectral norm of A may be found from the MATLAB command `sqrt(max(eig(A'*A)))` to be 2.6180 which is indeed not more than $\|A\|_F$ (in accordance with Eq. (6.2)). It is also the same as the spectral radius `max(eig(A))` of A in agreement with Eq. (6.1). (The **norm** command can be used to compute matrix norms.) Let $x = (5, -4, 3)^T$, for which $\|x\|_2 = [5^2 + (-4)^2 + (-3)^2]^{1/2} = \sqrt{50} \doteq 7.0711$. We could estimate $\|Ax\|_2$ by $\|A\|_F\|x\|_2 = \sqrt{8}\sqrt{50} = 20$ or by $\|A\|_s\|x\|_2 = 2.6180\sqrt{50} \doteq 18.5121$. In fact $Ax = (5, 0, -3)^T$ and so $\|Ax\|_2 = [5^2 + 0^2 + (-3)^2]^{1/2} = \sqrt{34} \doteq 5.8310$. $\square$

The matrix norm induced by the l_1 vector norm is $\|A\|_\gamma = \max_{j=1}^n \left\{ \sum_{i=1}^n |a_{ij}| \right\}$, that is, the maximum column sum. The matrix norm induced by the l_∞ vector norm is $\|A\|_\rho = \max_{i=1}^n \left\{ \sum_{j=1}^n |a_{ij}| \right\}$, that is, the maximum row sum. These are widely used because they are so simple to compute.

We almost always need consistent matrix norms because we are usually working with matrix-vector products (e.g., Ax and $A^{-1}b$), and whenever possible we choose natural norms because they give the best (that is, smallest) estimates of the sizes of these quantities. The spectral radius is not a norm so we cannot use it for this purpose.

We're now ready to apply the matrix norm to help us understand how the nature of the matrix A affects the quality of the computed solution $\widehat{x}$ to $Ax = b$, for which $x = A^{-1}b$ is the true solution. (We are assuming that A is nonsingular.) Consider the linear system

$$Ax = b.$$

No matter what method we use to solve it, we will find an approximation $\widehat{x}$ to x. We would like to say something about the absolute error

$$\alpha = \|x - \widehat{x}\|$$

or the relative error

$$\rho = \|x - \widehat{x}\|/\|x\|$$

(if x is nonzero) in our computed solution, but since we do not know x we cannot do this directly. An obvious substitute measure of error is the norm of the residual

$$r = b - A\widehat{x}$$

since $\|r\| = 0$ if $x = \hat{x}$ and, by continuity, $r \to 0$ as $x \to \hat{x}$. But now we must ask the following crucial question: Is it the case that if $\|r\|$ is small then the absolute error $\alpha = \|x - \hat{x}\|$ is small too? Remember, the residual measures the difference in the values of the function

$$y = Ax$$

at the true solution x (where we are given that $y = b$) and the computed solution (assuming that $y = A\hat{x}$ is computed accurately). In root-finding problems we stated that we could not safely use the reasoning that if $f(x_0)$ was nearly zero then x_0 was nearly a root of $f(x) = 0$. Is it different here, when we are solving the root-finding problem for the linear equation $Ax - b = 0$?

The answer is both yes and no. *Conditioning*–the sensitivity of the solution of a problem to small changes in the given data, leading to poorly behavior of numerical methods on those problems–can be defined for a very general class of problems. Nonlinear root-finding problems in particular can be well-conditioned or ill-conditioned (as discussed in Sec. 1.8); roots of multiplicity greater than one, for which the derivative of the function is zero at the root, are the most common example of ill-conditioned problems of this sort. In this sense there is nothing special about the case $Ax - b = 0$. Indeed, several of our methods for nonlinear root-finding were based on linearizing the function f to obtain a linear approximant. As a practical matter, however, the more useful answer is that the condition of A plays a special role in solving the *linear* system $Ax = b$.

We'll continue to focus on conditioning for square linear systems with A non-singular. If our computed solution is $\hat{x}$ then the residual

$$
\begin{aligned}
r &= b - A\hat{x} \\
&= Ax - A\hat{x} \\
&= A(x - \hat{x})
\end{aligned}
$$

is an easily computable measure of the error. Taking norms on both sides using any consistent vector and matrix norm gives

$$
\begin{aligned}
\|r\|_v &= \|A(x - \hat{x})\|_v \\
&\leq \|A\|_M \|x - \hat{x}\|_v
\end{aligned}
$$

(by consistency of the norms). But this inequality goes the wrong way as far as we are concerned; $\|r\|_v$ is what we can compute, and we want to use it to bound $\alpha = \|x - \hat{x}\|_v$ or $\rho = \|x - \hat{x}\|_v / \|x\|_v$. Let's try again, starting from the definition of the residual:

$$
\begin{aligned}
r &= b - A\hat{x} \\
&= Ax - A\hat{x} \\
A^{-1}r &= x - \hat{x} \\
\|A^{-1}r\|_v &= \|x - \hat{x}\|_v \\
\|A^{-1}\|_M \|r\|_v &\geq \|x - \hat{x}\|_v.
\end{aligned}
$$

This is better. If we know or can estimate $\|A^{-1}\|_M$ then we can put an upper bound on the absolute error α. For the relative error we have

$$(6.3) \qquad \|A^{-1}\|_M \|r\|_v / \|x\|_v \geq \|x - \widehat{x}\|_v / \|x\|_v$$

and from $Ax = b$ we have

$$\begin{aligned} \|Ax\|_v &= \|b\|_v \\ \|A\|_M \|x\|_v &\geq \|b\|_v \\ \|x\|_v &\geq \|b\|_v / \|A\|_M \end{aligned}$$

so Eq. (6.3) becomes

$$\|A^{-1}\|_M \|r\|_v / (\|b\|_v / \|A\|_M) \geq \|x - \widehat{x}\|_v / \|x\|_v$$

that is

$$(6.4) \qquad \frac{\|x - \widehat{x}\|_v}{\|x\|_v} \leq \|A\|_M \|A^{-1}\|_M \frac{\|r\|_v}{\|b\|_v}$$

which gives a bound on the relative error $\|x - \widehat{x}\|_v / \|x\|_v$ using the values of $\|r\|_v$ and $\|b\|_v$, which are easily computed, and the matrix norms of A and A^{-1}. Since we are usually more interested in ρ than in α, we define the **condition number** of A (with respect to $\|\cdot\|_M$) to be

$$\kappa(A) = \|A\|_M \|A^{-1}\|_M$$

(also written cond(A)). To simplify the notation, from here on we'll drop the subscripts on the norms and simply write $\kappa(A) = \|A\| \|A^{-1}\|$. It should be clear from context whether a vector or matrix norm is intended. (Typically we wish to use the Euclidean vector norm and the spectral matrix norm, if possible.) The condition number of a singular matrix is taken to be infinity. The condition number of a nonsingular matrix is always at least one, for

$$\begin{aligned} AA^{-1} &= I \\ \|AA^{-1}\| &= \|I\| \\ \|A\| \|A^{-1}\| &\geq \|I\| \end{aligned}$$

and the relation $\|A^2\| \leq \|A\|^2$ with $A = I$ gives $\|I\|^2 \geq \|I^2\| = \|I\|$ so that $\|I\| \geq 1$ (recall that only the zero matrix can have a zero norm). Hence $\kappa(A) = \|A\| \|A^{-1}\| \geq 1$. If $\kappa(A)$ is near one we say that A is well-conditioned and if $\kappa(A)$ is much larger than one we say that A is ill-conditioned; there is no specific cutoff value of $\kappa(A)$ for well-conditioned versus ill-conditioned matrices.

Consider again Eq. (6.4). We may write this equation in terms of the condition number as

$$\frac{\|x - \widehat{x}\|}{\|x\|} \leq \kappa(A) \frac{\|r\|}{\|b\|}$$

or

(6.5)
$$\rho \leq \kappa(A)\frac{\|r\|}{\|b\|}.$$

Suppose for simplicity that $\|b\| = 1$. Since $\kappa(A) \geq 1$, Eq. (6.5) states that the relative error in our computed solution $\hat{x}$ may be larger than the error as measured by the residual; in fact, it may be larger by a factor of $\kappa(A)$. If A is well-conditioned ($\kappa(A) \approx 1$) then it is valid to reason that a small residual implies a small relative error in $\hat{x}$; but if A is ill-conditioned ($\kappa(A) \gg 1$) then knowing that the residual is small tells us very little about the relative error in $\hat{x}$.

EXAMPLE 2: Let's look at the effect predicted by Eq. (6.5). Set $\epsilon = .1$ and $\varpi = 2\ln(\epsilon)$, and define

$$A = \begin{bmatrix} \cosh(\varpi) & \sinh(\varpi) \\ \sinh(\varpi) & \cosh(\varpi) \end{bmatrix}.$$

From MATLAB (using the command cond(A)) the condition number of A is 10^4. (The MATLAB functions sinh and cosh may be used to enter the hyperbolic trig functions.) Let's take $b = (1/\sqrt{2}, 1/\sqrt{2})^T$ (for which $\|b\| = 1$). The true solution is

$$\begin{aligned} x &= \begin{bmatrix} \cosh(\varpi) & \sinh(\varpi) \\ \sinh(\varpi) & \cosh(\varpi) \end{bmatrix}^{-1} \begin{pmatrix} 1/\sqrt{2} \\ 1/\sqrt{2} \end{pmatrix} \\ &= \frac{1}{\cosh^2(\varpi) - \sinh^2(\varpi)} \begin{bmatrix} \cosh(\varpi) & -\sinh(\varpi) \\ -\sinh(\varpi) & \cosh(\varpi) \end{bmatrix} \begin{pmatrix} 1/\sqrt{2} \\ 1/\sqrt{2} \end{pmatrix} \\ &= (1/\sqrt{2}) \begin{pmatrix} \cosh(\varpi) - \sinh(\varpi) \\ \cosh(\varpi) - \sinh(\varpi) \end{pmatrix} \\ &= (1/\sqrt{2}) \begin{pmatrix} \exp(-\varpi) \\ \exp(-\varpi) \end{pmatrix} \\ &= (1/\sqrt{2}) \begin{pmatrix} 1 \\ 1 \end{pmatrix} \epsilon^{-2} \end{aligned}$$

so that

(6.6)
$$x = \frac{1}{\sqrt{2}\epsilon^2} \begin{pmatrix} 1 \\ 1 \end{pmatrix}$$

and so $x = (70.71067811865474, 70.71067811865474)^T$ for $\epsilon = .1$. (Note the division by a small number in the formula for x.) From MATLAB, A\b is $\hat{x} = (70.71067811864299, 70.71067811864299)^T$. The relative error is $\|x - \hat{x}\|/\|x\|$ or about $1.66x10^{-13}$ and the norm of the residual $\|b - A\hat{x}\|$ is about $3.45x10^{-13}$. Despite a condition number of 10^4 the residual and the relative error are on the same order of magnitude. Note however that $\hat{x}$ is accurate to only 12 digits (70.7106781186) and not to the full 16 digits of x; we have lost about 4 digits of accuracy ($16 - 12 = 4$) in performing a calculation in which cond$(A) \approx 10^4$.

Let's try again with $\epsilon = .01$. We find that cond(A) is essentially 10^8 and $x = (7071.067811865475, 7071.067811865475)^T$. (Note that this is just the previous x rescaled by a factor of 100, as per Eq. (6.6).) From MATLAB, A\b is $\hat{x} = (7071.06791978500, 7071.06791978500)^T$. The relative error is $\|x - \hat{x}\|/\|x\|$ or about $2.16x10^{-8}$ and the norm of the residual $\|b - A\hat{x}\|$ is about $1.04x10^{-8}$. The relative error is larger than the norm of the residual, though only by a factor of about 2 (far from the worst case of 10^8). However, the entries of $\widehat{x}$ only agree with those of x to 7 digits (7071.067), though it's very nearly accurate to 8 digits; we have lost about 8 digits of accuracy $(16 - 7 = \text{about } 8)$ in performing a calculation in which $\operatorname{cond}(A) \approx 10^8$. $\square$

In both cases in Example 2 the norm of the residual is on the same order of magnitude as the relative error. As indicated by Eq. (6.5), this need not always be the case; between Eq. (6.5) and the subtraction of nearly equal quantities performed when computing r, we must be very careful in drawing conclusions about ρ based on it. Also in both cases in Example 2 we lost about p digits of accuracy in the solution of $Ax = b$ when the condition number of A was on the order of 10^p; this is typical. Although better results will be obtained for certain choices of b, both theory and experience show that:

> If the condition number of A is on the order of 10^p, about p significant figures will be lost in solving $Ax = b$.

(Commit the statement above to memory.) Note that this rule of thumb does *not* state that about p significant figures will be lost in solving $Ax = b$ by Gaussian elimination with partial pivoting; it states that about p significant figures will be lost by *any* method.

The reason is that Eq. (6.5) is a statement about the fundamental sensitivity of the solution x of the system $Ax = b$; in fact, it is more a statement about the sensitivity of the system $Ax = b$ to perturbations of A than it is a statement about numerical analysis. The very act of entering the matrix A on the computer will almost always cause some representation error, and so rather than solving $Ax = b$ we are in fact solving $A_1 x = b_1$ where $A \approx A_1$ and $b \approx b_1$. (These quantities are approximately equal in both the componentwise and normwise senses.) Hence the best we can possibly hope for is to solve

$$A_1 \xi = b_1$$

exactly for its solution ξ. How close is ξ to x? This is exactly the question that the condition number answers, and that answer is Eq. (6.5) and the guideline in italics above. Even supposing that $b_1 = b$, it is not hard to show that if A is nonsingular and the individual entries of A_1 differ by no more than ϵ from those of A then

$$(6.7a) \qquad\qquad \frac{\|\xi - x\|}{\|x\|} \le \kappa(A)\epsilon$$

for all $\epsilon > 0$ sufficiently small. (Recall that ξ is the *exact* solution of the perturbed system!) For example, if $\epsilon \approx 10^{-16}$ is the machine epsilon and $\kappa(A) \approx 10^p$ then

$$\frac{\|\xi - x\|}{\|x\|} \lesssim 10^p 10^{-16} = 10^{-16+p}$$

meaning that, generally, ξ and x may have components that differ in the $(16 - p)th$ place (e.g., about the tenth place if $p = 6$; the smaller entries of the vectors ξ and x may differ even sooner, but they contribute less to the norm.) We can't possibly hope to do better than get the exact solution of the closest system to $Ax = b$ that can be represented on the machine; we can only expect that we might do worse. Of course Eq. (6.7a) is an inequality, not an equality, so we will not get results this poor every time, but we certainly can and typically do get results like that, as seen in Example 2, and we should expect this.

It's worth repeating: The loss of about p significant figures when the condition number is roughly 10^p is a property of A, not a property of the numerical method used to solve $Ax = b$. It is not a failing of the method or the machine but rather an inherent property of the matrix–namely, its condition. In fact, we usually consider a method for solving $Ax = b$ "good" if it computes a solution $\widehat{x}$ with the property that

$$(6.8) \qquad\qquad (A + E)\widehat{x} = b$$

for some matrix E with small norm. This equation states that the computed solution $\widehat{x}$ of $Ax = b$ is the *exact* solution of a "nearby" problem involving the matrix $A + E$ that is close to A in the sense that $\|(A + E) - A\| = \|E\|$ is small. A result such as Eq. (6.8) is called a **backward error analysis** in that it refers the errors backward to the coefficient matrix; if the errors in $\widehat{x}$ may be accounted for by a slight perturbation of A, then unless we know A to great precision we cannot fault the algorithm for failing to find the desired solution x. A **forward error analysis** generally refers to a procedure for estimating the worst-case effects of round-off error as the computation proceeds step by step through the algorithm. Backward error analyses are preferred for a number of reasons; one is that the worst-case effects of round-off error are almost never seen in practice. Some errors are positive, some are negative, and they will largely cancel. For this reason forward error analyses often give extremely large bounds that are overly pessimistic for an actual computation.

A method that satisfies an equation such as Eq.(6.8), that is, one with a backward error analysis, is said to be **backwards stable**. Stability refers to the tendency of a method to produce answers that are about as good as the conditioning will allow; instability means that round-off errors are magnified to the point that the returned answer isn't even close to the true solution (measured with respect to the conditioning). The LU decomposition with complete pivoting is stable; the LU decomposition with partial pivoting is not stable, strictly speaking, but in practice it almost always behaves as though it were.

For matrices with a given condition number it is still true that some solution methods are superior to others, and there are techniques for improving an estimate $\widehat{x}$ of x (see Sec. 3.3). But when solving a linear system it is always useful to know the condition number of the coefficient matrix A. For some matrix norms this can be relatively easy to compute but for others, such as the spectral norm, it is hard to compute but not necessarily hard to approximate, and since its order of magnitude is all we really need, an estimate is usually sufficient. Using the Frobenius norm or some other norm will generally give an overestimate (that is, an upper bound) on the condition number with respect to the spectral norm but if that estimate is small enough then we need not compute $\kappa(A)$ more precisely.

We've covered a lot of ground in this section, and if matrix norms are new to you then it may take some time to adjust to them. Remember however that MATLAB can compute $\kappa(A)$ and the solution of $Ax = b$, and you now know how to interpret these quantities in an informed manner.

MATLAB

Let's start with some relevant MATLAB commands. The `norm` command computes vector and matrix norms. For example, enter:

```
» help norm
» A=rand(5);
» norm(A)            %Spectral norm.
» norm(A,'fro')      %Frobenius norm.
» norm(A,1)          %l1 (max. column sum) norm.
» norm(A,'inf')      %l-infinity (max. row sum) norm.
» v=1:10;p=3;
» norm(v,p)          %Holder p-norm for vectors.
```

(You may also use double quotes, e.g., `norm(A,"fro")`, in these commands.) For large matrices, the function `normest` may be used to estimate the spectral norm of a matrix; it is faster but less accurate.

The definition of the spectral norm is that it is equal to the square root of the largest eigenvalue of $A^T A$. Enter:

```
» sqrt(max(eig(A'*A)))
» norm(A)
```

The norm of A and the norm of A^{-1} may differ considerably in magnitude.

The command `cond` computes the condition number of A with respect to the spectral norm. Enter:

```
» cond(A)
» norm(A)*norm(inv(A))  %Definition of cond(A).
```

The `cond` command uses the singular value decomposition, which is more accurate and more efficient than the above approach. The square roots of the nonzero eigenvalues of $A^T A$ are by definition the singular values σ_i of A, which may be found by the `svd` command. Enter:

```
» sigma=svd(A)
» max(sigma)
```

to see this. As expected, $\|A\|_s$ is equal to the largest singular value. In fact, $\kappa_s(A)$ is the ratio of the largest and smallest singular values:

```
» cond(A), max(sigma)/min(sigma)
```

You can use `cond` to find the condition number w.r.t. other norms; for the Frobenius norm, use `cond(A,'fro')`. There is also a `condest` command that estimates the condition number of A with respect to the $\|\cdot\|_\gamma$ norm (induced by the l_1 vector norm; see Problem 2). The definition of condition number may be extended to nonsquare matrices and the `cond` command works for such matrices.

The **Hilbert matrix of order** n is the $n \times n$ matrix with entries $a_{ij} = 1/(i+j-1)$ and this family of matrices is known to be poorly conditioned. Enter:

```
» B=hilb(5)
» norm(B)
» norm(inv(B))
» cond(B)
```

for the 5×5 case. How badly conditioned is it? The condition number $\kappa(A)$ is sometimes called the condition of A with respect to matrix inversion because of its connection to the solution of $Ax = b$ which is (theoretically) equivalent to finding $A^{-1}b$. Enter:

```
» H=hilb(10);
```

```
» cond(H)
» inv(H)-invhilb(10)
» norm(ans)
```

The command `invhilb` produces the *exact* inverse of H (if n is sufficiently small, which it is here). The errors in the numerical computation of H^{-1} by `inv(H)` are *huge*; note the leading factor of 10^8, or type:

```
» format bank
» inv(H)-invhilb(10)
```

to see just how large these errors are. Enter:

```
» format short
» cond(invhilb(10))
» max(max(invhilb(10)))
```

From the size of the largest element of H^{-1} we see that the absolute errors in `inv(H)-invhilb(10)` are not quite as bad as they may seem at first. In fact, enter:

```
» format long
» inv(H)*H      %Should be the identity matrix, in principle.
» norm(ans)     %Should be 1.
» invhilb(10)*H
» norm(ans)     %Should be 1.
```

to see that the net effect of using the computed inverse is not as bad as expected; the errors in H^{-1} are worse than those seen in $H^{-1}H$. Enter:

```
» b=ones([10 1]);
» x1=inv(H)*b           %Computed solution of Hx=b.
» x=invhilb(10)*b       %Exact solution of Hx=b.
» norm(x-x1)/norm(x)    %Relative error.
» x2=H\b                %Solution by Gaussian elimination.
» norm(x-x2)/norm(x)    %Relative error.
» r=b-H*x2              %Residual.
» cond(H)*norm(r)/norm(b)   %Upper bound on relative error.
```

(see Eq. (6.5)). The relative error in $\hat{x}$ (x2) is nearly 10^6 times larger than the absolute error in $A\hat{x}$ (as measured by $\|r\|$). Problem 5 gives the bound $\|x - \hat{x}\| \leq \kappa(A)\|r\|/\|A\|$ on the absolute error; enter:

```
» cond(H)*norm(r)/norm(H)
» norm(x-x2)
```

and apparently the absolute error $\|x - \hat{x}\|$ is quite close to the limit. The norm of the residual is a terrible estimate of the size of the error in the computed solution for this matrix; the large condition number serves as our warning. Most importantly, with a condition number of about 10^{13} and starting from double-precision values with about 16 digits of precision, our final answer had a relative error of about 10^{-4}, matching our prediction that starting with 16 digits and losing 13 digits due to the ill-conditioning leaves us with about $16 - 13 \approx 4$ digits of accuracy in x_1.

Some matrices are artificially ill-conditioned. For example, the condition number of any matrix of the form $A = [1\ 0; 0\ \epsilon]$ is always $1/\epsilon$ (if $\epsilon > 0$) which may be made arbitrarily large by taking ϵ as small as desired; however, A will be well-behaved in almost any computation. This is technically an ill-conditioned matrix for small ϵ but the row equivalent matrix $[1\ 0; 0\ 1]$ is not.

Let's look again at the arrowhead matrix of the MATLAB subsection of Sec. 2.3. Enter:

```
» A=eye(10);A(1,:)=ones([1 10]);A(:,1)=ones([10 1])
» [L,U,P]=lu(A)
» cond(A)
» cond(L),cond(U)
```

to form A and find its LU decomposition. The condition numbers of L and U are each larger than that of A, though they would still be considered small. Enter:

```
» cond(P'*L*U)      %This should equal cond(A).
» cond(P*A)         %PA=LU.
```

We might have expected that since L and U have larger condition numbers that their product should too, but of course this cannot be the case since their product is A (actually, PA). This is related to a slightly subtle issue: When performing an LU decomposition, the error in the product LU tends to be much less than the errors in L and U individually. The errors are correlated in such a way that the product is a better approximation to PA than the individual factors are to L and U.

Problems

1.) a.) Given a norm $\| \cdot \|_v$ on $\mathbb{R}^n$ we call the pair $(\mathbb{R}^n, \| \cdot \|_v)$ a **normed space**. By definition, a circle in $(\mathbb{R}^n, \| \cdot \|_v)$ is the set of all $x \in \mathbb{R}^n$ such that $\|x\| = c$ for some $c > 0$. Sketch the unit circle $(c = 1)$ for $\mathbb{R}^2$ with the l_1, l_2, and l_∞ vector norms.

b.) We define distance in a normed space by $d(x,y) = \|x-y\|_v$. Find $d((2,1),(0,0))$ for $\mathbb{R}^2$ with the l_1, l_2, and l_∞ vector norms.

c.) The distance function on the normed space $(\mathbb{R}^n, \| \cdot \|_1)$ is also called the **taxicab distance** (or **Manhattan distance**). Explain why.

2.) a.) The matrix norm $\|A\|_\gamma = \max_{j=1}^n \{\sum_{i=1}^n |a_{ij}|\}$, that is, the maximum column sum, is the matrix norm induced by the l_1 vector norm. The matrix norm induced by the l_∞ vector norm is $\|A\|_\rho = \max_{i=1}^n \left\{\sum_{j=1}^n |a_{ij}|\right\}$, that is, the maximum row sum. For $A = [1\ \ 2\ \ 3; 4\ \ 5\ \ 6; 7\ \ 8\ \ 9]$, find (by hand) $\|A\|_F$, $\|A\|_\gamma$, $\|A\|_\rho$.

b.) Find $\|A\|_s$ by first finding the largest eigenvalue of $A^T A$ (use `max(eig(A'*A))`) and then taking the square root.

3.) a.) Show that if $\|A\| < 1$ in any matrix norm then $A^k \to 0$ (the zero matrix) as $k \to \infty$. (Note: A vector or matrix norm is always a continuous function of the elements of the vector or matrix.)

b.) Show that the spectral radius $\rho(A)$ is not a matrix norm. (Hint: Find a nonzero matrix for which $\rho(A) = 0$.)

c.) Show that if A is nonsingular then $\|A^{-1}\| \geq \|I\|/\|A\|$ in any matrix norm.

d.) Show that $\|I\| = 1$ for any natural norm.

4.) Show that $\kappa_s(A) \leq \kappa_F(A)$ for any nonsingular matrix A, where κ_s is the condition number with respect to the spectral norm and κ_F is the condition number with respect to the Frobenius norm.

5.) Use Eq. (6.3) to show that $\|x - \widehat{x}\| \leq \kappa(A)\|r\|/\|A\|$ is a bound on the absolute error in the computed solution $\widehat{x}$ of $Ax = b$.

6.) a.) Let $b - (1,-1,1,-1,1,-1,1,-1)^T$. Solve $Ax = b$ where A is the Hilbert matrix of order 8 using the LU decomposition (in MATLAB, this may be done by using `A\b`).

b.) What is $\kappa(A)$? Use the `invhilb` command to solve $Ax = b$. Is the relative error in your solution from part a. what you would expect?

c.) Let $b = (1, -1, 1, -1, 1, -1, 1, -1, 1, -1, 1, -1)^T$. Repeat parts a. and b. for the Hilbert matrix of order 12.

7.) For $A = [1 \quad 2 \quad 2; 2 \quad -1 \quad 1; 2 \quad 1 \quad -2]$ find $\|A\|_s$, $\|A\|_F$, $\|A\|_\gamma$, $\|A\|_\rho$, and $\|A\|_1 = \sum_{i=1}^n \sum_{j=1}^n |a_{ij}|$, and the corresponding condition numbers $\kappa_s(A)$, $\kappa_F(A)$, $\kappa_\gamma(A)$, $\kappa_\rho(A)$, and $\kappa_1(A)$. (See Problem 2.)

8.) a.) Show that $\kappa(A) = \kappa(cA)$ for any nonzero scalar c.

b.) Show that $\kappa(A) = \kappa(A^{-1})$ for any nonsingular matrix A.

9.) a.) Gaussian elimination with partial pivoting is generally well-behaved but can in principle give poor results for certain matrices. To construct one of the matrices for which it behaves poorly, set n=5 and use the command A=eye(n)-tril(ones([n n]),-1) followed by A(:,n)=ones([n 1]). Use MATLAB to find the LU decomposition of A and the condition numbers of A, L, and U.

b.) Set n=10 and repeat part a.

c.) Set n=80 and repeat part a.

d.) Use the matrix A from part c. and set x=ones([n 1]);b=A*x;. Solve $Ax = b$ using both x1=A\b and y=L\b;x2=U\y;. Look at x1 and x2; do they seem like good approximations of x (a vector of all ones)? Compute norm(x1-x) and norm(x2-x). How bad are these solutions?

e.) Compute norm(x1-x,1) and norm(x2-x,1). This is the total elementwise error, that is, norm(a-b,1) is equal to sum(abs(a-b)) (the l_1 norm $\|a-b\|_1 = \sum_{i=1}^n |a_i - b_i|$). How bad are these solutions?

f.) This matrix represents a worst case scenario for Gaussian elimination with partial pivoting. Set B=rand(80) and find the condition numbers of B, L, and U.

10.) a.) An orthogonal matrix satisfies $U^T U = I$. Show that $\|Ux\|_2 = \|x\|_2$ for all x if U is orthogonal. (Hint: Consider the quadratic form $x^T Ax$ with $A = U^T U$.)

b.) Show that an orthogonal matrix U has $\|U\|_s = 1$ and $\kappa_s(U) = 1$.

11.) Conduct an experiment in MATLAB that compares the determinant and the condition number as measures of near-singularity. Is the determinant a useful measure of near-singularity for numerical purposes?

12.) a.) Conduct an experiment in MATLAB to determine the expected condition number of an elementwise uniformly distributed (on $[0, 1]$) random $N \times N$ matrix (use rand(N) to form the matrices), using $N = 10$ and $N = 100$.

b.) Conduct an experiment in MATLAB to determine the expected condition number of an elementwise normally distributed (with $\mu = 0$, $\sigma^2 = 100$) random 10×10 matrix (use A=randn(N)/sqrt(N) to form the matrices) for $N = 10$ and $N = 100$.

13.) a.) Prove that $\|A\|_\gamma = \max_{j=1}^n \{\sum_{i=1}^n |a_{ij}|\}$ is the matrix norm induced by the l_1 vector norm.

b.) Prove that $\|A\|_\rho = \max_{i=1}^n \{\sum_{j=1}^n |a_{ij}|\}$ is the matrix norm induced by the l_∞ vector norm.

14.) Show that the Hölder p-norms $\|A\|_p = [\sum_{i=1}^n |a_{ij}|^p]^{1/p}$ are matrix norms if and only if $1 \le p \le 2$. (Hint: You may wish to use the **Hölder inequality** $\sum_{k=1}^n |x_k y_k| \le (\sum_{k=1}^n |x_k|^p)^{1/p} (\sum_{k=1}^n |y_k|^q)^{1/q}$ where $p \ge 1$, $q \ge 1$, and $1/p + 1/q = 1$.)

15.) a.) Show that $\max\{|a_{ij}|\}$ is not a matrix norm (where the maximum is taken over all elements of the matrix).

b.) Show that $n \max\{|a_{ij}|\}$ is a matrix norm on $n \times n$ matrices.

16.) Prove that for a fixed n, all vector norms on $\mathbb{R}^n$ are equivalent. Does your proof include matrix norms?

7. The QR Decomposition and Least Squares

The MATLAB backslash command \ (see `help slash`) generally uses Gaussian elimination with pivoting (in essence, the LU decomposition) to solve $Ax = b$ when A is square. For large square matrices we need another approach, namely, one that does not require $O(n^3)$ flops; we'll discuss such methods in Ch. 3. For now let us look at the problem of solving $Ax = b$ when A is not necessarily square. When A is $m \times n$ and $m \neq n$, the MATLAB backslash command generally uses the **QR decomposition**

$$A = QR$$

of A, where Q is an $m \times m$ orthogonal matrix (meaning that $Q^T Q = I$) and R is an $m \times n$ upper triangular matrix. For example, if A is 5×3 then

$$\begin{bmatrix} X & X & X \\ X & X & X \\ X & X & X \\ X & X & X \\ X & X & X \end{bmatrix} = \begin{bmatrix} X & X & X & X & X \\ X & X & X & X & X \\ X & X & X & X & X \\ X & X & X & X & X \\ X & X & X & X & X \end{bmatrix} \begin{bmatrix} X & X & X \\ 0 & X & X \\ 0 & 0 & X \\ 0 & 0 & 0 \\ 0 & 0 & 0 \end{bmatrix}$$

is the Wilkinson diagram of the QR decomposition. This decomposition is of great importance in numerical linear algebra.

Suppose that we already have a QR decomposition for a *square* matrix A. Then we can solve $Ax = b$ efficiently as follows:

$$\begin{aligned} Ax &= b \\ QRx &= b \\ Rx &= Q^T b \end{aligned}$$

and since R is a square upper triangular matrix this can be solved by back substitution, provided that no diagonal entry of R is zero. Since an orthogonal matrix has determinant ± 1, we have

$$\begin{aligned} \det(A) &= \det(QR) \\ &= \det(Q)\det(R) \\ &= \pm\det(R). \end{aligned}$$

Hence if A is nonsingular ($\det(A) \neq 0$) then $\det(R)$ is nonzero. But R is upper triangular, so $\det(R)$ is equal to the product of the diagonal entries of R; if $\det(R) \neq 0$ then $r_{ii} \neq 0$ for each $i = 1, ..., n$. This shows that the back substitution can be performed whenever A is nonsingular. In fact, if A is nonsingular then a QR decomposition of A must exist.

Let's consider the problem of solving $Ax = b$ when A is $m \times n$ and $m > n$, that is, when there are more equations than unknowns. The product Ax is always a linear combination of the columns of A:

$$Ax = \left[a_1 \vdots a_2 \vdots \cdots \vdots a_n \right] x$$
$$= x_1 a_1 + x_2 a_2 + \ldots + x_n a_n$$

where a_i is column i of A. For example,

$$\begin{pmatrix} 1 & 2 \\ 2 & 1 \end{pmatrix} \begin{pmatrix} 3 \\ -5 \end{pmatrix} = 3 \begin{pmatrix} 1 \\ 2 \end{pmatrix} + -5 \begin{pmatrix} 2 \\ 1 \end{pmatrix}$$
$$= \begin{pmatrix} -7 \\ 1 \end{pmatrix}$$

as you can check by computing the product in the usual row-dot-column manner. This means that the linear system $Ax = b$ can have a solution (that is, the linear system will be consistent) exactly when b may be written as some linear combination of the columns of A. Stated more formally, $Ax = b$ has a solution if and only if

$$b \in \mathrm{Col}(A)$$

where the vector space $\mathrm{Col}(A)$ is the column space of A, that is, the set of all linear combinations of the columns of A (considered as vectors in $\mathbb{R}^m$).

Looked at another way, $Ax = b$ has a solution exactly when the function $y = Ax$, viewed as a function mapping $\mathbb{R}^n$ to $\mathbb{R}^m$, can take on the value $y = b$ for some choice of $x \in \mathbb{R}^n$; that is, b must be in the range of A (the set of all $y \in \mathbb{R}^m$ such that $y = Ax$ for some $x \in \mathbb{R}^n$). If b is not in the range of A then the system is inconsistent, and so there is no solution x of $Ax = b$. For the function $y = Ax$, the range of the function is precisely the column space of the matrix.

But if $m > n$ then the range of $y = Ax$ is only a subset of the $\mathbb{R}^m$. (If A were square and nonsingular, the range would be equal to $\mathbb{R}^m$.) This means that $Ax = b$ will be inconsistent for many, and probably, most choices of $b \in \mathbb{R}^m$. Will we let the fact that no solution exists stop us from finding one?

EXAMPLE 1: Consider an experiment to determine the spring constant k of a spring using Hooke's law $F = k\Delta$ where F is the force applied to the spring and Δ is the resulting deflection from its equilibrium position. We attach an object of known mass to the spring $(F = -mg)$ and measure the resulting displacement Δ of the end of the spring (which will be negative). Then $k = F/\Delta$ is an estimate of the spring constant.

But using one measurement is clearly inferior to using several. Suppose we take three measurements and find the (Δ, F) data $(0.98, 2), (1.20, 2.5), (1.45, 3)$ in some units. We have three estimates $2/1 \doteq 2.0408$, $2.5/1.2 \doteq 2.0833$, and $3/1.6 \doteq 2.0690$ of k. Which is correct? Probably none of them. Averaging them is one way to handle the matter. But fundamentally what we have is the linear system

$$\begin{bmatrix} .98 \\ 1.20 \\ 2.5 \end{bmatrix} [k] = \begin{pmatrix} 2 \\ 2.5 \\ 3 \end{pmatrix}$$

of 3 equations in 1 unknown (namely, k). It is inconsistent and so technically it has no solution but we certainly want to find a value of k nonetheless. $\square$

The problem in Example 1 is much too simple. But problems with $m > n$ occur all the time and are said to be **overdetermined**. (If $m < n$ we say that the problem, or system, is **underdetermined**.) We have too much data, and, because of measurement or other error in the data, we cannot get an exact fit. As another example, if we were processing nuclear magnetic resonance (NMR) data then we might need to estimate the τ_1 relaxation time constant c from the relation

$$y = a + b\exp(-t/c)$$

given a long string of (t_i, y_i) data $(i = 1, ..., m)$. In other words, we would need to find the values of a, b, and c such that

$$
\begin{aligned}
a + b\exp(-t_1/c) &= y_1 \\
a + b\exp(-t_2/c) &= y_2 \\
&\vdots \\
a + b\exp(-t_m/c) &= y_m
\end{aligned}
$$

(7.1)

where typically $m \gg 3$. This is an overdetermined *nonlinear* system of m equations in 3 unknowns, and since no choice of a, b, and c will provide an exact solution for experimental data we would like to find the best (or most likely) values of a, b, and c for the given data. For a nonlinear system like this, we must use the nonlinear optimization methods of Ch. 7 to find the best choice of a, b, and c (that is, the choice that minimizes the error in some sense). This can be difficult.

Our focus in this section is on overdetermined linear systems $Ax = b$ $(m > n)$, which occur often. They can arise when we attempt to fit a polynomial

$$y = a_n t^n + a_{n-1} t^{n-1} + ... + a_1 t + a_0$$

to experimental data (m data pairs (t_i, y_i)) since the polynomial y is linear in its coefficients $a_n, ..., a_0$. It is common to define the "best" near-solution of $Ax = b$ in this case to be that choice of x which minimizes

(7.2) $\|b - Ax\|$

(the value of x for which the residual $r = b - Ax$ is as small as possible). For a variety of reasons we almost always use the Euclidean norm. This is called the **linear least squares solution** of the overdetermined system $Ax = b$. Strictly speaking it is a solution of the minimization problem (Eq. (7.2)) and not the (inconsistent) linear system but we refer to it as the solution of $Ax = b$ "in the sense of least squares".

EXAMPLE 2: Suppose we wish to fit the straight line $y = \alpha x + \beta$ to the data $(1, .9), (2, 2.1), (3, 2.9)$. This means we are seeking α and β such that

$$
\begin{bmatrix} 1 & 1 \\ 2 & 1 \\ 3 & 1 \end{bmatrix}
\begin{bmatrix} \alpha \\ \beta \end{bmatrix}
=
\begin{pmatrix} .9 \\ 2.1 \\ 2.9 \end{pmatrix}
$$

but no such α and β exist. The least squares solution is the value of $(\alpha, \beta)^T$ for which the residual

$$r = \begin{pmatrix} .9 \\ 2.1 \\ 2.9 \end{pmatrix} - \begin{bmatrix} 1 & 1 \\ 2 & 1 \\ 3 & 1 \end{bmatrix} \begin{bmatrix} \alpha \\ \beta \end{bmatrix}$$

$$= \begin{pmatrix} .9 - \alpha - \beta \\ 2.1 - 2\alpha - \beta \\ 2.9 - 3\alpha - \beta \end{pmatrix}$$

has the smallest possible norm

$$\|r\| = \left[(.9 - \alpha - \beta)^2 + (2.1 - 2\alpha - \beta)^2 + (2.9 - 3\alpha - \beta)^2\right]^{1/2}$$

$$= \left[13.63 - 27.6\alpha - 11.8\beta + 14\alpha^2 + 12\alpha\beta + 3\beta^2\right]^{1/2}$$

(simplified by a CAS). This will be minimal when

$$f(\alpha, \beta) = 13.63 - 27.6\alpha - 11.8\beta + 14\alpha^2 + 12\alpha\beta + 3\beta^2$$

is a minimum. Let's take the partial derivatives of f with respect to α and β and set them to zero in order to locate an extremum. From $\nabla f(\alpha, \beta) = 0$ we have the *square* linear system

$$28\alpha + 12\beta - 27.6 = 0$$
$$6\beta + 12\alpha - 11.8 = 0$$

that is

$$28\alpha + 12\beta = 27.6$$
$$12\alpha + 6\beta = 11.8$$

which determines the solution $(\alpha, \beta)^T$ of the least squares problem. That solution is

$$\alpha = 1$$
$$\beta = -1/30$$

so $y = x - 1/30$ is the best line (in the least squares sense). The norm of the residual is

$$\|r\| = \left[(.9 - 1 + 1/30)^2 + (2.1 - 2 + 1/30)^2 + (2.9 - 3 + 1/30)^2\right]^{1/2}$$

$$\doteq .1633$$

which should be the smallest possible. For example, if we were to take $(\alpha, \beta)^T = (1.1, -1/20)^T$ instead then we would have

$$\|r\| = \left[(.9 - 1.1 + 1/20)^2 + (2.1 - 2.2 + 1/20)^2 + (2.9 - 3.3 + 1/20)^2\right]^{1/2}$$

$$\doteq .3841$$

which is indeed larger (hence worse). Note that $Ax \doteq (0.9667, 1.9667, 2.9667)^T$ which is, as expected, different from b, since $Ax = b$ is insoluble. $\square$

How are we to find least squares solutions of linear systems in the general case? We'll sketch the details. It's known in linear algebra that the solution of

$$\min_{x \in \mathbb{R}^n} \|b - Ax\|$$

(Eq. (7.2)) is that value of x that satisfies $Ax = \Pi b$ where Π is the projection matrix from $\mathbb{R}^m$ to the range of A. (Hence Π maps vectors from $\mathbb{R}^m$ to vectors in $\mathrm{Col}(A) \subset \mathbb{R}^m$) This implies that the residual vector $r = b - Ax$ must be orthogonal to every vector in the range of A. Since the range of A is precisely the column space of A, some subset of the columns of A span it. This means that the dot product of r with each column of A must be zero:

$$r^T A = 0.$$

That's the key idea: $r^T A$ is the zero vector. It's more convenient for our purposes to write it in the form

$$
\begin{aligned}
A^T r &= 0 \\
A^T (b - Ax) &= 0 \\
A^T Ax &= A^T b
\end{aligned}
$$

and the square linear system $A^T Ax = A^T b$ is called the **normal equation** (or **normal equations**) for the system $Ax = b$. This is the same system derived when considering least squares from a statistical point of view.

Unless stated otherwise, we'll assume from here on that $m \geq n$ (the system is either square or overdetermined) and that A has full rank (meaning $\mathrm{rank}(A) = n$ since $m \geq n$). The underdetermined case may be handled in essentially the same manner as discussed below but using a QR decomposition of A^T rather than of A.

We now have a means of transforming an overdetermined linear system into a square linear system of the form $A^T Ax = A^T b$; in one sense, we have reduced the problem of solving a nonsquare system to that of solving a square system, and the solution of a square linear system is a problem that we have already addressed. It can be shown that if A has full rank then the $n \times n$ matrix $A^T A$ is nonsingular and hence it will be positive definite[12]. (Furthermore, if $m \gg n$ then the resulting matrix $A^T A$ is small relative to the size of A itself.) Hence to solve $Ax = b$ when $m > n$ we may form the normal equation $A^T Ax = A^T b$ and then find an LU decomposition of $A^T A$. Since $A^T A$ is positive definite it is possible to choose $L = U^T$ and to compute this accurately without pivoting (that is, the permutation matrix P may be taken to be I_n). This means that we may decompose $A^T A$ as

$$A^T A = L^T L$$

using the **Cholesky decomposition** (of $A^T A$). We then solve

$$
\begin{aligned}
A^T Ax &= A^T b \\
L^T Lx &= A^T b
\end{aligned}
$$

for $x = (A^T A)^{-1} A^T b$ in the usual two-step way:

---

[12]Given a matrix A, the matrix $A^T A$ is called its Gram matrix in some contexts.

$$L^T y = A^T b$$
$$Lx = y.$$

(we solve the first equation for y, then the second equation for x). This gives us a relatively straight-forward way to find the least-squares solution of an overdetermined system.

This method is widely used, especially when m is not too large. The good news is that the Cholesky decomposition may be computed using about half the flops needed for a general LU decomposition. But forming $A^T A$ can be inconvenient, and in addition, the solution of the normal equation is sensitive to round-off errors and this approach can exacerbate the problem. (Consider how much data gets packed into the 4 elements of a 2×2 matrix if we have $m = 1000$ data points and $n = 2$ unknowns, as in least-squares line-fitting with 1000 data points. That's a lot of floating point calculations.) This sensitivity is enough to force us to consider other methods.

Consider again the normal equation $A^T A x = A^T b$. If we had a QR decomposition of A then we could write this as

$$
\begin{aligned}
A^T A x &= A^T b \\
(QR)^T (QR)x &= (QR)^T b \\
R^T Q^T Q R x &= R^T Q^T b \\
R^T I R x &= R^T Q^T b \\
R^T R x &= R^T Q^T b
\end{aligned}
$$

since Q is orthogonal ($Q^T Q = I$). We'd like to conclude that the equation

$$Rx = Q^T b$$

must be satisfied by the solution of $A^T A x = A^T b$ but in general R is not even square so we can not use multiplication by $(R^T)^{-1}$ to arrive at this conclusion. In fact, it is not true in general that the solution of $Rx = Q^T b$ even exists; after all, $Ax = b$ is equivalent to $QRx = b$, that is, to $Rx = Q^T b$, so $Rx = Q^T b$ can have an actual solution x only if $Ax = b$ does. However, we are close. We need to find a way to simplify the expression $R^T R x = R^T Q^T b$.

The matrix R is upper triangular and since we have restricted ourselves to the case $m \geq n$ we may write the $m \times n$ matrix R as

$$R = \begin{pmatrix} R_1 \\ 0 \end{pmatrix}$$

in partitioned (block) form, where R_1 is an upper triangular $n \times n$ matrix and 0 represents an $(m - n) \times n$ zero matrix. Since $\text{rank}(R) = n$ it must be the case that $\text{rank}(R_1) = n$, so the square matrix R_1 is nonsingular. Hence every diagonal entry of R_1 must be nonzero. Now we may write $R^T R x = R^T Q^T b$ as

$$\begin{pmatrix} R_1 \\ 0 \end{pmatrix}^T \begin{pmatrix} R_1 \\ 0 \end{pmatrix} x = \begin{pmatrix} R_1 \\ 0 \end{pmatrix}^T Q^T b$$

$$\begin{pmatrix} R_1^T & 0^T \end{pmatrix} \begin{pmatrix} R_1 \\ 0 \end{pmatrix} x = \begin{pmatrix} R_1^T & 0^T \end{pmatrix} Q^T b$$

$$R_1^T R_1 x = \begin{pmatrix} R_1^T & 0^T \end{pmatrix} (Q^T b)$$

(verify the conformability of the matrix blocks). Note that multiplying by the block 0^T (an $n \times (m - n)$ zero matrix) on the RHS simply means that the last $(m - n)$ components of $Q^T b$ do not affect the computation. Since R_1 is nonsingular, we have

$$R_1 x = (R_1^T)^{-1} \begin{pmatrix} R_1^T & 0^T \end{pmatrix} (Q^T b)$$
$$= \begin{pmatrix} I_n & 0^T \end{pmatrix} (Q^T b).$$

The LHS $R_1 x$ is $(n \times n) \times (n \times 1) \to n \times 1$ and the RHS is $(n \times (n + (m - n)) \times (m \times m) \times (m \times 1) \to n \times 1$; this checks. In fact, this is the equation we want. If we define the vector q to be equal to the first n components of $Q^T b$ then this becomes

(7.3) $$R_1 x = q$$

which is a square linear system involving a nonsingular upper triangular $n \times n$ matrix. That's about as simple as we could hope for, and we refer to Eq. (7.3) as the **QR equation**. Its solution

$$x = R_1^{-1} q$$

(found by back substitution) is the least squares solution of $Ax = b$. We repeat that $Ax = b$, that is, $QRx = b$, will not usually have a solution in the strict sense when $m > n$.

This is important; (Eq. (7.3)) states that the least squares solution of $Ax = b$ is precisely the solution of $R_1 x = q$ as found by back substitution, where $A = QR$ is the QR decomposition of A and q is essentially $Q^T b$. Because we do not need to form $A^T A$, and because Q and R can be found accurately, the resulting computed solution is generally more accurate than the solution of the normal equation as found by Cholesky decomposition. There's actually a deeper connection; $R_1^T R_1$ actually *is* the Cholesky decomposition of $A^T A$ and so the QR decomposition approach may be viewed as simply a more accurate means of employing the Cholesky decomposition approach.

EXAMPLE 3: Consider again Example 2, where we wished to fit the straight line $y = \alpha x + \beta$ to the data $(1, .9), (2, 2.1), (3, 2.9)$. This led to the overdetermined system

$$\begin{bmatrix} 1 & 1 \\ 2 & 1 \\ 3 & 1 \end{bmatrix} \begin{bmatrix} \alpha \\ \beta \end{bmatrix} = \begin{pmatrix} .9 \\ 2.1 \\ 2.9 \end{pmatrix}$$

in which the coefficient matrix A clearly has full rank. Let's use the QR decomposition approach on this problem. If we enter the 3×2 matrix A as above then the MATLAB command [Q,R]=qr(A) produces:

```
Q =
    -0.2673 0.8729 0.4082
    -0.5345 0.2182 -0.8165
    -0.8018 -0.4364 0.4082
R =
    -3.7417 -1.6036
    0 0.6547
    0 0
```

so that $R_1 = [-3.7417 \ -1.6036; 0 \ 0.6547]$. We find

$$Q^T b = (-3.6882, -0.0218, -0.1633)^T$$

and so $q = (-3.6882, -0.0218)^T$. Hence we must solve $R_1 x = q$, that is,

$$(7.4) \qquad \begin{bmatrix} -3.7417 & -1.6036 \\ 0 & 0.6547 \end{bmatrix} \begin{bmatrix} \alpha \\ \beta \end{bmatrix} = \begin{pmatrix} -3.6882 \\ -0.0218 \end{pmatrix}$$

giving $\alpha \doteq 1.0000$, $\beta \doteq -0.0333$ as expected (where all computations were performed in double precision; solving Eq. (7.4) using only 4 digit arithmetic gives the less accurate $\alpha \doteq 0.9714$). $\square$

We still haven't discussed how to find a QR decomposition. We'll do that in the next section. For now we mention that, as the methods are usually implemented, finding the least squares solution of $Ax = b$ by Cholesky decomposition requires about half as many flops as using the QR decomposition when $m \gg n$; both methods require about $4n^3/3$ flops if $m = n$ (and n is large). Since $m \gg n$ is typical in cases of interest, the Cholesky decomposition approach will usually be faster, despite it having to first form $A^T A$. On the other hand, the QR decomposition is likely to be more accurate as it is less sensitive to rounding errors, and so a factor of 2 in computation time may be a reasonable price to pay for this increased accuracy; in fact, the QR decomposition is the preferred approach in practice.

EXAMPLE 4: Let H=hilb(20), A=H(1:20,1:10) and x=ones([10 1])-.5 and b=A*x (A is 20×10 and b is a 10×1 vector; we ignore the errors introduced into b by computing it from Ax). If we use the MATLAB command A\b, which uses the QR decomposition, the relative error in the computed solution is almost 10^{-6}. Since $\kappa(A) \approx 2.6x10^{11}$ we may have to live with this relatively poor quality (16 − 11 gives about 6 significant figures in this case). If we use the Cholesky decomposition approach (A'*A)\(A'*b) (MATLAB employs the Cholesky method automatically here) then the relative error in the computed solution is about 1.7. The absolute error in the l_∞ norm is about 2.4 meaning that at least one component of the computed solution differs from the correct solution x by nearly 2 units! This matches expectations in one sense, since $\kappa(A^T A) \approx 10^{17}$ meaning that we should expect to lose about 17 digits (out of a possible 16) in performing this computation. In numerical analysis and in scientific computation, 16 digits minus 17 digits equals 0 digits of accuracy remaining in the solution. $\square$

The case of Example 4 is extreme; still, the Cholesky decomposition approach can amplify errors beyond what the condition number of A would predict because it works with $A^T A$, not A, and as a rule of thumb,

$$\kappa\left(A^T A\right) \approx \left[\kappa\left(A\right)\right]^2$$

meaning that ill-conditioned matrices become much more so. But the QR decomposition of A generally gives more accurate results than use of the Cholesky decomposition of $A^T A$, even when we are not concerned about the additional round-off errors (and expense) associated with the formation of $A^T A$. In fact, if A is nonsingular then solving

$$Ax = b$$

by QR decomposition is generally a little more expensive, and a little more accurate, than using the LU decomposition of A.

The QR decomposition is not unique; after all, if $A = QR$ then $A = (-Q)(-R)$ also, and $-Q$ is orthogonal if Q is orthogonal. Under our assumptions ($m \geq n$ and rank$(A) = n$) there exists a unique QR decomposition of A for which the diagonal entries of R are positive. It's conventional to add the requirement that the diagonal entries of R must be positive to the definition of a QR decomposition (i.e., $A = QR$ where Q is orthogonal and R is upper triangular with positive diagonal entries) when A has full rank, but this convention isn't always used–in fact, MATLAB doesn't insist that the diagonal entries of R be positive.

Finally, note that the QR decomposition (or **QR factorization**) of A results in an $m \times n$ matrix R that will typically contain a large block of zeroes which will be removed when we actually use R; that is, if we know m and n then we really only need R_1 to reconstruct R if desired. Similarly, in forming q from the first n components of $Q^T b$ we are in effect partitioning the $m \times m$ matrix Q as

$$(7.5) \qquad\qquad Q = [Q_1 \ Q_0]$$

where Q_1 is an $m \times n$ matrix with orthonormal columns and Q_0 is an $m \times (m - n)$ matrix with orthonormal columns, and computing

$$\begin{aligned} Q^T b &= [Q_1 \ Q_0]^T b \\ &= [Q_1^T b \ Q_0^T b] \\ &= [q \ Q_0^T b] \end{aligned}$$

so that $q = Q_1^T b$. In addition,

$$\begin{aligned} QR &= [Q_1 \ Q_0] \begin{pmatrix} R_1 \\ 0 \end{pmatrix} \\ &= Q_1 R_1 + Q_0 0 \\ &= Q_1 R_1 \end{aligned}$$

(verify the conformability for each of these products); that is, $A = Q_1 R_1$. Hence we may write Eq. (7.3) as

$$R_1 x = Q_1^T b$$

which (under our assumptions) is a square system involving a nonsingular upper triangular matrix R_1. We say that $A = Q_1 R_1$ is the **reduced QR decomposition** (or **reduced QR factorization**) of A, and that $A = QR$ is the **full QR decomposition** (or **full QR factorization**) of A. The reduced QR factorization is the version that is commonly used in practice.

MATLAB

Type `help slash` to read a description of the slash command. Note that there is also a forward slash operator `/` that may be used to solve problems in the form $y^T A = b$. Both commands can also be used with matrices, e.g., if $AB = C$ then `B=A\C` (in the least squares sense if necessary).

The MATLAB command `qr` computes the full QR decomposition of a matrix. For example, enter:

```
» [Q,R]=qr(rand([7 4]))
```

The `qr` command does not force r_{ii} to be positive. The reduced QR factorization may be found using the `qr` command as well; enter:

```
» [Q1,R1]=qr(rand([7 4]),0)
```

to find the factors in the reduced form (or as MATLAB describes it, "economy size").

Let's compare the Cholesky and QR approaches. Enter:

```
» A=rand([5 3])
» M=A'*A
» X=chol(M)
» X'*X               %Should equal M.
» [Q,R]=qr(A)
» R'*R               %Should equal M.
» [Q1,R1]=qr(A,0)
» R1'*R1             %Should equal M.
» norm(M-X'*X)       %Error in M from Cholesky.
» norm(M-R'*R)       %Error in M from QR.
» norm(M-R1'*R1)     %Error in M from reduced QR.
```

For this small matrix all the errors will be in the level of noise. Note that R and X are the same save, possibly, for some minus signs that cancel out when $X^T X$ and $R^T R$ are formed.

Let's step through the details of Example 4 in MATLAB. (Look again at Example 4 first.) Enter:

```
» H=hilb(20);
» A=H(1:20,1:10)       %A is an ill-conditioned 20x10 matrix.
» cond(A)
» x=ones([10 1])-.5;
» b=A*x
» xqr=A\b              %Solve Ax=b by QR decomposition.
» xchol=(A'*A)\(A'*b)  %Solve Ax=b by Cholesky decomposition.
» norm(xqr-x)/norm(x)
» norm(xchol-x)/norm(x)
» norm(xchol-x,'inf')
» cond(A'*A)
```

The QR decomposition (used by the slash command when A is nonsquare) gives a much more accurate solution than solving the normal equation by Cholesky decomposition (used by the slash command for the square matrix $A^T A$) in this case; bear in mind however that this is an especially ill-conditioned example.

Let's find the least squares solution of an inconsistent overdetermined linear system by the reduced QR factorization step-by-step. (The slash command would do this for us automatically.) Enter:

```
» A1=magic(5)      %5x5 magic square.
» A2=pascal(5)     %5x5 Pascal's triangle as a matrix.
» A=[A1;A2]        %Concatenate along vertical axis.
» b=ones([10 1]);
» [Q1,R1]=qr(A,0)
» x=R1\(Q1'*b)     %Solve R1*x=Q1'*b.
» norm(b-A*x)      %Residual.
```

As we've mentioned before, a simple flops count between two methods can be very misleading. Memory management issues are often what dominate our concerns–how efficiently the algorithm implemented makes use of the memory architecture of the machine. More generally, even forming A itself is sometimes the longest part of the problem. Enter `type hilb` to see that forming the $n \times n$ Hilbert matrix in the manner used by MATLAB requires n^2 divisions and a like number of additions and subtractions (`H = E./(I+J-1)`, that is, $h_{ij} = 1/(i+j-1)$); this means that forming the matrix H is already an operation that requires $O(n^2)$ flops. If the entries a_{ij} of the matrix A in $Ax = b$ are given by a more complicated formula than this then the act of forming A may be what dominates the computation, as opposed to actually solving a system after A is formed. This is often true in data science applications, in which databases must be accessed and read to decide what to put into entry (i, j) of A (e.g., did customer i watch movie j?). In our case of comparing the Cholesky and QR approaches, note that $A^T A$ must be symmetric so if we code it up rather than using MATLAB's built-in multiplication operator then we can save half of the computations involved in forming it.

We mention a few other relevant MATLAB commands (see `help matfun`): The `rank` command determines the rank of a matrix (see `help rank` and Chapter 3 for details). The problem of determining the rank of a matrix numerically is more difficult than one might imagine at first. The command `orth(A)` returns an orthonormal matrix with the same range (column space) as `A`. The `lsqnonneg` command may be used to minimize Eq. (7.2) subject to a nonnegativity constraint, that is, to find $\min_{x \geq 0} \|b - Ax\|$.

Problems

1.) a.) If A is a 7×4 matrix with full rank, what are the sizes of x and b in $Ax = b$?
b.) What are the sizes of Q, R, Q_1, R_1, and q?
c.) What are the Wilkinson diagrams for the relations $A = QR$ and $A = Q_1 R_1$?
2.) a.) If A is a 4×7 matrix with rank 3, what are the sizes of x and b in $Ax = b$?
b.) What are the sizes of Q, R, Q_1, R_1, and q?
c.) What are the Wilkinson diagrams for the relations $A = QR$ and $A = Q_1 R_1$?
3.) Derive the QR equation $R_1 x = Q_1^T b$ by substituting the reduced QR factorization of A into the normal equation $A^T Ax = A^T b$.
4.) a.) What matrix A has $Q_1 = [1/\sqrt{2} \ \ 0; 0 \ \ 1; -1/\sqrt{2} \ \ 0]$ and $R_1 = [1 \ \ 2; 0 \ \ 3]$ as its reduced QR factorization?
b.) What is the full QR decomposition of A? (Hint: The only two 3×3 orthogonal matrices extending Q_1 have $\pm[1/\sqrt{2} \ \ 0 \ \ 1/\sqrt{2}]^T$ as their third column.)
c.) Use the reduced QR factorization of A to find the least squares solution of $Ax = (1, 1, 1)^T$. What is the norm of the residual (Eq. (7.2))?
d.) Verify that $R^T R = R_1^T R_1 = A^T A$.
e.) Find the least squares solution of $Ax = (1, 1, 1)^T$ by forming the normal equation and using the Cholesky decomposition.

5.) a.) Consider the orthogonal matrix $Q = [1/\sqrt{2}\ 1/\sqrt{2}\ 0; 0\ 0\ 1; 1/\sqrt{2}\ -1/\sqrt{2}\ 0]$ and the right triangular matrix $R = [1\ 2\ 2; 0\ 2\ -1; 0\ 0\ 2]$. Find the determinant of $A = QR$ (do not form A).
b.) Solve $Ax = (1, -1, 1)^T$ (do not form A).
c.) Find A.
6.) a.) What matrix A has $Q_1 = [1/\sqrt{3}\ \ 0; 1/\sqrt{3}\ \ 1/\sqrt{2}; -1/\sqrt{3}\ \ 1/\sqrt{2}]$ and $R_1 = [2\ 2; 0\ 1]$ as its reduced QR factorization?
b.) What is the full QR decomposition of A?
c.) Use the reduced QR factorization of A to find the least squares solution of $Ax = (1, -2, 1)^T$. What is the norm of the residual?
d.) Verify that $R^T R = R_1^T R_1 = A^T A$.
e.) Find the least squares solution of $Ax = (1, -2, 1)^T$ by forming the normal equation and using the Cholesky decomposition.
7.) a.) If A is an upper triangular matrix of full rank with at least as many rows as columns, what is its unique QR decomposition for which r_{ii} is positive for all i?
b.) If A is an orthogonal matrix, what is its unique QR decomposition for which r_{ii} is positive for all i?
8.) Show that if an $m \times n$ matrix A for which $m \geq n$ has a QR decomposition then it has a QR decomposition in which all diagonal entries of R are nonnegative.
9.) a.) Show that if Q is orthogonal then $\det(Q) = \pm 1$.
b.) Show that if Q is orthogonal then $-Q$ is orthogonal.
c.) Show that if Q_1 is an $m \times n$ matrix with orthonormal columns then $Q_1^T Q_1 = I_n$.
10.) As with the LU decomposition, it is not uncommon to find that the computed Q and R of a QR decomposition, call them Q^* and R^*, are relatively poor approximations in the sense that $\|Q - Q^*\|$ and $\|R - R^*\|$ are large but are good approximations in the sense that $\|A - Q^* R^*\|$ is small. (The errors in Q^* and R^* are correlated.) Construct an example that demonstrates this effect.
11.) A more careful operation count for Cholesky decomposition is approximately $mn^2 + \frac{1}{3}n^3$ flops (if the symmetry in $A^T A$ is exploited) and for the QR decomposition it is approximately $2mn^2 - \frac{2}{3}n^3$ flops, neglecting lower-order terms. For which choices $m \geq n$ is the Cholesky decomposition more efficient than the QR decomposition by this count?
12.) a.) Show that if u and v are orthogonal vectors then $\|u + v\|^2 = \|u\|^2 + \|v\|^2$ for the Euclidean norm. (Hint: Recall that $\|x\|^2 = x^T x$.)

b). Show that if the column vector v is partitioned as $v = (v_1^T : v_2^T)^T$ then $\|v\|^2 = \|v_1\|^2 + \|v_2\|^2$.
c). Consider the overdetermined linear system $Ax = b$. Since orthogonal matrices Q satisfy $\|Qx\| = \|x\|$ for the Euclidean norm, the norm of the residual $\|r\| = \|b - Ax\|$ satisfies $\|r\|^2 = \|b - Ax\|^2 = \|Q^T(b - Ax)\|^2$ (where $A = QR$ is the QR decomposition of A). Simplify $\|r\|^2 = \|Q^T(b - Ax)\|^2$ using Eq. (7.5) and part b. to the form $\|r\|^2 = \|q - R_1 x\|^2 + \|Q_0 b\|^2$.
d). Use the result of part c. to argue that the solution of the QR equation $R_1 x = q$ is the least squares solution of $Ax = b$. What formula does this give for the residual?
13.) a.) The MATLAB qr command does not enforce the convention $r_{ii} > 0$ in the QR decomposition $A = QR$. If M is the identity matrix save that wherever $r_{ii} < 0$ we set $m_{ii} = -1$ then how can we use M to obtain a QR decomposition $A = UV$ with U orthogonal and V right triangular and $v_{ii} > 0$?

b.) Write a MATLAB program that inputs a matrix, finds its QR decomposition via qr, and returns a factor R that follows the convention that $r_{ii} > 0$. Don't construct the matrix from part a. within your program.

14.) Write a program that accepts a positive definite matrix A and returns its reduced QR decomposition by using qr(A) within the program and then manipulating the results. (Your program is duplicating the functionality of qr(A,0).)

15.) Show that if A is positive definite then there is a positive definite matrix S such that $S^2 = A$. (This is called the matrix square root of A; see Sec. 2.5.) Use the MATLAB command sqrtm to compute the matrix square root of $A =$ [4 2 1;2 3 1;1 1 4].

16.) Suggest a method for solving the underdetermined linear system $Ax = b$ when it is consistent using the QR decomposition of A^T. (Consider making the unitary change of variables $y = Q^T x$.) Typically there are infinitely many solutions and we seek the one that minimizes $\|x\|$.

8. The QR Decomposition by Triangularization

In the last section we saw that the QR decomposition is a useful direct method for solving overdetermined linear systems; it is also used in algorithms to find eigenvalues. (See Ch. 3.) There are two major approaches to computing the QR decomposition, and we will describe one of them in this section. This is the more commonly employed approach; the second major approach, based on the Gram-Schmidt process, is discussed in the next section.

We first mention a few brief facts about orthogonal matrices: A $p \times q$ matrix M is said to be orthogonal if $M^T M = I_q$. Then $MM^T = I_p$, and the columns of M are orthonormal (not merely orthogonal). It follows that $\|Mx\| = \|x\|$ (the transformation by M preserves lengths). If $p = q$, M has determinant ± 1 and hence is nonsingular, all eigenvalues lie on the unit circle, $\|M\| = 1$ in any natural matrix norm, and $M^{-1} = M^T$.

We will continue to assume (unless stated otherwise) that A is an $m \times n$ matrix with $m \geq n$ and full rank n. Consider again the case of the LU decomposition where pivoting is not needed. We started with a matrix A that we wanted to reduce to triangular form using lower triangular matrices, so that we would have $MA = U$ when we were finished, with M lower triangular; we could then solve for $A = LU$ (where $L = M^{-1}$ was also lower triangular). The first step had the form

$$L_1 A = \begin{bmatrix} 1 & 0 & 0 \\ X & 1 & 0 \\ X & 0 & 1 \end{bmatrix} \begin{bmatrix} X & X & X & X \\ X & X & X & X \\ X & X & X & X \end{bmatrix} = \begin{bmatrix} X & X & X & X \\ 0 & X & X & X \\ 0 & X & X & X \end{bmatrix}$$

(previously we had restricted ourselves to square matrices but the LU decomposition may be applied to nonsquare matrices as well). At the next stage we really need only process the 2×3 block in the southeast using a 2×2 lower triangular matrix; we may represent this as

$$(8.1) \qquad \begin{bmatrix} 1 & {}_1 0_2 \\ {}_2 0_1 & L_2 \end{bmatrix} \begin{bmatrix} X & {}_1 X_3 \\ {}_2 0_1 & A_1 \end{bmatrix} = \begin{bmatrix} X & X & X & X \\ 0 & X & X & X \\ 0 & 0 & X & X \end{bmatrix}$$

where L_2 is the 2×2 unit lower triangular matrix $L_2 = [1 \ 0; -m_{32} \ 1]$ and A_2 is the southeast 2×3 block of the result $L_1 A$ of the first step. The matrix $_k 0_p$ represents a $k \times p$ zero matrix, and $_1 X_3 = [X \ X \ X]$.

Verify the conformability of the matrices in Eq. (8.1). It is important to be able to multiply matrices when they are partitioned into blocks of sub-matrices, as we have already seen in the previous section. Note how this emphasizes the iterative nature of Gaussian elimination–the same process is applied to smaller and smaller blocks of the current matrix.

In the more general case, a typical step of the LU decomposition on an $m \times n$ matrix with $m \geq n$ (for which pivoting is not needed) has the form

$$(8.2) \qquad \begin{bmatrix} I_r & _r 0_{m-r} \\ _{m-r} 0_r & L_{r+1} \end{bmatrix} \begin{bmatrix} X_a & X_b \\ 0 & X_c \end{bmatrix} = \begin{bmatrix} X_\alpha & X_\beta \\ _{m-r-1} 0_{r+1} & X_\gamma \end{bmatrix}.$$

where each entry is itself a matrix. We have processed r rows and are beginning to work on row $r + 1$ using the $(m - r) \times (m - r)$ unit lower triangular matrix L_{r+1}. Note that the entire matrix

$$\begin{bmatrix} I_r & _r 0_{m-r} \\ _{m-r} 0_r & L_{r+1} \end{bmatrix}$$

is unit lower triangular. The matrix to which it is applied,

$$\begin{bmatrix} X_a & X_b \\ _{m-r} 0_r & X_c \end{bmatrix}$$

is part-way to being upper triangular; the matrix X_a is $r \times r$ and is upper triangular (it and the zeroes below it represent the completed part of the matrix). The matrix X_b is $r \times (m - r)$ and represents the remaining entries of the first r rows that are no longer being used in the processing of A. The matrix X_c $((m - r) \times (m - r))$ is the part of the original matrix A, now changed by the row operations that brought us this far, that must still be triangularized. Performing the multiplication will give the result

$$\begin{bmatrix} X_\alpha & X_\beta \\ _{m-r-1} 0_{r+1} & X_\gamma \end{bmatrix}$$

where X_α is $(r + 1) \times (r + 1)$ (we have gained an additional column in upper triangular form, that is, that has all entries below the diagonal entry zero), X_β is $(r + 1) \times (m - r - 1)$, and X_γ is $(m - r - 1) \times (m - r - 1)$. The block X_γ is the part that must still be triangularized, and is one row and one column smaller than X_c was $((m - (r + 1)) \times (m - (r + 1))$ rather than $(m - r) \times (m - r))$.

Once again, please carefully verify the conformability of the matrices in Eq. (8.2). This block matrix form of the LU decomposition is useful for analyzing the method and for preparing to make use of options such as Level 3 BLAS (matrix-matrix operations). We have omitted pivoting only to emphasize the ideas without the details.

One of the key take-aways from this representation is this: Gaussian elimination (or the LU decomposition) may be viewed as a process of using successive (lower) triangular matrices to (upper) triangularize a matrix–it works because the process used to make A into the triangular matrix U is based on multiplications by matrices of exactly the form that we hope to obtain when we are done (lower triangular). The process may be described as one of *triangular triangularization,* that is, triangularization by triangular matrices.

What does this have to do with the QR decomposition? It suggests a way of finding a method for computing the decomposition. What if we could find an orthogonal matrix Q_1 that has an action similar to that of the unit lower triangular matrix in Eq. (8.2), in that the result $Q_1 A$ is one step closer to being upper triangular? Similar to Eq. (8.2) we would have an iterative process of the form

$$Q_{r+1} \begin{bmatrix} X & X & X & X & X \\ 0 & X & X & X & X \\ 0 & 0 & X & X & X \\ 0 & 0 & X & X & X \\ 0 & 0 & X & X & X \end{bmatrix} = \begin{bmatrix} X & X & X & X & X \\ 0 & X & X & X & X \\ 0 & 0 & X & X & X \\ 0 & 0 & 0 & X & X \\ 0 & 0 & 0 & X & X \end{bmatrix}$$

i.e.,

$$Q_{r+1} \begin{bmatrix} X_a & X_b \\ {}_{m-r+1}0_r & X_c \end{bmatrix} = \begin{bmatrix} X_\alpha & X_\beta \\ {}_{m-r}0_{r+1} & X_\gamma \end{bmatrix}$$

for some orthogonal matrix Q_{r+1}. (The matrix Q_{r+1} would likely have a block structure of its own.) If so then we might be able to find a sequence of such matrices that would triangularize A,

(8.3) $$Q_{m-1} \cdots Q_2 Q_1 A = R$$

in which case

$$\begin{aligned} A &= (Q_{m-1} \cdots Q_2 Q_1)^{-1} R \\ &= (Q_1^T Q_2^T \cdots Q_{m-1}^T) R \\ &= QR \end{aligned}$$

is a QR decomposition of A. (Transposes of square orthogonal matrices are themselves orthogonal.) This would be a process of *orthogonal triangularization* of A (that is, making A triangular using orthogonal matrices).

Can we carry out such a program? And if so, can we do it in a manner that does not amplify round-off errors to the extent that the method is useless on an actual machine? Yes.

EXAMPLE 1: Consider a 2×2 matrix with rank 2. Making it upper triangular means nothing more than eliminating the $(2,1)$ entry. If $A = [a \quad b; c \quad d]$ then we seek an orthogonal matrix $Q_1 = [q_{11} \quad q_{12}; q_{21} \quad q_{22}]$ such that $Q_1 A$ is upper triangular. The $(2,1)$ entry of this product is the dot product of row 2 of Q_1 with row 1 of A, namely $(q_{21}, q_{22}) \cdot (a, c) = a q_{21} + c q_{22} = 0$. If c is nonzero (the only case of interest) then we may write this as $q_{22} = -a q_{21}/c$. The only other condition that must be satisfied is the orthogonality of Q_1. It is a fact that every real 2×2 orthogonal matrix is either of the form $H(\theta) = [-\cos(\theta) \quad \sin(\theta); \sin(\theta) \quad \cos(\theta)]$ or $G(\theta) = [\cos(\theta) \quad -\sin(\theta); \sin(\theta) \quad \cos(\theta)]$ for some angle θ. The condition $q_{22} = -a q_{21}/c$ means $\cos(\theta) = -a \sin(\theta)/c$, that is, $\cot(\theta) = -a/c$ in either case.

So, in particular, if $A = [1 \quad 2; 3 \quad 4]$ then $H = [-\cos(\theta) \quad \sin(\theta); \sin(\theta) \quad \cos(\theta)]$ with $\theta = \cot^{-1}(-1/3) \doteq -1.2490$, so $H = [-0.3162 \quad -0.9487; -0.9487 \quad 0.3162]$. This gives $HA = [-3.1623 \quad -4.4272; 0.0000 \quad -0.6325]$ which is upper triangular as expected (call it R_H).

What if we use $G(\theta) = [0.3162 \; 0.9487; -0.9487 \; 0.3162]$? Then the resulting upper triangular matrix is $GA = [3.1623 \; 4.4272; 0.0000 \; -0.6325]$ (call it R_G).

Both $HA = R_H$ and $GA = R_G$ define QR decompositions $A = H^T R_H$ and $A = G^T R_G$ of A. Neither has r_{11} and r_{22} positive; in fact, R_H has both diagonal entries negative. By using $(-I)^2 = I$, the matrix analogue of $(-1)^2 = 1$, we may write

$$
\begin{aligned}
A &= H^T R_H \\
&= H^T I R_H \\
&= H^T (-I)(-I) R_H \\
&= H^T (-I)(-I) R_H \\
&= (-H^T)(-R_H)
\end{aligned}
$$

to obtain a QR decomposition of the desired form. The orthogonal matrix $-H^T$ corresponds to $[-\cos(\theta_1) \; \sin(\theta_1); \sin(\theta_1) \; \cos(\theta_1)]$ where $\theta_1 = \cot^{-1}(-1/3) + \pi$.

Nothing changes if the matrix to be decomposed is $2 \times n$; for example $(n = 3)$ if $B = [1 \; 2 \; -1; 3 \; 4 \; 2]$ then the same H as before gives $HB = [-3.1623 \; - 4.4272 \; - 1.5811; 0.0000 \; - 0.6325 \; 1.5811]$ which is upper triangular. Hence $B = (-H^T)(-HB)$ is a QR decomposition of B with positive diagonal entries in R. $\square$

We now have a means of finding a QR decomposition for 2x2 matrices; we need to generalize it to a method for finding the QR decomposition of $m \times n$ matrices. Methods based on generalizing the H matrices (which have $\det(H) = -1$ and represent a reflection in the plane) are different from methods based on generalizing the G matrices (which have $\det(G) = 1$ and represent a rotation in the plane). We'll generalize the H matrices first.

A **Householder matrix** (or **Householder reflector** or **Householder reflection** or **elementary reflector**) is a matrix of the form

$$
H_w = I_m - 2ww^T
$$

for some unit vector $w \in \mathbb{R}^m$. Householder matrices have a number of interesting properties: They are symmetric and orthogonal (which implies that $H_w^2 = I$, making H_w **idempotent**). In addition, if x is nonzero then $H_w x$ is the vector obtained by reflecting x across a certain hyperplane, namely, the hyperplane through the origin that is orthogonal to w. (Recall that a point–here the origin–and a nonzero vector determine a plane.) A Householder reflector algebraically performs the act of reflecting a vector across a plane.

We can choose a w that gives a desired result for the product $H_w x$. If u and v are linearly independent vectors in $\mathbb{R}^m$ and $\|u\|_2 = \|v\|_2$, and if we take w to be the normalized vector from u to v,

(8.4)
$$
w = \frac{v - u}{\|v - u\|_2}
$$

then

$$
H_w u = v.
$$

This is precisely the result we need. The first step of our scheme (Eq. (8.3)) should construct an orthogonal matrix Q_1 that zeroes out every entry below the $(1,1)$ entry of A:

$$\begin{bmatrix} X & X & X \\ X & X & X \\ X & X & X \end{bmatrix} \begin{bmatrix} X & X & X \\ X & X & X \\ X & X & X \end{bmatrix} = \begin{bmatrix} X & X & X \\ 0 & X & X \\ 0 & X & X \end{bmatrix}$$

(the right-hand side is Q_1A). Write A in the partitioned form $A = [a_1 \ A_{12}]$ where a_1 is the first column of A. If a_1 is the zero vector, we may take $Q_1 = I$ (that is, nothing needs to be done so we do nothing) and move on. Otherwise, we need Q_1 to be an orthogonal matrix that maps a_1 to $\alpha_1 e_1$ (for some scalar α_1, with e_1 the first column of I_m). Let's take $\alpha_1 = \pm\|a_1\|_2$, for then the vectors a_1 and $\alpha_1 e_1$ have the same length, and so from Eq. (8.4) a matrix that maps a_1 to $\alpha_1 e_1$ is H_w with

$$w = \frac{\alpha_1 e_1 - a_1}{\|\alpha_1 e_1 - a_1\|_2}$$

(for either choice of α_1). Then

$$\begin{aligned} H_w a_1 &= (I - 2ww^T)a_1 \\ &= \alpha_1 e_1 \end{aligned}$$

(and $H_w A_{12}$ is some matrix so that $H_w A = [\alpha_1 e_1 \ H_w A_{12}]$). We have constructed the desired orthogonal matrix Q_1 needed for Eq. (8.3).

EXAMPLE 2: Consider the matrix $A = [1 \ 2 \ 3; 4 \ 5 \ 6; 7 \ 8 \ 9]$. For the first step of QR decomposition using orthogonal triangularization by Householder reflectors, we set $a_1 = (1,4,7)^T$ for which $\alpha_1 = \pm[1^2 + 4^2 + 7^2]^{1/2} = \pm\sqrt{66} \doteq \pm 8.1240$. Since we must form the difference $\alpha_1 e_1 - a_1$ to find H_{w_1}, we may as well take $\alpha_1 < 0$ so that the only nonzero entry of $\alpha_1 e_1$, the first entry, is opposite in sign to the corresponding entry in a_1. This means that no subtraction (and hence potential cancellation of significant figures) can occur in that entry. We find

$$\begin{aligned} w &= [(-\sqrt{66},0,0)^T - (1,4,7)^T]/\|(-\sqrt{66},0,0)^T - (1,4,7)^T\| \\ &= -(\sqrt{66}+1,4,7)^T/\|-(\sqrt{66}+1,4,7)^T\| \\ &\doteq (-0.7494, -0.3285, -0.5749)^T \end{aligned}$$

and so taking this as w_1 gives

$$\begin{aligned} H_{w_1} &= I - 2(-0.7494, -0.3285, -0.5749)^T(-0.7494, -0.3285, -0.5749) \\ &\doteq \begin{bmatrix} -0.1231 & -0.4924 & -0.8616 \\ -0.4924 & 0.7841 & -0.3777 \\ -0.8616 & -0.3777 & 0.3389 \end{bmatrix} \end{aligned}$$

which is indeed an orthogonal matrix. Then

$$H_{w_1} A = \begin{bmatrix} -0.1231 & -0.4924 & -0.8616 \\ -0.4924 & 0.7841 & -0.3777 \\ -0.8616 & -0.3777 & 0.3389 \end{bmatrix} \begin{bmatrix} 1 & 2 & 3 \\ 4 & 5 & 6 \\ 7 & 8 & 9 \end{bmatrix}$$

$$\doteq \begin{bmatrix} -8.1240 & -9.6011 & -11.0782 \\ 0.0000 & -0.0860 & -0.1719 \\ 0.0000 & -0.9004 & -1.8009 \end{bmatrix}$$

which has the expected zeroes in the first column ($Q_1 = H_{w_1}$). Let's do another step. We want to operate on the southeast 2×2 block

$$A_{22} = \begin{bmatrix} -0.0860 & -0.1719 \\ -0.9004 & -1.8009 \end{bmatrix}.$$

The new a_1 is $(-0.0860, -0.9004)^T$ with $\alpha_1 = \pm 0.9045$. This time we should take $\alpha_1 > 0$ to avoid subtraction. We have

$$
\begin{aligned}
w &= [(0.9045, 0)^T - (-0.0860, -0.9004)^T]/\|(0.9045, 0)^T - (-0.0860, -0.9004)^T\| \\
&= (0.9045 + 0.0860, 0.9004)^T/\|(0.9045 + 0.0860, 0.9004)^T\| \\
&\doteq (0.9905, 0.9004)^T/1.3386 \\
&\doteq (0.7400, 0.6727)^T
\end{aligned}
$$

and taking this as w_2 we find

$$
\begin{aligned}
H_{w_2} &= I - 2(0.7400, 0.6727)^T (0.7400, 0.6727) \\
&\doteq \begin{bmatrix} -0.0951 & -0.9955 \\ -0.9955 & 0.0951 \end{bmatrix}
\end{aligned}
$$

and so

$$
\begin{aligned}
H_{w_2} A_{22} &= \begin{bmatrix} -0.0951 & -0.9955 \\ -0.9955 & 0.0951 \end{bmatrix} \begin{bmatrix} -0.0860 & -0.1719 \\ -0.9004 & -1.8009 \end{bmatrix} \\
&\doteq \begin{bmatrix} 0.9045 & 1.8091 \\ 0.0000 & -0.0001 \end{bmatrix}
\end{aligned}
$$

which has the expected form. Note that we can put this in the form of Eq. (8.3) by defining

$$
\begin{aligned}
Q_2 &= \begin{bmatrix} 1 & {}_1 0_2 \\ {}_2 0_1 & H_{w_2} \end{bmatrix} \\
&= \begin{bmatrix} 1 & 0 & 0 \\ 0 & -0.0951 & -0.9955 \\ 0 & -0.9955 & 0.0951 \end{bmatrix}
\end{aligned}
$$

because then the action of H_{w_2} on the sub-block A_{22} of $Q_1 A$ may be represented as $Q_2(Q_1 A)$. We have $Q_2 Q_1 A = R$ (in accordance with Eq. (8.3)) so that $A = QR$ where

$$R = \begin{bmatrix} -8.1240 & -9.6011 & -11.0782 \\ 0.0000 & 0.9045 & 1.8091 \\ 0.0000 & 0.0000 & -0.0001 \end{bmatrix}$$

and $Q = (Q_2 Q_1)^{-1} = Q_1^{-1} Q_2^{-1} = Q_1^T Q_2^T$, giving

$$Q = \begin{bmatrix} -0.1231 & -0.4924 & -0.8616 \\ -0.4924 & 0.7841 & -0.3777 \\ -0.8616 & -0.3777 & 0.3389 \end{bmatrix}^T \begin{bmatrix} 1 & 0 & 0 \\ 0 & -0.0951 & -0.9955 \\ 0 & -0.9955 & 0.0951 \end{bmatrix}^T$$

$$\doteq \begin{bmatrix} -0.1231 & 0.9046 & 0.4082 \\ -0.4924 & 0.3015 & -0.8165 \\ -0.8616 & -0.3015 & 0.4083 \end{bmatrix}$$

and it is easily checked that $Q^T Q = I_3$. This is a QR decomposition of A. (The $(3,3)$ entry of R would have been exactly zero in infinite-precision arithmetic because the original matrix is singular, however.) To make R have $r_{ii} > 0$ we could write $A = QJJR$ where $J = [-1 \ 0 \ 0; 0 \ 1 \ 0; 0 \ 0 \ -1]$; note that J satisfies $J^2 = I$. Then $A = (QJ)(JR)$, and where JR is of the desired form and $(QJ)^T(QJ) = J^T Q^T QJ = J^T IJ = J^T J = I$ so that QJ is orthogonal. $\square$

This approach is entirely general. Similar to Eq. (8.2) for the LU decomposition, a typical step of QR decomposition by Householder triangularization has the form

$$\begin{bmatrix} I_r & {}_r 0_{m-r} \\ {}_{m-r} 0_r & H_{r+1} \end{bmatrix} \begin{bmatrix} X_a & X_b \\ 0 & X_c \end{bmatrix} = \begin{bmatrix} X_\alpha & X_\beta \\ {}_{m-r-1} 0_{r+1} & X_\gamma \end{bmatrix}$$

where H_{r+1} is an $(m-r) \times (m-r)$ Householder reflector that will eliminate (make zero) all the entries in the first column of X_c save the first entry. The result of this process is R, and we can collect the matrices $Q_1, ..., Q_{m-1}$ to form Q. Since a zero column is not a problem–we simply take $Q_i = I$ for that step–this process may be applied to any $m \times n$ matrix, even if it does not have full rank. (Note that we never actually used the assumption that A had full rank.) However, if A does in fact have full rank then the QR decomposition with $r_{ii} > 0$ is unique.

Choosing the appropriate sign for the constant that multiplies e_1, as in Example 2, is more important than we have indicated; if cancellation of significant figures occurs then we may get very inaccurate results. Note that the computed R, even if inaccurate, will certainly be upper triangular (as we will set to zero those entries that we know must be zero), but this is not necessarily the case for the orthogonality property of Q. Choosing the sign of α_1 opposite to that of the first component of a_1 gives an algorithm that is well-behaved with respect to round-off errors (i.e., that is stable).

In many applications the matrix Q is not explicitly required and we can find a "Q-less" decomposition. Even if it is, constructing each Householder matrix is inefficient. Since

$$
\begin{aligned}
H_w x &= (I - 2ww^T)x \\
&= Ix - 2ww^T x \\
&= x - 2w(w^T x) \\
&= x - 2w(w^T x)
\end{aligned}
$$

we can compute the product $H_w x$ using only the vector w and further more we can do so efficiently: The factor $w^T x$ is just a dot product; for m-vectors, evaluating $H_w x$ in this way requires $O(m)$ flops as opposed to $O(m^2)$ for explicitly multiplying H_w into x, plus the advantage of not having to form a potentially large matrix. (Of course, $H_w^2 x = I$ also means $H_w = H_w^{-1}$.) This is a significant gain. In a similar way, if we store the w vector at each step then $Q^T b$ and Qy (if needed) may be formed without explicitly forming Q. The Q matrix may be formed by applying the technique for finding Qy successively to $e_1, ..., e_m$.

We haven't addressed the generalization of the matrix G of Example 1 yet. We'll do that in the MATLAB section. This will give a variant of the orthogonal triangularization approach to computing the QR decomposition that we have developed in this section. In the next section we will discuss a procedure, based on the Gram-Schmidt process, which finds the QR decomposition by *triangular orthogonalization* of A.

MATLAB

The Householder matrix $H_w = I - 2ww^T$ corresponding to a given w may be generated as follows; enter:

```
» w=[1 1 1]';w=w/norm(w);
» H=eye(3)-2*w*w'
```

Note that if w is a row vector then $w*w'$ will be a scalar (the dot, or inner, product) and MATLAB will automatically subtract $2w*w'$ from every entry of I_3; this will give incorrect results. The quantity $w*w'$ is a matrix, called the **outer product** (more generally, if x and y are column vectors then the matrix xy^T is their outer product). The use of eye(3) matters, because although MATLAB will compute the notionally nonconformable sum

```
» 1-2*w*w'
```

it does so by expanding the 1 to the equivalent of ones([3 3]).

Let's verify that the formula $Hx = x - 2w(w^T x)$ may be used to compute the product Hx without forming H. Enter:

```
» x=[2 1 2]'
» H*x              %Compute product using matrix.
» x-2*w*(w'*x)     %Compute product using vectors only.
```

How much more efficient is this? Let's use a much larger example. Matrices with thousands of rows are not uncommon in practice. Enter:

```
» w=rand([1000 1]);w=w/norm(w);x=rand([1000 1]);
» tic;H=eye(1000)-2*w*w';toc
» tic;H*x;toc
» tic;x-2*w*(w'*x);toc
```

The time to form H is already nontrivial. (The gallery command in MATLAB can be used to form Householder matrices.) Doing this 999 times while factoring a 1000×1000 matrix would be very time-consuming. (The size of the H_w matrices would be decreasing, of course, though not of the Q_i matrices in which they are embedded.) The time for performing Hx is variable but the flops count would be about 2 million $(O(n^2))$. The time for performing $x - 2w(w^T x)$ is almost certainly negligible and requires about 5000 flops $(O(n))$. This is a considerable difference; even if H is known, using it requires about 400 times as many floating point operations as using the alternative formula $x - 2w(w^T x)$, let alone the memory-access time.

Let's look at the generalization of the matrices G of Example 1. The Householder matrices H_w ($w \in \mathbb{R}^m$) represent reflections in $\mathbb{R}^m$ across $(m-1)$-dimensional hyperplanes defined by the 1-dimensional vector w; the matrices $G = [\cos(\theta) \ -\sin(\theta); \sin(\theta) \ \cos(\theta)]$ represent rotations through a counterclockwise angle θ in the plane, and are called **plane rotations** (or **plane rotation matrices**). If we had used the definition $[\cos(\theta) \ \sin(\theta); -\sin(\theta) \ \cos(\theta)]$ (equivalent to replacing θ by $-\theta$) instead then the direction of the rotation would be clockwise. Enter:

```
» clear all
» th=pi/2;
» G=[cos(th)  -sin(th);sin(th)  cos(th)]
» G*[1 0]'
» G*[1 1]'
» compass([1 0]',G*[1 0]'),grid
```

for a simple example (90° CCW rotation); the `compass` command gives a visual display of x and Gx. (See `help compass`.) Try another value of `th` and verify that the vector is rotated appropriately.

If x is a 2-vector with its second entry nonzero then clearly there exists a θ such that a rotation of x through the angle θ lies along the x-axis. For example, if we start with the vector `x=[-4 3]'` then there must be a rotation that gives the result $G_1 x = (5, 0)^T$ and another one that gives the result $G_2 x = (-5, 0)^T$ (recall that G_i preserves lengths since it is orthogonal). Enter:

```
» compass(-4,3,'b')
» hold on
» compass(5,0,'r')
» compass(-5,0,'y')
```

to visualize these choices; G_1 maps the blue vector to the red one, and G_2 maps the blue vector to the yellow one.

The MATLAB command `planerot` generates such plane rotations for a given x. (Some care must be taken in the computation of the elements of G, and `planerot` handles this issue; enter `type planerot` to see how the computation is made more accurate.). Enter:

```
» x=[-4 3]';
» G=planerot(x)
» G*x
```

to find a plane rotation G that maps x to a vector with second component zero and to verify that it is in fact G_1. Enter:

```
» [G,y]=planerot(x)
```

to obtain both G and $y = Gx$. Enter:

```
» [G,y]=planerot([1 0]')
» [G,y]=planerot([-1 0]')
```

to see that $G = I_2$ if the vector to be rotated already lies along the x-axis.

This means that if A is a 2×2 matrix then we can find a G to orthogonally triangularize. Enter:

```
» A=rand([2 2])
» G=planerot(A(:,1))     %Use first column of A.
» G'*G    %Check orthogonality.
» R=G*A
» G'*R    %Should be A.
```

This is a QR decomposition with $Q = G^T$, and it is truly different than the Householder approach (though it is still an orthogonal triangularization). We'd like to generalize it to $m \times n$ matrices. This is straight-forward, and we will demonstrate it using an example. Enter:

```
» clear all,clc
» A=rosser
```

The command **rosser** takes no input arguments and returns the Rosser test matrix, a symmetric 8×8 matrix used to test numerical eigenvalue routines. (We are just using it for convenience in this example.) We want to orthogonally triangularize it using rotations. Let's start by eliminating the $(2,1)$ entry 196. If we were only concerned with the upper 2×2 submatrix [611 196;196 899] then applying **planerot** would suffice; enter:

```
» A1=A(1:2,1)     %Entries (1,1) and (2,1) of A.
» G1=planerot(A1)
» G1*A1
```

to see the result. But if we simply embed G in an 8×8 identity matrix, similar to Eq. (8.2) but with G in the northwest corner, we will get the desired effect on the entire matrix A. Enter:

```
» I=eye(8);Q1=I;Q1(1:2,1:2)=G1
» Q1'*Q1     %Check orthogonality.
» R1=Q1*A
```

Note that the second entry of the first column of A is now zero. That's all we were trying to accomplish on this single step. Let's continue: We now want to eliminate (or annihilate) the $(3,1)$ entry of R1. Enter:

```
» A2=R1([1 3],1)     %Entries (1,1) and (3,1) of R1.
» G2=planerot(A2)
» Q2=I;Q2([1 3],[1 3])=G2
```

Look carefully at the northwest 3×3 submatrix of this matrix. The 4 elements of G2 have been inserted in the $(1,1)$, $(3,1)$, $(1,3)$, and $(3,3)$ entries of the current working matrix R1. What will happen when we apply Q2 to R1? Enter:

```
» Q2'*Q2     %Verify orthogonality.
» R2=Q2*R1
```

As desired, the effect of pre-multiplying R1 by Q2 is to eliminate the $(3,1)$ entry of R1. (Note that Q2 would not have this effect on other matrices–it was constructed to have this effect on this particular matrix.) Let's try again. Enter:

```
» A3=R2([1 4],1)     %Entries (1,1) and (4,1) of R2.
» G3=planerot(A3)
» Q3=I;Q3([1 4],[1 4])=G3
» R3=Q3*R2
```

(We should really be overwriting some of these variables, of course.) The matrix
R3 has zeroes in its $(2, 1)$, $(3, 1)$, *and* $(4, 1)$ entries; we've introduced a new desired
zero without having lost the previous one. Let's do one more; enter:

» Q4=I;Q4([1 5],[1 5])=planerot(R3([1 5],1)

» R4=Q4*R3

Success. Enter

» R1==R4

to see that a lot of the entries of the working matrix R_i are still unaffected by these
operations (as would be expected, since Q_i always has identity matrix blocks).
There is a much more efficient implementation available, of course.

A matrix such as the Q_i above, consisting of I_m with a plane rotation matrix
inserted in its (k, k), (k, p), (p, k), and (p, p) entries in the natural way, is called a
Givens matrix (or **Givens rotation**); the 2×2 matrix of Example 1 is a special
case. For an $m \times n$ matrix with $m \geq n$, we might in principle need $(m - 1) + (m -
2) + ... + 1 = m(m - 1)/2$ such matrices (working west to east across the matrix)
to introduce all the zeroes needed to drive A to triangular form. This will give
$Q_N Q_{N-1} \cdots Q_1 A = R$ $(N = m(m - 1)/2)$, or $A = QR$ with $Q = Q_1^T Q_2^T \cdots Q_N^T$
orthogonal and R upper triangular. This is the method of QR decomposition using
Givens rotations.

The traditional implementation of QR decomposition using Givens rotations is
more expensive than the QR decomposition using Householder reflections; however,
there is a fast implementation that is comparable in computational expense to the
Householder version. The error properties of the Givens and Householder schemes
are similar. The MATLAB qr command is based on Householder triangularization
(as is the QR decomposition used in the slash command, which also uses column
pivoting for improved accuracy), and the Householder triangularization approach is
used more often than the Givens rotation approach. The QR decomposition using
Givens rotations is the more efficient approach for upper Hessenberg matrices, which
have every entry with $i > j + 1$ zero), however, and this is an important subproblem
in certain computations. Givens rotations have other applications as well.

Problems

1.) a.) Show that the matrices $H(\theta)$ and $G(\theta)$ of Example 1 are orthogonal and
find their determinants.

b.) Show that every 2×2 orthogonal matrix corresponds to H or G for some angle
θ.

c.) Show that a Householder matrix H_w $(w \in \mathbb{R}^m)$ is symmetric and orthogonal,
and satisfies $H^2 = I_m$.

2.) Repeat Example 2 for the matrix $A = [2\ \ 2\ \ 3; 4\ \ 5\ \ 6; 7\ \ 8\ \ 9]$. (Supply
all intermediate details as in the example; use MATLAB for the multiplications.)
Include a check of your final result $(A = QR)$.

3.) a.) Suppose that in the process of performing QR decomposition by Householder
triangularization on some 8x6 matrix B we arrive at the matrix $\widetilde{B} = [I_5\ \ 0; 0\ \ A]$
after some step, where $A = [1\ \ 2\ \ 3; 4\ \ 5\ \ 6; 7\ \ 8\ \ 9]$ is the matrix of Example 2.
Write out explicitly (that is, not in block form) the 8×8 matrix Q_6 that would be
applied to $\widetilde{B}$ to find the next matrix $Q_6 \widetilde{B} = [I_6\ \ 0; 0\ \ A_{next}]$ in this process and the
matrix $Q_6 \widetilde{B}$.

b.) Find $Q_6 \tilde{B}$ using the formula $H_w x = x - 2w(w^T x)$ rather than matrix multiplication.

4.) Find the QR decomposition of `A=reshape(1:16,4,4)'` by the method of Householder triangularization (follow Example 2).

5.) a.) Partition the $n \times n$ matrix A into the block matrix $A = [A_{11} \ A_{12}; A_{21} \ A_{22}]$ where each $A_{i,j}$ is a submatrix of A and A_{11} is an $m \times m$ nonsingular matrix $(1 \leq m < n)$. Define the block matrix $E = [I_p \ 0; -A_{21}A_{11}^{-1} \ I_q]$ where I is an identity matrix of the appropriate order and 0 represents a zero matrix of the appropriate size. Give the sizes of A_{12}, A_{21}, A_{22}, I_p, the zero matrix 0, $-A_{21}A_{11}^{-1}$, and I_q.

b.) Show that $EA = [A_{11} \ A_{12}; 0 \ A_{22} - A_{21}A_{11}^{-1}A_{12}]$. What are the sizes of the zero matrix and the submatrix $A_{22} - A_{21}A_{11}^{-1}A_{12}$ (called the **Schur complement** of A_{11})?

c.) Draw the Wilkinson diagram for passing from A to EA if $n = 6$ and $m = 3$.

d.) What is the significance of this observation? Think in terms of Gaussian elimination and the use of block matrix-matrix operations.

6.) Find (step-by-step) the QR decomposition of $A = [0 \ 0 \ 1; 0 \ 1 \ 0; 1 \ 0 \ 0]$ using Householder reflectors.

7.) a.) What unit vector $w(\theta) \in \mathbb{R}^3$ corresponds to the 2×2 Householder matrix $H = [-\cos(\theta) \ \sin(\theta); \sin(\theta) \ \cos(\theta)]$?

b.) Let $w = (1, 0, 1)^T$ and $x = (1, 1, 1)^T$. Sketch w, the plane in $\mathbb{R}^3$ determined by w, x, and $H_w x$. Explain geometrically why $H_w x$ is the reflection of x in that plane.

8.) a.) Show that no 2×2 Householder reflector is also a Givens rotation.

b.) Show that a Givens rotation is orthogonal. Can it be symmetric?

c.) How many rows of an $m \times n$ matrix A might be affected by premultiplying it by a Householder reflector? How many rows of A might be affected by premultiplying it by a Givens rotation?

9.) Find the QR decomposition of `A=reshape(1:16,4,4)'` by Givens rotations. (See Problem 4.)

10.) a.) Find the QR decomposition of the Hilbert matrix of order $N = 4$. Use it to solve $QRx = b$ where b is a vector of ones. Find the relative error in your answer (use `invhilb(A)*b` to determine the true solution). Give the condition numbers of the Hilbert matrix, Q, and R.

b.) Repeat for $N = 5, 6, 7, 8, 9, 10, 11, 12$. Comment.

11.) a.) Find the QR decomposition of the Hadamard matrix of order $N = 4$. (Use the MATLAB command `A=hadamard(N)` to generate the matrix.) Use it to solve $QRx = b$ where b is a vector of ones. Find the relative error in your answer (use the fact that $A^T A = N I_N$ if A is a Hadamard matrix to determine the true solution). Give the condition numbers of the Hadamard matrix, Q, and R.

b.) Repeat for $N = 8, 12, 16, 20$. Comment.

12.) a.) Show that the Givens rotation that takes the nonzero vector $(x_1, x_2)^T$ to a vector on the positive x-axis may be found without computing θ using the formulas $\cos(\theta) = x_1/\sqrt{x_1^2 + x_2^2}$ and $\sin(\theta) = -x_2/\sqrt{x_1^2 + x_2^2}$.

b.) The formulas in part a. may produce an overflow in the computation of $x_1^2 + x_2^2$ even though the final result would not overflow. This is undesirable; underflows are also a concern for this method. One way to handle this is to set $s = |x_1| + |x_2|$ and then compute $\sqrt{x_1^2 + x_2^2}$ as $s\sqrt{(x_1/s)^2 + (x_2/s)^2}$ if s is nonzero, else take $\cos(\theta) = 1$,

$\sin(\theta) = 0$. (A technique such as this should always be used when forming Givens rotations.) Show that these expressions are mathematically equivalent.

c.) What problem is avoided by using $s\sqrt{(x_1/s)^2 + (x_2/s)^2}$ in the computation? Give a numerical example in which the use of this trick is beneficial.

d.) What problem is avoided by checking if $s = 0$ even though $(x_1, x_2)^T$ is known to be nonzero? Give a numerical example in which the use of this trick is beneficial.

e.) The MATLAB command `planerot` computes the quantity $\sqrt{x_1^2 + x_2^2}$ using the `norm` command. Does the `norm` command seem to have a safeguarding technique like that in part b. built into it? Justify your answer with the output of several MATLAB computations.

13.) Recall that a matrix is said to be upper Hessenberg if every entry below the first subdiagonal (that is, every entry with $i > j + 1$) is zero; such a matrix is "nearly" upper triangular. Show that if A is upper Hessenberg then the QR decomposition using Givens rotations is more efficient than the QR decomposition using Householder reflectors.

14.) Write a MATLAB program that performs the QR decomposition using Givens rotations. Compare the results of your program with those of the `qr` command.

15.) Let A be an $m \times n$ matrix with $m \geq n$. Prove that if $A = QR$ is a QR decomposition of A and A has full rank then the rows of Q form an orthonormal basis for Col(A) (the column space of A).

16.) Do the Household reflectors form a group for a fixed order n? Do the Givens rotations?

9. The QR Decomposition by Orthogonalization

We've been placing a greater emphasis on looking at numerical linear algebra algorithms in terms of sub-blocks of the matrices involved. This is important for many reasons: It allows for the efficient design and analysis of algorithms and the maximal use of the BLAS and/or the special architecture of the machine, for example. Often we wish to block problems into chunks of data that fit in the fast cache. In many cases, if we can analyze one step in the iterative process of reducing a matrix to a special form in terms of simple matrix algebra operations, then we can gain a great deal of understanding of the method. Some programming languages allow us to, in effect, lay out the matrix in blocks in memory rather than by rows or by columns.

Putting algorithms in blocked form is not as simple as dividing matrices into blocks. For example, the matrix

$$M = \begin{bmatrix} 1 & 0 & 0 & 0 \\ 0 & 0 & 0 & 1 \\ 0 & 0 & 1 & 0 \\ 0 & 1 & 0 & 0 \end{bmatrix}$$

certainly has an LU decomposition (after all, it's just a permutation of I_4), but if we block it in the form

$$
M = \begin{bmatrix}
1 & 0 & \vdots & 0 & 0 \\
0 & 0 & \vdots & 0 & 1 \\
\cdots & \cdots & \vdots & \cdots & \cdots \\
0 & 0 & \vdots & 1 & 0 \\
0 & 1 & \vdots & 0 & 0
\end{bmatrix}
$$

then every block is singular. This means that the block version of Gaussian elimination may fail even with complete pivoting. There is more to constructing a blocked algorithm than simply partitioning M into blocks! Indeed, given

$$
M = \begin{bmatrix}
A & \vdots & B \\
\cdots & \vdots & \cdots \\
C & \vdots & D
\end{bmatrix}
$$

the block form of Gaussian elimination generally proceeds via the **Schur comple-ment** $S = D - CA^{-1}B$ (of A in M),

$$
M \sim \begin{bmatrix}
I & \vdots & 0 \\
\cdots & \vdots & \cdots \\
CA^{-1} & \vdots & I
\end{bmatrix}
\begin{bmatrix}
A & \vdots & B \\
\cdots & \vdots & \cdots \\
C & \vdots & D
\end{bmatrix}
$$

$$
= \begin{bmatrix}
A & \vdots & B \\
\cdots & \vdots & \cdots \\
0 & \vdots & S
\end{bmatrix}
$$

assuming A is invertible. Of course, CA^{-1} is formed by solving $ZA = C$ for Z by Gaussian elimination (by reducing columns rather than rows, perhaps).

To a certain extent these ideas straddle numerical analysis and computer science, because the best way to block a matrix for a particular application will depend on software and hardware issues as well as mathematical ones. The term scientific computation is sometimes used to describe these practical issues which arise when one considers the prospect of coding a particular algorithm for use on a particular machine, whereas by computational science we mean the academic discipline that uses scientific computing to perform simulations for the purpose of scientific investigation. Commercial software is usually written and tested by teams of individuals with strengths in various aspects of computing (as well as in the scientific problems to which the programs will ultimately be applied), using a software engineering approach. The purpose of this text is not to make you a programmer of scientific software but rather an educated user of such programs who is also capable of understanding new methods and, if needed, modifying older methods as the need arises. An important part of this process is studying different algorithms for the same mathematical problem and learning why some methods are stronger in certain cases and others are stronger in other cases.

We've considered Householder triangularization as a means of computing a QR decomposition. There's a second major approach, based on the Gram-Schmidt process, which is in some ways more intuitive. The Householder triangularization (or Givens rotation) method of the previous section is usually superior to this approach, but there is at least one significant special case where the Gram-Schmidt approach can be the better choice.

We begin with a review of the Gram-Schmidt process from linear algebra. If $v_1, v_2, ..., v_k$ are linearly independent vectors then their span is some vector space $V = \text{span}(v_1, v_2, ..., v_k)$, for which $\{v_1, v_2, ..., v_k\}$ is a basis. We seek an orthonormal basis for V. The Gram-Schmidt process finds an orthonormal basis $\{u_1, u_2, ..., u_k\}$ for V from $\{v_1, v_2, ..., v_k\}$ as follows: First we set $u_1 = v_1/\|v_1\|$. Then we set $u_2' = v_2 - (v_2^T u_1)u_1$ before normalizing it as $u_2 = u_2'/\|u_2'\|$. Then we set $u_3' = v_3 - (v_3^T u_2)u_2 - (v_3^T u_1)u_1$ which we normalize to $u_3 = u_3'/\|u_3'\|$. Continuing in this way, we eventually reach $u_k' = v_k - (v_k^T u_{k-1})u_{k-1} - \cdots - (v_k^T u_1)u_1$ which we normalize $u_k = u_k'/\|u_k'\|$. If $\{v_1, v_2, ..., v_k\}$ is a linearly independent set then the process completes with $\{u_1, u_2, ..., u_k\}$ an orthonormal set with the same span. Note that the terms $(v_2^T v_1)$, etc., are inner products and the terms $(v_2^T u_1)u_1$, etc., are projections along the various unit vectors; the Gram-Schmidt process simply subtracts off the "nonorthogonal" parts of the current vector.

EXAMPLE 1: Suppose $\{v_1, v_2, v_3\} = \{(1,0,1)^T, (-1,1,1)^T, (0,1,1)^T\}$. Then we have $u_1 = v_1/\|v_1\|$, that is, $u_1 = (1,0,1)^T/\sqrt{2}$. Next, we need to form $u_2' = v_2 - (v_2^T u_1)u_1$. The dot product is easily seen to be zero in this case, however, so $u_2' = v_2$ and hence $u_2 = (-1,1,1)^T/\|(-1,1,1)^T$, which is $(-1/\sqrt{3}, 1/\sqrt{3}, 1/\sqrt{3})^T$. Lastly,

$$
\begin{aligned}
u_3' &= v_3 - (v_3^T u_2)u_2 - (v_3^T u_1)u_1 \\
&= \begin{pmatrix} 0 \\ 1 \\ 1 \end{pmatrix} - \left[\begin{pmatrix} 0 \\ 1 \\ 1 \end{pmatrix}^T \begin{pmatrix} -1/\sqrt{3} \\ 1/\sqrt{3} \\ 1/\sqrt{3} \end{pmatrix} \right] \begin{pmatrix} -1/\sqrt{3} \\ 1/\sqrt{3} \\ 1/\sqrt{3} \end{pmatrix} - \\
&\quad \left[\begin{pmatrix} 0 \\ 1 \\ 1 \end{pmatrix}^T \begin{pmatrix} 1/\sqrt{2} \\ 0 \\ 1/\sqrt{2} \end{pmatrix} \right] \begin{pmatrix} 1/\sqrt{2} \\ 0 \\ 1/\sqrt{2} \end{pmatrix} \\
&= \begin{pmatrix} 0 \\ 1 \\ 1 \end{pmatrix} - \left[2/\sqrt{3} \right] \begin{pmatrix} -1/\sqrt{3} \\ 1/\sqrt{3} \\ 1/\sqrt{3} \end{pmatrix} - \left[1/\sqrt{2} \right] \begin{pmatrix} 1/\sqrt{2} \\ 0 \\ 1/\sqrt{2} \end{pmatrix} \\
&= \begin{pmatrix} 0 \\ 1 \\ 1 \end{pmatrix} - \begin{pmatrix} -2/3 \\ 2/3 \\ 2/3 \end{pmatrix} - \begin{pmatrix} 1/2 \\ 0 \\ 1/2 \end{pmatrix} \\
&= \begin{pmatrix} 1/6 \\ 1/3 \\ -1/6 \end{pmatrix} \\
u_3 &= (1/6, 1/3, -1/6)^T/\|(1/6, 1/3, -1/6)^T\| \\
&= (1/6, 1/3, -1/6)^T/\sqrt{1/6} \\
&= (\sqrt{6}/6, \sqrt{6}/3, -\sqrt{6}/6)^T
\end{aligned}
$$

giving us an orthonormal set $\{u_1, u_2, u_3\}$ with the same span the original set had. (In this case, the span is $\mathbb{R}^3$.) To check this, we can form a matrix Q with columns u_1, u_2, u_3, that is,

(9.1)
$$Q = \begin{bmatrix} 1/\sqrt{2} & -1/\sqrt{3} & \sqrt{6}/6 \\ 0 & 1/\sqrt{3} & \sqrt{6}/3 \\ 1/\sqrt{2} & 1/\sqrt{3} & -\sqrt{6}/6 \end{bmatrix}$$

and see if it is orthogonal:

$$Q^T Q = \begin{bmatrix} 1/\sqrt{2} & 0 & 1/\sqrt{2} \\ -1/\sqrt{3} & 1/\sqrt{3} & 1/\sqrt{3} \\ \sqrt{6}/6 & \sqrt{6}/3 & -\sqrt{6}/6 \end{bmatrix} \begin{bmatrix} 1/\sqrt{2} & -1/\sqrt{3} & \sqrt{6}/6 \\ 0 & 1/\sqrt{3} & \sqrt{6}/3 \\ 1/\sqrt{2} & 1/\sqrt{3} & -\sqrt{6}/6 \end{bmatrix}$$

$$= \begin{bmatrix} 1 & 0 & 0 \\ 0 & 1 & 0 \\ 0 & 0 & 1 \end{bmatrix}.$$

This checks. We might have expected to find the standard basis for $\mathbb{R}^3$ but in general this will not be the case. $\square$

Let's return to the QR decomposition. Remember, this is to be a triangular orthogonalization approach. It's easiest to consider a square matrix; let A be an $m \times m$ matrix with $\text{rank}(A) = m$. Then $\text{Col}(A)$ is the span of the columns $a_1, ..., a_m$ of A, and $\{a_1, ..., a_m\}$ is a linearly independent set. (In fact, $\text{Col}(A) = \mathbb{R}^m$ in this case.) Hence the Gram-Schmidt process may be used to generate an orthonormal basis $\{u_1, u_2, ..., u_m\}$ for $\text{Col}(A)$. In fact, note from Example 1 that we can say something a bit stronger: The span of a_1 will be the same as the span of u_1 (they are multiples of each other, after all); the span of a_1 and a_2 will be the same as the span of u_1 and u_2 (u_1 and u_2 are linear combinations of a_1 and a_2), and so on.

Since the product Mx of a matrix and a vector is a linear combination of the columns of the matrix, we may express the Gram-Schmidt process relationship as follows: Let Q be the matrix with columns $u_1, u_2, ..., u_m$. Then

$$a_1 = Qr_1$$

where r_1 is a column vector that is nonzero only in its first entry (a_1 is a linear combination of u_1 only). Also

$$a_2 = Qr_2$$

where r_2 is a column vector that is nonzero only in its first and second entry (a_2 is a linear combination of u_1 and u_2 only). We could write

$$[a_1 \quad a_2] = Q[r_1 \quad r_2]$$

where $[a_1 \quad a_2]$ is an $m \times 2$ matrix consisting of the first two columns of A and $[r_1 \quad r_2]$ is an $m \times 2$ matrix that incorporates the information from the Gram-Schmidt process.

In general, then, the completed Gram-Schmidt process may be represented in the form $A = QR$ where R is an upper triangular matrix. The diagonal entries of R are nonzero; in fact, looking at the algorithm we see that they are the positive values $1/\|u_i'\|$ $(i = 1, ..., m)$.

This is the Gram-Schmidt orthogonalization approach to the QR decomposition, called the **classical Gram-Schmidt orthogonalization** method (or **classical Gram-Schmidt iteration**): We construct the orthogonal matrix Q column-by-column from the matrix A by the (inherently triangular) Gram-Schmidt process.

EXAMPLE 2: Refer back to Example 1; we're going to redo it in matrix form in order to show the close connection between the Gram-Schmidt process and the QR decomposition. Continue referring to Example 1 as you study this example. The vectors from Example 1 were the columns of the matrix

$$A = \begin{bmatrix} 1 & -1 & 0 \\ 0 & 1 & 1 \\ 1 & 1 & 1 \end{bmatrix}$$

(i.e., $A = [v_1|v_2|v_3]$). The first step in orthogonalizing A is to rescale its first column so that it is a column vector. Rescaling a column requires that we post-multiply the matrix by a diagonal matrix; we are performing an elementary column operation by using an elementary matrix. This gives

$$\begin{aligned} Q_1 &= AR_1 \\ &= \begin{bmatrix} 1 & -1 & 0 \\ 0 & 1 & 1 \\ 1 & 1 & 1 \end{bmatrix} \begin{bmatrix} 1/\sqrt{2} & 0 & 0 \\ 0 & 1 & 0 \\ 0 & 0 & 1 \end{bmatrix} \\ &= \begin{bmatrix} 1/\sqrt{2} & -1 & 0 \\ 0 & 1 & 1 \\ 1/\sqrt{2} & 1 & 1 \end{bmatrix} \end{aligned}$$

(note that Q_1 is not orthogonal; rather, it is the matrix that we are working on with the goal of *making* it orthogonal). Next we need to make column 2 of AR_1 be orthogonal to column 1 of AR_1. Referring to Example 1, this may be accomplished by multiplying column 1 by 0 (the columns are already orthogonal) and adding it to column 2–which, in this case, leaves column 2 unchanged as the columns are already orthogonal–and then normalizing by $\sqrt{3}$. Hence we need perform only the rescaling operation

$$\begin{aligned} Q_2 &= AR_1R_2 \\ &= \begin{bmatrix} 1/\sqrt{2} & -1 & 0 \\ 0 & 1 & 1 \\ 1/\sqrt{2} & 1 & 1 \end{bmatrix} \begin{bmatrix} 1 & 0 & 0 \\ 0 & 1/\sqrt{3} & 0 \\ 0 & 0 & 1 \end{bmatrix} \\ &= \begin{bmatrix} 1/\sqrt{2} & -1/\sqrt{3} & 0 \\ 0 & 1/\sqrt{3} & 1 \\ 1/\sqrt{2} & 1/\sqrt{3} & 1 \end{bmatrix} \end{aligned}$$

at this stage. Note that Q_2 now has two orthonormal columns. We have one last column to process. Referring to Example 1, we need to take the third column a_3 of Q_2 (which is also the third column of A) and subtract from it $1/\sqrt{2}$ times the first column u_1 of Q_2 and $2/\sqrt{3}$ times the second column u_2 of Q_2, and then divide the result by $1/\sqrt{6}$. That is, we need to form

$$\left(a_3 - 2/\sqrt{3}u_2 - 1/\sqrt{2}u_1\right)/\sqrt{1/6} \;=\; \sqrt{6}a_3 - 2\sqrt{6}/\sqrt{3}u_2 - \sqrt{6}/\sqrt{2}u_1$$
$$=\; \sqrt{6}a_3 - 2\sqrt{2}u_2 - \sqrt{3}u_1$$

and so we continue using triangular matrices to drive A to an orthogonal matrix by performing

$$
\begin{aligned}
Q_3 &= AR_1R_2R_3 \\
&= \begin{bmatrix} 1/\sqrt{2} & -1/\sqrt{3} & 0 \\ 0 & 1/\sqrt{3} & 1 \\ 1/\sqrt{2} & 1/\sqrt{3} & 1 \end{bmatrix} \begin{bmatrix} 1 & 0 & -\sqrt{3} \\ 0 & 1 & -2\sqrt{2} \\ 0 & 0 & \sqrt{6} \end{bmatrix} \\
&= \begin{bmatrix} 1/\sqrt{2} & -1/\sqrt{3} & \sqrt{6}/6 \\ 0 & 1/\sqrt{3} & \sqrt{6}/3 \\ 1/\sqrt{2} & 1/\sqrt{3} & -\sqrt{6}/6 \end{bmatrix}.
\end{aligned}
$$

where R_3 implements both the addition of the vectors and the rescaling. (In practice we would probably need to do this in two steps, since we need to know u'_3 before we can find its norm and rescale by it.) The matrix Q_3 is the desired orthogonal matrix Q (compare the matrix Q in Example 1, Eq. (9.1)), and the orthogonalizing matrix $\tilde{R}$ is the upper triangular matrix

$$
\begin{aligned}
\tilde{R} &= R_1R_2R_3 \\
&= \begin{bmatrix} 1/\sqrt{2} & 0 & 0 \\ 0 & 1 & 0 \\ 0 & 0 & 1 \end{bmatrix} \begin{bmatrix} 1 & 0 & 0 \\ 0 & 1/\sqrt{3} & 0 \\ 0 & 0 & 1 \end{bmatrix} \begin{bmatrix} 1 & 0 & -\sqrt{3} \\ 0 & 1 & -2\sqrt{2} \\ 0 & 0 & \sqrt{6} \end{bmatrix} \\
&= \begin{bmatrix} 1/\sqrt{2} & 0 & -\sqrt{6}/2 \\ 0 & 1/\sqrt{3} & -2\sqrt{6}/3 \\ 0 & 0 & \sqrt{6} \end{bmatrix}.
\end{aligned}
$$

We have $Q = A\tilde{R}$ (that is, postmultiplying by the triangular matrix $\tilde{R}$ orthogonalizes A), and this method guarantees that all diagonal entries of $\tilde{R}$ are strictly positive; hence the inverse of $\tilde{R}$ exists and $A = Q\tilde{R}^{-1}$. The inverse of an upper triangular matrix with positive diagonal entries is itself an upper triangular matrix with positive diagonal entries, so this is a QR decomposition of A with $R = \tilde{R}^{-1}$. Let's check:

$$QR = \begin{bmatrix} 1/\sqrt{2} & -1/\sqrt{3} & \sqrt{6}/6 \\ 0 & 1/\sqrt{3} & \sqrt{6}/3 \\ 1/\sqrt{2} & 1/\sqrt{3} & -\sqrt{6}/6 \end{bmatrix} \begin{bmatrix} \sqrt{2}/2 & 0 & -\sqrt{6}/2 \\ 0 & \sqrt{3}/3 & -2\sqrt{6}/3 \\ 0 & 0 & \sqrt{6} \end{bmatrix}^{-1}$$

$$= \begin{bmatrix} 1/\sqrt{2} & -1/\sqrt{3} & \sqrt{6}/6 \\ 0 & 1/\sqrt{3} & \sqrt{6}/3 \\ 1/\sqrt{2} & 1/\sqrt{3} & -\sqrt{6}/6 \end{bmatrix} \begin{bmatrix} \sqrt{2} & 0 & 1/\sqrt{2} \\ 0 & \sqrt{3} & -2/\sqrt{3} \\ 0 & 0 & 1/\sqrt{6} \end{bmatrix}$$

$$= \begin{bmatrix} 1 & -1 & 0 \\ 0 & 1 & 1 \\ 1 & 1 & 1 \end{bmatrix}$$

$$= A$$

(clearly, we would not compute the inverse matrix in an actual computation; a simple reorganization of the method above allows us to form R directly without ever forming $\tilde{R}$, similar to what we did with the LU decomposition). This checks. □

The classical Gram-Schmidt orthogonalization method is a natural way to find an orthogonal matrix Q such that $A = QR$, and as indicated we may view the effect of the R_i matrices as being elementary column operations that are applied to A (in general one R_i matrix will perform several elementary column operations). Note that the columns of Q are generated directly; we do not need to form it after the reduction from individual transformations. In fact the same is true of R, though we have not brought that fact out. The flops count is similar to that of the Householder triangularization approach (and if Q is needed explicitly the classical Gram-Schmidt orthogonalization method is more efficient). This is an excellent example of a simple, intuitive, straight-forward, and efficient method that is guaranteed to converge in infinite precision arithmetic (given our assumptions) and that *does not work*.

There are two principal problems with the classical Gram-Schmidt orthogonalization approach to the QR decomposition: Amplification of round-off errors (similar to unpivoted Gaussian elimination), and loss of the orthogonality property of Q. The problem is that in a computation such as

$$u_3' = v_3 - (v_3^T u_2)u_2 - (v_3^T u_1)u_1$$

it may be the case that v_3 is nearly in the span of u_1 and u_2, meaning that we will be subtracting nearly equal quantities and hence causing the cancellation of significant figures. For a large matrix, say 1000×1000, by the time you're subtracting 999 linearly independent vectors from a_{1000} there is a very good chance that you're going to lose significant precision. The classical Gram-Schmidt orthogonalization method can amplify errors significantly, and it should not be used to compute QR decompositions.

Was it a waste of time to introduce it? No; it's a useful theoretical framework for understanding both the QR decomposition and the Gram-Schmidt process itself, and as we shall see here and in Ch. 3 there are sometimes reasons why we would use a quick but relatively inaccurate method to solve a linear system. (See Sec. 3.4.) But more to the point, we can modify this method to get one with better error properties.

Let's look at the classical Gram-Schmidt orthogonalization method again. Let A be an $m \times n$ matrix with $m \geq n$ and $\text{rank}(A) = n$. We might write the algorithm more efficiently as follows (using MATLAB notation for the indexing):

CGS Algorithm:
1. Set $Q = A$.
2. Set $R(1,1) = \|Q(:,1)\|$.
3. Normalize column 1 of Q.
4. Begin loop ($k = 2$ to n):
5. Set $p = 1 : (k-1)$.
6. Set $R(p,k) = Q(:,p)^T Q(:,k)$.
7. Set $Q(:,k) = Q(:,k) - Q(:,p)R(p,k)$.
8. Normalize column k of Q; store normalization constant $\|Q(:,k)\|$ in $R(k,k)$.
9. End loop.

(Note that p is a vector representing the rows containing the superdiagonal entries of R for a given column k.) We interpret setting $Q = A$ to mean that A is to be overwritten by Q; we generally wouldn't want to make two copies of A to start things off. As with any algorithm discussed in this text, we are omitting myriad implementation details that a practical program must include in order to be efficient (and to safeguard against bad inputs, divisions by zeroes, etc.).

The **modified Gram-Schmidt orthogonalization method** (or **modified** or **stabilized Gram-Schmidt iteration**) for finding the QR decomposition of such a matrix has the following form:

MGS Algorithm:
1. Set $Q = A$.
2. Set $R(1,1) = \|Q(:,1)\|$.
3. Normalize column 1 of Q.
4. Begin loop ($k = 2$ to n):
5. Begin loop ($i = 1$ to $k - 1$):
6. Set $R(i,k) = Q(:,i)^T Q(:,k)$.
7. Set $Q(:,k) = Q(:,k) - R(i,k)Q(:,i)$.
8. End i loop.
9. Normalize column k of Q; store normalization constant $\|Q(:,k)\|$ in $R(k,k)$.
10. End k loop.

Rather than use the index vector p of the classical method, we step through p (using the i loop) in the updates of both R (step 5 of the classical method) and Q (step 6 of the classical method). In the classical method, step 5 completed before step 6, and we only changed column k of Q once (apart from the final normalization); in the modified method we change Q every time we compute an entry of R in step 5 (as an inner product, like $v_2^T u_1$ from Example 1). Nonetheless, the methods are equivalent in infinite precision arithmetic. (We omit the demonstration, which is based on viewing the calculation of a column of Q as a single projection in the classical method and as a series of projections in the modified method.) They also have similar flops counts (roughly $2mn^2$ for large m, n). However, they behave differently in finite precision. You'll be asked in the Problems to discuss why this is so.

The modified Gram-Schmidt orthogonalization method has round-off error properties similar to those of Householder triangularization; in this sense it is a considerable improvement over the classical Gram-Schmidt orthogonalization method.

(These comments also apply to the problem of making a basis into an orthonormal basis; this problem is simply that of finding a QR decomposition where we do not care about R.) However, loss of the orthogonality property of Q can still occur in the modified method, though it is not as severe as in the classical method. If an accurate orthogonal matrix Q is needed, Householder triangularization is usually preferable. It is possible however to use a "reorthogonalization" procedure with either of the Gram-Schmidt methods in order to improve the quality of Q.

The modified Gram-Schmidt orthogonalization method as we have presented it is sometimes called the column version of the method since it fills in R by columns. There is a row version that fills in R by rows and that is more amenable to the use of higher-level BLAS; note that whereas our version of the classical Gram-Schmidt orthogonalization method was written using matrix-vector products such as $Q(:,p)^T Q(:,k)$ and $Q(:,p)R(p,k)$ (recall that p is a vector of indices, so $Q(:,p)$ and $R(p,k)$ are matrix blocks), they do not appear in our modified Gram-Schmidt orthogonalization algorithm. We said early in this section that there was a special circumstance under which we might prefer the classical method, and this is it: The classical method is vector-oriented, and if we are using a parallel machine then the speed benefits of using the classical method may make up for the loss of accuracy in some circumstances (particularly if we only need R, not Q, say because we wish to compute $\det(A)$ or $\operatorname{rank}(A)$). We can apply a reorthogonalization procedure if necessary.

We'll see this happen again as we continue studying numerical linear algebra in Ch. 3: Advanced architecture machines can make older methods of computation that are less accurate or less efficient (in the serial machine sense) become attractive options once again. For most purposes however the modified Gram-Schmidt orthogonalization method should be used rather than the classical version.

We make one final point about the QR decomposition. As with the LU decomposition, the errors in the computed Q and R matrices do not lead, generally, to comparably sized errors in their product QR; the errors are somehow "correlated". That is to say, if $A = QR$ in infinite precision arithmetic with all diagonal entries of R positive, and Q_c and R_c are the corresponding finite precision QR factors (computed by any method), then $\|Q - Q_c\|$ and $\|R - R_c\|$ might be relatively large while $\|A - Q_c R_c\|$ is not. This phenomenon is not fully understood.

MATLAB

The `orth` command finds an orthonormal basis for the column space of a matrix (that is, the space spanned by its columns), and the `null` command finds an orthonormal basis for its null space (that is, the set of all x such that $Ax = 0$). Enter:

» `type orth`
» `type null`

to see that each uses the `svd` command (see the next section). Previous versions used the `qr` command for this.

Let's look at a rank-deficient case. We'll make a random matrix that has its first and third column the same. Enter:

» `A=rand(3);A(:,3)=A(:,1)`

Since the range of A is a two-dimensional space, $Ax = b$ will be inconsistent for most $b \in \mathbb{R}^3$. Enter:

» `[Q,R]=qr(A)`

Notice that R, like A, is singular (since $R(3,3)$ is zero, or very nearly so). Let's check the decomposition; enter:

```
» A,Q*R
» norm(A-Q*R)
```

The first two columns of Q span the same space as the first two columns of A. Let's do a check; enter:

```
» w=cross(A(:,1),A(:,2))
```

to generate a vector orthogonal to the column space of A (which is a plane in $\mathbb{R}^3$ spanned by $A(:,1)$ and $A(:,2)$). Enter:

```
» dot(w,Q(:,1))
» dot(w,Q(:,2))
```

These are essentially (perhaps exactly) zero, as expected; that checks (vectors orthogonal to the span of the first two columns of A are orthogonal to the span of the first two columns of Q). Enter:

```
» dot(w,Q(:,3))
```

This is not zero. Recall that the third column of Q is an arbitrary vector orthonormal to the first two columns of Q. The reduced factorization is:

```
» [Q1,R1]=qr(A,0)
```

This is the same as the QR decomposition since A is square. Now enter:

```
» clear all
» A=rand([6 3])
» A(:,4)=A(:,1)      %Create fourth column (equal to first).
» [Q,R]=qr(A)
» [Q1,R1]=qr(A,0)
```

Once again R is singular. In the square case we said that R was singular because A was singular but we now see that the real reason R is singular in each case is because A does not have full rank. In the current case, A is a 6×4 matrix with rank$(A) = 3$. The reduced QR factorization matrix Q_1 is a nonsquare matrix with orthonormal columns; the full QR decomposition matrix Q agrees with Q_1 for the first four columns, and then has two added orthonormal columns.

The first 3 columns of Q_1 form an orthonormal basis for the column space of A; all the other columns are just there to fill out the Q_1 (or Q) matrix. The fourth column of Q_1 is just there for multiplicative conformability; it is an arbitrary extension of the orthonormal basis for Col(A) contained in $Q_1(:,1:3)$. Similarly for the fifth and sixth columns of Q.

Because of the column-by-column way in which the QR decomposition is computed, given the QR decomposition of A the QR decomposition of the matrix formed by deleting a column from A may be found efficiently. (This is true whether the matrices Q and R are found by orthogonal triangularization or triangular orthogonalization.) The QR decomposition is used in certain data science applications in which addition/deletion of rows/columns of A is common, because A is a **term-by-document matrix**, that is, its columns represent documents (books, web pages, etc.) and its rows represent words appearing in them, or vice versa. As web pages disappear or are newly created, the matrix must be updated.

The MATLAB command qrdelete may be used to perform a deletion. Let's look at an example first. Enter:

```
» clear all
» format compact
```

```
» A=pascal(6); B=A
» [Q,R]=qr(A);
» B(:,3)=[]      %Eliminate column 3 of B using empty vector [].
» [QB,RB]=qr(B);
» Q,QB           %Compare the entries of Q and QB.
» Q==QB          %Compare the entries of Q and QB.
» R,RB           %Compare the entries of R and RB.
» R(:,1:5)==RB   %Compare the entries of R and RB.
```

(We shouldn't really be testing for exact equality here, of course.) We recomputed the QR decomposition of B in this case, but the qrdelete command is more efficient. Enter:

```
» [QC,RC]=qrdelete(Q,R,3);
» abs(QC-QB)<1E-10      %Compare the entries of QC and QB.
» abs(RC-RB)<1E-10      %Compare the entries of RC and RB.
» norm(QC-QB)
» norm(RC-RB)
```

It appears that the matrices QB and QC and the matrices RB and RC are essentially the same. Enter:

```
» B
» QB*RB
» QC*RC
```

to see that these are both QR decompositions of B. If we already have a QR decomposition of A, the qrdelete approach is much more efficient. There is also a qrinsert command; these both use Givens rotations to find the new QR decomposition. There is also a qrupdate command that performs a similar function. There are many techniques for efficiently updating decompositions of a given matrix when the matrix undergoes a change of a proscribed form.

The MATLAB command B(:,k)=[] is an efficient way to delete a column of a matrix in MATLAB. We can also represent such deletions using matrix multiplications for theoretical purposes. For example, enter:

```
» A=reshape(1:16,4,4)'
» F=eye(4);E(3,:)=[]
» F*A
» A*F'
```

The matrix F is sometimes called a **deletion matrix**; it deletes a row if used to pre-multiply a matrix or a column if its transpose is used to post-multiply a matrix. This is, of course, not an elementary operation and hence F is not an elementary matrix.

Problems

1.) a.) Use the Gram-Schmidt process (as in Example 1) to find an orthonormal basis for the space spanned by $\{(1,0,1,0,1)^T, (2,2,1,2,2)^T, (1,-1,1,0,0)^T\}$.

b.) Use the Gram-Schmidt process (as in Example 1) to find an orthonormal basis for the space spanned by $\{(1,-1,1,0,0)^T, (2,2,1,2,2)^T, (1,0,1,0,1)^T\}$.

2.) a.) Apply the classical Gram-Schmidt method (as in Example 2) to find a QR decomposition of the matrix $A = [2\ 1\ 1; 1\ 2\ 1; 1\ 1\ 2]$.

b.) Apply the CGS Algorithm (by hand) to find a QR decomposition of A.

c.) Apply the MGS Algorithm (by hand) to find a QR decomposition of A.

d.) Explain why the MGS Algorithm is more accurate than the CGS Algorithm.

3.) Apply the CGS Algorithm (by hand) to find an orthonormal basis for the space spanned by $\{(1,0,1,0,1)^T, (2,2,1,2,2)^T, (1,-1,1,0,0)^T\}$.

4.) a.) Write a MATLAB program that implements the CGS Algorithm. Demonstrate it on the matrix $A = [2 \ 1 \ 1; 1 \ 2 \ 1; 1 \ 1 \ 2]$.

b.) Find a matrix A for which your program returns a Q that demonstrates the loss of orthogonality.

5.) a.) Write a MATLAB program that implements the MGS Algorithm. Demonstrate it on the matrix $A = [2 \ 1 \ 1; 1 \ 2 \ 1; 1 \ 1 \ 2]$.

b.) Demonstrate your program on the example you found for Problem 4(b). Does the MGS Algorithm perform better?

6.) a.) Apply the CGS Algorithm (by hand) to find a QR decomposition of the matrix $A = [1 \ 2 \ 3 \ 4; 5 \ 6 \ 7 \ 8; 9 \ 10 \ 11 \ 12; 13 \ 14 \ 15 \ 16]$.

b.) Apply the MGS Algorithm (by hand) to find a QR decomposition of A.

7.) a.) Apply the CGS Algorithm (by hand) to find a QR decomposition of the matrix $A = [1 \ 0 \ 1 \ 0; 0 \ 1 \ 0 \ 1; 1 \ 0 \ 0 \ 1; 0 \ 0 \ 1 \ 1]$. Use it to find $\det(A)$.

b.) Apply the MGS Algorithm (by hand) to find a QR decomposition of A.

8.) a.) If the $m \times n$ matrix A ($m \geq n$) is of full rank then we define its **Moore-Penrose pseudo-inverse** to be the square matrix $A^+ = (A^T A)^{-1} A^T$. (The MATLAB command pinv computes the pseudo-inverse.) Show that the least squares solution of the over-determined linear system $Ax = b$ is $x = A^+ b$ if A has full rank.

b.) Show that $A^+ A = I$ and $AA^+ y = y$ for all conformable vectors y.

c.) Show that $(A^+)^+ = A$. Why is the term pseudo-inverse appropriate for A^+?

d.) Suppose $A = QR$ is a QR decomposition of A. Express A^+ in terms of Q and R.

e.) What is the Moore-Penrose pseudo-inverse of the zero matrix?

f.) Show that if A is symmetric then A^+ is symmetric.

g.) Show that $(A^k)^+ = (A^+)^k$.

h.) If $A = [1 \ 2; 2 \ 1; -1 \ 2]$, find A^+ and use it to find the least squares solution of $Ax = (2,2,2)^T$. Also, find the condition number of A, defined by $\kappa(A) = \|A\|\|A^+\|$.

9.) a.) If the $m \times n$ matrix A ($m \leq n$) is of full rank then we define its Moore-Penrose pseudo-inverse to be the square matrix $A^+ = A^T (AA^T)^{-1}$. Show that the least squares solution of $Ax = b$ is given by $x = A^+ b$ (when the underdetermined system is consistent its least squares solution is the one for which $\|x\|$ is smallest).

b.) Repeat Problem 8 parts b. through g. for this case, if possible. Not all cases will be true! For example, $A^+ A$ need not equal I, though $AA^+ A = A$ and $A^+ AA^+ = A^+$.

10.) Prove that the MGS Algorithm is mathematically equivalent to the CGS Algorithm.

11.) a.) Using your CGS Algorithm program from Problem 4 and your MGS Algorithm program from Problem 5, demonstrate experimentally the numerical superiority of the MGS Algorithm. Explain the reasoning behind your design of the experiment.

b.) Using your MGS Algorithm program from Problem 5, compare the accuracy of the MGS Algorithm to that of the Householder triangularization approach (as implemented in the MATLAB qr command.) Explain the reasoning behind your design of the experiment.

12.) a.) Prove that if the $m \times n$ matrix A $(m \geq n)$ is of full rank then the CGS Algorithm will produce an orthogonal Q and an upper triangular R with positive entries on the main diagonal.

b.) Prove that if the $m \times n$ matrix A $(m \geq n)$ is of full rank then the MGS Algorithm will produce an orthogonal Q and an upper triangular R with positive entries on the main diagonal.

13.) a.) Assuming A is nonsingular, express $\|A\|_s$ in terms of the spectral norms of its QR factors Q and R.

b.) Assuming A is nonsingular, express $\kappa_s(A)$ in terms of the condition numbers of its QR factors Q and R.

c.) Do the results of parts a. and b. hold when the quantities are computed in MATLAB? Experiment to determine this.

14.) Develop an RQ decomposition of A by applying the Gram–Schmidt process to the rows of A, beginning with the last row.

15.) Give an exact floating point operations count for the CGS Algorithm and for the MGS Algorithm.

16.) Prove that the Gram-Schmidt process transforms a linearly independent set of vectors to an orthonormal set of vectors spanning the same vector space.

10. The Singular Value Decomposition

We have considered two principal techniques for solving $Ax = b$: The LU decomposition and the QR decomposition. In the next chapter we will discuss iterative methods for this problem. But there is one other major matrix decomposition technique: The **singular value decomposition** (**SVD**)

$$(10.1) \qquad A = U\Sigma V^T$$

of an $m \times n$ matrix A. Here U is an $m \times m$ orthogonal matrix, V is an $n \times n$ orthogonal matrix, and Σ is an $m \times n$ diagonal matrix with nonnegative entries on its main diagonal:

$$\begin{bmatrix} X & X & X \\ X & X & X \\ X & X & X \\ X & X & X \\ X & X & X \end{bmatrix} = \begin{bmatrix} X & X & X & X & X \\ X & X & X & X & X \\ X & X & X & X & X \\ X & X & X & X & X \\ X & X & X & X & X \end{bmatrix} \begin{bmatrix} X & 0 & 0 \\ 0 & X & 0 \\ 0 & 0 & X \\ 0 & 0 & 0 \\ 0 & 0 & 0 \end{bmatrix} \begin{bmatrix} X & X & X \\ X & X & X \\ X & X & X \end{bmatrix}$$

(a 5×3 example). The diagonal entries of Σ are called the **singular values** of A. The singular values are denoted

$$\sigma_i = \Sigma_{ii}$$

and are ordered so that $\sigma_1 \geq \sigma_2 \geq \cdots \geq \sigma_p$ (where there are $p = \min(m,n)$ singular values). There are exactly $r = \text{rank}(A)$ positive singular values, and any additional diagonal entries of Σ are zero. It is a fact that the nonzero singular values of A are the square roots of the nonzero eigenvalues of AA^T, which are the same as the nonzero eigenvalues of $A^T A$.

Every matrix has a singular value decomposition, and this decomposition occurs in many applications so it is important to have some general familiarity with it. We will only be able to present a brief overview of it as a detailed discussion of the SVD requires knowledge of linear algebra beyond what is assumed here.

Suppose that A is square and nonsingular. Then if we wish to solve $Ax = b$ and know an SVD of A, we can write

$$
\begin{aligned}
U\Sigma V^T x &= b \\
U^T U\Sigma V^T x &= U^T b \\
\Sigma V^T x &= U^T b \\
V^T x &= \Sigma^{-1} U^T b \\
VV^T x &= V\Sigma^{-1} U^T b \\
x &= V\Sigma^{-1} U^T b
\end{aligned}
$$

(since $U^T U = I$ and $VV^T = I$ by orthogonality). Because A has full rank all eigenvalues of $A^T A$ are nonzero and hence all diagonal entries of Σ are nonzero as well, so the square diagonal matrix Σ is nonsingular and Σ^{-1} is trivial to compute. Hence if we can compute the SVD accurately then we can solve $Ax = b$ very efficiently.

Similarly, consider the problem of finding the least squares solution of the overdetermined linear system $Ax = b$. The least squares solution of $Ax = b$ is the solution of $A^T Ax = A^T b$ (see Sec. 2.7), that is, the solution of

$$
\begin{aligned}
(U\Sigma V^T)^T U\Sigma V^T x &= (U\Sigma V^T)^T b \\
V\Sigma^T U^T U\Sigma V^T x &= V\Sigma^T U^T b \\
V\Sigma^T \Sigma V^T x &= V\Sigma^T U^T b \\
\text{(10.2)} \qquad V^T V\Sigma^T \Sigma V^T x &= V^T V\Sigma^T U^T b \\
\Sigma^T \Sigma V^T x &= \Sigma^T U^T b \\
\Sigma V^T x &= U^T b \\
V^T x &= \Sigma^{-1} U^T b \\
x &= V\Sigma^{-1} U^T b
\end{aligned}
$$

(some additional justification is needed to remove the factor Σ^T in going from $\Sigma^T \Sigma V^T x = \Sigma^T U^T b$ to $\Sigma V^T x = U^T b$). This is the same formal solution as we found for the linear system $Ax = b$, but recall that A is no longer a square matrix.

One advantage of the SVD is clear: We can compute the (least squares) solution $x = V\Sigma^{-1} U^T b$ efficiently once the SVD is known. In addition, since the condition number of an orthogonal matrix is always unity for any natural norm, we can do so very accurately.

Unfortunately, the SVD is significantly more expensive to compute than the QR factorization when m and n are of similar size. (If $m \gg n$ then the work required is comparable.) The QR factorization in turn is more expensive to compute than the LU decomposition. If m and n are equal then solving a least squares problem by the SVD is about an order of magnitude more costly than using the QR factorization. For many applications the LU decomposition has perfectly acceptable accuracy, but for least squares problems it is generally advisable to use the QR factorization unless the problem is known to be a difficult[13] one, in which case using the SVD is probably justified. (Recall that the condition number for solving $A^T A x = A^T b$ is roughly the square of that for solving $Ax = b$ so extra care is needed for these problems.) As always, the choice will depend on the user's needs as well as available hardware and software. The SVD is typically the most expensive, but most accurate, way to solve these types of problems. Careful monitoring of the smallest nonzero singular values can allow for even better results.

Software packages often return the **reduced singular value decomposition** $A = U_1 \Sigma_1 V^T$ with U_1 an $m \times n$ matrix with orthonormal columns, V an $n \times n$ matrix with orthonormal columns, and Σ_1 a diagonal $n \times n$ matrix. (The SVD in Eq. (10.1) is called the **full singular value decomposition**.) The reduced SVD removes the zero rows at the bottom of Σ, if any, as they do not affect any computations.

We briefly mention a few interesting facts about the SVD: If $r = \text{rank}(A)$ is positive then the first r columns of U form a basis for the column space of A and the first r columns of V form a basis for the row space of A. The Frobenius norm of a square matrix A is equal to $\sqrt{\sigma_1^2 + \ldots + \sigma_p^2}$ and the spectral norm is given by $\|A\|_2 = \sigma_1$ (the largest singular value of A). These ideas are further explored in the Problems.

There are multiple ways to compute the SVD, but the details are beyond the scope of this text. A typical algorithm uses Householder reflectors and Givens rotations to obtain an upper bidiagonal matrix B, that is, a matrix that is nonzero only on its diagonal and immediate superdiagonal, as an intermediate step. This gives an equation similar to Eq. (10.1) but with the middle matrix bidiagonal before proceeding to find the actual SVD. Other algorithms are based on the relationship between the singular values of A and the eigenvalues of $A^T A$.

The SVD is important in many applications in scientific computing, including finding the rank of a matrix, rank-deficient least squares problems, improving the conditioning of an ill-conditioned matrix, information retrieval by search engines (where the term-by-document matrix A may be on the order of millions by millions), image processing, and many others. We might have a term-by-document matrix with movies as rows and users as columns, say (with 8 movies and 6 users)

$$M = \begin{pmatrix} 2 & 3 & 0 & 0 & 0 & 3 \\ 0 & 0 & 0 & 1 & 0 & 0 \\ 0 & 0 & 1 & 0 & 2 & 0 \\ 4 & 0 & 0 & 1 & 0 & 0 \\ 0 & 2 & 0 & 0 & 0 & 0 \\ 0 & 0 & 0 & 0 & 3 & 0 \\ 0 & 0 & 3 & 0 & 0 & 2 \\ 0 & 0 & 0 & 1 & 0 & 0 \end{pmatrix}$$

indicating that, for example, user 2 has watched movie 1 exactly 3 times. We expect a matrix like this to be large and very sparse. One way of formulating recommendation for users is to find the SVD $U\Sigma V^T$ of this matrix, reduce the smallest positive singular values to zero in Σ, giving Σ_R, and then form a rank-reduced version of M using Σ_R with the same U and V, giving

---

[13]Here difficult generally means that A is rank-deficient or nearly so.

$$M_R = U\Sigma_R V^T$$

the rank of which has been decreased by the number of singular values zeroed out. (This technique is called **rank reduction** and appears in a variety of contexts in computational matrix algebra.) This can be used to cluster users together into groups with similar interests by using M_R as a projection matrix; the lower rank effectively compels M_R to bring like-minded users, as well as movies, into clusters. (The same thing can be done with the QR decomposition, by changing some nonzero entries on the main diagonal of R to zero.) This idea leads to a data-mining technique called **latent semantic indexing**.

Unfortunately, going into these matters in more detail would require familiarity with linear algebra material just beyond what we have assumed as prerequisite. But the SVD is so important in scientific computing that omitting it entirely would have been an oversight. We will focus below on gaining facility with using the SVD in MATLAB so that you are sufficiently knowledgeable to be able to make use of it when needed.

MATLAB

The MATLAB command for computing the SVD of a matrix is [U,S,V]=svd(A), which returns the SVD factors U, Σ, and V. The command [U,S,V]=svd(A,0) gives the reduced SVD of A. Note that the U of the SVD is different from the U of the LU decomposition.

Let's solve a linear least squares problem using the SVD. We'll use a 100×10 matrix (100 equations in 10 unknowns). Enter:

```
» A=rand([100 10]);
» b=rand([100 1]);
» [U,S,V]=svd(A);
» x=V*(S\(U'*b));          %Solution.
» norm(A*x-b)              %Residual.
» [U0,S0,V0]=svd(A,0);     %Reduced form.
» x0=V0*(S0\(U0'*b));      %Solution.
» norm(A*x0-b)             %Residual.
```

Let's check the SVD itself. Enter:

```
» norm(A-U*S*V')
```

to check how well USV^T reconstructs A.

One of the many computational applications of the SVD is in finding the rank of a matrix numerically. (Defining the numerical rank of a matrix is an issue in itself.) Enter:

```
» type rank
```

to see that MATLAB computes the rank of a matrix A by finding its SVD and counting the number of singular values that are significantly greater than zero. For example, enter:

```
» rank(A)
» diag(S)       %Singular values of A.
» S1=S;S1(10,10)=0;diag(S1)
» A1=U*S1*V';
» rank(A1)
```

Setting a singular value to zero reduces the rank of the reconstructed matrix. This is used to produce lower-rank approximations of A, which are often needed in applications.

Problems

1.) Find an SVD of I_4. Is it unique?

2.) What matrix has $U = [1/\sqrt{2}\ 0\ 1/\sqrt{2}; 0\ 1\ 0; 1/\sqrt{2}\ 0\ -1/\sqrt{2}]$, $V = [1/\sqrt{2}\ -1/\sqrt{2}; 1/\sqrt{2}\ 1/\sqrt{2}]$, and $\Sigma = [2\ 0; 0\ 1; 0\ 0]$ as the factors in its SVD? What are its singular values? What is its rank?

3.) Carefully justify the derivation in Eq. (10.2). Pay attention to the sizes of the matrices involved.

4.) Prove that the rank of a matrix is equal to the number of nonzero singular values it has.

5.) Prove that the nonzero singular values of A are the square roots of the nonzero eigenvalues of AA^T and of $A^T A$.

6.) If A is an 6×4 matrix, what are the sizes of U, Σ, and V in its full SVD? What are the sizes of U_1, Σ_1, and V in its reduced SVD?

7.) a.) Write a MATLAB program that finds the rank of a matrix by finding its SVD (use the MATLAB svd command) and then counting the number of singular values that exceed a user-supplied tolerance.

b.) Write a MATLAB program that computes the rank of a matrix by finding its QR decomposition (use the MATLAB qr command) and finding the number of diagonal entries of R having magnitudes that exceed a user-supplied tolerance.

c.) Compare your programs on several test matrices. Can you explain why the SVD is so often used in practice rather than the QR decomposition?

8.) a.)Construct the matrix for interpolating a degree 15 polynomial to $\sin(x)$ on 200 equally spaced nodes on $[0, 1]$. Solve for the coefficients using the LU decomposition, the QR decomposition, and the SVD. Give the residual errors.

b.) Construct the matrix for interpolating a degree 20 polynomial to $\sin(x)$ on 400 equally spaced nodes on $[0, 1]$. Solve for the coefficients using the LU decomposition, the QR decomposition, and the SVD. Give the residual errors.

9.) If A has full rank must $A^T A$ and AA^T have full rank too?

10.) Prove that $\det(A) = \pm\sigma_1 \cdots \sigma_p$.

11.) a.) Prove that if $r = \text{rank}(A)$ is positive then the first r columns of U form a basis for the column space of A.

b.) Prove that if $r = \text{rank}(A)$ is positive then the first r columns of V form a basis for the row space of A.

12.) Prove that the Frobenius norm of a square matrix A is equal to $\sqrt{\sigma_1^2 + \dots + \sigma_p^2}$.

13.) a.) It is a fact that $\|A\|_2 = \sigma_1$ (the largest singular value of A). What is $\kappa(A)$ in terms of the singular values of A? What does this imply about the numerical significance of the sizes of the singular values of A? Demonstrate this by using the SVD to create matrices with large condition numbers by intentionally setting the largest singular value to a large number.

b.) Write a MATLAB program that computes the condition number of a matrix using this fact and the svd command. (This is not an efficient approach to finding $\kappa(A)$.)

14.) How could the SVD be used to find A^{-1}? Is this practical computationally?

15.) Prove that every matrix has a singular value decomposition.

16.) a.) It is a fact that $A_k = \sum_{i=1}^{i=k} \sigma_i u_i v_i^T$, where u_i and v_i are the columns of U and V, respectively, is equal to A if $k = \text{rank}(A)$. Verify this for 3 matrices.

b.) Let $r = \text{rank}(A)$. Verify for 3 matrices that if $A_k = \sum_{i=1}^{i=k} \sigma_i u_i v_i^T$ $(k = 1, ...r)$ then $\|A - A_k\|$ is a monotonically decreasing function of k. (Recall that $\sigma_1 \geq \sigma_2 \geq \cdots \geq \sigma_r$, with all other singular values zero, by convention.)

c.) Prove that A_k has rank k for $k \leq r$. What happens if $k > r$?

d.) The **Eckart-Young Theorem** states that A_k is the best approximation to A among all matrices of rank k in the sense that $\left\| A - \tilde{A} \right\|_F$ is minimized w.r.t. the Frobenius norm when $\tilde{A} = A_k$, where $\tilde{A}$ is an arbitrary rank k matrix, and that $\|A - A_k\|_F^2 = \sum_{i=k+1}^{r} \sigma_i^2$. Verify this experimentally.

e.) The **Eckart-Young-Mirsky Theorem** states that A_k is the best approximation to A among all matrices of rank k in the sense that $\left\| A - \tilde{A} \right\|_s$ is minimized w.r.t. the spectral norm when $\tilde{A} = A_k$, where $\tilde{A}$ is an arbitrary rank k matrix, and that $\|A - A_k\|_s = \sigma_{k+1}$. Verify this experimentally.

f.) Select an image that is represented as a single matrix A. (You can find demo images in MATLAB.) Decompose it as $A = U\Sigma V^T$ and then form A_k where k is about 90% of $\text{rank}(A)$. Display the image A_k. How does it compare to A? Depending on the image, it might look better as this process can smooth out unwanted noise. Repeat for smaller and smaller k. Then repeat this process for two more images. Comment on the effect of this process of smoothing via rank reduction.

CHAPTER 2 BIBLIOGRAPHY:

Applied Numerical Linear Algebra, William W. Hager, Prentice-Hall 1988
An undergraduate-level introductory textbook.
Numerical Linear Algebra, Lloyd N. Trefethen and David Bau, III, SIAM 1997
An informal graduate-level applied introduction to numerical linear algebra.
Matrix Algorithms Volume I Basic Decompositions, G. W. Stewart, SIAM 1998
A detailed, modern book on matrix decompositions.
Matrix Computations, 3rd ed., Gene H. Golub and Charles F. Van Loan, The
Johns Hopkins University Press 1996
The classic reference on the subject.
Applied Numerical Linear Algebra, James W. Demmel, SIAM 1997
A comprehensive graduate-level text.
Numerical Methods for Least Squares Problems, Åke Björck, SIAM 1996
A detailed graduate-level reference.
Numerical Analysis, 6th ed., Richard L. Burden and J. Douglas Faires, Brooks/Cole
Publishing Company 1997
A traditional introductory undergraduate level textbook.
*Introduction to Scientific Computing: A Matrix-Vector Approach Using MAT-
LAB*, 2nd ed., Charles F. Van Loan, Prentice-Hall 2000
An introductory survey with an emphasis on computing in MATLAB.
*Evaluating Derivatives: Principles and Techniques of Algorithmic Differentia-
tion*, Andreas Griewank, SIAM 2000
A comprehensive introduction to automatic differentiation.
Data Science for Mathematicians, Nathan Carter (ed.), CRC Press 2020

CHAPTER 3

Iterative Methods

1. Jacobi and Gauss-Seidel Iteration

In the previous chapter we discussed direct methods for the solution of linear systems. These are the methods of choice for moderate size problems. Computing the LU decomposition of an $n \times n$ matrix, for example, requires about $\frac{2}{3}n^3$ floating point operations. If $n = 100$ this is about 6.7×10^5 floating point operations which takes well under a second on a standard desktop computer; if $n = 1000$ however this is about 6.7×10^8 flops which takes considerably longer to finish (due to memory management issues in addition to the thousand-fold extra flops required). Since matrices as large as tens of thousands by tens of thousands are commonly encountered in practice (in problems involving the numerical solution of partial differential equations, such as in fluid dynamics) and matrices as large as hundreds of thousands by hundreds of thousands are not uncommon (for example, applications in data science, or the analysis of genetic data), more efficient methods are clearly needed.

We will have little to say about large, dense matrices. If approximating such a matrix with a simpler matrix is not acceptable, the computation will take a long time. Moving entries of the matrix from memory to the processor(s) will likely be more time-consuming than the actual computations. These are extremely challenging problems—consult an expert.

Fortunately, it is common for large matrices occurring in practice to be sparse. For large, sparse matrices there are a number of iterative methods that—in combination with a smart storage system for the sparse matrix—can lead to considerably shorter computation times. As a rule, direct methods are $O(n^3)$ and iterative methods are, in the cases in which they are appropriate, $O(n^2)$. The goal for sparse matrices is always to get a method that is $O(N)$ where N is the number of nonzero entries in a typical row of the matrix.

In this section we will discuss two classical methods for the solution of linear systems by iterative methods. Remember, the implicit assumption of sparsity is standard when we consider these methods for solving $Ax = b$–they may not be very efficient otherwise.

Consider a square linear system $Ax = b$. We seek an iteration of the form $x^{k+1} = F(x^k)$ where an initial guess $x^0 \in \mathbb{R}^\times$ is given and F is simple to compute. One way to achieve this is based on additively decomposing A into its upper triangle, lower triangle, and diagonal parts, called a **matrix splitting**:

(1.1) $$A = U + L + D$$

DOI: 10.1201/9781003042273-3

where U is a square matrix with the strict upper triangle of A in its upper triangle and is zero otherwise, L is a square matrix with the strict lower triangle of A in its lower triangle and is zero otherwise, and D is a diagonal matrix containing the main diagonal of A. By "strict" upper or lower triangle, we mean that the diagonal is excluded. For example,

$$
\begin{pmatrix} 1 & 2 & 3 \\ 4 & 5 & 6 \\ 7 & 8 & 9 \end{pmatrix} = \begin{pmatrix} 0 & 2 & 3 \\ 0 & 0 & 6 \\ 0 & 0 & 0 \end{pmatrix} + \begin{pmatrix} 0 & 0 & 0 \\ 4 & 0 & 0 \\ 7 & 8 & 0 \end{pmatrix} + \begin{pmatrix} 1 & 0 & 0 \\ 0 & 5 & 0 \\ 0 & 0 & 9 \end{pmatrix}
$$
$$
= \quad U + L + D
$$

is a splitting of the form of Eq. (1.1). Now we can write

$$
\begin{aligned}
Ax &= b \\
(U + L + D)x &= b \\
Dx &= -(U + L)x + b \\
x &= -D^{-1}(U + L)x + D^{-1}b
\end{aligned}
$$
(1.2)

(if no diagonal element of A is zero, which we will assume throughout this section). This has the form $x = g(x)$ of a fixed point iteration; perhaps

(1.3) $$ x^{k+1} = -D^{-1}(U + L)x^k + D^{-1}b $$

will be a useful iteration. Indeed, Eq. (1.3) is called **Jacobi iteration**. Note that D is diagonal so the computation of D^{-1} is trivial.

EXAMPLE 1: Let's apply Jacobi iteration to the system $Ax = (3, -1, 4)^T$ with the matrix $A = [4\ 2\ 1; 1\ 3\ 1; 1\ 1\ 4]$. From Eq. (1.2) we form $Dx = -(U + L)x + b$ which is equivalent to moving the nondiagonal entries to the RHS; that is,

$$
\begin{aligned}
4x_1 + 2x_2 + x_3 &= 3 \\
x_1 + 3x_2 + x_3 &= -1 \\
x_1 + x_2 + 4x_3 &= 4
\end{aligned}
$$

becomes

$$
\begin{aligned}
4x_1 &= 3 - 2x_2 - x_3 \\
3x_2 &= -1 - x_1 - x_3 \\
4x_3 &= 4 - x_1 - x_2
\end{aligned}
$$

corresponding to the matrix-vector form $Dx = -(U + L)x + b$. We obtain Eq. (1.3)

$$
\begin{aligned}
x_1^{k+1} &= \frac{1}{4}(3 - 2x_2^k - x_3^k) \\
x_2^{k+1} &= \frac{1}{3}(-1 - x_1^k - x_3^k) \\
x_3^{k+1} &= \frac{1}{4}(4 - x_1^k - x_2^k)
\end{aligned}
$$

by dividing out the diagonal entries and indexing the two sides appropriately with $k = 0, 1, 2, \ldots$ as our iteration counter. Let's take $x^0 = (0, 0, 0)^T$ as the initial guess. Then we have

$$
\begin{aligned}
x_1^1 &= \frac{1}{4}(3 - 2 \cdot 0 - 0) \\
&= \frac{3}{4} \\
x_2^1 &= \frac{1}{3}(-1 - 0 - 0) \\
&= -\frac{1}{3} \\
x_3^1 &= \frac{1}{4}(4 - 0 - 0) \\
&= 1
\end{aligned}
$$

$$
\begin{aligned}
x_1^2 &= \frac{1}{4}(3 - 2 \cdot \frac{-1}{3} - 1) \\
&\doteq .6667 \\
x_2^2 &= \frac{1}{3}(-1 - \frac{3}{4} - 1) \\
&\doteq -.9167 \\
x_3^2 &= \frac{1}{4}(4 - \frac{3}{4} - \frac{-1}{3}) \\
&\doteq .8958
\end{aligned}
$$

which seems to be slowly converging to the true solution $x_1 = 1$, $x_2 = -1$, $x_3 = 1$. The method does not recognize that x_3^1 was exact or that x_1^1 was a better estimate than x_1^2; but the errors for these estimates are

$$
\begin{aligned}
\alpha_1 &= \|x^1 - x\| \doteq .7120 \\
\alpha_2 &= \|x^2 - x\| \doteq .3590 \\
\rho_1 &= \|x^1 - x\| \doteq .7120/\sqrt{3} \\
&\doteq .4111 \\
\rho_2 &= \|x^2 - x\| \doteq .3590/\sqrt{3} \\
&\doteq .2073
\end{aligned}
$$

so the method is certainly reducing the overall error. We would continue iterating until some convergence criterion was met. $\square$

We used the formula for Jacobi iteration in component form in Example 1. If x_i^k is used to denote the ith component of x^k, then Jacobi iteration corresponds to solving equation i for the diagonal entry x_i,

$$a_{ii}x_i = b_i - \sum_{\substack{j=1 \\ j \neq i}}^{n} a_{ij}x_j$$

$$x_i = \frac{b_i}{a_{ii}} - \frac{1}{a_{ii}} \sum_{\substack{j=1 \\ j \neq i}}^{n} a_{ij}x_j$$

then iterating

$$x_i^{k+1} = \frac{b_i}{a_{ii}} - \frac{1}{a_{ii}} \sum_{\substack{j=1 \\ j \neq i}}^{n} a_{ij}x_j^k$$

until some convergence criterion is met. The convergence criterion used for Jacobi iteration may be that the absolute error, relative error, or residual error

$$r_k = \|b - Ax_k\|$$

be less than some tolerance or any similar rule. We generally will use the approximate relative error

$$\rho_k = \frac{\left\|x^k - x^{k-1}\right\|}{\|x^k\|}$$

as the convergence criterion. Often the tolerance τ is set as some small percentage of $\|x^0\|$ (but not smaller than some $\tau_\alpha > 0$; see Sec. 2.4).

We now ask the standard questions: Does Jacobi iteration (Eq. (1.3)) converge for all matrices and all initial guesses x^0? How rapidly does it converge? Is the method stable? And can we accelerate the convergence?

Experience quickly shows that Jacobi iteration can diverge for some initial guesses for some matrices. In order to explore the convergence properties of this method, let's write Eq. (1.3) as

$$x^{k+1} = Mx^k + \beta$$

where the matrix $M = -D^{-1}(U+L)$ and vector $\beta = D^{-1}b$ are known. Any iteration that can be written in the form $x^{k+1} = Mx^k + \beta$ is said to be a **stationary method** (meaning that the parameters M and β do not depend on k). Note that

$$\begin{aligned} \left\|x^{k+1}\right\| &= \left\|Mx^k + \beta\right\| \\ &\leq \|M\|\left\|x^k\right\| + \|\beta\| \end{aligned}$$

for the natural norm induced by the chosen vector norm (see Sec. 2.6). So

$$\begin{aligned}
\left\|x^{k+1}\right\| &\leq \|M\|\left\|x^k\right\| + \|\beta\| \\
&\leq \|M\|\left(\|M\|\left\|x^{k-1}\right\| + \|\beta\|\right) + \|\beta\| \\
&= \|M\|^2\left\|x^{k-1}\right\| + \|M\|\|\beta\| + \|\beta\| \\
&= \|M\|^2\left\|x^{k-1}\right\| + (\|M\|+1)\|\beta\| \\
&\leq \|M\|^2\left(\|M\|\left\|x^{k-2}\right\| + \|\beta\|\right) + (\|M\|+1)\|\beta\| \\
&= \|M\|^3\left\|x^{k-2}\right\| + (\|M\|^2 + \|M\| + 1)\|\beta\| \\
&\vdots \\
&\leq \|M\|^{k+1}\left\|x^0\right\| + (\|M\|^k + ... + \|M\| + 1)\|\beta\|
\end{aligned}$$

and so if $\|M\| < 1$ then

$$(1.4) \qquad \left\|x^{k+1}\right\| \leq \|M\|^{k+1}\left\|x^0\right\| + \frac{1}{1-\|M\|}\|\beta\|$$

since $\|M\|^k + ... + \|M\| + 1$ is a partial sum of the geometric series

$$(1.5) \qquad \sum_{k=0}^{\infty}\|M\|^k = \frac{1}{1-\|M\|}$$

which contains only positive terms. Since $\|M\|^{k+1} \to 0$ as $k \to \infty$ if $\|M\| < 1$, we suspect from Eq. (1.4) that $\|M\| < 1$ will give convergence. Unfortunately, all we've shown is that the norm of the iterates is bounded, and that in the limit of large k the bound is essentially

$$\left\|x^{k+1}\right\| \leq \frac{1}{1-\|M\|}\|\beta\|$$

(plus the increasingly small contribution of $\|M\|^{k+1}\left\|x^0\right\|$ in Eq. (1.4)) if $\|M\| < 1$. This isn't enough to guarantee convergence of x^k.

We can obtain a more precise result by looking at the eigenvalues of M and in particular its largest eigenvalue. Recall from Sec. 2.6 that the spectral radius of a matrix M is

$$\rho(M) = \max_{i=1}^{n}(|\lambda_i|)$$

where $\lambda_1, ..., \lambda_n$ are the eigenvalues of M. It is a fact that if $\rho(M) < 1$ then M is **convergent,**, meaning that $M^k \to 0$ as $k \to \infty$ (that is, every entry of M^k tends to the the scalar zero so that M^k itself tends to the zero matrix). If M is convergent then

$$\sum_{k=0}^{\infty}M^k = (I-M)^{-1}$$

(an analogue of Eq. (1.5)). The quantity $(I-M)^{-1}$ is called the **resolvent** of M; note that M itself need not be invertible for the resolvent to exist. Now, consider again the linear fixed point iteration

$$
\begin{aligned}
x^{k+1} &= Mx^k + \beta \\
&= M(Mx^{k-1} + \beta) + \beta \\
&= M^2 x^{k-1} + (M + I)\beta \\
&= M^2(Mx^{k-2} + \beta) + (M + I)\beta \\
&= M^3 x^{k-2} + (M^2 + M + I)\beta \\
&\quad\vdots \\
&= M^{k+1}x^0 + (M^k + \ldots + M + I)\beta
\end{aligned}
$$

in terms of x^0. If M is convergent,

$$
\begin{aligned}
\lim_{k\to\infty} x^{k+1} &= \lim_{k\to\infty}(M^{k+1}x^0 + (M^k + \ldots + M + I)\beta) \\
&= 0 + (I - M)^{-1}\beta \\
&= (I - M)^{-1}\beta
\end{aligned}
$$

for any initial guess x^0. In fact, the iteration $x^{k+1} = Mx^k + \beta$ converges to this unique value for every initial guess if and only if M is convergent. (If $\rho(M) > 1$ the method can still converge for some x^0, in principle, but not for every x^0. For such an M the method is unstable however. The case $\rho(M) = 1$ requires special care.) Indeed, $x = (I - M)^{-1}\beta$ is a fixed point of the method, for

$$
\begin{aligned}
Mx + \beta &= M(I - M)^{-1}\beta + \beta \\
&= M\sum_{k=0}^{\infty} M^k\beta + \beta \\
&= \sum_{k=0}^{\infty} M^{k+1}\beta + \beta \\
&= \sum_{k=1}^{\infty} M^k\beta + I\beta \\
&= \sum_{k=0}^{\infty} M^k\beta \\
&= (I - M)^{-1}\beta \\
&= x
\end{aligned}
$$

(as $M^0 = I$ by convention). In summary, if M is convergent then $x = Mx + \beta$ has a unique fixed point at $x = (I - M)^{-1}\beta$ and fixed point iteration always converges to this value (compare the Contraction Mapping Theorem).

Hence, Jacobi iteration, or any other matrix splitting technique, will converge if the matrix $A = U + L + D$ leads to an M that is a convergent matrix. Since it will converge for *any* initial guess, stability is immediate (a perturbation of a given x^k will necessarily perturb it to another value from which the method converges). The speed of convergence can be shown to be like $\rho(M)^k$, that is,

$$\frac{\|x^k - x\|}{\|x^0 - x\|} = O(\rho(M)^k)$$

as $k \to \infty$. (This is linear convergence.) Hence a small spectral radius is desirable, and if $\rho(M)$ is near unity then we must expect very slow convergence.

Clearly, it will not always be the case that A is such that $M = -D^{-1}(U + L)$ is convergent. However, there is an important class of matrices for which this will always be the case. We say that a square matrix A is **diagonally dominant** if

$$|a_{ii}| \geq \sum_{\substack{j=1 \\ j \neq i}}^{n} |a_{ij}|$$

$(i = 1, ..., n)$ that is, if the diagonal element of each row equals or exceeds, in absolute value, the sum of the absolute values of all other entries in that row. If the inequality is strict we say that the matrix is **strictly diagonally dominant** and if it is weak we say that the matrix is **weakly diagonally dominant**. If A is strictly diagonally dominant then $M = -D^{-1}(U + L)$ is convergent and Jacobi iteration will converge. In fact, the method will frequently converge if A is weakly diagonally dominant. Diagonally dominant matrices occur often in applications involving the numerical solution of partial differential equations. Symmetric diagonally dominant matrices are positive semi-definite, and symmetric strictly diagonally dominant matrices are positive definite.

If A is not diagonally dominant then in principle we must check $\rho(M)$ to see if the method is applicable. This is usually too expensive however and so we use the fact that

$$\rho(M) \leq \|M\|$$

for any natural norm. If $\|M\| < 1$ in some natural norm then M is convergent and we may use Jacobi iteration with confidence that it will succeed.

EXAMPLE 2: The matrix $A = [1\ 1\ 3; 1\ 3\ 1; 3\ 1\ 1]$ is not diagonally dominant, but interchanging the first and last rows gives $A_1 = [3\ 1\ 1; 1\ 3\ 1; 1\ 1\ 3]$ which is strictly diagonally dominant. Jacobi iteration will converge if applied to A_1.

The matrix $M = [.3\ .2\ .1; .2\ .2\ -.2; .4\ -.5\ 0]$ has Frobenius norm (see Sec. 2.6) $\|M\| = \sqrt{.3^2 + .2^2 + ... + (-.5)^2 + 0^2} \doteq .8185$ so $\rho(M) < 1$. (In fact, $\rho(M) \doteq .4531$.) Hence, iteration of $x^{k+1} = Mx^k + c$ will converge for any c and any x^0. $\square$

The easiest norm to use in checking whether $\|M\| < 1$ is the matrix norm that is induced by the l_∞ vector norm (see Sec. 2.6),

$$\|M\|_\rho = \max_{i=1}^{n} \left\{ \sum_{j=1}^{n} |m_{ij}| \right\}$$

that is, the maximum row sum of M. This matrix norm is called the **max norm** or **inf norm**. The spectral norm $\|M\|_s$ will be closer to $\rho(M)$ but is harder to compute. For the matrix M in Example 2, the row sums are .6, .6, and .9, so $\|M\| = .9$. This is larger than the Frobenius norm $\|A\|_F \doteq .8185$ and the spectral norm $\|A\|_s \doteq .6419$ and all exceed the spectral radius $\rho(M) \doteq .4531$, as they must. Based only on the max norm we would estimate that the error decreases to about

90% of its previous level each iteration in the limit (that is, the error only decreases by about 10%); based on the Frobenius norm, to about 82% of its previous level; the spectral norm, about 64%; and the spectral radius gives the correct value, about 45% (the error decreases by slightly more than half). The spectral radius may be estimated if it is truly needed.

If $\rho(M)$ is near unity, convergence will be slow. Can we speed it up? One possibility is suggested by taking another look at what we did in Example 1.

EXAMPLE 3: Consider again the system $Ax = (3, -1, 4)^T$ with the matrix $A = [4\,2\,1; 1\,3\,1; 1\,1\,4]$. We have from Eq. (1.3) that

$$
\begin{aligned}
x_1^{k+1} &= \frac{1}{4}(3 - 2x_2^k - x_3^k) \\
x_2^{k+1} &= \frac{1}{3}(-1 - x_1^k - x_3^k) \\
x_3^{k+1} &= \frac{1}{4}(4 - x_1^k - x_2^k)
\end{aligned}
$$

and we will again take $x^0 = (0, 0, 0)^T$ as the initial guess. Then we have

$$
\begin{aligned}
x_1^1 &= \frac{1}{4}(3 - 2 \cdot 0 - 0) \\
&= \frac{3}{4} \\
x_2^1 &= \frac{1}{3}(-1 - x_1^0 - x_3^0)
\end{aligned}
$$

but now it may occur to us that we have a presumably improved estimate of the true value of x_1, namely x_1^1, and we may wish to use this better estimate in place of x_1^0. Let's do that:

$$
\begin{aligned}
x_2^1 &= \frac{1}{3}(-1 - \frac{3}{4} - 0) \\
&= -\frac{7}{12}
\end{aligned}
$$

a better estimate of $x_2 = -1$ than the previous $x_2^1 = -1/3$. Let's use the same trick for x_3^1:

$$
\begin{aligned}
x_3^1 &= \frac{1}{4}(4 - \frac{3}{4} - \frac{-7}{12}) \\
&= \frac{23}{24} \\
&\doteq .9583
\end{aligned}
$$

which is very close to the previous estimate of 1. Continuing in this way, using new estimates as soon as they become available, we have

$$x_1^2 = \frac{1}{4}\left(3 - 2 \cdot \frac{-7}{12} - \frac{23}{24}\right)$$

$$\doteq .8021$$

$$x_2^2 = \frac{1}{3}\left(-1 - .8021 - \frac{23}{24}\right)$$

$$\doteq -.9201$$

$$x_3^2 = \frac{1}{4}(3 - .8021 - -.9201)$$

$$\doteq .7795$$

which is superior to the estimate $x^2 = (.7292, -.8333, .6458)^T$ from Jacobi iteration in every component (recall that the solution is $x = (1, -1, 1)^T$). $\square$

The method of Example 3 is referred to as **Gauss-Seidel iteration**. It simply uses newer information as soon as it becomes available (a simple idea that is widely applicable in scientific computing). Gauss-Seidel iteration corresponds to the matrix splitting

$$Ax = b$$
$$(U + L + D)x = b$$
$$(L + D)x = -Ux + b$$
$$x = -(L + D)^{-1}Ux + (L + D)^{-1}b$$

(i.e., $M = -(L + D)^{-1}U$ if $(L + D)^{-1}$ exists). This gives the iteration

(1.6)
$$x^{k+1} = -(L + D)^{-1}Ux^k + (L + D)^{-1}b.$$

Of course, we would not form $(L + D)^{-1}$ explicitly but would instead proceed as in the example above. In component form,

$$x_i^{k+1} = \frac{b_i}{a_{ii}} - \frac{1}{a_{ii}}\sum_{j=1}^{i-1} a_{ij}x_j^{k+1} - \frac{1}{a_{ii}}\sum_{j=i+1}^{n} a_{ij}x_j^k$$

where by convention an empty sum is zero. Again, we are assuming throughout that the diagonal entries a_{ii} are nonzero $(i = 1, ..., n)$.

Typically, if Jacobi iteration converges for a given matrix A then Gauss-Seidel iteration converges also and is faster (that is, the spectral radius of its iteration matrix M is smaller); however, this is not always the case. If A is strictly diagonally dominant or positive definite then Gauss-Seidel iteration will converge for any initial guess. There are block forms of both algorithms.

Gauss-Seidel iteration is generally superior to Jacobi iteration. One exception might be on a parallel machine: If $x \in \mathbb{R}^n$ and there are n processors available, Jacobi iteration can be performed very efficiently (processor i computes the updated x_i) but Gauss-Seidel does not gain much benefit. However, usually the number p of processors is much less than n so the speed-up is not as great as indicated. In that case we might have each processor compute n/p of the entries of x using the idea of Gauss-Seidel iteration (that is, making use of any updated variables available to it) but might or might not try to pass updated values between processors. There are a number of strategies in common use for performing a slightly modified version of Gauss-Seidel iteration on a parallel machine. While the theoretical convergence results no longer apply, in practice convergence is usually achieved with these modified methods.

Note that the particular iterates found using Jacobi and Gauss-Seidel iteration depend on the ordering of the unknowns. If we write the same linear system in a different order we will generate a different Jacobi or Gauss-Seidel sequence.

These methods were used for the solution of large linear systems for many years. Today however the Jacobi and Gauss-Seidel iterations are most often used in conjunction with some other iterative method. An example of such a technique will be discussed in Sec. 3.3.

MATLAB

In principle we could implement Jacobi iteration using Eq. (1.3), $x^{k+1} = -D^{-1}(U+L)x^k + D^{-1}b$, and Gauss-Seidel iteration using Eq. (1.6), $x^{k+1} = -(L+D)^{-1}Ux^k + (L+D)^{-1}b$. But n is generally large in these applications so even solving the triangular system $(L+D)x^{k+1} = -Ux^k + b$ for Gauss-Seidel iteration may be too time-consuming. In fact, it is often the case that we avoid ever forming A, L, U, or D explicitly due to memory concerns and the fact that A is typically sparse. Instead we write a program that returns selected entries of A or that computes Ax given x. We then implement the updates as separate formulas for $x_1^{k+1}, ..., x_n^{k+1}$.

For convenience however let us use the matrix forms for these iterations here. Enter:

» A=4*eye(10);for i=2:9;A(i,[i-1 i+1])=[1 1];end
» A(10,1)=1;A(1,10)=1;A(1,2)=1;A(10,9)=1

Matrices more-or-less like this occur frequently in applications involving partial differential equations. Enter:

» D=diag(diag(A));L=tril(A,-1);U=triu(A,1);

to split the matrix. Let's compare Jacobi and Gauss-Seidel iteration to solve $Ax = b$. Enter:

» b=6*ones([10 1]);

which makes the true solution a vector of all ones. Enter:

```
» x0j=zeros([10 1]);x0gs=x0j;          %Initial guess.
» xj=D\(-(U+L)*x0j+b);rj=b-A*xj        %Jacobi iteration.
» xgs=(L+D)\(-U*x0gs+b);rgs=b-A*xgs    %Gauss-Seidel iteration.
» norm(rj),norm(rgs)             %Residual norms.
» alpha_j=norm(xj-ones([10 1]))   %Absolute error.
» alpha_gs=norm(xgs-ones([10 1]))    %Absolute error.
» x0j=xj;x0gs=xgs;
```

Note that Gauss-Seidel iteration has reduced the residual and absolute errors more rapidly than Jacobi iteration; note also that rgs has a zero component. Compare the residuals and absolute errors for several more iterations. Generally alpha_gs should be smaller than alpha_j but a smaller residual does not guarantee this.

We can accelerate the convergence of Jacobi or Gauss-Seidel iteration using a technique similar to the damping used in Newton's method (see Sec. 1.5). We rewrite the iteration as $x^{k+1} = x^k + \Delta^k$ where the vector Δ^k is the difference between x^k and x^{k+1}. Then we introduce a new parameter $\varpi > 0$, called the **relaxation factor**, and write $x^{k+1} = x^k + \varpi\Delta^k$. If $\varpi = 1$ we have the standard method. If $0 < \varpi < 1$ we say that we are using **under-relaxation** and if $\varpi > 1$ we say that we are using **over-relaxation**; the term **successive over-relaxation** (or **SOR**) is used to refer to both cases. Under-relaxation may be used to force the iteration to converge in some cases where it is diverging. More commonly over-relaxation ($\varpi > 1$) is used to accelerate convergence. The idea is that if Δ^k is a step from x^k toward x^{k+1} then in all likelihood it's a step in the right direction and so we should probably take a larger one. The right value of ϖ can be difficult to determine, but a good choice of ϖ can give much faster convergence. (A poor choice of ϖ may cause divergence.) The SOR method for Gauss-Seidel iteration can be written in the form $x^{k+1} = Mx^k + c$ where $M = (D + \varpi L)^{-1}((1 - \varpi)D - \varpi U)$ and $c = \varpi(D + \varpi L)^{-1}b$.

For example, let's use SOR on the system we have been solving. Enter:

» x0gs=zeros([10 1]);x0SOR=x0gs;w=1.5;

» xgs=(L+D)\(-U*x0gs+b);rgs=b-A*xgs;   %Gauss-Seidel iteration.

» xSOR=(D+w*L)\((1-w)*D-w*U)*x0SOR+w*(D+w*L)\b;rSOR=b-A*xSOR;   %SOR.

» norm(rgs),norm(rSOR)

» x0gs=xgs;x0SOR=xSOR;

Repeat for several iterations; the SOR method seems to be inferior to the Gauss-Seidel iteration. Let's check the spectral radius of the two iteration matrices. Enter:

» max(abs(eig((L+D)\(-U))))   %Gauss-Seidel.

» max(abs(eig((D+w*L)\((1-w)*D-w*U))))   %SOR.

to find the values of $\rho(M)$ in each case. We have $\rho(M) \doteq .3093$ for Jacobi iteration and $\rho(M) \doteq .6135$ for SOR, so it is not surprising that SOR performed as it did. Enter:

» bestz=Inf;bestw=0;

» for w=1:.005:2;z=max(abs(eig((D+w*L)\((1-w)*D-w*U))));

»   if z<bestz;bestz=z;bestw=w;end;

» end

» bestw

to see that $\varpi \approx 1.07$ appears to be the optimal choice in this case. Enter:

» w=1.07;

» max(abs(eig((D+w*L)\((1-w)*D-w*U))))

to see that $\rho(M) \doteq .2335$ for this choice of ϖ. This is a considerable improvement over Gauss-Seidel iteration. Compare Gauss-Seidel iteration to SOR with this value of ϖ.

It can be shown that a necessary condition for the SOR method to converge for all x^0 is that no diagonal entry of A be zero and $0 < \varpi < 2$. If A is positive definite and $0 < \varpi < 2$ then the SOR method will converge for any x^0. (This justifies the claim that standard Gauss-Seidel iteration converges for a positive definite matrix, as it corresponds to $\varpi = 1$.) In practice we almost always let $\varpi = \varpi_k$ with $\varpi_k \to 1$ as $k \to \infty$. The best value of the relaxation factor ϖ can be determined for certain special matrices but in general it must be found by trial and error. This is worthwhile when the same matrix will be used repeatedly for different choices of b, which is a common occurrence–possibly solving the same system $Ax = b$ over several months during a complicated engineering design.

Problems

1.) a.) Perform another 3 iterations of the method in Example 1. Compute the relative error of your answer.

b.) Perform another 3 iterations of the method in Example 3. Compute the relative error of your answer.

2.) a.) Find the spectral radius of the matrix in Example 1 and use it to estimate the size of the error after 5 iterations. Compare this to the estimate found by using the max, Frobenius, and spectral norms.

b.) Find the spectral radius of the matrix in Example 3 and use it to estimate the size of the error after 5 iterations. Compare this to the estimate found by using the max, Frobenius, and spectral norms.

c.) Compare your estimates with the results from Problem 1.

3.) a.) Prove that Jacobi iteration must converge for a strictly diagonally dominant matrix. (Hint: Show that $\|M\| < 1$ for some convenient norm.)

b.) Prove that Gauss-Seidel iteration must converge for a strictly diagonally dominant matrix.

4.) a.) Verify that Eq. (1.6) is equivalent to the scalar form of Gauss-Seidel iteration as described in Example 3.

b.) Jacobi iteration is also known as the **method of simultaneous displacements**, while Gauss-Seidel iteration is also known as the **method of successive displacements**. Why are these descriptions appropriate?

5.) a.) Prove that $\rho(A) \leq \|A\|$ for any natural matrix norm.

b.) Show that if $\rho(M) < 1$ then the resolvent $(I - M)^{-1}$ of M exists.

6.) Write a MATLAB program that performs the SOR method on a given system. The user should provide A, b, x^0, ϖ, and any additional parameters needed (such as tolerances) and the program should return an estimate of the solution.

7.) a.) Perform 5 iterations of SOR on the system in Example 1 using $\varpi = 1.1, 1.3, 1.5$. Compare your results to your results from Problem 1.

b.) Perform 5 iterations of SOR on the system in Example 3 using $\varpi = 1.1, 1.3, 1.5$. Compare your results to your results from Problem 1.

8.) Construct a chart that shows the advantage of a method with $\rho(M) = .3$ over a method with $\rho(M) = .5$.

9.) a.) Compute (by hand) the spectral, Frobenius, and max norms of the $n \times n$ identity matrix, and its spectral radius.

b.) Compute (by hand) the spectral, Frobenius, and max norms of the matrix $A = [2\ 1; 1\ 2]$, and its spectral radius.

10.) Let A be a 10×10 tridiagonal matrix with 4 on the main diagonal and -1 on the superdiagonal and subdiagonal, and let b be a vector of all ones. Solve $Ax = b$ to a relative error of 10^{-6} using Jacobi iteration, Gauss-Seidel iteration, and SOR using an appropriate ϖ (determined by trial-and-error). Comment.

11.) a.) Verify experimentally that $\rho(M) \geq |\varpi - 1|$ for SOR. Why does this imply that $0 < \varpi < 2$ is a necessary condition for convergence of SOR?

b.) If A is a positive definite tridiagonal matrix then the optimal ϖ for SOR has $\rho(M) = \varpi - 1$. Verify this experimentally (as in the MATLAB subsection).

c.) Prove that if the main diagonal of A contains no zero entries then $\rho(M) \geq |\varpi - 1|$ for the iteration matrix M of the SOR method.

12.) Generate 20 random 100×100 matrices and compare $\rho(M)$ for both Jacobi iteration and Gauss-Seidel iteration. (Use a script file to do this.) Comment.

13.) Consider the formula $\sum_{k=0}^{\infty} M^k = (I - M)^{-1}$. Comment on the efficiency of using partial sums of this series to approximate the resolvent $(I - M)^{-1}$.

14.) a.) The simplest possible matrix splitting method for $Ax = b$ is the **Richardson iteration** $x^{k+1} = (I - A)x^k + b$. Under what conditions on A will it converge?

b.) Give a direct proof that $\|M\| < 1$ implies convergence of $x^{k+1} = Mx^k + c$ for any c and any x^0 by taking the norm on both sides of $x^{k+1} - x = Mx^k + c - x$ where x is the solution. (Hint: Note that $x = Mx + c$ and that $\alpha_k = \|x^k - x\|$ is the absolute error.)

15.) a.) A commonly occurring system of equations in the numerical study of heat flow takes the form $u_{ij} = \frac{1}{4}\left(u_{i-1,j} + u_{i+1,j} + u_{i,j-1} + u_{i,j+1}\right)$ where the u_{ij} are the unknowns ($i = 1, ...n$, $j = 1, ...m$). Write this as a linear system in matrix form. Is the matrix sparse?

b.) This system could be reindexed $x_1 = u_{11}, x_2 = u_{12}, ...$, but it is convenient to retain the u_{ij}. Explain how the system could be solved by Jacobi or Gauss-Seidel iteration in place by treating the unknowns u_{ij} as a matrix and never forming the coefficient matrix of the system in part a.). Is this more efficient?

c.) A strategy for partially parallelizing the Gauss-Seidel iteration is **red-black Gauss-Seidel iteration**, where we imagine the entries u_{ij} with $i + j$ even to be "red" and the entries u_{ij} with $i + j$ odd to be "black" (as if the matrix U were written on a checker board). We compute u_{ij}^{k+1} from u_{ij}^k by first updating all red entries, then updating all black entries. Discuss how this compares to standard Gauss-Seidel iteration.

d.) Write a MATLAB program to perform red-black Gauss-Seidel iteration for this system. Test your program.

16.) Verify experimentally that the solution of $x^{k+1} = Mx^k + \beta$ is $(I - M)^{-1}\beta$ when $\rho(m) = 1$.

2. Sparsity

The iterative methods discussed in the previous section will only be sufficiently fast for large matrices if the matrices are sparse. One commonly encountered type of sparse matrix is a tridiagonal matrix. Such a matrix is essentially composed of three vectors–the vector $d = (a_{ii})_{i=1}^n$ of diagonal elements, the vector $s = (a_{i+1,i})_{i=1}^{n-1}$ of subdiagonal elements, and the vector $t = (a_{i,i+1})_{i=1}^{n-1}$ of superdiagonal elements:

$$A = \begin{pmatrix} d_{11} & t_{12} & & & & \\ s_{21} & d_{22} & t_{23} & & & \\ & s_{32} & d_{33} & t_{34} & & \\ & & s_{43} & d_{44} & t_{45} & \\ & & & s_{54} & d_{55} & t_{56} \\ & & & & s_{65} & d_{66} \end{pmatrix}.$$

(The double indices are used here for clarity–actually the d, s, and t vectors would be singly-indexed.) A matrix such as this can have at most $(n-1)+n+(n-1) = 3n-2$ nonzero entries, which is considerably less than the n^2 possible nonzero entries for a dense matrix. How can we realize storage and speed gains for matrices such as this?

For such a specialized case we could write a special-purpose matrix-vector multiplication program that inputs a vector x and returns Ax, computed efficiently. We could make similar programs for other computations involving A. Indeed, if we knew that our problem would involve a tridiagonal matrix we would want to do this in our code.

In fact, let's look at such a special case: Performing Gaussian elimination when the matrix is known to be tridiagonal. In the very first step, assuming we don't need to pivot, the multiplier will be $m_{21} = s_{21}/d_{11}$ and when we perform $R_2 \leftarrow R_2 - m_{21}R_1$ we only need to do one computation: After setting the $(2,1)$ entry to zero (no computing involved), we need to make the new $(2,2)$ entry be

$$\delta_{22} = d_{22} - (s_{21}/d_{11})\,t_{12}.$$

That's it; the step only involves a single computation, giving

$$\begin{pmatrix} d_{11} & t_{12} & & & & \\ 0 & \delta_{22} & t_{23} & & & \\ & s_{32} & d_{33} & t_{34} & & \\ & & s_{43} & d_{44} & t_{45} & \\ & & & s_{54} & d_{55} & t_{56} \\ & & & & s_{65} & d_{66} \end{pmatrix}.$$

and now the first column has been cleared: It would be senseless to "eliminate" the known zeroes in the $(3,1)$ through $(6,1)$ positions. But clearly, the next stage–clearing column 2 below the $(2,2)$ position–will also involve only a single computation, $\delta_{33} = d_{33} - (s_{32}/d_{22})\,t_{23}$.

The formalization of this observation is known as the Thomas algorithm, which is simply an efficient form of Gaussian elimination for solving $Ax = b$ when A is known to be a tridiagonal matrix. Let's go back to singly-indexing the diagonal entries d_i $(i = 1,...n)$, subdiagonal entries s_i $(i = 2,...n)$, and superdiagonal entries t_i $(i = 1,...n-1)$. We'll assume for simplicity that no pivoting is needed in order to illustrate the main idea. Then the algorithm is:

Thomas Algorithm:
1. Set $\varpi_1 = d_1$, $\rho_1 = t_1/\varpi_1$, $\varsigma_1 = b_1/\varpi_1$.
2. Begin loop $(i = 2$ to $n)$:
3. Set $\varpi_i = d_i - s_i\rho_{i-1}$.
4. Set $\rho_i = t_i/\varpi_i$.
5. Set $\varsigma_i = (b_i - s_i\varsigma_{i-1})/\varpi_1$.
6. End loop.
7. Set $x_n = \varsigma_n$.
8. Begin loop $(k = n-1$ to $1)$:
9. Set $x_k = \varsigma_k - \rho_k x_{k+1}$
10. End loop

This algorithm uses about $8n - 6$ flops, including the back substitution phase, and so is certainly $O(n)$ (as desired for sparse matrices). It requires 3 temporary vectors ϖ, ρ, and ς. It will be used automatically by MATLAB if it detects a tridiagonal form.

The Thomas algorithm as stated does not benefit significantly from parallelization. A variant[1] that is slightly more parallel-friendly reduces the matrix starting from both the northwest and southeast corners simultaneously. In essence, we proceed as follows:

---

[1]This is related to the so-called **twisted factorization** methods that are crosses of LU and UL factorizations.

$$
\begin{pmatrix}
d_1 & t_1 & & & & \\
s_2 & d_2 & t_2 & & & \\
& s_3 & d_3 & t_3 & & \\
& & s_4 & d_4 & t_4 & \\
& & & s_5 & d_5 & t_5 \\
& & & & s_6 & d_6
\end{pmatrix}
\sim
\begin{pmatrix}
d_1 & t_1 & & & & \\
0 & \delta_2 & t_2 & & & \\
& s_3 & d_3 & t_3 & & \\
& & s_4 & d_4 & t_4 & \\
& & & s_5 & d_5 & t_5 \\
& & & & 0 & \delta_6
\end{pmatrix}
$$

where $\delta_2 = d_2 - (s_2/d_1)\, t_1$ and $\delta_6 = d_5 - (s_6/d_5)\, t_5$. Those two computations may be done on two separate processors (or, more likely, in two separate threads on a single processor). At the next stage we have

$$
\begin{pmatrix}
d_1 & t_1 & & & & \\
0 & \delta_2 & t_2 & & & \\
& 0 & \delta_3 & t_3 & & \\
& & s_4 & d_4 & t_4 & \\
& & & 0 & \delta_5 & t_5 \\
& & & & 0 & \delta_6
\end{pmatrix}
$$

and then finally we eliminate s_4 and compute δ_4 to complete the decomposition. If n were odd there wouldn't be a need for a single serial clean-up step like this at the end.

A generalization of this idea called Cyclic Reduction (or **odd-even reduction**) may be a better choice when a truly parallel device is available. The serial form of Cyclic Reduction is equivalent to re-ordering the rows *and the columns* of the matrix to bring all odd-indexed rows to the top and to enforce a certain blocked form, namely

(2.1)
$$
\begin{pmatrix}
d_1 & t_1 & & & & \\
s_2 & d_2 & t_2 & & & \\
& s_3 & d_3 & t_3 & & \\
& & s_4 & d_4 & t_4 & \\
& & & s_5 & d_5 & t_5 \\
& & & & s_6 & d_6
\end{pmatrix}
\sim
\begin{pmatrix}
d_1 & & & t_1 & & \\
& d_3 & & s_3 & t_3 & \\
& & d_5 & & s_5 & t_5 \\
s_2 & t_2 & & d_2 & & \\
& s_4 & t_4 & & d_4 & \\
& & s_6 & & & d_6
\end{pmatrix}
$$

which can be achieved by the operation[2] PAP^T where P is the permutation matrix

(2.2)
$$
\begin{pmatrix}
1 & 0 & 0 & 0 & 0 & 0 \\
0 & 0 & 1 & 0 & 0 & 0 \\
0 & 0 & 0 & 0 & 0 & 1 \\
0 & 1 & 0 & 0 & 0 & 0 \\
0 & 0 & 0 & 1 & 0 & 0 \\
0 & 0 & 0 & 0 & 0 & 1
\end{pmatrix}
$$

which is equivalent to performing elementary row operations of interchanging rows via the premultiplication, and then elementary column operations of interchanging columns via the postmultiplication. (Of course, all this is simulated in software rather than being actually performed.) But Eq. (2.1) is ideally prepared for a single step of block Gaussian elimination $L\left(PAP^T\right)$ to reduce it to

---

[2]This is both a congruence and, as $P^T = P^{-1}$, a similarity.

$$\begin{pmatrix} d_1 & & & t_1 & & \\ & d_3 & & s_3 & t_3 & \\ & & d_5 & & s_5 & t_5 \\ & & & d_2^{(1)} & t_2^{(1)} & \\ & & & s_4^{(1)} & d_4^{(1)} & t_4^{(1)} \\ & & & & s_6^{(1)} & d_6^{(1)} \end{pmatrix}$$

(assuming nonzero odd-indexed diagonal entries within A). The result is a more nearly upper triangular matrix that is, in effect, halfway to an LU decomposition of A. In particular we can already solve for the odd-indexed unknowns, using the above matrix and the appropriate parts of the lower triangular matrix L used to form it.

Furthermore, the southeast 3×3 block is a new tridiagonal matrix–the Schur complement of the block Gaussian elimination operation involving the upper bidiagonal (hence tridiagonal) southwest 3×3 block of the right-hand side matrix in Eq. (2.1) –and so we can now repeat the same odd-even reduction on it. (Sometimes the first and last rows of A are handled separately as special cases and the odd-even reduction idea is only applied to the remaining $n-2$ rows.) That's the key idea: Mimic the permutation within the code, implement the Gauss transformation L based on the simulated block of odd-indexed entries to get a transformed version of the even-indexed entries (the forward reduction phase), and solve for the odd-indexed entries using this information (the back substitution phase); then, repeat on the even-indexed entries after re-indexing them. Obviously, this is easiest when n is a power of two but this is not necessary; we can handle any straggling rows as needed.

This serial version of the method does create more fill-in than the Thomas algorithm and requires about twice as many operations, but this can be more than compensated for by modifying the idea slightly on a machine with appropriate architecture. The specific variation of the algorithm that would be used would depend on the machine, but the essential insight in **Parallel Cyclic Reduction** is that repeated forward reduction phases, with the backward substitution phases deferred, allow us to send smaller subproblems of the same type to different processors until either we run out of processors or reduce the problem to a trivial size.

The ideas of both the Thomas algorithm and Cyclic Reduction are easily generalized to pentadiagonal matrices (which have two possibly nonzero entries immediately to either side of the main diagonal, that is, the first two subdiagonals and the first two superdiagonals may have nonzero entries) as well as other matrices that are banded, meaning that all of their nonzero entries are clustered near the main diagonal of the matrix. In fact, both of these methods for tridiagonal matrices extend to block tridiagonal matrices of the form

$$M = \begin{pmatrix} A_1 & B_1 & & & & \\ C_1 & A_2 & B_2 & & & \\ & C_2 & A_3 & B_3 & & \\ & & C_3 & A_4 & B_4 & \\ & & & C_4 & A_5 & B_5 \\ & & & & C_5 & A_6 \end{pmatrix}.$$

where M is $n \times n$ and each of the A_i, B_i, C_i are $p \times p$ matrices (not necessarily themselves tridiagonal), with all other entries of M being null. These occur commonly in, for example, the numerical solution of partial differential equations.

But what if the sparse matrix does not have a simple, easily recognizable structure? In the general case we might store a sparse matrix using three vectors: A vector r of row indices, a vector c of column indices, and a vector v of the nonzero (double precision) values. For example, for the unstructured sparse matrix

$$A = \begin{pmatrix} 8 & 7 & 0 & 0 & 0 & 8 \\ 0 & 0 & 0 & 0 & 0 & 7 \\ 0 & 9 & 0 & 0 & 8 & 0 \\ 8 & 0 & 7 & 0 & 0 & 7 \\ 0 & 0 & 0 & 9 & 0 & 8 \\ 0 & 7 & 9 & 0 & 0 & 0 \end{pmatrix}$$

we would have

$$
\begin{aligned}
r &= (1,1,1,2,3,3,4,4,4,5,5,6,6) \\
c &= (1,2,6,6,2,5,1,3,6,4,6,2,3) \\
v &= (8,7,8,7,9,8,8,7,7,9,8,7,9)
\end{aligned}
$$

(2.3)

as our representation of the matrix, called the **coordinate scheme**. A representation such as this is called a **packed** form of the matrix, and the operation of putting it into such a form is called a **gather** operation.

To understand how to interpret the packed form in Eq. (2.3), look at the fourth entry in each vector: The (r_4, c_4) entry of the matrix (that is, the $(2,6)$ entry) is $v_4 = 7$. Similarly, the $(5,4)$ entry is 9. All unlisted entries are understood to be zero. Since entirely zero rows or columns are a possibility we need to store separately the total number of rows and of columns; for example, we would also get the coordinate representation in Eq. (2.3) if we had the 8×6 matrix that consisted of the same A as before but with two entirely null columns at the bottom.

As each of these three vectors has 13 entries, we are using 39 entries to store the nonzero entries of a 6×6 matrix. Such a matrix has only 36 entries (including the zeroes) and so this is not desirable[3]. However, suppose that the matrix A was 1000×1000 and had only 3000 nonzero entries. Then this scheme uses nine thousand entries whereas storing the entire matrix requires one million entries. That is a considerable savings of memory, and of time required to access it.

There are more efficient ways to store matrices in packed form. Most such schemes use linked lists. What form is best may depend on the application; for example, the coordinate scheme of Eq. (2.3) makes it relatively easy to access the matrix by rows (e.g., if we needed to find all entries of row 4) but would be very inconvenient for an algorithm that accesses matrices by columns, as a search of c would be required to find all entries in a given column. However, we will continue to use this form for purposes of illustration.

How do we perform operations involving sparse matrices (and vectors)? Suppose A is stored in packed form and we wish to compute Ax where x is also packed. (Packing a vector x requires only a single index vector.) We might first expand

---

[3]Of course, 26 of the 39 entries here are indices and so must be small integers, which can be represented more compactly than a floating point number.

x into a full vector with zeroes in appropriate places, called a **scatter** operation. Then we could, for each row of A, multiply the nonzero entries of A, as found from the vector c, by the corresponding entries in x and sum them:

$$(2.4) \qquad\qquad (Ax)_i = \sum v_j x_j$$

where the sum is over all indices j found in c that correspond to row i (from r). After all rows have been processed the resulting vector may be gathered into packed form. If it is convenient to access the packed form of A by columns then it may be easier to use the formula

$$(2.5) \qquad\qquad Ax = \sum_{i=1}^{n} x_i a_i$$

where a_i represents column i of A. (We are using the fact that a matrix-vector product is a linear combination of the columns of the matrix). Of course, we skip columns of A that are entirely full of zeroes.

While we benefit from manipulating many fewer entries, finding those entries requires searching, and indexing entries out of, several vectors. This is a relatively slow operation. We will generally benefit from using the sparse form only for matrices that are significantly sparse; a common rule of thumb is that the matrix should have at least 95% null entries. Treating a matrix as a sparse matrix can make a considerable difference in speed of execution; that difference may be a speed-up if the technique is used appropriately but could also be a slow-down if it is used for a matrix that might better be treated as if it were full.

There is much more that could be said about the storage and manipulation of sparse matrices and vectors. We will explore the issue further in MATLAB rather than theoretically. Many times, however, you will be working in Fortran, Python, or C++, say. Then, to perform an LU decomposition, for example, you will generally need a routine that packs the matrix into sparse storage and a routine that operates on the matrix to find its LU factors given that particular packed form. A Fortran routine must respect the fact that Fortran stores matrices in column-major form; Python and C++ default to row-major, but you may be able to change this if desired. You might also need a routine to scatter the matrix back into unpacked form. Coordinating these steps may make for a programming challenge even if the subprograms to perform the individual steps are readily available.

MATLAB

Sometimes a routine for storing matrices in packed form must be written because a matrix with a very specialized structure is being manipulated. However, there are programs available (at *www.netlib.org*, for example) that are appropriate for general use. In MATLAB sparse matrix support is provided. Enter:

» `help sparfun`

for a listing of sparse matrix functions. Enter:

» `A=[1 0 2 0 0]'*[1 0 2 0 0]`

» `S=sparse(A)`

» `whos`

to enter a full matrix A and then convert it to sparse form (S); note the difference in bytes used from the `whos` command. Note also that MATLAB displays the matrix using the coordinate scheme, sorted by columns. In fact, we could have used that scheme to define it; enter:

```
» sparse([1 3 1 3],[1 1 3 3],[1 2 2 4])
```

(this has the form sparse(r,c,v) in the notation of Eq. (2.3)). Enter:

```
» full(ans)
```

to convert it back to full form. Of course, MATLAB has no way of knowing that this should be a 5×5 matrix padded by zero rows and columns.

There are a number of useful commands for sparse matrices (by which we mean matrices stored as though they were sparse, whether or not they are truly sparse). Enter:

```
» nnz(S)          %Number of nonzero entries in S.
» spy(S)          %Wilkinson diagram.
» nonzeros(S)     %The vector v of nonzero values.
» S(1,3)
» S(1,3)=0        %Note that r, c, and v are updated automatically.
» S(1,3)=2;       %Reset this entry.
» issparse(S)     %Check for sparse storage.
» issparse(A)     %Check for sparse storage.
```

There are other MATLAB functions for sparse matrices; see help sparfun.
We are focusing on iterative methods in this chapter, but sparse matrix techniques are also used in conjunction with direct methods. Enter:

```
» clear all
» S=speye(10)     %Sparse form identity matrix.
» full(S)
» S(1,1:10)=ones([1 10])
» S(1:10,1)=ones([10 1])
» full(S)
» [L,U]=lu(S)
» nnz(S),nnz(L),nnz(U)
» spy(S)
» spy(L)
» spy(U)
```

to create a sparse arrowhead matrix and find its LU decomposition. Note that the factors L and U are not truly sparse, although they are stored as though they were. Enter:

```
» full(U)
» whos
```

to see that there is still some advantage in storing U as though it were sparse. (The full form of U appears as ans.) This phenomenon is known as **fill-in**: A sparse matrix typically fill in the zero entries when Gaussian elimination is performed on it, so that sparsity is not preserved by Gaussian elimination. This is unfortunate and problematic in practice. But, enter:

```
» S1=S([10,2:9,1],:)
» full(S1)
» [L1,U1]=lu(S1)
» nnz(S1),nnz(L1),nnz(U1)
» whos          %Compare L and L1, U and U1.
» spy(S1)
```

```
» spy(L1)
» spy(U1)
```
to see that a simple row interchange operation can improve the degree of fill-in considerably. The reordering algorithms listed under **help sparfun** are used to find an ordering of the rows and columns of the matrix that will (hopefully) reduce fill-in, based on the pattern of the nonzero elements of the matrix. For example, enter:
```
» p=symrcm(S)    %New ordering.
» S2=S(p,p);     %Same rows, different order.
» [L2,U2]=lu(S2);
» whos           %Compare L, L1, and L2, then U, U1, and U2.
```
to see that this ordering, generated by a method known as the **reverse Cuthill-McKee algorithm**, further reduces the amount of storage required by L. Of course, a price is paid in searching and then reordering the matrix.

If it is known that a sparse matrix will have new nonzero elements added to it at a later point, memory can be allocated for these future entries using the **sparse** command or the **spalloc** command. There are also sparse versions of **rand** (**sprand**), **randn** (**sprandn**), and other similar commands. See also the **spones** command.

We emphasize that storing a non-sparse matrix in sparse form may mean using more storage than the full form and may slow rather than speed-up computations. It's not always clear when sparse form is better, but when it is it can make for a significant decrease in memory usage and computing time.

To see the effect of Eq. (2.2) on re-ordering a matrix via the transformation PAP^T, enter:
```
» A=reshape(1:36,6,6)
» P=eye(6); P(1:6,:)=P([1 3 5 2 4 6],:)
» P*A*P'    %Note re-ordering of both rows and cols.
```
This is the form of the P matrix used in the derivation of the Cyclic Reduction algorithm, though of course not within actual code.

Problems

1.) a.) Write the matrix $A = [1\ 0\ 1\ 0; -1\ 0\ 0\ 0; 0\ 5\ 0\ 0; 0\ 0\ 0\ -2]$ in packed form using the coordinate scheme. Is this an efficient representation of A?

b.) If $r = (1, 1, 2, 2, 2, 3, 3, 4, 5, 5, 5, 6, 8)$, $c = (1, 3, 5, 6, 8, 2, 3, 1, 4, 6, 8, 6, 1)$, and $v = (-1, 7, 2, 3, 4, -5, 8, -7, -1, -5, 3, -4, 6)$ for a matrix stored in the coordinate scheme, what is a_{32}? What is a_{61}? What row of A contains the most entries? What column of A contains the most entries? Are there any zero rows or columns? What can you say about the size of A? Is this an efficient representation of A?

2.) Write a detailed algorithm (pseudo-code) for changing a full matrix to packed form using the coordinate scheme.

3.) a.) Write a detailed algorithm (pseudo-code) for performing the multiplication of a packed matrix times a packed vector using Eq. (2.4). You may assume that there is a command such as the MATLAB **zeros** command that creates a full vector of all zeroes. Do not assume that the entries in the r vector are ordered.

b.) Write a detailed algorithm (pseudo-code) for performing the multiplication of a packed matrix times a packed vector using Eq. (2.5).

4.) If entries of A, r, c, and v all require the same amount of memory, when is it storage-efficient to use the coordinate scheme? What if the entries of r and c only require half as much memory as an entry of v (or A)? What if the entries of r and c only require one-quarter as much memory as an entry of v (or A)?

5.) Compare the time and/or flops required for MATLAB to solve a 1000×1000 linear system stored in sparse form to the same system stored in full form when the matrix is approximately .1% full; .5% full; 1% full; 2% full; 5% full; and then so on in increments of 5 percentage points to 50% full. (You may wish to use the sprandn command to create the matrices.) Solve the systems using the slash command.

6.) a.) Derive the Thomas algorithm by applying Gaussian elimination to a tridiagonal matrix and then writing the method in the form of the algorithm given in the text.

b.) How would you extend the idea of the Thomas algorithm to a method for block tridiagonal matrices? Do not assume the blocks themselves are tridiagonal, but you may assume that there is a routine to perform Gaussian elimination on the blocks (or a Schur complement routine, if you wish).

7.) a.) Write out the algorithm for performing (serial) Cyclic Reduction.

b.) Suggest an algorithm for Parallel Cyclic Reduction under the assumption that the matrix is $n \times n$ with $n = 2^N$ and that there are n processors available. What if there are only $n/2$ processors?

8.) a.) One strategy to decrease fill-in in the LU decomposition of a sparse matrix A is to use as a pivot that value a_{ij} in the active submatrix with the smallest **Markowitz count** $(r_i - 1)(c_j - 1)$ where r_i is the number of nonzero entries in row i, called the **row count** and c_j is the number of nonzero entries in column j, called the **column count**. Write a MATLAB program that performs Gaussian elimination on a matrix with the pivots chosen to minimize the Markowitz count. Test your program; does it decrease fill-in?

b.) In practice this method cannot be stable since the size of the pivot is not taken into account. For this reason it is used in conjunction with **threshold pivoting**, where we insist that the new pivot not only minimize the Markowitz count but that it also has magnitude at least $\alpha \max(|a_{ij}|)$ where $0 < \alpha < 1$ and the max is taken over all entries in the active submatrix. In this way the new pivot may not be the largest possible pivot but if α is not too small it should not amplify errors significantly. If no potential pivot satisfies both the thresholding and Markowitz count criteria we allow the count to be larger than the minimal value. Modify your MATLAB program to use threshold pivoting. Test your program; comment.

c.) For a large matrix this method requires too much searching and comparing, and so it is common to use heuristics such as searching only for k rows and columns from the current position or taking the first entry that meets the thresholding requirement and has a suitably small Markowitz count. Suggest a modification along these general lines and implement it as an option in your program. Test your program; comment.

9.) a.) Write a detailed algorithm (pseudo-code) for updating r, c, and v to reflect the deletion of a single entry from the matrix.

b.) Write a detailed algorithm (pseudo-code) for updating r, c, and v to reflect the deletion of a single row from the matrix.

10.) a.) Another scheme for storing a sparse matrix is to store its elements as a sparse vector. The entries of A are read into the vector v by rows, first all entries of row 1, then all entries of row 2, and so on. A second vector c contains the column in which the corresponding entry in v occurs. A third vector l contains the length of the rows, in order; note that this vector may be shorter than c and v. Alternatively (or, for convenience, additionally) a vector s may be stored that indicates where in v each row starts. For example, if the first row of A has four nonzero entries and the second row has three, then $s(1) = 1$, $s(2) = 5$ (the values for row 2 start in entry 5 of v), $s(3) = 8$, and $l(1) = 4$, $l(2) = 3$. Rewrite the matrix of Eq. (2.3) and the arrowhead matrix of the MATLAB subsection in this format.

b.) What are the advantages and disadvantages of this scheme?

11.) a.) A common storage scheme for sparse matrices is the **row-linked list**, where we store the values in a vector v and the columns of those values in a vector c of the same length. In addition, a vector s stores the start of the ith row in its ith entry. (Compare Problem 10.) However, rather than listing the entries in v by rows, we store a fourth vector λ of links. The vector λ has the same length as v and $\lambda(i)$ points to the next entry in row i (or is zero to indicate the end of a row). For example, if $v = (-1, -2, -3, -4, -5)$, $c = (2, 3, 3, 1, 3)$, $s = (1, 3, 5)$, and $\lambda = (4, 0, 0, 2, 0)$, then we reconstruct the matrix A as follows: Row t1 starts at $s(1) = 1$ for which $c(1) = 2$, $v(1) = -1$, so $a_{12} = -1$. The link points to $\lambda(1) = 4$ for which $c(4) = 1$, $v(4) = -4$, so $a_{11} = -4$. The link points to $\lambda(4) = 2$ for which $c(2) = 3$, $v(2) = -2$, so $a_{13} = -2$. The link points to $\lambda(2) = 0$ so we have reached the end of row 1. Similarly, row 2 starts at $s(2) = 3$ giving $a_{23} = -3$ and as $\lambda(3) = 0$ this is the end of the row; finally, $a_{33} = -5$. An advantage of the row-linked list is that it is easy to add new entries to the matrix. Rewrite the matrix of Eq. (2.3) and the arrowhead matrix of the MATLAB subsection in this format.

b.) Rewrite the matrix of Eq. (2.3) and the arrowhead matrix of the MATLAB subsection using the **column-linked list** format, where we store links to the next entries in columns rather than rows.

c.) If we may need to access a packed matrix both by its rows and its columns then we sometimes store both the row and column links. Comment on the efficiency of this scheme with respect to storage.

12.) a.) Create a 100×100 arrowhead matrix such as the one in the MATLAB subsection and compare the sparsity of its LU and QR factors as given and after reordering using the results of the `symrcm` and `colamd` commands.

b.) Repeat part a. using several tridiagonal matrices.

c.) Repeat part a. using several matrices created by `sprandn`.

13.) a.) Write a detailed algorithm (pseudo-code) for performing Gaussian elimination on a sparse matrix using the coordinate scheme. (Do not use the MATLAB sparse matrix functions; manipulate r, c, and v directly.) Assume the matrix is already packed.

b.) Implement your program and test it.

14.) a.) Consider the arrowhead matrix of the MATLAB subsection. What is the optimal reordering of the rows and columns so as to minimize fill-in when Gaussian elimination is performed on it?

b.) Give several examples of matrices whose structures ensure that Gaussian elimination will create no fill-in no matter what nonzero values occupy those positions.

15.) a.) Write a MATLAB program that performs Jacobi iteration on a sparse matrix given in the coordinate scheme.

b.) Write a MATLAB program that performs Gauss-Seidel iteration on a sparse matrix.

c.) Test your programs on several large sparse matrices. Compare your results to the use of the MATLAB slash command. Comment.

16.) a.) Extend the idea of the Thomas algorithm to a method for pentadiagonal matrices.

b.) What is the operations count for your method?

c.) Implement and test your method.

3. Iterative Refinement

The Jacobi, Gauss-Seidel, and SOR methods of Sec. 3.1 were previously widely employed for the solution of large sparse systems. The modern approach is different: We use such a method to prime another iterative technique. This priming can take one of two forms, and we explore them in this section and the next.

The technique we will discuss here is based on the observation that if x_0 is an estimate of the true solution x of $Ax = b$ then the residual r satisfies

$$\begin{aligned} r &= b - Ax_0 \\ &= Ax - Ax_0 \\ &= A(x - x_0) \end{aligned}$$

(since $Ax = b$). But $e = x - x_0$ is precisely the error in estimating x using x_0; we have found a linear system satisfied by the error vector,

(3.1) $$Ae = r$$

and in principle we can solve Eq. (3.1) for e and set $x = x_0 + e$ to determine x. In practice of course we will have error in e as well but even still $x_1 = x_0 + e$ may be an improved estimate of x. This technique is called **iterative refinement** (or **iterative improvement**). It uses the current x_0 to predict a correction e to be applied to it.

We could use iterative refinement simply to recover some of the accuracy that is lost in solving $Ax = b$ by some other method. For example, it is reasonable to take the approximate solution of $Ax = b$ as found by LU decomposition and perform one or two iterations of iterative improvement on it in order to clean it up. As the LU decomposition of A is already known from solving $Ax = b$ to get the approximate solution, this can be done efficiently by simply using it in Eq. (3.1). Since iterative improvement requires only $O(n^2)$ operations if we save the LU factors of A and reuse them, whereas solving $Ax = b$ in the first place requires $O(n^3)$ operations, this is an inexpensive measure to employ to improve a solution. For well-conditioned matrices one iteration will likely suffice. On the other hand, if we have spent $O(n^3)$ operations and gained an inaccurate solution due to ill-conditioning, then spending a mere $O(n^2)$ additional operations to improve it may be wise (we are "saving" the computation, hopefully).

However, as indicated at the start of this section, we can also use it as our solution method. We generate an initial guess x_0 that is suitably close to the true solution–say, by performing several iterations of Gauss-Seidel iteration to generate this x_0–and perform iterative improvement on it until the error ceases to be reduced. In fact, it is not uncommon to use Gaussian elimination *without pivoting* to generate the initial guess! Iterative refinement will take the inaccurate but relatively rapidly generated result and improve upon it.

EXAMPLE 1: Let $A = [4\ 2\ 1; 1\ 3\ 1; 1\ 1\ 4]$ and let $b = (3, -1, 2)^T$ be the system from Example 1 of Sec./ 3.1, for which $x = (1, -1, 1)^T$ is the true solution. Two iterations of Jacobi iteration gave the approximation $(.6667, -.9167, .8958)^T$ which we take as the initial guess x_0 for iterative improvement. We have $r_0 = b - Ax_0 \doteq (1.2708, .1876, .6668)^T$ for the residual error ($\|r_0\| \doteq 1.4473$). Applying iterative refinement, we solve $Ae_1 = r_0$ giving $e_1 = (.3333, -.0833, .1042)^T$ and hence $x_1 = x_0 + e_1 = (1, -1, 1)^T$ to the precision available. The residual error is now $r_1 = b - Ax_1 = 0$ to the precision available ($\|r_1\| = 0.0$ to about 16 decimal places). This reduction in the residual and absolute errors cannot be improved upon in double precision. $\square$

For a well-conditioned matrix it is not unreasonable to expect that the absolute error α may be reduced to the order of machine precision by repeated application of this technique; for ill-conditioned matrices we expect less dramatic but still significant improvement, unless $\kappa(A)$ is on the order of 10^p or larger, where p is the number of digits of precision. However, to achieve this the computation of the residual r, which involves extreme cancellation of significant figures, must be done with extra precision. The small matrix in Example 1 had small integer entries and so is not typical of matrices to which this method would be applied.

If the other computations involved in the computation are being done in single precision, then the evaluation of

$$(3.2) \qquad\qquad\qquad r_k = b - Ax_k$$

(which should be nearly zero if x_k is a good estimate of x) should be in done in double precision; if we are using double precision for the other computations then we should use **extended precision**, which in this case means the use of at least 63 bits for the mantissa (and at least 15 bits for the exponent) with a total of 80 bits, for the computation of r_k. (Conceivably, quadruple precision will be available, but this is uncommon when not on a supercomputer.) Availability of extended precision varies from machine to machine, as does the precise allocation of those 80 (or possibly more) bits. If needed it can always be simulated in software. It is not uncommon to have the facility to compute inner products of vectors in extended precision readily available.

In addition to the extra care used in computing the residuals we should use the original A in the calculation of Ax_k in Eq. (3.2), as opposed to LUx_k (which would introduce additional errors). This means that we cannot overwrite A with its LU factors. This technique is referred to as **mixed precision iterative improvement**. Without the extra precision iterative improvement will be much less successful, though for moderately ill-conditioned matrices ($\kappa(A)$ less than order of 10^p) it may still yield a worthwhile gain in accuracy.

In using this method and comparing the (measurable) norm of the residual error $\|r_k\|$ to the (unknown) relative error ρ_k it is worthwhile to recall the formula

$$\rho \le \kappa(A)\frac{\|r\|}{\|b\|}$$

(Eq. (2.6.5)), that is,

(3.3)
$$\frac{\|x_k - x^*\|}{\|x^*\|} \leq \kappa(A)\frac{\|r_k\|}{\|b\|}$$

which bounds the relative error in the approximation x_k of x in terms of the condition number of A, the norm of the residual, and the norm of the constant vector b. If $\kappa(A)$ can be estimated then this formula may be used to check the relative error in our approximate solution.

At this point it may be worthwhile to consider that many problems of practical interest involve inexact data, and so it may not be sensible to seek great accuracy in their solution. In solving a partial differential equation numerically it is common to end up with a linear system $Ax = b$ where A is known exactly (essentially just a sparse matrix of small integers, in some cases) but b reflects imprecisely known boundary data from physical measurements. Given that we are therefore solving the equation

$$Ax = b + \delta b$$

(where δb is the error in our knowledge of the true b vector), rather than the desired problem $Ax = b$, going to extra effort to obtain a more accurate solution of this inaccurate system of equations may not be justified. Users of numerical software commonly want "more power" in the form of higher precision used in the computations (meaning they will take longer to complete), more accuracy in the solution (meaning it will take longer to achieve a desired tolerance), more detailed models, and so on, but often a single precision computation to just a few digits of precision is all that is justified by the data and model.

MATLAB

We will not use variable precision arithmetic in MATLAB at this point. However, enter:

```
» help vpa
» vpa(2.814+2.814,3)    %Mimics fl(x)+fl(y) in 3-digit arithmetic.
» vpa(vpa(2.814,3)+vpa(2.814,3),3)    %Mimics fl(fl(x)+fl(y)).
» whos    %Note that ans has type 'sym' (symbolic).
```

For numbers that exceed 15 or 16 digits it's necessary to use the **sym** command to prevent them from being stored in double precision format, but the symbolic computations involved will be *very* slow compared to the use of standard floating-point arithmetic in single or double precision. Of course, if you have access to extended or quadruple precision via hardware, then you do have a good option for getting higher precision.

Let's use mixed precision iterative improvement to recover some of the accuracy that is lost in solving $Ax = b$ when A is the Hilbert matrix of order 6. Enter:

```
» help single
» b=ones([6 1],'single'); H=hilb(6,'single')
» x=invhilb(6,'single')*b    %True solution.
» whos
```

We have created two single precision objects, b and H, explicitly, and MATLAB has then created x as single by using the rule that it attempts to preserve the precision level of the objects involved in a computation when able to do so. Not every command accepts `'single'` as an optional extra argument, but many do. Enter:

```
» x0=H\b      %Approximate solution.
» norm(x-x0)/norm(x)   %Check the rel. error.
```

In this case we know H to double precision so we can use that; in other cases we may only ever have H to single precision. Enter:

```
» r=single(double(b)-hilb(6,'double')*double(x0))
» e=H\r;
» x1=x0+e;    %Apply the correction.
» whos('x1')    %Single precision, as expected.
» r1=single(double(b)-hilb(6,'double')*double(x1))
» norm(r1)
» norm(x-x1)/norm(x)
```

The residual error $\|r\|$ has actually increased slightly but the relative error is almost an order of magnitude smaller. We have gained about one more decimal digit of accuracy for the single application of mixed precision iterative improvement. Let's try one more step; enter:

```
» e=H\r1;
» x2=x1+e;    %Apply the correction.
» r2=single(double(b)-hilb(6,'double')*double(x2));
» norm(r2)
» norm(x-x2)/norm(x)
```

This is the smallest residual error $\|r\|$ yet, but we know that the residual error can be misleading. The relative error, however, has once again decreased by about an order of magnitude, meaning we've gained about one additional digit of accuracy. This is slow progress, but it's still progress. With a condition number of roughly 10^7 and working in single precision, this may not be a surprise. Whether it's worth the effort or not depends on the user's needs. You can see the improvement:

```
» x(1)     %True value
» x0(1),x1(1),x2(1)   %Gaining a new accurate sig. fig. each time.
```

Sometimes the limitations of the computing environment mean that we may have to use **fixed precision iterative improvement**, where the same precision is used for all computations. Mixed precision iterative improvement is the better technique, but it's not always easy to do so. In the previous example, if we could have done all our computations in double precision from the outset, we would have–that example mimics a case where we are limited to single precision for most computations (probably due to storage concerns[4]), but could use double precision occasionally.

Let's try the fixed precision technique in double precision with a slightly better conditioned matrix, say, the Hilbert matrix of order 5 that has a condition number on the order of 10^5. Enter:

```
» clear all
» A=hilb(5)
» cond(A)
» b=[1 -1 1 -1 1]';
```

---

[4]Increasingly in supercomputing environments, the use of lower precision formats is driven by concerns over the amount of much energy that is being used for the computation. It can be surprisingly high.

```
» x0=A\b
» x=invhilb(5)*b    %Exact soln.
» norm(x-x0)/norm(x)    %Relative error.
```

The relative error in x1 is already on the order of 10^{-12}, which is as expected (16 digits of precision minus 5 lost because of a condition number on the order of 10^5 means that we expect there to be roughly 11 digits of accuracy left, which is about what we found).

Now let's apply the improvement technique. Enter:

```
» r=b-A*x0
» e=A\r
» x1=x0+e
» norm(x-x1)/norm(x)
```

This gave an extremely modest improvement that was probably not worth the effort. (If you're using a different version of MATLAB, you might even see the opposite result!) Unless we can compute r more accurately we probably won't do much better. If you try this in a CAS or if your version of MATLAB has access to variable precision arithmetic then you may be able to experiment with this by mimicking extended precision. A typical extended precision format beyond double precision has about 19 decimal digits of accuracy.

Having to retain A to compute r (and not recompute it from L and U, which will be used to solve $Ae = r$ in the form $L(Ue) = r$), and needing extended precision for the computation of r, represent significant drawbacks of this method. On the other hand, it can in principle give good improvements at relatively small computational cost. If we are working in single, or even half, precision, the "extended" precision needed is almost surely available. Although MATLAB defaults to double precision, a great many computations can be done faster in single precision while giving answers of the quality needed. It's not good practice to use higher precision than the problem merits; we seek to use smarter algorithms instead. Iterative improvement is a general purpose tool for improving the quality of a solution.

Problems

1.) Solve $Ax = b$ where $A = [2 \; 1; 1 \; 1]$, $b = (.15, .15)^T$ using mixed precision iterative improvement. Use 3 digit floating point arithmetic as single precision and 6 digit floating point arithmetic as double precision. Generate an initial guess using a single Jacobi iteration step starting from $(1, 1)^T$.

2.) Solve $Ax = b$ where $A = [20 \; 1; 1 \; .1]$, $b = (.15, .15)^T$ using mixed precision iterative improvement. Use 3 digit floating point arithmetic as single precision and 6 digit floating point arithmetic as double precision. Generate an initial guess using a single Jacobi iteration step starting from $(1, 1)^T$.

3.) a.) Solve the system in Problem 1 using Jacobi iteration with the given initial guess (in 3 digit floating point arithmetic). How many iterations are required to achieve the same accuracy as in Problem 1?

b.) Repeat part a. using Gauss-Seidel iteration.

4.) a.) Solve the system in Problem 2 using Jacobi iteration with the given initial guess (in 3 digit floating point arithmetic). How many iterations are required to achieve the same accuracy as in Problem 2?

b.) Repeat part a. using Gauss-Seidel iteration.

5.) Construct a sequence of increasingly ill-conditioned matrices in MATLAB, with condition numbers ranging up to at least 10^{20}, and solve systems involving them (for which you know the true solution) using two iterations of Jacobi iteration followed by as many iterations of iterative improvement as seem justified. Record the relative errors. Comment.

6.) Write a MATLAB program that accepts a square matrix A, a conformable vector b, and an optional positive integer k and returns the solution of $Ax = b$ as found by MATLAB's backslash command followed by k iterations of iterative improvement (use $k = 1$ if k is not provided by the user). Demonstrate your program on well-conditioned and ill-conditioned matrices.

7.) Let $A = [2\ 1\ 1; 1\ 2\ 1; 1\ 1\ 2]$ and $b = (4, 4, 4)^T$ so that $Ax = b$ has the solution $x = (1, 1, 1)^T$. Use iterative improvement, mixing single and double precision, to solve $Ax = b$ starting from an initial guess of $x_0 = 10^m (1, -2, 3)^T$ for $m = 1, 2, ...,$ until m is so large that convergence is no longer achieved. Comment.

8.) Write a MATLAB program that solves $Ax = b$ using Gaussian elimination to a desired tolerance using Eq. (3.3). Have your program apply iterative improvement to insure that the tolerance is met and return both the solution and the final upper bound on ρ.

9.) Write a MATLAB program that solves $Ax = b$ using QR decomposition to a desired tolerance using Eq. (3.3). Have your program apply iterative improvement to insure that the tolerance is met and return both the solution and the final upper bound on ρ.

10.) Write a MATLAB program that solves $Ax = b$ using Gauss-Seidel iteration to a desired tolerance using Eq. (3.3). Have your program apply one round of iterative improvement after the tolerance is met and return both the solution and the final upper bound on ρ (after the iterative improvement step).

11.) Using arbitrary precision arithmetic (such as in a CAS, or via MATLAB's vpa command), repeat Problem 5 using very high precision. Comment on your results, including the price paid in execution speed for using higher precision.

12.) a.) Under reasonable assumptions, fixed precision iterative improvement reduces the relative error $\|x_k - x\|_\infty / \|x\|_\infty$ at each step by a factor of about $2n\epsilon\kappa_\infty(A)$ where ϵ is the machine epsilon. Verify this experimentally.

b.) For mixed precision iterative improvement the relative error is reduced by a factor of about ϵ at each step. Verify this experimentally.

c.) How much better is mixed precision iterative improvement than fixed precision iterative improvement?

13.) A common means of solving a moderately large sparse system is to use Gaussian elimination with thresholding (see Problem 8 of Sec. 3.2) and then improve the result using iterative improvement or another iterative technique. The threshold is chosen so as to pivot rarely. In fact, one might skip pivoting entirely, thereby generating a relatively inaccurate result relatively quickly. Write a MATLAB program that performs Gaussian elimination without pivoting followed by iterative improvement; test it on several ill-conditioned matrices.

14.) For a large sparse matrix, preserving the sparsity structure is often important. This is all the more so if the scheme used to store the sparse matrix makes adding a new nonzero entry time-consuming. For this reason one might design a method as follows: Perform Gaussian elimination with or without pivoting, but whenever an operation would create fill-in in the matrix the result of that operation will be ignored and the entry will remain zero. Hence we annihilate nonzero entries but whenever this would create a new nonzero entry elsewhere in that row we simply leave that result zero. This technique is known as **incomplete Gaussian elimination**. Write a MATLAB program that implements incomplete Gaussian elimination without pivoting followed by iterative improvement. Test it on several large sparse matrices.

15.) Using arbitrary precision arithmetic (such as in a CAS, or via MATLAB's vpa command), write a program that implements mixed precision iterative improvement mimicking the use of double and extended precision arithmetic. Test your program.

16.) Write a MATLAB program that inputs a single precision matrix A, vector b, and vector x_0 that solves $Ax = b$ via mixed precision iterative improvement starting with the initial guess x_0, with double precision as the higher level of precision. You may choose whether you wish to use an initial Jacobi or Gauss-Seidel step(s), or simply go to iterative improvement. You should return a single precision answer and an optional status variable indicating failure to converge.

4. Preconditioning

Iterative improvement is one way in which we can prime an iterative method using the ideas of the Jacobi iteration

$$(4.1) \qquad x^{k+1} = -D^{-1}(U + L)x^k + D^{-1}b$$

or the Gauss-Seidel iteration

$$(4.2) \qquad x^{k+1} = -(L + D)^{-1}Ux^k + (L + D)^{-1}b$$

(where $A = U + L + D$).R. However, there is another way that is standard nowadays (and is an area of active research): **Preconditioning**, where we take a linear system $Ax = b$ and modify it to an equivalent system of the form

$$M_1 A M_2 y = \tilde{b}$$

(where $y = M_2^{-1}x$ and $\tilde{b} = M_1 b$) for which

$$\kappa(M_1 A M_2) << \kappa(A)$$

so that $M_1 A M_2$ is significantly better conditioned than A itself[5]. We refer to M_1 as a **left preconditioner** and M_2 as a **right preconditioner**. If the costs involved in finding and using M_1 and M_2 are not too severe compared to what is gained from the reduction in the condition number, this can be very beneficial.

Preconditioning followed by an iterative method is the standard approach for large sparse linear systems. In fact, an iteration or two of Jacobi or Gauss-Seidel iteration followed by iterative improvement (see Sec. 3.3) is often considered an example of this approach, with the Jacobi or Gauss-Seidel iteration being roughly like a form of preconditioning.

In some cases it is important to use both a right preconditioner and a left preconditioner; for example, if A is positive definite then $M_1 A$ may not be but $M_1 A M_2$ can be made to be positive definite, and this is likely to be desirable.

---

[5]In some applications we might want $\rho(I - M_1 A M_2) << \rho(I - A)$, or some other requirement involving the eigenvalues, and we use the term preconditioning to refer to this as well.

However, for simplicity we will look only at left preconditioners (so that $M_2 = I$). It is convenient to write the preconditioner as $P = M_1^{-1}$ so that the linear system $Ax = b$ becomes

(4.3) $$P^{-1}Ax = P^{-1}b$$

after preconditioning by P. Of course, we will not actually be inverting P at any point unless doing so is trivial.

One of the simplest preconditioners is also one of the most widely used. It is motivated in part by Eq. (4.1), Jacobi iteration, where D^{-1} is used to rescale all the nondiagonal entries of the matrix A. Perhaps $P^{-1} = D^{-1}$, that is, $P = D$, will yield a useful preconditioner? Indeed,

$$D^{-1}Ax = D^{-1}b$$

where D is the matrix of the diagonal entries of A, is often a good preconditioner and is called the **Jacobi preconditioner**. (We say that we are using **Jacobi preconditioning**.) That is, it is frequently the case that if A is ill-conditioned then $D^{-1}A$ is better conditioned than A.

EXAMPLE 1: Consider $A = [2\ 1; .1\ .01]$. Then $\kappa(A) \doteq 62.6103$ but $\kappa(D^{-1}A) \doteq 25.5233$ ($D = [2\ 0; 0\ .01]$ so $D^{-1}A = [1\ .5; 10\ 1]$). Jacobi preconditioning has reduced the condition number to less than half of what it was. Of course, we need to reduce it by an order of magnitude to see a meaningful difference in the accuracy of the solution of $Ax = b$ (recall that if $\kappa(A) \approx 10^k$ then we expect to lose about k digits of accuracy in solving $Ax = b$). If $A = [2\ 1; .01\ .01]$ then $\kappa(A) \doteq 500.0180$ but $\kappa(D^{-1}A) \doteq 6.3423$ ($D = [2\ 0; 0\ .01]$ so $D^{-1}A = [1\ .5; 1\ 1]$), a considerable improvement; if $A = [2\ 1; .0001\ .0001]$ then $\kappa(A) \doteq 50000.0$ ($5.0E4$) but again $\kappa(D^{-1}A) \doteq 6.3423$ ($D = [2\ 0; 0\ .0001]$ so $D^{-1}A = [1\ .5; .01\ .01]$). In the last case Jacobi preconditioning could gain us 4 digits of accuracy! $\square$

The Jacobi preconditioner is a type of **diagonal preconditioner** (in which P is a diagonal matrix). Sometimes we use an even simpler diagonal preconditioner, with $P = cI$ for some nonzero constant c. We might also use a diagonal preconditioner designed to **balance** (or **equibalance**) the matrix, meaning that we attempt to make all row vectors of the matrix have about the same length with respect to some vector norm. This frequently improves the conditioning, but it is not always obvious how best to choose the entries of P to achieve the desired effect.

The simple and inexpensive nature of Jacobi iteration makes it worthwhile to try it in many cases. However, there are a number of other preconditioners in common use. One is the **Gauss-Seidel preconditioner** $P = (L + D)$ (from Eq. (4.2); the minus sign is irrelevant here), and similarly for SOR. This preconditioning matrix is not quite as simple as before, though at least it is lower triangular. As we look at more computationally expensive preconditioners, note from Eq. (4.3)

$$P^{-1}Ax = P^{-1}b$$

that the trivial preconditioner $P = I$ has no effect, while the choice $P = A$ (so that $P^{-1} = A^{-1}$) solves the system immediately but is of course prohibitively expensive. (Recall that sparse matrices typically have full inverses.) We would like to choose a P that is simple to use, like a diagonal matrix–in particular, systems $Px = c$ should be easy to solve–but that is somehow "near" A^{-1} so that $P^{-1}A$ is nearer to

I, which has unit condition number for natural norms and is easy to manipulate. (This is sometimes called an **approximate inverse** preconditioner, as opposed to preconditioning for, say, some change in the eigenvalues of the preconditioned matrix.) These two conditions are in tension; except in exceptional circumstances, we can't have both P and $P^{-1}A$ be simple.

Suppose that A is large and sparse. What is it that keeps us from solving $Ax = b$ by the LU decomposition? It isn't just that the method is $O(n^3)$ because if A is very sparse we have relatively few nonzero entries to handle and should be able to write a program that makes use of this fact. But we also have fill-in–the LU factors of A are apt to be dense even if A is sparse. What if we were to form the LU decomposition of A approximately, so that $A \approx \tilde{L}\tilde{U}$? We could then use the approximate LU decomposition $P = \tilde{L}\tilde{U}$ as a preconditioner. If the approximation is good enough, perhaps $P^{-1}A$ might be near enough to I to represent an improvement.

One way to do this is to simply prohibit fill-in. We allow an entry of $\tilde{L}$ or $\tilde{U}$ to be nonzero only if the corresponding entry of A is nonzero. More generally, we might specify a range of entries of $\tilde{L}$ and $\tilde{U}$ that may be filled in, e.g., up to the first three subdiagonals and up to the first three superdiagonals. Any such factorization is known as **incomplete LU factorization**, leading to an **incomplete LU preconditioner** $P = \tilde{L}\tilde{U}$ (often referred to simply as an **ILU** method). The corresponding technique for positive definite matrices is **incomplete Cholesky factorization** and hence an **incomplete Cholesky preconditioner**.

EXAMPLE 2: Consider

$$A = \begin{pmatrix} 10 & .1 & .1 & .1 \\ .1 & .1 & 0 & 0 \\ .1 & 0 & .1 & 0 \\ .1 & 0 & 0 & .1 \end{pmatrix}$$

which has condition number $\kappa(A) \doteq 103.1553$. Its LU decomposition is, to the precision indicated,

$$L = \begin{pmatrix} 1 & 0 & 0 & 0 \\ .1 & 1 & 0 & 0 \\ .1 & -.0101 & 1 & 0 \\ .1 & -.0101 & -.0102 & 1 \end{pmatrix}$$

$$U = \begin{pmatrix} 10 & .1 & .1 & .1 \\ 0 & .099 & -.001 & -.001 \\ 0 & 0 & .099 & -.001 \\ 0 & 0 & 0 & .099 \end{pmatrix}$$

and we see that both L and U have completely filled in. The incomplete LU factorization, allowing no nonzero entries where A has zero entries, is

$$\widetilde{L} = \begin{pmatrix} 1 & 0 & 0 & 0 \\ .1 & 1 & 0 & 0 \\ .1 & 0 & 1 & 0 \\ .1 & 0 & 0 & 1 \end{pmatrix}$$

$$\widetilde{U} = \begin{pmatrix} 10 & .1 & .1 & .1 \\ 0 & .099 & 0 & 0 \\ 0 & 0 & .099 & 0 \\ 0 & 0 & 0 & .099 \end{pmatrix}$$

(notice that the replaced entries were all relatively small in magnitude). Then

$$\widetilde{L}\widetilde{U} \doteq \begin{pmatrix} 10 & .1 & .1 & .1 \\ .1 & .1 & .001 & .001 \\ .1 & .001 & .1 & .001 \\ .1 & .001 & .001 & .1 \end{pmatrix}$$

and so

$$(\widetilde{L}\widetilde{U})^{-1}A \doteq \begin{pmatrix} 1 & .0002 & .0002 & .0002 \\ 0 & 1 & -.0101 & -.0101 \\ 0 & -.0101 & 1.0001 & -.0101 \\ 0 & -.0101 & -.0101 & 1.0002 \end{pmatrix}$$

is the ILU preconditioned matrix, which has $\kappa((\widetilde{L}\widetilde{U})^{-1}A) \doteq 1.0310$. This is a considerable improvement but at a nontrivial cost. $\square$

Incomplete LU factorization with thresholding is the technique of discarding small elements in L and U. (Thresholding can also refer to a pivoting strategy, as described in Sec. 3.2.) This gives a decomposition $A = \widehat{L}\widehat{U} + E$ where the entries of E, the discarded entries, are small in magnitude, and hence $\|E\|$ is small. We then use $P = \widehat{L}\widehat{U}$ as the preconditioner. If the entries of E are truly negligible then $P^{-1}A$ should be close to the identity matrix[6].

Clearly, we will not form $P^{-1}A$ for a preconditioner like this. We will need to restructure our algorithms so that we solve a matrix involving P to simulate the effect of forming $P^{-1}A$. For example, suppose we need to form the residual of a preconditioned system $P^{-1}Ax = P^{-1}b$. We have

$$\begin{aligned} r &= P^{-1}b - P^{-1}Ax \\ r &= P^{-1}(b - Ax) \\ Pr &= b - Ax \end{aligned}$$

so we find r by computing the unpreconditioned residual $\rho = b - Ax$ and then solving $Pr = \rho$ for r. Although the preconditioner is nominally P^{-1}, we will be solving systems involving P. An example of this is given in the MATLAB subsection of Sec. 3.5.

What method should be applied to the preconditioned system $P^{-1}Ax = \widetilde{b}$? In practice we will usually have an iterative method in mind and will choose a preconditioner that is appropriate for that method. Commonly used iterative methods

---

[6]A statement like this always begs the question "Close in what sense?"; normwise ($\|P^{-1}A - I\|$) is the obvious sense but not the only one that might be of interest.

include iterative refinement (Sec. 3.3), GMRES (Sec. 3.5), conjugate gradients (in the positive definite case; see Sec. 7.7), and many others. In principle one could precondition adaptively, that is, use a different preconditioner P_i at each step of the main method, but this is rarely done. For some algorithms it will be something other than a reduction in the condition number that will do the most to improve the speed and accuracy of the computation (namely, the location of its eigenvalues or singular values).

Preconditioners are also used for other problems in numerical linear algebra, most notably eigenvalue problems (to be discussed later in this chapter). Efficient and effective preconditioning algorithms are extremely important in practice[7].

MATLAB

We'll look at the effects of preconditioning in this subsection but will leave the details concerning its practical implementation, avoiding inverses, until the MATLAB subsection of Sec. 3.5. Let's look again at the first matrix of Example 1. Enter:

```
>> A=[2 1;.1 .01]
>> cond(A)
>> AJ=inv(diag(diag(A)))*A
>> cond(AJ)
```

to duplicate the results in the example. Now let's try Gauss-Seidel preconditioning. Enter:

```
>> AGS=inv(tril(A))*A
>> cond(AGS)
```

Not only is the condition number for the Gauss-Seidel preconditioner smaller than that for the Jacobi preconditioner, but the resulting matrix is more nearly equibalanced: The max norms of the rows are closer and the entries of the matrix vary less in magnitude.

Let's try this with a bigger matrix. We'll use the Hilbert matrix of order 10. Enter:

```
>> H=hilb(10);
>> cond(H)
>> HJ=(diag(diag(H)))\H;
>> cond(HJ)
>> HGS=tril(H)\H;        %Better approach than inv(tril(H))*H.
>> cond(HGS)
```

The Jacobi preconditioner reduces the condition number by about an order of magnitude and the more expensive Gauss-Seidel preconditioner reduces it by about another order of magnitude.

There are commands in MATLAB for incomplete LU and incomplete Cholesky factorization factorizations (ilu and ichol, respectively).

```
>> S=sprand(100,100,.2);
>> S=S+(speye(size(S)));  %Ensure nonzero main diagonal for LU.
>> cond(S)       %Ignore the warning about condest.
>> [L1,U1,P]=ilu(S);   %ILU factorization with no fill-in.
```

[7]Trefethen and Bau write, "The name of the new game [in numerical linear algebra] is *iteration with preconditioning*. Increasingly often it is not optimal to solve a problem exactly in one pass; instead, solve it approximately, then iterate."

```
>> spones(L1)~=spones(tril(P*S)) %Where do L and A patterns differ?
>> spones(U1)~=spones(triu(P*S)) %Where do U and A patterns differ?
```
The ILU factors should only differ in their patterns of nonzero elements in places
where L or U has gained an additional zero entry by cancellation (which is unlikely
but possible). Very likely the matrices returned in the last two steps were entirely
zero; there was no fill-in (and no new zeros created in either the L or U factors).
Enter:
```
>> norm(full(P'*L1*U1-S))  %How well does L1*U1 approximate PS?
>> nnz(inv(P'*L1*U1))  %How sparse is the approx. inverse?
```
The approximate inverse is certainly not sparse–in all likelihood, it was completely
nonzero (10000 nonzero entries). Clearly we do not want to form $(P^T LU) \approx A^{-1}$.
We will see in the MATLAB subsection of Sec. 3.5 how the ILU approach can be
used effectively and efficiently.

Problems

1.) a.) Create 3 different 3×3 matrices for which Jacobi preconditioning improves
the condition number. Also compare $\|A - I\|$ to $\|P^{-1}A - I\|$ for each case.
b.) Create 3 different 3×3 matrices for which Gauss-Seidel preconditioning improves
the condition number. Also compare $\|A - I\|$ to $\|P^{-1}A - I\|$ for each case.
2.) Create a large ($n \geq 100$) matrix for which ILU preconditioning improves the
condition number. Simulate the incomplete LU factorization by performing the LU
decomposition, then explicitly making appropriate entries of the matrix zero.
3.) Write a complete algorithm (pseudo-code) for *efficiently* performing incomplete
LU factorization with no fill-in allowed.
4.) Suggest a method for using SOR as a preconditioner.
5.) a.) Find a diagonal preconditioner that minimizes the condition number of
$P^{-1}A$ if $A = [10 \ 1; .1 \ .2]$.
b.) Let S=sprandn(100,100,.2). Use trial and error to find a good diagonal
preconditioner for S.
6.) a.) The MATLAB command cgs(A,b,[],[],P) attempts to solve $P^{-1}Ax =
P^{-1}b$ by an iterative method. Use this to experiment with various preconditioners
P; comment.
b.) The MATLAB command bicg(A,b,[],[],P) attempts to solve $P^{-1}Ax =
P^{-1}b$ by an iterative method. Use this to experiment with various preconditioners
P; comment.
c.) The MATLAB command bicgstab(A,b,[],[],P) attempts to solve $P^{-1}Ax =
P^{-1}b$ by an iterative method. Use this to experiment with various preconditioners
P; comment.
7.) Solve $Hx = b$, where H is the Hilbert matrix of order 10 and b is a vector of all
ones, using the LU decomposition. Repeat using Jacobi preconditioning and then
again using Gauss-Seidel preconditioning. Compare the errors and the efficiency.
8.) Write a MATLAB program that computes an ILU factorization with thresholding for a sparse matrix. Compare your approach to that in luinc (with drop
tolerance).
9.) Could you use a unitary matrix as a preconditioner? Why or why not?
10.) Solve $Ax = b$ with A=gallery('dramadah',N) for increasing N, with b is
a vector of all ones, using the LU decomposition. (These are the anti-Hadamard
matrices: They have only 0 or 1 as entries and their inverses have large, but integer,

entries.) Repeat using Jacobi preconditioning and then again using Gauss-Seidel preconditioning. Comment on the errors and efficiency.

11.) a.) Solve $Ax = b$, where A is the prolate matrix of order N (use the MATLAB command `A=gallery('prolate',N)`) and b is a vector such that the true solution x is known, using the LU decomposition for $N = 5, 6, ..., 15$. Repeat using Jacobi preconditioning and then again using Gauss-Seidel preconditioning. Comment on the errors.

b.) The **prolate matrices** are **Toeplitz matrices**, meaning that every diagonal is constant. A common preconditioner for Toeplitz matrices is an appropriate **circulant matrix**, that is, a Toeplitz matrix for which each row is found from the row above it by pushing the entries one entry to the right, with wrap-around. The MATLAB command `gallery('circul',V)` creates the circulant matrix with first row V. Use trial and error to find a useful circulant preconditioner for the prolate matrix of order 8.

12.) a.) Write a MATLAB program that computes the incomplete Cholesky factorization with no fill-in allowed for a sparse positive definite matrix. Compare your approach to that in `ichol`.

b.) Write a MATLAB program that computes the incomplete Cholesky factorization with thresholding for a sparse positive definite matrix. Compare your approach to that in `ichol` (with drop tolerance).

13.) The MATLAB command `gallery('dorr',N)` creates the **Dorr matrix** of order N, an ill-conditioned sparse tridiagonal matrix. Compare Jacobi, Gauss-Seidel, and ILU preconditioning of the Dorr Matrix for $N = 10, 50, 100$.

14.) Conduct an experiment to compare the estimates returned by `cond(A,1)` and `condest(A)` for both accuracy and speed.

b.) Do the same for `norm(A,1)` and `normest1(A)`.

15.) Let H be the Hilbert matrix of order N. Using both a left and a right preconditioner M_1 and M_2, respectively, try to find a combination of M_1 and M_2 that gives the lowest condition number for $M_1 A M_2$. Explain your reasoning regarding the choice of M_1 and M_2.

16.) The command `W=gallery('wathen',5,5)` creates a matrix that arises in the numerical solution of partial differential equations by the finite element method. Use `spy(W)` to see its banded structure. What is its bandwidth, that is, how far from the main diagonal does a nonzero entry appear? What would be an appropriate technique for solving a system involving this matrix?

5. Krylov Space Methods

Many iterative methods for solving linear systems $Ax = b$ and for finding eigenvalues and eigenvectors of a matrix A are based on the **Krylov space**

$$(5.1) \qquad K_k = \text{span}\{w, Aw, A^2w, ..., A^{k-1}w\}$$

associated with A and a given vector w, which is a subspace of $\mathbb{R}^n$. We call it the **Krylov subspace** generated by A and w. We'll focus on using it for the solution of linear systems rather than eigenvalue problems. The general idea is this: Given the linear system $Ax = b$ we first try to find an approximate solution w_1 that is a multiple of w. We then look for a better approximate solution w_2 that was a linear combination of w and Aw, that is, one that is an element of K_2. We continue in this way, so that the kth approximation w_k is an element of K_k. Since

$$K_n = \mathbb{R}^n$$

(for a typical A and w) it appears that if a solution exists, then it is in K_n. We hope to find a good approximate solution in K_k for $k \ll n$ because in the cases of interest n will be very large (and A will be sparse).

There are many ways to build such a method. The typical choice for w is either b or the residual $b - Aw_0$ for some initial guess w_0. We'll look at a method that uses $w = b$, so that Eq. (5.1) becomes

$$K_k = \text{span}\{b, Ab, A^2 b, ..., A^{k-1} b\}.$$

Choose an initial guess w_0 as to the solution of $Ax = b$. Then define

$$V_k = w_0 + K_k$$

by which we mean that V_k is the set of all vectors of the form $w_0 + w$, where w_0 is the initial guess and w is any element of K_k. (Frequently we take $w_0 = 0$, in which case $V_k = K_k$.) We say that V_k is an **affine space**, meaning a vector space shifted by a particular vector (here, w_0). Note that if $w_0 \notin K_k$ then V_k is not a vector space.

At each stage of the iteration we will choose the new approximation w_k from this space. The natural way to define an approximate solution w_k of $Ax = b$ drawn from V_k is to let w_k be the value that minimizes the norm of the residual

$$\|b - Ax\|$$

over all vectors in V_k. We work with the residual squared for convenience and write this as

$$(5.2) \qquad\qquad w_k = \min_{w \in V_k} \|b - Aw\|^2$$

that is, w_k solves a least squares problem over a restricted part of the entire space $\mathbb{R}^n$, namely the affine space V_k. But the change of variables $z_k = w_k - w_0$ gives

$$
\begin{aligned}
z_k &= \min_{z \in K_k} \|b - A(z + w_0)\|^2 \\
&= \min_{z \in K_k} \|b - Aw_0 - Az\|^2 \\
(5.3) \qquad &= \min_{z \in K_k} \|r_0 - Az\|^2
\end{aligned}
$$

(where $r_0 = b - Aw_0$ is the initial residual), because $w_k - w_0$ is an element of the Krylov subspace K_k. Since K_k is a vector space, unlike V_k, this form is often easier to work with: We now have a standard linear least squares problem $Az = r_0$, and we know how to solve those.

If we solve for $k = 1, 2, ..., n$ we will eventually find the least squares solution of $Ax = b$, which is the true solution x if A is nonsingular; this would be grossly inefficient. The **GMRES** (for *generalized minimum residual*) method succeeds because it is easier to solve Eq. (5.2) over a smaller-dimensional space than a larger-dimensional one, and because it is frequently the case that an acceptable solution can be found for $k << n$. This justifies thinking of the method as an iterative method even though k cannot possibly exceed n; n will be so large in practice that k will never be nearly as big, and we imagine iterating for $k = 1, 2, 3 \ldots$ until some convergence criterion is met. We will solve Eq. (5.3) for a single vector K_1, then over a two-dimensional space K_2, and so on, until we stop. Since

$$K_1 \subset K_2 \subset \cdots \subset K_n$$

the previous solution is always included in the new space, and hence our new solution cannot possibly be worse than the current one (and should be better).

We may use any of our usual termination criteria with GMRES. A commonly employed criterion is that the **relative residual**

$$\frac{\|r_k\|}{\|b\|}$$

be less than some specified tolerance, where as usual $r_k = b - Aw_k$ denotes the residual after the kth iteration. It can be shown that if A is diagonalizable and

$$\mu = \|A - I\|$$

is less than unity then

$$\|r_k\| \leq \mu^k \|r_0\|$$

so that we may estimate the number of iterations required to achieve a desired decrease in the residual. Since $\mu < 1$ will not be true of a general matrix we will certainly want to try to find a preconditioner P such that

$$\left\|P^{-1}A - I\right\| < 1$$

(we might use both a left and a right preconditioner) when using GMRES, though preconditioning is not strictly necessary for convergence.

The algorithm in outline, then, is to choose a w_0 and successively solve Eq. (5.2) or Eq. (5.3) until convergence is achieved. How shall we solve the least squares problem at each iteration problem? It's natural to think of using the QR decomposition. Define

$$\Gamma_k = \left[b \mid Ab \mid A^2b \mid \cdots \mid A^{k-1}b\right]$$

called the kth **Krylov matrix**. Every vector in K_k is some linear combination of the k columns of Γ_k, and every linear combination of the columns of Γ_k is in K_k, so there must be some vector $y \in \mathbb{R}^k$ such that

$$w_k - w_0 = \Gamma_k y$$

and in terms of y Eq. (5.3) becomes the problem of finding the solution y of

(5.4) $$\min_{y \in \mathbb{R}^k} \|r_0 - A\Gamma_k y\|^2$$

i.e.

$$\min_{y \in \mathbb{R}^k} \|r_0 - M_k y\|^2$$

where $M_k = A\Gamma_k$ is an $n \times k$ matrix. We first find a $y \in \mathbb{R}^1$, then a $y \in \mathbb{R}^2$, and so on, and use them to construct our approximations $w_k \in \mathbb{R}^n$.

This is a series of standard least squares problems and each could be solved using the QR decomposition of M_k (see Sec. 2.7). However, this approach is both unstable and inefficient. Part of the problem is that the Krylov matrix Γ_k tends to be very ill-conditioned, because eventually $A^{p-1}b$ and $A^p b$ tend to be nearly linearly dependent, or at the very least they nearly lie in the span of the previous vectors. This problem is inherited by the R factor in the QR factorization of Γ_k. We will use a closely related approach based on using the Gram-Schmidt process (see Sec. 2.9) to find an orthonormal basis–the columns of Q–for the Krylov space K_k. This will allow us to avoid the formation, and use, of R. The Gram-Schmidt process in this setting is referred to as the **Arnoldi process**, and it may be summarized as follows:

Arnoldi Process Algorithm:
1. Set $q_1 = (b - Ax_0)/\|b - Ax_0\|$.
2. Begin loop ($m = 1$ to $k - 1$):
3. Set $v_{m+1} = Aq_m - \sum_{i=1}^m ((Aq_m)^T q_i)q_i$.
4. Set $q_{m+1} = v_{m+1}/\|v_{m+1}\|$.
5. End loop.

(This is essentially just the Modified Gram-Schmidt algorithm from Sec. 2.9. We should only compute Aq_m once in step 3.) If it should happen that $\|q_m\| = 0$ for some m then the process cannot be completed; however, it is a fact that if $\|q_m\| = 0$ then the true solution x of $Ax = b$ lies in $V_m = w_0 + K_m$, and hence we have already found the true solution. Otherwise, $\{q_1, ..., q_k\}$ is the desired orthonormal basis for K_k. Let

$$Q_k = [q_1 \mid q_2 \mid q_3 \mid \cdots \mid q_k]$$

be the corresponding matrix. This is the Q that would be found in the reduced QR factorization of Γ_k; again, the columns of Q_k form an orthonormal basis for K_k that has been found by applying the Gram-Schmidt process to Γ_k, the matrix associated with K_k.

By the same reasoning that led us to Eq. (5.4) we can now rephrase our least squares problem in terms of Q_k as

(5.5) $$\min_{y \in \mathbb{R}^k} \|r_0 - AQ_ky\|^2$$

and this will be much better behaved because Q_k has orthonormal columns and hence is well conditioned whereas K_k in Eq. (5.4) was likely to be ill-conditioned. We will solve Eq. (5.5) for y_k, then set $w_k = w_0 + Q_ky_k$.

EXAMPLE 1: Let $A = [2\ 1\ -1; 0\ 2\ 2; -2\ 1\ 2]$ and $b = (2, 4, 1)^T$. To solve $Ax = b$ we'll take $w_0 = 0$ and use Eq. (5.5). We have $r_0 = b - Aw_0 = b$, $\Gamma_1 = b$, and so the Arnoldi process gives $q_1 = r_0/\|r_0\| \doteq (.4364, .8729, .2182)^T$. Hence we must solve

$$
\begin{aligned}
y_1 &= \min_{y \in \mathbb{R}^k} \|r_0 - AQ_1 y\|^2 \\
&= \min_{y \in \mathbb{R}^k} \|r_0 - Aq_1 y\|^2 \\
&\doteq \min_{y \in \mathbb{R}^k} \left\|(2,4,1)^T - (1.5275, 2.1822, .4364)^T y\right\|^2
\end{aligned}
$$

for $y_1 \in \mathbb{R}^1$. We need the least squares solution of

$$
\begin{pmatrix} 1.5275 \\ 2.1822 \\ .4364 \end{pmatrix} y_1 = \begin{pmatrix} 2 \\ 4 \\ 1 \end{pmatrix}
$$

and we use the reduced QR decomposition of $q_1 \doteq (1.5275, 2.1822, .4364)^T$,

$$
(5.6) \qquad\qquad
\begin{aligned}
Q &\doteq \begin{pmatrix} -.5659 \\ -.8085 \\ -.1617 \end{pmatrix} \\
R &\doteq -2.6992
\end{aligned}
$$

(from MATLAB, using `[Q,R]=qr(A*q1,0)`) to find

$$
\begin{aligned}
y_1 &= R^{-1} Q^T r_0 \\
&\doteq 1.6773
\end{aligned}
$$

(see Sec. 2.7). Our current estimate of the solution of $Ax = b$ is

$$
\begin{aligned}
w_1 &= Q_1 y_1 \\
&= q_1 y_1 \\
&\doteq \begin{pmatrix} .7320 \\ 1.4641 \\ .3660 \end{pmatrix}
\end{aligned}
$$

for which $r_1 = b - Aw_1 \doteq (-.5621, .3399, .2680)^T$ ($\|r_1\| \doteq .7094$, compared to $\|r_0\| \doteq 4.5826$). This isn't an awful approximation of the true solution, which is $(1, 1, 1)^T$.

Let's do a second iteration of the method. First we need v_2. From the Arnoldi process, we have

$$v_2 = Aq_1 - \sum_{i=1}^{1} ((Aq_1)^T q_i) q_i$$

$$= \begin{pmatrix} 1.5275 \\ 2.1822 \\ .4364 \end{pmatrix} - \left[\begin{pmatrix} 1.5275 \\ 2.1822 \\ .4364 \end{pmatrix}^T \begin{pmatrix} .4364 \\ .8729 \\ .2182 \end{pmatrix} \right] \begin{pmatrix} .4364 \\ .8729 \\ .2182 \end{pmatrix}$$

$$\doteq \begin{pmatrix} .3637 \\ -.1455 \\ -.1455 \end{pmatrix}$$

$$q2 = \frac{\begin{pmatrix} .3637 \\ -.1455 \\ -.1455 \end{pmatrix}}{\left\| \begin{pmatrix} .3637 \\ -.1455 \\ -.1455 \end{pmatrix} \right\|}$$

$$\doteq \begin{pmatrix} .8704 \\ -.3482 \\ -.3482 \end{pmatrix}$$

so

$$Q_2 = [q_1 \mid q_2]$$

$$= \begin{pmatrix} .4364 & .8704 \\ .8729 & -.3482 \\ .2182 & -.3482 \end{pmatrix}$$

is the orthonormal basis of the Krylov subspace K_2. We must solve

$$y_2 = \min_{y \in \mathbb{R}^2} \| r_0 - AQ_2 y \|^2$$

for $y_2 \in \mathbb{R}^2$. (Note that we always use r_0 in the formula determining y_k, not r_k.)
Once again we will solve this using the reduced QR factorization of AQ_2,

(5.7)
$$Q \doteq \begin{pmatrix} -.5659 & .5899 \\ -.8085 & -.2600 \\ -.1617 & -.7645 \end{pmatrix}$$

$$R \doteq \begin{pmatrix} -2.6992 & .5911 \\ 0 & 3.5182 \end{pmatrix}$$

(compare the Q and R factors in Eq. (5.6) and in Eq. (5.7)). Solving the system
$AQ_2 y = r_0$, that is, $QRy = r_0$, gives

$$y_2 = R^{-1} Q^T r_0$$

$$\doteq \begin{pmatrix} 1.6384 \\ -0.1776 \end{pmatrix}$$

(note that the first component of y_2 is not equal to the sole component of y_1, though it is close). We have

$$w_2 = Q_2 y_2$$
$$\doteq \begin{pmatrix} .5605 \\ 1.4919 \\ .4194 \end{pmatrix}$$

for which $r_2 = b - Aw_2 \doteq (-.1935, .1774, -.2097)^T$ ($\|r_2\| \doteq .3360$, compared to $\|r_1\| \doteq .7094$ and $\|r_0\| \doteq 4.5826$). The absolute errors are $\alpha_0 \doteq 1.7321$, $\alpha_1 \doteq .8302$, $\alpha_2 \doteq .8788$. It might seem unsettling to see the absolute error increase as we go from w_1 to w_2, but it serves as a reminder that GMRES minimizes the residual error, not the absolute error. $\square$

At stage $k + 1$ of GMRES we are searching for an approximate solution over a subspace K_{k+1} (or more generally an affine space V_{k+1}) that includes the subspace K_k over which we searched at the previous stage, so the approximation can only improve or stay the same, not get worse. If there were no better approximation in K_{k+1}, after all, then we could use the approximation previously found in K_k. Hence, the residuals $\{r_k\}$ must be nonincreasing

$$\|r_{k+1}\| \le \|r_k\|$$

(recall that, as per Eq. (5.2), it is the residual r_k that we are minimizing at each step even though it is r_0 that appears explicitly in Eq. (5.5)). If A is nonsingular then GMRES will converge in at most n steps but again, this theoretical guarantee is useless for the cases in which GMRES is applied. We need convergence in many fewer iterations than n.

As suggested by comparing the two Q factors found in the two stages shown in the example, the implementation of GMRES can be made significantly more efficient than what we have described. We can only sketch the details here. At each iteration, there is a $(k + 1) \times k$ upper Hessenberg matrix H_k such that

(5.8) $$AQ_k = Q_{k+1} H_k.$$

and it can be found efficiently as part of the Arnoldi process. Putting this in Eq. (5.5) gives

$$y_k = \min_{y \in \mathbb{R}^k} \|r_0 - Q_{k+1} H_k y\|^2$$

which can be simplified as follows:

$$y_k = \min_{y \in \mathbb{R}^k} \|r_0 - Q_{k+1} H_k y\|^2$$

(5.9) $$= \min_{y \in \mathbb{R}^k} \left\| Q_{k+1} (Q_{k+1}^T r_0 - H_k y) \right\|^2$$

$$= \min_{y \in \mathbb{R}^k} \left\| Q_{k+1}^T r_0 - H_k y \right\|^2$$

$$= \min_{y \in \mathbb{R}^k} \|\rho e_1 - H_k y\|^2$$

where $\rho = \|r_0\|$ and e_1 is the first standard basis vector. In GMRES we solve the least squares problem

$$(5.10) \qquad\qquad \min_{y \in \mathbb{R}^k} \|\rho e_1 - H_k y\|^2$$

or equivalently

$$\min_{y \in \mathbb{R}^k} \|\rho e_1 - H_k y\|$$

at each iteration, using QR factorization (on H_k) in the usual way. (Recall from Sec. 2.8 that the Givens rotations method is especially efficient on upper Hessenberg matrices, so this will be a relatively fast process.) It is a fact that

$$r_k = Q_{k+1}(\rho e_1 - H_k y_k)$$

so we need not form $w_k = w_0 + Q_k y_k$ at each iteration to check the residual but can wait and only form the final approximation w_k.

But better yet, there is an efficient way to find the QR factors of H_{k+1} from those of H_k, as you might have guessed from comparing Eq. (5.6) and Eq. (5.7). This is called an **updating** procedure as it updates the old Q and R.

If you look carefully at what we've discussed and at Example 1 you'll see that we need not perform the entire Arnoldi procedure first before going to the minimization step. We can find an orthonormal basis for K_k, solve Eq. (5.10), then return to the Arnoldi process to get the next needed vector (for the orthonormal basis for K_{k+1}), and then solve Eq. (5.10) with H_{k+1}, and so on until the convergence criterion is met. In broad outline we have:

GMRES Algorithm (Outline):
1. Compute ρ.
2. Begin loop ($k = 1$ to n):
3. Perform a step of the Arnoldi process.
4. Form Q_k and H_k.
5. Solve $\min_{y \in \mathbb{R}^k} \|\rho e_1 - H_k y\|$ using the QR decomposition.
6. If convergence criterion is met, set $w_k = w_0 + Q_k y_k$ and terminate.
7. End loop.

In step 5 we can use the updating procedure to get the QR factors of H_k from the QR factors of H_{k-1} (don't confuse this Q with the matrix Q_k that is a factor of the Krylov matrix Γ_k). An algorithm that implements GMRES in this way is:

GMRES Algorithm:
1. Set $r_0 = b - A w_0$, $\rho = \|r_0\|$, $q_1 = r_0 / \rho$.
2. Begin loop ($k = 1$ to n):
3. Begin loop ($p = 1$ to k):
4. Set $h_{p,k} = q_k A^T q_p^T$.
5. End loop.
6. Set $t_{k+1} = A q_k - \sum_{p=1}^{k} h_{p,k} q_p$.
7. Set $h_{k+1,k} = \|v_{k+1}\|$.
8. Set $q_{k+1} = t_{k+1}/h_{k+1,k}$.
9. Set Q and R to be the reduced QR factors of H_k.
10. Set $y_k = \rho R^{-1} Q^T e_1$.

11. If convergence criterion is met or maximum iteration count is exceeded, set $w_k = w_0 + Q_k y_k$ and terminate.

12. End loop.

(We have left out the updating procedure in step 9 for clarity.) If the convergence criterion we are using is based on the residual r_k, which is almost certainly the case, we may use

$$\|r_k\| = \|\rho e_1 - H_k y_k\|$$

to compute it after y_k has been found, without the need to find $r_k = b - Aw_k = b - AQ_k y_k$. Note that while we need not retain the values of the temporary vectors t_k, for example, we must retain $\{q_i\}$ throughout the iteration. This can use a lot of memory.

The ill-conditioning of Γ_k means that we may lose the orthogonality of the basis vectors. Because of this we periodically **reorthogonalize** the basis for K_k (by applying the modified Gram-Schmidt process to the current basis). There are heuristics for detecting loss of orthogonality.

The advantages of GMRES are principally that it is applicable to matrices without special properties[8] (like positive definiteness, etc.) and that it depends on A only through matrix-vector products such as the computation of Aw_0 in computing the residual and the computation of products like Aq_m in the Arnoldi process. This fact can be useful when A is so large that storing it and working with it directly is impractical, but we have an explicit formula for computing a_{ij}. (This is a common case, especially in the numerical solution of partial differential equations.) We may write a special program that computes Ax from the entries of A without ever creating a copy of A. For this reason the method is said to be **matrix-free**. Matrix-free algorithms are often convenient for large, sparse systems, but to use them the method must access A only in this way. This means that products such as $A^T y$ cannot be used (the program that computes Ax, which may be supplied by someone else, will not compute $A^T y$), and so an inner product such as

$$h_{pk} = q_k A^T q_p$$

must be computed as

$$h_{pk} = (Aq_k^T)^T q_p$$

in order to keep the method matrix-free. This represents no real inconvenience, but we must be alert for opportunities to rewrite our algorithms so as to take advantage of efficiencies like this. Although it may seem unsettling to be attempting to solve $Ax = b$ when you can't actually "see" A, this situation is common in practice.

The disadvantages of GMRES include its computational expense (all the more so if the matrix-vector products are expensive) and the need to store a basis for K_k. The latter is usually the biggest concern: If A is very large then we may be able to store it efficiently as a sparse matrix or avoid storing it at all by using a program to compute the matrix-vector products. The entries in an orthonormal basis for K_k, however, form an $n \times k$ matrix that is likely to have no zero entries. As k increases the storage cost to store Q_k, needed to find w_k from

---

[8]However, it is a fact that GMRES works particularly well for normal matrices.

$$w_k = w_0 + Q_k y_k$$

will increase, often to the point where it is impractical to continue; maintaining orthogonality will also become more and more difficult.

For these reasons it is common to use a **restarting** provision, where we periodically stop the iteration and begin it again using the most current estimate, w_k, as the new initial guess. This means that we are starting over again with a one-dimensional subspace and building up a new K_1, K_2, ..., until convergence or the next restart. A common version of this restarted GMRES algorithm is known as **GMRES(m)**, which restarts after every m iterations. (The parameter m is usually based on a combination of available memory, and experience with the particular problem.) A restarted version of GMRES will generally have worse convergence properties but will require less memory.

The theory of GMRES and other Krylov space methods is related to the theory of polynomials in a fascinating way. We quote one theorem among many to give an idea of the flavor of these results:

THEOREM 1: If A is a nonsingular matrix then the residuals generated by the GMRES method satisfy

$$\|r_k\| = \min_{p \in P_k^1} (\|p(A)r_0\|)$$

where P_k^1 is the set of all polynomials of degree at most k such that $p(0) = 1$.

The set P_k^1 is sometimes called the set of **residual polynomials** of degree k in this context; $p(A)$ is the polynomial $p(z)$ evaluated at $z = A$ in the obvious manner. It may be helpful to use the Cayley-Hamilton theorem, which states that every matrix satisfies its own characteristic equation, to generate a residual polynomial to use to estimate $\|r_k\|$; a polynomial with roots at the eigenvalues of A is also frequently useful. When we use polynomials in this way to derive bounds on $\|r_k\|$ using Theorem 1 or a similar theorem we say that we are using a **test polynomial**.

We mention again that GMRES is typically used with preconditioning. The GMRES algorithm tends to perform better when the eigenvalues of A are **clustered** in the complex plane, that is, when they tend to group around some (non-zero) point, and so clustering the eigenvalues is a goal of preconditioning. This requires different preconditioners than those in Sec. 3.4.

MATLAB

For the very large, sparse matrices we're talking about one would ordinarily use (or write) a special purpose program in C++, Fortran, Python, or the like, which would call subroutines from Netlib or a similar source of high-quality numerical subroutines. In fact, MATLAB started as an interactive way of using such Fortran subroutines, and Fortran is still popular in large measure because of the large number of scientific computing subroutines created in it in the 1970s and 1980s.

Let's step through Example 1 in MATLAB. Compare the results to those in the example. Enter:

```
>> A=[2 1 -1;0 2 2;-2 1 2],b=[2;4;1]
>> r0=b                  %Recall that w0=0 here.
>> q1=r0/norm(r0)    %This is also Q1.
>> AQ=A*q1
>> [Q,R]=qr(AQ,0)
```

```
>> y1=R\Q'*b                %Solution of minimization problem.
>> w1=q1*y1
>> r1=b-A*w1
>> norm(r1)
>> norm(w1-[1;1;1])     %Abs. error; true solution is [1;1;1].
>> clear Q R            %Start the second iteration:
>> v2=A*q1-(q1'*A'*q1)*q1;q2=v2/norm(v2)
>> Q2=[q1,q2]
>> H=Q2'*A*q1     %Nonsquare upper Hessenberg, as expected.
>> AQ2=A*Q2
>> [Q,R]=qr(AQ2,0)
>> y2=R\Q'*b
>> w2=Q2*y2
>> r2=b-A*w2
>> norm(r2)
>> norm(w2-[1;1;1])     %Abs. error.
```

The absolute error has increased slightly, even though the residual has decreased considerably. Unfortunately in a real application of the method we cannot monitor the absolute error, only the residual. We must rely on formulas like $\rho_k \leq \kappa(A) \|r_k\| / \|b\|$ from Sec. 2.6 to bound the relative error ρ_k in terms of the residual, or the related result $\alpha_k/\alpha_0 \leq \kappa(A) \|r_k\| / \|r_0\|$ for the absolute error.

We did not need the variable H in order to work the problem in this way. We computed it merely to see the 2×1 upper Hessenberg matrix promised by Eq. (5.5).

How would we implement preconditioning? Suppose we were using a left preconditioner P so that the system we wanted to solve was now $P^{-1}Ax = P^{-1}b$. If P is the Jacobi preconditioner we might simply perform the multiplication and then use GMRES on $P^{-1}A$. For most other preconditioners this would not be feasible, so we simulate it. Note that for the preconditioned system the residuals are of the form $r_k = P^{-1}b - P^{-1}Ax_k = P^{-1}(b - Ax_k)$. Hence, $Pr_k = b - Ax_k$ which can be solved for r_k (remember that one criterion for a good preconditioner is that systems involving P should be easy to solve). Whenever we need b we use $P^{-1}b$ (which we might simply solve for once and store), and whenever we need A we use $P^{-1}A$ (which we might be able to compute and store or might need to solve for each time).

There is a GMRES command in MATLAB, gmres. Enter help gmres and note that you may enter a program that returns Ax rather than a matrix A. Note that you may also pass a preconditioner to gmres and may have it restart periodically (that is, it incorporates GMRES(k) as well). The routine also accepts an initial guess w_0 and a maximum number of iterations. The help suggests the following example using GMRES(10); enter:

```
>> A=gallery('wilk',21); b=sum(A,2);   %Makes soln. all ones.
>> tol=1e-12;maxit=15;M1=diag([10:-1:1 1 1:10]);
>> x=gmres(A,b,10,tol,maxit,M1,[],[]);
>> norm(b-A*x)/norm(b)     %Relative residual.
>> norm(b-A*x)             %Residual.
>> norm(x-ones(size(x)))   %Absolute error.
```

The preconditioner M1 used in this example is essentially the Jacobi preconditioner, with 1 substituted where the diagonal of A has a zero entry. Let's try again with the Gauss-Seidel method, again adjusting the zero diagonal entry to one. Enter:

```
>> M1=tril(A);M1(11,11)=1;
>> x=gmres(A,b,10,tol,maxit,M1);
>> norm(b-A*x)/norm(b)    %Relative residual for original system.
>> norm(b-A*x)            %Residual for original system.
>> norm(x-ones(size(x)))  %Absolute error.
```

It isn't actually fair to compare the errors because we asked gmres to compute until it hit its tolerance of 10^{-12} and that's what it did in both cases. The exact errors when it crossed that threshold aren't necessarily indicative of which approach is better.

What happens if we use no preconditioning? Enter:

```
>> z=gmres(A,b,10,tol);
>> norm(b-A*z)/norm(b)    %Relative residual.
>> norm(z-ones(size(x)))  %Absolute error.
```

The routine indicates that convergence was not achieved. The relative residual when it stopped was on the order of 10^{-9} and the absolute error was on the order of 10^{-8}.

In fact, if you look at the help for gmres you will see that it is set up to use the ILU method in an efficient manner. Enter:

```
>> H=sparse(hilb(8));  %(This isn't really very sparse.)
>> xtrue=ones([8 1]);b=H*x;
>> [L,U,P]=ilu(H);
>> x=gmres(A,b,10,tol,maxit,P'*L,U);
>> norm(x-xtrue)
```

to have gmres use the same preconditioner but without actually forming $P^T LU$. Instead, it solves systems such as $P^T LUx = c$ in the usual LU decomposition two-step manner. Using these two triangular solves is more efficient (and more accurate, although of course the decomposition itself is already intentionally inaccurate).

Repeat this experiment using A=sprandn(100,100,.2); solve using no preconditioner, the Jacobi preconditioner, and the ILU method. Repeat for a few different matrices A. Your results will likely indicate that preconditioning is more effective on structured matrices than unstructured ones, as a rule. There are many other preconditioners that are specialized to highly structured problems.

Enter type gmres and see how this method is implemented. Enter help minres and help qmr to see details of two methods related to GMRES.

Problems

1.) Perform one more iteration in Example 1 and verify that you get the true solution.

2.) Let $A = [-2\ 1\ 2\ 4; 1\ 0\ -1\ 2; -3\ 2\ -1\ -1; 0\ 1\ -2\ 1]$ and $b = (1,1,1,1)^T$. Approximate the solution of $Ax = b$ using three iterations of GMRES (as in Example 1) starting from $x_0 = (1,0,-1,0)^T$.

3.) a.) Write a MATLAB program that implements GMRES.
b.) Write a MATLAB program that implements GMRES(m).

4.) a.) Let $A = [2\ 2\ 1\ -1\ 3; 1\ 0\ -1\ -3\ 1; 0\ -1\ 3\ 2\ -1; 1\ 0\ -1\ 2\ 1]$ and $b = (1,1,1,1,1)^T$. Approximate the solution of $Ax = b$ using four iterations of GMRES.

b.) Approximate the solution of $Ax = b$ using two iterations of GMRES(2) (that is, perform two iterations of GMRES, restart, then perform two more iterations of GMRES). Compare with your answer from part a.

5.) If A is diagonalizable, nonsingular, and has $p < n$ distinct eigenvalues, then GMRES will converge in at most p iterations. Demonstrate this for two full 4×4 matrices, having 1 and 2 distinct eigenvalues, respectively.

6.) a.) Let $A = [-2 \ 1 \ 2 \ 4; 0 \ 1 \ -1 \ 2; -3 \ 2 \ -1 \ -1; 0 \ 1 \ -2 \ 1]$, $b = (1, 1, 1, 1)^T$, and $x_0 = (1, 0, -1, 0)^T$ as in Problem 2. Approximate the solution of $Ax = b$ using three iterations of preconditioned GMRES using the Jacobi preconditioner. Do not use `gmres`; step through the computation.

b.) Repeat part a. with the Gauss-Seidel preconditioner.

7.) Use `sprandn` to generate a 100×100 sparse matrix A that is about 10% dense. Pick a solution x and generate the corresponding b. Approximate the solution using: Four iterations of GMRES; two iterations of GMRES(2); four iterations of preconditioned GMRES using the Jacobi preconditioner; and four iterations of preconditioned GMRES using the Gauss-Seidel preconditioner. Use $x_0 = 0$ in each case. Compare your answers and also the efficiency of each method.

8.) a.) Use `sprandn` to generate a 500×500 sparse matrix A that is about 5% dense. Pick a solution x and generate the corresponding b. Approximate the solution using: Ten iterations of GMRES; two iterations of GMRES(5); ten iterations of preconditioned GMRES using the Gauss-Seidel preconditioner; and ten iterations of preconditioned GMRES using the ILU preconditioner. Use $x_0 = 0$ in each case. Compare your answers and also the efficiency of each method.

b.) Repeat part a. but rather than limiting the number of iterations continue until a relative residual of less than 10^{-8} is achieved.

9.) Carefully justify the steps in Eq. (5.9).

10.) It is possible to compute the QR factors of H_{k+1} from those of H_k using one Givens matrix. Suggest a method for doing so (for a hint, enter `type qrinsert` in MATLAB).

11.) a.) Write a MATLAB program for performing GMRES using the modified Gram-Schmidt method (from the MGS algorithm in Sec. 2.9).

b.) Compare the performance of GMRES with the modified Gram-Schmidt method to GMRES as given in the text. Find a matrix for which the former gives noticeably better results.

12.) If A is nonsingular and normal, meaning that $A^T A = A A^T$, and b is a linear combination of m linearly independent eigenvectors of A, then GMRES will converge in at most m iterations. Verify this for three different (nontrivial) normal matrices. Note: Symmetric matrices are normal, but there are nonsymmetric normal matrices.

13.) Verify Theorem 1 for the iteration in Example 1 for $k = 1, 2$.

14.) Prove Theorem 1. Hint: Show that if $x_k \in x_0 + K_k$ then $x_k = x_0 + \sum_{i=0}^{k} c_i A^i r_0$, then use this in $r_k = b - Ax_k$.

15.) Write an efficient MATLAB program that implements GMRES with preconditioning. If a preconditioner is not supplied by the user the program should use a reasonable default preconditioner.

16.) The `gallery` command in MATLAB can be used to generate a Krylov matrix. Write a MATLAB program that uses this to create such a matrix and then re-orthogonalizes it using MGS.

6. Numerical Eigenproblems

We've focused on solving the linear system $Ax = b$ in this and the previous chapter. There are other types of matrix computations that are of interest, however. The two primary types of matrix problems are those of solving the linear system $Ax = b$ and of finding the eigenvalues of A, and most other matrix problems are solved in the service of one of these two goals.

Recall that the spectrum of A, $\mathrm{sp}(A)$, is the set of all eigenvalues of A, that is, the set of all complex numbers λ such that

$$Ax = \lambda x$$

for some nonzero vector x, called an eigenvector of A associated with λ. These are of importance in many physical and mathematical applications, such as determining whether a control system is stable, or finding the singular values and hence condition number of a matrix (since the nonzero singular values of A are the square roots of the eigenvalues of $A^T A$; see Sec. 2.10).

All methods for finding the eigenvalues of a general matrix will be iterative methods, for finding the eigenvalues of an $n \times n$ matrix is equivalent to finding the roots of its characteristic polynomial

$$c(\lambda) = a_n \lambda^n + a_{n-1} \lambda^{n-1} + ... + a_1 \lambda + a_0$$

(where a_n is nonzero), and there is no way to find the roots of a general polynomial of degree 5 or higher without using iterative methods[9]. Since finding eigenvalues is equivalent to finding the roots of $c(\lambda)$, the same must be true of eigenvalue problems. (In fact, often we find the zeroes of a polynomial by constructing a matrix that has it as characteristic polynomial and then finding the eigenvalues of the matrix.). Hence, we will not have a finite-step method like Gaussian elimination for numerical eigenproblems; unlike the case for linear systems, we can not choose between direct methods and iterative methods.

Suppose that A has a full set of linearly independent eigenvectors[10] $\{x_1, ... x_n\}$ associated, in that order, with the eigenvalues $\lambda_1, ..., \lambda_n$, where $|\lambda_1| \geq |\lambda_2| \geq \cdots \geq |\lambda_n|$. Then

$$v_0 = \alpha_1 x_1 + ... + \alpha_n x_n$$

for some coefficients $\alpha_1, ..., \alpha_n$, because the eigenvectors form a basis for $\mathbb{R}^n$. Then

$$
\begin{aligned}
A v_0 &= A\left(\alpha_1 x_1 + ... + \alpha_n x_n\right) \\
&= \alpha_1 A x_1 + ... + \alpha_n A x_n \\
&= \alpha_1 \lambda_1 x_1 + ... + \alpha_n \lambda_n x_n
\end{aligned}
$$

because $A x_i = \lambda_i x_i$ $(i = 1, ..., n)$. So

---

[9]For every nonzero polynomial there is a matrix that has it as its characteristic polynomial.
[10]Recall that this is necessarily so in the commonly occurring case that A is symmetric.

$$
\begin{aligned}
A^2 v_0 &= A(Av_0) \\
&= Av_1 \\
&\quad \alpha_1 \lambda_1 A x_1 + \ldots + \alpha_n \lambda_n A x_n \\
&= \alpha_1 \lambda_1^2 x_1 + \ldots + \alpha_n \lambda_n^2 x_n
\end{aligned}
$$

and in general

$$
\begin{aligned}
A^k v_0 &= \alpha_1 \lambda_1^k x_1 + \ldots + \alpha_n \lambda_n^k x_n \\
&= \lambda_1^k \left(\alpha_1 x_1 + \alpha_2 \frac{\lambda_2^k}{\lambda_1^k} x_2 + \ldots + \alpha_n \frac{\lambda_n^k}{\lambda_1^k} x_n \right) \\
&= \lambda_1^k \left(\alpha_1 x_1 + \alpha_2 \left(\frac{\lambda_2}{\lambda_1} \right)^k x_2 + \ldots + \alpha_n \left(\frac{\lambda_n}{\lambda_1} \right)^k x_n \right)
\end{aligned}
$$

assuming $\lambda_1 \neq 0$. If it happens that $|\lambda_1| > |\lambda_i|$ for $i \neq 1$ we say that the matrix has a **dominant eigenvalue** λ_1. Then $A^k v_0 \approx \alpha_1 \lambda_1^k x_1$ for large k; $A^k v_0$ is tending toward a vector in the direction of an eigenvector associated with the largest eigenvalue of A (in magnitude), and the ratio of the max norms $\left\| A^k v_0 \right\|_\infty$ and $\left\| A^{k+1} v_0 \right\|_\infty$ is $|\lambda_1|$, the size of that eigenvalue.

The **power method** uses this idea to find the dominant eigenvalue of a matrix. It proceeds as follows: Pick an estimate v_0 of an eigenvector of A associated with a largest eigenvalue (in modulus) of A. Normalize it so that $\|v_0\|_\infty = 1$. Then compute

$$
\begin{aligned}
(6.1) \qquad v_k &= \frac{Av_{k-1}}{\mu_k} \\
&= \frac{A^k v_0}{\mu_k \mu_{k-1} \cdots \mu_1}
\end{aligned}
$$

($k = 1, 2, \ldots$), where μ_k is an entry of the vector in the numerator Av_{k-1} that has maximum modulus, insuring that $\|v_k\|_\infty = 1$ at each step. (There are other ways to normalize this iteration.) You'll be asked to show in the Problems that the normalization by μ_k in Eq. (6.1) is just what is needed to insure that

$$
\begin{aligned}
(6.2) \qquad v_k &\to x_1 \\
\mu_k &\to \lambda_1
\end{aligned}
$$

as $k \to \infty$. Clearly, the convergence is linear, with rate $|\lambda_2/\lambda_1|$; the more widely separated the top two eigenvalues, the smaller this ratio, and the faster the convergence. We speak of the **eigengap** $|\lambda_i - \lambda_{i+1}|$ between any two successive eigenvalues of a matrix; here we hope for a large eigengap.

What could go wrong? If A does not have a dominant eigenvalue it is still possible to extract some information from the iterates v_k but fundamentally the method fails. Fortunately in many applications it is known or exceedingly probable that there is a dominant eigenvalue. The other potential problem is that α_1 is zero, that is, that we have the incredibly bad luck to choose an initial vector v_0 that is orthogonal to x_1. Even if this should happen–and it's very unlikely–round-off errors will quickly introduce a component along x_1 and the method will recover. (Note, round-off error is actually *beneficial* here!) Usually, little effort is made to insure that the initial guess v_0 is a good guess at x_1; we choose any vector with no zero entries as the initial guess.

If we are principally interested in the eigenvalue, not the eigenvector, then we may compute the **Rayleigh quotient**

$$R_k = \frac{v_k^T A v_k}{v_k^T v_k}$$

which can be shown to converge to the eigenvalue quadratically. Other acceleration techniques for linearly convergent sequences could also be applied.

EXAMPLE 1: Consider the matrix $A = [1 \ 2; 2 \ 1]$. The eigenvalues are -1 and 3 so there is a dominant eigenvalue. Let's use $v_0 = (2, 1)^T$. We have

$$\begin{aligned} v_1 &= \frac{A v_0}{\mu_1} \\ &= \frac{1}{\mu_1} \begin{pmatrix} 1 & 2 \\ 2 & 1 \end{pmatrix} \begin{pmatrix} 2 \\ 1 \end{pmatrix} \\ &= \frac{1}{\mu_1} \begin{pmatrix} 4 \\ 5 \end{pmatrix} \\ &= \begin{pmatrix} 4/5 \\ 1 \end{pmatrix} \end{aligned}$$

$(\mu_1 = 5)$. Then

$$\begin{aligned} v_2 &= \frac{A v_1}{\mu_2} \\ &= \frac{1}{\mu_2} \begin{pmatrix} 1 & 2 \\ 2 & 1 \end{pmatrix} \begin{pmatrix} 4/5 \\ 1 \end{pmatrix} \\ &= \frac{1}{\mu_2} \begin{pmatrix} 14/5 \\ 13/5 \end{pmatrix} \\ &= \begin{pmatrix} 1 \\ 13/14 \end{pmatrix} \end{aligned}$$

$(\mu_2 = 14/5)$. Since the dominant eigenvalue is 3, $\mu_2 = 2.8$ is already a fair approximation; the eigenvector associated with the dominant eigenvalue is $(1, 1)^T$ and $(1, 13/14)^T \doteq (1, .9286)^T$ is a decent approximation to it for only two iterations. The Rayleigh quotients are

$$\begin{aligned} R_1 &= \frac{v_1^T A v_1}{v_1^T v_1} \\ &= \frac{\begin{pmatrix} 4/5 \\ 1 \end{pmatrix}^T \begin{pmatrix} 1 & 2 \\ 2 & 1 \end{pmatrix} \begin{pmatrix} 4/5 \\ 1 \end{pmatrix}}{\begin{pmatrix} 4/5 \\ 1 \end{pmatrix}^T \begin{pmatrix} 4/5 \\ 1 \end{pmatrix}} \\ &\doteq 2.9512 \end{aligned}$$

and

$$R_2 = \frac{v_2^T A v_2}{v_2^T v_2}$$

$$= \frac{\begin{pmatrix} 1 \\ 13/14 \end{pmatrix}^T \begin{pmatrix} 1 & 2 \\ 2 & 1 \end{pmatrix} \begin{pmatrix} 1 \\ 13/14 \end{pmatrix}}{\begin{pmatrix} 1 \\ 13/14 \end{pmatrix}^T \begin{pmatrix} 1 \\ 13/14 \end{pmatrix}}$$

$$\doteq 2.9945$$

and these are indeed much better approximations of $\lambda_1 = 3$, as expected (the relative error in R_2 is .0018 or about 2/10 of one percent). $\square$

A simple trick allows us to find the smallest eigenvalue (in modulus) of a non-singular matrix A. Note that if (λ_i, x_i) are an eigenpair of A,

$$Ax_i = \lambda_i x_i$$

and no eigenvalue of A is zero, then

$$\frac{1}{\lambda_i} x_i = A^{-1} x_i$$

that is, the eigenvalues of A^{-1} are the reciprocals of those of A. Hence, the largest eigenvalue of A^{-1} is the smallest eigenvalue of A, and so applying the power method to A^{-1} will locate it. Of course, as should go without saying at this point, we do not actually compute A^{-1} but rather compute the quantity

$$(6.3) \qquad\qquad v_k = \frac{A^{-k} v_{k-1}}{\mu_k}$$

by solving $Ay = v_k$ for the next numerator in Eq. (6.3) at each step. Note that we will be solving a system involving the same A many times, so it would be sensible to use a decomposition of it, or possibly even to transform A to a similar[11] matrix B such that linear systems involving B are easier to solve (e.g., if B was sparse or otherwise structured). Using Eq. (6.3) to find the smallest eigenvalue of A is known as the **inverse power method**.

The power method is useful if we only need the largest eigenvalue of A. This would allow us, for example, to determine the spectral radius

$$\rho(A) = \max\{|\lambda_i|\}$$

of the matrix. It is not too uncommon to need *only* the largest eigenvalue of a matrix. As an example, Google's PageRank algorithm for determining search engine results uses the power method in a case where a dominant eigenvalue $\lambda_1 = 1$ with a large eigengap is expected, meaning that the method requires at most a few iterations of an incredibly large matrix (representing the World Wide Web). The ranking is determined by x_1, the eigenvector associated with λ_1.

In many problems involving physical systems the largest *several* eigenvalues of a matrix are needed. For example, many applications involving systems of differential

---

[11]Recall that matrices A and B are said to be *similar* if $B = P^{-1}AP$ for some matrix P; similar matrices have the same spectrum.

equations ask for only the largest three or so eigenvalues. How can we find more eigenvalues? Deflation is an obvious approach (see Sec. 1.8)[12]. Note that if λ is an eigenvalue of A with associated eigenvector x then $(0, x)$ is an eigenpair of

$$A - \lambda I_n$$

for,

$$
\begin{aligned}
(A - \lambda I_n)x &= Ax - \lambda I_n x \\
&= \lambda x - \lambda x \\
&= 0 \\
&= 0 \cdot x
\end{aligned}
$$

and so if λ_1 was the largest eigenvalue in modulus of A, it is essentially turned into a null eigenvalue of the deflated matrix

$$M_1 = A - \lambda_1 I_n.$$

If M_1 has a dominant eigenvalue then we would apply the power method to it. If μ_1 is the dominant eigenvalue of M_1, with eigenvector z_1, then

$$
\begin{aligned}
M_1 z_1 &= \mu_1 z_1 \\
(A - \lambda_1 I_n) z_1 &= \mu_1 I_n z_1 \\
A z_1 &= (\lambda_1 I_n + \mu_1 I_n) z_1 \\
&= (\lambda_1 + \mu_1) z_1
\end{aligned}
$$

that is, $\lambda_1 + \mu_1$ is an eigenvalue of A, with eigenvector z_1. Since $|\mu_1|$ must be positive if it is dominant within M_1, this is truly a different eigenvalue than λ_1. This process could then be repeated to find more eigenvalues.

We won't discuss deflation in any detail. It has the usual drawbacks; it produces relatively low quality estimates, with error growing as the process continues, and its results should be cleaned up by another method. It is usually used not with the identity matrix but instead in the form

$$(6.4) \qquad\qquad\qquad A - \lambda_1 \Lambda$$

where Λ is a rank one matrix that is formed as the outer product of (the approximation to) x_1 and another vector that determines the method of deflation used. That is, we take $\Lambda = x_1 c^T$ for some vector c that is specified by some particular variation of deflation.

A related idea is to note that, if σ is a given value, then the eigenvalues of $A - \sigma I_n$ are related to the eigenvalues λ_i of A as follows:

$$
\begin{aligned}
(A - \sigma I_n)x &= Ax - \sigma I_n x \\
&= \lambda x - \sigma x \\
&= (\lambda - \sigma)x
\end{aligned}
$$

---

[12]The term deflation is also used to refer to a method for splitting an eigenvalue problem into smaller problems. We will not discuss this sense of deflation.

i.e., $\tau_i = \lambda_i - \sigma$ are the eigenvalues of $A - \sigma I_n$. Applying the power method to $A - \sigma I_n$ to find the largest eigenvalue of the shifted matrix is known as the **shifted power method** (with shift σ.).

Most often the power method is applied in the following manner: We obtain an initial estimate of the eigenvalue λ_i that we seek, which may or may not be the largest or smallest eigenvalue of A. We then apply the inverse power method to $A - \sigma I_n$. Since the eigenvalues of this matrix are

$$\frac{1}{\lambda_i - \sigma}$$

(assuming $\sigma \notin \mathrm{sp}(A)$), the largest eigenvalue of $(A - \sigma I_n)^{-1}$ corresponds to the eigenvalue λ_i of A that is closest to σ. Hence, this approach, called the **inverse shifted power method**[13] or **inverse iteration**, allows us to find the eigenvalue of A closest to a given value σ. This is the form in which the power method is most commonly seen, unless it is being used to get an initial estimate of an eigenvalue for some other method to improve. Note that the power method with deflation may be used to get an estimate of the second-largest, third-largest, etc., eigenvalues for use in this method.

The inverse iteration method typically converges much more rapidly than the power method (per iteration, but of course iterations are now more expensive as we must solve a linear system at each step). Of course the choice of shift affects this; if σ is very close to the eigenvalue then $1/(\lambda_i - \sigma)$ is likely to be very large and hence far from the next nearest eigenvalue, so rapid convergence of the method can be expected due to the sizable eigengap.

In the spirit of the Gauss-Seidel iteration (see Sec. 3.1), a natural idea is to update the shift at each step of the shifted or inverse shifted power method with the estimate provided by the Rayleigh quotient, that is, to set $\sigma_k = R_{k-1}$ and use $A - \sigma_k I_n$ at each step. This is sensible since R_k is our best available estimate of the eigenvalue. The inverse shifted power method with this modification is called **Rayleigh quotient iteration** and yields very rapid convergence to the eigenvalue.

We might worry for all shifted methods that as $\sigma \to \lambda$ the matrix $A - \sigma I_n$ may become ill-conditioned with respect to the solution of linear systems. This is so, but the net effect on the resulting eigenvalue estimates is less than would be expected. The reason is that the errors tend to align with the direction of the approximate eigenvector–but since any nonzero multiple of an eigenvector is itself an eigenvector, the resultant is just another approximate eigenvector. The condition number of a matrix with respect to inversion, $\kappa(A)$, isn't necessarily a good notion of the conditioning of the eigenvalue problem for A, and there is a notion of condition number with respect to eigenvalues (see the MATLAB subsection). Ill-conditioning in this context is usually due to the matrix having repeated eigenvalues, which have a zero eigengap.

Inverse iteration is a good technique for improving the accuracy of an eigenvalue or eigenvector, for finding an eigenvector corresponding to a known eigenvalue, or for finding a single eigenvalue near a given value σ. If more than just a few eigenpairs of A are sought, however, there are significantly more efficient methods.

MATLAB

[13]Many authors simply call this the inverse power method.

We have already seen the `eig` command. If A is a square matrix, `[V,D]=eig(A)` finds matrices such that `A*V=V*D`, that is, V is the matrix of eigenvectors and D is a diagonal matrix containing the corresponding eigenvalues on the main diagonal. Enter:
```
>> help eig
```
to see the options; in particular, note that by default a balancing procedure is used to help improve the condition of the matrix, and this can be turned off if there is concern that it may be making matters worse.

As mentioned in this section, although we have been focusing on the condition number of a matrix with respect to the solution of a linear system $Ax = b$ involving it, the idea of a condition number is more general and can be discussed with respect to other problems. In Sec. 1.8, ill-conditioned nonlinear root-finding problems are discussed; there, ill-conditioning is associated with multiple roots. An eigenvalue problem is ill-conditioned when A is near a matrix with multiple eigenvalues (which would correspond to multiple roots of the characteristic polynomial of the matrix). The MATLAB command `condeig` computes the condition number of a matrix's eigenvalues. Enter:
```
>> H=hilb(10);
>> cond(H)
>> condeig(H)
```
Each eigenvalue of the 10×10 Hilbert matrix is very well-conditioned with respect to the eigenvalue problem, even though the matrix is very poorly conditioned with respect to the solution of linear systems. Enter:
```
>> A=[2 1;0 2]
>> cond(A)
>> condeig(A)
```
This matrix is well-conditioned with respect to the solution of linear systems, but since it is defective (as it only has one linearly independent eigenvector) with repeated eigenvalues 2 and 2, it is very poorly conditioned with respect to the eigenvalue problem. The situation is not nearly as bad for $2I_2$. Enter:
```
>> condeig(2*eye(2))
```
Despite this matrix also having repeated eigenvalues 2 and 2, it is well-conditioned. The defective nature of the previous matrix makes it especially poorly conditioned despite its trivial size.

The eigenvalue condition number depends on the matrix of eigenvectors and its inverse, similar to the way in which the condition number with respect to the solution of linear systems depends on A and A^{-1}. Recall that if A is symmetric then the matrix of eigenvectors is orthogonal and hence its inverse is easy to compute. Enter:
```
>> type condeig
```
to see how the eigenvalue condition number is computed in MATLAB. Note from the help for `condeig` that you may use it to compute the eigenvalues, eigenvectors, and condition numbers all at once.

The MATLAB command `poly` may be used to find the characteristic polynomial of A. However, this is a slow process. Enter:
```
>> type poly
```
and note that MATLAB finds the characteristic polynomial by using the `eig` command to find its roots, then constructing the coefficients of the polynomial. To

actually find the coefficients in $c(\lambda)$ without knowing the roots of the characteristic equation is prohibitively expensive in the general case. Even if we could do that, the corresponding root-finding problem $c(\lambda) = 0$ can be ill-conditioned (viewed as a root-finding problem) even when the eigenvalue problem itself is well-conditioned. Using techniques intended to find eigenvalues of matrices directly is the best approach in essentially all cases.

Problems

1.) a.) Apply three iterations of the power method to estimate the largest eigenvalue of $A = [2 \ 1; 1 \ 2]$.

b.) Compute the three Rayleigh quotients.

2.) Apply three iterations of the shifted power method to estimate an eigenvalue of $A = [2 \ 1; 1 \ 2]$ using the shift $\sigma = 1.25$.

3.) Apply three iterations of the inverse power method to estimate the smallest eigenvalue of $A = [2 \ 1; 1 \ 2]$.

4.) a.) Why must a dominant eigenvalue be real?

b.) What happens if the largest eigenvalue of A is a multiple eigenvalue, but all other eigenvalues of A are smaller in modulus?

c.) If x is an eigenvector of A, what can you say about the Rayleigh quotient $R(x) = x^T A x / x^T x$?

5.) Verify that the results in Eq. (6.2) are correct.

6.) a.) Apply three iterations of the inverse shifted power method to estimate the eigenvalue of $A = [2 \ 1; 1 \ 2]$ nearest to 1.2.

b.) Apply three iterations of the inverse shifted power method to estimate the eigenvalue of $A = [2 \ 1; 1 \ 2]$ nearest to 1.1.

c.) Apply three iterations of the inverse shifted power method to estimate the eigenvalue of $A = [2 \ 1; 1 \ 2]$ nearest to 1.05. Comment on the effect of the shift on the convergence of the method.

7.) What is the formula for the eigenvalue condition number used in the `condeig` program?

8.) Write a MATLAB program for performing the power method using the Rayleigh quotient. Attempt to detect failure to converge (due to the lack of a dominant eigenvalue). The user should be asked to supply only the matrix. Return an estimate of both the eigenvalue and an associated eigenvector.

9.) Apply the power method to the matrix $A = [-5 \ -4; 4 \ 5]$. Explain your results.

10.) a.) Apply five iterations of the power method to estimate the largest eigenvalue of $A = [3 \ 1 \ 0; 1 \ 3 \ 0; 0 \ 0 \ 1]$.

b.) Compute the Rayleigh quotient of the last estimate of the eigenvalue.

c.) Use deflation in the form $A - \lambda I_3$ to approximate the next smallest eigenvalue of A using five iterations of the power method.

d.) Compute the Rayleigh quotient of the last estimate of this eigenvalue.

11.) The method of **Wielandt deflation** uses Eq. (6.4) with $\Lambda = x_1 c^T$ where $c = A_i / (\lambda_1 \xi_i)$. Here A_i is the ith row of A, written as a column vector; λ_1 is the estimate of the largest eigenvalue of A; and ξ_i is the ith entry of x_1, the estimate of the associated eigenvector. Repeat Problem 10 part c and 10 part d . using Wielandt deflation. Is it a better technique than using $\Lambda = I_n$?

12.) Write a MATLAB program for performing the inverse shifted power method using the Rayleigh quotient. Attempt to detect failure to converge (due to the lack of a dominant eigenvalue). The user should be asked to supply only the matrix. Return an estimate of both the eigenvalue and an associated eigenvector.

13.) a.) Use the power method with deflation to find all eigenvalues of the Hilbert matrix of order 10 to an accuracy of at least 10^{-4}.

b.) Repeat with the inverse shifted power method, choosing shifts in a reasonable way based on the information available (perhaps by using the Gershgorin Circle Theorem). Comment.

14.) What happens in the power method if the largest eigenvalue of A is complex?

15.) a.) Prove that the Rayleigh quotient converges to the dominant eigenvalue.

b.) Prove that this convergence is quadratic.

16.) a.) The **numerical range** (or **field of values**) of a (possibly complex) $n \times n$ matrix A is defined to be the set $W(A)$ of all values that the Rayleigh quotient $R = v^H A v / v^H v$ can take on, for all nonzero $v \in \mathbb{C}^n$. (Note the use of the Hermitian transpose.) In this sense it is the usual sense of the term range for the Rayleigh quotient when viewed as a function $R(v)$ from $\mathbb{C}^n$ to $\mathbb{C}$. The largest entry of $W(A)$, in modulus, is called the **numerical radius** $r(A)$ of the matrix. (The numerical radius always exists.) Prove that $r(A)$ is a matrix norm and then conclude that $\rho(A) \leq r(A)$ (that is, the numerical radius is an upper bound on the spectral radius).

b.) Show that $W(A)$ can equally well be defined as the set of all $v^H A v$ over all unit vectors in $\mathbb{C}^n$.

c.) When does the numerical range contain the zero vector, that is, when is $0 \in W(A)$?

7. The QR Method of Francis

Inverse iteration finds only one eigenpair at a time. There is a Krylov subspace generalization of the power method that locates several eigenpairs of a matrix. The method generates a set of vectors called a **Krylov sequence**

$$\{v_0, A v_0, A^2 v_0, ..., A^{k-1} v_0\}$$

in which we effectively perform the power method computations but retain previously computed vectors. Next, we find an orthonormal basis $\{q_1, ..., q_k\}$ for the Krylov subspace corresponding to this Krylov sequence, that is, we form the vector space

$$K_k = \text{span}\{v_0, A v_0, A^2 v_0, ..., A^{k-1} v_0\}$$

then find an orthonormal basis such that

$$K_k = \text{span}\{q_1, ..., q_k\}$$

which is called **Arnoldi's method** (or **Arnoldi iteration**) in this context. This is followed by a matrix version of the Rayleigh quotient. This is primarily just an efficient implementation of the power method, but well worth the effort.

There is also the method of **simultaneous iteration** (or the **block power method**) that applies the power method to several vectors at once; that is, it computes

$$A^k[v_0 \quad w_0 \ldots z_0]$$

where the vectors are normalized appropriately. Once again an orthogonalization procedure must be applied. This method yields the first several eigenvectors of A (that is, those associated with the eigenvalues of largest magnitude). For many physical applications, where the largest eigenvalues indicate the speed at which the solution to a differential equation is decreasing (because the solution is a combination of exponentials of the form $\exp(c_i\lambda_i t)$ for some c_i, possibly all unity, and where the λ_i are eigenvalues), it's enough to know what the largest few are as the remaining eigenvalues give negligible contributions to the sum.

But what if we want *all* the eigenvalues of a matrix? To take just one example, the roots command uses the eigenvalues of a matrix to find all roots of a polynomial that is the characteristic polynomial of that matrix. Certainly we want all of the eigenvalues in this case. There are many others.

The natural idea, at least in the case where A is diagonalizable (meaning that its eigenvectors span $\mathbb{R}^n$), is to look at the eigenvalue decomposition

$$A = PDP^{-1}$$

where D is a diagonal matrix with the eigenvalues of A on the main diagonal and P is a nonsingular matrix with the corresponding eigenvectors as its columns. Recall that in general the diagonal elements of D and corresponding columns of P may be complex even though A is real.

If A is normal, meaning that $A^T A = A A^T$, then we may take P to be unitary (meaning that P is equal to its conjugate transpose). The most important case of this is when A is symmetric, in which case we may take P to be orthogonal. (Symmetric matrices are trivially normal, and they have real eigenvalues and real eigenvectors.) The symmetric eigenvalue problem is perhaps the most commonly encountered situation.

However, not all matrices have an eigenvalue decomposition. For this and other reasons we are inclined to seek another decomposition. All square matrices have a **Jordan decomposition** (or **Jordan normal form** or **Jordan canonical form**)

$$A = PJP^{-1}$$

where J is diagonal if A is diagonalizable and a specific type of upper triangular matrix otherwise. But all matrices also have a **Schur decomposition** (or **Schur form**)

$$\begin{aligned} A &= QTQ^{-1} \\ &= QTQ^T \end{aligned}$$

where Q is orthogonal, so that $Q^{-1} = Q^T$, and T is upper triangular. We will use the Schur form in our algorithm.

Why is the Schur form more attractive than the other forms? Note that all of the forms above are similarity transformations[14], that is, they are of the form $A = SBS^{-1}$. Recall that all matrices related by similarity transformations have the same eigenvalues. Hence in each case (the eigenvalue decomposition $A = PDP^{-1}$,

---

[14]The Schur form is also a congruence; two matrices A and B are congruent if $A = M^T B M$ with M nonsingular.

the Jordan normal form $A = PJP^{-1}$, and the Schur form $A = QTQ^T$ where $Q^T = Q^{-1}$), the matrices A, D, J, and T all have the same eigenvalues. Also, each of D, J, and T are upper triangular so their eigenvalues are simply their diagonal entries.

But unlike the eigenvalue decomposition, the Schur form always exists; and unlike the eigenvalue and Jordan decompositions, the similarity transformation to it is always orthogonal. This is very desirable for numerical computation because orthogonal matrices are so well-behaved (despite almost always being dense). Hence we are inclined to seek an *orthogonal triangularization* to Schur form (see Sec. 2.8), that is, we want to develop an algorithm intended to process A as follows:

$$
\begin{aligned}
T_0 &= A \\
T_1 &= Q_1^T A Q_1 \\
T_2 &= Q_2^T Q_1^T A Q_1 Q_2 \\
T_3 &= Q_3^T Q_2^T Q_1^T A Q_1 Q_2 Q_3 \\
&\vdots
\end{aligned}
$$

(7.1)

with each T_i more nearly upper triangular, until convergence to the Schur form $T = Q^T A Q$ is reached (to within some tolerance). Each Q_i will be orthogonal, which has numerical benefits (the condition number is $\kappa(Q_i) = 1$ if using a natural norm), and each step is a similarity transformation, so the eigenvalues are preserved. We will need to use inverse iteration after finding the eigenvalues of the main diagonal of T if the eigenvectors are also desired. In many cases we seek only the eigenvalues.

How shall we find the similarity transformations Q_i? We'll omit the detailed derivation and go directly to the surprising result: We can obtain the Schur form by decomposing $T_0 = A$ into its QR decomposition

$$A = Q_0 R_0$$

(see Sec. 2.7) then forming the product of the QR factors in the reverse order

$$T_1 = R_0 Q_0$$

and repeating. The algorithm, in its simplest form, is no more complicated than that[15]. We implement the process in Eq. (7.1) as follows:

$$
\begin{aligned}
Q_0 R_0 &= A \\
T_1 &= R_0 Q_0 \\
Q_1 R_1 &= T_1 \\
T_2 &= R_1 Q_1 \\
Q_1 R_1 &= T_2 \\
T_3 &= R_2 Q_2 \\
&\vdots
\end{aligned}
$$

---

[15]In fact, we could use the LU decomposition in a similar way to find eigenvalues; this is the LR algorithm.

and repeat until the entries in the lower triangle of T_i are sufficiently small. This is the **QR method**, for which the QR decomposition of Sec. 2.7 was actually first developed.

Is this really a similarity transformation at each step? Yes. Let's look at the first step. We have, from the QR method,

$$
\begin{aligned}
T_1 &= R_0 Q_0 \\
T_1 Q_0^T &= R_0 \\
Q_0 T_1 Q_0^T &= Q_0 R_0 \\
Q_0 T_1 Q_0^T &= T_0
\end{aligned}
$$

so $T_0 = A$ and T_1 are related by a similarity transformation. The same is true for any other step, and so the end result is an approximately upper triangular matrix T_N that is similar to A

$$
\begin{aligned}
T_N &= \left(Q_N^T Q_{N-1}^T \cdots Q_0^T \right) A \left(Q_0 Q_1 \cdots Q_N \right) \\
&= Q_a^T A Q_a
\end{aligned}
$$

where $Q_a = Q_0 Q_1 \cdots Q_N$ is the approximate Q of the Schur decomposition. It can be shown that if T_N is nearly upper triangular then its eigenvalues are nearly the entries on the main diagonal–as would be expected; this is a continuity result–so we take the diagonal entries as the eigenvalues and then use inverse iteration to get the eigenvectors if they are desired.

EXAMPLE 1: Consider the matrix $A = [3\ 1\ 0; 1\ 3\ 0; 0\ 0\ 1]$. The eigenvalues of A are 4, 2, 1. The QR method gives

$$
\begin{aligned}
A &= Q_0 R_0 \\
&\doteq \begin{pmatrix} -.9487 & -.3162 & 0 \\ -.3162 & .9487 & 0 \\ 0 & 0 & 1.0000 \end{pmatrix} \begin{pmatrix} -3.1623 & -1.8974 & 0 \\ 0 & 2.5298 & 0 \\ 0 & 0 & 1.0000 \end{pmatrix} \\
T_1 &= R_0 Q_0 \\
&\doteq \begin{pmatrix} 3.6000 & -0.8000 & 0 \\ -0.8000 & 2.4000 & 0 \\ 0 & 0 & 1.0000 \end{pmatrix}
\end{aligned}
$$

(all computations are done in MATLAB). The diagonal entries already appear to be approaching the eigenvalues. Note also that the symmetry of A is preserved, although we are not taking advantage of it here. Decomposing T_1 into its QR factors and then multiplying them in the reverse order gives

$$T_1 = Q_1 R_1$$

$$\doteq \begin{pmatrix} -.9762 & .2169 & 0 \\ .2169 & .9762 & 0 \\ 0 & 0 & 1.0000 \end{pmatrix} \begin{pmatrix} -3.6878 & 1.3016 & 0 \\ 0 & 2.1693 & 0 \\ 0 & 0 & 1.0000 \end{pmatrix}$$

$$T_2 = R_1 Q_1$$

$$\doteq \begin{pmatrix} 3.8824 & .4706 & 0 \\ 0.4706 & 2.1176 & 0 \\ 0 & 0 & 1.0000 \end{pmatrix}$$

and again the symmetry has been preserved (and though it's less obvious, so have the eigenvalues). The only nonzero value in T_i is the $(2,1)$ entry, the magnitude of which has decreased from 1 in T_0 to 0.8000 in T_1 to .4706 in T_2; this is the entry we are trying to drive to zero. The diagonal of T_i is even closer to the eigenvalues 4, 2, and 1. In fact

$$T_3 \doteq \begin{pmatrix} 3.9692 & -.2462 & 0 \\ -.2462 & 2.0308 & 0 \\ 0 & 0 & 1.0000 \end{pmatrix}$$

$$T_4 \doteq \begin{pmatrix} 3.9922 & .1245 & 0 \\ .1245 & 2.0078 & 0 \\ 0 & 0 & 1.0000 \end{pmatrix}$$

in fair agreement with $\mathrm{sp}(A) = \{4, 2, 1\}$. In fact, the Gershgorin Circle Theorem indicates that the larger two eigenvalues have been located to within a radius of .1245 (in the complex plane). $\square$

Note how the $(2,1)$ entry of T_i is reduced by about one-half at each step. There is a reason for this slow convergence–the QR method can be viewed as a reformulation of the block power method mentioned at the beginning of this section, and so the slowest-fading entry below the main diagonal goes to zero like $|\lambda_2/\lambda_1|$ in general. (In the example, $|\lambda_2/\lambda_1| = |2/4|$, and hence we see a reduction of about $1/2$ at each step.) The method is stable and converges under fairly general conditions, but this slow (linear) convergence is problematic.

The speed can be improved with shifts. The **QR method with shifts**, also called the **QR method of Francis**[16], picks a shift σ_k at each step and proceeds as follows:

$$\begin{aligned} Q_0 R_0 &= A - \sigma_0 I_n \\ T_1 &= R_0 Q_0 + \sigma_0 I_n \\ Q_1 R_1 &= T_1 - \sigma_1 I_n \\ T_2 &= R_1 Q_1 + \sigma_1 I_n \\ Q_1 R_1 &= T_2 + \sigma_2 I_n \\ T_3 &= R_2 Q_2 - \sigma_2 I_n \end{aligned}$$

(7.2)

$$\vdots$$

---

[16]John G. F. Francis of the U.K. published his version of the algorithm shortly before Vera Kublanovskaya of the U.S.S.R. independently published her version.

until the subdiagonal elements of T_i are sufficiently small (or some other convergence criterion is met). How shall we choose the shifts? There are several possibilities. One method in common use is based on the Rayleigh quotient. Another is the **Wilkinson shift**, defined as the eigenvalue of the southeast 2×2 submatrix of T_i that is closest to the southeastmost entry in T_i. That is, if we represent T_i with a Wilkinson diagram like

$$T_i = \begin{pmatrix} X & X & X & X & X \\ \vdots & \vdots & \vdots & \vdots & \vdots \\ X & X & X & X & X \\ X & X & X & \alpha & \beta \\ X & X & X & \gamma & \delta \end{pmatrix}$$

then the Wilkinson shift σ_i is the eigenvalue of

$$(7.3) \qquad \begin{pmatrix} \alpha & \beta \\ \gamma & \delta \end{pmatrix}$$

that is closest to δ (with either eigenvalue chosen in the event of a tie). With the Wilkinson shift (or the shift based on the Rayleigh quotient), the method always converges at least quadratically and in most cases cubically. Hence, the QR method with an appropriate choice of shifts converges very rapidly, *per iteration*.

A generalization called the **implicit QR algorithm** effectively makes use of multiple shifts simultaneously. The simplest version of the implicit QR algorithm is the **double-shift** method that uses both eigenvalues of Eq. (7.3).

The reasoning for emphasizing the fact that the convergence is rapid per iteration (which, after all, is what we always mean by convergence) is that the method is $O(n^4)$ for a general $n \times n$ matrix. That's very likely to be too high a price to pay. For this reason, there is a preprocessing step that is *always* done first: We reduce the matrix A to upper Hessenberg form

$$MAM^{-1} = \begin{pmatrix} X & X & X & X & X & X & X \\ X & X & X & X & X & X & X \\ 0 & X & X & X & X & X & X \\ 0 & 0 & X & X & X & X & X \\ 0 & 0 & 0 & X & X & X & X \\ 0 & 0 & 0 & 0 & X & X & X \\ 0 & 0 & 0 & 0 & 0 & X & X \end{pmatrix}$$

by a similarity transformation (based for example on Householder reflectors; see Sec. 2.8). Only after this is completed do we begin performing the QR method. The matrix M can be made to be orthogonal, and since orthogonal matrices have nice properties that is what we do. Note that if A is symmetric then the upper Hessenberg matrix is actually tridiagonal. Sometimes Givens rotations are used instead of Householder reflectors, and sometimes we reduce the matrix to **bidiagonal** form (where only the main diagonal and the subdiagonal or only the main diagonal and the superdiagonal are nonzero) instead. In the common case that A is symmetric we can realize some additional savings.

The QR method of Francis may be applied to an upper Hessenberg matrix at a cost of $O(n^2)$ operations–very reasonable when compared to the cost of simply solving a linear system. The reduction to upper Hessenberg form is $O(n^3)$ and so the whole process is $O(n^3)$. If A is symmetric so that its upper Hessenberg form is already tridiagonal, so much the better.

We have omitted a great many important practical details of implementation. The subdiagonal elements of the T_i tend to zero in a way that can be described in much greater detail and used to specify convergence criteria. Once an eigenvalue is located to the desired precision we may split[17] the matrix into separate pieces, which results in a speed gain.

If A is symmetric, all eigenvalues are real. Otherwise, there may be complex conjugate pairs of eigenvalues. We must choose between using complex arithmetic–which is usually undesirable because of the resulting slowdown of arithmetic computations, or possibly even because of software limitations–or using some artifice to avoid the complex arithmetic. There are two (related) ways to dodge the need for complex arithmetic. One is a sort of double-iteration of the method. The other is the **real Schur form**, which gives a block upper triangular matrix with 2×2 real matrices on the main diagonal in place of the corresponding pair of complex conjugate eigenvalues. For example, the Schur decomposition of a matrix with eigenvalues $\pm i$ would lead to entries of i and $-i$ on the main diagonal of T, while the real Schur form would place a block having eigenvalues $\pm i$, say

$$\begin{pmatrix} 0 & 1 \\ -1 & 0 \end{pmatrix}$$

on the main diagonal of T. At the final stage of the method we would find the eigenvalues of the 2×2 blocks, but before then we would work with entirely real calculations. Note that a block upper triangular matrix is not, in general, upper triangular. Nonethless, generalizations of algorithms like the Thomas algorithm may be applied to them.

There's a lot more that could be said about this powerful and sophisticated method, but we will stop here. The QR method with shifts is one of the principal numerical eigenproblem algorithms, but not the only one. As always, it is best to use a professionally written version of it if one wishes to use the method.

MATLAB

There is a MATLAB command for the Jordan decomposition but it is a symbolic method and hence its utility is limited. Enter:

```
>> help jordan
>> type jordan
```

to see information on this command. It's unlikely that you will need the Jordan normal form for a practical application.

The command **hess** may be used to convert a matrix to upper Hessenberg form. Recall that the **qr** command may be used to find the QR decomposition needed in the QR method (see Sec. 2.7).

The **eigs** command attempts to find a few eigenvalues of a matrix (typically a large, sparse matrix). Enter:

```
>> help eigs
```

to see its options. By default it returns the 6 largest eigenvalues but can return more or fewer and can also return the k smallest eigenvalues. If you look carefully in its

---

[17]Called deflation, but not in the sense used in Sec. 1.8 or Sec. 3.6.

documentation (doc eigs) you'll see it can accept a function handle for a function that computes Ax and hence can be used in matrix-free mode. The svds command does the same thing for singular values. This is extremely important in practice, where large, sparse matrices are often given this way (e.g., in the numerical solution of partial differential equations). The eigs command uses the ARPACK package of Fortran[18] routines for solving large eigenvalue problems by Arnoldi methods.

The eig command may use any of a number of different LAPACK routines depending on the structure of the matrix it is given; enter:

```
>> doc eig
```

and look under the heading *Algorithm*. The most general case is handled by the QR method. Incidentally, it's possible to call other ARPACK and LAPACK routines within MATLAB, just as one can with any compiled routine.

The schur command computes the real Schur form of a matrix. (This is the form that is block triangular.) Enter:

```
>> A=gallery('hanowa',6)
>> [U,T]=schur(A)
>> U*T*U'        %Should match A.
```

The Schur decomposition is block triangular. The three blocks have the same eigenvalues as the matrix; for example, note that one block is $[-1 \ 1; -1 \ -1]$ and enter:

```
>> eig(A)
>> eig(T)
>> eig([-1 1;-1 -1])
```

The matrices A and T have the same eigenvalues, and the submatrix $[-1 \ 1; -1 \ -1]$ represents two of them by a real block. Enter:

```
>> [U,C]=rsf2csf(U,T);C
```

to convert the real Schur form to the Schur form. Note that the Schur form is triangular, not merely block triangular, but that it is complex-valued.

Remember, MATLAB was originally just a user-friendly interface to LIN-PACK, a predecessor of LAPACK, and other FORTRAN77 routines. While it has grown, many of its most important commands remain easy-to-use interfaces to packages such as LAPACK and ARPACK. These packages and a great many others (EISPACK for eigenvalues, QUADPACK for quadrature, etc.) are available at *www.netlib.org* and you may download and use them free of charge. Many are available in both Fortran and C or C++. All may be called directly from MATLAB, though preparing them for this purpose takes a bit of effort.

These packages provide a great resource. If your problems should outgrow the capabilities of MATLAB, or you need greater control of the parameters, or you need a fully compiled version, you may need to use them. If you're working in Fortran, C++, Python (which has numerical libraries called NumPy and SciPy available), etc., and you can't find what you need within them, you should absolutely be making heavy use of these libraries and, if possible, the BLAS. Browse *www.netlib.org* now to see what is available. (There is a naming convention, and learning it will make it easier to find what you're looking for there.) While MATLAB is unparalleled for rapid prototyping of ideas, after determining that an algorithm will work in MATLAB you'll often need to recode it. (Sometimes an executable MATLAB .mex

---

[18]In fact ARPACK is largely based around Fortran77 routines, from the 1978 standard. Old Fortran routines are still in surprisingly widespread use!

file will suffice; see `help mex`.) There are many resources available to you as you write your own scientific code.

Problems

1.) a.) Verify that the matrices T_3 and T_4 in Example 1 are correct.

b.) Compute T_5 and T_6. What are the bounds from the Gershgorin Circle Theorem? What are the actual absolute and relative errors in the eigenvalues? Do the diagonal elements seem to be converging to the eigenvalues as rapidly as the $(2,1)$ entry is converging to zero, or faster?

b.) Compute T_7 and T_8. Comment on the quality of the solution at this stage.

2.) a.) Apply 10 iterations of the QR method (without shifts) to approximate the eigenvalues of $A = [2 \ 1; 1 \ 2]$. Use the MATLAB `qr` command to find the decompositions.

b.) Repeat for $A = [9 \ 1; 1 \ 9]$. Comment on the speed of convergence of the method for the two matrices and explain the difference.

c.) Approximate the eigenvectors of the matrix in part a. using inverse iteration.

d.) Approximate the eigenvectors of the matrix in part b. using inverse iteration.

3.) Apply 10 iterations of the QR method (without shifts) to approximate the eigenvalues of $A = [2 \ 2; 2 \ 2]$. Use the MATLAB `qr` command . Explain the rapid convergence of the method.

4.) Show that if A is symmetric and $H = MAM^{-1}$ is upper Hessenberg, with M orthogonal, then H is in fact tridiagonal.

5.) Write a detailed algorithm (pseudo-code) for performing the QR method with Wilkinson shifts on an upper Hessenberg matrix.

6.) Repeat Example 1 using the QR method with Wilkinson shifts.

7.) Repeat Problem 2 using the QR method with Wilkinson shifts.

8.) Conduct a numerical experiment to compare the efficiency of the QR method (without shifts) on a matrix with and without an initial reduction to upper Hessenberg form (using the command `hess`). Discuss your results.

9.) Show that the QR method with shifts is a similarity transformation at each step (see Eq. (7.2)).

10.) Let A be symmetric. The Wilkinson shift corresponding to the submatrix in Eq. (7.3) is often computed using the formula $\sigma = \delta - \text{sgn}(\Delta)\beta^2 / \left(|\Delta| + \sqrt{\Delta^2 + \beta^2} \right)$, where $\Delta = (\alpha - \delta)/2$, for reasons of stability. Discuss this formula. What potential problems does it address? What formula would you suggest using to compute σ if A were not symmetric?

11.) Write a MATLAB program that performs the QR method with Wilkinson shifts using an initial reduction to upper Hessenberg form (for which you may use the command `hess`). Use any reasonable convergence criterion. Your program need only return eigenvalues.

12.) The **Rayleigh quotient shift** is $\sigma = q^T A q / q^T q$ where q is the last column of Q at each step. As mentioned previously, it gives at least quadratic and typically cubic convergence. Repeat Problem 2 with this shifting strategy; compare your results to those in Problem 7.

13.) Conduct a numerical experiment to demonstrate the cubic convergence of the QR method with Wilkinson shifts.

14.) a.) Develop a method of transforming a square matrix A to upper Hessenberg form using Householder reflectors. The resulting upper Hessenberg matrix must be orthogonally similar to A, that is, it must equal MAM^{-1} where M is an orthogonal matrix ($M^{-1} = M^T$). See Sec. 2.8. Explain clearly why your method works.

b.) Implement and test your method.

15.) a.) Develop a method of transforming a square matrix A to upper Hessenberg form using Givens rotations. The resulting upper Hessenberg matrix must be orthogonally similar to A, that is, it must equal MAM^{-1} where M is an orthogonal matrix ($M^{-1} = M^T$). See Sec. 2.8. Explain clearly why your method works.

b.) Implement and test your method.

16.) Write a pair of MATLAB programs: One that implements the functionality of rsf2csf, and one that performs the reverse function of taking a (complex) Schur form to the real Schur form. For the latter you will need a formula for finding a 2×2 matrix having specified eigenvalues $\alpha \pm i\beta$, $\beta \neq 0$.

CHAPTER 3 BIBLIOGRAPHY:

Applied Numerical Linear Algebra, William W. Hager, Prentice-Hall 1988
An undergraduate-level introductory textbook.
Numerical Linear Algebra, Lloyd N. Trefethen and David Bau, III, SIAM 1997
An informal graduate-level applied introduction to numerical linear algebra.
Iterative Methods for Solving Linear Systems, Anne Greenbaum, SIAM 1997
A modern graduate-level survey of iterative methods.
Direct Methods for Sparse Matrices, I.S. Duff, A. M. Erisman, and J. K. Reid,
Oxford University Press 1986
Emphasizes the LU decomposition for sparse systems.
Iterative Methods for Linear and Nonlinear Equations, C. T. Kelley, SIAM 1995
Emphasizes the conjugate gradient method and variants.
Matrix Computations, 3rd ed., Gene H. Golub and Charles F. Van Loan, The
Johns Hopkins University Press 1996
The classic reference on the subject.
Matrix Algorithms Volume II Basic Eigensystems, G. W. Stewart, SIAM 2001
A detailed, modern book on the eigenvalue problem.
The Algebraic Eigenvalue Problem, J. H. Wilkinson, Oxford University Press
1965
The classic reference on the eigenvalue problem.
Applied Numerical Linear Algebra, James W. Demmel, SIAM 1997
A comprehensive graduate-level text.
The Symmetric Eigenvalue Problem, Beresford N. Parlett, SIAM 1998
The classic reference for the symmetric case.
A Multigrid Tutorial, 2nd ed., William L. Briggs, Van Emden Henson, and
Steve F. McCormick, SIAM 2000
An introduction to multigrid methods.

CHAPTER 4

Polynomial Interpolation

1. Lagrange Interpolating Polynomials

So far we've discussed two of the three most commonly occurring problems in numerical analysis: Root-finding for nonlinear equations and the solution of linear systems. These problems are each important in their own right but in addition it is not uncommon for another type of problem to require the solution of a problem from one of these classes at each step; an example is Newton's method for systems (Sec. 2.4), which requires the solution of a linear system at each step.

Polynomial interpolation is of interest in and of itself, but it is also of interest as a theoretical tool to devise and analyze numerical methods. We've already seen how linear interpolation (for example, in the method of false position, Sec. 1.1) and quadratic interpolation (for example, in Brent's method, Sec. 1.6), were useful tools for deriving methods, and cubic interpolation will be used in Ch. 7. Interpolation has also been useful for arguing that those methods should work well for sufficiently smooth functions. We will start by developing this powerful tool somewhat more formally, and then look at related techniques for finding the equation of a smooth curve through a given set of points. These will be applied in later chapters.

Suppose we have a set of $n + 1$ points $x_0, x_1, ..., x_n$, called **nodes** (or **breakpoints**), ordered so that $x_0 < x_1 < \cdots < x_n$, and a set of associated y-values $y_0, y_1, ..., y_n$. Possibly the y-values are found from a function $f(x)$ defined on $[x_0, x_n]$ by $y_i = f(x_i)$, or possibly they're just measured data. The polynomial interpolation problem is to find a polynomial $p(x)$ of degree at most n that interpolates the data $(x_0, y_0), (x_1, y_1), ..., (x_n, y_n)$, by which we mean that

$$
\begin{aligned}
y_0 &= p(x_0) \\
y_1 &= p(x_1) \\
&\ \ \vdots \\
y_n &= p(x_n)
\end{aligned}
$$

(1.1)

and we say that p interpolates f (or the data $y_0, y_1, ..., y_n$) at $x_0, x_1, ..., x_n$, and that p is an interpolant. We stress that an interpolant must agree with the function f or the values $y_0, y_1, ..., y_n$ at the corresponding points $x_0, x_1, ..., x_n$: An interpolant, unlike a least squares curve, must go through every data point.

The obvious technique for finding the coefficients in a polynomial interpolant is as follows: Let $x_0 < x_1 < \cdots < x_n$ and $y_0, y_1, ..., y_n$ be given, and let

$$p(x) = a_n x^n + a_{n-1} x^{n-1} + \cdots + a_1 x + a_0$$

DOI: 10.1201/9781003042273-4

represent the interpolant. We must find $a_n, ..., a_1, a_0$. We can represent the interpolation requirements of Eq. (1.1) as

$$a_n x_0^n + a_{n-1} x_0^{n-1} + \cdots + a_1 x_0 + a_0 = y_0$$
$$a_n x_1^n + a_{n-1} x_1^{n-1} + \cdots + a_1 x_1 + a_0 = y_0$$
$$\vdots$$
$$a_n x_n^n + a_{n-1} x_n^{n-1} + \cdots + a_1 x_n + a_0 = y_0$$

that is

$$\begin{bmatrix} 1 & x_0 & x_0^2 & \cdots & x_0^n \\ 1 & x_1 & x_1^2 & \cdots & x_1^n \\ 1 & x_2 & x_2^2 & \cdots & x_2^n \\ \vdots & \vdots & \vdots & \ddots & \vdots \\ 1 & x_n & x_n^2 & \cdots & x_n^n \end{bmatrix} \begin{bmatrix} a_0 \\ a_1 \\ \vdots \\ a_n \end{bmatrix} = \begin{pmatrix} y_0 \\ y_1 \\ \vdots \\ y_n \end{pmatrix}$$

where the matrix

$$V = \begin{bmatrix} 1 & x_0 & x_0^2 & \cdots & x_0^n \\ 1 & x_1 & x_1^2 & \cdots & x_1^n \\ 1 & x_2 & x_2^2 & \cdots & x_2^n \\ \vdots & \vdots & \vdots & \ddots & \vdots \\ 1 & x_n & x_n^2 & \cdots & x_n^n \end{bmatrix}$$

contains the known x-values. A matrix with the form of V, where each row has entries $\alpha^0, \alpha^1, ..., \alpha^n$ for some α, is called a **Vandermonde matrix**[1] and we say that a system involving it is a **Vandermonde system**. If the x_i are distinct then it is nonsingular, which implies that the polynomial interpolant exists and is unique. The coefficients of the interpolant are then given by the solution of

$$V \begin{bmatrix} a_0 \\ a_1 \\ \vdots \\ a_n \end{bmatrix} = \begin{pmatrix} y_0 \\ y_1 \\ \vdots \\ y_n \end{pmatrix}$$

and we may solve this linear system by any method desired. (Not surprisingly, there are special methods for solving systems involving Vandermonde matrices, and these should be used.) Notice that we arrive at a linear system because a polynomial in the variable x of degree at least 2 is *nonlinear* in x but is always *linear* in its coefficients, and it is the coefficients that we seek.

EXAMPLE 1: Suppose we wish to interpolate a polynomial to the data points $(0,0), (\pi/2, 1), (\pi, 0), (3\pi/2, -1)$ from the sine curve. We have $(x_0, x_1, x_2, x_3) = (0, \pi/2, \pi, 3\pi/2)$, $(y_0, y_1, y_2, y_3) = (0, 1, 0, -1)$, and so we must solve

---

[1] Not all authors require that a Vandermonde matrix be square.

$$\begin{bmatrix} 1 & 0 & 0 & 0 \\ 1 & \frac{\pi}{2} & \frac{\pi^2}{4} & \frac{\pi^3}{8} \\ 1 & \pi & \pi^2 & \pi^3 \\ 1 & \frac{3\pi}{2} & \frac{9\pi^2}{4} & \frac{27\pi^3}{8} \end{bmatrix} \begin{bmatrix} a_0 \\ a_1 \\ a_2 \\ a_3 \end{bmatrix} = \begin{pmatrix} 0 \\ 1 \\ 0 \\ -1 \end{pmatrix}$$

for a_0, a_1, a_2, a_3. Using V\y in MATLAB® gives

$$\begin{bmatrix} a_0 \\ a_1 \\ a_2 \\ a_3 \end{bmatrix} \doteq \begin{bmatrix} 0 \\ 1.6977 \\ -0.8106 \\ 0.0860 \end{bmatrix}$$

that is,

$$p(x) = .0860x^3 - .8106x^2 + 1.6977x$$

is the interpolant. It is easily checked that Eq. (1.1) is satisfied by $p(x)$ (up to round-off error). □

Why might we want to use a polynomial interpolant to a function? We might need to find a simple approximation for the function. After all, to evaluate

$$f(x) = \sin(x)$$

on a computer we must approximate it by a function using only arithmetic operations, and a polynomial interpolant is one possibility. (See Ch. 8 for details on how the sine function is actually implemented within a computing environment, which does involve interpolation.) Another possibility is to use the Taylor series, but this is rarely done; while the Taylor series is of great use in deriving and analyzing numerical methods, the series itself usually provides a poor approximation for the function except in a very small interval about its center, and computing it requires derivative information that is usually inconvenient to access or even unavailable.

But polynomial interpolation is usually a poor way to approximate a function. The Vandermonde matrix is frequently ill-conditioned. Even if we solve it accurately, however, if n is large and x is large then error in evaluating the polynomial

$$p(x) = a_n x^n + a_{n-1} x^{n-1} + \cdots + a_1 x + a_0$$

can be considerable. (If n is large and x is large then any error in x will be amplified greatly in raising x to the nth power.) Cancellation could be an issue as well, particularly if the signs of the coefficients alternate. There are tricks for addressing these issues, at least in part, but there is another issue: Polynomial interpolants of high degree tend to be oscillatory, that is, they often must make large twists and turns to go through the nodes. An example is provided by the **Runge function** (Fig. 1)

(1.2) $$f(x) = \frac{1}{1 + 25x^2}$$

($-1 \le x \le 1$). In the Problems you are asked to create and plot several interpolants to it. Large oscillations develop near the endpoints. This type of behavior is typical; the polynomial interpolant must often make a large excursion in order to be able to turn smoothly and ultimately interpolate the function at the next node. For the

Runge function, the amplitude of the turns actually increases as n increases–that is, the approximation gets worse as n increases, in the sense that $\max |f(x) - p(x)|$ grows and in fact diverges. This possible failure of polynomial interpolants to converge as $n \to \infty$ is known as the **Runge phenomenon**.

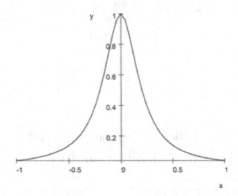

FIGURE 1. Runge function.

The situation can be improved by using nodes that are not equally spaced, when we are free to choose them, but the fact of the matter is that interpolation by high-degree polynomials is rarely a useful tool in practice. Its importance for us will be in deriving other numerical methods, as we did with the low-order interpolants for root-finding methods in Ch. 1. (They'll be essential in the numerical integration procedures to be considered in the next chapter.) Deriving and understanding other methods is certainly an important application.

Suppose we wished to interpolate a line to two distinct points x_a, x_b. Define the functions $L_a(x)$ and $L_b(x)$ by

$$
\begin{aligned}
L_a(x) &= \frac{x - x_b}{x_a - x_b} \\
L_b(x) &= \frac{x - x_a}{x_b - x_a}
\end{aligned}
$$

and note that

$$
\begin{aligned}
L_a(x_a) &= 1, L_a(x_b) = 0 \\
L_b(x_a) &= 0, L_b(x_b) = 1
\end{aligned}
$$

and that $L_a(x)$ and $L_b(x)$ are each linear functions of x. Given $y_a = f(x_a)$ and $y_b = f(x_b)$ we see that

(1.3) $$\ell(x) = y_a L_a(x) + y_b L_b(x)$$

has the property that

$$\ell(x_a) \; = \; y_a L_a(x_a) + y_b L_b(x_a)$$
$$= \; y_a \cdot 1 + y_b \cdot 0$$
$$= \; y_a$$
$$\ell(x_b) \; = \; y_a L_a(x_b) + y_b L_b(x_b)$$
$$= \; y_a \cdot 0 + y_b \cdot 1$$
$$= \; y_b$$

that is, $\ell(x)$ interpolates f at (x_a, y_a) and (x_b, y_b). Since $\ell(x)$ is the sum of two linear functions it is also linear (possibly constant) and since there is a unique line between any two points with distinct abscissae, $y = \ell(x)$ must be the equation of that line. The functions $L_a(x)$ and $L_b(x)$ appearing in Eq. (1.3) would be convenient to use if we knew that x_a and x_b were likely to stay the same while the y-values varied.

Let's generalize this technique to an arbitrary list of $n + 1$ distinct nodes $x_0, x_1, ..., x_n$ (ordered so that $x_0 < x_1 < \cdots < x_n$). Define the functions

$$L_{n,i}(x) \; = \; \frac{(x - x_0)(x - x_1) \cdots (x - x_{i-1})(x - x_{i+1}) \cdots (x - x_n)}{(x_i - x_0)(x_i - x_1) \cdots (x_i - x_{i-1})(x_i - x_{i+1}) \cdots (x_i - x_n)}$$
$$= \; \prod_{\substack{k=0 \\ k \neq i}}^{n} \frac{(x - x_k)}{(x_i - x_k)}$$

$(i = 0, ..., n)$. These are polynomials of precise degree n with the property that

$$L_{n,i}(x_j) = 0$$

if $j \neq i$ and

$$L_{n,i}(x_i) = 1$$

(see Fig. 2.) These functions are called the **Lagrange interpolating polynomials**). Note that they depend not only on n and $i \in \{0, 1, ..., n\}$ but also on the nodes $x_0, x_1, ..., x_n$.

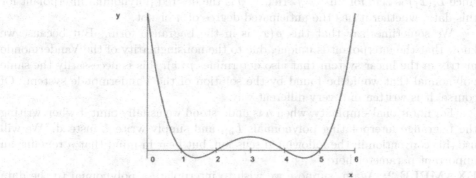

FIGURE 2. Lagrange polynomial $L_{4,3}$.

Since a sum of polynomials of degree n is a polynomial of degree at most n, any linear combination of Lagrange interpolating polynomials $L_{n,i}$ must be a polynomial of degree at most n. From this we can find the interpolating polynomial $p(x)$ of degree at most n determined by the $(n+1)$ data points

$$(x_0, y_0), (x_1, y_1), ..., (x_n, y_n)$$

(which we presume to have distinct abscissal values) very easily in terms of the Lagrange interpolating polynomials determined by the nodes $\{x_i\}_{i=1}^n$, simply by summing up each $L_{n,i}$ times its corresponding ordinate value y_i. That is to say, we form the linear combination

$$\begin{aligned} p(x) &= y_0 L_{n,0}(x) + y_1 L_{n,1}(x) + \cdots + y_n L_{n,n}(x) \\ &= \sum_{i=0}^{n} y_i L_{n,i}(x) \end{aligned}$$

of the $L_{n,i}$, and the resulting function $p(x)$ is a polynomial of degree n (or possibly less, if the points lie along a lower-order polynomial curve).

This polynomial has the expected property that when it is evaluated at one of the nodes, say $x = x_j$, it returns the appropriate y-value, namely

$$\begin{aligned} p(x_j) &= \sum_{i=0}^{n} y_i L_{n,i}(x_j) \\ &= y_0 L_{n,0}(x_j) + y_1 L_{n,1}(x_j) + \cdots + y_n L_{n,n}(x_j) \\ &= y_j \cdot 1 \\ &= y_j \end{aligned}$$

since $L_i(x_j)$ is zero for all $i \neq j$. Hence, p is the desired polynomial interpolant for this data, whether it has the anticipated degree of n or not.

We sometimes say that this $p(x)$ is in the Lagrange form. But because we know that the interpolant is unique, due to the nonsingularity of the Vandermonde matrix in the linear system that also determines $p(x)$, this is necessarily the same polynomial that would be found by the solution of the Vandermonde system. Of course, it is written in a very different way.

For notational simplicity, when n is understood we usually omit it when writing the Lagrange interpolating polynomials $L_{n,i}$ and simply write L_i instead. We will use this convention in the following discussion, but bear in mind that n remains an important parameter here.

EXAMPLE 2: Again, suppose we wish to interpolate a polynomial to the data points $(0,0), (\pi/2, 1), (\pi, 0), (3\pi/2, -1)$ from the sine curve. We have $(x_0, x_1, x_2, x_3) = (0, \pi/2, \pi, 3\pi/2)$, $(y_0, y_1, y_2, y_3) = (0, 1, 0, -1)$, and so

$$L_0(x) = \frac{(x - x_1)(x - x_2)(x - x_3)}{(x_0 - x_1)(x_0 - x_2)(x_0 - x_3)}$$

$$= \frac{(x - \pi/2)(x - \pi)(x - 3\pi/2)}{(0 - \pi/2)(0 - \pi)(0 - 3\pi/2)}$$

$$= -\frac{4}{3\pi^3}(x - \pi/2)(x - \pi)(x - 3\pi/2)$$

$$L_1(x) = \frac{(x - x_0)(x - x_2)(x - x_3)}{(x_1 - x_0)(x_1 - x_2)(x_1 - x_3)}$$

$$= \frac{(x - 0)(x - \pi)(x - 3\pi/2)}{(\pi/2 - 0)(\pi/2 - \pi)(\pi/2 - 3\pi/2)}$$

$$= \frac{4}{\pi^3}x(x - \pi)(x - 3\pi/2)$$

$$L_2(x) = \frac{(x - x_0)(x - x_1)(x - x_3)}{(x_2 - x_0)(x_2 - x_1)(x_2 - x_3)}$$

$$= \frac{(x - 0)(x - \pi/2)(x - 3\pi/2)}{(\pi - 0)(\pi - \pi/2)(\pi - 3\pi/2)}$$

$$= -\frac{4}{\pi^3}x(x - \pi/2)(x - 3\pi/2)$$

$$L_3(x) = \frac{(x - x_0)(x - x_1)(x - x_2)}{(x_3 - x_0)(x_3 - x_1)(x_3 - x_2)}$$

$$= \frac{(x - 0)(x - \pi/2)(x - \pi)}{(3\pi/2 - 0)(3\pi/2 - \pi/2)(3\pi/2 - \pi)}$$

$$= \frac{4}{3\pi^3}x(x - \pi/2)(x - \pi)$$

are the Lagrange polynomials. Hence

$$p(x) = 0 \cdot L_0(x) + 1 \cdot L_1(x) + 0 \cdot L_2(x) + (-1) \cdot L_3(x)$$

$$= L_1(x) - L_3(x)$$

$$= \frac{4}{\pi^3}x(x - \pi)(x - 3\pi/2) - \frac{4}{3\pi^3}x(x - \pi/2)(x - \pi)$$

$$\doteq .0860x^3 - .8106x^2 + 1.6977x$$

is the interpolating polynomial, in agreement with the result of Example 1. $\square$

It is trivial to find the coefficients needed for the Lagrange form of the polynomial interpolant; they are the given values $y_0, y_1, ..., y_n$, or the values of a given function f at the given nodes $x_0, x_1, ..., x_n$. It is harder to find the coefficients for the polynomial written in the natural form $a_n x^n + a_{n-1}x^{n-1} + \cdots + a_1 x + a_0$ since we must solve a linear system involving a dense and frequently ill-conditioned linear system. On the other hand, if we have the polynomial in the form $a_n x^n + a_{n-1}x^{n-1} + \cdots + a_1 x + a_0$ then it is easy to evaluate $p(x)$ at a given x, but in order to evaluate $p(x)$ in the Lagrange form $p(x) = y_0 L_0(x) + y_1 L_1(x) + \cdots + y_n L_n(x)$ we must find the Lagrange polynomials

$$L_0, L_1, ..., L_n$$

for the given nodes $x_0, x_1, ..., x_n$, and this involves some work (in addition to concerns about overflow and underflow in intermediate calculations). The natural form is easy to evaluate but hard to find, and the Lagrange form is just the opposite.

There is another commonly encountered form, the **Newton form**, that is based on writing the polynomial interpolant as

$$(1.4) \quad p(x) = b_n(x-x_{n-1})\cdots(x-x_1)(x-x_0)+\cdots+b_2(x-x_1)(x-x_0)+b_1(x-x_0)+b_0$$

and for which computation of the coefficients $b_0, b_1, ..., b_n$ is easier than for the natural form while the evaluation of the resulting interpolant is easier than and as accurate as for the Lagrange form. This makes it a nice compromise between the two in situations in which we will be making heavy use of the interpolant for its most obvious purpose–to find approximate values of $f(x)$ at points lying between the nodes. Our primary interest in polynomial interpolation will be for deriving other numerical methods with it, however, for which the Lagrange and standard forms are most convenient.

The terms *natural form*, *Lagrange form*, and *Newton form* are based on the notion that the interpolating polynomial $p(x)$ is an element of the linear (vector) space P_n of all polynomials of degree at most n^2, for which

$$\{1, x, x^2, ..., x^n\}$$

is called the natural basis, but for which

$$\{L_0, L_1, ..., L_n\}$$

and

$$\{1, (x - x_0), (x - x_1)(x - x_0), ..., (x - x_{n-1})\cdots(x - x_1)(x - x_0)\}$$

are also bases. (That is, every polynomial of degree at most n may be written as a linear combination of the elements of these sets, and that combination is unique up to order.) The different ways of writing the interpolating polynomial are based on different choices of a basis for the space P_n, leading to different coefficients and easier or harder ways of finding these coefficients. We sometimes refer to the problem of interpolating a polynomial to a function, or to data, as the Lagrange interpolation problem (regardless of the form in which we write that interpolant).

Let's look at a theorem concerning the quality of a polynomial interpolant as well as a theorem concerning polynomial approximation. It's convenient to use the notation $C^k(I)$ for the set (in fact, linear space) of all functions defined on an interval I that are k-times continuously differentiable on that interval; often we write this as $C^k(\alpha, \beta)$ or $C^k[\alpha, \beta]$ to indicate the interval $I = (\alpha, \beta)$ or $I = [\alpha, \beta]$, respectively. The case $k = 0$ is also written $C(I)$ and represents the set of all continuous function on I.

THEOREM 1 (CAUCHY REMAINDER THEOREM FOR POLYNO- MIAL INTERPOLATION): Let $a \le x_0 < x_1 < \cdots < x_n \le b$ and f in $C^{n+1}[a, b]$ be given. Then the polynomial interpolant $p(x)$ to f at $x_0, x_1, ..., x_n$ satisfies

---

[2]Including the zero polynomial, even though strictly speaking it has *no* degree.

$$f(x) - p(x) = \frac{f^{(n+1)}(\xi)}{(n+1)!}(x - x_0)(x - x_1) \cdots (x - x_n)$$

for any x, for some $\xi = \xi(x)$ in $(\min(x, a), \max(x, b))$.

PROOF: Fix $x \notin \{x_0, x_1, ..., x_n\}$ and define $\varpi(x) = (x - x_0)(x - x_1) \cdots (x - x_n)$. Let

$$\Phi(x) = \frac{f(x) - p(x)}{\varpi(x)}$$

$$\Omega(y; x) = f(y) - p(y) - \varpi(y)\Phi(x)$$

and note that $\Omega(y; x) = 0$ if $y \in \{x_0, x_1, ..., x_n\}$ or if $y = x$. Hence by the generalized Rolle's theorem there exists a ξ in $(\min(x, a), \max(x, b))$ such that $\Omega^{(n+1)}(\xi; x) = 0$. But

$$\frac{d^{n+1}}{dy^{n+1}}\Omega(y; x) = \frac{d^{n+1}}{dy^{n+1}}\left(f(y) - p(y) - \varpi(y)\Phi(x)\right)$$

$$= f^{(n+1)}(y) - 0 - \Phi(x)\frac{d^{n+1}}{dy^{n+1}}\varpi(y)$$

$$= f^{(n+1)}(y) - (n+1)!\Phi(x)$$

and so $\Omega^{(n+1)}(\xi; x) = 0$ means that

$$0 = f^{(n+1)}(\xi) - (n+1)!\Phi(x)$$

$$\Phi(x) = \frac{f^{(n+1)}(\xi)}{(n+1)!}$$

$$\frac{f(x) - p(x)}{\varpi(x)} = \frac{f^{(n+1)}(\xi)}{(n+1)!}$$

$$f(x) - p(x) = \varpi(x)\frac{f^{(n+1)}(\xi)}{(n+1)!}$$

for any $x \notin \{x_0, x_1, ..., x_n\}$. But if $x \in \{x_0, x_1, ..., x_n\}$ then $f(x) - p(x) = 0$ and $\varpi(x) = 0$, so the result holds for any x. $\square$

Note that this theorem may be used to estimate the error in the interpolating polynomial within the range of its data $x_0 < x_1 < \cdots < x_n$ and also outside this range. The error bound grows rapidly outside $[x_0, x_n]$, and it is therefore highly inadvisable to use an interpolant to approximate values of the function outside of the range of the data (that is, for $x < x_0$ or $x > x_n$). (Compare this error term with that from the Taylor series with remainder.) Interpolants, polynomial or otherwise, should only be used to approximate values within the range of the data on which they were based if possible; extrapolated values are extremely suspect.

THEOREM 2 (WEIERSTRASS APPROXIMATION THEOREM): If f is in $C[a, b]$ then for every $\epsilon > 0$ there exists a polynomial $p(x)$ such that

$$|f(x) - p(x)| \leq \epsilon$$

for every $x \in [a, b]$.

The Weierstrass Approximation Theorem is one of the most important theorems in approximation theory. It asserts that every continuous function can be approximated arbitrarily well by a polynomial of sufficiently high degree. Of course, it need not be the case that $p(x)$ is equal to $f(x)$ for some x: This is a theorem about approximation, not interpolation. Still, it justifies the belief that approximation by a polynomial is a viable approach.

A final point on interpolation. It is common in applications to need only the values of the interpolant at some point or points. In such a case we do not need the coefficients at all, in principle. There are significantly more efficient techniques available for computing the value of $p(x)$ at a point x without first finding an explicit expression for $p(x)$. As with any computation, it pays to ask at the start what we actually *need* and to only compute that if possible. The traditional algorithm for this purpose is Neville's algorithm, which is closely related to the Newton form. Suppose we want to find the value of $p(z)$ at some point z, where $p(x)$ is the (unknown) polynomial interpolant of degree at most n to the $n+1$ distinct points (x_i, y_i), $i = 1, 2, ..., n$.

For $j > i$, let $p_{i,j}(x)$ represent the polynomial interpolant of degree at most $j - i$ through the $(j - i + 1)$ points (x_k, y_k) for $k = i, i+1, \ldots, j$; for $j = i$, we'll use the convention that $p_{i,i}(z) = y_i$ for $i = 0, 1, ..., n$. Then the $p_{i,j}$ satisfy

$$(1.5) \qquad p_{i,j}(z) = \frac{(z - x_j) p_{i,j-1}(z) - (z - x_i) p_{i+1,j}(z)}{x_i - x_j}$$

which is a recurrence relation involving the *values* of various interpolating polynomials at the common point z. The zeroth stage is simply the specification of the initial data as $p_{i,i}(z)$ (that is, all cases where $j - i = 0$). At the first stage we compute

$$p_{0,1}(z) = \frac{(z - x_1) p_{0,0}(z) - (z - x_0) p_{1,1}(z)}{x_0 - x_1}$$

$$p_{1,2}(z) = \frac{(z - x_2) p_{1,1}(z) - (z - x_1) p_{2,2}(z)}{x_1 - x_2}$$

and so on, up to $p_{n-1,n}(z)$. This gives all defined cases where $j - i = 1$. The second stage gives us

$$p_{0,2}(z) = \frac{(z - x_2) p_{0,1}(z) - (z - x_0) p_{1,2}(z)}{x_0 - x_2}$$

and all the remaining $p_{i,j}(z)$ where $j - i = 2$. Eventually, at stage n we reach the single value

$$p_{0,n}(z)$$

found from $p_{0,n-1}(z)$ and $p_{1,n}(z)$, which is the desired value of the polynomial interpolant $p(x)$ at $x = z$. (By uniqueness, $p_{0,n}(x) = p(x)$ for all x.) We have found a value of $p(x)$ without having found an expression for it. It's not difficult to obtain $p'(z)$ too by a formula similar to that in Eq. (1.4).

The algorithm is much simpler to use than it may appear and traditionally leant itself to by-hands computation by human computers, typically aided by mechanical or, later, electronic adding machines. A pre-printed tableau (i.e., table) indicated the 'program' to be followed to complete the computation, with blank slots for the various data that would be provided and values generated and arrows or other indicia serving to suggest the flow of the program to be followed, would be used by the person tasked to perform the computations. The basic idea for Neville's algorithm is to use a tableau like:

$$
\begin{array}{c|cccc}
x_i & y_i & 1 & 2 & 3 \\
\cdots & \cdots & \cdots & \cdots & \cdots \\
\underline{} & \underline{} & & & \\
\underline{} & \underline{} & \underline{} & \underline{} & \underline{} \\
\underline{} & \underline{} & \underline{} & & \\
\underline{} & \underline{} & \underline{} & &
\end{array}
$$

and the computer would fill in the slots column-by-column, from left-to-right, until the desired value was reached on the right. In fact, the use of a program of instructions performed on a tableau like this as a practical means of using an algorithm is part of the reason software today is referred to by the term 'program'.

The $p_{i,j}$ are sometimes called **partial interpolating polynomials** in this context, as they interpolate only part of the data. Sometimes we even stop the algorithm early if the values of the partial interpolating polynomials are starting to closely agree with one another, as this can avoid the problems inherent in the use of a needlessly high-degree interpolant. Neville's algorithm is a method of building up the degree n interpolant's value–or, with slight modification, the polynomial itself–from degree 0 polynomials $p_{i,i}$, then degree 1 polynomials $p_{i,i+1}$, and so on. It is $O\left(n^2\right)$ and more efficient than finding $p(x)$ first and then evaluating it when a single value is all that is needed.

Neville's algorithm is important, but the bigger take-away for us is: Don't assume that you need to perform task A to get result B in scientific computing simply because you were taught to do so in algebra, calculus, etc. Just as when someone says "I need to know how to find M^{-1}" when in fact they're simply trying to solve $Mx = b$ and don't realize that they don't need, and should not want, the inverse, it pays to ask yourself (or those seeking your advice) what is *really* needed, and check whether there isn't a better way to get at it.

MATLAB

There are a number of MATLAB commands for performing interpolation. Let's re-do Example 1 in MATLAB. Enter:

```
» x=[0 pi/2 pi 3*pi/2]';y=[0 1 0 -1]';
» V=vander(x)
» V=fliplr(V)
```

The MATLAB command `vander` sets up a Vandermonde matrix, but the MATLAB convention differs from ours. (That's not uncommon. The row of ones in a Vandermonde-style matrix might be in either the first or last column, or even the first or last row, depending on the source consulted.) Enter:

```
» p=V\y
```

to find the coefficients of the polynomial. Note that p(1) is the constant term and p(4) is the coefficient of x^3; again, the MATLAB convention is the opposite, so let's flip p around. Enter:

» p=flipud(p)'

to put p in the appropriate form for use in other MATLAB commands. (Had we set up the problem so as to conform with MATLAB's notion of a Vandermonde matrix this would not be necessary.) The MATLAB command polyval may be used to evaluate a polynomial given its coefficients. Enter:

» z=rand;p(1)*z^3+p(2)*z^2+p(3)*z+p(4)

» polyval(p,z)

The values should be the same. In fact we may use the polyval command to evaluate the polynomial at a vector. Enter:

» z=linspace(0,2*pi,100);

» zy=polyval(p,z)   %Polynomial evaluated at z.

» plot(z,zy,'g',z,sin(z),'r'),grid

The actual curve is in red and the approximant is in green. Enter:

» z=linspace(0,4*pi,200);   %Only this line changes.

» zy=polyval(p,z);.

» plot(z,zy,'g',z,sin(z),'r'),grid

The interpolant can only be expected to give reasonable results where it had data (in our case, 0 to $3\pi/2$); extrapolating outside of this range by evaluating on $[0, 4\pi]$ will give poor results, as predicted by Theorem 1.

There is a built-in command for performing the interpolation and it should generally be used if a polynomial interpolant is desired. Enter:

» x,y   %Data.

» pnew=polyfit(x,y,3)

» p   %Should be the same.

The form of the polyfit command is polyfit(x,y,n) where x and y are the data to be interpolated and n is the degree of the polynomial. The assumption throughout this section has been that n is equal to length(x)-1; if this is not so then the polyfit command will give a least squares fit. (See Sec. 2.7.) For another interpolation command, see help interp1.

Poor interpolation behavior is not uncommon for Lagrange interpolation using equally spaced nodes. Enter:

» x=linspace(0,2*pi);n=9;

» plot(x,abs(sin(x)),'b')   %Note the plot's inaccuracy near pi.

» nodes1=linspace(0,2*pi,n+1);   %Choose 10 equally spaced nodes.

» ynodes1=abs(sin(nodes1));   %Corresponding y-values.

» p=polyfit(nodes1,ynodes1,n);

» y1=polyval(p,x);   %Evaluate p(x) on a finer grid.

» hold on;plot(x,y1,'r')

This isn't very good, though we are only using 10 points. Let's see if we can improve it. Enter:

» nodes1   %Current nodes.

» plot(nodes1,zeros([1 n+1]),'go') %Locate nodes as green 'o's.

» cheby=cos((2*(n-(0:n))+1)*pi/(2*n+2)) %Chebyshev nodes on [-1,1].

» nodes2=cheby*pi+pi;   %Rescale and shift to [0,2*pi].

» plot(nodes2,zeros([1 n+1]),'co') %Locate nodes as cyan 'x's.

Notice that the second set of nodes on the x-axis is not equally spaced; in particular, notice how they are more closely packed near the endpoints. Enter:

```
» ynodes2=abs(sin(nodes2));     %Corresponding y-values.
» p2=polyfit(nodes2,ynodes2,n);
» y2=polyval(p2,x);      %Evaluate p2(x) on a finer grid.
» plot(x,y2,'k')     %Plot in black.
```

The approximation is worse in the very middle but is considerably better everywhere else. Try again with a larger value of n. Equally spaced nodes are much easier to use but unequally spaced nodes typically give considerably better results. The **Chebyshev nodes** $x_i = \cos((2(n-i)+1)\pi/(2n+2))$ $(i = 0, 1, ..., n)$ are often a good choice on $[-1, 1]$ (rescaled to the appropriate interval) but there is no guarantee that they will give good results.

We should address the issue of accurate evaluation of polynomials. For a polynomial of degree n, we must compute x^n when using the standard form, and if n is large and $|x| > 1$ then errors could be magnified considerably. For example, enter:

```
» clear all
» N=20;y=10^N,y1=(10*(1+2*eps))^N
» a=abs(y1-y)     %Absolute error.
» r=a/y          %Relative error.
```

The absolute error is quite large though the relative error is small. Larger perturbations and larger exponents would give worse results. Cancellation is also an issue; enter:

```
» x=linspace(.995,1.005,200);
» plot(x,(x-1).^8)
```

to plot $f(x) = (x-1)^8$ for $x \in [.995, 1.005]$. Then enter:

```
» g=x.^8-8*x.^7+28*x.^6-56*x.^5+70*x.^4-56*x.^3+28*x.^2-8*x+1;
» plot(x,g),grid
```

to plot $g(x) = x^8 - 8x^7 + 28x^6 - 56x^5 + 70x^4 - 56x^3 + 28x^2 - 8x + 1$ over the same range. It is a fact that $f(x) = g(x)$ for all x; $g(x)$ is just the expanded version of $f(x)$. However, cancellation of significant figures caused by the subtractions lead to a very noisy function near $x = 1$. (Imagine trying to find the sole root $x^* = 1$ of this function using a root-finding method!) There is a way to address this issue. One aspect of it is buried in the Lagrange and Newton forms; if we know that we will be evaluating a polynomial $p(x)$ near $x = x_0$ then it usually makes sense to write $p(x) = b_n(x - x_0)^n + \cdots + b_1(x - x_0) + b_0$ rather than using powers of x since then we will be raising the smaller quantity $(x - x_0)$ to the nth power rather than the larger quantity x. This is the same form as the Taylor series of p centered at x_0, strictly speaking. The other aspect of evaluating a polynomial accurately is **nested evaluation** (also called **Horner's rule**, which is inherent in **Horner's method**). We illustrate it by example. If $g(z) = 3z^3 - 6z^2 + 4z - 5$ (possibly $z = x - x_0$) then rather then evaluate $g(z)$ directly, as in:

```
» z=pi;
» 3*z^3-6*z^2+4*z-5
```

we rewrite $g(z)$ in the mathematically equivalent form $g(z) = ((3z - 6)z + 4)z - 5$ and evaluate it as follows:

```
» ((3*z-6)*z+4)*z-5
```

The nested form uses fewer flops[3]. This is true in general, and it typically gives a more accurate result as well. Horner's rule should generally be used to evaluate polynomials whenever it is feasible to do so. If you have the MATLAB Symbolic Toolbox, type `help horner` for a useful command.

We mention that the `conv` command (convolution) may be used to find polynomial products; that is, if `p1` and `p2` are vectors representing coefficients of the polynomials $p_1(x)$ and $p_2(x)$, then `conv(p1,p2)` is a vector that contains the coefficients of the polynomial $p_1(x)p_2(x)$. More generally `conv(x,y)` convolves the vectors `x` and `y`. The `deconv` (deconvolution) command may be used to divide polynomials, while the `polyder` command differentiates polynomials (again, in the form of a vector of coefficients).

Problems

1.) Use the Vandermonde matrix approach and the Lagrange interpolating polynomial approach to find the equation $y = \alpha x^2 + \beta x + \gamma$ of the quadratic through the points $\{(0,1),(1,2),(2,1)\}$.

2.) a.) Approximate $f(x) = e^{x/2}$ over $[1,9]$ by a fourth degree polynomial in two ways: Using the Lagrange form of the interpolating polynomial with the nodes $1,3,5,7,9$, and using a Taylor polynomial centered at $x = 5$.
b.) Plot the error estimate for these approximants (from Theorem 1 and from the remainder form of Taylor series) for $x \in [0,12]$.
c.) Plot the actual error for these approximants for $x \in [0,12]$.
d.) Find the actual error for both approximants at $x = 1.5 : .5 : 8.5$. Comment.
e.) Find the actual error for both approximants at $x = 4.7 : .1 : 5.3$. Comment.

3.) a.) Prove that if $p(x)$ and $q(x)$ are polynomials of degree at most n that interpolate a function f at the distinct points $x_0, x_1, ..., x_n$ then $p = q$. Use the function $h(x) = p(x) - q(x)$ (not the nonsingularity of the corresponding Vandermonde matrix).
b.) Show that a polynomial of degree m is its own polynomial interpolant of degree at most n ($n \geq m$) no matter how the distinct points $x_0, x_1, ..., x_n$ are chosen.
c.) Argue that a Vandermonde matrix is nonsingular if the points $x_0, x_1, ..., x_n$ on which it is based are distinct by showing that it maps only the zero vector to the zero vector.

4.) A sine table is desired. The table will be for equally spaced angles in $[0, \pi/2]$ and will give sine values correct to 10 decimal places. It is to be such that linear interpolation ($n = 1$) between neighboring entries will always give a result that is correct to at least 4 decimal places. Use the error formula from Theorem 1 to suggest a suitable spacing between the nodes.

5.) a.) Find the interpolating polynomial for $f(x) = x^4 - 2x^3 + 2x - 1$ based on the nodes $\{0,1,2,3\}$. Use the Lagrange method but write your final answer in the natural form.
b.) Write your answer in the Newton form (Eq. (1.4)).

6.) a.) Consider the Runge function (Eq. (1.2)). Find the polynomial of degree $n = 10$ that interpolates it at equally spaced points in $[-1,1]$. Plot the function together with its interpolant. Approximate the maximum error by evaluating the

---

[3]Comparisons can be difficult as a power like z^3 might be computed as $z \cdot z \cdot z$ (two flops) or by something like $\exp(3 \cdot \ln z)$, which involves two functions that each require many flops, depending on the software environment.

Runge function and the interpolant on a fine grid and finding the largest absolute difference.

b.) Repeat for $n = 15$.

c.) Repeat for $n = 20$.

d.) Repeat for $n = 50$. Does raising the degree of the interpolating polynomial give a better approximation?

e.) Find, by trial and error, a polynomial of degree $n = 10$ that interpolates the Runge function reasonably well over $[-1, 1]$ by using unequally spaced nodes. (Hint: Place more nodes near the endpoints ± 1 of the interval.)

7.) a.) Using `polyfit` and `polyval`, find a good approximation for $\sin(x)$ on $[0, 2\pi]$ by choosing an interpolating polynomial of degree at most 8. The nodes need not be equally spaced. Plot the polynomial and $\sin(x)$ on a single graph.

b.) Use Theorem 1 to estimate the worst-case error in your approximation for $x \in [0, 2\pi]$.

c.) Plot $\sin(x)$ and your approximant for $x \in [-2\pi, 4\pi]$.

8.) A sine table is desired. The table will be for equally spaced angles in $[0, \pi/2]$ and will list the sine values to 10 decimal places. Quadratic interpolation ($n = 2$) between neighboring entries must always give a result that is correct to at least 4 decimal places. Use the error formula from Theorem 1 to suggest a suitable spacing between the nodes. Compare your answer to Problem 4.

9.) a.) Use Neville's algorithm to compute $p(3)$ if the x-values are $0, 1, 2, 4, 5$ and the corresponding y-values are $3, 1, 1, 3, 2$, respectively, where $p(x)$ is the degree 4 interpolant to the data.

b.) Repeat for $p(-1)$, which lies outside the range of the abscissal data.

c.) Write an algorithm (pseudo-code) for efficiently performing Neville's algorithm.

d.) Construct a tableau, with blank spaces and with arrows indicating the left-to-right flow, for Neville's algorithm when there will be 5 data points and the value of the degree 4 interpolant at an as-yet-unspecified point is sought.

10.) Write a MATLAB program that accepts an n-vector and produces labeled plots of $L_{n,0}, ..., L_{n,n}$. Each plot should have its own figure window; use `pause` to display the separate graphs briefly as they are produced. The program should return some appropriate quantity.

11.) Derive Neville's algorithm.

12.) a.) Write a MATLAB program that finds the inverse interpolant (see Sec. 1.6) for given data consisting of two vectors of length $n + 1$.

b.) Write a MATLAB program that performs root-finding by inverse interpolation based on $n + 1$ nodes, that is, given an initial bracket, the method chooses $n - 1$ additional points within the bracket, performs inverse interpolation with a polynomial of degree at most n, and uses that as a new point, discarding an older one, and then performs inverse interpolation again on the new set of $n + 1$ points until some convergence criterion is met.

13.) a.) Show that Eq. (1.5) is correct.

b.) Derive a similar formula for the derivative of $p(x)$ and then algorithm for finding the value of $p'(z)$ by a similar method.

14.) a.) For $n = 4$, use Eq. (1.4) and the conditions of Eq. (1.1) to write a lower triangular matrix system for the coefficients $b_0, ..., b_4$ of the Newton form of the interpolating polynomial.

b.) Write out the solution of the system found in part a. using forward substitution.

c.) Note that b_0 may be found directly in terms of $y_0 = f(x_0)$. Write b_1 as a difference of two y-values (or f values) divided by a difference of two x-values. Then write b_2 as a difference of two terms of the form $(\Delta y / \Delta x)$ divided by a difference of x-values. Continue for b_3 and b_4. This is known as the **divided differences** technique for finding the Newton form.

d.) Write a MATLAB program that computes the interpolatory polynomial of degree at most n for given data consisting of two vectors of length $n + 1$ in the Newton form by the divided differences technique. (The `diff` command may be useful.) Demonstrate your program.

e.) How do divided differences appear in Neville's algorithm?

15.) a.) In some cases we are given data of the form (x_i, y_i, y_i') $(i = 0, ..., n+1)$ where $y_i = f(x_i)$ and $y_i' = f'(x_i)$ and we seek a polynomial interpolant of degree at most $2n + 1$ to the data (x_i, y_i) which also has the prescribed derivatives; that is, $p(x_i) = y_i$ (Eq. (1.1)) and also $p'(x_i) = y_i'$. This is called the **Hermite interpolation** problem. If the x_i are distinct, must such a polynomial exist for arbitrary choices of y_i and y_i'? If so, is it unique?

b.) Solve the Hermite interpolation problem for the sine function using the nodes $\{x_0, x_1, x_2, x_3\} = \{0, \pi/2, \pi, 3\pi/2\}$. Compare your result to Example 1.

16.) a.) What is the change-of-basis matrix for going from the natural basis of the space P_n to the Lagrange form basis?

b.) What is the change-of-basis matrix for going from the natural basis of the space P_n to the Newton form basis?

2. Piecewise Linear Interpolation

There are many reasons that we might need to approximate a function. On most standard machines, anything that is to be computed must be broken down, at some level, to addition, subtraction, multiplication, or division (plus book-keeping operations)[4]. Hence when a calculator returns a value for $\sin(3)$ it is only because the function $y = \sin(x)$ has been approximated by a **rational function** (that is, a ratio of polynomials). At a higher level, we often need to approximate special functions that occur in particular problems. In these cases it is far from clear that we need an interpolant per se, which is required to pass through certain points; we really just need to control the error in our approximation. We'll discuss approximation in more generality later.

Another common example of a situation that calls for approximation of a function is the numerical solution of an ODE (or the numerical integral of a function), which yields a discrete set of points through which we would like to draw a smooth curve. We might also have measured data through which we would like to pass a curve (say, from a digitized graph or from tracking a moving object). In these cases it is usually desirable to pass the curve through the known data points and so an interpolant, rather than just an approximant, is needed.

Polynomial interpolation has many uses. As we have indicated in the previous section, however, polynomial interpolants are usually not the right choice for approximating a function. The oscillatory nature of high-order polynomial interpolants, plus the errors in evaluating x^n for large n, prevent them from being useful

---

[4]Of course, we are over-simplifying matters.

as a general-purpose tool. An oscillatory interpolant will be a perfect interpolant $(p(x_i) = y_i)$ but will be a very poor approximation of the curve between those points, and we always want to get a good approximation of the underlying function. Taylor polynomials are worse; they are guaranteed to interpolate only at a single point, their center, and while they are an excellent approximation there the quality of the approximation drops off rapidly. In addition, they require derivative information about the function that is rarely available. Polynomial interpolants and Taylor polynomials are of great use in numerical analysis but they are principally used to derive and analyze methods, not as methods in and of themselves. We'll see examples of this in the next two chapters (numerical integration and the numerical solution of ODEs).

One way to improve the fidelity of the approximant is to require that the polynomial not only interpolate the function but that the derivative of the polynomial interpolate the derivative of the function as well; that is, we could seek an interpolant $p(x)$ such that

$$
\begin{aligned}
p(x_i) &= f(x_i) \\
p'(x_i) &= f'(x_i)
\end{aligned}
$$

$(i = 0, 1, ..., n)$. This is called the **Hermite interpolation** problem and, similar to Lagrange interpolation, there are special polynomials that may be used to solve it. Derivative information is often not available, however, and the degree of p in Hermite interpolation must be about twice the degree of the corresponding Lagrange interpolant since we must satisfy twice as many conditions. This is inconvenient, even though we expect better accuracy, both because we have twice as much information (at each point not only $f(x_i)$ but also $f'(x_i)$ is given) and because making the derivatives of $p(x)$ match those of $f(x)$ forces the shape of the interpolant to more nearly match that of the function.

We should mention an important point that was alluded to in the previous section: The difficulties with polynomial interpolation are made that much worse by using equally spaced points. Choosing the nodes intelligently can give a much less oscillatory fit. Of course, we would like to automate the process as much as possible so that a program may be written that does not prompt for help from the user in the middle of its run (especially since sooner or later some other program will certainly be calling it to perform some computation), and so we would need a technique for choosing the points. These exist; but oscillations can still occur, evaluating x^n for large n is still an issue, an error in any coefficient affects the values of the polynomial everywhere, and quite simply there are better and more efficient methods.

How shall we find such a method? We must give up either "polynomial" or "interpolation" and we know that interpolation is sometimes desired. Evidently we will have to look to non-polynomial functions. Indeed, there are trigonometric and rational interpolants, for example. Still, polynomials were attractive in the first place because they are simple, familiar, easily computed, easily differentiated and integrated, etc., and it would be nice to be able to use them. Fortunately, there is a compromise—and we have been using it since Sec. 1.1. Rather than use functions that are polynomials in the strict sense, we will use piecewise-defined functions that are polynomials on subintervals of the interval of interest. This is called **piecewise**

polynomial interpolation, and the subintervals are often referred to as **elements** or **panels**.

The simplest approach is **piecewise linear interpolation**; in which each polynomial piece is linear. This is what the MATLAB `plot` function does: Given a set of points, it "connects the dots" using straight line segments. Between any two points the curve is a (linear) polynomial, while over an entire interval it is not a polynomial. It takes on the correct values at the nodes (that is, it interpolates). We've already seen that if we choose sufficiently many points then this gives a nice, smooth-looking curve that approximates the function; every time we've plotted $\sin(x)$, for example, we have been using this technique and have been viewing a piecewise linear interpolant. This simple method bears closer attention.

It's convenient to define an analogue of the Lagrange polynomials for this case. There are two ways we might proceed: Given the $n+1$ nodes $x_0, x_1, ..., x_n$ (ordered so that $x_0 < x_1 < \cdots < x_n$), we might define a separate linear function on each of the n subintervals $[x_i, x_{i+1}]$ $(i = 0, ..., n-1)$ by

$$m_i(x) = s_i x + b_i$$

on the subinterval $[x_i, x_{i+1}]$, and $m_i(x) = 0$ outside of this subinterval. We then choose the constants s_i and b_i on each subinterval according to the interpolation condition

$$(2.1) \qquad \begin{aligned} s_i x_i + b_i &= y_i \\ s_i x_{i+1} + b_i &= y_{i+1} \end{aligned}$$

$(i = 0, ..., n-1)$. This gives a system of $2n$ equations (2 per element, n elements) in $2n$ unknowns (2 per element, n elements), and this system is trivially broken down into n blocks, each 2×2 (Eq. (2.1)).

A more natural analogue of the Lagrange polynomials, however, is obtained by choosing a basis $\Lambda_0, ..., \Lambda_{n-1}$ of functions with the property that $\Lambda_i(x)$ is a piecewise linear function on $[x_0, x_n]$ satisfying

$$\begin{aligned} \Lambda_i(x_j) &= \begin{cases} 1 & \text{if} \quad j = i \\ 0 & \text{if} \quad j \neq i \end{cases} \\ &= \delta_{ij} \end{aligned}$$

$(i = 0, ..., n-1)$ at the nodes. Here δ_{ij} is the **Kronecker delta function** which is equal to 1 if $i = j$ and is equal to 0 otherwise. This means that the $\Lambda_i(x)$ must be the **hat functions** (or **chapeau functions**), which are zero everywhere except on the subintervals to the immediate left and right of x_i, where $\Lambda_i(x)$ goes from 0 at x_{i-1} to 1 at x_i and back down to 0 at x_{i+1} (see Fig. 1, based on the nodes $\{1, 2, 3, 4, 5\}$). For the leftmost subinterval $[x_0, x_1]$ and the rightmost subinterval $[x_{n-1}, x_n]$ we omit the line segments that would fall outside of the interval $[x_0, x_n]$.

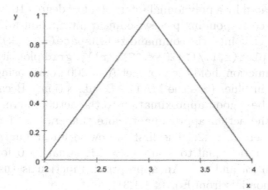

FIGURE 1. Hat function ($x_i = 3$).

How shall we use the functions $\{\Lambda_0, ..., \Lambda_{n-1}\}$ (called the **hat basis** or **chapeau basis**)? In exactly the same way as we did with the Lagrange polynomials for (true) polynomial interpolation. The piecewise linear polynomial interpolant with the nodes $x_0, x_1, ..., x_n$ and corresponding function values $y_0, y_1, ..., y_n$ is

(2.2) $$p(x) = y_0\Lambda_0 + \cdots + y_n\Lambda_n.$$

(We call $\Lambda_0, ..., \Lambda_{n-1}$ the **piecewise linear Lagrange interpolating polynomials** for these nodes.) Clearly, the interpolant is unique, so this is exactly the same solution as would be found from Eq. (2.1), but it is easier to compute the coefficients in this case; on the other hand, evaluating $p(x)$ by Eq. (2.2) within a panel requires that we evaluate 2 piecewise Lagrange interpolating polynomials, the one from the node on the left and the one from the node on the right. (See Fig. 2.) Hence 2 linear functions are summed to find the value on the linear segment for that subinterval. This is twice the work required to evaluate the equivalent interpolant determined by the solution of Eq. (2.1).

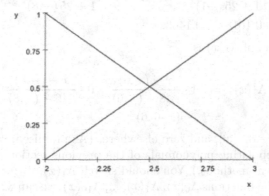

FIGURE 2. Hat functions over $[2, 3]$.

EXAMPLE 1: Recall from the previous section that the Runge function

$$f(x) = \frac{1}{1 + 25x^2}$$

is poorly approximated by a polynomial interpolant of degree 10 (which uses $N = 11$ nodes). To see the corresponding piecewise linear interpolant all we need do is plot this function using 11 points; the command `x=linspace(-1,1,N);` gives the desired x-values, and then `plot(x,1./(1+25*x.^2),'r')`, `grid` plots it. To compare it to the actual Runge function, hold the plot, get $N = 200$ points using `linspace` again, and overlay a plot in blue. (See the MATLAB subsection.) Because of Theorem 1 below, this should be a good approximation to the actual curve.

To construct the actual approximant, note that $x_0 = -1$ and $x_i = x_0 + ih$ ($i = 0, 1, ..., 10$) where $h = .2$. The first piecewise linear Lagrange interpolating polynomial Λ_0 must be equal to 1 at $x_0 = -1$, equal to 0 for $x \geq x_1 = -.8$, and linear between x_0 and x_1. An appropriate function is (using the Lagrange interpolating polynomials from Eq. (4.1.1.3))

$$\begin{aligned} \Lambda_0(x) &= 1 \cdot \frac{x - (-.8)}{-1 - (-.8)} + 0 \cdot \frac{x - (-1)}{-.8 - (-1)} \\ &= -5(x + .8) \end{aligned}$$

for $x \in [x_0, x_1]$ (and zero elsewhere). (You should sketch $\Lambda_0(x)$.) Similarly, again using the Lagrange interpolating polynomials from Eq. (4.1.1.3), we can find the left-hand part of $\Lambda_1(x)$ as

$$\begin{aligned} \Lambda_1(x) &= 0 \cdot \frac{x - (-.8)}{-1 - (-.8)} + 1 \cdot \frac{x - (-1)}{-.8 - (-1)} \\ &= 5(x + 1) \end{aligned}$$

for $x \in [x_0, x_1] = [-1, -.8]$. This means that the actual interpolant over $[x_0, x_1]$ is

$$\begin{aligned} m_0 &= f(x_0)\Lambda_0(x) + f(x_1)\Lambda_1(x) \\ &= f(-1) \cdot (-5(x + .8)) + f(-.8) \cdot (5(x + 1)) \\ &= \frac{1}{1 + 25(-1)^2}(-5(x + .8)) + \frac{1}{1 + 25(-.8)^2}(5(x + 1)) \\ &\doteq 0.1018x + 0.1403. \end{aligned}$$

The right-hand part of $\Lambda_1(x)$ is

$$\begin{aligned} \Lambda_1(x) &= 1 \cdot \frac{x - (-.6)}{-.8 - (-.6)} + 0 \cdot \frac{x - (-.8)}{-.6 - (-.8)} \\ &= -5(x + .6) \end{aligned}$$

for $x \in [x_1, x_2] = [-.8, -.6]$, and zero elsewhere. (As promised, we are finding that the Lagrange interpolating polynomials of the previous section are useful tools in other, more practical methods.) You should sketch $\Lambda_1(x)$.

Hence, using the functions $\Lambda_0(x), \Lambda_1(x), ..., \Lambda_{10}(x)$, we can write the interpolant in the form

$$p(x) = f(x_0)\Lambda_0(x) + f(x_1)\Lambda_1(x) + \cdots + f(x_{10})\Lambda_{10}(x)$$

which, being piecewise defined, is equal to 10 different linear segments on 10 different subintervals. We emphasize that $p(x)$ is a single, continuous function defined over the entire interval. $\square$

Notice that equally spaced nodes are not a problem in this approach; in addition, an error in one piece of the interpolant has only a local effect (it does not degrade the quality of the interpolant over the entire data range, as happens for polynomial interpolation). It is a fact that any continuous function on a closed and bounded interval may be approximated arbitrarily well by a piecewise linear interpolant on a sufficiently fine grid. In fact, we have the following theorem:

THEOREM 1: Let $x_0 < x_1 < \cdots < x_n$ and f in $C^2[x_0, x_n]$ be given. Then the piecewise linear interpolant $p(x)$ to f at $x_0, x_1, ..., x_n$ satisfies

$$(2.3) \qquad |f(x) - p(x)| \leq \frac{Mh^2}{8}$$

for any $x \in [a, b]$, where M is an upper bound on $|f''(x)|$ for $x \in [a, b]$ and $h = \max(x_{i+1} - x_i)$ $(i = 0, ..., n - 1)$.

This gives an estimate of the worst-case error in the approximation; it may be employed over any smaller interval within $[x_0, x_n]$ by taking M to be a bound for $|f''(x)|$ on just that interval. Note that the larger the second derivative $f''(x)$ is, the worse the linear approximation might be–as would be expected, as this is an indication of curvature. Looked at another way, the smaller M is, the closer f'' is to being zero, and hence the closer f is to being locally linear (and thus the closer it is to being well-approximated by a line segment). The theorem follows from Theorem 1 of the previous section.

Often we must use the nodes we are given. This is especially true when we do not have an explicit formula for $f(x)$. The nodes may be the x-values at which data was measured (as with experimental data) and it is often infeasible to specify the desired x-values or to get additional data, which often presents the problem of prohibitive expense and delay. In other cases, however, we *can* select the nodes–for example, if we are writing a subprogram to approximate a special function $f(x)$ of interest. It seems clear from Eq. (2.2) that we should pick more points where the curvature of $f(x)$ is greater in order to get a smaller h in that region and hence a lower overall error bound; this is called selecting the nodes adaptively. If we have a formula for f then we may do so by a trial-and-error approach, time permitting; if not, f'' can be estimated by finite differences (see Sec. 6.1). The key idea is that we use some heuristic to detect regions where the function is changing more rapidly and place relatively more points there.

EXAMPLE 2: Let's look at plotting $f(x) = \sin(x)$ over $[0, \pi]$. We know that $f''(x) = -\sin(x)$. If we use, say, 10 equally spaced points in the first subinterval $0 \leq x \leq .1$ then the error bound $Mh^2/8$ from Eq. (2.3) gives ($M = \sin(.1)$, $h = .01$) $E \leq \sin(.1)(.01)^2/8 \doteq 1.2479E - 6$ on this interval. Note that this means that the approximant is good to within about $1.25E - 6$ for *every* $x \in [0, .1]$. How many equally spaced points would we have to put in the subinterval $[1.4, 1.5]$ to get the same accuracy? Here $M = \sin(\pi/2) = 1$ so we need to choose h such that

$$Mh^2/8 \leq 1.25E - 6$$
$$h^2 \leq 8\left(\frac{5}{4}\right)10^{-6}$$
$$= 10^{-5}$$
$$h \leq .0032$$

meaning that in this interval of width $w = .1$ we need about $w/.0032 \approx 316.2$ points in this subinterval to get the same guaranteed accuracy. (We'd need to round this up to 317 to be sure.) It takes 30 times as many points to track the function around the curve at the maximum at $x = \pi/2$! It's unfortunate that we need so many nodes but at least the error bound gives us a means of guaranteeing that a specified accuracy is achieved. □

An adaptive routine for constructing a piecewise linear interpolant would typically take a step of a certain size, use some heuristic to test the error or the second derivative, and then refine the grid if necessary by relaxing it if it seems that a larger step could be taken, and refining it if it seems a smaller step is necessary. (Just as we want a finer grid for improved accuracy when needed, we want to use a coarser grid where possible for efficiency. This efficiency is often what allows us to place more nodes in regions where the function is changing rapidly, because there is often an effective limit on the number of function evaluations allowed.) We'll consider this idea in more detail when we consider numerical integration and the numerical solution of ODEs (in Ch. 5 and Ch. 6, respectively).

Notice that if we have constructed a piecewise linear interpolant for a function and we decide to add a few more points then it requires relatively little work to construct the new interpolant; the new points only affect the function in the subintervals in which they lie, and we only need to recompute the interpolant in these regions. Compare this to a polynomial interpolation problem when we add a single node: Every coefficient of the polynomial will most likely change.

There is an evaluation issue that we have overlooked until now. If the nodes are irregularly spaced, then given an x we must determine in which element $[x_i, x_{i+1}]$ it lies before we can evaluate $p(x)$. There are a number of ways to handle this efficiently; the most easily implemented method is to treat it in a manner similar to a root-finding problem and apply the idea of bisection (binary search) as a search method. For example, if $n = 101$ and the nodes $x_0 < x_1 < \cdots < x_{100}$ are stored in a list then to determine the subinterval containing a given $x \in [x_0, x_{100}]$ we might check if $x < x_{50}$; if so, we might then check whether $x < x_{25}$ and if not then we know that $x_{25} \leq x < x_{50}$. Our next test might be $x < x_{37}$, and so on until we determine the interval in which x lies.

MATLAB

Let's try the Runge function plot mentioned in the text. Enter:

```
» x=linspace(-1,1,11); plot(x,1./(1+25*x.^2),'r'), grid
» hold on; y=linspace(-1,1,200); plot(y,1./(1+25*y.^2),'b')
```

to plot a piecewise linear approximation to the Runge function and the Runge function together. Of course, they're really both piecewise linear plots, but the second one is at a resolution too fine for the eye to see the difference at this scale.

We've been focusing on using MATLAB as a tool for performing numerical experiments; the emphasis in MATLAB and other computation-oriented languages on vectorization actually affects the development of numerical algorithms as well. We haven't focused as much on some of the other tools available in MATLAB. Let's take this opportunity to explore the plotting capabilities in greater detail. It is not uncommon for those who work in other languages to import the results of their calculations into MATLAB for display (we speak of "visualization" of data).

We've used the **plot** command often. Let's look at a more sophisticated application. Enter:

```
» close all   %Make sure all figures are closed.
» th=linspace(0,2*pi);
» fig1=plot(th,sin(th),'g',th,cos(th),'b:')
» grid on
```

Note that the `plot` command returned a vector; each entry is called a handle and it identifies an object within the plot. We can use these handles to determine properties of these objects. Enter:

```
» get(fig1(1))
```

to see the properties of the first line. The color is [0 1 0], that is, no Red, some Green, and no Blue in the RGB color scheme. Enter:

```
» get(fig1(2))
```

to see the properties of the second (dashed) line. The color is [0 0 1], that is, no Red, no Green and some Blue, and the LineStyle is dashed. What about the figure itself? It's the first figure, and its handle is therefore 1. Enter:

```
» get(1)
```

to see some of its properties. For example, if the color is [0 0 0], this means a black background; if it's something like [0.9400 0.9400 0.9400], this means a white background (equal amounts of Red, Green, and Blue). Enter:

```
» set(1)
```

to see a display of possible settings for this figure window. Let's make the background Red. Enter:

```
» set(1,'color',[1 0 0])
» figure(1)
```

That's too Red. Let's try again. Enter:

```
» set(1,'color',[.6 0 0])
» figure(1)
```

We can mix colors too. Enter:

```
» set(1,'color',[.8 .6 .4])
» figure(1)
```

Maybe a white background would be better. Enter:

```
» set(1,'color',[1 1 1])
» figure(1)
```

We can adjust the colors and thickness of the plotted lines too. Enter:

```
» set(fig1(1),'color',[0 .5 0])
» set(fig1(1),'LineWidth',8)
» figure(1)
```

to change the shade of green used on the first curve and to thicken it. You may enter `set(fig1(1))` to see what else about the sine curve's plot may be changed using `set`. We can delete a curve as well; enter:

```
» delete(fig1(2))
» figure(1)
```

to delete just the cosine curve from the plot. The `refresh` command may be used to refresh a plot.

Let's get a new figure window. We'll leave the previous one where it is. Enter:

```
» H=figure
```

to open a new figure window and store its handle in H. This window will remain the active plot unless we select the previous figure window (or a new one) by a MATLAB command or by clicking on it. For example, enter:

» gcf   %Get current figure's handle.

(this should be 2) then click on the first figure window (containing the sine curve) to make it active. Enter:

» gcf

again; it should be 1. Let's make the empty figure window active again. Enter:

» figure(2)

» gcf

Now let's plot a circle in it. Enter:

» plot(sin(th),cos(th),'y-',sin(th),cos(th),'wo')

» figure(H)

The discrete points are plotted as white circles, and the yellow line interpolates those circles. (The circle is a function of neither x nor y so this is a more general piecewise linear interpolation than we have considered in the reading.) The aspect ratio is off. To make it actually look circular, enter:

» axis equal

and look at it again. (Note that the **axes** command is different from the **axis** command. We want the latter.) Let's enlarge the box around the circle a bit; enter:

» axis([-1.1 1.1 -1.1 1.1]),figure(H)

Now let's title and label it. You can use some TEX syntax in MATLAB. For example, enter:

» title('Plot of x=sin(\theta), y=cos(\theta)')

» figure(H)

and note that the \theta is displayed as the Greek letter θ. Similarly for the other Greek letters; for those for which the capital letter is also available, it may be obtained as \Theta (etc.). There are many other TEX formatting options that may be used; enter:

» title('Plot of {\itx}=sin(\theta), {\ity}=cos(\theta)')

» figure(H)

to have the x and y written in italics (the braces limit the scope of the italic switch). Other TEX commands may be used to give other common mathematical symbols (e.g., \infty for ∞, \int for $\int$) and effects (e.g., subscripts and superscripts). Enter:

» xlabel('{\itx} axis','color','red')

» ylabel('{\ity} axis','color','blue')

» figure(H)

to label the axes in red and blue. Other properties of the title and labels may be set as well, including the style and size of font used.

The **text** command may be used to place text within a plot. Enter:

» htext=text(-.6,0,'This is a circle.')

» get(htext)

» figure(H)

to write the indicated text starting at the point $x = -.6$, $y = 0$ on the current axes. Everything listed by the **get** command is something we can change; enter:

» set(htext,'color',[0 0 1],'FontName','Courier')

» figure(H)

to make some changes. Let's write it smaller and better centered. Enter:

» delete(htext)

```
» figure(H)      %Text should be gone
» text(-.1,0,'This is a circle.','color',...
[0 0 1],'FontSize',6)
```
(recall that the ellipsis ... means continuation of the current line onto the next). We did not ask for a handle this time so none is returned.

The properties we have discussed (color, FontSize, etc.), and others, may also be used with `title`, `xlabel`, `ylabel`, and `text`. More than one piece of text may be written. To highlight interesting aspects of a plot; enter:
```
» text(0,.9,'\uparrow','color',[1 0 0])
» text(-.05,.8,'Top','color',[1 0 0])
```
to point to the top of the circle, from within it. To add text by the mouse, use the `gtext` command. Enter:
```
» gtext('Hello world!')
```
then click on some point within the figure window. The text should appear at the point where you clicked the mouse. The `gtext` command returns a handle (after the click) if one is requested. The `ginput` command may be used to collect the location of a mouse click; enter:
```
» [x,y]=ginput(2)
```
then click twice in the figure window. The `x` and `y` vectors collect the x and y coordinates of those mouse clicks. If used in the form `[x,y]=ginput` an unlimited number of points may be gathered (until the return key is pressed).

We can also add a legend to the plot that indicates which line type corresponds to which curve. Enter:
```
» clf     %Clear figure 2.
    » x=0:.1:pi;
    » y=[sin(x);cos(x)];
    » legend('sine','cosine','Location','SouthEastOutside')
```
to add the legend on the lower right. The legend may be dragged to a new location using the mouse, including within the figure. (See `help legend` for more options.) The command `legend off` (or `legend('off')`) removes the legend from the plot.

There is much more that can be done; see a MATLAB manual and a TEX manual for details. It is possible to make very professional looking plots (two dimensional and three dimensional) in MATLAB. Plots may be rotated, etc., using the mouse.

We briefly mention some other plot-related commands; you may look up the details using `help` if you need these capabilities. The `reset` command resets properties set by `set`. The `clf` and `cla` commands clear figures and axes, respectively. (Many of these commands are most useful when embedded in a MATLAB program that makes many plots, as for a presentation.) The `axes` command positions the axes. The `axis` command has more options than we have discussed. The `subplot` command may be used to put more than one plot in a single figure window; for example, enter:
```
» clf     %Clear figure 2.
» subplot(2,1,1),plot(th,sin(th))
» subplot(2,1,2),plot(th,cos(th))
```
The figure window may be subdivided further than this. The `semilogx`, `semilogy`, and `loglog` commands may be used for logarithmic plots. The `polar` command may be used for polar plots. The `line` command may be used to draw lines. The

`fplot` command is essentially an interface to `plot` but can be convenient for plotting a function; see `help fplot`. The `plot3` command is a three-dimensional analogue of `plot`; for example, enter:

» `t=0:pi/50:10*pi;`
» `plot3(sin(t),cos(t),t);`

(This example is from the `plot3` help.) The `view` command may be used to view a plot from another angle. Enter:

» `view([10 100 10])`
» `plot(t,sin(t))`
» `view(10,20)`

See `help view` for details. The `patch` and `fill` commands may be used to create filled-in polygons in a plot. Enter:

» `close all`
» `t=t/max(t);`
» `plot(t,sin(t)),hold on`
» `fill([0 .5 .5 0],[0 0 .5 .5],'g')`

to draw a filled-in green polygon on this plot, with successive vertices at $(0,0)$, $(.5,0)$, $(.5,.5)$, $(0,.5)$ (in that order). See also `help colormap`.

The `print` command may be used to print a figure from MATLAB (rather than using the figure's drop-down commands), either to the printer or to a file. The print command has many options; see also `printopt`, `orient`, and `saveas`.

There's a lot more plotting power in MATLAB; we haven't even touched on three-dimensional plots in this section and have only hinted at the color options. See `help graphics` and `help graph2D`, and `help graph3d` for more details.

Problems

1.) a.) Sketch the hat functions $\Lambda_0, ...\Lambda_4$ for the nodes $\{1,3,4,6,7\}$.

b.) Find the formulas for these piecewise Lagrange interpolating polynomials $(\Lambda_0, ...\Lambda_4)$ using the Lagrange interpolating polynomials (as in Example 1). Be sure to indicate where the functions are zero on $[1,7]$ as well as where they are nonzero.

c.) Use the hat basis to express the piecewise linear interpolant to $\sin(x)$ at these points.

2.) a.) What spacing (assuming equally spaced nodes) should be used to plot $2\sin(x)$ over the interval $[0,\pi]$ in order to ensure that the error in the piecewise linear interpolant is no more than .01?

b.) What selection of nodes should be used to plot $2\sin(x)$ over the interval $[0,\pi]$ in order to ensure that the error in the piecewise linear interpolant is no more than .01? Use the following procedure to answer this question: Set $x_0 = 0$ and $h = .1$. Generate a test point by taking a step of size h into the interval. Use Theorem 1 to check the error over that interval. If the error is too large, cut h in half, and try again; if the predicted error is less than .005 (half the tolerance), double h and try again. Repeat until you find an acceptable node, then use that value of h in the next step. You may write a program or script file to do this if you wish.

3.) a.) Recreate the interpolation plot from Example 1 using the sequence of MAT-LAB commands `x=linspace(-1,1,11);` followed by `plot(x,1./(1+25*x.^2)`, `grid, hold on` then `y=linspace(-1,1,200);` and finally `plot(y,1./(1+25*y.^2))`. Note how the "true" curve is simply a piecewise linear curve with very many nodes.

b.) Repeat with only 10 equally spaced points.

c.) How many points must you use before the curve looks like a smooth sine curve and not a piecewise linear curve?

4.) Use Theorem 4.1.1 to prove Theorem 1. Note: The theorem is correct as stated, but getting the denominator down to 8 requires some insight.

5.) Consider the nodes $\{0, \pi/2, \pi\}$ and the function $y = \sin(x)$. Fit a piecewise cubic interpolant $c_1(x)$ to this function by choosing a cubic for the subinterval $[0, \pi/2]$ that not only interpolates y at the endpoints but also has the property that the derivative of the cubic interpolates y'; that is, the cubic $c_1(x)$ should satisfy $c_1(0) = y(0)$, $c_1(\pi/2) = y(\pi/2)$, $c_1'(0) = y'(0)$, $c_1'(\pi/2) = y'(\pi/2)$. Then do the same with a second cubic $c_2(x)$ on the subinterval $[\pi/2, \pi]$. Plot the resulting approximant.

6.) (Refer to Problem 3. a.) Plot the graph from Example 1. Include an appropriate title and axis labels and a legend. Use the `text` or `gtext` command to note on the graph the point or points where the approximation is worst.

7.) Write a MATLAB program to perform binary search: Given a vector of nodes $x_1 < x_2 < x_3 < \cdots < x_{n+1}$ and a scalar value x, the program should return the lowest index i such that $x_i \le x \le x_{i+1}$, the vector $[x_i, x_{i+1}]$, and a status variable s indicating success if an interval is found ($s = 1$), and failure if $x < x_1$ ($s = -1$), $x > x_{n+1}$ ($s = -2$), or x not of the appropriate form (e.g., x is a vector, or NaN, or a string variable, etc.; $s = -3$).

8.) Choose an efficient selection of nodes to plot the **error function erf(x)** over the interval $[0, 1]$ with error in the piecewise linear interpolant no more than .01. Give the nodes and a plot of `erf(x)` and the approximant.

9.) Choose an efficient selection of nodes to plot the function `sin(12x)` over the interval $[0, 1]$ with error in the piecewise linear interpolant no more than .01. Give the nodes and a plot of `sin(12x)` and the approximant.

10.) Write a MATLAB program that accepts a function as an input argument and solicits nodes from the user using `ginput`, then plots the corresponding hat basis functions. Plot 5 hat functions per figure window, each a different color, and pause between opening new windows. The windows should be titled and labeled appropriately; you may use a command like `title(['Numbers ',num2str(N),' to ',num2str(N+4),'.'])`, for example.

11.) Write a MATLAB program that accepts a function as an input argument and solicits nodes from the user using `ginput`, then plots the piecewise linear interpolant for the function using those nodes. (Ignore the y-values collected by `ginput`.) Note that you may use the figure command to open a window and `text` or `title` to display a request that the user enter points. Your function should plot the function and the approximant on a single graph; plot the function using 5 times as many points as there are nodes. Return the nodes as output. Demonstrate your program on the Runge function.

12.) Write a detailed algorithm (pseudo-code) for evaluating a piecewise linear interpolant by performing binary search. Assume the ordered list of nodes, the corresponding function values, and a point x are given; perform the search, evaluate the interpolant at x, and return the value of the interpolant or an error message if appropriate.

13.) Write a MATLAB program that accepts a function and a number of subintervals and plots a piecewise linear interpolant with that number of equally spaced subintervals. The program should then prompt the user to specify a subinterval by clicking on the plot; the region beneath the linear segment on the selected subinterval should then be filled-in. If an output argument is requested then the number of the selected subinterval should be returned, counting from zero.

14.) Write a MATLAB program that performs piecewise linear interpolation using the technique of Eq. (2.1). Your program should take as inputs the nodes and the function or a vector of y-values (use some test to decide which has been supplied), and return the coefficients in a 2 column matrix. Write a second program that evaluates the interpolant at a given point.

15.) Write a MATLAB program that accepts a vector of nodes the length of which is a multiple of 3 and a function or a vector of y-values (use some test to decide which has been supplied), and returns the coefficients of a piecewise quadratic interpolant as a 3 column matrix. Determine the quadratics by fitting the first quadratic through the first three points x_0, x_1, x_2, the next quadratic through the points x_2, x_3, x_4, the next quadratic through the points x_4, x_5, x_6, and so on. Your program should have an optional input argument that requests a plot of the interpolant.

16.) a.) Write a MATLAB program that approximates the definite integral of a function using a piecewise linear interpolant. Your program should input the function, a desired number of nodes, and limits of integration, and should output an approximation of the integral. Your program should have an optional input argument that requests a plot of the function (you may wish to use `fplot`) and the interpolant on the same graph, with the area under the interpolant filled-in. Choose colors and ordering so that the function is visible over the filled-in region.
b.) Experiment with your program; how many nodes are needed to get an error of no more than 1% for some common functions?

3. Cubic Splines

Piecewise linear interpolation is a useful tool. However, while the resulting interpolant is necessarily continuous, it is not smooth (differentiable). For many applications we would like to have a smooth interpolant. (Think of video game graphics for which the characters have angular, gem-faceted faces–piecewise planar–versus more sophisticated graphics that produce smooth, rounded, natural-looking features.) Taking a finer and finer grid will improve the visual appearance of a piecewise linear interpolant but this is computationally expensive, not always possible–if the nodes are measured data, for example–and still doesn't produce an approximating function that is truly differentiable. We want to be able to produce a smooth interpolant using a reasonable number of nodes.

The idea of piecewise linear interpolation may be generalized to piecewise polynomial interpolation using quadratics, cubics, and so on, as the pieces. For example, given the nodes $\{0, \pi/2, \pi\}$ and the function $y = \sin(x)$, we could get a differentiable interpolant by fitting a quadratic to the three points (in essence, treating the interval from 0 to π as a single element containing three nodes), or by fitting a quadratic on $\{0, \pi/2\}$ and another quadratic on $\{\pi/2, \pi\}$ and making each have the correct derivative in the middle. That is, if $q_1(x) = a_1 x^2 + b_1 x + c_1$ and $q_2(x) = a_2 x^2 + b_2 x + c_2$ are the two quadratics, we want to choose the coefficients so that

$$q_1(0) = 0$$
$$q_1(\pi/2) = 1$$
$$q_1'(\pi/2) = 0$$
$$q_2(\pi/2) = 1$$
$$q_2'(\pi/2) = 0$$
$$q_2(\pi) = 0$$

that is

$$c_1 = 0$$
$$a_1(\pi/2)^2 + b_1(\pi/2) + c_1 = 1$$
$$2a_1(\pi/2) + b_1 = 0$$
$$a_2(\pi/2)^2 + b_2(\pi/2) + c_2 = 1$$
$$2a_2(\pi/2) + b_2 = 0$$
$$a_2\pi^2 + b_2\pi + c_2 = 0$$

and the solution is

$$a_1 = -0.4053$$
$$b_1 = 1.2732$$
$$c_1 = 0$$
$$a_2 = -0.4053$$
$$b_2 = 1.2732$$
$$c_2 = 0$$

(see Fig. 1, where the interpolant is the thinner curve). The quadratics happen to be the same due to the underlying symmetry; in general they would not be. We have forced them to give a continuous and differentiable curve on $[0, \pi]$ by requiring that the derivatives have the same value where the quadratics meet–namely, the true value of $y'(x)$ at that point. Fitting a quadratic between each pair of nodes is called **piecewise quadratic interpolation**.

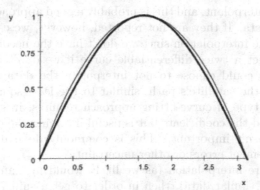

FIGURE 1. Fitting a quadratic.

If there were another node to the right of π, (for example, if the nodes were augmented to become $\{0, \pi/2, \pi, 3\pi/2\}$), then we wouldn't have another free coefficient to match the derivative at the node at π. The obvious thing to do is to use cubics on each section, for then we have four freely selectable coefficients on each subinterval. We write

$$(3.1) \qquad c_i(x) = \alpha_i x^3 + \beta_i x^2 + \gamma_i x + \delta_i$$

on subinterval i, and require

$$(3.2) \qquad \begin{array}{rcl} c_i(x_i) & = & y_i \\ c_i'(x_i) & = & y_i' \\ c_i(x_{i+1}) & = & y_{i+1} \\ c_i'(x_{i+1}) & = & y_{i+1}' \end{array}$$

at the left and right endpoints x_i and x_{i+1}, respectively, of the subinterval. This gives four equations in the four unknowns $\alpha_i, \beta_i, \gamma_i, \delta_i$ for each of the n subintervals, and if the nodes are distinct then there is a unique solution for every choice of function and derivative values. Fitting a cubic between each pair of nodes is called **piecewise cubic interpolation**.

More generally, we could use a polynomial of any degree on a given element, or even of different degrees on different elements; this is called **piecewise polynomial interpolation**. If we require that an appropriate number of derivative values be matched, as above, then it is called **piecewise polynomial Hermite interpolation**. For the reasons indicated above, polynomials of odd degree are most commonly employed (especially linear, cubic, and quintic).

Piecewise cubic Hermite interpolation (using Eq. (3.1) and Eq. (3.2)) has an attractive property: The resulting interpolant is differentiable. In general it will not be twice differentiable. For many problems, however, having a twice differentiable interpolant is important. If we have discrete observations of a moving particle (say, a satellite that reports its position once per second) and we wish to recreate its orbit, then interpolating the data is a reasonable approach; however, the velocity and acceleration of the particle are certainly continuous functions as well, so on physical grounds we would like our interpolant to be at least twice continuously differentiable. If velocity and acceleration data are also reported, we could use a piecewise quintic interpolant, and this is probably a good approach in that case since it uses all of the data. If they are not reported, however, we cannot use piecewise polynomial Hermite interpolation since we don't have the needed derivative data.

How can we get a twice differentiable curve if we don't have the derivative values, then? We could choose to not interpolate the data and instead find a best-fit ellipse for the satellite's path, similar to the least-squares lines discussed in Sec. 2.7. This type of curve-fitting approach requires, in general, a nonlinear optimization to find the coefficients but is useful if forcing the curve to go through every data point isn't important. This is commonly done if the data contains significant measurement errors or other uncertainties.

If we do require interpolation (as we likely would in many cases involving a moving particle or similar data), then in order to get a sufficiently smooth interpolant we have two options: Lagrange interpolation, which as we have seen can give a very strange-looking path (think of the Runge function), or piecewise polynomial interpolation where we select the values for the derivatives so as to give a smooth curve. Let's explore the latter approach.

EXAMPLE 1: Suppose we wish to fit a twice continuously differentiable curve to the data $\{(-1, 0.3679), (0, 1), (2, 7.3891)\}$. In order to do so, we need to choose polynomials $p_1(x)$ (on $[-1, 0]$) and $p_2(x)$ (on $[0, 2]$) that will satisfy the conditions

$$
\begin{aligned}
p_1(-1) &= 0.3679 \\
p_1(0) &= 1 \\
p_1'(0) &= p_2'(.75) \\
p_1''(0) &= p_2''(.75) \\
p_2(0) &= 1 \\
p_2(2) &= 7.3891
\end{aligned}
$$

at the nodes -1, 0, and 2. The first two conditions and the last two conditions ensure interpolation, and the middle two ensure smoothness at the middle node 0, where the polynomials meet. We have a total of six conditions. We should be able to use two quadratics $p_1(x) = ax^2 + bx + c$ and $p_2(x) = \alpha x^2 + \beta x + \gamma$ since this gives a total of six unknowns $a, b, c, \alpha, \beta, \gamma$. We have the requirements

$$
\begin{aligned}
a - b + c &= 0.3679 \\
c &= 1 \\
b &= \beta \\
2a &= 2\alpha \\
\gamma &= 1 \\
4\alpha + 2\beta + \gamma &= 7.3891
\end{aligned}
$$

upon substituting in the equations for p_1 and p_2. Clearly, $a = \alpha$, $b = \beta$, and $c = \gamma = 1$, so $p_1 = p_2$. The first and last equation then give

$$
\begin{aligned}
a - b + 1 &= 0.3679 \\
4a + 2b + 1 &= 7.3891
\end{aligned}
$$

and the solution is $a = 0.8542$, $b = 1.4862$. Hence $p_1(x) = p_2(x) = .8542x^2 + 1.4862x + 1$ defines the interpolant.

We never actually set the values for the derivatives at the node 0; we just required that they be equal. They have taken the values

$$
\begin{aligned}
p_1'(0) &= 1.4862 \\
p_1''(0) &= 2(.8542) \\
&= 1.7083
\end{aligned}
$$

in order to meet this requirement. In fact, the data was from the function $y = e^x$ for which $y'(0) = y''(0) = 1$. The derivative is not well approximated but the interpolant is twice continuously differentiable. Indeed, since $p_1(x) = p_2(x)$, it is infinitely differentiable, but typically we only get as much derivative agreement as we explicitly require. □

We now have a means of taking data $\{(x_0, y_0), ..., (x_n, y_n)\}$ and fitting a smooth curve to it. The data may be measured data (possibly provided directly to the computer by another device, e.g., a digitizer reading in discrete points along the boundary of some object of interest like the body of an automobile) or may be the output of another program that provides such points (say, a program that numerically solves an ODE by finding the solution only on a discrete grid). Consider again Example 1. If we had had 4 grid points instead of 3, then we would have had 3 elements. Nothing changes for the first and last element, but the middle element has to match derivatives across two nodes rather than just one. A quadratic won't work here; we need more parameters in the polynomial for that section. This will be true for most subintervals unless n, the number of subintervals, is trivially small.

The usual approach is to use a piecewise cubic interpolant on every subinterval (even the ones at the ends). Each cubic has 4 coefficients for a total of $4n$ parameters over the entire interpolant. How many conditions are there that must be satisfied? Each of the n cubics must interpolate at 2 points (its left and right endpoints), giving $2n$ conditions; in addition, at each of the $n - 1$ internal nodes we specify 2 constraints, namely, first differentiability and second differentiability. This gives a total of $2n + 2(n - 1) = 4n - 2$ equations in the $4n$ parameters. We need two additional conditions to have a uniquely determined interpolant.

Using quadratics on the end subintervals would reduce the number of unknowns to $4n - 2$, but it's convenient to have a piecewise cubic interpolant rather than a mixed-degree interpolant. Instead we add two conditions at the end subintervals, called **boundary conditions**. There are a number of standard boundary conditions. Let $p_0(x)$ and $p_{n-1}(x)$ be the leftmost and rightmost cubic. The **free boundary conditions** are

$$
\begin{aligned}
p_0''(x_0) &= 0 \\
p_{n-1}''(x_n) &= 0
\end{aligned}
$$

(also called the **natural boundary conditions**). The **clamped boundary conditions** are

$$
\begin{aligned}
p_0'(x_0) &= s_0 \\
p_{n-1}'(x_n) &= s_n
\end{aligned}
$$

where s_0 and s_1 are given values; typically we take s_0 and s_1 to be the derivatives $s_0 = f'(x_0)$ and $s_n = f'(x_n)$ of the function $y = f(x)$ or approximations of them. Of course, if the data $\{(x_0, y_0), ..., (x_n, y_n)\}$ is random, say due to additive error, there might not even *be* a deterministic functional relationship $y = f(x)$ between x and y.

Let $S(x)$ be the interpolant; that is, $S(x)$ is equal to $p_0(x)$ on $[x_0, x_1]$, $p_1(x)$ on $[x_1, x_2]$,..., $p_{n-1}(x)$ on $[x_{n-1}, x_n]$. The **not-a-knot boundary conditions** are that $p_0(x) = p_1(x)$ for all x and $p_{n-2}(x) = p_{n-1}(x)$ for all x (almost as though the nodes x_1 and x_{n-2} were not being used, though note that $S(x)$ must still interpolate at x_1 and x_{n-2}); an equivalent condition is that S is three times continuously differentiable at the knots x_1 and x_{n-2}. If the derivative values at the endpoints are not available or easily approximated, this condition is commonly employed.

EXAMPLE 2: Let's find a twice continuously differentiable interpolant of this form for $y = e^x$ using the nodes $\{(-1, 0.3679), (0, 1), (2, 7.3891), (4, 54.5982)\}$. Since we know the function, we'll use the clamped boundary conditions. We need to find 3 cubics $p_0(x) = a_0 x^3 + b_0 x^2 + c_0 x + d_0$, $p_1(x) = a_1 x^3 + b_1 x^2 + c_1 x + d_1$, and $p_2(x) = a_2 x^3 + b_2 x^2 + c_2 x + d_2$. The conditions are:

$$
\begin{array}{llll}
p_0(-1) = 0.3679 & p_0(0) = 1 & p_1(2) = 7.3891 & p_2(4) = 54.5982 \\
p_0'(-1) = 0.3679 & p_1(0) = 1 & p_2(2) = 7.3891 & p_2'(4) = 54.5982 \\
& p_0'(0) = p_1'(0) & p_1'(2) = p_2'(2) & \\
& p_0''(0) = p_1''(0) & p_1''(2) = p_2''(2) &
\end{array}
$$

or, in terms of the 12 unknown coefficients,

$$
\begin{aligned}
-a_0 + b_0 - c_0 + d_0 &= 0.3679 \\
3a_0 - 2b_0 + c_0 &= 0.3679
\end{aligned}
$$

at the node at -1,

$$
\begin{aligned}
d_0 &= 1 \\
d_1 &= 1 \\
c_0 &= c_1 \\
2b_0 &= 2b_1
\end{aligned}
$$

at the node at 0,

$$
\begin{aligned}
8a_1 + 4b_1 + 2c_1 + d_1 &= 7.3891 \\
8a_2 + 4b_2 + 2c_2 + d_2 &= 7.3891 \\
12a_1 + 4b_1 + c_1 &= 12a_2 + 4b_2 + c_2 \\
12a_1 + 2b_1 &= 12a_2 + 2b_2
\end{aligned}
$$

at the node at 2, and

$$
\begin{aligned}
64a_2 + 16b_2 + 4c_2 + d_2 &= 54.5982 \\
48a_2 + 8b_2 + c_2 &= 54.5982
\end{aligned}
$$

at the node at 4. This gives a linear system involving a 12×12 coefficient matrix, though we could clearly reduce the order by inspecting the above relations and using $d_0 = 1$, $d_1 = 1$, $c_0 = c_1$, and $b_0 = b_1$ to eliminate four of the unknowns. The solution of this system is

$$
\begin{aligned}
a_0 &= 0.1803, b_0 = 0.6249, c_0 = 1.0766, d_0 = 1.0000 \\
a_1 &= 0.2170, b_1 = 0.6249, c_1 = 1.0766, d_1 = 1.0000 \\
a_2 &= 3.3924, b_2 = -18.4275, c_2 = 39.1813, d_2 = -24.4031
\end{aligned}
$$

(using MATLAB); we'll plot this in the MATLAB subsection. □

Notice that the equations are truly coupled; c_0 is equal to c_1, but another equation couples c_1 and c_2, for example. The differentiability conditions are not local to an element; every value affects every other one, in the general case. This is different from the piecewise cubic interpolant with known values of the derivatives, where each cubic depended only on data from the nodes at either end of its subinterval. We refer to any twice continuously differentiable piecewise cubic interpolant to the data $\{(x_0, y_0), ..., (x_n, y_n)\}$ as a **cubic spline**. We call it a **natural** (or **free**), **clamped** (or **complete**), or **not-a-knot cubic spline** according to the boundary conditions used.

Note that a piecewise cubic interpolant that is used to interpolate both f and f' at some nodes (a piecewise cubic Hermite interpolant) uses the same number of coefficients as a cubic spline for f at those nodes but requires about twice as much data, namely, the values of both $f(x_i)$ and $f'(x_i)$, and is only *once* continuously differentiable. On the other hand, since the piecewise cubic Hermite interpolant uses more data we expect it to give a better fit in the quantitative sense of deviating less from the true function even though it may be worse in the qualitative sense of not being *twice* differentiable.

Splines of other orders may be defined as well. The piecewise linear interpolants of the previous sections are considered **linear splines** and in those cases in which a thrice continuously differentiable interpolant is needed a **quintic spline**, with degree five polynomial sections interpolated to the data $\{(x_0, y_0), ..., (x_n, y_n)\}$, may be used. However, linear and cubic splines are by far the most commonly used types of splines. We may also use other boundary conditions (for example, if the function is known to have some property such as periodicity).

It appears that we will need to solve a $4n \times 4n$ system in order to find a cubic spline for $n + 1$ data points. This is not the case, however. By arranging the computation carefully we can reduce the $4n \times 4n$ system to a much smaller one. We'll discuss this in the next section.

MATLAB

Let's look again at Example 2, where we found a cubic spline interpolant for $y = \exp(x)$ at the nodes $\{-1, 0, 2, 4\}$. Enter the coefficients as a vector:

```
» p=[0.1803 0.6249 1.0766 1.0000 0.2170 0.6249 1.0766 1.0000 ...
   3.3924 -18.4275 39.1813 -24.4031]
```

The first four entries of p are the coefficients in $p_0(x) = a_0 x^3 + b_0 x^2 + c_0 x + d_0$, and so on. Let's plot the interpolant. The first element is $[-1, 0]$. Enter:

```
» x=linspace(-1,0);y=polyval(p(1:4),x);
» xtemp=linspace(0,2);y=[y,polyval(p(5:8),xtemp)];x=[x,xtemp];
» xtemp=linspace(2,4);y=[y,polyval(p(9:12),xtemp)];x=[x,xtemp];
```

to form a vector of x-values from -1 to 4 and corresponding y-values. (We are using piecewise linear interpolation to plot the cubic spline.) Enter:

```
» plot(x,y,'y',x,exp(x),'g'),grid
```

to plot the interpolant and the function. This is a good fit (recall that we used only 4 nodes). Do the derivatives truly match? Let's check at $x = 2$. Enter:

```
» p1d=polyder(p(5:8)),p2d=polyder(p(9:12))
```

to find the first derivatives there. Enter:

```
» exp(2)
» polyval(p1d,2),polyval(p2d,2)
```

to verify that the spline's one-sided derivatives are equal, though they are not equal to the actual derivative value $y'(2) = \exp(2)$. (The piecewise cubic Hermite interpolant would have the correct value here, since it would have taken that as data.) Similarly, enter:

```
» p1dd=polyder(p1d),p2dd=polyder(p2d)
» polyval(p1dd,2),polyval(p2dd,2)
```

to verify that the second derivatives match across this node, as expected, although they do not match $y''(2) = \exp(2)$. The piecewise cubic Hermite interpolant would not have had a second derivative here, in general.

There are a number of built-in spline functions in MATLAB, as well as an optional Spline Toolbox. The `spline` command takes as arguments a vector `x` of nodes, a vector `y` of corresponding y-values, and a finer grid `xi` of values, and returns the values of the spline function at the points in `xi`. The not-a-knot boundary condition is used by default. Enter:

```
» x=[0 1 2 3];y=cos(x);    %Nodes and y-values.
» xi=linspace(min(x),max(x),300);
» yi=spline(x,y,xi);
» plot(x,y,'ro',xi,yi,'b'),grid
```

This evaluates the spline at the 300 points in `xi` and uses those points to plot the spline; the nodes are plotted as red circles. Note that the coefficients of the cubics of which the spline is comprised are not returned, just the values of the spline interpolant. The spline interpolates at the nodes, as expected, and has a smooth appearance. Enter:

```
» hold on
» plot(xi,cos(xi),'k')    %Compare to true function.
» plot(x,y,'g')           %Compare piecewise linear interp.
```

Note how well the piecewise linear interpolant fits in the middle section, which stretches across $\pi/2$; this is because the function $y = \cos(x)$ being interpolated has $y''(\pi/2) = -\cos(\pi/2) = 0$ in the center of this section and hence small second derivative near that centerpoint, so a line is a good approximation in this region.

What if we want the coefficients of the spline? The `spline` command may be used to obtain this information, but in a special form called the pp-representation (piecewise polynomial representation) that we will not discuss. See `help` for `spline`, `mkpp`, `unmkpp`, and `ppval` for more information; see also `interp1` for more spline functionality.

One of the biggest uses of splines is to draw a smooth (and hence physically plausible) curve through measured data, or discrete data generated by some other program. Let's simulate measured data for a trajectory. Enter:

```
» t=1+sort(rand([1 100]));   %Random nodes on [1,2], sorted.
» y=-(t-1).*(t-2)+(rand([1 100])-.5)/50;  %Trajectory plus noise.
» plot(t,y,'g',t,-(t-1).*(t-2),'r')  %Linear spline.
» hold on
» tt=linspace(1,2,600);yy=spline(t,y,tt);
» plot(tt,yy,'b')  %Cubic spline.
```

The cubic spline is worse than the linear spline! Even splines can suffer from some of the difficulties associated with high-order polynomial interpolation. Because of the additive noise, which is not small compared to the data (especially near $t = 1$ and $t = 2$, where y is zero), the values of y do not form a smooth curve. In fact, the relation $y(t) = -(t - 1)(t - 2) + e(t)$ where $e(t)$ is uniformly distributed noise on $[-.5, .5]$ does not even represent a deterministic functional relationship if the values are independent, and the resulting curves are nondifferentiable. Unless the noise is sufficiently small, fitting a twice continuously differentiable function may not be wise.

Often the noise is not so large as to make this a problem. As another example, let's fit a cubic spline to data generated from the numerical solution of an ODE. The MATLAB command ode23 may be used to numerically solve ODEs, as we'll discuss in Ch. 6. Enter:

```
» clear all
» close all
» g=@(t,y) exp(y).*sin(t.*y.^2)
» [t,y]=ode23(g,[0 3],1);
» plot(t,y,'b'),grid
```

to solve the ODE IVP $y' = g(t, y) = e^y \sin(ty^2)$, $y(0) = 1$ numerically and plot that solution using a linear spline. The ode23 command returns a vector of nodes t and the approximate values y of $y(t)$ at those nodes. Look at the sharp peak near $t = .7$; as plotted, the curve looks nondifferentiable. Now enter:

```
» tt=linspace(0,3,600);yy=spline(t,y,tt);
» figure(2);plot(tt,yy,'g'),grid
```

to plot the cubic spline interpolant to the numerical solution of the ODE IVP. Compare the plots; the cubic spline is smoother, especially near the origin and the hump on the graph. This gives a more natural-looking fit. Of course, without knowing the true solution $y(x)$, we can not assert that it is in fact a better approximation to the solution. The cubic spline doesn't use more data than the linear spline, it just uses it in a different way.

Problems

1.) a.) Fit a piecewise quadratic interpolant to $y = \cos(x)$ at the nodes $\{0, 1, 2\}$. (Match the derivative of $y(x)$ at the center node.)
b.) Fit a piecewise cubic Hermite interpolant to $y = \cos(x)$ at the nodes $\{0, 1, 2\}$. (Match the derivatives of $y(x)$ at each node.) Is it a spline (that is, does the second derivative of the interpolant exist at $x = 1$)?
c.) Fit a natural cubic spline interpolant to $y = \cos(x)$ at the nodes $\{0, 1, 2\}$. Plot it, $\cos(x)$, and your piecewise cubic interpolant from part b. on a single graph.
2.) Let $f(x) = \exp(x)$. Find the piecewise cubic interpolant which interpolates both f and f' at $\{0, .1, .2, .3\}$ and the natural cubic spline that interpolates f. Plot them on a single graph along with $f(x)$.
3.) Find the natural and clamped cubic splines for $f(x) = \log(x)$ at the nodes $\{1, 2, 3\}$.
4.) Can a piecewise quadratic interpolant be used to find a differentiable interpolant to a function at an arbitrary list of nodes? If so, it is called a **quadratic spline** (or **parabolic spline**). What boundary conditions might need to be imposed?
5.) Suppose a quintic spline $S(x)$ is to be fit to the function $f(x)$ at the nodes $\{x_0, x_1, ..., x_n\}$. What conditions must S satisfy? Use the natural boundary conditions.
6.) a.) Repeat Example 2 using the natural boundary conditions. Compare your result to the clamped spline found in Example 2.
b.) Repeat Example 2 using the not-a-knot boundary conditions.

7.) If a function is known to be periodic on $[x_0, x_n]$ then we might use the **cyclic** (or **periodic**) **boundary conditions** $p_0'(x_0) = p_{n-1}'(x_n)$, $p_0''(x_0) = p_{n-1}''(x_n)$. Create a cyclic spline for $\sin(2x)$ using the nodes $\{0, 1, 2, \pi\}$. List the three cubics and plot the interpolant and $\sin(2x)$ on a single graph.

8.) a.) Write a MATLAB program that performs piecewise cubic Hermite interpolation. The input arguments should be a vector of nodes, a vector of corresponding y-values, and a vector of corresponding y'-values. The output should be a matrix with one row of 4 coefficients for each element.

b.) Write a MATLAB program that inputs the coefficients from your program of part a. and the nodes plus a vector of x-values and returns the y-values of the spline at those x-values (similar to the MATLAB `spline` command).

9.) a.) Write a MATLAB program that performs piecewise quadratic interpolation using elements based on 3 nodes; that is, fit a quadratic to x_0, x_1, x_2, then another quadratic to x_2, x_3, x_4, and so on. The input arguments should be a vector of $n + 1$ nodes containing where n is even and a vector of corresponding y-values. The output should be a matrix with one row of 3 coefficients for each of the $n/2$ elements.

b.) Write a MATLAB program that inputs the coefficients from your program of part a. and the nodes plus a vector of x-values and returns the y-values of the spline at those x-values (similar to the MATLAB `spline` command).

10.) a.) The **quadratic boundary conditions** for a cubic spline require the leftmost cubic $p_0(x)$ and the rightmost cubic $p_{n-1}(x)$ to be quadratics. Repeat problem 7. using this boundary condition.

b.) Experiment with the natural, clamped (with the correct derivative values), not-a-knot, and quadratic boundary conditions for a variety of functions. Comment on their relative merits and weaknesses.

c.) Experiment with the natural, not-a-knot, and quadratic boundary conditions for randomly generated data. Comment on their relative merits and weaknesses.

11.) Prove that the condition that the spline S is three times continuously differentiable at the knots x_1 and x_{n-2} is equivalent to the not-a-knot requirement that $p_0(x) = p_1(x)$ for all x and $p_{n-2}(x) = p_{n-1}(x)$ for all x.

12.) a.) Find the clamped cubic spline interpolant $S(x)$ for $y = e^{-x} \cos(2x)$ at the nodes $\{0, .1, .2, .3\}$. Plot $S(x)$ and $y(x)$ on the same graph.

b.) Plot $S'(x)$ and $y'(x)$ on the same graph.

c.) Plot $S''(x)$ and $y''(x)$ on the same graph.

13.) a.) Write a MATLAB program that performs piecewise quintic Hermite interpolation. The input arguments should be a vector of nodes, a vector of corresponding y-values, a vector of corresponding y'-values, and a vector of corresponding y''-values. The output should be a matrix with one row of 6 coefficients for each element.

b.) Write a MATLAB program that inputs the coefficients from your program of part a. and the nodes plus a vector of x-values and returns the y-values of the spline at those x-values (similar to the MATLAB `spline` command).

14.) Use the MATLAB `spline` command to plot an approximation of the function $y = e^{2x} \cos(\ln(x))$ for $x \in [1, 7]$. Try to use as few nodes as possible to get a good fit.

15.) Prove that if $y(x)$ is a cubic then it is equal to its piecewise cubic Hermite interpolant and to its clamped cubic spline. Will it equal its natural and not-a-knot cubic splines? Will it equal its piecewise quintic Hermite interpolant and clamped quintic spline?

16.) a.) It's occasionally best to treat the endpoints x_0 and x_n as being so special that they essentially lay outside the splined region. In this case we could spline $[x_1, x_{n-1}]$ with our two "boundary conditions" being the additional requirements that the leftmost cubic also interpolate at x_0 and the rightmost cubic also interpolate at x_n. (Regrettably, this is sometimes called an "interpolating spline"–despite the fact that by definition, all splines interpolate.) We often do not use the cubic on $[x_0, x_1]$ or $[x_{n-1}, x_n]$ in this case. Write a MATLAB program that implements this method, then choose several test functions and plot the resulting curves. Compare them to the corresponding clamped splines on $[x_0, x_n]$. Does either approach seem better over $[x_1, x_{n-1}]$?

b.) A variant of this idea uses all $n + 1$ nodes on $[x_0, x_n]$ but generates its own values for the boundary conditions by using a quadratic interpolant on $[x_0, x_1, x_2]$ to estimate a y-value y_{-1} at some point x_{-1} lying to the left of x_0, say $x_{-1} = x_0 - (x_1 - x_0)$, and similarly for y_{n+1} at x_{n+1}, where x_{n+1} is to the right of x_n but at a distance that is again determined by the scale of the size of the elements. A cubic spline is then constructed on $[x_0, x_n]$ with the two estimated values used as the needed boundary conditions, with the resulting spline not evaluated on $[x_{-1}, x_0]$ or $[x_n, x_{n+1}]$. This differs from part a. in that there we were provided with actual data points, and here we are creating "fictitious points" (or "faux points"). Discuss this idea. Does it seem reasonable for the method to extrapolate to find its own boundary values? When might this be a good, or bad, approach?

4. Computation of the Cubic Spline Coefficients

Cubic splines are widely used. When data from the profile of a physical object like the hood of a car is scanned in, it consists of a discrete set of points that must be smoothly interpolated. The fact that cubic splines are twice differentiable ensures that the resulting curve has a realistic, natural look[5].

We need to consider the accuracy of the spline approximant and also discuss how to compute the spline coefficients efficiently. This is going to involve a lot of little details, but it will be worth it in the end. Let's start with an improved representation of the cubic pieces. As before, let $S(x)$ be the spline and let $p_0(x), p_1(x), ..., p_{n-1}(x)$ be the cubics on $[x_0, x_1]$, $[x_1, x_2]$, ..., $[x_{n-1}, x_n]$. Rather than writing $p_i(x) = a_i x^3 + b_i x^2 + c_i x + d_i$, let's write

$$(4.1) \qquad p_i(x) = \alpha_i(x - x_i)^3 + \beta_i(x - x_i)^2 + \gamma_i(x - x_i) + \delta_i$$

($i = 0, 1, ..., n - 1$). (The Newton form of the Lagrange interpolant, Eq. (4.1.1.4), uses a somewhat similar approach.) There are two immediate advantages to using this form. One is that if x is large then a small error in x will be magnified considerably in the computation of the x^3 term in $a_i x^3 + b_i x^2 + c_i x + d_i$, but if the subintervals are not too large then the term $(x - x_i)^3$ in the equivalent form of Eq. (4.1) will be relatively small and hence the effect of an error in x will be lessened.

---

[5]For the profile of a car, say, we might use three splines so that we could get the hood and windshield to meet an angle (non-matching first derivatives) and similarly the trunk and the back windshield.

(In a similar way, any error in a_i would be multiplied by the large value x^3 but an error in α_i would be multiplied by the smaller value $(x - x_i)^3$.) Hence, this form should result in more accurate evaluations of the spline.

The second benefit of using this form is that the interpolation condition $p_i(x_i) = y_i$ (where $y_i = f(x_i)$ are given values) immediately gives

$$
\begin{aligned}
y_i &= p_i(x_i) \\
&= \alpha_i(x_i - x_i)^3 + \beta_i(x_i - x_i)^2 + \gamma_i(x_i - x_i) + \delta_i \\
&= \delta_i
\end{aligned}
$$

so that the values of $\delta_0, ..., \delta_{n-1}$ are trivially determined. Recall that in the previous section we found the spline coefficients by solving a system of $4n$ equations in $4n$ unknowns; this representation of the cubics reduces the number of unknowns to $3n$ while at the same time using a representation that is more accurate when evaluated at a given x. The direct solution methods we've discussed take $O(N^3)$ floating point operations to solve an $N \times N$ system; for $N = 4n$ this means about $c(4n)^3$ flops for some $c > 0$ (if N is large), while for $N = 3n$ this means about $c(3n)^3$ flops for some c. Hence solving the reduced system takes about $3^3/4^3$ or $27/64$ as many flops (plus the memory management benefits). We've saved over half the computational effort at no cost. In graphics applications it isn't uncommon to have hundreds of thousands of nodes $\{x_i\}_{i=0}^{n}$, so this is a big deal.

We can do even better. Let's suppose that the spline exists (we'll justify this assumption later). Then $S''(x)$ is a well-defined piecewise linear interpolant to the unknown values $S''(x_0), S''(x_1), ..., S''(x_n)$ at the nodes $x_0, x_1, ..., x_n$. Let's write $\sigma_i = S''(x_i)$ for $i = 0, 1, ..., n$. In terms of these unknown values, it must be the case that

$$
(4.2) \qquad S''(x) = \sum_{i=0}^{n} \sigma_i \Lambda_i(x)
$$

in terms of the piecewise linear Lagrange interpolating polynomials. (This is just Eq. (4.2.2.2).) If we can determine $\sigma_0, \sigma_1, ..., \sigma_n$, then we can find an expression for $S''(x)$. From Eq. (4.1),

$$
\begin{aligned}
p_i''(x) &= 6\alpha_i(x - x_i) + 2\beta_i \\
p_i''(x_i) &= 2\beta_i
\end{aligned}
$$

but $p_i''(x_i)$ is just $S''(x_i)$, which by definition is σ_i. In other words, if we can find $\sigma_0, \sigma_1, ..., \sigma_n$, then we can find $\beta_0, \beta_1, ..., \beta_{n-1}$ trivially from

$$
\beta_i = \sigma_i/2.
$$

Can we also relate the values of the remaining coefficients α_i and γ_i ($i = 0, 1, ..., n - 1$) to the values $\sigma_0, \sigma_1, ..., \sigma_n$ (and possibly the known x-values and y-values)? If so, then we would only need to solve a matrix system for the $n + 1$ unknowns $\sigma_0, \sigma_1, ..., \sigma_n$.

It will be convenient to write $h_i = x_{i+1} - x_i$. Consider Eq. (4.2) on a single element $[x_i, x_{i+1}]$; proceeding as in Example 4.2.1, we have

$$
\begin{aligned}
S''(x) &= \sigma_i \Lambda_i(x) + \sigma_{i+1}\Lambda_{i+1}(x) \\
&= \sigma_i \frac{x - x_{i+1}}{x_i - x_{i+1}} + \sigma_{i+1}\frac{x - x_i}{x_{i+1} - x_i} \\
&= -\sigma_i \frac{x - x_{i+1}}{h_i} + \sigma_{i+1}\frac{x - x_i}{h_i}
\end{aligned}
$$

on this element. Integrating,

$$
(4.3) \qquad S'(x) = -\sigma_i \frac{(x - x_{i+1})^2}{2h_i} + \sigma_{i+1}\frac{(x - x_i)^2}{2h_i} + \tau_i
$$

for some constant of integration τ_i, and

$$
(4.4) \qquad S(x) = -\sigma_i \frac{(x - x_{i+1})^3}{6h_i} + \sigma_{i+1}\frac{(x - x_i)^3}{6h_i} + \tau_i(x - x_i) + \kappa_i
$$

for some constants of integration τ_i and κ_i. (You'll be asked to verify this formula in the Problems. You may wish to complete the squares in the numerators, after integrating, to get the form in Eq. (4.3).) But from Eq. (4.4), if we take $x = x_i$ we have

$$
\begin{aligned}
S(x_i) &= -\sigma_i \frac{(x_i - x_{i+1})^3}{6h_i} + \sigma_{i+1}\frac{(x_i - x_i)^3}{6h_i} + \tau_i(x_i - x_i) + \kappa_i \\
&= -\sigma_i \frac{(x_i - x_{i+1})^3}{6h_i} + \kappa_i \\
&= \sigma_i \frac{h_i^3}{6h_i} + \kappa_i \\
&= \sigma_i \frac{h_i^2}{6} + \kappa_i \\
&= y_i
\end{aligned}
$$

by the interpolation condition. That is,

$$
\kappa_i = y_i - \sigma_i h_i^2/6
$$

so that if we can find the σ_i then we can find the κ_i. Using Eq. (4.4) again, if we take $x = x_{i+1}$ we have

$$
\begin{aligned}
S(x_{i+1}) &= -\sigma_i \frac{(x_{i+1} - x_{i+1})^3}{6h_i} + \sigma_{i+1}\frac{(x_{i+1} - x_i)^3}{6h_i} + \tau_i(x_{i+1} - x_i) + \kappa_i \\
&= \sigma_{i+1}\frac{(x_{i+1} - x_i)^3}{6h_i} + \tau_i(x_{i+1} - x_i) + \kappa_i \\
&= \sigma_{i+1}\frac{h_i^{\;3}}{6h_i} + \tau_i h_i + \kappa_i \\
&= \sigma_{i+1}\frac{h_i^2}{6} + \tau_i h_i + \kappa_i \\
&= y_{i+1}
\end{aligned}
$$

by the interpolation condition. That is,

$$\sigma_{i+1}\frac{h_i^2}{6} + \tau_i h_i + \kappa_i = y_{i+1}$$

or

$$
\begin{aligned}
\tau_i h_i &= y_{i+1} - \sigma_{i+1}h_i^2/6 - \kappa_i \\
\tau_i &= (y_{i+1} - \sigma_{i+1}h_i^2/6 - \kappa_i)/h_i \\
&= (y_{i+1} - \sigma_{i+1}h_i^2/6 - (y_i - \sigma_i h_i^2/6))/h_i \\
&= \frac{y_{i+1} - y_i}{h_i} - \frac{h_i}{6}(\sigma_{i+1} - \sigma_i)
\end{aligned}
$$

(4.5)

$(i = 0, 1, ..., n - 1)$ which gives the values of the τ_i in terms of the σ_i, the known y-values, and the known spacings h_i between the nodes.

Notice from Eq. (4.4) that the cubic polynomial on the element i is completely determined by knowledge of the σ_i, τ_i, and κ_i (plus the known grid spacings), and we now have formulas for τ_i, and κ_i in terms of the unknowns σ_i plus the known y and h values; we only need to find the $n + 1$ values $\sigma_0, \sigma_1, ..., \sigma_n$ in order to completely determine the spline. It's a trivial matter to compare Eq. (4.1) and Eq. (4.4) in order to find the needed expressions for the α_i and γ_i (recall that we already know that $\delta_i = y_i$ and $\beta_i = \sigma_i/2$). In fact,

(4.6)
$$
\begin{aligned}
\alpha_i &= \frac{\sigma_{i+1} - \sigma_i}{6h_i} \\
\gamma_i &= \frac{y_{i+1} - y_i}{h_i} - \frac{h_i}{6}(\sigma_{i+1} + 2\sigma_i)
\end{aligned}
$$

as you'll be asked to verify in the Problems. We now have expressions for all four of the coefficients $\alpha_i, \beta_i, \gamma_i, \delta_i$ in terms of the σ_i.

All we need now is a way to determine $\sigma_0, \sigma_1, ..., \sigma_n$. We have ensured that the second derivative of $S(x)$ exists by piecewise linear interpolation (Eq. (4.2)) and have been using the interpolation conditions that S must satisfy; we haven't really made use of $S'(x)$. The requirement that $S'(x)$ be continuous means that at $x_1, ..., x_{n-1}$ we must have

$$p'_{i-1}(x_i) = p'_i(x_i)$$

(convince yourself that the subscripts are correct); that is,

$$3\alpha_{i-1}(x_i - x_{i-1})^2 + 2\beta_{i-1}(x_i - x_{i-1}) + \gamma_{i-1} = 3\alpha_i(x_i - x_i)^2 + 2\beta_i(x_i - x_i) + \gamma_i$$

or

$$3\alpha_{i-1}h_i^2 + 2\beta_{i-1}h_i + \gamma_{i-1} = \gamma_i$$

for the $n - 1$ interior nodes $i = 1, ..., n - 1$. Substituting from Eq. (4.6) and using and $\beta_i = \sigma_i/2$ gives

(4.7)
$$3\frac{\sigma_i - \sigma_{i-1}}{6h_{i-1}}h_i^2 + 2\sigma_{i-1}h_i/2 = \frac{y_{i+1} - y_i}{h_i} - \frac{h_i}{6}(\sigma_{i+1} + 2\sigma_i) - \left[\frac{y_i - y_{i-1}}{h_{i-1}} - \frac{h_{i-1}}{6}(\sigma_i + 2\sigma_{i-1})\right]$$

or

$$3\frac{\sigma_i - \sigma_{i-1}}{h_{i-1}}h_i^2 + 6\sigma_{i-1}h_i = 6\frac{y_{i+1} - y_i}{h_i} - h_i(\sigma_{i+1} + 2\sigma_i) -$$
$$\left[6\frac{y_i - y_{i-1}}{h_{i-1}} - h_{i-1}(\sigma_i + 2\sigma_{i-1})\right]$$

or, after considerable simplification,

$$(4.8) \qquad \varpi_i\sigma_{i-1} + 2\sigma_i + (1 - \varpi_i)\sigma_{i+1} = r_i$$

where

$$(4.9) \qquad \varpi_i = \frac{h_{i-1}}{h_{i-1} + h_i}$$
$$r_i = \frac{6}{h_i + h_{i-1}}\left(\frac{y_{i+1} - y_i}{h_i} - \frac{y_i - y_{i-1}}{h_{i-1}}\right)$$

(again, you'll be asked to fill in the details in the Problems). Note the similarity of the equation determining r_i to a difference equation for the second derivative (it looks like a difference of two approximations of a first derivative). Our linear system now consists of the $n - 1$ equations represented by Eq. (4.8) for $i = 1, ..., n - 1$, plus the 2 boundary conditions for the spline, for a total of $n + 1$ equations in $n + 1$ unknowns. For example, note that the natural boundary conditions (second derivatives constrained to be zero at the endpoints of the spline) correspond to the two additional conditions $\sigma_0 = 0$, $\sigma_n = 0$.

We've gone through a lot of detailed algebra to get here, and we're not done yet. Let's take stock. We started with a $4n \times 4n$ linear system (from the end of the previous section). We've come to an $(n+1) \times (n+1)$ linear system (plus the small cost of using Eq. (4.6) to find α_i and γ_i plus the comparatively negligible cost of forming δ_i and β_i). This reduction of a $4n \times 4n$ system to an $(n+1) \times (n+1)$ system represents a considerable benefit with respect to the amount of work that we must perform to find the spline interpolant, at the cost of our having had to wade through a modest amount of algebra–just once, during the derivation of the algorithm. This is a really good thing.

EXAMPLE 1: Consider the data $\{(-1, -1), (0, 1), (1, 1)\}$. Let's interpolate a natural spline to it using the method we've just developed. There are 3 nodes ($n = 2$), so we need to find the 3 values σ_0, σ_1, and σ_2. The natural boundary conditions give $\sigma_0 = 0$, $\sigma_2 = 0$, and Eq. (4.8) gives

$$\varpi_1\sigma_0 + 2\sigma_1 + (1 - \varpi_1)\sigma_2 = r_1$$
$$\frac{h_0}{h_0 + h_1}\sigma_0 + 2\sigma_1 + \left(1 - \frac{h_0}{h_0 + h_1}\right)\sigma_2 = \frac{6}{h_0 + h_1}\left(\frac{y_2 - y_1}{h_1} - \frac{y_1 - y_0}{h_0}\right)$$
$$\frac{1}{2}\sigma_0 + 2\sigma_1 + \left(1 - \frac{1}{2}\right)\sigma_2 = \frac{6}{2}\left(\frac{1-1}{1} - \frac{1--1}{1}\right)$$
$$\frac{1}{2}\sigma_0 + 2\sigma_1 + \frac{1}{2}\sigma_2 = -6$$

(using Eq. (4.9) and the data). Hence we must solve the system

$$
\begin{pmatrix} 1 & 0 & 0 \\ 1/2 & 2 & 1/2 \\ 0 & 0 & 1 \end{pmatrix} \begin{pmatrix} \sigma_0 \\ \sigma_1 \\ \sigma_2 \end{pmatrix} = \begin{pmatrix} 0 \\ -6 \\ 0 \end{pmatrix}
$$

for σ_0, σ_1, and σ_2. (We're not using the fact that σ_0 and σ_2 are already determined.) This is our $(n+1)\mathrm{x}(n+1)$ system. The solution is easily found to be $\sigma_0 = 0$, $\sigma_1 = -3$, and $\sigma_2 = 0$.

To actually construct the spline, we need to find the coefficients. Since there are two elements we are looking for two cubics. We have

$$
\begin{aligned}
\delta_0 &= y_0 \\
&= -1
\end{aligned}
$$

$$
\begin{aligned}
\delta_1 &= y_1 \\
&= 1
\end{aligned}
$$

$$
\begin{aligned}
\beta_0 &= \sigma_0/2 \\
&= 0
\end{aligned}
$$

$$
\begin{aligned}
\beta_1 &= \sigma_1/2 \\
&= -3/2
\end{aligned}
$$

and from Eq. (4.6),

$$
\begin{aligned}
\alpha_0 &= \frac{\sigma_1 - \sigma_0}{6h_0} \\
&= (-3 - 0)/(6 \cdot 1) \\
&= -1/2
\end{aligned}
$$

$$
\begin{aligned}
\alpha_1 &= \frac{\sigma_2 - \sigma_1}{6h_1} \\
&= (0 - -3)/(6 \cdot 1) \\
&= 1/2
\end{aligned}
$$

$$
\begin{aligned}
\gamma_0 &= \frac{y_1 - y_0}{h_0} - \frac{h_0}{6}(\sigma_1 + 2\sigma_0) \\
&= (1 - -1)/1 - (1/6)(-3 + 2 \cdot 0) \\
&= 5/2
\end{aligned}
$$

$$
\begin{aligned}
\gamma_1 &= \frac{y_2 - y_1}{h_1} - \frac{h_1}{6}(\sigma_2 + 2\sigma_1) \\
&= (1 - 1)/1 - (1/6)(0 + 2 \cdot -3) \\
&= 1
\end{aligned}
$$

so that

$$p_0(x) = \alpha_0(x - x_0)^3 + \beta_0(x - x_0)^2 + \gamma_0(x - x_0) + \delta_0$$
$$= -\frac{1}{2}(x + 1)^3 + 0(x + 1)^2 + \frac{5}{2}(x + 1) - 1$$
$$= -\frac{1}{2}(x + 1)^3 + \frac{5}{2}(x + 1) - 1$$

is the cubic on the first element $[-1, 0]$, and

$$p_1(x) = \alpha_1(x - x_1)^3 + \beta_1(x - x_1)^2 + \gamma_1(x - x_1) + \delta_1$$
$$= \frac{1}{2}(x - 0)^3 - \frac{3}{2}(x - 0)^2 + 1(x - 0) + 1$$
$$= \frac{1}{2}x^3 - \frac{3}{2}x^2 + x + 1$$

is the cubic on the second element $[0, 1]$. Verify that $p_0(-1) = -1$, $p_0(0) = 1$, $p_1(0) = 1$, and $p_1(1) = 1$ (interpolation). Note also that at the node $x_1 = 0$ we have $p_0'(0) = 1$, $p_1'(0) = 1$ (these match as expected) and $p_0''(0) = -3$, $p_1''(0) = -3$ (these match and furthermore are equal to σ_1, as expected). $\square$

We still must put Eq. (4.8) in matrix form and discuss the other boundary conditions. The natural spline is easy; the linear system has the form

(4.10)
$$\begin{bmatrix} 1 & 0 & 0 & 0 & 0 & 0 \\ \varpi_1 & 2 & 1-\varpi_1 & 0 & 0 & 0 \\ 0 & \varpi_2 & 2 & 1-\varpi_2 & 0 & 0 \\ 0 & 0 & \varpi_3 & 2 & 1-\varpi_3 & 0 \\ 0 & 0 & 0 & \varpi_4 & 2 & 1-\varpi_4 \\ 0 & 0 & 0 & 0 & 0 & 1 \end{bmatrix} \begin{pmatrix} \sigma_0 \\ \sigma_1 \\ \vdots \\ \sigma_{n-1} \\ \sigma_n \end{pmatrix} = \begin{pmatrix} 0 \\ r_1 \\ \vdots \\ r_{n-1} \\ 0 \end{pmatrix}$$

where the coefficient matrix T is tridiagonal, that is, it has nonzero entries only on the main diagonal, the first superdiagonal, and the first subdiagonal. (The first and last equations are trivial and we would rewrite this as an $(n-1) \times (n-1)$ system in $\sigma_1, ..., \sigma_{n-1}$ in practice, but we'll leave it in the form of Eq. (4.10).) In addition, note from Eq. (4.9) that $0 < \varpi_i < 1$ and hence $0 < 1 - \varpi_i < 1$ as well. This means that the matrix T has the property that the diagonal entry of each row is greater than the sum $\varpi_i + (1 - \varpi_i) = 1$ of the off-diagonal entries in that row. Since

$$|a_{ii}| > \sum_{\substack{j=1 \\ j \neq i}}^{n} |a_{ij}|$$

for each row i the matrix is strictly diagonally dominant matrix (see Sec. 3.1). Strictly diagonally dominant matrices are nonsingular, and (naive) Gaussian elimination may be performed on them (pivoting is not necessary, neither due to the possibility of a zero pivot nor for controlling the growth of errors). The matrix T has a number of other desirable properties as well, for example, it has the positive definiteness property $x^T T x > 0$ if $x \neq 0$ (though it is not necessarily symmetric and hence is not, in general, a positive definite matrix).

So the linear system in Eq. (4.10) has a unique solution (the natural spline exists and is unique) and we may find it accurately using a direct method. The fact that T is tridiagonal means we have available a very efficient way of solving the system (the $O(n)$ Thomas algorithm).

The term *spline* is a drafting term that originally referred to a thin, pliable strip of wood that was passed through pins stuck through paper to allow a smooth curve to be drawn through the points marked by the pins (by tracing along the spline). Clamps or weights (called *ducks*) were used to adjust the angle at which the spline passed through certain points, if desired. If no clamps were used the spline would be free of the pins and take on a straight shape to the left of the leftmost pin and to the right of the rightmost pin, where it was no longer subject to the forces exerted by the pins; the free (a.k.a. natural) boundary conditions $S''(x_0) = 0$, $S''(x_n) = 0$ simulate this case (the second derivative being zero corresponds to linearity). The clamped spline, where a derivative is specified at the endpoints, corresponds to clamping the spline at the leftmost and rightmost pins so as to force it to enter and leave at a prescribed angle. Clearly, this will be more accurate in general, since it uses additional information about the desired shape of the spline. Of course, this information is not always available; on the other hand, the natural boundary conditions impose two conditions ($S''(x_0) = 0$, $S''(x_n) = 0$) that are almost certainly incorrect. For this reason the not-a-knot boundary conditions are usually preferred when the data needed for the clamped spline is not available (or cannot be accurately estimated).

For the clamped boundary conditions, all that changes in Eq. (4.10) are the first and last rows of T and the vector on the right-hand side. The conditions $S'(x_0) = s_0$, $S'(x_n) = s_n$ may be applied to Eq. (4.3); at $x = x_0$, for example, we have

$$S'(x_0) = -\sigma_0 \frac{(x_0 - x_1)^2}{2h_0} + \sigma_1 \frac{(x_0 - x_0)^2}{2h_0} + \tau_0$$

$$s_0 = -\sigma_0 h_0/2 + \left(y_1 - \sigma_1 h_0^2/6 - \kappa_0\right)/h_0$$

(using Eq. (4.5) to find τ_0). Substituting in the expression for κ_0 gives

$$
\begin{aligned}
s_0 &= -\sigma_0 h_0/2 + \left(y_1 - \sigma_1 h_0^2/6 - \kappa_0\right)/h_0 \\
&= -\sigma_0 h_0/2 + \left(y_1 - \sigma_1 h_0^2/6 - (y_0 - \sigma_0 h_0^2/6)\right)/h_0 \\
&= -\sigma_0 h_0/2 + (y_1 - y_0)/h_0 + (\sigma_0 - \sigma_1)h_0/6 \\
&= (y_1 - y_0)/h_0 - \sigma_0 h_0/3 - \sigma_1 h_0/6
\end{aligned}
$$

which is a single equation in the 2 unknowns σ_0 and σ_1. Writing it in a form similar to Eq. (4.8) gives

$$2\sigma_0 + \sigma_1 = 6(y_1 - y_0)/h_0^2 - 6s_0/h_0$$

which will become the new first row of Eq. (4.10) for this boundary condition. In a similar way we find the equation

$$\sigma_{n-1} + 2\sigma_n = 6s_n/h_{n-1} - 6(y_n - y_{n-1})/h_{n-1}^2$$

from Eq. (4.3) evaluated at $x = x_n$. This gives a linear system $As = w$ where

$$(4.11) \quad A = \begin{bmatrix} 2 & 1 & 0 & 0 & 0 & 0 \\ \varpi_1 & 2 & 1-\varpi_1 & 0 & 0 & 0 \\ 0 & \varpi_2 & 2 & 1-\varpi_2 & 0 & 0 \\ 0 & 0 & \varpi_3 & 2 & 1-\varpi_3 & 0 \\ 0 & 0 & 0 & \varpi_4 & 2 & 1-\varpi_4 \\ 0 & 0 & 0 & 0 & 1 & 2 \end{bmatrix}$$

$$b = \begin{pmatrix} \sigma_0 \\ \sigma_1 \\ \vdots \\ \sigma_{n-1} \\ \sigma_n \end{pmatrix}$$

$$w = \begin{pmatrix} 6(y_1 - y_0)/h_0^2 - 6s_0/h_0 \\ r_1 \\ \vdots \\ r_{n-1} \\ 6s_n/h_{n-1} - 6(y_n - y_{n-1})/h_{n-1}^2 \end{pmatrix}$$

for the complete spline. Once again, the coefficient matrix is tridiagonal and diagonally dominant, and though non-symmetric does possess the positive definiteness property, giving us many opportunities to increase the efficiency and accuracy of our solution.

We could also put the linear system for the not-a-knot spline in matrix form, but this has been enough detailed derivation for now. It will likely pay to re-read this section. The big point is: A small amount of time spent setting up a problem and formulating it in a proper manner can lead to big gains in efficiency. In this case, not only have we reduced the system from a $4n \times 4n$ system that would require $O(N^3)$ flops for solution ($N = 4n$) to an $(n+1) \times (n+1)$ structured system with many special properties that may be handled in $O(n)$ flops, but recall that Eq. (4.1) will also be more accurate when we actually need to evaluate $S(x)$ at some x that is not a node.

We can define basis functions for splines as we did with Lagrange interpolating polynomials in Sec. 4.1 and piecewise Lagrange interpolating polynomials in Sec. 4.2. In fact, the hat basis of Sec. 4.2 is a set of basis functions for linear splines.

We close with two theorems. The notation is as used elsewhere in this section.
THEOREM 1: If $f \in C^4[x_0, x_n]$ and there is an $M > 0$ such that $\left| f^{(iv)}(x) \right| \leq M$ for $x \in [x_0, x_n]$ then

$$\max_{x \in [x_0, x_n]} \{|f(x) - S(x)|\} \leq \frac{5M}{384} \max_{i=0}^{n-1} \{|h_i^4|\}$$

for the clamped cubic spline interpolant $S(x)$ to $f(x)$.

Note that the error in the clamped cubic spline decreases as $O(h^4)$ where h is the maximum distance between nodes. In particular, the spline converges to the function as $h \to 0$ if the function has a bounded fourth derivative (justifying refinement of the grid as a means of obtaining an improved approximant). There is an $O(h^3)$ error formula for $\max\{f'(x) - S'(x)\}$, and so on, justifying the use of a splines for numerical differentiation. Note also that there is no assumption of equal spacing, and as usual equal spacing is not likely to be the best choice. Adaptive selection of knots is useful but we do not always have the freedom to choose these values.

THEOREM 2 (Minimum Curvature Property of Natural Splines): If $S(x)$ is the free cubic spline interpolant to $f(x)$ at the nodes $x_0, ..., x_n$ and if $g \in C^2[x_0, x_n]$ is any other function that interpolates $f(x)$ at the same nodes $x_0, ..., x_n$ then

$$\int_{x_0}^{x_n} [S''(x)]^2 \, dx < \int_{x_0}^{x_n} [g''(x)]^2 \, dx.$$

Note that g need not be a spline (or even a polynomial) and that no assumption is made about $f(x)$ other than that it is defined at the nodes. Theorem 2 states that, amongst all sufficiently smooth interpolants to a given function, the natural spline has minimum total curvature (as measured by the second derivative). It can be shown that this means that it has approximately minimum strain energy of all such curves through those nodes. In this sense the mathematical spline mimics the physical spline, which assumes the least strain shape. Although this is an interesting property, it does not alter the fact that the clamped and not-a-knot boundary conditions are generally preferable (or perhaps other boundary conditions if we know something special about the function f, e.g., that it is periodic).

MATLAB

The matrices in Eq. (4.10) and Eq. (4.11) are sparse; they have $(n+1) + 2n = 3n + 1$ nonzero entries out of a total of n^2 entries, so roughly $3/n$ of the entries are nonzero. It's wasteful to form an $n \times n$ matrix to hold roughly $3n$ entries–not only in terms of memory usage but also in terms of the wasted time forming the matrix initially and then retrieving rows from memory as needed. If we only stored the $3n + 1$ needed entries that might all fit in fast memory.

For this reason the tridiagonal matrices in Eq. (4.10) and Eq. (4.11) would never be formed at all. We could write a program that uses Eq. (4.8) directly and never works explicitly with a matrix, or at least store it in sparse form. In either case we obtain speed and storage benefits. The MATLAB backslash command recognizes certain special matrices and uses a more efficient method for them (for example, matrices that are tridiagonal). Nonetheless, if we wanted to handle large tridiagonal matrices we should write our own code.

The sparse command can be used to form a matrix by row and column coordinates. For example, enter:

```
» rows=[1 1 2 2 2 3 3 3 4 4 4 5 5 5 6 6];
» cols=[1 2 1 2 3 2 3 4 3 4 5 4 5 6 5 6];
» T=sparse(rows,cols,.5*ones([16 1]))
» full(T)    %Standard form.
» T=T+diag(1.5*ones([6 1]))
» T(1,2)=1;T(6,5)=1;
» full(T)
```

to create the matrix T of Eq. (4.11) if $\varpi_i = 1/2$. Note that MATLAB knows to retain the sparse form when adding matrices, if possible.

Let's consider how we might present a table of interpolated values using MAT-LAB. If we had the data $\{(0,0), (.1, .0998), (.2, .1987)\}$ from the sine function and wished to interpolate values at $x_i = .01i$ $(i = 0, 1, ..., 20)$ then we could do the following; enter:

```
» x=(0:.01:.2)'
» y=spline([0 .1 .2],sin([0 .1 .2]),x)   %List y values.
» disp('        x              y'),disp([x y])  %9 spaces before x,y.
```

The display of the y-values is adequate. What if we want 5 places after the decimal point, though? Enter:

```
» s_table=sprintf('%5.5f %5.5f\n',[x y])
» disp(s_table)
```

The `sprintf` command formats a matrix according to the conventions of the C++ programming language. We won't detail these. For another example, enter:

```
» format long
» z=exp(10)
» disp(sprintf('The value of z is %5.3f',z))    %Floating pt. format.
» disp(sprintf('The value of z is %5.2f',z))
» disp(sprintf('The value of z is %5.1f',z))
» disp(sprintf('The value of z is %5.0f',z))
» disp(sprintf('The value of z is %20.8f',z))
» disp(sprintf('The value of z is %5.3e',z))    %Exponential format.
» disp(sprintf('The value of z is %5.2e',z))
» disp(sprintf('The value of z is %5.0e',z))
» disp(sprintf('z=%5.3f units and z^2=%5.3f units.',z,z^2))
» disp(sprintf('z=%5.3f units\nz^2=%5.3f units.',z,z^2))
```

The \n formats a newline. A specification such as %5.3f represents a format of 5 total spaces for the number, of which 3 are allotted to the values after the decimal point (hence the extra spaces when %20.8f was used); the total number of spaces used may be more if the format does not allow for the non-fractional part of the number. Use of the `sprintf` command (or the `fprintf` command if writing to a file) allows careful control over the format of your output; this is useful for creating table displays. Note that the `sprintf` command actually returns a character variable (see `help char` and `help string`, as MATLAB does draw a distinction), so we use `disp` to display it in order to suppress the MATLAB "ans =" formatting of displayed values.

There are a great many generalizations of the notion of splines. Multivariate versions of these generalizations are used in computer graphics, for example. There is an extensive optional Spline Toolbox for MATLAB.

Problems

1.) Fill in the details in going from Eq. (4.2) to Eq. (4.4).

2.) Derive Eq. (4.6) by comparing Eq. (4.4) and Eq. (4.1).

3.) Derive Eq. (4.8) and Eq. (4.9) from Eq. (4.7).

4.) a.) Fit a natural cubic spline to the data $\{(0,1), (1,2), (2,1), (3,2)\}$ using Use Eq. (4.10). Form and solve the 4×4 matrix by hand. Write the three cubics (Eq. (4.1)) explicitly.

b.) Use your spline to interpolate values at $x = .5, 1.5, 2.5$.

5.) a.) Use Eq. (4.10), Eq. (4.11), and MATLAB to fit a natural cubic spline and a clamped cubic spline to the function $y = e^{-x} \cos(2x)$ at the nodes $\{0, .1, .2, .3, .4, .5\}$. Write the five cubics (Eq. (4.1)) explicitly. Plot the function and the splines on a single graph. Comment.

b.) Repeat part a. with the nodes $\{0, .05, .18, .32, .45, .5\}$.

c.) Use Theorem 1 to bound the error in the clamped splines in parts a. and b.

6.) a.) Use Eq. (4.10) to fit a natural cubic spline to the function $y = \sin(x^2)$ using the nodes $0, .1, ..., 1$. Give the coefficients and plot your result.

b.) Use Eq. (4.11) to fit a clamped cubic spline to the function $y = \sin(x^2)$ using the nodes $0, .1, ..., 1$. Give the coefficients and plot your result.

7.) a.) Use Eq. (4.10) to fit a natural cubic spline to the function $y = x^9 - 5x^7 + x^4 - 2x^2 + 1$ using the nodes $0, .1, ..., 1$. Give the coefficients and plot your result.

b.) Use Eq. (4.11) to fit a clamped cubic spline to the function $y = x^9 - 5x^7 + x^4 - 2x^2 + 1$ using the nodes $0, .1, ..., 1$. Give the coefficients and plot your result.

8.) If a function is periodic on $[x_0, x_n]$ then we may use the cyclic boundary conditions $p_0'(x_0) = p_{n-1}'(x_n)$, $p_0''(x_0) = p_{n-1}''(x_n)$. Give the matrix form for the cyclic spline equations.

9.) a.) Write a MATLAB program that approximates the definite integral of a given function between given endpoints using a cubic spline based on n equally spaced nodes, where n is an optional input argument. Demonstrate your program and discuss its accuracy.

b.) Write a MATLAB program that approximates the derivative of a given function using a cubic spline based on n equally spaced nodes, where n is an optional input argument; a second optional argument should be used to request a plot of the function and its derivative. Your output should be the approximate derivatives at the interior nodes. Demonstrate your program and discuss its accuracy.

10.) Write a MATLAB program that inputs a function and a vector of nodes and displays a 6-digit table of the function values at the nodes and at the midpoints of the subintervals between the nodes. Return the y-values from the table as your output.

11.) Give the matrix form for the not-a-knot spline equations.

12.) Derive the systems corresponding to Eq. (4.10) and Eq. (4.11) for quintic splines.

13.) Is there a more efficient algorithm than the Thomas algorithm for solving a *positive definite* tridiagonal system?

14.) Show that a symmetric diagonally dominant matrix that has positive diagonal entries is necessarily positive definite. (Hint: Use Gershgorin's Theorem.) What happens if the matrix is not symmetric?

15.) a.) Verify Theorem 1 using MATLAB experiments.

b.) Verify Theorem 2 using MATLAB experiments.

16.) Prove Theorem 2.

5. Rational Function Interpolation

Splines are widely used in engineering and in mathematics. They are among the most important tools in approximation theory, and are often used in the finite element method for solving partial differential equations. They're important in computer graphics. Their utility goes well beyond making a smooth plot out of discrete data. However, we now turn to other techniques.

We came to splines as we sought to improve upon the weaknesses of polynomial interpolation. We noted that we had to either give up "polynomial" or give up "interpolation" to get a better approximant. Techniques that give up interpolation are can be used to get good approximations. Splines gave up polynomial in favor of piecewise polynomial. Another approach is to turn to **rational functions**, which are defined as functions that are ratios of polynomials

$$(5.1) \qquad \frac{P_n(x)}{Q_d(x)} = \frac{p_n x^n + p_{n-1} x^{n-1} + \ldots + p_1 x + p_0}{q_d x^d + q_{d-1} x^{d-1} + \ldots + q_1 x + q_0}$$

where $Q_d(x)$ is not the zero polynomial. Hence the two polynomials have no roots in common. Note that every polynomial is a rational function, as we can take $Q_d(x) = 1$.

The **degree** of a rational function is defined to be $n + d$ and its **degree type** is the pair (n, d). Interpolating a rational function to a set of data is known as **rational function interpolation**. Typically n and d are equal or nearly so, and are chosen in advance by the user. If $n \leq d$ we say that the rational function is **proper**.

An advantage of rational function interpolation is that a function $R(x) = P_n(x)/Q_d(x)$ may have a pole, that is, a point x_0 where $Q_d(x_0) = 0$, leading to a singularity in the function at that point (possibly cancellable by corresponding zeros of $P_n(x)$). This allows us to model a broader class of functions. Additionally, if we attempt to approximate a real function $f(x)$ by a polynomial interpolant, poles of $f(x)$ off the real line (in the complex plane) can lead to a poor approximation on the real line, as they do for power series; rational functions avoid these issues and can give good approximations under such circumstances. In fact, rational function interpolation and approximation are commonly employed for theoretical work in complex analysis.

Rational function approximation is also a common method for the approximate evaluation of special functions (like $\cos(x)$, $\Gamma(x)$, etc.) on the computer, which after all can only add, subtract, multiply, and divide at the base level–that is, it computes polynomial quantities. The methods for computing named functions like these are built into a chip or compiler. Rational function interpolants are used in some extrapolation techniques for improving the speed of convergence (similar to Aitken's Δ^2 method) and hence are a useful theoretical tool in numerical analysis.

We assume that the coefficients $p_0, \ldots, p_n, q_0, \ldots, q_d$ are to be found. If only the values of the interpolant, and not its coefficients, are needed, then an approach similar to Neville's method can be used. This is often the case and it is not only wasteful to compute the coefficients when they are not needed but can also be less accurate. For our purposes we will need to be able to find the coefficients so that they can be used in deriving numerical methods, but don't compute them if you don't need them!

How shall we interpolate a function of the form of Eq. (5.1) to a set of data $\{(x_i, y_i)\}_{i=0}^{N}$? First, note that we have $n + d + 2$ unknowns in our model and so in order to interpolate, not just approximate, we must have this many conditions on the interpolant. However, we can not hope to get a truly unique interpolant in general, for if

$$(5.2) \qquad \frac{P_n(x)}{Q_d(x)} = \frac{\Pi_\nu(x)}{K_\delta(x)}$$

then it too interpolates. As a simple example, anything interpolated by

$$\frac{x^2 - 1}{2x + 3}$$

is also interpolated by

(5.3)
$$\frac{2x^2 - 2}{4x + 6} = \frac{2\left(x^2 - 1\right)}{2\left(2x + 3\right)}$$

or by

$$\frac{x^3 + 2x^2 - x - 2}{2x^2 + 7x + 6} = \frac{\left(x^2 - 1\right)\left(x + 2\right)}{\left(2x + 3\right)\left(x + 2\right)}$$

(assuming we evaluate this as its limit at $x = 2$ rather than interpreting it as the undefined quantity 0/0). For this reason we reconsider Eq. (5.2) and define two rational functions to be **equivalent** if

$$P_n(x)K_\delta(x) = \Pi_\nu(x)Q_d(x)$$

that is, if they represent the same rational function. With this definition, if a rational function interpolant exists for a certain data set then it is unique, up to equivalence. When the numerator and the denominator polynomials have no roots in common, so that there are no common factors $(x - z_0)$ that could be divided out, we say that they are **relatively prime**. When $P_n(x)$ and $Q_d(x)$ are relatively prime we say that the rational function $P_n(x)/Q_d(x)$ is relatively prime or in **reduced form**.

Unfortunately, we see from Eq. (5.3) that even if two rational functions are in reduced form they can still differ by a multiplicative constant. For that reason we need a normalization criterion to achieve uniqueness of the interpolant. In some cases we require that both the numerator and denominator polynomials be monic, but for interpolation we typically require that $q_0 = 1$ instead.

How many conditions do we need to impose on Eq. (5.1)? There were $n + d + 2$ unknown coefficients, but insisting that $q_0 = 1$ reduces it to $n + d + 1$ constraints. The number of data points $N + 1$ must equal $n + d + 1$.

EXAMPLE 1: Let us look for a rational function interpolant of degree type at most $(1, 1)$, $R(x) = P_1(x)/Q_1(x)$, for the data $\{(0, 1), (1, 1), (2, 3)\}$. Our interpolant has the form

(5.4)
$$R(x) = \frac{p_1 x + p_0}{q_1 x + 1}$$

and interpolation means that we must have

$$1 = \frac{p_0}{1}$$
$$1 = \frac{p_1 + p_0}{q_1 + 1}$$
$$3 = \frac{2p_1 + p_0}{2q_1 + 1}$$

and so

$$p_0 = \frac{p_1 + p_0}{q_1 + 1}$$

(from the first two equations), that is,

$$
\begin{aligned}
p_0 (q_1 + 1) &= p_1 + p_0 \\
p_0 q_1 + p_0 &= p_1 + p_0 \\
p_0 q_1 &= p_1.
\end{aligned}
$$

Hence $p_0 = p_1/q_1$. Using this in Eq. (5.4) gives

$$
\begin{aligned}
R(x) &= \frac{p_1 x + p_1/q_1}{q_1 x + 1} \\
&= \frac{p_1}{q_1} \frac{q_1 x + 1}{q_1 x + 1} \\
&= \frac{p_1}{q_1}
\end{aligned}
$$

(the last two rational functions are equivalent), so that $R(x)$ is a constant. This constant must be 1 because we have interpolated at $x = 0$ and $x = 1$ where $f(x) = 1$, but then $R(2) = 1$ and the function doesn't interpolate at $x = 2$. The assumption that we could divide by q_1, that is, that $q_1 \neq 0$, must have been incorrect. Taking $q_1 = 0$ gives

$$
\begin{aligned}
R(x) &= \frac{p_1 x + p_0}{1} \\
&= p_1 x + p_0
\end{aligned}
$$

which is a linear function. Since it interpolates $(0, 1)$ and $(1, 1)$ it must once again be the identity function $R(x) = x$, but once again it does interpolate at $x = 2$. There is no rational function interpolant for this data set. $\square$

Hence while a given data set has at most one rational function interpolant (up to equivalence), it might not have even that, despite the number of unknowns matching the number of constraints in our system. This is unlike the case for polynomial interpolation.

If the interpolant exists, it is the solution of an underconstrained $N + 1$ by $n + d + 2$ linear system of the form

(5.5)
$$
\begin{aligned}
Q_d(x_0)y_0 - P_n(x_0) &= 0 \\
Q_d(x_1)y_1 - P_n(x_1) &= 0 \\
&\vdots \\
Q_d(x_N)y_0 - P_n(x_N) &= 0
\end{aligned}
$$

(an $(N + 1) \times (N + 2)$ homogeneous linear system, before we impose the additional requirement that $q_0 = 1$). Such a system always has a nontrivial solution. All non-trivial solutions of this system can be shown to define equivalent rational functions, but the solution of Eq. (5.5) is *not* always an interpolant as it doesn't always go through every data point. We say that the points that are not interpolated by this rational function are **unattainable** by a rational function of the specified degree type, and that the points at which it does interpolate are **attainable**. (The terms **inaccessible** and **accessible** are also used.) When there are unattainable points, the degree $n + d$ of the interpolant $P_n(x)/Q_d(x)$ will be less than that predicted by the relation $N + 1 = n + d + 1$ (i.e., $n + d < N$ in such a case).

EXAMPLE 2: Let us look for a rational function interpolant $P_1(x)/Q_1(x)$ for the data $\{(0,1),(1,2),(2,1)\}$. Our interpolant has the form

$$R(x) = \frac{p_1 x + p_0}{q_1 x + q_0}$$

and Eq. (5.5) yields

$$(q_1 x_0 + 1)\, y_0 - (p_1 x_0 + p_0) = 0$$
$$(q_1 x_1 + 1)\, y_1 - (p_1 x_1 + p_0) = 0$$
$$(q_1 x_2 + 1)\, y_2 - (p_1 x_2 + p_0) = 0$$

i.e.,

$$q_0 - p_0 = 0$$
$$2\,(q_1 + q_0) - (p_1 + p_0) = 0$$
$$(2q_1 + q_0) - (2p_1 + p_0) = 0$$

or

$$\begin{pmatrix} -1 & 0 & 1 & 0 \\ -1 & -1 & 2 & 2 \\ -1 & -2 & 1 & 2 \end{pmatrix} \begin{pmatrix} p_0 \\ p_1 \\ q_0 \\ q_1 \end{pmatrix} = \begin{pmatrix} 0 \\ 0 \\ 0 \end{pmatrix}$$

the solution of which is $q_1 = -q_0$, $p_1 = -q_0$, and $p_0 = q_0$, upon taking q_0 as the free parameter. Setting $q_0 = 1$ gives

$$R(x) = \frac{-x + 1}{-x + 1}$$
$$= 1$$

so once again there is an unattainable point (here, $x_1 = 1$) that is missed by the interpolant. $\square$

In Example 1 the degree of the rational function was 1 rather than the 2 that would have been expected for a data set with no unattainable points; in Example 2 the degree was 0. We might have expected the degree to be just one less than the expected value because only one point wasn't interpolated. However, it can be shown that if there are unattainable points, then the attainable points can be interpolated by a rational function that has a lesser degree than would be expected based on the number of attainable points. Whenever a set of m points can be interpolated by a rational function with a degree lower than $m - 1$ we say that the data set is **degenerate** (or a **degenerate configuration**); hence, when there is an unattainable point or points, the attainable points necessarily form a degenerate configuration.

In Example 2 we found the solution as $R(x) = (-x+1)/(-x+1)$ and simplified it, taking its value at $x = 1$ to be defined by the obvious limit. The original form, as given by the solution of Eq. (5.5), was not relatively prime. In fact, it can be shown that there are no unattainable points exactly when Eq. (5.5) has a solution that is relatively prime. Equivalently, there are no unattainable points exactly when the denominator $Q_d(x)$ from Eq. (5.5) satisfies

(5.6) $$Q_d(x_i) \neq 0$$

($i = 0, 1, ..., n$), that is, when the denominator polynomial never has a root at a knot x_i. Compare these two characterizations of failure to interpolate to the cases considered in Example 1 and Example 2.

There is another way to solve for the coefficients rather than using Eq. (5.5). Note that if the denominator $Q_d(x)$ were known then we could use Eq. (5.6) to verify that an interpolant did indeed exist. If we found that the problem was solvable we could use Eq. (5.1) and the interpolation condition to write

$$P_n(x_i) = y_i Q_d(x_i)$$

which has the form

$$P_n(x_i) = v_i$$

($i = 0, 1, ..., n$) where the v_i are known values. But this is just the polynomial interpolation problem of Sec. 4.1 with the data $\{x_i, y_i Q_d(x_i)\}_{i=0}^{N}$ and may be solved by any convenient means, such as the Lagrange interpolating polynomials, giving

$$
\begin{aligned}
P_n(x) &= \varpi(x) \sum_{i=0}^{n} \frac{v_i}{\varpi'(x_i)(x - x_i)} \\
&= \varpi(x) \sum_{i=0}^{n} \frac{y_i Q_d(x_i)}{\varpi'(x_i)(x - x_i)}
\end{aligned}
$$

(5.7)

(this is a slightly different version of the formula for Lagrange interpolation given in Sec. 4.1, using the function $\varpi(x) = (x - x_0)(x - x_1) \cdots (x - x_n)$ from the proof of Theorem 4.1.1). Note that we are interpolating a polynomial of degree n to $N + 1$ data points where typically $n \approx N/2$ (because $n \approx d$, the numerator and denominator polynomials are of about the same order) and so when a solution of the rational function interpolation problem exists it must be the case that the transformed data $\{x_i, y_i Q_d(x_i)\}_{i=0}^{N}$ is interpolable by a polynomial of lower degree than would otherwise have been expected.

Hence we have reduced the rational function interpolation problem to that of finding the denominator polynomial $Q_d(x)$. How shall we find it? Although the derivation is not difficult it is somewhat tedious so we shall simply note that, starting from a formula like Eq. (5.7), we can show that the coefficients $q_0, ..., q_d$ of $Q_d(x)$ satisfy

(5.8) $$\sum_{k=0}^{d} c_{mk} q_k = 0$$

($m = 0, 1, ..., d - 1$), where

$$c_{mk} = \sum_{i=0}^{N} \frac{x_i^{m+k} y_i}{\varpi'(x_i)}$$

$(m = 0, 1, ..., d - 1,\ k = 0, 1, ..., d)$. The coefficients c_{mk} are determined by the data $\{x_i, y_i\}_{i=0}^n$. This gives a linear system (Eq. (5.8))

$$
\begin{aligned}
c_{0,0}q_0 + \cdots + c_{0,d}q_d &= 0 \\
c_{1,0}q_0 + \cdots + c_{1,d}q_d &= 0 \\
&\ \ \vdots \\
c_{d-1,0}q_0 + \cdots + c_{d-1,d}q_d &= 0
\end{aligned}
$$

i.e., $Cq = 0$ where $C = (c_{mk})_{m,k=0,0}^{m,k=d-1,d}$ and $q = (q_0, ..., q_d)^T$. Because this is a $(d-1) \times d$ homogeneous linear system, it has a nontrivial solution, which we may normalize by taking $q_0 = 1$.

This method is **Jacobi's method for rational interpolation**. In summary, we form the c_{mk} from the data; solve the linear system $Cq = 0$; verify that $Q_d(x_i) \neq 0$ for all $i = 0, 1, ..., N$; then solve the polynomial interpolation problem with the data $\{x_i, y_i Q_d(x_i)\}_{i=0}^N$ for $P_n(x)$, expecting that a polynomial of degree n will suffice despite the fact that typically $n < N$.

Round-off errors may force us to use least squares to fit $P_n(x)$ to the data in order to insure that we get a $P_n(x)$ of the correct degree. But since Eq. (5.5) is an $(N+1) \times (N+2)$ linear system, and Eq. (5.8) is a $(d-1) \times d$ linear system where typically $d \approx N/2$, Jacobi's method for rational interpolation has advantages when N is large.

It's worth repeating: If you only need values of the interpolant in an application, this is the wrong approach. We want a method that computes the coefficients because it can be used to derive other numerical methods, much as linear interpolation was used to develop the method of false position in Sec. 1.1. But it's a poor way to proceed if what we have is experimental or tabulated data from which we wish to interpolate missing values, say. A commonly used algorithm when just the values are needed is the **Bulirsch-Stoer algorithm**, which creates a tableau of values using a recurrence relation based on differences of lower-degree rational function interpolants. Other approaches use continued fractions.

MATLAB

We have focused on square systems throughout this text and to a lesser extent on overdetermined systems (in the context of least squares problems). The two linear systems mentioned in this section, Eq. (5.5) and Eq. (5.8), are undetermined systems. Since they are homogeneous they must have infinitely many solutions.

One way to get a solution for this type of system is to use the `rref` command to find the reduced row echelon form of the matrix. Enter:
```
>> help rref
>> A=rand([3 4])
>> rref(A)
```
We may use this form to solve for the first 3 variables in terms of the 4th. Enter:
```
>> [U,R]=rref(A)
>> length(R)
```
The length of R is the rank of the matrix. The calculation of the rank is of course subject to the effects of representation error (the matrix we want is not stored exactly in the machine) and round-off error.

For larger matrices we do not want to have to interpret the reduced row echelon form ourselves. Enter:

```
>> b=[0 0 0]';A\b
```

This gives *a* solution, but not *all* solutions. Unfortunately, we want any nonzero solution and it gives only the zero solution. (See `help slash`. Another method for undetermined systems is the pseudo-inverse; see `help pinv`.) The most efficient approach may well be to write a program that uses Gaussian elimination to find any nontrivial solution using backward substitution. (Recall that the reduced row echelon form uses the less efficient Gauss-Jordan elimination.) However, using the tools from MATLAB and results from linear algebra, we can do it more easily (but less computationally efficiently). Enter:

```
>> null(A)
```

The columns of the result of the `null` command give a basis for the null space of the matrix. But the solution set of $Ax = 0$ is precisely null(A), so any element of the null set is a solution of the equation. In particular, any basis vector is a nontrivial solution. (If A is square and nonsingular, null(A) contains only the zero vector and hence has no basis.) Hence to solve Eq. (5.8), say, we can find null(C), which must have at least one basis vector (that is, dim null$(C) \geq 1$) and we may use that basis vector as our nontrivial solution q to $Cq = 0$.

We emphasize that using the `null` command, which relies on the SVD, is efficient only in the sense that it is supplied by MATLAB and so minimizes your programming time; it's very inefficient with respect to flops used. For small problems there is a version based on the result of `rref`; enter:

```
>> B=[1 0 1 0;1 1 2 -1;2 1 3 -1]
>> null(B)
>> null(B,'r')
>> type null
```

This only works for small matrices with small integer entries and hence is of very limited utility. Enter:

```
>> null(A,'r')
```

to see what happens when the matrix is small but does not have integer entries. Again, the `null` command is convenient but inefficient and if we wish to solve these systems regularly we should write a program using Gaussian elimination or some other method to handle such systems.

Problems

1.) Use Eq. (5.5) to find a rational function interpolant of the form $P_1(x)/Q_1(x)$ to $\sin(x)$ at $x = 0, \pi/2, \pi$. Are there any unattainable points?

2.) Attempt to use Jacobi's method for rational interpolation on the problem in Example 2. Does it succeed?

3.) a.) Use Eq. (5.5) to interpolate a rational function of the form $P_2(x)/Q_2(x)$ to the data $\{(-2, 2), (-1, 1), (0, 0), (1, 1), (2, 2)\}$.

b.) Repeat part a. using Jacobi's method for rational interpolation.

c.) Attempt to interpolate a rational function of the form $P_3(x)/Q_1(x)$ to the same data. Are any points unattainable? If so, find the interpolant for the attainable points.

4.) Show that if a rational function interpolant exists (i.e., there are no unattainable points) then it is unique, up to equivalence.

5.) Show that Eq. (5.7) is a valid representation of the interpolating polynomial through the points $\{x_i, y_i Q_d(x_i)\}_{i=0}^n$.

6.) a.) Use a rational function interpolant of degree 6 and degree type $(3,3)$ to approximate the tangent function over $[0, \pi]$. Attempt to avoid having unattainable points.

b.) Repeat with a polynomial interpolant of the same degree. Comment.

c.) Repeat with a rational function interpolant of degree 6 and degree type $(4,2)$.

d.) Repeat with a rational function interpolant of degree 6 and degree type $(2,4)$.

e.) Repeat with a rational function interpolant of degree 6 and degree type $(5,1)$.

f.) Repeat with a rational function interpolant of degree 6 and degree type $(1,5)$.

7.) Construct, by trial-and-error, a rational function interpolant to $\sin(x)$ on $[0, \pi/2]$ that has error at most 10^{-6} at x=0:.01:pi/2 (in MATLAB notation). Try to find the simplest rational function you can for this purpose.

8.) a.) Develop a root-finding method, similar to inverse linear interpolation, that uses a rational function interpolant of degree type $(1,1)$. Test your method against other one-dimensional root-finding methods.

b.) Why is a root-finding method based on a rational function interpolant a natural choice for a function such as that in Fig. 1 from Sec. 1.4? Test your method on such a function and compare its performance to that of other methods.

9.) Perform a numerical experiment to test whether rational function interpolation performs poorly on equally spaced nodes, as polynomial interpolation does.

10.) Write a MATLAB program for efficiently solving an underdetermined system. You may assume that the system is $n \times (n+1)$ and that any one nontrivial solution is desired, and you may use the lu command or any other MATLAB command.

11.) a.) Write the Runge function as a function of a complex variable z. Where are its poles?

b.) Perform a numerical experiment to compare polynomial interpolation near and far from complex poles. Pick several functions that are well-behaved on the real line but that have poles near it. Interpolate the real function both near to and far from the complex poles. Does the polynomial interpolant appear to feel the effect of the poles that are lying off the real line?

12.) Perform a numerical experiment to compare polynomial interpolation and rational function interpolation. Discuss their strengths and weaknesses.

13.) Write a MATLAB program for performing rational function interpolation using Eq. (5.5). The user should specify the data and the desired n and d.

14.) Write a MATLAB program for performing rational function interpolation using Jacobi's method for rational interpolation. The user should specify the data and the desired n and d.

15.) Prove that if there are unattainable points then the attainable points form a degenerate configuration.

16.) a.) When the coefficients and the variable are complex-valued, a rational function of degree type $(1,1)$ that is nonconstant is called a **Möbius transformation**. (The domain is the extended complex plane $\mathbb{C} \cup \{\infty\}$.) We usually write the transformation as $f(z) = (az + b) / (cz + d)$. What condition on the coefficients is needed to insure that it is not in fact a constant function, that is, that it is not reducible to degree type $(0,0)$?

b.) Find the fixed points of a Möbius transformation in terms of its coefficients. You may need to consider separate cases.

c.) Find the zeroes (also called the holes in this context) and the poles of a Möbius transformation in terms of its coefficients. You may need to consider separate cases.
d.) Find the inverse transformation.
e.) Show that knowing the value of $f(z)$ at three distinct points in $\mathbb{C} \cup \{\infty\}$ determines the transformation uniquely.

6. The Best Approximation Problem

Let's discuss the approximation problem in a more general sense than just interpolation. What is the best way to approximate a function f by a simpler function g? There is no one answer. First we must specify what we mean by a "simple" function–say, P_2, the set of all polynomials of degree at most two. But even then we must address the question: "Best" in what sense? Best might mean easiest to manipulate, easiest to calculate, closest to the function, or some compromise between these goals, depending on the application.

Even something as seemingly simple as asking that the approximant g give the closest approximation to f of all functions in the set of simple functions from which g is drawn doesn't define best unambiguously. By closest do we mean minimizing the L_∞ norm of the difference $f - g$

$$(6.1) \qquad\qquad \max|f(x) - g(x)|$$

over the interval of interest $[a, b]$, or minimizing the L_1 norm of the difference

$$(6.2) \qquad\qquad \int_a^b |f(x) - g(x)|\, dx$$

or minimizing the L_2 norm of the difference

$$(6.3) \qquad\qquad \left(\int_a^b (f(x) - g(x))^2 dx \right)^{1/2}$$

or perhaps something else? (These Hölder norms on function spaces are defined in analogy with the corresponding vector norms.) In many areas of scientific computing we are asked by users for the best way to do something, and can spend a lot of time getting at what the user really means by *best*.

For example, consider the functions $f_1(x) = x^{-3}$ and $f_2(x) = 0$ on $[.1, 1]$. With respect to the L_∞ norm of Eq. (6.1), the functions are far apart, for

$$\max|f_1(x) - f_2(x)| = 1000$$

over $[.1, 1]$; but with respect to the L_1 norm of Eq. (6.2), they are much closer,

$$\int_a^b |f_1(x) - f_2(x)|\, dx \doteq 49.5$$

(the area between the curves is only about 50 units). If we hope to use f_2 to approximate the *values* of f_1 then we must expect errors on the order of 10^3, but if we hope to use f_2 to approximate *integrals* of f_1 then we should expect errors on the order of 10^1; f_2 is a much better approximant in the L_1 sense than in the

L_∞ sense. If we were interested in RMS quantities, such as the difference between two waveforms, then we would consider L_2, which is in fact the most commonly employed norm.

We call the quantity $\max |f(x) - g(x)|$ the L_∞ distance between the functions f and g, and $\int_a^b |f(x) - g(x)|\, dx$ and $\left(\int_a^b (f(x) - g(x))^2 dx\right)^{1/2}$ the L_1 and L_2 distances, respectively, between f and g. (Compare the similar terminology for distances between vectors in Sec. 2.6.) Of course, especially in the L_1 and L_2 cases, these do not correspond to distances as we usually think of them, but the concept is still useful. The distance between f and g is taken to be a norm of the function $h = f - g$.

Here's the key point: When we speak of our desire to find the best approximant in a certain situation, we must be prepared to define *best* by specifying: a.) the class of simple functions to be used as candidate approximants, and b.) the notion of distance between functions that is to be minimized[6]. We then minimize the distance between f and g which is, in every case of interest to us, given by

$$d(f, g) = \|f - g\|$$

for some norm. This is called the **best approximation problem**. For example, asking that we find the polynomial that best approximates $\sin(x)$ is insufficiently specific for us to proceed, but asking for the element of P_3 that minimizes the L_2 distance between it and $\sin(x)$ on the interval $[0, 1]$ means solving the problem

$$\min_{g \in P_3} \left(\int_0^1 (\sin(x) - g(x))^2 dx \right)^{1/2}$$

that is, finding the values of a, b, c, and d that make

$$\int_0^1 (\sin(x) - (ax^3 + bx^2 + cx + d))^2 dx$$

as small as possible. This particular problem could be solved either analytically by partially differentiating w.r.t. a, b, c, and d, or numerically by a minimization routine.

The best approximation problem with the L_∞ norm is sometimes called **minimax approximation** because it seeks to minimize the maximum distance between the two functions, that is, it finds the g that minimizes

$$\max |f(x) - g(x)|$$

over some interval, with g from some defined class of functions. Using the L_1 norm means minimizing the area between the curves (that is, the area of the function $h(x) = |f(x) - g(x)|$). Using the L_2 norm is called **least squares approximation** because we are trying to find the g that minimizes Eq. (6.3)

$$\left(\int_a^b (f(x) - g(x))^2 dx \right)^{1/2}$$

which is equivalent to minimizing

---

[6]We want the class of simple functions to be a subspace of a normed linear space that also includes f, with the distance induced by that norm.

$$\int_a^b (f(x) - g(x))^2 dx$$

which is, in essence, a sum (integral) of squares, similar to linear least squares. (The L_2 norm is also called the least squares or Euclidean norm. There is a close connection to Fourier series here.) Least squares and minimax approximation are widely employed.

Another norm that is used in some circumstances is the **Sobolev r-norm**, which is defined (for sufficiently differentiable functions) by

$$\|f\|_{S,r} = \left(\int_a^b \left(f^2(x) + [f'(x)]^2 + [f''(x)]^2 + ...[f^{(r)}(x)]^2 \right) dx \right)^{1/2}$$

for some integer $r \geq 0$. Note that two functions may be close in the L_∞ norm but far in the Sobolev r-norm if the functions remain close to one another but one wiggles, creating a large first derivative, while the other does not wiggle. In Fig. 1, the functions $f(x) = x$ and $g(x) = \sin(\varpi x)/10$ are plotted for $\varpi = 20$. They are close in the L_p norms but not in the Sobolev 1-norm, because

$$
\begin{aligned}
f'(x) &= 1 \\
g'(x) &= \frac{\varpi}{10} \cos(20x) \\
&= 2 \cos(20x)
\end{aligned}
$$

and as ϖ increases the L_p norms would remain small but the Sobolev 1-norm could be made arbitrarily large. A good approximation of f need not be a good approximation of f'. If it is important not only that the function values be close but also that the overall shape of the approximant be similar to that of the function being approximated then it may be useful to consider using a Sobolev norm.

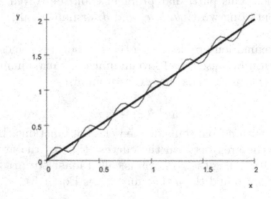

FIGURE 1. Differing Sobolev norms.

Note carefully that we are *approximating*, not *interpolating*. An approximant need never intersect the function it approximates (or, in a discrete case like linear least squares, the data set it approximates); it need only be near the function, in the sense of some distance function.

The best approximation problem for a normed linear space $f \in V$ with approximants to be chosen from a subspace $W \subset V$ can be viewed as the problem of finding an approximation operator $P : V \to W$ such that if $f \in V$ then $Pf \in W$ is its best approximation (which we assume is unique). When working on the computer we may need to use an approximate approximation operator $\tilde{P}$. We will not consider the problem at that level of abstraction however.

There is a theorem that states that, for reasonable choices of the class of "simple" functions and the norm defining the distance, a solution to the best approximation problem always exists. This means that there is a specific function that achieves the minimum. In the case of the L_2 norm it is also unique if the class of approximants is a linear space; this need not be the case for the L_1 and L_∞ norms[7] (though it frequently is true in cases of interest).

Let's look in more detail at minimax approximation–the best approximation problem with respect to the L_∞ norm. This is also called **best uniform approximation** when the space is (a subspace of) the space $C(I)$ of all continuous functions on some closed interval I, because the L_∞ norm

$$\|f\|_\infty = \max_{x \in I} |f(x)|$$

is also called the **uniform norm**. One reason that best uniform approximation is of interest is that a good L_∞ approximant is always a good L_1 and L_2 approximant, though the converse is not true in general. For instance, we have a special case of the equivalence of norms for $C(I)$:

THEOREM 1: Let f, g in $C(I)$ be given where I is a closed interval of length λ. Then

$$\|f - g\|_1 \le \lambda^{1/2} \|f - g\|_2 \le \lambda \|f - g\|_\infty .$$

PROOF: Let $h = f - g$ and $I = [a, b]$. Then the Euclidean norm of h can be estimated by

---

[7]In fact, if we use any Hölder p-norm with $1 < p < \infty$ then we will have a unique solution. Note that p need not be an integer.

$$
\begin{aligned}
\|h\|_2 &= \left(\int_a^b h^2(\xi) d\xi \right)^{1/2} \\
&\leq \left(\int_a^b \max_{x \in I} \{ h^2(\xi) \} d\xi \right)^{1/2} \\
&= \left(\int_a^b \left(\max_{x \in I} \{ |h(\xi)| \} \right)^2 d\xi \right)^{1/2} \\
&= \left(\int_a^b (\|h\|_\infty)^2 d\xi \right)^{1/2} \\
&= \left(\|h\|_\infty^2 \int_a^b d\xi \right)^{1/2} \\
&= \|h\|_\infty (b-a)^{1/2} \\
&= \lambda^{1/2} \|h\|_\infty
\end{aligned}
$$

so that $\lambda^{1/2} \|h\|_2 \leq L \|h\|_\infty$, which is the right-hand inequality. Let's look at the left-hand inequality. By definition,

$$
\begin{aligned}
\|h\|_1 &= \int_a^b |h(\xi)| \, d\xi \\
&= \int_a^b |h(\xi)| \cdot 1 d\xi \\
&\leq \left(\int_a^b h^2(\xi) d\xi \right)^{1/2} \left(\int_a^b 1^2 d\xi \right)^{1/2}
\end{aligned}
$$

where we have used the Cauchy-Schwarz Theorem. But this is just

$$
\begin{aligned}
\|h\|_1 &\leq \|h\|_2 \|1\|_2 \\
&= (b-a)^{1/2} \|h\|_2 \\
&= \lambda^{1/2} \|h\|_2
\end{aligned}
$$

giving the desired result, $\|h\|_1 \leq \lambda^{1/2} \|h\|_2 \leq \lambda \|h\|_\infty$. $\square$

Hence if the error $e_\infty = \|f - g\|_\infty$ in an approximation of f by g is small with respect to the uniform norm on a certain interval, then

$$e_1 \leq \lambda e_\infty$$
$$e_2 \leq \lambda^{1/2} e_\infty$$

so that the errors with respect to the L_1 and L_2 norms are at most a constant multiple of e_∞. As e_∞ is decreased, so are (the bounds on) e_1 and e_2, and hence as $e_\infty \to 0$ so do e_1 and e_2. Perhaps surprisingly, though, it is not always the case that as $e_1 \to 0$ or $e_2 \to 0$, so does e_∞. This is one motivation for studying minimax approximation.

The best uniform approximation problem is also called the problem of **Chebyshev approximation** when the class of approximants is the space P_n of all polynomials of degree at most n. Typically when we perform best uniform approximation the approximants are either polynomials or rational functions. Splines and trigonometric polynomials are also used. Rational approximants generally give better results than polynomials and are widely used for the implementation of special functions in software but we will limit ourselves to polynomial approximants for now.

Let's consider then the problem of best approximation with respect to the uniform norm, using polynomials of degree at most n as our approximants[8]. Why should we look at this problem? We will want to find approximations to special functions such as the error function so that we can evaluate these approximations on a computer. The use of the uniform norm insures that if we set an error tolerance– say, that evaluations of f must be accurate to within some given tolerance τ–then we can guarantee that the tolerance will be met by requiring that our approximation g satisfy

$$\max |f - g| \leq \tau$$

over the interval of interest. (This neglects the effects of round-off error.) We know from Theorem 1 that if we have a good uniform approximation then we'll have a good L_1 and L_2 approximant as well.

Another use of best uniform approximation is to find a smooth approximation g to a discontinuous or nondifferentiable function f when we'd like to apply an algorithm to f that requires continuity or differentiability. If the approximation is good we can use it in the algorithm in place of f. In this case we may also be concerned about how well g' approximates f'.

Best uniform approximations have a number of interesting properties. The following characterization theorem is important in practice:

THEOREM 2 (MINIMAX CHARACTERIZATION THEOREM): If f is in $C[a, b]$ and p is a polynomial of degree n then p is the unique best uniform approximation to f from polynomials of degree at most n if and only if there are distinct points $\xi_0, \xi_1, ..., \xi_{n+1}$ in $[a, b]$ such that the maximum error is achieved at each ξ_i, $i = 0, 1, ..., n + 1$ and the error alternates in sign at the ξ_i.

---

[8]The Weierstrass Approximation Theorem states that we can approximate a continuous function on $[a, b]$ as closely as we like (in the uniform norm) by a polynomial, but the degree of the polynomial may be arbitrarily high.

That is, if p is a Chebyshev approximation to f of degree n then there are $n+2$ points[9] ξ_i at which the worst-case error is actually achieved,

$$(6.4) \qquad \|f - p\|_\infty = |f(\xi_i) - p(\xi_i)|$$

and furthermore the errors

$$e_i = f(\xi_i) - p(\xi_i), e_{i+1} = f(\xi_{i+1}) - p(\xi_{i+1})$$

are equal in magnitude but alternate in sign, that is,

$$(6.5) \qquad f(\xi_i) - p(\xi_i) = -\big(f(\xi_{i+1}) - p(\xi_{i+1})\big).$$

Furthermore, any polynomial of degree n that has $n+2$ points satisfying Eq. (6.4) and Eq. (6.5) is necessarily a best uniform approximation to f. (We refer to Eq. (6.5) as the **alternating sign condition**.) The theorem also asserts that the Chebyshev approximation is unique.

This leads to a class of algorithms, called **exchange algorithms**, that approximate the best uniform approximant by choosing an initial set of points $\{\xi_i\}_{i=0}^{n+1}$, called a **reference**, and then finding the element $p \in P_n$ that minimizes

$$(6.6) \qquad \max_i \{f(\xi_i) - p(\xi_i)\}$$

by some means. If the error is not sufficiently small, we exchange one or more points of the current reference with maxima of

$$e(x) = |f(x) - p(x)|$$

(considered over all of $[a, b]$, not just on the reference) to get a new reference. We then repeat the process. The best known of these methods is the **Remez exchange algorithm**[10]. The Remez algorithm is actually meant for best uniform approximation by rational functions, where the presence of degenerate points can complicate matters.

How well do the Chebyshev approximations approximate the function for large n? Must they converge to the function that is being approximated in the limit? Again, we have a theorem:

THEOREM 3 (JACKSON'S THEOREM): If f is in $C^k[a, b]$ and p_n is the Chebyshev approximation of f of degree n then

$$\|f - p_n\|_\infty \leq \frac{c_k \left\|f^{(k)}\right\|_\infty}{n^k}$$

for some constant c_k.

Importantly, c_k does not depend on the function f nor on the degree n of the approximant, and so we have the result that $\|f - p_n\|_\infty$ is $O(n^{-k})$ as $n \to \infty$. In particular, the Chebyshev approximations converge to f as the degree n increases.

---

[9]Of course, if we attempt to approximate a linear function f by a polynomial of degree at most 10 then we'll get a degree 1 approximant with infinitely many points of intersection and hence zero error. We are focusing on the case where the polynomial approximant has maximal degree.

[10]*Remez* is sometimes seen as *Remes*.

MATLAB

The MATLAB `norm` command finds the norms of vectors, not functions, so we must use the `int` or `quad` command to compute the L_p norms of functions. For example, enter:

```
>> integral(@(x) abs(sin(x)-(x+x.^3)),0,pi)
```

to find the L_1 distance between $\sin(x)$ and $x + x^3$ over $[0, \pi]$.

Methods for evaluating the standard special functions fall in several major categories, each of which has subcategories: Series expansions, approximation using rational functions, and special case formulae–things that just happen to work for that one function–are the most common. Numerical integration, numerical solution of an ODE, and stochastic simulation are all methods of last resort.

In some cases we perform an argument reduction first. For example, to evaluate $\sin(x)$ we relate it to a value of the sine function in $[0, \pi/8]$, say, and use a routine for evaluating sine in that interval (which might be based on a best uniform approximation). In many cases we use a transformation to simplify the problem or a relationship to another special function we have already approximated well.

The gamma function $\Gamma(x)$ is implemented in MATLAB by the `gamma` command. Enter:

```
» doc gamma
```

and scroll down to the section labeled Algorithms. The `gamma` command uses one of several best uniform approximants depending on the argument ranges; the approximants are rational functions. It sometimes happens that a special routine is needed for a function related to a special function. Enter:

```
» help gammaln
```

This function computes $\ln(\Gamma(x))$ in a way that is often more accurate than using $\ln(\cdot)$ composed with $\Gamma(\cdot)$. Enter:

```
» exp(gammaln(50))-gamma(50)    %Should be equal.
» ans/gamma(50)    %Relative error, if gamma(50) is accurate.
```

The absolute error is very large, though the relative error is small. Typical algorithms for $\ln(\Gamma(x))$ do not utilize a routine for $\Gamma(x)$.

The error function and its variants are usually computed by minimax or nearly minimax rational approximations; the inverse error function (`erfinv` in MATLAB) can be found by a rational approximation, improved by a single step of a high-order root-finding algorithm. These functions are of great importance in statistics[11]. Enter:

```
» doc erf
```

and scroll down to the section labeled Tips. This suggests using `erfc` for the complementary error function $\mathrm{erfc}(x) = 1 - \mathrm{erf}(x)$ rather than computing it by the definition in order to maintain accuracy. This is a general rule: It's usually worth directly approximating an often-used special function rather than losing accuracy by basing it off a related one. (When a related function is used within a provided routine, the error has been controlled for in advance.)

Other cases that often require a special routine in order to get the desired accuracy are the functions $\exp(x) - 1$ and $\ln(x + 1)$ for small x. Use of $\exp(\cdot)$ and $\ln(\cdot)$ may give poor results; e.g., consider the severe cancellation of significant

---

[11]One of the most important early applications of computers was the creation of large, accurate tables of special functions for applications in statistics and engineering.

figures in computing $\exp(x) - 1$ for x very near zero. A rational approximation or series approximation may give better results.

Problems

1.) Find the L_1, L_2, and L_∞ distances between $\sin(x)$ and $\cos(x)$ over $[0, \pi]$.

2.) Find the L_1, L_2, and L_∞ distances between $\sin(x)$ and its Taylor series of orders 1, 3, and 5 over $[0, \pi]$.

3.) Find the L_1, L_2, and L_∞ norms of $\exp(x)$ over $[0, 1]$.

4.) Prove directly from the definition (that is, without using Theorem 2) that a Chebyshev approximation of a function in $C[a, b]$ from P_1 must have at least three points at which the error is maximal.

5.) Find the element of P_3 that minimizes the L_2 distance between it and $\sin(x)$ on the interval $[0, 1]$ (that is, find the best approximation from P_3 for $\sin(x)$ on the interval $[0, 1]$ using the standard norm).

6.) Find the element of P_3 that minimizes the L_2 distance between it and $\sin(x)$ on the interval $[0, \pi]$ (that is, find the best approximation from P_3 for $\sin(x)$ on the interval $[0, \pi]$ using the standard norm). Compare your answer for this problem with your answer for Problem 5.

7.) Find the best L_2 approximation for $\exp(x)$ on $[-1, 1]$ using linear combinations of 1, $\sin(x)$, and $\cos(x)$. d.) Compare your answer to the Fourier series expansion of $\exp(x)$ on $[-1, 1]$.

8.) Construct an example of two functions such that their difference over $[0, 1]$ has L_∞ norm less than a given $\epsilon > 0$ but Sobolev 1-norm greater than a given $\delta > 0$.

9.) Compute the Sobolev 1-norm and the Sobolev 2-norm of the sine and cosine functions over $[0, \pi]$.

10.) a.) Find the best approximation for $\exp(x)$ on $[0, 1]$ from P_5 with respect to the L_2 norm. What is the distance of $\exp(x)$ from the approximant? What is the distance of $\exp(x)$ from its Taylor series of order 5?

b.) Repeat using the L_∞ norm.

c.) Repeat using the L_1 norm.

11.) Show that the L_2 norm is a norm on the linear space $C[a, b]$ of all functions continuous on the finite interval $[a, b]$; that is, show that it satisfies the properties of positive definiteness, homogeneity ($\|\alpha f\| = |\alpha|\|f\|$ for all scalars α), and the triangle inequality (see Sec. 2.6).

12.) Show that the L_1 norm is a norm on the linear space $C[a, b]$ of all functions continuous on the finite interval $[a, b]$.

13.) Show that the L_∞ norm is a norm on the linear space $C[a, b]$ of all functions continuous on the finite interval $[a, b]$.

14.) Show that the Sobolev r-norm is a norm on the linear space $C^r[a, b]$ of all functions r times continuously differentiable on the finite interval $[a, b]$.

15.) Write a MATLAB program that uses `integral` to approximate the distance between two functions with respect to the L_1 or L_2 norms. Input the functions, the interval, and an indicator variable showing which norm is desired.

16.) a.) If $p \geq 1$ and $q \geq 1$ then we say that p and q are **Hölder conjugates** if $1/p + 1/q = 1$. Prove Young's inequality for products: If p and q are Hölder conjugates then $ab \leq a^p p + b^q / q$ for all nonnegative a and b, with equality only when $a^p = b^q$.

b.) Suppose p and q are Hölder conjugates. Prove the Hölder inequality for vectors, $\sum_{k=1}^{n} |x_k y_k| \leq \left(\sum_{k=1}^{n} |x_k|^p\right)^{1/p} \left(\sum_{k=1}^{n} |y_k|^q\right)^{1/q}$ (i.e., $\|x \cdot y\|_1 \leq \|x\|_p \|y\|_q$). When is equality obtained?

c.) Suppose p and q are Hölder conjugates. Prove Hölder's inequality for functions, $\|fg\|_1 \leq \|f\|_p \|g\|_q$). When is equality obtained?

CHAPTER 4 BIBLIOGRAPHY:

A Practical Guide to Splines, Carl de Boor, Springer-Verlag New York 1978
A classic text on splines.
Multivariate Splines, Charles K. Chui, SIAM 1988
A graduate-level monograph on splines in several dimensions.
Interpolation and Approximation, Philip J. Davis, Dover Publications 1975
A classic text on interpolation and approximation in a general setting.
Spline Functions: Computational Methods, Larry L. Schumaker
A computational approach.

CHAPTER 5

Numerical Integration

1. Closed Newton-Cotes Formulas

Many definite integrals of interest cannot be evaluated analytically. Perhaps the best known example is the area under the bell-shaped curve (standard normal distribution), for which we define for $x > 0$ the error function

$$(1.1) \qquad \text{erf}(x) = \frac{2}{\sqrt{\pi}} \int_0^x e^{-t^2/2} dt$$

and then the probability that a standard normal random variable takes a value in $[-x, x]$ is

$$\frac{1}{\sqrt{2}} \text{erf}(x).$$

The integral defining the error function appears very frequently in probability and statistics and is extensively tabulated. It is tabulated, or approximated numerically in software, because it can be shown that there is no way to express the antiderivative of $\exp\left(-z^2\right)$ in terms of elementary functions. Hence the need for methods to perform approximate integration is unavoidable. Unless otherwise mentioned, we assume throughout this chapter that the integrand f is continuous on the finite interval $[a, b]$ (where $a < b$).

There are a great many analytical techniques for approximating integrals. We might imagine approximating the definite integral

$$I = \int_a^b f(x) dx$$

by expanding $f(x)$ in a Taylor or other series then integrating that series term-by-term to find a series for I. Most of the methods we will consider, however, will be numerical methods that have the form of a **weighted average rule**

$$(1.2) \qquad \int_a^b f(x) dx \approx w_0 f(x_0) + w_1 f(x_1) + \ldots + w_n f(x_n)$$

for some weights $w_0, w_1, \ldots, w_n$ and some nodes $x_0, x_1, \ldots, x_n$ (which need not lie in $[a, b]$). The selection of weights and nodes defines the method. Techniques for performing numerical integration are also called **numerical quadrature** methods.

For example, in the calculus you likely encountered several numerical integration techniques, including the **midpoint rule**

DOI: 10.1201/9781003042273-5

$$\int_a^b f(x)dx \approx (b-a)f(\frac{a+b}{2})$$

($w_1 = b-a$, $x_1 = (a+b)/2$), which approximates f by a constant, namely, its value at the midpoint of the interval of integration, and the **trapezoidal rule**

$$(1.3) \qquad \int_a^b f(x)dx \approx \frac{b-a}{2}(f(a)+f(b))$$

($w_0 = w_1 = (b-a)/2$, $x_0 = a$, $x_1 = b$) which approximates f by linear interpolation over the interval. Geometrically, we can think of the midpoint rule as approximating the area under f by the area of a rectangle and the trapezoidal rule as approximating the area under f by the area of a trapezoid.

EXAMPLE 1: Consider Eq. (1.1). We do not necessarily need to perform numerical quadrature; there are ways to approximate the function $erf(x)$ analytically (say, by a truncated series). However, let's try using the midpoint and trapezoidal rules to approximate $erf(1)$. We have

$$
\begin{aligned}
erf(1) &= \frac{2}{\sqrt{\pi}} \int_0^1 e^{-z^2} dz \\
&\approx \frac{2}{\sqrt{\pi}} \exp(-(1/2)^2) \\
&\doteq .8788
\end{aligned}
$$

from the midpoint rule, since $x = 1/2$ is the midpoint of the interval, and

$$
\begin{aligned}
erf(1) &= \frac{2}{\sqrt{\pi}} \int_0^1 e^{-z^2} dz \\
&\approx \frac{2}{\sqrt{\pi}} \left(\frac{1-0}{2}\right) \left(\exp(-(0)^2 + \exp(-(1)^2)\right) \\
&\doteq .7717
\end{aligned}
$$

from the trapezoidal rule. The actual value (using the MATLAB® `erf` function) is .8427. The trapezoidal rule gives an underestimate in this case because the linear interpolant always lies below the function over the interval of integration. □

There are many approximate integration rules beyond these two and in fact there are a number of families of such rules. The midpoint and trapezoidal rules are members of a family of quadrature rules that approximate $f(x)$ by an interpolating polynomial, then integrate that polynomial. That is, these rules form a polynomial interpolant

$$p(x) = c_0 + c_1 x + ... + c_n x^n$$

and reason that since $p(x) \approx f(x)$ over the interval of integration,

$$\int_a^b f(x)dx \approx \int_a^b p(x)dx$$

$$= c_0 x + \frac{c_1}{2}x^2 + ... + \frac{c_n}{(n+1)!}x^{n+1}\Big|_a^b$$

$$= c_0(b-a) + \frac{c_1}{2}(b^2-a^2) + ... + \frac{c_n}{(n+1)!}\left(b^{n+1}-a^{n+1}\right)$$

where the constants $c_0, ..., c_n$ typically depend on the nodes $x_0, ..., x_n$, the endpoints a and b, and the values of f at certain points. Rules of this type–formed by polynomially interpolating f, then integrating that interpolant exactly–are called **interpolatory rules**[1], and if the nodes are equally spaced then they are called **Newton-Cotes rules**.

For example, the trapezoidal rule is the Newton-Cotes rule found by using an interpolant of degree 1 with the nodes $x_0 = a$, $x_1 = b$. Recall from Sec. 4.1 that we may write this interpolant as

$$p(x) = f(a)L_a(x) + f(b)L_b(x)$$

where

$$L_a(x) = \frac{x-b}{a-b}$$

$$L_b(x) = \frac{x-a}{b-a}$$

are the appropriate Lagrange interpolating polynomials. Then

$$\int_a^b p(x)dx = \int_a^b \left(f(a)L_a(x) + f(b)L_b(x)\right)dx$$

$$= f(a)\int_a^b \frac{x-b}{a-b}dx + f(b)\int_a^b \frac{x-a}{b-a}dx$$

$$= f(a)\left(\frac{b-a}{2}\right) + f(b)\left(\frac{b-a}{2}\right)$$

$$= \frac{b-a}{2}(f(a)+f(b))$$

in agreement with Eq. (1.3). Again, with respect to Eq. (1.2) we have weights $w_0 = w_1 = (b-a)/2$ and nodes $x_0 = a$, $x_1 = b$.

Let's explore the Newton-Cotes rules in rather more detail. Notice that there is no reason that we must have $x_0 = a$ and $x_n = b$, or even that all nodes must be contained in $[a, b]$ (though usually we do not assume that f is known outside of this integral). If it is the case that $x_0 = a$ and $x_n = b$ we say that the rule is a **closed Newton-Cotes rule** (based on $(n+1)$ points). The trapezoidal rule is closed but the midpoint rule is not; its only node in $[a, b]$ is $x_0 = (a+b)/2$. Note that the $n+1$ points of a closed Newton-Cotes rule divide $[a, b]$ into n subintervals.

We can easily derive the formula for the closed Newton-Cotes rule based on $(n+1)$ points using the methods of Sec. 4.1. The interpolant is

---

[1]We could of course interpolate with some function other than polynomials, but this is less common.

5. NUMERICAL INTEGRATION

$$p(x) = f_0 L_{n,0}(x) + f_1 L_{n,1}(x) + \cdots + f_{n+1} L_{n,n}(x)$$

where $f_i = f(x_i)$, $i = 0, ..., n$. Hence

$$
\begin{aligned}
\int_a^b p(x)dx &= \int_a^b \left(f_0 L_{n,1,0}(x) + f_1 L_{n,1}(x) + \cdots + f_{n+1} L_{n,n}(x) \right) dx \\
&= \sum_{i=0}^n f_i \left(\int_a^b L_{n,i}(x)dx \right) \\
&= \sum_{i=0}^n B_{n,i} f_i
\end{aligned}
$$

where the weights

$$
\begin{aligned}
B_{n,i} &= \int_a^b L_{n,i}(x)dx \\
&= \int_a^b \frac{(x - x_0)(x - x_1) \cdots (x - x_{i-1})(x - x_{i+1}) \cdots (x - x_n)}{(x_i - x_0)(x_i - x_1) \cdots (x_i - x_{i-1})(x_i - x_{i+1}) \cdots (x_i - x_n)} dx
\end{aligned}
$$

do not have a simple formula despite being merely the definite integrals of polynomials. Typically we compute the $B_{n,i}$ for $a = 0$, $b = 1$ and then use a change of variables to map that to a general $[a, b]$.

The trapezoidal rule on $[a, b]$, for example, has weights $B_{1,0} = B_{1,1} = (x_1 - x_0)/2$ (with $x_0 = a$, $x_1 = b$). We let

(1.4)
$$h = \frac{b - a}{n}$$

be the width of each of the n subintervals of $[a, b]$ so that the trapezoidal rule weights may be written as $B_{1,0} = B_{1,1} = h/2$.

There are tables of values of $\{B_{n,i}\}_{i=0}^{n+1}$ for various n. Some of the more frequently used closed Newton-Cotes rules are **Simpson's rule** for $n + 1 = 3$,

$$\int_a^b f(x)dx \approx \frac{h}{3} \left(f(x_0) + 4f(x_1) + f(x_2) \right)$$

Simpson's 3/8 rule for $n + 1 = 4$,

$$\int_a^b f(x)dx \approx \frac{3h}{8} \left(f(x_0) + 3f(x_1) + 3f(x_2) + f(x_3) \right)$$

and **Boole's rule** (or **Milne's rule**) for $n + 1 = 5$,

$$\int_a^b f(x)dx \approx \frac{2h}{45} \left(7f(x_0) + 32f(x_1) + 12f(x_2) + 32f(x_3) + 7f(x_4) \right)$$

where in each case the value of h is given by Eq. (1.4). Although all the coefficients of $f(x_i)$ are positive in these formulae, for sufficiently high n they are of mixed sign and increasingly large.

What is the approximation error involved in using one of these rules? Consider Theorem 1 of Sec. 4.1, which states that if f is suitably differentiable then the error in the interpolant is

$$f(x) - p(x) = \frac{f^{(n+1)}(\xi)}{(n+1)!}(x - x_0)(x - x_1) \cdots (x - x_n)$$

for any $x \in [a, b]$ and for some $\xi = \xi(x)$ in (a, b). Since this formula is valid for any x in the interval, we have

$$(1.5) \qquad E = \int_a^b f(x)dx - \int_a^b p(x)dx$$

$$= \int_a^b \frac{f^{(n+1)}(\xi(x))}{(n+1)!}(x - x_0)(x - x_1) \cdots (x - x_n)dx$$

for the approximation error. Since ξ is an unknown function of x this cannot be integrated exactly, but nonetheless it can be shown that for n odd

$$(1.6) \qquad E = \frac{h^{n+2}f^{(n+1)}(\gamma)}{(n+1)!} \int_0^n \varpi(x)dx$$

while for n even,

$$(1.7) \qquad E = \frac{h^{n+3}f^{(n+2)}(\gamma)}{(n+2)!} \int_0^n x\varpi(x)dx$$

where

$$(1.8) \qquad \varpi(x) = x(x - 1)(x - 2) \cdots (x - n)$$

and $\gamma \in (a, b)$. Note from the expressions for E that the error for $n = 2$ (that is, Simpson's rule) is $O(h^5)$, and that the error for $n = 3$ (Simpson's 3/8 rule) is the same, $O(h^5)$. In fact, all the closed Newton-Cotes rules based on an odd number of points (even number n of subintervals) have the same order of error as the rule that follows them (based on an even number of points and odd number n of subintervals).

But there is another interesting consequence of Eq. (1.7) and Eq. (1.6). Since we are approximating $f(x)$ by a polynomial interpolant of degree n and then integrating that polynomial *exactly*, if $f(x)$ should happen to be a polynomial of degree at most n then $f(x)$ and its interpolant $p(x)$ are precisely the same function, and so

$$\int_a^b f(x)dx = \int_a^b p(x)dx$$

apart from rounding error. This is reflected in Eq. (1.6) which shows that if n is even and

$$f^{(n+1)}(x) = 0$$

over $[a, b]$ then in particular $f^{(n+1)}(\gamma) = 0$ and so $E = 0$. A polynomial of degree n or less has $f^{(n+1)}(x)$ identically zero so it is integrated exactly, just as expected. But now consider the case of n even; Eq. (1.7) shows that $E = 0$ if

$$f^{(n+2)}(x) = 0$$

over $[a, b]$, so if n is even then the rule integrates polynomials of degree at most $n + 1$ exactly! This is quite a surprise–Simpson's rule, which fits a quadratic to three points, nonetheless integrates cubics exactly (up to round-off error).

EXAMPLE 2: Let $f(x) = 2x^2 - 3x + 2$ and consider $\int_1^2 f(x)dx$. Simpson's rule gives

$$
\begin{aligned}
\int_1^2 f(x)dx &\approx \frac{h}{3}(f(x_0) + 4f(x_1) + f(x_2)) \\
&= \frac{1/2}{3}(f(1) + 4f(1.5) + f(2)) \\
&= \frac{1}{6}(1 + 4*2 + 4) \\
&= \frac{13}{6}
\end{aligned}
$$

which is easily verified to be the correct answer. Let $g(x) = 3x^3 - 4x^2 + 2x + 1$ and consider $\int_1^2 g(x)dx$. Then Simpson's rule gives

$$
\begin{aligned}
\int_1^2 g(x)dx &\approx \frac{h}{3}(g(x_0) + 4g(x_1) + g(x_2)) \\
&= \frac{1/2}{3}(g(1) + 4g(1.5) + g(2)) \\
&= \frac{1}{6}(2 + 4*\frac{41}{8} + 13) \\
&= \frac{71}{12}
\end{aligned}
$$

which again is easily verified to be the correct answer. $\square$

We say that a numerical quadrature rule has **precision** or **degree of accuracy** n if it integrates the monomial x^i exactly for $i = 0, ..., n$ but does not integrate x^{n+1} exactly. If a rule has precision n then it integrates all polynomials of degree at most n exactly (up to round-off error). If n is odd, the closed Newton-Cotes rule based on $n + 1$ points has precision n and is of order $O(h^{n+2})$. But $n - 1$ is even in this case, and so the rule based on just $n - 1$ points has the same precision n and the same order $O(h^{n+2})$ but uses one fewer point, and has smaller coefficients (which will amplify errors less). Hence rules based on n even are usually preferred in applications, especially if f is costly to evaluate. In fact, it will be shown in the next section that for Simpson's rule

$$(1.9) \qquad\qquad E = -\frac{h^5}{90}f^{(4)}(\gamma_1)$$

and it can also be shown that for Simpson's 3/8 rule we have the exceedingly similar error term

$$E = -\frac{3h^5}{80} f^{(4)}(\gamma_2)$$

and so even the size of the asymptotic error constant $-1/90$ of Simpson's rule is smaller than that of the asymptotic error constant $-3/80$ of Simpson's 3/8 rule. Of course, the specific points γ_1 and γ_2 will almost certainly differ for any given function f and interval $[a, b]$, thus affecting the fourth derivative values between these two formulas.

What will we do if we need to integrate a function over a large interval of integration? We cannot expect Simpson's rule, based on three points, to do well over an interval like $[0, 10]$; the behavior of Eq. (1.9) won't become apparent until $h = (b - a)/2$ is less than 1. On the other hand, we know that interpolating by a high-degree polynomial is generally a bad idea; the high oscillations would introduce a great deal of error. In fact, closed Newton-Cotes rules need not converge to the value of the integral as $n \to \infty$ even if the integrand is analytic on $[a, b]$.

The obvious solution is to use the fact that we may break any integral down into a sum of integrals over smaller regions using the relation

$$\int_a^b f(x)dx = \int_a^c f(x)dx + \int_c^b f(x)dx$$

and apply the rule to each of the integrals over the smaller intervals. This is called the **composite** (or **compound**) form of the rule, as opposed to the **simple** form we have used until now. We divide the interval $[a, b]$ into nodes $x_0 = a < x_1 < ... < x_N = b$ as before but now we apply a simple rule over every several points. For example, the composite trapezoidal rule is

$$\begin{aligned}
\int_a^b f(x)dx &\approx \frac{x_1 - x_0}{2}(f(x_0) + f(x_1)) + \frac{x_2 - x_1}{2}(f(x_1) + f(x_2)) + ... \\
&\quad + \frac{x_N - x_{N-1}}{2}(f(x_{N-1}) + f(x_N)) \\
&= \frac{h}{2}(f(x_0) + f(x_1)) + \frac{h}{2}(f(x_1) + f(x_2)) + ... + \\
&\quad \frac{h}{2}(f(x_{N-1}) + f(x_N)) \\
&= \frac{h}{2}(f(x_0) + f(x_1) + f(x_1) + f(x_2) + ... + f(x_{N-1}) + f(x_N)) \\
&= \frac{h}{2}(f(x_0) + 2f(x_1) + 2f(x_2) + ... + 2f(x_{N-1}) + f(x_N)) \\
&= \frac{h}{2}\left(f(a) + 2\sum_{i=1}^{N-1} f(x_i) + f(b)\right)
\end{aligned}$$

$(h = (b - a)/N)$, and the composite Simpson's rule is

$$\int_a^b f(x)dx \quad \approx \quad \frac{(x_2 - x_0)/2}{3}(f(x_0) + 4f(x_1) + f(x_2)) + \dots$$

$$+\frac{(x_N - x_{N-2})/2}{3}(f(x_{N-2}) + 4f(x_{N-1}) + f(x_N))$$

$$= \quad \frac{h}{3}(f(x_0) + 4f(x_1) + f(x_2)) + \frac{h}{3}(f(x_2) + 4f(x_3) + f(x_4)) + \dots$$

$$+\frac{h}{3}(f(x_{N-2}) + 4f(x_{N-1}) + f(x_N))$$

$$= \quad \frac{h}{3}(f(x_0) + 4f(x_1) + f(x_2)) + f(x_2) + 4f(x_3) + f(x_4) + \dots$$

$$+f(x_{N-2}) + 4f(x_{N-1}) + f(x_N))$$

$$= \quad \frac{h}{3}(f(x_0) + 4f(x_1) + 2f(x_2) + 4f(x_3) + 2f(x_4) + \dots$$

$$+2f(x_{N-2}) + 4f(x_{N-1}) + f(x_N))$$

$$= \quad \frac{h}{3}\left(f(a) + 4\sum_{\substack{i=1 \\ i\ odd}}^{N-1} f(x_i) + 2\sum_{\substack{i=1 \\ i\ even}}^{N-1} f(x_i) + f(b) \right)$$

$$= \quad \frac{h}{3}\left(f(a) + 4\sum_{k=1}^{N/2} f(x_{2k-1}) + 2\sum_{k=1}^{N/2-1} f(x_{2k}) + f(b) \right)$$

if N is even. Note how both of these rules give less weight to the values of f at the endpoints a, b and more to the interior points, and that the composite Simpson's rule gives even less weight to the endpoints than does the composite trapezoidal rule. The endpoint information is less reliable–we do not know what the function is doing to the left of x_0, though we have some idea what it is doing to the left and right of x_1 because we have data there–so we weight these points less.

Clearly we will typically use composite rules, not simple rules, in practice. What is the error associated with a composite rule? Let's look at Simpson's rule; assume we have an even number of panels N, so that the simple Simpson's rule is being applied $N/2$ times. The error of the simple rule is

$$E = -\frac{h^5}{90}f^{(4)}(\gamma)$$

(Eq. (1.9)). In the case of the composite rule we have $h = (b-a)/N$ so

$$E_i = -\frac{(b-a)^5}{N^5 90}f^{(4)}(\gamma_i)$$

on the ith subinterval. We will be adding up $N/2$ terms (because Simpson's rule uses 2 panels at a time), yielding

$$E_1 + \dots + E_{N/2} = -\frac{(b-a)^5}{N^5 90}f^{(4)}(\gamma_1) + \dots + -\frac{(b-a)^5}{N^5 90}f^{(4)}(\gamma_N)$$

or, if $f^{(4)}(x)$ is roughly constant on $[a,b]$, say $f^{(4)}(x) \approx K$,

$$E_1 + \ldots E_{N/2} \approx -\frac{(b-a)^5}{N^5 90}K + \ldots + -\frac{(b-a)^5}{N^5 90}K$$

$$= -N\frac{(b-a)^5}{N^5 90}K$$

$$= -\frac{(b-a)^5}{N^4 90}K$$

$$= -\frac{(b-a)h^4}{180}K$$

so that we expect the order of the method to be reduced to $O(h^4)$. Indeed, this is what happens, and it can be shown that for the composite trapezoidal rule

(1.10) $$E = -\frac{(b-a)h^2}{12}f''(\gamma)$$

(compare Eq. (1.7) for $n = 1$), and for the composite Simpson's rule

(1.11) $$E = -\frac{(b-a)h^4}{180}f^{(4)}(\gamma).$$

where in each case γ is an unknown point within the interval $[x_0, x_N]$. Note that the degree of accuracy is not affected, only the order of convergence (as $h \to 0$). The formulae in Eq. (1.10) and Eq. (1.11) are the ones that will be used to estimate errors in any practical case, not the error formulae for the simple versions of the rules.

EXAMPLE 3: How large must N be if we want to use the composite trapezoidal rule to approximate the integral of $f(x) = \ln(x)$ over $1 \leq x \leq 2$ with an error of no more than $.5 \cdot 10^{-6}$? Using Eq. (1.10) we have

$$|E| \leq .5 \cdot 10^{-6}$$

$$\frac{(b-a)h^2}{12}M \leq .5 \cdot 10^{-6}$$

where M is a bound on $|f''(x)|$ over $[1,2]$. Since $f''(x) = -x^{-2}$ we can take $M = 1$, so we have

$$\frac{(2-1)h^2}{12} \cdot 1 \leq .5 \cdot 10^{-6}$$

$$h^2 \leq 6 \cdot 10^{-6}$$

$$h \leq .0025$$

(rounded up from $h \leq .002449\ldots$). We need h to be $.0025$ or smaller, that is, $N \geq 1/h = 400$ panels ($N + 1 = 401$ points). Fortunately function evaluations are not overly expensive for this function. For the composite Simpson's rule we have (with $M = 6$ now a bound on $|f^{(4)}(x)|$)

$$\frac{(b-a)h^4}{180}M \;\; \leq \;\; .5 \cdot 10^{-6}$$

$$h^4 \;\; \leq \;\; 90 \cdot 10^{-6}/6$$

$$h \;\; \leq \;\; .0039$$

(rounded up again), or $N \geq 1/h \doteq 258.1989$ giving $N = 259$. Since we must evaluate f at $N+1$ points using either method, this is much more efficient if the cost of function evaluations is our biggest concern. $\square$

We can apply these methods to any continuous function but the error estimates will not apply unless we have sufficient differentiability. We should expect worse performance of the methods for functions that have fewer continuous derivatives.

We have been talking about the approximation error E of closed Newton-Cotes quadrature rules, defined in Eq. (1.5) as

$$E = \int_a^b f(x)dx - \int_a^b p(x)dx$$

which is also called the **truncation error** since it arises from truncating a term from the exact formula

$$\int_a^b f(x)dx = \int_a^b p(x)dx + \int_a^b \frac{f^{(n+1)}(\xi(x))}{(n+1)!}(x-x_0)(x-x_1)\cdots(x-x_n)dx$$

for $\int_a^b f(x)dx$. We also need to consider round-off error, that is, the issue of stability. We define the **total error** to be the sum

$$E_{Total} = E_{Tr} + E_R$$

of the error due to truncation (that is, the quality of our method from an analytical perspective) and the error due to the effects of using finite precision arithmetic. Note that the act of truncation implies that there will be some error in our computations so we must make sure that it doesn't grow too large.

We know how to bound the error in a single arithmetic computation from Sec. 1.7. However, f may be supplied by another program or may be provided as data with no functional relationship specified. This is a common type of numerical integration problem: Measured data is provided as a list of x and y values, and so exact integration is not an option. The data may be the result of a laboratory experiment, for example. Whether the values of $f(x_i)$ are found by computation or measurement, we assume that the x_i are known exactly and that the error in the values of $f(x_i)$ is bounded,

$$|f(x_i) - f_i| \leq \varepsilon$$

where f_i is the computed/measured value of $f(x_i)$ and $\varepsilon > 0$ is a constant. Consider a weighted average rule as in Eq. (1.2),

$$\int_a^b f(x)dx = w_0 f(x_0) + w_1 f(x_1) + ... + w_n f(x_n) + E_{Tr}$$

where E_{Tr} represents the truncation error. The (simple) rule itself is

$$A_f = w_0 f(x_0) + w_1 f(x_1) + ... + w_n f(x_n)$$

but we will be computing

(1.12) $$\tilde{A}_f = w_0 f_0 + w_1 f_1 + ... + w_n f_n$$

instead. The multiplications $w_i f_i$ introduce a small error as do the additions but usually they are small with respect to the errors in f so we will neglect them. We have, for one application of the simple rule,

$$
\begin{aligned}
E_R^1 &= \left| A_f - \tilde{A}_f \right| \\
&= \left| w_0 f(x_0) + w_1 f(x_1) + ... + w_n f(x_n) - (w_0 f_0 + w_1 f_1 + ... + w_n f_n) \right| \\
&= \left| w_0 (f(x_0) - f_0) + w_1 (f(x_1) - f_1) + ... + w_n (f(x_n) - f_n) \right| \\
&\leq |w_0| \, e + |w_1| \, e + ... + |w_n| \, \varepsilon \\
&= (|w_0| + |w_1| + ... + |w_n|) \varepsilon
\end{aligned}
$$

which depends on the coefficients of the simple rule. Let

$$\Omega_n = \sum_{i=0}^{n} |w_i|$$

which is, in fact, a notion of condition number for this problem. We know that Ω_n generally depends linearly on $h = (b - a) / N$ where N is the number of panels in the composite rule. Suppose $\Omega_n \leq \alpha h$ for some $\alpha > 0$. Then

$$
\begin{aligned}
E_R &\leq \Omega_n \varepsilon \cdot N \\
&\leq \alpha h \varepsilon \cdot \frac{b - a}{h} \\
&= \alpha \varepsilon (b - a)
\end{aligned}
$$

which is, crucially, independent of the number of panels N to which the composite rule is applied! This is very stable behavior.

We said that the computation of Eq. (1.12) itself was unlikely to be a significant source of error. This is true if the weights w_i are not of mixed sign. However, for $n = 9$ and $n \geq 11$ the closed Newton-Cotes rules have weights of mixed sign. In this case the computation of

$$\tilde{A}_f = w_0 f_0 + w_1 f_1 + ... + w_n f_n$$

will likely involve the subtraction of nearly equal quantities, since $f_{i+1} \approx f_i$ by continuity. Cancellation of significant figures will be a serious problem. Additionally, in the closed Newton-Cotes rules the size of

$$\sum_{i=0}^{n} |w_i| = |w_0| + |w_1| + ... + |w_n|$$

grows without bound as n increases, so the errors can be large even for small data. What we see from this is that the closed Newton-Cotes formulas cannot be stable for large n. However, weighted average rules like Eq. (1.2) *do* have desirable round-off error properties if the weights w_i are nonnegative.

The composite versions of closed Newton-Cotes rules for $n \leq 8$ may be safely applied. We expect more accuracy at a slightly higher computational cost for the higher-degree rules. Note also that Simpson's rule requires an even number of panels (odd number of data points), Simpson's 3/8 rule requires that the number of panels be divisible by 3 (the number of data points must be $3n + 1$), and so on. We will soon discuss ways to circumvent these limitations.

MATLAB

The basic MATLAB command for numerical integration is `integral`. (Older versions of MATLAB used `quad`, as in 'quadrature'; `integral` is a vectorized version of `quadgk`.) Enter:

```
>> help integral
```

The `integral` command is designed to work well on vectorized functions. It has a default absolute error tolerance of 10^{-10}. Enter:

```
>> format long
>> f=@(x) 1./x
>> f([1:3])    %Test vectorization.
>> q=integral(f,1,10)
>> ql=quadl(f,1,10,1E-10)
>> qq=quad(f,1,10,1E-10)    %Adaptive Simpson's rule
>> a=log(10)          %Answer.
>> norm(q-a),norm(ql-a),norm(qq-a)
```

(Note: You may find that the quad commands are no longer be available.) The result from `integral` is best, but each method has met or exceeded its default error tolerance ($1E - 10$) so each routine has succeeded. Let's try another function. Enter:

```
>> g=@(x) abs(x-1)+abs(x+1)+sin(x)
>> q=integral(g,1,10)
>> ql=quadl(g,1,10,1E-10)
>> qq=quad(g,1,10,1E-10)
```

If you have the Symbolic Toolbox, enter:

```
>> syms x
>> G=int(abs(x-1)+abs(x+1)+sin(x),x)
```

to find the symbolic antiderivative of $g(x)$. This allows us to find the analytical value of the integral very accurately.

Error estimates as given in the text are bounds on the worst case and typically the methods do much better than indicated. For example, in Example 3 we estimated that to use the composite trapezoidal rule to approximate the integral of $f(x) = \ln(x)$ over $1 \leq x \leq 2$ with an error of no more than $.5 \cdot 10^{-6}$ would require $N = 400$ panels. Let's see. If you have the Symbolic Toolbox, enter:

```
>> syms z
>> int(log(z),z)
```

to see that the antiderivative is $z \ln(z) - 1$, so the correct value of the integral is $2 \ln(2) - 1$. Enter:

```
>> A=2*log(2)-1
>> integral(@log,1,2)         %Check.
```

Let's try some simpler methods. Enter:

```
>> a=1;b=2;f=@log
>> T1=.5*(b-a)*(f(a)+f(b))    %Simple trap.  rule.
```

```
>> x1=1.5;
>> T2=.5*(x1-a)*(f(a)+f(x1))+.5*(b-x1)*(f(x1)+f(b)) %Two panels.
>> N=3;h=(b-a)/N;
>> S=2*sum(f((a+h):h:(b-h)));T3=h*(f(a)+f(b)+S)/2 %Comp.  trap.
>> abs(A-T3)
```

The results are increasingly accurate. Let's try $N = 400$ as per Example 3. Enter:

```
>> N=400;h=(b-a)/N;
>> S=2*sum(f((a+h):h:(b-h)));T400=h*(f(a)+f(b)+S)/2
>> norm(A-T400)
```

The error is about $2.6E - 7$, about half the guarantee provided by the error bound. Since the error bound reflects the worst case over *all* sufficiently differentiable functions, any given function will likely have an error that is rather smaller. Enter:

```
>> (h^2)/12    %Error is O(h^2).
```

The error is roughly $h^2/12$, as predicted. In fact, the error being roughly $O(h^2)$ means that, for large N, $E \approx ch^2$ for some c (treating $f''(x)$ as roughly constant). Looked at another way, $c \approx E/h^2$. Enter:

```
>> c1=norm(A-T400)/h^2
>> N=800;h=(b-a)/N;
>> S=2*sum(f((a+h):h:(b-h)));T800=h*(f(a)+f(b)+S)/2
>> norm(A-T800)
>> c2=norm(A-T800)/h^2
```

to estimate c; these two estimates are in close agreement at about $1/24$, suggesting that in Eq. (1.10) we must have $f''(\gamma) \approx 1/2$. That means that we can estimate the error for $N = 1600$. Enter:

```
>> N=1600;h=(b-a)/N;
>> c2*h^2                    %Estimate.
>> S=2*sum(f((a+h):h:(b-h)));T1600=h*(f(a)+f(b)+S)/2
>> norm(A-T1600)
```

This is excellent agreement with our prediction! Using this idea we could make an estimate, tailored to this function (rather than using the general purpose Eq. (1.10)), as to how many more subintervals would be needed to achieve a desired error tolerance. Indeed, this is the sort of thing done in adaptive methods such as quad that attempt to locally estimate the error and choose more subintervals if needed. (There are other ways of doing so.) Enter:

```
>> c3=norm(A-S)/h^2
```

to see that the estimate of the asymptotic error constant $-(b-a)f''(\gamma)/12$ of Eq. (1.10) has changed little.

What do we do if we want to use Simpson's rule but the number of panels is odd? There are several possible approaches. If you have the Symbolic Toolbox, enter:

```
>> syms x
>> int(x.^8-3*x.^5+4*x.^2+2+sin(x),x)
```

The antiderivative is $F(x) = x^9/9 - x^6/2 + 4x^3/3 + 2x - \cos(x)$ so the integral of the function from 1 to 2 is $F(2) - F(1)$. Enter:

```
>> clear all
>> x=1;y1=1/9*x^9-1/2*x^6+4/3*x^3+2*x-cos(x);
>> x=2;y2=1/9*x^9-1/2*x^6+4/3*x^3+2*x-cos(x);
>> A=y2-y1   %Value of definite integral.
```

```
>> f=@(x) x.^8-3*x.^5+4*x.^2+2+sin(x)
>> N=9;h=1/N;           %N=9 panels.
>> xval=1:h:2           %Nodes.
>> fval=f(xval);        %Function evaluated at nodes.
>> Simp=dot([1 4 2 4 2 4 2 4 1],fval(1:9))*h/3
```
This gives the Simpson's rule estimate over the first 8 panels. Let's check its error; enter:
```
>> x=xval(end-1);y2=1/9*x^9-1/2*x^6+4/3*x^3+2*x-cos(x);A2=y2-y1
>> norm(A2-Simp)
```
The absolute error in estimating the integral up to $x = 1.8$ is about .006. We might imagine using the trapezoidal rule over the last panel; enter:
```
>> Trap=h*(fval(end-1)+fval(end))/2
>> Aest=Simp+Trap                  %Estimated area.
>> norm(A-Aest)
```
This is about .3! We had an estimate that was good to about .006 over $[1, 1.8]$ and ended up with one that was good only to about .3 over $[1, 2]$ because we contaminated our earlier $O(h^4)$ results with an $O(h^2)$ estimate. As you might guess, this is definitely a bad idea. Instead, let's try an approach that maintains the same order of the error. Enter:
```
>> Simp2=dot([1 4 2 4 2 4 1],fval(1:7))*h/3
```
This is the Simpson's rule estimate over the first 6 of the 9 panels. Now let's use the (simple) Simpson's 3/8 rule, which is $O(h^4)$, over the last 3 panels. Enter:
```
>> Simps38=dot([1 3 3 1],fval((end-3):end))*3*h/8
>> Aest2=Simp2+Simps38
>> norm(A-Aest2)
```
While this isn't as small as we'd like, it's an order of magnitude better than before. The step size h isn't small enough yet for the $O(h^4)$ behavior to fully kick in. Enter:
```
>> z=1:.01:2;plot(z,f(z),'r'),grid
>> line([2-h 2],[f(2-h) f(2)])
```
to see why the trapezoidal rule had difficulty getting an accurate estimate at this step size (note the vertical scale); enter:
```
>> line([1 1+h],[f(1) f(1+h)])
```
to see that the fitted line would have been a much better approximant near $x = 1$. Of course, since we have stored the f values in fval we could get a better estimate by taking additional nodes at the midpoints of each panel then using those points together with those in fval to get an estimate at resolution $h/2$. This saves us recomputing f at the original nodes, which is important if computing f is expensive. If f is measured data there may be a literal expense involved in repeating the experiment or requesting that additional field measurements be made.

The trapezoidal rule is provided in the command trapz and the cumulative trapezoidal rule (simulating indefinite integration) in cumtrapz. Enter:
```
>> f=@(x) x.^2        %Integral from 0 to 1 is 1/3.
>> x=0:.1:1;
>> trapz(x,f(x))      %Quadrature from 0 to 1.
>> cumtrapz(x,f(x))   %Cumulative quadrature from 0 to 1.
>> plot(x,ans)        %Numerical antiderivative of f.
```
The trapezoidal rule is provided because of its use in the numerical solution of ODEs and also its high accuracy for certain special quadrature problems: The trapezoidal

rule becomes more accurate if $f(a) = f(b)$, and more so if $f^{(k)}(a) = f^{(k)}(b)$ for $k = 0, 1, ..., p$. The more nearly periodic the function is on the interval, the better it performs, and even the simple trapezoidal rule will often outperform much "better" methods for such functions.

Problems

1.) Estimate erf(1) using the simple Simpson's, Simpson's 3/8, and Boole's rules. Construct a simple graph showing the error vs. the number of points $n + 1$ used (include the midpoint and trapezoidal rule errors from Example 1).

2.) a.) Estimate the number of subintervals needed to approximate $\int_0^1 \sin(x)dx$ to within 10^{-3} using the trapezoidal rule, then apply the trapezoidal rule with that number of subdivisions. What is the actual error in your method?
b.) Repeat with Simpson's rule.

3.) Compare the errors in using Simpson's rule and Simpson's 3/8 rule to approximate $\int_0^1 \cos(x)dx$, $\int_1^{10} \ln(x)dx$, and $\int_{-1}^1 (x^7 - 2x^3 + 1)dx$ with $N = 6, 12, 18$ panels.

4.) The error formula for the trapezoidal rule assumes that f is at least twice continuously differentiable. Use the trapezoidal rule to approximate $\int_{-1}^1 |x^3|dx$ and $\int_{-1}^1 g(x)dx$ where $g(x) = 0$ if $x < 0$ and $g(x) = x^2$ if $x \geq 0$. Use $N = 5, 9, 15$. Comment.

5.) Write a MATLAB program that accepts a function, an interval, a number of subintervals, and a choice of method (trapezoidal rule, Simpson's rule, Simpson's 3/8 rule, or Boole's rule) and returns the appropriate approximate integral of the function over the interval.

6.) a.) Create a 4 digit table of values of erf(x) for $x = 0, .1, .2, ..., 3$ using Simpson's rule. Your table entries should be accurately rounded to the 4 digits listed. Determine the appropriate number of panels to use from Eq. (1.11).
b.) How many panels would have been needed to create the table using the trapezoidal rule?

7.) a.) Approximate $\int_0^{20} x^5 \ln(x + 1)dx$ to within 10^{-6} using the trapezoidal, Simpson's, and Simpson's 3/8 rules. Comment on the relative efficiency of the methods.
b.) Approximate $\int_0^{20} (|x - 2| + |x - 5| + |x - 18|) \ln(x + 1)dx$ to within 10^{-6}. Use the trapezoidal, Simpson's, and Simpson's 3/8 rules. Comment on the relative efficiency of the methods.
c.) Approximate $\int_0^{20} x \ln(\sin(x) + 1.01)dx$ to within 10^{-6}. Use the trapezoidal, Simpson's, and Simpson's 3/8 rules. Comment on the relative efficiency of the methods.

8.) a.) What is the degree of accuracy of the rule $\int_{-1}^1 f(x)dx = f(-1/\sqrt{3}) + f(1/\sqrt{3})$?
b.) Experiment with this rule; compare it to the trapezoidal rule, which is also based on two points. Which seems to perform better?
c.) How could you use this rule to approximate $\int_a^b \phi(x)dx$?

9.) a.) Derive Simpson's rule and Simpson's 3/8 rule by performing the appropriate integrations $B_{n,i} = \int_0^1 L_{n,i}(x)dx$ analytically.
b.) What are the closed Newton-Cotes rules based on 6 and 7 points? (The rule based on 7 points is called **Weddle's rule**. There isn't a standard name for the

other rule.) You may integrate numerically or by using a CAS to find the coefficients, but may need to transform $\int_a^b L_{n,i}(x)dx$ to $\int_0^1 L_{n,i}(z)dz$ first, then transform back, to find the $B_{n,i}$ in terms of a and b.

10.) a.) Cubic splines interpolate a function with a piecewise cubic function. Does interpolating a function with a cubic spline then integrating the interpolant give the same result as the Simpson's 3/8 rule, which also corresponds to cubic interpolation?

b.) Perform an experiment to compare the results of interpolating free, clamped, and not-a-knot cubic splines and integrating the splines exactly. Which seems to work best? How do they compare with Simpson's rule and Simpson's 3/8 rule?

11.) a.) Approximate $\pi = 4\arctan(1)$ using the fact that $\arctan(x) = \int_0^x (1 + z^2)^{-1}dz$ ($x \geq 0$). Use `quad` and `quadl`.

b.) Find the positive solution of $\text{erf}(x) = x$ using Newton's method. Evaluate $\text{erf}(x)$ by using Simpson's rule applied to $\text{erf}(x) = \frac{2}{\sqrt{\pi}} \int_0^x e^{-z^2} dz$.

12.) a.) Use the trapezoidal rule to estimate $\int_0^1 \sqrt{x}\sin(x)dx$. What does the error estimate suggest for this integrand?

b.) Make the change of variables $x = z^2$ to get $\int_0^1 2z^2 \sin(z^2)dz$. Use the trapezoidal rule to estimate the value of this integral. What does the error estimate suggest for this integrand?

13.) a.) The function $\varpi(x)$ in Eq. (1.8) appears because of the following formula for the Lagrange interpolating polynomials with respect to the nodes $x_0, ..., x_n$: If $\varpi(x) = \prod_{i=0}^n (x - x_0)$ then $L_{n,i}(x) = \varpi(x)/[\varpi'(x)(x - x_i)]$. Prove this.

b.) Prove that the unique rule of the form of Eq. (1.2) with $a < x_0 < x_1 < \cdots < x_n$ that has the property that it integrates the monomials $f(x) = 1, x, x^2, ..., x^n$ exactly is the interpolatory rule.

14.) Use Simpson's rule to approximate $\iint_R \sin(x^3 + y^2)dA$, where R is the unit square $0 \leq x \leq 1$, $0 \leq y \leq 1$, by treating it as an iterated integral. For fixed values of y, find $\int_{x=0}^{x=1} \sin(x^3 + y^2)dx$ numerically to get a series of values of $F(y) = \int_{x=0}^{x=1} \sin(x^3 + y^2)dx$. Then find $\int_{y=0}^{y=1} F(y)dy$.

15.) For the simple Boole's rule, $E = (-8/945) h^7 f^{(6)}(\gamma)$. Verify that the error is $O(h^7)$ experimentally.

16.) Construct an experiment to verify that the trapezoidal rule has $O(h^2)$ error and Simpson's rule has $O(h^4)$ error. Produce appropriate plots.

2. Open Newton-Cotes Formulas

We have discussed the closed Newton-Cotes formulae. The other type of Newton-Cotes formulae are the **open Newton-Cotes rules** (based on $n+1$ points), which approximate the value of $\int_a^b f(x)dx$ by interpolating a polynomial of degree n to the points

$$x_0 = a + h, x_1 = a + 2h, ..., x_n = a + (n+1)h$$

(note that $x_n = b-h$) and then integrating that polynomial over $[a, b]$. The methods use only interior points, not the endpoints, with spacing

$$h = \frac{b - a}{n + 2}$$

(note that this is slightly smaller than the corresponding h for a closed Newton-Cotes rule based on $n + 1$ points, which was $h = (b - a)/n$). Examples include the midpoint rule

$$\int_a^b f(x)dx \approx 2hf(x_0)$$

($x_0 = (a + b)/2$), which corresponds to $n = 0$ and has error

$$E = \frac{h^3}{3}f''(\gamma)$$

($\gamma \in (a, b)$), the $n = 1$ case

$$\int_a^b f(x)dx \approx \frac{3h}{2}\left(f(x_0) + f(x_1)\right)$$

with error

$$E = \frac{3h^3}{4}f''(\gamma)$$

and the $n = 2$ case

$$\int_a^b f(x)dx \approx \frac{4h}{3}\left(2f(x_0) - f(x_1) + 2f(x_2)\right)$$

with error

$$E = \frac{14h^5}{45}f^{(4)}(\gamma)$$

and of course the corresponding composite rules are $O(h^2)$, $O(h^2)$, and $O(h^4)$, respectively. The weights w_i are given once again by

$$(2.1) \qquad\qquad w_i = \int_a^b L_{n,i}(x)dx$$

($i = 0, 1, ..., n$), though note that the nodes themselves are different (and $L_{n,i}(x)$ depends on the locations of the nodes) so these will not be numerically the same as the closed Newton-Cotes weights. The weights w_i in a Newton-Cotes rule are also called the **Cotes numbers**.

Open Newton-Cotes formulae share the properties of the closed formulae; those for which n is even are more accurate than might be expected, for example. There are special situations where methods that do not use the endpoints are needed but generally speaking the open formulae are used much less often than the closed formulae.

In the previous section we gave a formula for the coefficients in the closed Newton-Cotes rule based on $n + 1$ points. This formula, which is the same as that for the open rules (Eq. (2.1)), involves the integrals of the Lagrange interpolating polynomials. Let's look at a different approach that is applicable to both the open and closed rules. One characterization of the Newton-Cotes rules is that they are the unique rules based on $n + 1$ equally spaced points in $[a, b]$ for which

$$\int_a^b x^i dx$$

is found exactly for $i = 0, 1, ..., n$. Let's use this fact to find the coefficients in Simpson's rule. Simpson's rule is a weighted average rule

$$\int_a^b f(x)dx \approx w_0 f(x_0) + w_1 f(x_1) + w_2 f(x_2)$$

based on the three points

$$x_0 = a, x_1 = \frac{a+b}{2}, x_2 = b$$

and it must be exact for each of the monomials 1, x, x^2. So we need:

$$w_0 \cdot 1 + w_1 \cdot 1 + w_2 \cdot 1 = \int_a^b 1 dx$$
$$= b - a$$
$$w_0 \cdot a + w_1 \cdot \frac{a+b}{2} + w_2 \cdot b = \int_a^b x dx$$
$$= \frac{1}{2}(b^2 - a^2)$$
$$w_0 \cdot a^2 + w_1 \cdot \left(\frac{a+b}{2}\right)^2 + w_2 \cdot b^2 = \int_a^b x^2 dx$$
$$= \frac{1}{3}(b^3 - a^3)$$

which, for fixed a and b, is a linear system in the unknown weights w_0, w_1, w_2. In particular,

$$\begin{pmatrix} 1 & 1 & 1 \\ a & \frac{a+b}{2} & b \\ a^2 & \left(\frac{a+b}{2}\right)^2 & b^2 \end{pmatrix} \begin{pmatrix} w_0 \\ w_1 \\ w_2 \end{pmatrix} = \begin{pmatrix} b - a \\ \frac{1}{2}(b^2 - a^2) \\ \frac{1}{3}(b^3 - a^3) \end{pmatrix}$$

which is essentially a Vandermonde system (see Sec. 4.1), though rotated from how we defined previously. Nonetheless it is still nonsingular if $a \neq b$. For fixed a and b it is easy to solve this system numerically; for general a and b the analytical solution is

(2.2) $$w_0 = \frac{b-a}{6}, w_1 = \frac{2(b-a)}{3}, w_2 = \frac{b-a}{6}$$

in agreement with Simpson's rule. Of course, although we required that this rule be exact for 1, x, and x^2, we know that it is also exact for x^3.

This technique is known as the **method of undetermined coefficients**. Let's apply it to a specific problem. Suppose we are using Simpson's rule to approximate the integral of some function $f(x)$ for which we have only measured values $y_i = f(x_i)$ at equally spaced points $x_0, ..., x_n$. If n is even we can approximate

$$\int_{x_0}^{x_{n-1}} f(x)dx$$

using Simpson's rule, but to find

$$\int_{x_0}^{x_n} f(x)dx$$

we will need to approximate

$$\int_{x_{n-1}}^{x_n} f(x)dx$$

as well. We know that using the trapezoidal rule is a bad idea as it will contaminate our approximation of $\int_{x_0}^{x_{n-1}} f(x)dx$ with its $O(h^2)$ error. We'd like a $O(h^4)$ rule for the single panel $[x_{n-1}, x_n]$ and in fact one can be found in the following form:

$$\int_{x_{n-1}}^{x_n} f(x)dx \approx w_{n-2}f(x_{n-2}) + w_{n-1}f(x_{n-1}) + w_n f(x_n).$$

A rule like this, intended to take care of the extra panel for Simpson's rule, is called a **semi-simp rule** (or **half-simp rule**). It uses the value of f to the immediate left of the point x_{n-1} to help build a model for f and hence get a better approximation over the interval $[x_{n-1}, x_n]$. Let's ask that it be exact for 1, x, and x^2 again. We have

$$
\begin{aligned}
w_0 \cdot 1 + w_1 \cdot 1 + w_2 \cdot 1 &= \int_{x_{n-1}}^{x_n} 1 dx \\
&= x_n - x_{n-1} \\
w_0 \cdot x_{n-2} + w_1 \cdot x_{n-1} + w_2 \cdot x_n &= \int_{x_{n-1}}^{x_n} x dx \\
&= \frac{1}{2}(x_n^2 - x_{n-1}^2) \\
w_0 \cdot x_{n-2}^2 + w_1 \cdot (x_{n-1})^2 + w_2 \cdot x_n^2 &= \int_{x_{n-1}}^{x_n} x^2 dx \\
&= \frac{1}{3}(x_n^3 - x_{n-1}^3)
\end{aligned}
$$

giving the Vandermonde system

$$
\begin{pmatrix}
1 & 1 & 1 \\
x_{n-2} & x_{n-1} & x_n \\
x_{n-2}^2 & (x_{n-1})^2 & x_n^2
\end{pmatrix}
\begin{pmatrix}
w_0 \\
w_1 \\
w_2
\end{pmatrix}
=
\begin{pmatrix}
x_n - x_{n-1} \\
\frac{1}{2}(x_n^2 - x_{n-1}^2) \\
\frac{1}{3}(x_n^3 - x_{n-1}^3)
\end{pmatrix}
$$

to be solved for w_0, w_1, w_2. The solution may be found using the fact that the determinant of a Vandermonde matrix

$$(2.3) \qquad A = \begin{pmatrix} 1 & 1 & 1 \\ \alpha & \beta & \gamma \\ \alpha^2 & \beta^2 & \gamma^2 \end{pmatrix}$$

is $\det(A) = (\beta - \alpha)(\gamma - \alpha)(\gamma - \beta)$, and Cramer's rule. (Cramer's rule is not useful as a numerical method but can be useful for symbolic solution of linear systems.) Recall that Cramer's rule asserts that if A is nonsingular then the solution of $Ax = b$ is

$$x_i = \frac{\det(A_i)}{\det(A)}$$

($i = 1, ..., n$) where A_i is equal to A with its ith column replaced by b. In this case we have

$$
\begin{aligned}
\det(A) &= (x_{n-1} - x_{n-2})(x_n - x_{n-2})(x_n - x_{n-1}) \\
&= h \cdot 2h \cdot h \\
&= 2h^3
\end{aligned}
$$

($h = x_n - x_{n-1}$). Also

$$
\begin{aligned}
\det(A_1) &= \begin{vmatrix} x_n - x_{n-1} & 1 & 1 \\ \frac{1}{2}(x_n^2 - x_{n-1}^2) & x_{n-1} & x_n \\ \frac{1}{3}(x_n^3 - x_{n-1}^3) & (x_{n-1})^2 & x_n^2 \end{vmatrix} \\
&= -\frac{1}{6}(x_{n-1} - x_n)^4 \\
&= -\frac{1}{6}h^4
\end{aligned}
$$

$$
\begin{aligned}
\det(A_2) &= \begin{vmatrix} 1 & x_n - x_{n-1} & 1 \\ x_{n-2} & \frac{1}{2}(x_n^2 - x_{n-1}^2) & x_n \\ x_{n-2}^2 & \frac{1}{3}(x_n^3 - x_{n-1}^3) & x_n^2 \end{vmatrix} \\
&= \frac{1}{6}(x_n - x_{n-2})(x_{n-1} - x_n)^2 (2x_{n-1} + x_n - 3x_{n-2}) \\
&= \frac{1}{6} \cdot 2h \cdot h^2 \cdot (2(x_{n-2} + h) + (x_{n-2} + 2h) - 3x_{n-2}) \\
&= \frac{1}{6} \cdot 2h \cdot h^2 \cdot (2h + 2h) \\
&= \frac{4}{3}h^4
\end{aligned}
$$

$$
\begin{aligned}
\det(A_3) &= \begin{vmatrix} 1 & 1 & x_n - x_{n-1} \\ x_{n-2} & x_{n-1} & \frac{1}{2}(x_n^2 - x_{n-1}^2) \\ x_{n-2}^2 & (x_{n-1})^2 & \frac{1}{3}(x_n^3 - x_{n-1}^3) \end{vmatrix} \\
&= \frac{1}{6}(x_{n-1} - x_{n-2})(x_{n-1} - x_n)^2 (x_{n-1} - 3x_{n-2} + 2x_n) \\
&= \frac{1}{6} \cdot h \cdot h^2 ((x_{n-2} + h) - 3x_{n-2} + 2(x_{n-2} + 2h)) \\
&= \frac{1}{6} \cdot h \cdot h^2 \cdot (h + 4h) \\
&= \frac{5}{6}h^4
\end{aligned}
$$

(using a CAS). So

$$w_0 = \frac{-\frac{1}{6}h^4}{2h^3}$$

$$= -\frac{1}{12}h$$

$$w_1 = -\frac{\frac{4}{3}h^4}{2h^3}$$

$$= \frac{2}{3}h$$

$$w_2 = \frac{\frac{5}{6}h^4}{2h^3}$$

$$= \frac{5}{12}h$$

are the coefficients, and the half-simp rule is

$$\int_{x_{n-1}}^{x_n} f(x)dx \approx \frac{h}{12}(-f(x_{n-2}) + 8f(x_{n-1}) + 5f(x_n)).$$

We can now employ Simpson's rule and if we encounter an unmatched last panel we can use the half-simp rule. This is especially useful if the data is being integrated in real time.

EXAMPLE 1: Let's use Simpson's rule and the half-simp rule to approximate $\int_0^{2.5} \exp(x)dx$ using $N = 5$ panels. We have $x_0 = 0$, $x_1 = .5$, $x_2 = 1$, $x_3 = 1.5$, $x_4 = 2$, $x_5 = 2.5$, $x_6 = 3$. Simpson's rule gives

$$\int_0^2 \exp(x)dx \approx \frac{.5}{3}(f(0) + 4f(.5) + 2f(1) + 4f(1.5) + f(2))$$

$$\doteq 6.3912$$

(the correct value is $\exp(2) - \exp(0) \doteq 6.3891$). The half-simp rule gives

$$\int_2^{2.5} \exp(x)dx \approx \frac{.5}{12}(-f(1.5) + 8f(2) + 5f(2.5))$$

$$\doteq 4.8143$$

(the correct value is $\exp(2.5) - \exp(2) \doteq 4.7934$). Hence

$$\int_0^{2.5} \exp(x)dx \approx 6.3912 + 4.8143$$

$$\doteq 11.1846$$

(the correct value is $\exp(2.5) - \exp(0) \doteq 11.1825$). This is an error of about .002, which isn't bad for using only 7 points. □

We could create similar special methods for other Newton-Cotes formulae (such as Simpson's 3/8 rule). The trapezoidal rule and Simpson's rule are by far the most popular of the closed Newton-Cotes rules however.

We need to know how accurate the half-simp rule is. In order to determine that it's convenient to first consider another method of deriving the simple Simpson's rule. We'll expand the integrand $f(x)$ in a Taylor series about the leftmost point–assuming, as usual, as much differentiability as needed[2]–and try to identify something on the RHS that looks like Simpson's rule. We start by replacing the integrand by its Taylor series expansion about the left-hand endpoint x_0, giving

$$
\int_{x_0}^{x_2} f(x)dx = \int_{x_0}^{x_2} \left(f(x_0) + f'(x_0)(x - x_0) + \frac{f''(x_0)}{2!}(x - x_0)^2 + \right.
$$
$$
\left. \frac{f'''(x_0)}{3!}(x - x_0)^3 + \frac{f^{(4)}(x_0)}{4!}(x - x_0)^4 + \ldots \right) dx
$$

and then integrate the power series term-by-term,

$$
\int_{x_0}^{x_2} f(x)dx = \left(xf(x_0) + \frac{f'(x_0)}{2}(x - x_0)^2 + \frac{f''(x_0)}{3 \cdot 2!}(x - x_0)^3 + \right.
$$
$$
\left. \left. \frac{f'''(x_0)}{4 \cdot 3!}(x - x_0)^4 + \frac{f^{(4)}(x_0)}{5 \cdot 4!}(x - x_0)^5 + \ldots \right) \right|_{x_0}^{x_2}
$$
$$
= (x_2 - x_0)f(x_0) + \frac{f'(x_0)}{2}(x_2 - x_0)^2 + \frac{f''(x_0)}{6}(x_2 - x_0)^3 +
$$
$$
\frac{f'''(x_0)}{24}(x_2 - x_0)^4 + \frac{f^{(4)}(x_0)}{120}(x_2 - x_0)^5 + \ldots
$$

after evaluating this integral. We now substitute for the difference in terms of the step size h, giving

$$
\int_{x_0}^{x_2} f(x)dx = 2hf(x_0) + 2h^2 f'(x_0) + 8h^3 \frac{f''(x_0)}{6} + 16h^4 \frac{f'''(x_0)}{24} +
$$
$$
32h^5 \frac{f^{(4)}(x_0)}{120} + \ldots
$$
$$
= 2hf(x_0) + 2h^2 f'(x_0) + \frac{4}{3}h^3 f''(x_0) + \frac{2}{3}h^4 f'''(x_0) +
$$
$$
\frac{4}{15}h^5 f^{(4)}(x_0) + \ldots
$$

which is a relatively simple expression for the definite integral, given that it is represented as an infinite sum. But we now need to collect terms on the right in such a way as to identify Simpson's rule lurking within this expression. This is going to involve a heavy dose of algebra, so we need to keep in mind that our goal is identifying that $h/3, 4h/3, h/3$ pattern. We have

---

[2]We're actually assume the existence of the full Taylor's series, which is a more powerful assumption than even infinite differentiability.

$$\int_{x_0}^{x_2} f(x)dx = \frac{h}{3}\left(f(x_0) + 4f(x_0) + 1f(x_0)\right) + \frac{h}{3}\left(4f'(x_0) + 1\cdot 2f'(x_0)\right)h +$$

$$\frac{h}{3}\left(4\cdot\frac{1}{2}f''(x_0) + 1\cdot 2f''(x_0)\right)h^2 +$$

$$\frac{h}{3}\left(4\cdot\frac{1}{6}f'''(x_0) + 1\cdot\frac{4}{3}f'''(x_0)\right)h^3 +$$

$$\frac{h}{3}\left(4\cdot\frac{1}{24}f^{(4)}(x_0) + 1\cdot\frac{2}{3}f^{(4)}(x_0)\right)h^4 - \frac{1}{90}f^{(4)}(x_0)h^5 + \dots$$

$$= \frac{h}{3}f(x_0) +$$

$$\frac{4h}{3}\left(f(x_0) + f'(x_0)h + \frac{1}{2}f''(x_0)h^2 + \frac{1}{6}f'''(x_0)h^3 +\right.$$

$$\left.\frac{1}{24}f^{(4)}(x_0)h^4 + \dots\right) +$$

$$\frac{h}{3}\left(f(x_0) + 2f'(x_0)h + 2f''(x_0)h^2 + \frac{4}{3}f'''(x_0)h^3 +\right.$$

$$\left.\frac{2}{3}f^{(4)}(x_0)h^4 + \dots\right) +$$

$$-\frac{1}{90}f^{(4)}(x_0)h^5 + \dots$$

$$= \frac{h}{3}f(x_0) + \frac{4h}{3}f(x_1) +$$

$$\frac{h}{3}(f(x_0) + f'(x_0)(2h) + \frac{1}{2}f''(x_0)(2h)^2 +$$

$$\frac{1}{6}f'''(x_0)(2h)^3 + \frac{1}{24}f^{(4)}(x_0)(2h)^4 + \dots) +$$

$$-\frac{1}{90}f^{(4)}(x_0)h^5 + \dots$$

$$= \frac{h}{3}f(x_0) + \frac{4h}{3}f(x_1) + \frac{h}{3}f(x_2) - \frac{1}{90}f^{(4)}(x_0)h^5 + \dots$$

$$= \frac{h}{3}\left(f(x_0) + 4f(x_1) + f(x_2)\right) - \frac{1}{90}f^{(4)}(x_0)h^5 + \dots$$

(note that in the last several lines the terms indicated by +... within parentheses represent terms in h^6 and higher since they are multiplied by the h outside the parentheses). So the truncation error in the simple Simpson's rule is

(2.4) $$E = -\frac{1}{90}f^{(4)}(x_0)h^5 + \dots$$

in agreement with Eq. (5.1.1.9), where a different analysis has yielded

(2.5) $$E = -\frac{1}{90}f^{(4)}(\gamma)h^5$$

for some unknown $\gamma \in (x_0, x_2)$. We have shown that the simple Simpson's rule is indeed $O(h^5)$ as $h \to 0$.

Think again about what we just did in deriving Eq. (2.4): We expanded the integrand in a Taylor series; integrated that series term-by-term; grouped terms so as to identify the Newton-Cotes rule on the RHS; then identified the error term. We can use the same technique even if the function $f(x)$ is not real analytic as assumed here, that is, even if it does not have a Taylor series over the interval of integration, but we must take care regarding how we integrate the error term (because it contains an unknown point of evaluation, which will vary with h). We would reach the same conclusion, assuming we have at least as many orders of differentiability as are needed to reach the result in Eq. (2.4).

Now let's use this same approach with the half-simp rule in order to determine its accuracy. We need it to be at least as accurate as the composite Simpson's rule. (The half-simp rule effectively only has a simple form.) Remember in this case that the integral is taken over the interval from x_{n-1} to x_n, even though the rule itself also uses the point x_{n-2} that lies to the left of the left-hand endpoint of the interval of integration.

We begin once again by expanding the integrand in its Taylor series about the left-hand endpoint, giving the series representation

$$
\int_{x_{n-1}}^{x_n} f(x)dx = \int_{x_{n-1}}^{x_n} \left(f(x_{n-1}) + f'(x_{n-1})(x - x_{n-1}) + \right.
$$
$$
\frac{f''(x_{n-1})}{2!}(x - x_{n-1})^2 + \frac{f'''(x_{n-1})}{3!}(x - x_{n-1})^3 +
$$
$$
\left. \frac{f^{(4)}(x_{n-1})}{4!}(x - x_{n-1})^4 + ... \right) dx
$$

and so, integrating term-by-term, we have

$$
\int_{x_{n-1}}^{x_n} f(x)dx = \left(xf(x_{n-1}) + \frac{f'(x_{n-1})}{2}(x - x_{n-1})^2 + \frac{f''(x_{n-1})}{3 \cdot 2!}(x - x_{n-1})^3 + \right.
$$
$$
\left. \frac{f'''(x_{n-1})}{4 \cdot 3!}(x - x_{n-1})^4 + \frac{f^{(4)}(x_{n-1})}{5 \cdot 4!}(x - x_{n-1})^5 + ... \right)\Bigg|_{x_{n-1}}^{x_n}
$$
$$
= (x_n - x_{n-1})f(x_{n-1}) + \frac{f'(x_{n-1})}{2}(x_n - x_{n-1})^2 +
$$
$$
\frac{f''(x_{n-1})}{6}(x_n - x_{n-1})^3 + \frac{f'''(x_{n-1})}{24}(x_n - x_{n-1})^4 +
$$
$$
\frac{f^{(4)}(x_{n-1})}{5 \cdot 4!}(x_n - x_{n-1})^5 + ...
$$
$$
\frac{5h}{12}\left(f(x_{n-1}) + f'(x_{n-1})h + \frac{1}{2}f''(x_{n-1})h^2 + \frac{1}{6}f'''(x_{n-1}) + \right.
$$
$$
\left. \frac{1}{24}f^{(4)}(x_{n-1}) + ... \right) +
$$
$$
\frac{1}{12}f'''(x_{n-1})h^4
$$

which lacks any reference to the point x_{n-2} lying outside the interval, which we need in order to recognize and isolate the rule. We will have to re-group the terms on the RHS in such a way as to introduce an expression for the Taylor series for $f(x)$ expanded about x_{n-2} so that we can introduce that point:

$$
\begin{aligned}
\int_{x_{n-1}}^{x_n} f(x)dx &= hf(x_{n-1}) + \frac{f'(x_{n-1})}{2}h^2 + \frac{f''(x_{n-1})}{6}h^3 + \\
&\quad \frac{f'''(x_{n-1})}{24}h^4 + \frac{f^{(4)}(x_{n-1})}{120}h^5 + \dots \\
&= \frac{h}{12}(-f(x_{n-1}) + 8f(x_{n-1}) + 5f(x_{n-1})) + \\
&\quad \frac{h}{12}(-1\cdot -1f'(x_{n-1}) + 5f'(x_{n-1}))h + \\
&\quad \frac{h}{12}\left(-1\cdot\frac{1}{2}f''(x_{n-1}) + 5\cdot\frac{1}{2}f''(x_{n-1})\right)h^2 + \\
&\quad \frac{h}{12}\left(-1\cdot\frac{-1}{6}f'''(x_{n-1}) + 5\cdot\frac{1}{6}f'''(x_{n-1})\right)h^3 + \\
&\quad \frac{1}{12}f'''(x_{n-1})h^4 + \\
&\quad \frac{h}{12}\left(-1\cdot\frac{1}{24}f^{(4)}(x_{n-1}) + 5\cdot\frac{1}{24}f^{(4)}(x_{n-1})\right)h^4 + \\
&\quad -\frac{19}{120}f^{(4)}(x_{n-1})h^5 + \dots \\
&= \frac{-h}{12}\left(f(x_{n-1}) - f'(x_{n-1})h + \frac{1}{2}f''(x_{n-1})h^2 - \frac{1}{6}f'''(x_{n-1}) + \right. \\
&\quad \left. \frac{1}{24}f^{(4)}(x_{n-1}) - \dots\right) + \\
&\quad \frac{8h}{12}f(x_{n-1}) + \\
&\quad \frac{5h}{12}\left(f(x_{n-1}) + f'(x_{n-1})h + \frac{1}{2}f''(x_{n-1})h^2 + \frac{1}{6}f'''(x_{n-1}) + \right. \\
&\quad \left. \frac{1}{24}f^{(4)}(x_{n-1}) + \dots\right) + \\
&\quad \frac{1}{12}f'''(x_{n-1})h^4 \\
&= \frac{h}{12}(-f(x_{n-2}) + 8f(x_{n-1}) + 5f(x_n)) + \frac{1}{12}f'''(x_{n-1})h^4 + \dots
\end{aligned}
$$

where all neglected terms are $O(h^5)$ or higher. Hence the truncation error for the half-simp rule is $O(h^4)$, and so we are justified in using it with the $O(h^4)$ composite Simpson's rule.

The method of undetermined coefficients is useful for deriving rules but is less useful for computing the needed coefficients for large n. For one thing, note that in the 3×3 case the determinant of the Vandermonde matrix is h^3, which for small h is near zero; this is just one symptom of the fact that Vandermonde matrices based on equally spaced points are apt to be ill-conditioned. The closed Newton-Cotes formulae form a very popular family of quadrature rules but finding efficient means of generating the weights requires some ingenuity. We have given the weights as integrals and also as solutions of a linear system, but there are faster ways of evaluating them numerically.

We have been discussing approximate integration, but there is another computational tool available for integration. Symbolic integration, as provided by a CAS, is a powerful tool for those integrals which can be solved by formula. Even then, however, we may get an answer like

$$\int_a^b \frac{1}{x} dx = \ln(b) - \ln(a)$$

and $\ln(x)$ is itself approximated in the software. Since $\ln(x)$ is generally a provided function it is convenient to use it but we should remember that its values are generated by a numerical method and hence subject to approximation error as well as the rounding error of the arithmetic operations involved in using that approximation. Thus for a more complicated integral with a non-polynomial integrand we might consider simply using numerical quadrature in the first place rather than finding an antiderivative which must then be evaluated approximately. But for many integrals, and for all cases involving experimental data or function values generated by another program (so that f itself is unavailable), we need an approximate integration rule.

The methods we have been discussing paralellize easily, as we can divide the number of panels by the number of processors and ship out to each processor an appropriately sized region on which it will perform the quadrature. There is little need for inter-processor communication.

MATLAB

If there is a singularity at an endpoint then the open formulae can be very inaccurate. On the other hand, if some derivative of the integrand is singular at the endpoint, open rules are usually preferred to closed rules as they avoid this singularity.

Let's look at a function with an interesting singularity, namely the function $f(x) = 3x^2 + \ln((\pi - x)^2)/\pi^4 + 1$. Enter:

```
>> f=@(x) 3*x.^2+log((pi-x).^2)/pi^4+1
>> x=0:.00001:5;
>> y=f(x);
>> plot(x,y),grid
```

Where is the singularity that should occur when $x = \pi$ (giving $\ln(0)$)? Let's check for it; enter:

```
>> f(pi)
>> min(y)
```

The singularity is there, but even at this fine resolution the smallest value we see is greater than 1 even though f is not bounded from below on this interval! (The smallest value near $x = \pi$ is even larger.) All the negative values must lie in an extremely small interval about $x = \pi$. Let's try plotting it again. Enter:

```
>> x=(pi-1E-8):1E-10:(pi+1E-8);
>> y=f(x);
>> plot(x,y),grid
>> min(y)
>> s=y(isfinite(y));          %Remove the infinite value.
>> min(s)
```

This is a little better, but still no (finite) negative values. Let's try one more time. Enter:

```
>> x=(pi-10*eps):eps:(pi+10*eps);
>> y=f(x);
>> plot(x,y),grid
>> min(y)
>> s=y(isfinite(y));
>> min(s)
```

The plot is not terribly informative except to show that the function is still not very small. In fact, $f(x) = 3x^2 + \ln((\pi - x)^2)/\pi^4 + 1$ is positive for every IEEE standard double precision floating point number $x > 0$. We have seen this in our `f(x)` (which differs from $f(x)$ because, for example, `pi` is not equal to π and `log(x)` is not exactly equal to $\ln(x)$). This singularity is not detectable numerically. Enter:

```
>> integral(f,2,4)
```

In fact $f(x)$ is integrable, and the MATLAB routines cannot see the singularity anyway so they do not notice any difficulty (we say that we are "ignoring the singularity"). On the other hand, enter:

```
>> g=@(x) x.^(-.99)
>> integral(g,0,1)               %Should be about 1.
```

This function is also singular and integrable, and all the routine correctly recognizes this and terminates with a warning and highly inaccurate answers. Let's try another singular function. Enter:

```
>> h=@(x) sin(1./x)./x
>> integral(h,0,1)               %Should be about .6247.
```

Perhaps this isn't so surprising since $h(0)$ is undefined. Let's try the idea of "avoiding the singularity", as an open rule would do automatically; enter:

```
>> integral(h,eps,1)             %Should be about .6247.
```

Maybe ϵ is too small. Enter:

```
>> integral(h,1E-4,1)            %Should be about .6247.
>> ezplot(h,0,1)
```

In fact, $h(x)$ is much more ill-behaved near $x = 0$ than a plot could ever show. But, a simple change of variables shows that $h(x) = \int_1^\infty \mathrm{sinc}(x)dx$ so if we have a routine for integrals on unbounded domains, we may be able to evaluate this. Enter:

```
>> integral(@(x) sin(x)./x,1,1000)
```

to see that we get plausible answers from this. We could continue to increase the upper limit until the values ceased to differ appreciably. An initial change of variables can sometimes help you get the most out of built-in numerical integration routines.

A function like $\mathrm{sinc}(x) = \sin(x)/x$ may appear to be singular (in this case at $x = 0$) but in fact the limit as x approaches the apparent singularity the function may be well-defined ($\mathrm{sinc}(0) = 1$) and hence not singular. Of course, the naive evaluation of $\sin(x)/x$ at $x = 0$ would not give its correct value.

Although weighted average rules are stable, subdividing the interval too finely means we are doing many operations, which is inefficient. Additionally, when performing many arithmetic and other operations for a large sum and a small h we run the risk of eventually adding small numbers (from the last few intervals) to a large number (the area up to that point) if we do not take some care in how we form the sum. Taking too many panels ("overcomputing") will increase the error; stability means it won't grow too large too fast, not that it won't happen. Enter:

```
>> k=0;
>> f=@(x) sin(2*pi*x)
>> k=k+1;h=10^-k;x=0:h:1;
>> trapz(x,f(x))        %Should be zero.
>> k=k+1;h=10^-k;x=0:h:1;
>> trapz(x,f(x))
>> k=k+1;h=10^-k;x=0:h:1;     %Re-enter previous two lines.
>> trapz(x,f(x))
>> k=k+1;h=10^-k;x=0:h:1;     %Re-enter previous two lines.
>> trapz(x,f(x))
>> k=k+1;h=10^-k;x=0:h:1;     %Re-enter previous two lines.
>> trapz(x,f(x))
>> k=k+1;h=10^-k;x=0:h:1;     %Re-enter previous two lines.
>> trapz(x,f(x))
```

The error is remaining on the order of 10^{-17}, which is certainly acceptable given that the machine epsilon is about 10^{-16}, but still it has started increasing even though h keeps decreasing. Because the rule is stable, round-off error is growing slowly, but even still it is there and it is degrading our results.

Note how small the error was for relatively large h (at $k = 1$); the trapezoidal rule is giving especially good results because of the periodicity of the function on the interval. Even at $k = 1$, the error is essentially zero.

Problems

1.) Use Cramer's rule to derive Eq. (2.2) and verify that these are the weights of Simpson's rule as given in the previous section.

2.) Use Simpson's rule with the half-simp rule to approximate $\int_0^1 f(x)dx$ for $f(x) = \sin(x^2)$ and $f(x) = x^7$ using $N = 9$ and $N = 21$ panels. Compare your answers to those found by using Simpson's 3/8 rule over the last three panels rather than the half-simp rule over the last panel.

3.) What is the half-simp rule for integrating over the first (not last) panel? Use the method of undetermined coefficients.

4.) a.) What are the degrees of accuracy of the first three open Newton-Cotes formulae $(n = 0, 1, 2)$? Directly show that they are exact for the appropriate monomials.

b.) Directly show that Simpson's rule and the half-simp rule are exact for constants, linear functions, and quadratics. What about cubics? Quartics?

5.) Prove that Eq. (2.1) gives the correct Cotes numbers for the open Newton-Cotes rules.

6.) a.) Derive the half-simp rule in the following way: Find a rule of the form $\int_{-1}^{1} f(x)dx \approx w_0 f(x_0) + w_1 f(x_1) + w_2 f(x_2)$ that is exact for $f(x) = 1, x, x^2$ by the method of undetermined coefficients and then use a transformation that allows you to apply it to integrals of the form $\int_a^b \phi(x)dx$.

b.) Derive a rule that could be used with Simpson's 3/8 rule if there is a single panel left.

c.) What rule would you use for Simpson's 3/8 rule with two panels left?

7.) Why do all the integration formulas for approximating $\int_a^b f(x)dx$ that we've seen have the property that $\sum_{i=0}^n w_i = b - a$?

8.) Construct an experiment to compare the closed Newton-Cotes rules based on $n+1$ points with the open Newton-Cotes rules based on $n+1$ points. Is one type consistently better than the other?

9.) It was stated in the text that Cramer's rule isn't useful as a numerical method, and this is so. However, one can imagine highly specialized cases where the structure of A makes it easy to accurately find the determinants, and only a few components x_i of x are needed. Such a situation is very rare in practice; but suppose that A is a large nonsingular upper triangular matrix and that $b = e_1$. If only the first component x_1 of $Ax = b$ is needed, is Cramer's rule more efficient than backward solution of the triangular system? What numerical issues arise in computing $\det(A) = \prod_{i=1}^n a_{ii}$ by such a method? Suggest a method for scaling in this computation to avoid underflow and overflow and to maintain accuracy.

10.) a.) Derive the error term for the midpoint and trapezoidal rules using Taylor series.

b.) Derive the error term for the Simpson's 3/8 rule using Taylor series.

c.) Derive the error term for Simpson's rule using Taylor series with remainder and an appropriate mean value theorem.

11.) Show that Eq. (2.4) can be written in the form Eq. (2.5).

12.) a.) Consider evaluating $\ln(x)$ for $x \geq 1$ using the rule $\int_1^x x^{-1}dx = \ln(x)$. How many points would you need to evaluate $\ln(2)$ to within 10^{-15} using Simpson's rule? How about with Boole's rule?

b.) The closed Newton-Cotes rule based on 7 points has an error term of the form $E = -9h^9 f^{(8)}(\gamma)/1400$. How many points would be required for this rule?

13.) a.) Show that the determinant of the matrix in Eq. (2.3) is $(\beta-\alpha)(\gamma-\alpha)(\gamma-\beta)$.

b.) Generalize this to a formula for the determinant of a nxn Vandermonde matrix. If you can't prove it correct, verify it experimentally.

14.) a.) Write a MATLAB program that performs approximate integration for measured data (that is, the input will be a vector y of function values and a step size h) using Simpson's rule and, when needed, the half-simp rule.

b.) Evaluate $\int_0^1 \sin(x)dx$ using Simpson's rule and, when necessary, the half-simp rule, using $n = 5, 6, ...16$ panels. Comment on the differences, if any, between using an odd and even number of panels.

15.) a.) Derive the open Newton-Cotes rule based on 4 points using the method of undetermined coefficients.

b.) Show that the error term is $E = 95h^5 f^{(4)}(\gamma)/144$ using Taylor series.

16.) a.) Derive the closed Newton-Cotes rule based on 7 points (**Weddle's rule**) using the method of undetermined coefficients.

b.) Show that the error term is $E = -9h^9 f^{(8)}(\gamma)/1400$ using Taylor series.

3. Gaussian Quadrature

The Newton-Cotes formulae use a set of equally spaced nodes to which they interpolate a polynomial. The composite rules may be thought of as using a piecewise polynomial interpolant. These rules are widely used, though what we see employed most commonly from this class of methods in numerical integration routines are closed Newton-Cotes rules of low order. (The open rules are seen more often in the numerical solution of ordinary differential equations.) But we know that interpolating to equally spaced nodes can be problematic. Can we do better with unequally spaced nodes? Yes, we can, if what we mean by better is a higher degree of accuracy. Here's an example: The rule

$$(3.1) \qquad \int_{-1}^{1} f(x)dx = f(-1/\sqrt{3}) + f(1/\sqrt{3})$$

has degree of accuracy 3, whereas the Newton-Cotes rules based on 2 points both have degree of accuracy 1. Think about it: This is a rule based on only 2 points that evaluates all cubics correctly. That's pretty impressive. Since

$$(3.2) \qquad \int_{a}^{b} g(t)dt = \frac{b-a}{2} \int_{-1}^{1} g(t(x))dx$$

where $t(x) = ((b-a)x + b + a)/2$ we see that this technique is not restricted to the interval $[-1,1]$.

EXAMPLE 1: Let's try this method on $\int_{-1}^{1} x^6 dx = 2/7$. The trapezoidal rule uses 2 points and gives $\int_{-1}^{1} x^6 dx \approx f(-1) + f(1) = 2$ $(E \doteq 1.9973)$. The open Newton-Cotes rule for $n = 1$ uses 2 points and gives $\int_{-1}^{1} x^6 dx \approx f(-1/3) + f(1/3) = .0027$ $(E \doteq 1.7143)$. But Eq. (3.1) also uses 2 points and gives $\int_{-1}^{1} x^6 dx \approx f(-1/\sqrt{3}) + f(1/\sqrt{3}) \doteq .0741 (E \doteq .2116)$. In each of these three cases the area under the curve $f(x) = x^6$ is approximated by the area under a straight line, but the quality of the estimates differs considerably. $\square$

How do we find rules like Eq. (3.1)? Suppose we decide that in the general weighted average rule

$$(3.3) \qquad \int_{a}^{b} f(x)dx \approx w_0 f(x_0) + w_1 f(x_1) + ... + w_n f(x_n)$$

we are free to choose not only the weights w_i but also the locations of the nodes x_i. Then we have $2n + 2$ degrees of freedom rather than $n + 1$. We will try to choose these quantities so as to maximize the degree of accuracy of the method.

It's clear that any interpolatory rule based on $n+1$ points must have degree of accuracy at least n because we are interpolating a polynomial

$$p(x) = a_n x^n + a_{n-1} x^{n-1} + ... + a_1 x + a_0$$

to the function; if the function is itself a polynomial of degree n or less, we'll have a perfect fit. In the other direction we have the following theorem:

THEOREM 1: A weighted average rule based on $n + 1$ points that has degree of accuracy at least n is an interpolatory rule.

PROOF: The proof is essentially the method of undetermined coefficients. If a rule of the form of Eq. (3.3) has degree of accuracy at least n then we know that, by definition,

$$w_0 \cdot 1 + w_1 \cdot 1 + \cdots + w_n \cdot 1 \;=\; \int_a^b 1\,dx$$

$$w_0 \cdot x_0 + w_1 \cdot x_1 + \cdots + w_n \cdot x_n \;=\; \int_a^b x\,dx$$

$$w_0 \cdot x_0^2 + w_1 \cdot x_1^2 + \cdots + w_n \cdot x_n^2 \;=\; \int_a^b x^2\,dx$$

$$\vdots$$

$$w_0 \cdot x_0^n + w_1 \cdot x_1^n + \cdots + w_2 \cdot x_n^n \;=\; \int_a^b x^n\,dx$$

and as the x_i are distinct we recognize this as a square and in fact nonsingular Vandermonde system. Hence there is one and only one solution. But we have seen already that any interpolatory rule based on $n+1$ points has degree of accuracy at least n, and so the interpolatory rule based on $x_0, x_1, ..., x_n$ is a solution of this system. Hence, it is the unique solution. $\square$

This is a useful theorem. We now know that when we seek to find a rule based on $n+1$ points that is to have degree of accuracy greater than n, we will necessarily need to use an interpolatory rule. We know what the weights must be for an interpolatory rule; hence the only question that remains before us is how to position the nodes so as to maximize the degree of accuracy. Once those have been determined, the coefficients in an interpolatory rule must necessarily be of the form we have seen previously, based on quadratures of the Lagrange interpolating polynomials:

$$w_i = \int_a^b L_{n,i}(x)\,dx$$

$(i = 0, 1, ..., n)$. This is true no matter where the nodes are placed. (Of course, we should bear in mind that $L_{n,i}(x)$ itself depends on the locations of those nodes.) Computing the w_i will not necessarily be easy, but there's no freedom concerning what values the weights in the rule must take.

How shall we find $x_0, x_1, ..., x_n$? We could use the method of undetermined coefficients. Let's try this for $n = 1$ (a method based on 2 points). Since there are 4 degrees of freedom (w_0, w_1, x_0, x_1) in the rule

$$\int_a^b f(x)\,dx \approx w_0 f(x_0) + w_1 f(x_1)$$

we will ask that the method be exact for the monomials $1, x, x^2, x^3$. This gives the requirements

$$w_0 \cdot 1 + w_1 \cdot 1 \;=\; \int_a^b 1 \, dx$$

$$=\; b - a$$

$$w_0 \cdot x_0 + w_1 \cdot x_1 \;=\; \int_a^b x \, dx$$

$$=\; \frac{1}{2}(b^2 - a^2)$$

$$w_0 \cdot x_0^2 + w_1 \cdot x_1^2 \;=\; \int_a^b x^2 \, dx$$

$$=\; \frac{1}{3}(b^3 - a^3)$$

$$w_0 \cdot x_0^3 + w_1 \cdot x_1^3 \;=\; \int_a^b x^3 \, dx$$

$$=\; \frac{1}{4}(b^4 - a^4)$$

which is a nonlinear system of 4 equations in the 4 unknowns w_0, w_1, x_0, x_1, namely, the system

$$(3.4) \qquad \begin{aligned} w_0 + w_1 &= b - a \\ w_0 x_0 + w_1 x_1 &= \frac{1}{2}(b^2 - a^2) \\ w_0 x_0^2 + w_1 x_1^2 &= \frac{1}{3}(b^3 - a^3) \\ w_0 x_0^3 + w_1 x_1^3 &= \frac{1}{4}(b^4 - a^4) \end{aligned}$$

which we must solve. For $a = -1$, $b = 1$ it is not hard to show that the solution is $w_0 = 1$, $w_1 = 1$, $x_0 = -1/\sqrt{3}$, $x_1 = 1/\sqrt{3}$ as in Eq. (3.1). However, for larger n it's clear that this approach, based on solving an $n \times n$ nonlinear system, will be difficult to apply. Fortunately there's another way to find the locations of the nodes.

It's easiest to describe the approach in terms of **product integration rules**, which are weighted average rules of the form

$$(3.5) \qquad \int_a^b w(x) f(x) \, dx \approx w_0 f(x_0) + w_1 f(x_1) + \ldots + w_n f(x_n)$$

for some **weight function** $w(x)$ that is nonnegative on $[a, b]$[3]. Every rule we've seen so far has been of this form with $w(x) = 1$. We say that a rule of the form of Eq. (3.5) is a **Gaussian quadrature rule** (or **Gauss rule**) if it integrates all monomials of degree at most $2n + 1$ exactly, that is, if

$$\int_a^b w(x) p(x) \, dx = w_0 p(x_0) + w_1 p(x_1) + \ldots + w_n p(x_n)$$

---

[3]We will also require that $\int_a^b w(x) x^k \, dx$ converge for $k = 0, 1, 2, \ldots$.

for all polynomials p of degree less than or equal to $2n + 1$. Note that it's actually the integral of the function $w(x)p(x)$ that is found by the product rule but we say that it integrates polynomials exactly, not that it integrates the weight function times polynomials exactly (though the latter is precisely what it does).

EXAMPLE 2: Let $w(x) = (1 - x^2)^{-1/2}$. This weight function, called the **Chebyshev weight of the first kind**, is positive over $(-1, 1)$ and integrable over $[-1, 1]$ but is singular at $x = \pm 1$. The product integration rule

$$\int_{-1}^{1} w(x)f(x)dx \approx \frac{\pi}{n+1} \sum_{k=0}^{n} f(x_{k,n})$$

where

$$x_{k,n} = \cos\left(\frac{(k + \frac{1}{2})\pi}{n + 1}\right)$$

is a Gauss rule, discussed in detail in the next section. Hence it should integrate

$$\frac{x^2}{\sqrt{1 - x^2}}$$

(corresponding to $f(x) = x^2$) exactly. Let's take $n = 1$, which should be exact for polynomials of degree at most $2n + 1 = 3$. We have $x_{0,1} = \cos(\pi/4) = 1/\sqrt{2}$ and $x_{1,1} = \cos(3\pi/4) = -1/\sqrt{2}$, so

$$
\begin{aligned}
\int_{-1}^{1} \frac{x^2}{\sqrt{1 - x^2}} dx &\approx \frac{\pi}{2}(f(x_{0,1}) + f(x_{1,1})) \\
&= \frac{\pi}{2}(f(\cos(\pi/4)) + f(\cos(3\pi/4))) \\
&= \frac{\pi}{2}\left(\cos^2(\pi/4) + \cos^2(3\pi/4)\right) \\
&= \frac{\pi}{2}\left(\frac{1}{2} + \frac{1}{2}\right) \\
&= \frac{\pi}{2}
\end{aligned}
$$

and indeed it can be shown that

$$\int_{-1}^{1} \frac{x^2}{\sqrt{1 - x^2}} dx = \frac{\pi}{2}$$

so that the rule has integrated $f(x) = x^2$ exactly using only 2 points and without the need to manipulate the singular function $w(x)$. $\square$

This shows one advantage of writing a rule as a product integration rule: If we need to integrate $g(x) = w(x)f(x)$ and $w(x)$ is poorly behaved (highly oscillatory, unbounded at some point, expensive to compute, etc.) then we may be able to handle its difficulties analytically by burying its effect in the design of the method, that is, in the weights w_i, and only have to evaluate $f(x)$ numerically. In the example above, the Chebyshev weight function did not need to be evaluated during the quadrature.

Product integration rules are useful when $w(x)f(x)$ is hard to approximate but $f(x)$ isn't. After all, if a polynomial of low-to-moderate order can't approximate

$w(x)f(x)$ well, then none of the methods we've seen so far are likely to be successful. We could in principle approximate $w(x)f(x)$ by something other than an interpolant, such as a Taylor or Fourier series, but the product rule approach is generally simpler.

All Gauss rules are open rules, that is, they do not use the endpoints of the interval of integration. But to describe them will require a digression. We define the **inner product** of two real functions f, g with respect to the weight $w(x)$ over the interval $[a, b]$ by

$$(3.6) \qquad\qquad (f, g) = \int_a^b f(x)g(x)w(x)dx.$$

This has the usual properties of an inner product: Positive definiteness,

$$(f, f) \geq 0 \text{ and } (f, f) = 0 \text{ iff } f = 0$$

symmetry,

$$(f, g) = (g, f)$$

and linearity

$$(\alpha f + \beta g, h) = \alpha(f, h) + \beta(g, h)$$

($\alpha, \beta \in \mathbb{R}$), where f, g, and h are integrable functions over $[a, b]$. Two functions are said to be orthogonal with respect to this inner product if their inner product is zero.

Let P_n be the set of all polynomials in x of degree at most n[4]. Then P_n is a linear space, for the sum of two polynomials in x of degree at most n is a polynomial in x of degree at most n, and similarly for a scalar multiple of a polynomial in x of degree at most n. Every element of P_n is integrable over any interval $[a, b]$ so Eq. (3.6) defines an inner product on P_n, making it an **inner product space** (a linear space together with an inner product).

Since P_n is a linear space, it has a basis. The obvious basis is the set $M = \{1, x, x^2, ..., x^n\}$ of monomials, but it's always convenient to have an orthonormal basis. The Gram-Schmidt process can be used in an inner product space to make a linearly independent set of vectors into an orthonormal set. Let's skip the normalization for now and just look for an orthogonal basis. We start with

$$\begin{aligned} p_0(x) &= m_0(x) \\ &= 1 \end{aligned}$$

and apply the Gram-Schmidt process (see Sec. 2.9) but do not normalize at each step:

---

[4]Including, of course, the zero polynomial.

$$p_1(x) = m_1(x) - \frac{(p_0, m_1)}{(p_0, p_0)}p_0(x)$$

$$= x - \frac{1}{b-a}\left(\int_a^b 1 \cdot x \cdot w(x)dx\right) \cdot 1$$

(3.7)
$$= x - \frac{1}{b-a}\int_a^b xw(x)dx$$

$$p_2(x) = m_2(x) - \frac{(p_0, m_2)}{(p_0, p_0)}p_0(x) - \frac{(p_1, m_2)}{(p_1, p_1)}p_1(x)$$

(3.8)
$$= x^2 - \frac{1}{b-a}\left(\int_a^b 1 \cdot x^2 \cdot w(x)dx\right) \cdot 1 -$$

(3.9)
$$\frac{1}{(p_1, p_1)}\left(\int_a^b p_1(x) \cdot x^2 \cdot w(x)dx\right) \cdot p_1(x)$$

and so on. For a specific choice of $w(x)$ each of the definite integrals is just a scalar and so, for example, Eq. (3.7) shows that

$$p_1(x) = x - c_0$$

for some constant c, and so

$$p_2(x) = x^2 - c_1 x - c_2$$

(for some c_1, c_2) by Eq. (3.8). These polynomials are **monic**, that is, the coefficient of their leading terms is always unity, and by construction they are orthogonal,

$$(p_k, p_l) = \int_a^b p_k(x)p_l(x)w(x)dx = 0$$

if $k \neq l$. In fact, note that if $g(x)$ is a polynomial of degree at most m then

$$g(x) = c_0 p_0(x) + ... + c_n p_m(x)$$

by the fact that $\{p_0, ..., p_m\}$ is a basis for P_m, and so

$$\begin{aligned}(p_{m+1}, g) &= (p_{m+1}, c_0 p_0(x) + ... + c_n p_m(x)) \\ &= c_0(p_{m+1}, p_0(x)) + ... + c_n(p_{m+1}, p_m(x)) \\ &= 0 + ... + 0 \\ &= 0\end{aligned}$$

since $\{p_0, ..., p_m, p_{m+1}\}$ is orthogonal. That is, the polynomial p_{m+1} is orthogonal to *every* polynomial of degree at most m, not just to $p_0, ..., p_m$.

This procedure can be continued indefinitely, getting an orthogonal basis for P_n for larger and larger values of n, and hence can be applied to $\{1, x, x^2, ...\}$ to define an orthogonal family of polynomials

$$\{p_i(x)\}_{i=0}^{\infty}$$

(for a fixed weight function $w(x)$ that determines the specific family, and on a fixed interval $[a, b]$). In summary, we have an infinite sequence $p_0, p_1, p_2, \ldots$ of monic, pairwise orthogonal polynomials. For some applications being monic is more convenient than being orthonormal, but it's easy enough to form

$$\left\{ \frac{p_i(x)}{\sqrt{(p_i(x), p_i(x))}} \right\}_{i=0}^{\infty}$$

if we want to have an orthonormal family of polynomials w.r.t. the vector norm—where the "vectors" now are functions–determined from the inner product as

$$\|f\| = \sqrt{(f, f)}$$

which does satisfy the properties of an inner product, as given previously for vectors in $\mathbb{R}^n$ (see Sec. 2.6).

So what? Well, it turns out that a great deal is known about such families, and they occur in several areas of applied mathematics. The zeroes of any member of such a family are real and distinct and located in $[a, b]$. That is, p_{n+1} has $n+1$ zeroes $a < x_0 < x_1 < \cdots < x_n < b$. Our Gauss rule

$$\int_a^b w(x)f(x)dx = w_0 f(x_0) + w_1 f(x_1) + \ldots + w_n f(x_n)$$

will use the zeroes of p_{n+1} as its nodes and will have the weights

$$w_i = \int_a^b L_{n,i}(x)w(x)dx$$

$(i = 0, 1, \ldots, n)$. Let's prove this fact:

THEOREM 1: Let $\{p_i\}_{i=0}^{\infty}$ be a family of orthogonal polynomials with respect to the weight function $w(x)$ over $[a, b]$. Let $x_0, x_1, \ldots x_n$ be the zeroes of p_{n+1}. Then the quadrature rule

$$\int_a^b w(x)f(x)dx \approx w_0 f(x_0) + w_1 f(x_1) + \ldots + w_n f(x_n)$$

where

$$w_i = \int_a^b L_{n,i}(x)w(x)dx$$

$(i = 0, 1, \ldots, n)$ integrates $f(x) = x^k$ exactly for $k = 0, 1, \ldots, 2n+1$.

PROOF: Since the weights w_i are the weights found by replacing f by an interpolating polynomial of degree n the method must integrate $f(x) = x^k$ exactly for $k = 0, 1, \ldots, n$. Suppose that $f(x)$ is any polynomial of degree at most $2n+1$. Then

$$f(x) = Q(x)p_{n+1}(x) + R(x)$$

where $Q(x)$ is the quotient upon (synthetic) division by $p_{n+1}(x)$ and $R(x)$ is the remainder. Both $Q(x)$ and $R(x)$ must have degree at most n because $p_{n+1}(x)$ has degree $n+1$. Then applying the quadrature rule gives

$$w_0 f(x_0) + w_1 f(x_1) + \ldots + w_n f(x_n) = \sum_{i-0}^{n} w_i (Q(x_i) p_{n+1}(x_i) + R(x_i))$$

$$= \sum_{i-0}^{n} w_i R(x_i)$$

because we chose the nodes so that $p_{n+1}(x_i) = 0$ $(i = 0, 1, \ldots, n)$. But the rule is exact for polynomials of degree at most n, so

$$\sum_{i-0}^{n} w_i R(x_i) = \int_a^b w(x) R(x) dx$$

$$= \int_a^b w(x) R(x) dx + \int_a^b w(x) Q(x) p_{n+1}(x) dx$$

since $(p_{n+1}(x), Q(x)) = 0$ due to the fact that $Q(x)$ is of degree at most n. Pulling this all together we have

$$w_0 f(x_0) + w_1 f(x_1) + \ldots + w_n f(x_n) = \int_a^b w(x) R(x) dx +$$

$$\int_a^b w(x) Q(x) p_{n+1}(x) dx$$

$$= \int_a^b w(x) (Q(x) p_{n+1}(x) R(x)) dx$$

$$= \int_a^b w(x) f(x) dx$$

if $f(x)$ is a polynomial of degree at most $2n + 1$. $\square$

This shows that the degree of accuracy of a Gauss rule is at least $2n + 1$, and in fact no weighted average rule

$$\int_a^b w(x) f(x) dx \approx w_0 f(x_0) + w_1 f(x_1) + \ldots + w_n f(x_n)$$

can have degree of accuracy $2n + 2$, so Gauss rules are optimal in this sense. It is also a fact that the weights w_i satisfy $w_i \in [0, 1]$ which is beneficial with respect to round-off error (in particular, $w_i > 0$). Gauss rules are not strictly optimal with respect to absolute error but do have excellent error properties. The formula for the simple rules is

(3.10)
$$E = \frac{f^{(2n+2)}(\gamma)}{(2n+2)!} (p_{n+1}, p_{n+1})$$

for some $\gamma \in (a, b)$. Of course, these rules can also be used in composite form, as with the Newton-Cotes rules.

We have reduced the problem of finding rules with degree of accuracy as high as possible to choosing a weight $w(x)$, generating the relevant orthogonal polynomials, finding their roots, and then finding the weights

$$w_i = \int_a^b L_{n,i}(x)w(x)dx$$

($i = 0, 1, ..., n$). But we know that finding all roots of a polynomial usually isn't easy; even in a case like this, where the roots are real, simple, and located in a known interval, for large n the roots near the endpoints of the interval may not be well separated[5]. We'll discuss more efficient techniques for generating these values.

The case $w(x) = 1$ is obviously of special interest. The orthogonal polynomials in this case are known as the **Legendre polynomials** if $[a,b] = [-1,1]$. (See Fig. 1 for the first four.) More general intervals may be handled by a simple linear transformation between $[-1,1]$ and $[a,b]$.

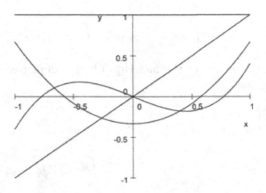

FIGURE 1. Legendre polynomials.

The quadrature rule derived from this family is called **Gauss-Legendre quadrature**. The formula Eq. (3.1) is the Gauss-Legendre rule for $n = 1$, and the midpoint rule is the case $n = 0$.

The MATLAB `integral` command is based on a variation of the Gaussian quadrature idea. Its predecessor, the `quad` command, used Simpson's rule.

MATLAB

The Gauss-Legendre rule based on 1 point is the midpoint rule and the Gauss-Legendre rule based on 2 points is Eq. (3.1). Let's find the rule based on 3 points. We take $p_0(x) = 1$; from Eq. (3.7), $p_1(x) = x$, and from Eq. (3.8), $p_2(x) = x^2 - 1/3$. If you have the Symbolic Toolbox, enter:

```
>> syms x p2
>> p2=x^2-1/3;
>> p3=x^3-int(1*x^3,x,-1,1)/int(1,x,-1,1)-...
   int(x*x^3,x,-1,1)*x/int(x^2,x,-1,1)-...
   int(p2*x^3,x,-1,1)*x/int(p2^2,x,-1,1)
```

to see that $p_3(x) = x^3 - \frac{3}{5}x$. Let's check:

```
>> clear all
>> p=@(x) x.^3-.6*x
```

---

[5]"We did a fair amount of work on Gaussian integration, which was a very hard thing to do in the pre computer days." –Philip J. Davis, on computations done with Philip Rabinowitz at what is now NIST in the mid-1950s.

```
>> integral(p,-1,1)
>> integral(@(x) x.*p(x),-1,1)
>> integral(@(x) (x.^2).*p(x),-1,1)
```
Yes, $p_3(x)$ appears to be orthogonal to $p_0(x)$, $p_1(x)$, and $p_2(x)$. Enter:
```
>> nodes=roots([1 0 -.6 0])
```
to get the nodes ($x_0 = -\sqrt{3/5}, x_1 = 0, x_2 = \sqrt{3/5}$). Now let's generate the weights using the Lagrange interpolating polynomials. Enter:
```
>> s=sqrt(3/5)
>> w0=integral(@(x) x.*(x-s)./((-s)*(-s-s)),-1,1)
>> w1=integral(@(x) (x+s).*(x-s)./(s*(-s)),-1,1)
>> w2=integral(@(x) x.*(x+s)./(s*(s+s)),-1,1)
>> format rat   %Display approximately as ratios of small integers.
>> w0,w1,w2
>> format      %Reset.
```
The weights appear to be 5/9, 8/9, and 5/9. (Since the functions are just polynomials it would have been easy to integrate them analytically or symbolically and verify this.) Let's check; we have used $n = 2$ so we should be able to integrate polynomials of degree at most 5 exactly. Enter:
```
>> nodes=sort(nodes)           %Reorder the nodes.
>> g=@(x) 3*x.^5-7*x.^4+2*x.^3-5*x.^2-8*x+3
>> w0*g(nodes(1))+w1*g(nodes(2))+w2*g(nodes(3)) %Gauss-Legendre.
>> integral(g,-1,1)
```
The answers agree, so this checks. The Gauss-Legendre rule based on 3 points is $\int_a^b f(x)dx \approx \frac{5}{9}f(-\sqrt{3/5}) + \frac{8}{9}f(0) + \frac{5}{9}f(\sqrt{3/5})$.

Let's plot the first three Legendre polynomials. Some patterns already seem clear (every other term is missing, for example). Enter:
```
>> z=-1:.01:1;
>> plot(z,1,'b'),hold on
>> plot(z,z,'r')
>> plot(z,z.^2-1/3,'g')
>> plot(z,z.^3-3*z/5,'k')
>> axis([-1 1 -1.1 1.1])
>> grid
```
The polynomials appear to be constrained to take on values that remain within $[-1, 1]$. Indeed, this is the case, and for this reason the change of variables $x = \cos(\phi)$ is often used to simplify formulas involving Legendre polynomials. The orthogonality is difficult to see from the plot as it represents the fact that the innder product $\int_a^b p_k(x)p_l(x)dx$ is zero if $k \neq l$ and not anything about right angles, but you may be able to imagine how the areas cancel.

The Legendre polynomials as we have defined them are orthogonal but not orthonormal on $[a, b]$. You may see the Legendre polynomials defined in other texts as normalized in more than one way. This does not affect the location of the roots.

Problems

1.) Show that Eq. (3.2) is correct. Use it to approximate $\int_2^4 \sin(x^2)dx$ using Eq. (3.1).

2.) Show that every interpolatory rule for $\int_a^b f(x)w(x)dx$ has the weights $w_i = \int_a^b L_{n,i}(x)w(x)dx$.

3.) Show that for $a = -1$, $b = 1$ the solution of Eq. (3.4) is $w_0 = 1$, $w_1 = 1$, $x_0 = -1/\sqrt{3}$, $x_1 = 1/\sqrt{3}$.

4.) Using the norm $\|f\| = (f, f)^{1/2}$, with $w(x) = 1$ and $[a, b] = [-1, 1]$, find $\|1\|$, $\|x\|$, and $\|\sin(x)\|$.

5.) What are the first three Legendre polynomials? Show that the midpoint rule is the Gauss-Legendre rule for $n = 0$ and that Eq. (3.1) is the Gauss-Legendre rule based on $n + 1 = 2$ points.

6.) Prove that no weighted average formula for $\int_a^b w(x)f(x)dx$ can have degree of accuracy $2n + 2$. Hint: Consider the polynomial $(x - x_0)^2(x - x_1)^2 \cdots (x - x_n)^2$.

7.) a.) Find the Gauss-Legendre rules based on 2 and 3 points.

b.) Use them to approximate $\int_0^2 (1 + e^x)^{-2}dx$. Bound the errors using Eq. (3.10).

8.) a.) Determine the Gauss-Legendre rule based on 4 points.

b.) Determine the Gauss-Legendre rule based on 5 points. Did the work done in part a. simplify these calculations?

c.) Use the Gauss-Legendre rules based on N points, $N = 1, 2, 3, 4, 5$, to approximate $\int_{-1}^1 x^3 \ln(x)dx = -20 \ln(2) + 14$.

d.) Repeat part c. for $\int_{-1}^1 x^{m/2}dx$, $m = 1, 3, 5$.

e.) Bound the error in your answers to part c. using Eq. (3.10). Compare to the actual errors.

9.) Conduct an experiment to compare the open Newton-Cotes, closed Newton-Cotes, and Gauss-Legendre rules based on n points for all n for which you have formulae. Discuss your results.

10.) a.) Use the method of undetermined coefficients to derive a product integration rule of the form $\int_0^{2\pi} f(x)\cos(x)dx \approx w_0 f(0) + w_1 f(\pi) + w_2 f(2\pi)$. This is the first **Filon cosine formula**. For which monomials is it exact?

b.) Compare this rule to Simpson's rule, which is also based on 3 points, for several choices of f.

c.) Use the method of undetermined coefficients to derive a product integration rule of the form $\int_0^{2\pi} f(x)\sin(x)dx \approx w_0 f(0) + w_1 f(\pi) + w_2 f(2\pi)$. This is the first **Filon sine formula**. For which monomials is it exact?

d.) Compare this rule to Simpson's rule and the Gauss-Legendre rule based on 3 points for several choices of f.

11.) Let $x = \cos(\phi)$ and write $p_n(x) = p_n(\cos(\phi))$ for the nth Legendre polynomial. Plot $p_n(\cos(\phi))$ as a function of ϕ for $n = 0, 1, ..., 10$. Comment on the qualitative properties of these graphs.

12.) a.) Let $w(x) = \exp(-x^2)$ and $a = -\infty$, $b = \infty$. The orthogonal polynomials generated by the inner product $(f, g) = \int_{-\infty}^{\infty} f(x)g(x)\exp(-x^2)dx$ are the **Hermite polynomials**. Find the first 4 Hermite polynomials.

b.) The quadrature rule for $\int_{-\infty}^{\infty} f(x)\exp(-x^2)dx$ which these polynomials define is called **Gauss-Hermite quadrature**. Find the first 4 Gauss-Hermite quadrature rules.

c.) Use the first 4 Gauss-Hermite quadrature rules to approximate the improper integral $\int_{-\infty}^{\infty} \cos(x)\exp(-x^2)dx$.

13.) Use composite Gauss-Legendre integration based on 3 points to approximate the indefinite integral $\int_0^x \sin(t)dt$ at $x = 0, .1, .2, ..., 6.4$. Plot your integral. Use linear interpolation of these values to estimate the value at $x = 2\pi$.

14.) For $n = 0, 1, 2, 3$, show that the Legendre polynomial $y = p_n(x)$ satisfies **Legendre's differential equation** $(1 - x^2)y'' - 2xy' + n(n+1)y = 0$. (This is actually true for all $n = 0, 1, 2, ...$) Is this true even at ± 1?

15.) Normalize the Legendre polynomials p_0, p_1, p_2, p_3 by dividing them by their respective norms $(p_i, p_i)^{1/2}$.

16.) What polynomials are generated by $w(x) = 1$ on $[0, 2]$? How do they relate to the Legendre polynomials?

4. Gauss-Chebyshev Quadrature

Gauss-Legendre rules are the most popular Gauss rules because $w(x) = 1$ is so convenient; in fact, the term Gauss rule is often used to mean Gauss-Legendre rule. For a general-purpose routine like the `integral` command in MATLAB this is clearly a good choice: A routine is needed that will work on as broad a class of integrands as possible, because it's intended to be used by a wide class of users, not all of whom will have training in numerical analysis.

However, it's important to have special rules for special cases such as noisy functions, nonsmooth functions, singular at points within the interval, highly oscillatory integrands, infinite intervals of integration, and so on. There are many such rules that may be used for difficult quadrature problems. Many Gauss rules are of this sort, because they use an interval of integration that is unbounded, as with **Gauss-Laguerre quadrature** which is based on the **Laguerre polynomials**, generated by the weight function

$$w(x) = \exp(-x)$$

on the interval $[0, \infty)$. This provides a rule for approximating $\int_0^\infty f(x) \exp(-x) dx$ that could be used for the numerical calculation of Laplace transforms, for example.

Let's continue to explore Gaussian quadrature. We say that a weight function is **admissible** if it is nonnegative on $[a, b]$ (possibly equal to positive infinity at some points) and

$$\int_a^b w(x)x^k dx$$

converges for $k = 0, 1, 2, ...$ Often we normalize $w(x)$ so that

$$\int_a^b w(x)dx = 1$$

but this is not essential. Every admissible weight function that is positive on (a, b) generates a sequence of orthogonal polynomials and hence a Gauss rule[6]. The orthogonal polynomials can be generated by the Gram-Schmidt process (with normalization optional) as described in the previous section:

---

[6]Sometimes we allow $w(x)$ to be of mixed sign. Such rules, if they exist for a given $w(x)$, can have degree of accuracy even greater than Gauss rules.

$$p_1(x) = m_1(x) - \frac{(p_0, m_1)}{(p_0, p_0)} p_0(x)$$

$$p_2(x) = m_2(x) - \frac{(p_0, m_2)}{(p_0, p_0)} p_0(x) - \frac{(p_1, m_2)}{(p_1, p_1)} p_1(x)$$

$$p_3(x) = m_3(x) - \frac{(p_0, m_3)}{(p_0, p_0)} p_0(x) - \frac{(p_1, m_3)}{(p_1, p_1)} p_1(x) - \frac{(p_2, m_3)}{(p_2, p_2)} p_2(x)$$

$$\vdots$$

where $m_i(x) = x^i$ ($i = 0, 1, 2, ...$) and $p_0(x) = 1$. Notice that these equations are of the form

$$p_1(x) = m_1(x) - c_0 p_0(x)$$

$$p_2(x) = m_2(x) - c_1 p_0(x) - c_2 p_1(x)$$

$$p_3(x) = m_3(x) - c_3 p_0(x) - c_4 p_1(x) - c_5 p_2(x)$$

$$\vdots$$

for some constants c_i formed from inner products of monomials with orthogonal polynomials. We might expect some of these constants to be zero by orthogonality and indeed, it is always possible to generate the orthogonal polynomials corresponding to a given weight function $w(x)$ and interval by

(4.1) $$p_{n+1}(x) = x p_n(x) - \alpha_{n+1} p_n(x) - \beta_{n+1} p_{n-1}(x)$$

called a **three-term recurrence relation** because the three terms p_{n+1}, p_n, and p_{n-1} appear in it. All orthogonal polynomials satisfy a three-term recurrence relation of this form with

$$\alpha_{n+1} = \frac{(p_n, x p_n)}{(p_n, p_n)}$$

$$= \frac{\int_a^b x p_n^2(x) w(x) dx}{\int_a^b p_n^2(x) w(x) dx}$$

$$\beta_{n+1} = \frac{(p_n, x p_{n-1})}{(p_{n-1}, p_{n-1})}$$

$$= \frac{\int_a^b x p_n(x) p_{n-1}(x) w(x) dx}{\int_a^b p_{n-1}^2(x) w(x) dx}$$

($n = 0, 1, ...$) where we define $p_{-1}(x) = 0$ so that $\beta_1 = 0$. Every polynomial generated in this way, for every choice of $w(x)$, has zeroes that are real, simple, and lie entirely in (a, b). If the weight function is symmetric with respect to the midpoint m of the interval $[a, b]$ then the roots are also located symmetrically about m and the Gauss rule weights for the symmetrically placed points $\pm x_i = m \pm \epsilon_i$ are the same.

In fact, one can give a formula for the weights: The weights in a Gauss rule based on $n + 1$ points are given by

$$(4.2) \qquad w_i = -\frac{(p_{n+1}, p_{n+1})}{p_{n+2}(x_i)p'_{n+1}(x_i)}$$

$(i = 0, 1, ..., n)$, where the x_i are the zeroes of $p_{n+1}(x)$ There are other formulas for the weights of a Gauss rule.

EXAMPLE 1: The first two Legendre polynomials are $p_0(x) = 1$ and $p_1(x) = x$ so the next Legendre polynomial is, by Eq. (4.1),

$$\begin{aligned} p_2(x) &= xp_1(x) - \alpha_2 p_1(x) - \beta_2 p_0(x) \\ &= x \cdot x - \frac{\int_{-1}^{1} x \cdot x^2 dx}{\int_{-1}^{1} x^2 dx} x - \frac{\int_{-1}^{1} x \cdot x \cdot 1 dx}{\int_{-1}^{1} 1^2 dx} 1 \\ &= x^2 - 0 \cdot x - \frac{2/3}{2} \cdot 1 \\ &= x^2 - \frac{1}{3} \end{aligned}$$

$(x_0 = -1/\sqrt{3}, \; x_1 = 1/\sqrt{3})$ and the corresponding weights in the Gauss-Legendre rule are given by Eq. (4.2),

$$\begin{aligned} w_0 &= -\frac{(p_2, p_2)}{p_3(x_0)p'_2(x_0)} \\ &= -\frac{\int_{-1}^{1} \left(x^2 - \frac{1}{3}\right)^2 dx}{p_3(-1/\sqrt{3})p'_2(-1/\sqrt{3})} \\ &= -\frac{8/45}{-8/45} \\ &= 1 \end{aligned}$$

(using $p_3(x) = x^3 - \frac{3}{5}x$), and $w_1 = w_0$ because $p_2(x) = x^2 - \frac{1}{3}$ is symmetric with respect to the origin. The weights are $w_0 = 1, w_1 = 1$. $\square$

Unlike the simple examples we have seen so far, the weights are in general irrational numbers that do not have simple expressions. The nodes and weights may be explicitly coded into a program that computes with a fixed number of points $n + 1$ or may be generated by a numerical method (e.g., Newton's method to find the roots and a formula like Eq. (4.2) for the weights). It's common to see a program listing for a numerical integration routine that begins with a listing of the nodes and weights for, say, $n = 2$ to $n = 20$ so that the user may select the desired order. Hence we do not expect to pay the computational cost of generating the nodes and weights at runtime.

Another important class of Gaussian quadrature rules is based on the interval $[-1, 1]$ and either the weight function

$$w(x) = \frac{1}{\sqrt{1 - x^2}}$$

which gives rise to the **Chebyshev**[7] **polynomials of the first kind** $T_n(x)$ and hence the **Gauss-Chebyshev rule of the first kind**, or the weight function

---

[7]You will see this name transliterated in different ways in different texts.

$$w(x) = \sqrt{1 - x^2}$$

which gives rise to the **Chebyshev polynomials of the second kind** $U_n(x)$ and hence the **Gauss-Chebyshev rule of the second kind**. These are called the **Gauss-Chebyshev rules**. The term Gauss-Chebyshev rule alone generally refers to the rule of the first kind, and similarly for Chebyshev polynomials. We'll focus on Chebyshev polynomials of the first kind.

The first two Chebyshev polynomials are $T_0(x) = 1$ and $T_1(x) = x$. Others may be found from the three-term recurrence relation (Eq. (4.1)). A convenient representation is given by

$$(4.3) \qquad\qquad T_n(x) \;=\; \frac{1}{2^{n-1}}\cos(n\arccos(x))$$

$$=\; \frac{1}{2^{n-1}}\cos(n\theta)$$

$(n \geq 1)$ using the change of variables $\theta = \arccos(x)$. Clearly, $T_n(x)$ takes on values in $[-1, 1]$ (in fact, in $[-2^{-(n-1)}, 2^{-(n-1)}]$). It is not hard to show that

$$(4.4) \qquad\qquad T_{n+1}(x) = xT_n(x) - \frac{1}{4}T_{n-1}(x)$$

for $n \geq 2$ (and $T_2(x) = xT_1(x) - \frac{1}{2}T_0(x)$), and that the roots of $T_n(x) = 0$ are given by

$$x_k = \cos\left(\frac{2k+1}{2n}\pi\right)$$

$(n \geq 1, k = 0, 1, ..., n - 1)$, called the **Chebyshev nodes** (or **Chebyshev points**). A Gauss-Chebyshev rule based on $n + 1$ points would use the Chebyshev points that are the $n + 1$ roots of $T_{n+1}(x) = 0$. Note that this formula does not give the points ordered as $x_0 < x_1 < \cdots < x_{n-1}$.

It's apparent from the figure that the Chebyshev points tend to bunch up near the endpoints. This is entirely general, and becomes more pronounced as $n \to \infty$. In fact, the Chebyshev points are the points at which one would generally wish to interpolate a function defined on $[-1, 1]$ (or, with a linear change of variable, any finite interval) because of the following result:

THEOREM 1: If $n \geq 1$ and $q(x)$ is a monic polynomial of degree n

$$\max_{x \in [-1,1]} (|T_n(x)|) < \max_{x \in [-1,1]} (|q(x)|)$$

unless $q(x) = T_n(x)$.

This states that $T_n(x)$ is the monic polynomial of degree n that is "smallest" on $[-1, 1]$ in the sense of having the least deviation from 0 (in the max norm; plot a few cases to see this). Recall Theorem 4.1.1 regarding polynomial interpolation: Let $a \leq x_0 < x_1 < \cdots < x_n \leq b$ and f in $C^{n+1}[a, b]$ be given. Then the polynomial interpolant $p(x)$ to f at $x_0, x_1, ..., x_n$ satisfies

$$(4.5) \qquad f(x) - p(x) = \frac{f^{(n+1)}(\xi)}{(n+1)!}(x - x_0)(x - x_1)\cdots(x - x_n)$$

for any $x \in [a,b]$, for some $\xi = \xi(x)$ in (a,b). We cannot control $\xi(x)$ but in all likelihood $f^{(n+1)}(\xi(x))$ does not vary much over the interval. We can however control the term

$$\varpi(x) = (x - x_0)(x - x_1)\cdots(x - x_n)$$

(see also Eq. (6.1.1.8)). By Theorem 1, this quantity is minimized over $[-1,1]$ by choosing the nodes to be the $n+1$ zeroes of $T_{n+1}(x)$, namely

$$(4.6) \qquad x_k = \cos\left(\frac{2k+1}{2n+2}\pi\right)$$

($k = 0, 1, ..., n$). While this may not strictly minimize the RHS of Eq. (4.5) because of the presence of the unknown $\xi(x)$ it is very likely be very close to the minimum, and is the best we can do with the available information. Hence, as a rule interpolation at the Chebyshev points is to be preferred when we are not forced to use other points.

This also explains why Gauss-Chebyshev quadrature is the basis for a number of popular rules. All Gauss rules are interpolatory rules; they interpolate the function $f(x)$ in

$$\int_a^b w(x)f(x)dx$$

with a polynomial. The use of the Chebyshev points gives the polynomial interpolant of $f(x)$ that minimizes the maximum deviation $|f(x) - p(x)|$ over the interval of integration, which certainly should reduce the error. The error bound for Gauss-Chebyshev quadrature of the first kind is

$$E = \frac{\pi}{2^{2n+1}(2n+2)!}f^{(2n)}(\gamma_1)$$

for some $\gamma_1 \in (-1,1)$. The error bound for Gauss-Chebyshev quadrature of the second kind is

$$E = \frac{\pi}{2^{2n+3}(2n+2)!}f^{(2n)}(\gamma_2)$$

for some $\gamma_2 \in (-1,1)$. The bound for in the second case is smaller by a factor of 4 (assuming $f^{(2n)}(\gamma_1) \approx f^{(2n)}(\gamma_2)$) but the rule of the first kind is easier to apply because of the surprising fact that

$$w_i = \frac{\pi}{n+1}$$

($i = 0, 1, ..., n$) are the weights of the Gauss-Chebyshev rule of the first kind. Since w_i doesn't depend on i, just n, we can save work in generating the weights and in computing the weighted average.

EXAMPLE 2: Let's approximate $\int_{-1}^{1} \cos(\pi x)dx$ using Gauss-Chebyshev quadrature. We must first write this as

$$\int_{-1}^{1} \cos(\pi x)dx = \int_{-1}^{1} \left(\cos(\pi x)\sqrt{1-x^2}\right)\frac{1}{\sqrt{1-x^2}}dx$$

$$= \int_{-1}^{1} f(x)w(x)dx$$

where $f(x) = \cos(\pi x)\sqrt{1-x^2}$. Using $n+1 = 5$ points gives

$$\int_{-1}^{1} \cos(\pi x)dx \approx \frac{\pi}{n+1}\sum_{i=0}^{n} f(x_i)$$

$$= \frac{\pi}{5}\left(f(x_0) + f(x_1) + f(x_2) + f(x_3) + f(x_4)\right)$$

$$= \frac{\pi}{5}[f\left(\cos\left(\pi/10\right)\right) + f\left(\cos\left(3\pi/10\right)\right) + f\left(\cos\left(5\pi/10\right)\right) +$$

$$f\left(\cos\left(7\pi/10\right)\right) + f\left(\cos\left(9\pi/10\right)\right)]$$

$$\doteq -.0323$$

and the correct value is $\int_{-1}^{1} \cos(\pi x)dx = 0$. If we use instead $n+1 = 21$ points we get $-.0019$. These results are not very impressive, but we have traded a nice integrand like $\cos(\pi x)$ for the much less nice integrand $f(x) = \cos(\pi x)\sqrt{1-x^2}$. Let's try another example. We'll approximate $\int_{-1}^{1} \cos(\pi x)/\sqrt{1-x^2}dx$. We have

$$\int_{-1}^{1} \cos(\pi x)/\sqrt{1-x^2}dx \approx \frac{\pi}{n+1}\sum_{i=0}^{n} \cos(\pi x_i)$$

$$\doteq -.9557$$

while the actual value is $-.9558$. This is excellent agreement for only 5 points. If $f(x)$ is already an ugly function then writing $\int_{-1}^{1} f(x)dx$ as

$$\int_{-1}^{1} \left(f(x)\sqrt{1-x^2}\right)/\sqrt{1-x^2}dx$$

and applying Gauss-Chebyshev quadrature to $g(x) = f(x)\sqrt{1-x^2}$ is likely to be fine; if $f(x)$ is very well-behaved this may not be advisable. $\square$

All weighted average rules that have equal weights $w_0 = w_1 = \cdots = w_n$ for a given n, that is, all rules of the form

$$\int_{a}^{b} w(x)f(x)dx \approx c_{n+1}\sum_{i=0}^{n+1} f(x_i)$$

($w_i = c_{n+1}$ for all $i = 0,1,...,n$) are called **Chebyshev-type rules**. The use of equal weights reduces the effects of round-off error, which can be especially useful for noisy functions where f is not known very accurately.

In light of Theorem 1 we might try using an interpolatory rule for $\int_{-1}^{1} f(x)dx$ based on interpolating $f(x)$ at the Chebyshev nodes but without using a weight function: That is, we might use the rule

$$\int_{-1}^{1} f(x)dx \approx \sum_{i=0}^{n+1} w_n f(x_i)$$

where the x_i are the Chebyshev points and the weights w_i are determined in the usual manner,

$$w_i = \int_{-1}^{1} L_{n,i}(x)dx.$$

Mapping the Chebyshev nodes from $[-1, 1]$ to $[a, b]$ and applying this idea gives a rule known as the **classical Clenshaw-Curtis method** (or **Fejér formula**); interpolating $f(x)$ at the Chebyshev *extrema* mapped from $[-1, 1]$ to $[a, b]$ gives a rule known as the **practical Clenshaw-Curtis method**. The practical Clenshaw-Curtis formulae have certain computational advantages over, and similar performance to, the classical Clenshaw-Curtis formulae, which in turn are generally superior to Newton-Cotes formulae though with about the same degree of accuracy. Applying the Clenshaw-Curtis idea with a weight function other than $w(x) = 1$ gives a **modified Clenshaw-Curtis method**. A Clenshaw-Curtis method is the default numerical integrator in MAPLE.

MATLAB

Let's plot the first several Chebyshev polynomials of the first kind. We'll enter them explicitly rather than generate them by the recurrence relation. Enter:

```
>> x=-1:.01:1;T0=ones(size(x));
>> T1=x;
>> T2=x.^2-.5;
>> T3=x.^3-.75*x;
>> T4=x.^4-x.^2+1/8;
>> T5=x.^5-.8*x.^3+(5/16)*x;
>> plot(x,[T0;T1;T2;T3;T4;T5]),grid
```

You can tell which polynomial is which by the number of roots (n for the degree n polynomial) or extrema ($n - 1$), and by the fact that the higher the degree the smaller the maximum deviation of the polynomial from 0.

Let's look at the interpolation properties of the Chebyshev points. The points are given by $x_k = \cos\left(\frac{2k+1}{2n+2}\pi\right)$ for $k = 0, 1, ..., n$. We'll use the Runge function (see Sec. 5.1) which we know is a difficult function for equally spaced points. Enter:

```
>> clear all
>> f=@(x) 1./(1+25*x.^2)
>> n=8;k=0:n;
>> x=-1:.01:1;
>> xc=cos(pi*(2*k+1)/(2*n+2))        %Chebyshev points.
>> xl=-1:.25:1                        %Equally spaced points.
>> xr=2*rand([1 9])-1                 %Randomly chosen points.
>> xu=[-.95 -.8 -.6 -.3 0 .3 .6 .8 .95] %Unequally spaced points.
>> pc=polyfit(xc,f(xc),8);
>> pl=polyfit(xl,f(xl),8);
>> pr=polyfit(xr,f(xr),8);
>> pu=polyfit(xu,f(xu),8);
>> yc=polyval(pc,x);
```

```
>> yl=polyval(pl,x);
>> yr=polyval(pr,x);
>> yu=polyval(pu,x);
>> plot(x,yc,'b',x,yl,'c',x,yr,'r',x,yu,'g',x,f(x),'k'),grid
```
The blue curve should be the best fit to the black curve. Enter:
```
>> max(abs(yc-f(x)))
>> max(abs(yl-f(x)))
>> max(abs(yr-f(x)))
>> max(abs(yu-f(x)))
```
The error is minimized by the Chebyshev nodes. Repeat this for a higher n (recall that interpolation by high-degree polynomials is usually a bad idea, especially with equally spaced nodes).

Of course, the Runge function is a particularly difficult function for interpolation; try this experiment again using $f(x) = \sin(x)$. Make a reasonable choice for xu (by eye).

Problems

1.) a.) Use the Gauss-Chebyshev rules based on $1, 2, 3, 4, 5$ points to approximate $\int_{-1}^{1} \sin(x) / \left(1 - x^2\right)^{1/2} dx$. Explain your results.

b.) Repeat part a. for $\int_{0}^{5} x \cos(3x) dx$. You will need to make a change of variables and then write the integrand as $f(x) / \left(1 - x^2\right)^{1/2}$ for some $f(x)$.

2.) Use Eq. (4.2) to derive the weights for the Gauss-Legendre rule based on 3 points.

3.) Which rules that we have seen so far are Chebyshev-type rules?

4.) The Chebyshev polynomials of the second kind $U_{n+1}(x)$ have zeroes $x_k = \cos((k + 1)\pi/(n + 2))$, $k = 0, 1, ..., n$, and the Gauss-Chebyshev rule of the second kind has weights $w_k = (\pi/(n + 2)) \sin^2((k + 1)\pi/(n + 2))$, $k = 0, 1, ..., n$. Use the Gauss-Chebyshev rule of the second kind based on $1, 2, 3, 4, 5$ points to approximate $\int_{0}^{5} x \cos(3x) dx$.

5.) a.) Find the first 5 Laguerre polynomials using Eq. (4.1) then use Eq. (4.2) to derive as many Gauss-Laguerre rules as you can from these 5 polynomials. State the degrees of accuracy of each rule.

b.) Use your Gauss-Laguerre rules to approximate $\int_{0}^{\infty} \sin(x) \exp(-x) dx$.

6.) Show that the Chebyshev polynomials are defined by Eq. (4.3) by proving directly that $\int_{-1}^{1} T_k(x) T_l(x) w(x) dx = 0$ if $k \neq l$ $(w(x) = \left(1 - x^2\right)^{-1/2})$, then showing that $T_n(x)$ is monic.

7.) a.) Show that the nth Legendre polynomial only contains even powers of x and is an even function if n is even, and only contains odd powers of x and is an odd function if n is odd.

b.) Show that the nth Chebyshev polynomial of the first kind only contains even powers of x and is an even function if n is even and only contains odd powers of x and is an odd function if n is odd.

8.) a.) Prove that the three-term recurrence relation Eq. (4.1) is correct by first arguing that it is possible to write $p_{n+1}(x) = xp_n(x) - \alpha_{n+1}p_n(x) - \beta_{n+1}p_{n-1}(x) - ... - \gamma_{n+1}p_0(x)$ and then considering (p_{n+1}, p_n).

b.) Prove that Eq. (4.4) is correct.

9.) Use the Gauss-Chebyshev rules of the first and second kind based on $5, 10, 15, 20$ points to estimate the area under the Runge function. Do the same with the closed and open Newton-Cotes formulae based on $5, 10, 15, 20$ points.

10.) a.) Prove that the Chebyshev nodes are given by Eq. (4.6).

b.) Prove that the extrema of $T_n(x)$ are given by $y_k = \cos(k\pi/n)$, $k = 0, 1, ..., n$. What is $T_n(y_k)$?

11.) The Chebyshev polynomials of the second kind $U_n(x)$ satisfy $U_n(x) = \sin((n+1)\arccos(x))/\left(1 - x^2\right)^{1/2}$. What properties of $U_n(x)$ can you deduce from this relation?

b.) Verify the recurrence relations $U_{n+1}(x) - 2xU_n(x) + U_{n-1}(x) = 0$, $T_n(x) = U_n(x) - xU_{n-1}(x)$, $(1 - x^2)U_{n-1}(x) = xT_n(x) - T_{n+1}(x)$, and $T_n'(x) = nU_{n-1}(x)$.

12.) Write a MATLAB program that accepts a function, a finite interval of integration, and a positive integer N and returns the integral of the function as approximated by the Gauss-Legendre rule based on N points. Test it on several functions.

13.) a.) Write a MATLAB program that accepts a function, a finite interval of integration, and a positive integer N and returns the integral of the function as approximated by the Gauss-Chebyshev rule of the first kind based on N points. Test it on several functions.

b.) Extend your program so that it accepts an additional positive integer m and performs N-point composite Gauss-Chebyshev quadrature on m panels. Test it on several functions.

14.) Write a MATLAB program that accepts a function, an interval of integration of the form $(-\infty, b]$ or $[a, \infty)$, and a positive integer N and returns the integral of the function as approximated by the Gauss-Laguerre rule of the first kind based on N points.

15.) The orthogonal polynomials generated by $w(x) = \exp(-x^2)$ and the interval $(-\infty, \infty)$ are the Hermite polynomials. Write a MATLAB program that accepts a function and a positive integer $N \leq 3$ and returns the doubly improper infinite integral of the function as approximated by the Gauss-Hermite rule of the first kind based on N points. Also return a status variable to indicate failure to converge or invalid inputs. Test it on several functions.

16.) The orthogonal polynomials generated by $w(x) = \exp(-x)$ and the interval $[0, \infty]$ are the Laguerre polynomials. Write a MATLAB program that accepts a function and a positive integer N and returns the integral of the function as approximated by the Gauss-Hermite rule of the first kind based on $N \leq 3$ points. Also return a status variable to indicate failure to converge or invalid inputs. Test it on several functions.

5. Adaptivity and Automatic Integration

We've skipped over some very simple quadrature rules. The composite midpoint rule may be modified to allow for unequally spaced points: Let $x_0, x_1, ..., x_n$ be ordered points in (a, b) and define $x_{-1} = a$, $x_{n+1} = b$. Let $h_i = x_{i+1} - x_i$. Then we can center a rectangle at x_i with width

$$w_i = \frac{h_i}{2} + \frac{h_{i+1}}{2}$$

and use the rule

$$\int_a^b f(x)dx \approx \sum_{i=0}^n w_i f(x_i)$$

to approximate the integral. (Note that the nodes are no longer truly midpoints of the interval; we could use the midpoints of the intervals defined by $x_0, x_1, ..., x_n$ if we have f but not if we have only a list of values $f(x_0), f(x_1), ..., f(x_n)$.) We could use a similar approach for the trapezoidal rule with unequally spaced nodes.

The midpoint rule is one of several rules that explicitly form a finite Riemann sum to approximate an integral and so are called **Riemann sum rules.** (Many other rules can be shown to be equivalent to a Riemann sum.) Other examples include the **left endpoint rule**, where we choose a number of panels and approximate the area under f over panel $[x_i, x_{i+1}]$ by the area of the rectangle with height $f(x_i)$ and width $h_i = x_{i+1} - x_i$, the **right endpoint rule**, where we approximate the area under f over panel $[x_i, x_{i+1}]$ by the area of the rectangle with height $f(x_{i+1})$ and width h_i, and Monte Carlo methods where we generate a list of nodes $x_0, x_1, ..., x_n$ pseudo-randomly and then apply any of these methods. All these rules are exact for constants.

For a simple rule, if we seek greater accuracy by increasing the number of points $n + 1$ in the weighted average in the hopes that

$$\sum_{i=0}^n w_i f(x_i) \to \int_a^b f(x)w(x)dx$$

as $n \to \infty$, we must know that the rule converges to the integral as n increases. If the interval of integration is finite then every interpolatory rule that has nonnegative weights (for all n) converges for all Riemann integrable functions. (In particular, it converges for all continuous and piecewise continuous functions.) Hence, all Gauss rules converge. The simple closed and open families of Newton-Cotes rules can diverge.

Instead of letting the number of points used in a simple rule increase, though, we're more likely to take a composite rule approach by applying a fixed simple rule (that is, with a fixed n) on smaller and smaller subintervals. Composite schemes rely on the following theorem to justify their application:

THEOREM 1: Let $h = b - a$. A weighted average rule for $\int_a^b w(x)f(x)dx$ that has degree of accuracy $d \geq 0$ satisfies

$$E = O(h^{d+2})$$

as $h \to 0$ for any function f that is at least $d + 1$ times continuously differentiable on $[a, b]$.

(Hence we expect $O(h^{d+1})$ behavior for the composite form.) This means that if f is sufficiently differentiable and the rule integrates constants correctly then the error in using the simple rule on a smaller and smaller subinterval goes to zero like h^{d+2}, which is at least h^2. It follows that composite Riemann sum rules are $O(h)$; the midpoint rule is an exception ($O(h^2)$).

A surprising amount can be said about rules that are exact for constants. We also have the theorem:

THEOREM 2: A weighted average rule for $\int_a^b f(x)dx$ with degree of accuracy $d \geq 0$ has absolute error E that satisfies

$$E \leq (b - a + K) \min \left(\max_{x \in [a,b]} (|p(x) - f(x)|) \right)$$

where the minimum is taken over all polynomials with degree at most d.

PROOF: It's convenient to write $Qf = \sum_{i=0}^{n} w_i f(x_i)$ and $\int f = \int_a^b f(x)dx$. The absolute error then satisfies

$$
\begin{aligned}
E &= \left| Qf - \int f \right| \\
&= \left| Qf + Qp - Qp + \int p - \int p - \int f \right| \\
&= \left| Qf - Qp + \int p - \int f \right|
\end{aligned}
$$

for any polynomial $p(x)$ of degree at most d because the rule is exact for such polynomials. So

$$
\begin{aligned}
E &\leq |Qf - Qp| + \left| \int p - \int f \right| \\
&= |Q(f - p)| + \left| \int (p - f) \right|
\end{aligned}
$$

by the linearity of both the quadrature rule and integration. But

$$
\begin{aligned}
|Q(f - p)| &= \left| \sum_{i=0}^{n} w_i \left(f(x_i) - p(x_i) \right) \right| \\
&\leq \sum_{i=0}^{n} |w_i| \, |f(x_i) - p(x_i)| \\
&\leq K \max_{x \in [a,b]} (|p(x) - f(x)|)
\end{aligned}
$$

and

$$
\begin{aligned}
\left| \int (p - f) \right| &= \left| \int_a^b (p - f)dx \right| \\
&\leq \int_a^b |p - f| \, dx \\
&\leq \int_a^b \max_{x \in [a,b]} (|p(x) - f(x)|) \, dx \\
&= (b - a) \max_{x \in [a,b]} (|p(x) - f(x)|)
\end{aligned}
$$

so

$$E \leq K \max_{x \in [a,b]} (|p(x) - f(x)|) + (b - a) \max_{x \in [a,b]} (|p(x) - f(x)|)$$

$$= (b - a + K) \max_{x \in [a,b]} (|p(x) - f(x)|)$$

for *every* polynomial $p(x)$ of degree at most d. In particular, it is true for that polynomial that minimizes

$$\max_{x \in [a,b]} (|p(x) - f(x)|)$$

which gives the desired result. $\square$

This theorem is easily modified to allow for a weight function in the integral. Clearly a higher degree of accuracy will mean a better result if K is bounded, as it is for the Gauss-Chebyshev rules and the Gauss-Legendre rule, since

$$\min \left(\max_{x \in [a,b]} (|p(x) - f(x)|) \right) = \min_{x \in P_d} (\|p(x) - f(x)\|_\infty)$$

can only decrease as d increases, because the space of all polynomials of degree at most d, P_d, is a subset of the space of all polynomials of degree at most m, P_m, if $m > d$.

We now have theoretical justification for improving the accuracy of our approximate integrations by increasing the degree of accuracy of our method (say, by using the Gauss-Legendre family of rules based on an increasing number of points), due to Theorem 1, or by using a fixed method in a composite manner and relying on the $O(h^{d+1})$ behavior promised by Theorem 2 for sufficiently differentiable functions. The latter approach is usually more efficient–do we really want to have to find the 101 nodes for a Gauss rule based on $n + 1 = 101$ points by repeated application of Newton's method? Furthermore, if time constraints limit us to about 101 function evaluations, wouldn't we be better off choosing them adaptively, that is, putting more where the function is changing more rapidly and fewer where it is changing more slowly, as a recursively called composite rule could do?

We've discussed several special-purpose routines for special situations. When you have to write your own code (which is generally not advisable if a professionally produced routine will serve your purposes), and you know that you will be handling certain types of integrals with certain properties, you'll want to choose a rule that fits your needs. This may mean recognizing that a weight function appears naturally in your integrands, or that your problems always involve unbounded intervals or singular integrands, or utilizing the availability of advanced computer architectures, or making speed versus accuracy trade-offs. Fortunately there are many schemes from which to choose.

But how do we choose a rule to implement in a general purpose integration routine like `integral`, where f may or may not be well behaved? To put it another way, why did the makers of MATLAB and Maple choose variations of the Gauss quadrature ideas as their general-purpose routine? What qualities should we be looking for in an integration routine that attempts to use adaptive methods in order to achieve a desired error tolerance as efficiently as possible? And exactly how does it check to see if the error is as small as desired?

A low-order Newton-Cotes rule such as Simpson's rule has the advantages of simplicity and of requiring less smoothness of the function. Simpson's rule has an error term involving $f^{(4)}(x)$ and so it expects a function with at least 4 continuous derivatives. The Gauss rules based on the same number of points have error terms involving $f^{(6)}(x)$ and so they expect a function with at least 6 continuous derivatives. This difference becomes more pronounced as n increases; a Gauss rule will require about twice the number of continuous derivatives of a Newton-Cotes rule. Hence Newton-Cotes rules may make sense for less smooth functions and Gauss rules may make sense for very smooth functions. It's common to see a variety of options available in a software environment, some recommended for low accuracies with nonsmooth functions, and others recommended for higher accuracies with smooth functions. The Gauss and Gauss-type rules give better accuracy but require additional differentiability.

We say that a method or program performs **automatic integration** if it accepts an integrand, interval, and tolerance, then returns either the value of the integral to within that tolerance or an error message indicating that the tolerance was not met (possibly because the maximum number of function evaluations, or maximum recursion level if recursive, *was* met). Every program you are likely to encounter for performing integration will be an automatic integrator. Of course, the program may erroneously indicate that the tolerance was met on rare occasions—perfection is an unrealistic goal.

Let's discuss how we can estimate the error for the purposes of automatic integration. There are different approaches for different methods so we'll focus on composite Newton-Cotes formulae, though it should be clear how the idea generalizes. Consider the trapezoidal rule

$$\int_\alpha^\beta f(x)dx = T_h f - \frac{(\beta - \alpha)h^2}{12}f''(\gamma_h)$$

where $T_h f$ is the result of applying the trapezoidal rule with spacing h. Let's assume that f is indeed twice continuously differentiable so that this error estimate is valid. Then

$$
\begin{aligned}
\int_\alpha^\beta f(x)dx &= T_{h/2}f - \frac{(\beta - \alpha)(h/2)^2}{12}f''(\gamma_{h/2}) \\
&= T_{h/2}f - \frac{(\beta - \alpha)h^2}{48}f''(\gamma_{h/2})
\end{aligned}
$$

is the error at the spacing $h/2$. That is,

$$
\begin{aligned}
T_h f - \frac{(b - a)h^2}{12}f''(\gamma_h) &= T_{h/2}f - \frac{(\beta - \alpha)h^2}{48}f''(\gamma_{h/2}) \\
T_h f - T_{h/2}f &= \frac{(\beta - \alpha)h^2}{12}f''(\gamma_h) - \frac{(\beta - \alpha)h^2}{48}f''(\gamma_{h/2}) \\
&= \frac{(\beta - \alpha)h^2}{48}(4f''(\gamma_h) - f''(\gamma_{h/2}))
\end{aligned}
$$

(notice that this is an equality, not an approximation). Now suppose that the interval $[\alpha, \beta]$ is not very large, presumably because we are applying the trapezoidal rule on just one subinterval of the larger interval in which we're interested. Then we expect that

(5.1) $f''(\gamma_h) \approx f''(\gamma_{h/2})$

and so

$$T_h f - T_{h/2} f \;\approx\; 3\frac{(\beta - \alpha)h^2}{48}f''(\gamma)$$
$$= \; -3E_{h/2}$$

where $E_{h/2}$ is the error in the approximation $T_{h/2}f$. Hence by computing with two rules with known error formulae we can approximate the error. If the absolute value of

(5.2) $E_{h/2} \approx -\dfrac{1}{3}\left(T_h f - T_{h/2} f\right)$

is sufficiently small then we accept $T_{h/2}f$ as our approximation (or possibly use $T_{h/2}f$ and $T_h f$ to extrapolate to an even better solution); otherwise we subdivide the interval $[\alpha, \beta]$ and use the rule at two different resolutions again.

EXAMPLE 1: Let $f(x) = \exp(x)$, $[\alpha, \beta] = [0, .1]$. Then with $h = .1$ we have

$$\int_0^{.1} \exp(x)dx \;\approx\; T_{.1}f$$
$$\doteq\; .10525854590378$$

and with $h = .05$ we have

$$\int_0^{.1} f(x)dx \;\approx\; T_{.05}f$$
$$\doteq\; .10519282777069$$

(note that we could have reused function values from the computation of $T_{.1}f$ here). Hence Eq. (5.2) gives

$$-\frac{1}{3}(T_{.1}f - T_{.05}f) \;\doteq\; \frac{1}{-3}\cdot 6.5718E-5$$
$$= \; -2.19060E-5$$

as the estimate of $E_{.05}$. In fact,

$$\int_0^{.1} \exp(x)dx - T_{.05}f \;\doteq\; .10517091807565 - .10519282777069$$
$$\doteq\; -2.19097E-5$$

so that $-2.19060E-5$ is a very good estimate of the true error $-2.19097E-5$. The approximation $T_{.05}f$ may be used with some confidence that the error is within a tolerance of about $2.2E-5$; we might also try to improve the estimate by adding back in the (signed) error estimate, that is,

$$T = T_{.05}f - \frac{1}{3}\left(T_{.1}f - T_{.05}f\right)$$

should cancel out most of the error (compare the idea behind the method of iterative refinement from Sec. 3.3). Indeed, the error in T has magnitude about $3.65E - 9$, a considerable improvement. However, note that we do not have an error estimate for T if we use this approach, just the estimate for $T_{h/2}f$ which should be a rather pessimistic estimate for T. $\square$

The method works because we know something specific about the form of the error for the trapezoidal method. In fact, we have an important theorem from approximation theory that has applications in several areas of applied mathematics:

THEOREM 3 (EULER-MACLAURIN SUMMATION FORMULA): If f is $2k + 1$ times continuously differentiable on $[a, b]$ then

$$
\begin{aligned}
\int_a^b f(x)dx &= T_hf - \frac{B_2}{2!}h^2(f'(b) - f'(a)) - \frac{B_4}{4!}h^4(f^{(3)}(b) - f^{(3)}(a)) \\
&\quad - \frac{B_6}{6!}h^6(f^{(5)}(b) - f^{(5)}(a)) - \cdots - \\
&\quad \frac{B_{2k}}{(2k)!}h^{(2k-1)}(f^{(2k-1)}(b) - f^{(2k-1)}(a)) + R_{2k+1} \\
&= T_hf - \sum_{m=1}^k \frac{B_{2m}}{(2m)!}h^{2m}\left(f^{(2m-1)}(b) - f^{(2m-1)}(a)\right) + R_{2k+1}
\end{aligned}
$$

where $h = b - a$,

$$
B_p = \sum_{i=0}^p \sum_{j=0}^q (-1)^j \binom{q}{j} \frac{j^m}{q+1}
$$

for $p \geq 0$, and

$$
\begin{aligned}
R_{2k+1}(z) &= \int_a^b P_{2k+1}(1 - t) f^{(2k+1)}(t)dt \\
P_{2k+1}(z) &= (-1)^{k-1} \sum_{m=1}^\infty \frac{2\cos(2m\pi z)}{(2m\pi)^{2k+1}}
\end{aligned}
$$

for $k \geq 0$.

The numbers B_p are called the **Bernoulli numbers** and satisfy $B_0 = 1$, $B_1 = -1/2$, and $B_{2i+1} = 0$ if $i \geq 1$. (A different convention uses $B_1 = 1/2$.) They may also be found recursively. The polynomials $P_{2k+1}(z)$ arise from a transformation of a different formula for R_{2k+1} that is based on the **Bernoulli polynomials**

$$
B_n(x) = (-1)^{k-1} \sum_{m=0}^n B_{n-k} \binom{n}{m} x^k
$$

where the B_{n-k} are the Bernoulli numbers. In practice, the Bernoulli numbers are time-consuming to compute.

The importance of this theorem for us is multi-fold. Although questions of convergence in the Euler-Maclaurin summation formula can be difficult, if f is sufficiently differentiable then the remainder term is $O\left(h^{2k+1}\right)$ and so the truncation error of the trapezoidal rule T_hf can be written in the form

$$
E = c_2h^2 + c_4h^4 + c_6h^6 + \cdots + c_{2k}h^{2k} + O(h^{2k+1})
$$

and if f is infinitely differentiable then we can write it as

$$E = c_2 h^2 + c_4 h^4 + c_6 h^6 + \dots$$

indicating the $O\left(h^2\right)$ error that we expect. But since we now have explicit formulas for the c_{2m} we can use this expression to find any number of formulas analogous to Eq. (5.2). Additionally, we note that if f is such that $f'(a) = f'(b)$ then

$$E = c_4 h^4 + c_6 h^6 + \dots + c_{2k} h^{2k} + O(h^{2k+1})$$

(the error is $O\left(h^4\right)$). The more consecutive odd-order derivatives of f that have the same values at the endpoints of the interval of integration, starting from the first derivatives, the higher the order of the error. Clearly, if we know our function has this sort of near-periodicity on the interval, we should choose the (simple!) trapezoidal rule over other methods. But this observation also leads to a whole class of quadrature methods that take an integral $\int_a^b f(x)dx$ and transform it to an equivalent integral

$$\int_a^b f(x)dx = \int_\alpha^\beta g(t)dt$$

for which $g(t)$ satisfies $g^{(m)}(\alpha) = g^{(m)}(\beta)$ for $m = 1, 2, \dots, M$ (or at least for odd m). If M is large enough, then applying the trapezoidal rule to $\int_\alpha^\beta g(t)dt$ should give extremely good results. The efficiency of such a technique would depend on the difficulty of finding and using the transformation, of course. This technique is referred to as the use of a **periodizing transformation** (see Sec. 5.8).

The corresponding automatic integration method with Simpson's rule is widely used as the basis of an automatic integrator. We could use any rule for which the error E has a tractable form and for which we can make the statement that E_h is approximately equal to a fixed and known multiple of $E_{h/k}$ for some $k > 1$. This is true of a great many rules.

For any particular choice of h it could certainly happen that the estimate based on step size h is better than the one based on $h/2$. There are heuristics for checking whether a function or its derivatives are sufficiently smooth in some region so that this is unlikely to happen but if h is small this is not usually necessary.

There are other error estimation techniques, some of which are specific to certain methods. If we can estimate the error, however, we can build an **adaptive numerical integrator**. Such a routine accepts a function f, an interval $[a,b]$, and a tolerance τ, and attempts to return

$$\int_a^b f(x)dx$$

to within the tolerance specified by adaptively refining the grid. Let Q_h be a quadrature rule based on a spacing of h (possibly $Q_h = T_h$) and k an integer greater than 1. We proceed as follows: Evaluate $Q_{b-a}f$ and $Q_{(b-a)/k}f$. Estimate the error; if it is less than τ, return $Q_{(b-a)/k}f$. Otherwise, divide $[a,b]$ into k equal subintervals and repeat the above process on each subinterval with tolerance τ/k. When the integral on each subinterval has been successfully approximated, sum up the contributions and return that result.

This is an adaptive, recursive algorithm. It is adaptive because it chooses the intervals to focus on based on how difficult it finds the integrand to be, and recursive because it calls itself on these problematic regions, then calls itself again on the problematic subregions of these problematic regions, continually parceling out proportional amounts of the tolerance τ, until convergence is achieved or some maximum number of function evaluations or recursive calls is reached. See Fig. 1.

$$\underbrace{[-1,1]}$$
$$\underbrace{[-1,0]} \qquad [0,1]^*$$
$$\underbrace{[-1,-.5]} \qquad [-.5,0]^*$$
$$[-1,-.75]^* \quad [-.75,-.5]^*$$

Figure 1 Possible path of a recursive integrator.

(* = convergence achieved on that subinterval)

If we detect that an estimate on a given subinterval is well below the tolerance then we may be able to "trade" back the "excess" tolerance to a subinterval on which the routine is struggling to meet its tolerance.

This approach uses the same error tolerance τ/λ on every subinterval of $[a,b]$ of length λ. This may be inefficient. If we know in advance that the function has rough areas and smooth areas and we know where they are then we might divide the tolerance τ unequally, allowing more error where the function is poorly behaved and less error where it is well-behaved. For an automatic integrator, of course, we cannot make any such assumptions about the function to be integrated.

MATLAB

Let's check the error estimate $E_{h/2} \approx -\frac{1}{3}\left(T_h f - T_{h/2} f\right)$ from Eq. (5.2). We'll use a function with a known integral of course. Enter:

```
>> f=@(x) 6*x.^5
>> h=.1;
>> x=0:h:1;
>> T1=trapz(x,f(x))
>> x=0:h/2:1;
>> T2=trapz(x,f(x))
>> format long
>> E_est=-(T1-T2)/3        %Estimated error.
>> E=1-T2                  %Actual error.
>> abs(E_est-E)
```

The error is on the order of 10^{-3} and we have an estimate of it that's accurate to about 10^{-5}. That's an excellent approximation of the error. Let's try with a much smaller h; enter:

```
>> h=.0001;
>> x=0:h:1;
>> T1=trapz(x,f(x))
>> x=0:h/2:1;
>> T2=trapz(x,f(x))
>> E_est=-(T1-T2)/3        %Estimated error.
>> E=1-T2                  %Actual error.
>> abs(E_est-E)
```

The error is on the order of 10^{-9} and we have an estimate of it that's good to about 10^{-17}. The error estimate becomes better as h decreases. Enter:

```
>> T2_c=T2+E_est          %Correct T2 using estimated error.
>> abs(T2_c-1)
>> format short
```

If we try to "correct" the estimate $T_{h/2}$ by adding the error estimate from Eq. (5.2) to it then we get an answer that, while not exact, is very good. This is also an efficient way to improve the accuracy. Enter:

```
>> 1/h+1/(h/2)            %Number of function evals. used to find T2_c.
>> xx=linspace(0,1,30000);
>> T3=trapz(xx,f(xx))
>> abs(T3-1)
```

The error is on the order of 10^{-9}, as it was for T2; using Eq. (5.2) with spacings h and $h/2$ is more effective than using the same total number of function evaluations in a trapezoidal rule with a smaller h. This is a good example of working smarter, not harder. Unfortunately we do not have an error estimate for the "corrected" T2. Still, the formula $T_c = T_2f - \frac{1}{3}(T_1f - T_2f)$ seems useful. If we use it then we typically report the error estimate for T_2 even though it is likely to be a gross overestimate of the true error. The corresponding correction formula for Simpson's rule is $T_c = T_2f - \frac{1}{3}(T_1f - T_2f)$.

Let's look at another Riemann sum rule. Monte Carlo methods make use of pseudo-random numbers such as those generated by rand. A simple Monte Carlo quadrature method would be the composite midpoint rule or left or right endpoint rule with randomly selected nodes. Enter:

```
>> f=@sin
>> x=pi*rand([1 100]);     %100 randomly selected pts. in [0,pi].
>> x=sort(x);              %Get ordered nodes.
>> h=diff(x);              %Panel widths.
>> A=dot(h,f(x(1:end-1)))  %Left endpoint rule.
```

(The correct answer is 2.) As a rule, the convergence of these methods is $O(1/\sqrt{n})$ when n pseudo-randomly chosen points are used, independent of the degree of smoothness of the integrand. This order of convergence isn't very impressive, though the convergence of the corresponding Monte Carlo methods for multiple integrals is also $O(1/\sqrt{n})$ no matter the number of variables with respect to which we are integrating, so there can be advantages in higher dimensions. Indeed, high-dimension multiple integrals are generally evaluated in this way.

Problems

1.) a.) Use the adaptive recursive trapezoidal rule to estimate $\int_0^1 (x + \sin(x^2))dx$ with an error tolerance of .01.

b.) Use the adaptive recursive trapezoidal rule to estimate $\int_0^1 \exp(-x)dx$ with an error tolerance of .001.

2.) a.) What is the error estimate corresponding to Eq. (5.2) for Simpson's rule?

b.) Write a complete algorithm (pseudo-code) for an adaptive recursive automatic integrator based on Simpson's rule. The method should make efficient use of previously computed values.

3.) a.) Write a MATLAB program for performing adaptive recursive automatic integration based on the trapezoidal rule. Test your program.

b.) Modify your program to add back in the estimated error. Test your program.

4.) Would it be feasible to build an adaptive recursive method based on Gauss-Legendre quadrature?

5.) a.) What is the error estimate corresponding to Eq. (5.2) for Simpson's 3/8 rule?

b.) Write a MATLAB program for performing adaptive recursive automatic integration based on Simpson's 3/8 rule. Test your program.

6.) Prove that all Riemann sum rules based on N points are at least $O(1/N)$.

7.) a.) Use the trapezoidal rule to approximate $\int_{-1}^{1} \sin(\pi x)dx$, $\int_{-1}^{1} \cos(\pi x)dx$, and $\int_{0}^{1}(x^5 - 2x^3 + 1)dx$.

b.) Repeat part a. using Simpson's rule. Comment.

c.) Use the trapezoidal rule to approximate $\int_{-1}^{1} \cos(\pi x)dx$ by integrating from 0 to 1 and doubling the result. Is your answer as accurate as the corresponding result from part a.? Why?

8.) a.) Use the adaptive recursive trapezoidal rule to estimate $\int_{.01}^{1} \sin(x^{-2})dx$. Create a diagram similar to Fig. 1 showing how the interval of integration was subdivided.

b.) Use the adaptive recursive Simpson's rule to estimate $\int_{.01}^{1} \sin(x^{-2})dx$. Create a diagram similar to Fig. 1 showing how the interval of integration was subdivided.

9.) Verify Theorem 2 for a particular integral $\int_{a}^{b} f(x)dx$ and method and a variety of polynomials $p(x)$.

10.) Choose a variety of functions and test whether Eq. (5.1) is a reasonable approximation.

11.) Modify Theorem 2 to allow for a weight function in the integral.

12.) If endpoint information about the first derivatives of a function f is available then the Euler-Maclaurin summation formula implies that $\int_{a}^{b} f(x)dx \approx T_h f - \frac{B_2}{2}h^2(f'(b) - f'(a))$ is an $O(h^4)$ approximation, called the **trapezoidal rule with endpoint correction** (or sometimes the **corrected trapezoidal rule**, but see Problem 13). Note that $B_2 = -1/6$. Use this rule to approximate $\int_{-1}^{1} \exp(x)dx$, $\int_{-1}^{1} \sin(x^2)dx$, and $\int_{1}^{3} x^3 \cos(x \ln(x))dx$.

13.) The technique of adding back in the error in the trapezoidal rule to find $T_c = T_{h/2}f - \frac{1}{3}\left(T_h f - T_{h/2}f\right)$, as used in the MATLAB subsection, is called the **trapezoidal rule with error correction** (or sometimes the **corrected trapezoidal rule**, but see Problem 12). What can we say the size of the error in T_c? Can we further "correct the correction" by generating a sequence of approximations like $T_h, T_{h/2}, T_{h/4}, T_{h/8}, ...$?

14.) a.) What is the error estimate corresponding to Eq. (5.2) for Boole's rule?

b.) Write a MATLAB program for performing adaptive recursive automatic integration based on Boole's rule. Test your program.

15.) Write a MATLAB program for performing Monte Carlo integration as a Riemann sum rule with points selected pseudo-randomly. Test your program.

16.) Prove Theorem 1.

6. Automatic Integration with Gauss Rules

There is a sense of conditioning in the context of quadrature, just as there was for the solution of linear systems. The condition number of a weighted average rule is

$$K = \sum_{i=0}^{n} |w_i|$$

and the larger this value is, the worse the results we expect. The condition number in this case bounds the effect of a perturbation $\tilde{f}$ of the integrand f (say, due to evaluation error when attempting to compute $f(x)$) as

(6.1) $$E \le K \max \left| \tilde{f} - f \right|$$

where E measures the absolute difference of the quadrature rule applied to the perturbed function $\tilde{f}$ and the quadrature rule applied to f itself. Small errors in evaluating f can be multiplied by a factor as large as the sum of the absolute values of the weights.

Which of the quadrature rules that we've considered are well-conditioned? For the Newton-Cotes rules we know that

$$\sum_{i=0}^{n} |w_i| \to \infty$$

as $n \to \infty$, so they are ill-conditioned for large n. For the Gauss-Chebyshev rule, however,

$$\begin{aligned} \sum_{i=0}^{n} |w_i| &= \sum_{i=0}^{n} \frac{\pi}{n+1} \\ &= \pi \end{aligned}$$

so this method is well-conditioned for all n, as is the Gauss-Chebyshev of the second kind, which has a slightly smaller condition number. For the Gauss-Legendre rule

$$\sum_{i=0}^{n} |w_i| = b - a$$

so this rule is well-conditioned unless the interval of integration is very large. In general, Gauss rules end to be well-conditioned[8].

How can we build an automatic integrator based on a Gauss rule? Using it recursively or as a composite rule is a possibility, but there is another approach that is widely used and is the basis of several methods. The idea stems from asking this question: Suppose we want to use an interpolatory rule based on $n+1$ points of which $m < n+1$ have already been chosen (called the **pre-assigned nodes** or **pre-assigned abscissas**); how should we choose the remaining $n+1-m$ points so as to maximize the degree of accuracy of the resulting method? Gauss rules answer the question for $m = 0$, but now we are imagining a case where we have

---

[8]Quadrature rules with nonnegative weights generally have better conditioning than those with weights of mixed sign.

used a Gauss rule based on m points and wish to improve upon the approximation by using a rule based on $n+1 > m$ points. We're going to give up the $2n+1$ degree of accuracy we could obtain by using a Gauss rule in favor of the efficiency of being able to reuse the values

$$f(x_0), f(x_1), ..., f(x_{m-1})$$

that have already been computed (and retained) from the initial rule. The degree of accuracy of the new rule will be roughly the degree of accuracy of the old rule ($2m - 1$ for a Gauss rule based on m points) plus the number of points added ($n + 1 - m$).

An automatic integrator generally monitors the error by computing the integral twice, at two different resolutions, and using these values to generate an estimate of the absolute error. This will be much more efficient if function values can be reused. We'll try to find a way to extend the Gauss rule selected as our main method to another Gauss-like rule that we can use to check our error.

Suppose that we were going to use a Gauss rule recursively. For the sake of definiteness, let $[a, b] = [-1, 1]$. We might first apply the rule with $n+1$ points over $[-1, 1]$ and then apply it with $2n + 1$ points over the same interval, allowing us to estimate the error, even if only by accepting the estimate based on $2n + 1$ points as essentially accurate. If the error is acceptable, we could stop and take the estimate based on more points as our result; if it isn't, we could apply both rules over $[-1, 0]$ and $[0, 1]$ and see which of these is the problematic interval. Since the interval over which we are applying the rule is smaller each time the result should be better.

But if we use a Gauss rule on $[-1, 1]$ then none of the function evaluations will be reusable for the application of that same rule on $[-1, 0]$! All Gauss rules are open–we won't even be able to reuse the endpoint evaluations. What if we used an "almost" Gauss rule that had pre-assigned abscissa $a = -1$, $b = 1$ and that also used an odd number of points. Then when we applied the rule the first time to the whole interval of integration $[-1, 1]$ we'd need to evaluate $f(-1)$, $f(0)$, and $f(1)$. But when we applied it to the subinterval $[-1, 0]$ we'd be able to reuse $f(-1)$ and $f(0)$, and similarly for $f(0)$ and $f(1)$ on the subinterval $[0, 1]$.

The weighted average rule based on $n + 1$ points for the interval $[-1, 1]$ having the pre-assigned abscissas $x_0 = -1$, $x_n = 1$ and the maximum degree of accuracy achievable is called the **Lobatto rule** (or sometimes the **Gauss-Lobatto rule**) and is given by

$$\int_{-1}^{1} f(x)dx \approx \frac{2}{n(n+1)}(f(-1) + f(1)) + \sum_{i=1}^{n-1} w_i f(x_i)$$

where the nodes $x_1, ..., x_{n-1}$ are the $n - 1$ zeroes of $p'_n(x)$, the derivative of the Legendre polynomial of degree n, and the weights are given by

(6.2) $$w_i = \frac{2}{n(n+1)k_n^2 p_n^2(x_i)}$$

($i = 1, 2, ..., n - 1$). No zero of $p'_n(x)$ is also a zero of $p_n(x)$, so this is well-defined. Here

$$k_n = \frac{(2n)!}{2^n (n!)^2}$$

is the leading coefficient of the nth Legendre polynomial when it is normalized in a different way than the requirement that each polynomial be monic. In particular, the three-term recursion relation for the Legendre polynomials with leading coefficient k_n is $P_0(x) = 1$, $P_1(x) = x$, and

$$P_{n+1}(x) = \frac{2n+1}{n+1} x P_n(x) - \frac{n}{n+1} P_{n-1}(x).$$

In this form every polynomial P_n satisfies $P_n(1) = 1$ but the leading coefficients are no longer unity[9].

The Lobatto rule allows us to save 2 function evaluations on each subinterval. What have we given up? The error for the Lobatto rule has the form

(6.3) $$E = c_n f^{(2n)}(\gamma)$$

(for an especially complicated c_n), and so the method has degree of accuracy $2n-1$. This is 2 less than the maximum possible degree of accuracy $2n+1$ that is attained when we are free to choose the locations of all nodes, which makes intuitive sense as we have given up two degrees of freedom in insisting upon $x_0 = -1$ and $x_n = 1$.

The weighted average rule based on $n+1$ points for the interval $[-1, 1]$ that has the single pre-assigned abscissa $x_0 = -1$ and the maximum degree of accuracy achievable is called the **Radau rule** and is given by

$$\int_{-1}^{1} f(x)dx \approx \frac{2}{(n+1)^2} f(-1) + \sum_{i=1}^{n} w_i f(x_i)$$

where the nodes $x_1, ..., x_n$ are the n zeroes of

$$\frac{k_n p_n(x) + k_{n+1} p_{n+1}(x)}{x - 1}$$

(with $p_n(x)$ the monic Legendre polynomial of degree n) and the weights are given by

(6.4a) $$w_i = \frac{1 - x_i}{(n+1)^2 k_n^2 p_n^2(x_i)}$$

($i = 1, 2, ..., n$). The error has the form

$$E = d_n f^{(2n+1)}(\gamma)$$

for a complicated d_n, and so the method has degree of accuracy $2n$. This is 1 less than the maximum possible degree of accuracy $2n + 1$ that is attained when we are free to choose the locations of all nodes, which again makes intuitive sense as we have given up one degree of freedom in insisting upon $x_0 = -1$. A Radau rule may also be defined with $x_n = 1$ pre-assigned and x_0 free (the **Radau right-hand formula** as opposed to the **Radau left-hand formula** described above).

The Radau quadrature rule is useful in the numerical solution of ordinary differential equations. The Lobatto quadrature rule is used for automatic integration, for example, in the MATLAB `quadl` routine. Rules of this general form, having some pre-assigned abscissas with the remaining abscissas chosen so as to have the

---

[9]Note that we are distinguishing between $p(x)$ for the monic Legendre polynomials and $P(x)$ for those with $P(1) = 1$.

maximal degree of accuracy, are called **Gauss-type rules with pre-assigned abscissas**.

Clearly, if we have to find the nodes by solving for the zeroes of $p'(x)$ numerically and then finding the weights using Eq. (6.2) or Eq. (6.4a) then this could be a time-consuming and error-prone undertaking, although Newton's method tends to work well for finding the nodes for moderate n. Typically a program does not perform Lobatto quadrature for an arbitrary n but rather performs automatic integration using the Lobatto rule for two fixed numbers of points, both of which have been hard-coded in advance, or a Lobatto rule followed by some other rule. The nodes and weights appear explicitly in the program[10]. Routines are available that perform Lobatto quadrature for arbitrary n but they usually use different approaches for the weights; one method is based on forming a matrix that has the nodes as its eigenvalues and that forms the weights using the eigenvectors.

EXAMPLE 1: Let's determine the Lobatto rule based on $n + 1 = 4$ points. Two of those points are $x_0 = -1$ and $x_3 = 1$. The remaining points x_1 and x_2 are the zeroes of

$$p_3'(x) = \left(x^3 - \frac{3}{5}x\right)'$$
$$= 3x^2 - \frac{3}{5}$$

so $x_1 = -1/\sqrt{5}$, $x_2 = 1/\sqrt{5}$. The weights are

$$w_1 = \frac{2}{12k_3^2 p_3^2(x_1)}$$
$$= \frac{1}{6\left(\frac{6!}{8(3!)^2}\right)^2 \left(x_1^3 - \frac{3}{5}x_1\right)^2}$$
$$= \frac{5}{6}$$
$$w_2 = \frac{2}{12k_3^2 p_3^2(x_2)}$$
$$= \frac{5}{6}$$

and $w_0 = w_3 = 2/12 = 1/6$. Thus the weights are

$$w \doteq (.1667, .8333, .8333, .1667)^T$$
$$= \frac{1}{6}(1, 5, 5, 1)^T$$

(you can find the line defining these weights in `quadl`). Note once again that the endpoints are given lesser weight than the interior points. The nodes are always symmetric about the origin for Lobatto quadrature and the weights of symmetric points are equal.

---

[10]The "*Handbook of Mathematical Functions*" edited by Abramowitz and Stegun (and since updated to the online "*Digital Library of Mathematical Functions*") lists the nodes and weights for Lobatto quadrature through 10 points, to 10 decimal places.

Let's use this rule on $\int_{-1}^{1} \frac{1}{e^x+1} dx = 1$. We already have the weights and the nodes, so we compute

$$\int_{-1}^{1} \frac{1}{e^x+1} dx \approx \frac{1}{6}f(-1) + \frac{5}{6}f\left(-\frac{1}{\sqrt{5}}\right) + \frac{5}{6}f\left(\frac{1}{\sqrt{5}}\right) + \frac{1}{6}f(1)$$

which is exact to the precision available. This is pretty impressive for only 4 points. If we use the 4 point Lobatto rule on $\int_{-1}^{1} \sin(x)dx$ we have $\int_{-1}^{1} \sin(x)dx \approx 2.7756E - 17$, also an excellent approximation. $\square$

With the 4 point Lobatto quadrature algorithm we could approximate the integral $\int_{-1}^{1} f(x)dx$ using $f(-1), f(-1/\sqrt{5}), f(1/\sqrt{5}), f(1)$ and then refine our estimate by computing $\int_{-1}^{0} f(x)dx$ and $\int_{0}^{1} f(x)dx$ using the same rule. We would only need to compute 2 new function values in $[-1,0]$ and 2 new function values in $[0,1]$ to get a new estimate of $\int_{-1}^{1} f(x)dx = \int_{-1}^{0} f(x)dx + \int_{0}^{1} f(x)dx$, which is the same as the cost in function evaluations for the initial estimate. A true Gauss rule would require 6 new function evaluations if 0 was a node (odd number of points) and 8 new function evaluations otherwise (even number of points).

For a rule like 4 point Lobatto quadrature that has degree of accuracy 7, Theorem 1 of the previous section predicts that the error will decay like h^9. In other words, if the error in approximating $\int_{-1}^{1} f(x)dx$ is about ch^9 then the error in the approximation of $\int_{-1}^{0} f(x)dx$ or $\int_{0}^{1} f(x)dx$ is about $ch^9/2^9$ (neglecting round-off error).

EXAMPLE 2: Let's integrate $f(x) = \sin^2(x)$ over $[-1,1]$. The correct value is .54535128658716. The 4 point Lobatto rule gives .54771978564889 for an absolute error of $\alpha_0 \doteq .00236849906173$. Now we'll repeat over two subintervals, using Eq. (6.3.3.2),

$$\int_{a}^{b} g(t)dt = \frac{b-a}{2} \int_{-1}^{1} g(t(x))dx$$

where $t(x) = ((b-a)x + b + a)/2$. In our case this gives

$$\int_{-1}^{0} \sin^2(t)dt = \frac{0-(-1)}{2} \int_{-1}^{1} \sin^2(x)dx$$

$$= \frac{1}{2} \int_{-1}^{1} \sin^2((x-1)/2)dx$$

(we could have also mapped the nodes in $[-1,1]$ to nodes in $[-1,0]$ instead). The Lobatto rule now gives

$$\int_{-1}^{0} \sin^2(x)dx \approx \frac{1}{2}\left(\frac{1}{6}f_{1,1}(-1) + \frac{5}{6}f_{1,1}\left(-\frac{1}{\sqrt{5}}\right) + \frac{5}{6}f_{1,1}\left(\frac{1}{\sqrt{5}}\right) + \frac{1}{6}f_{1,1}(1)\right)$$
$$\doteq .27268670270676$$

(where $f_{1,1}(x) = \sin^2((x-1)/2)$). Note that $f_{1,1}(-1) = f(-1) = \sin^2(-1)$ and $f_{1,1}(1) = f(0) = \sin^2(0)$ so we can reuse two function values here. Since $\int_{-1}^{0} \sin^2(x)dx \doteq .27267564329358$ the absolute error is $\alpha_{1,1} \doteq 1.105941318502168E - 5$. Similarly,

$$\int_0^1 \sin^2(t)dt = \frac{1-0}{2}\int_{-1}^1 \sin^2((x+1)/2)dx$$

$$\approx \frac{1}{2}\left(\frac{1}{6}f_{1,2}(-1) + \frac{5}{6}f_{1,2}\left(-\frac{1}{\sqrt{5}}\right) + \frac{5}{6}f_{1,2}\left(\frac{1}{\sqrt{5}}\right) + \frac{1}{6}f_{1,2}(1)\right)$$

$$\doteq .27268670270676$$

(where $f_{1,2}(x) = \sin^2((x+1)/2)$), and again note that $f_{1,2}(-1) = f(0)$ and $f_{1,2}(1) = f(1)$ have already been computed. We have

$$\alpha_{1,2} = \alpha_{1,1} \doteq 1.105941318502168E - 5$$

because $f(x)$ is an even function. Our new estimate is

$$\int_{-1}^1 \sin^2(x)dx = .27268670270676 + .27268670270676$$

$$= .54537340541353$$

with error $\alpha_1 \doteq 2.211882637004337E - 5$. Compared with the original error $\alpha_0 \doteq 2.368499061730000E - 3$ this is a considerable improvement.

From Theorem 1 of the previous section, if $\alpha_0 \doteq 2.368499061730000E - 3$ then we expect that $\alpha_{1,1} \approx \alpha_0/2^9 = 4.6260E - 6$. We didn't get results that good but then we have only taken one step and h has only gone from $h = 2$ to $h = 1$. □

More generally, we might apply a Gauss rule based on m points to an integral on $[-1, 1]$ and then wish to approximate the same integral with a Gauss rule based on $n + 1$ points, $n + 1 > m$. Because of efficiency concerns we would probably use a Gauss-type rule (based on $n + 1$ points) instead, with the first m abscissas pre-assigned so that we could reuse the corresponding function values. A special case of this is **Gauss-Kronrod quadrature** (or **Kronrod extension**), which uses $n = 2m$ and generates a Gauss-type rule based on $2m + 1$ points using the m pre-assigned abscissas. (The m previous weights must be discarded however.) The new nodes interlace with the old nodes, and the resulting rule has degree of accuracy $3m + 1$ if m is even and $3m + 2$ if m is odd.

Kronrod extensions also exist for the Radau and Lobatto rules. If you look carefully at the MATLAB routine `quadl` you'll see the following lines:

```
% Four point Lobatto quadrature.
Q1 = (h/6)*[1 5 5 1]*y(1:2:7)';
% Seven point Kronrod refinement.
Q2 = (h/1470)*[77 432 625 672 625 432 77]*y';
```

The routine is using the same 4 point Lobatto rule we have been using, which uses four values of the previously computed vector y of function values (y_1, y_3, y_5, and y_7), followed by a Gauss-type rule which reuses those values and also uses the remaining three values of y (y_2, y_4, and y_5). This doesn't take a 4 point rule to a 9 point rule because when applied to the Radau or Lobatto rules we don't consider the pre-assigned abscissas; that is, a Kronrod extension of a Lobatto rule based on m points gives a new rule based on $2(m - 2) + 1 + 2$ points (double the number $m - 2$ of points that were not pre-assigned in the original rule, add 1, then add back in the 2 pre-assigned abscissas). For the 4 point Lobatto rule $m - 2 = 2$ points are not pre-assigned, leading to a $2(m - 2) + 1 + 2 = 7$ point Kronrod extension.

The MATLAB `integral` command is its main automatic numerical integrator. It's a vectorized version of the Gauss-Kronrod method. (See also its predecessor `quadqk`. All of the 'quad' commands are expected to be removed in a future release of MATLAB.) It incorporates a wide variety of heuristics to handle challenging cases.

Both the Lobatto and Radau rules can be generalized to give, for any admissible weight function $w(x)$, a rule of the form

$$\int_{-1}^{1} w(x)f(x)dx \approx \sum_{i=0}^{n} w_i f(x_i)$$

with similar properties. It is not possible to generalize the Gauss-Kronrod rules for all admissible weight functions, though it can be done for some choices of $w(x)$.

MATLAB

The method of undetermined coefficients, also called the **method of moments** because the integrals $\int_a^b w(x)x^k dx$ are called **moments**, can be used to find the nodes and weights of Radau, Lobatto, and similar rules. Let's use this technique to find the 4 point Lobatto rule. The nodes $x_0 = -1$ and $x_3 = 1$ are pre-assigned, and we need to find x_1, x_2, w_0, w_1, w_2, and w_3. The rule with $n + 1 = 4$ points should have degree of accuracy 5. The equations are

$$
\begin{aligned}
w_0 \cdot 1 + w_1 \cdot 1 + w_2 \cdot 1 + w_3 \cdot 1 &= \int_{-1}^{1} 1 dx \\
&= 2 \\
w_0 \cdot -1 + w_1 \cdot x_1 + w_2 \cdot x_2 + w_3 \cdot 1 &= \int_{-1}^{1} x dx \\
&= 0 \\
w_0 \cdot 1 + w_1 \cdot x_1^2 + w_2 \cdot x_2^2 + w_3 \cdot 1 &= \int_{-1}^{1} x^2 dx \\
&= 2/3 \\
w_0 \cdot -1 + w_1 \cdot x_1^3 + w_2 \cdot x_2^3 + w_3 \cdot 1 &= \int_{-1}^{1} x^3 dx \\
&= 0 \\
w_0 \cdot 1 + w_1 \cdot x_1^4 + w_2 \cdot x_2^4 + w_3 \cdot 1 &= \int_{-1}^{1} x^4 dx \\
&= 2/5 \\
w_0 \cdot -1 + w_1 \cdot x_1^5 + w_2 \cdot x_2^5 + w_3 \cdot 1 &= \int_{-1}^{1} x^5 dx \\
&= 0
\end{aligned}
$$

that is

$$w_0 + w_1 + w_2 + w_3 \; = \; 2$$
$$-w_0 + w_1 x_1 + w_2 x_2 + w_3 \; = \; 0$$
$$w_0 + w_1 x_1^2 + w_2 x_2^2 + w_3 \; = \; 2/3$$
$$-w_0 + w_1 x_1^3 + w_2 x_2^3 + w_3 \; = \; 0$$
$$w_0 + w_1 x_1^4 + w_2 x_2^4 + w_3 \; = \; 2/5$$
$$-w_0 + w_1 x_1^5 + w_2 x_2^5 + w_3 \; = \; 0$$

which is a nonlinear system of 6 equations in 6 unknowns. Let's see if MATLAB can solve this (using the Symbolic Toolbox). Enter:

```
>> help solve
```

for information on the symbolic solution command. Enter:

```
>> syms w0 w1 w2 w3
>> syms x1 x2
>> S=solve(w0+w1+w2+w3-2,-w0+w1*x1+w2*x2+w3,...
   w0+w1*x1^2+w2*x2^2+w3-2/3,-w0+w1*x1^3+w2*x2^3+w3,...
   w0+w1*x1^4+w2*x2^4+w3-2/5,-w0+w1*x1^5+w2*x2^5+w3)
```

and be patient. A symbolic answer is indicated. It is in the form of a structure (see `help struct`), so to get the individual components we need to get them from the proper fields. Enter:

```
>> fieldnames(S)
```

to see the fields associated with this structure. Then enter:

```
>> S.w0
>> S.w1
>> S.w2
>> S.w3
>> S.x1
>> S.x2
```

The first entries in each field correspond to one solution, and the second entries correspond to a second solution. This is because we did not indicate that $x_1 < x_2$. The solution is correct. Symbolic computing rarely solves the problem we want solved, but can help us to find useful formulas for working on problems numerically. If `solve` had failed we might have needed to find a numerical solution via `fsolve`.

Many MATLAB routines allow for an optional structure argument to specify options for the command. Enter:

```
>> help fzero
```

We can specify an optional structure that sets options for this command, and there is a command `optimset` that will help us do so. Enter:

```
>> fzero(@sin,3)
>> options = optimset('fzero');  %Initialize the structure options.
>> options = optimset(options,'Display','iter');
>> fzero(@sin,3,options)
```

Setting options can allow you to get a great deal more value out of many standard commands.

Problems

1.) Show that the 4 point Lobatto rule is exact for odd functions. What other rules is this true of?

2.) Use the 4 point Lobatto rule to estimate $\int_{-1}^{1} \cos(x^2)dx$ and $\int_{-1}^{1} \sin^2(x)dx$.

3.) Use the 4 point Lobatto rule recursively to estimate $\int_{-1}^{1} \exp(x)dx$ to within $1E-6$.

4.) a.) What are the nodes and weights in the 3 point Lobatto rule? Use the Legendre polynomial to determine the nodes and Eq. (6.2) to determine the weights. Use your result to provide an explanation of the fact that Simpson's rule has a higher degree of accuracy than would otherwise be expected of a quadratic interpolatory rule.

b.) What are the nodes and weights in the 5 point Lobatto rule?

5.) Find the first three Radau quadrature rules, and use them to estimate the value of $\int_{-1}^{1} \exp(x)dx$.

6.) Compare the 4 point Radau and Lobatto rules experimentally. Comment.

7.) Write a MATLAB program that accepts a function, an interval, and a tolerance and returns an estimate of the integral of the function over that interval. Your program should use the 4 point Lobatto rule followed by the 7 point Kronrod extension to get two estimates of the integral and terminate if the absolute difference between these two estimates is less than the tolerance; otherwise it should subdivide the interval into 6 new panels defined by the 7 points you've already computed and repeat this procedure. (This is the idea of `quadl`.) Make efficient use of previously computed function values.

8.) The constant c_n in Eq. (6.3) is negative. Compare Eq. (6.3) based on $n+1$ points to Eq. (6.3.3.10) based on n points; show that if $f^{(2n)}(x)$ is of one sign on the interval of integration then the Gauss-Legendre based on n points and the Lobatto rule based on $n+1$ points bracket the true value of the integral (neglecting round-off error). Since the constants in each of the error terms are comparable in magnitude, the weighted average of $(n+1)/(2n+1)$ times the Gauss-Legendre based on n points and $n/(2n+1)$ times the Lobatto rule based on $n+1$ points is typically a more accurate estimate of the integral than either rule alone. Demonstrate this experimentally.

9.) a.) Argue that defining the condition number of a numerical integration problem to be the smallest value K that satisfies Eq. (6.1) is reasonable.

b.) Note that K depends on both f and the quadrature rule, in general; show that the definition implies that $K = \sum_{i=0}^{n} |w_i|$ by taking appropriate norms on both sides of Eq. (6.1).

10.) Derive the 3 point Lobatto rule by fixing $x_0 = a$ and $x_2 = b$ then using the method of undetermined coefficients.

11.) Derive the 3 point Radau rule by fixing $x_0 = a$ then using the method of undetermined coefficients.

12.) Create a table of the condition numbers of all rules for which you have the explicit coefficients. Comment.

13.) Write a MATLAB program that performs Radau quadrature for an arbitrary n. Use the formulas given in this section for the nodes (you may use the MATLAB `roots` command) and the weights.

14.) a.) Find the first three right-hand Radau formulas.

b.) What are the first three right-hand Radau formulas transformed to an arbitrary finite interval $[a, b]$?

15.) What are the first three Lobatto rules with respect to the first Chebyshev weight function $w(x) = 1/\sqrt{1-x^2}$?

16.) What are the first three Lobatto rules with respect to the second Chebyshev weight function $w(x) = \sqrt{1 - x^2}$?

7. The Trapezoidal Rule with Extrapolation

We can extract a lot of useful information from the Euler-Maclaurin summation formula. Using the trapezoidal rule formula

$$\int_\alpha^\beta f(x)dx = T_h f - \frac{(\beta - \alpha)h^2}{12} f''(\gamma)$$

we derived the estimate

$$E_{h/2} \approx -\frac{1}{3}\left(T_h f - T_{h/2} f\right)$$

(Eq. (5.6.5.2)). This led to the idea of adding this (signed) error back into our estimate of the error to get an improved or "corrected" estimate

(7.1) $$T_c = T_{h/2}f - \frac{1}{3}\left(T_h f - T_{h/2} f\right)$$

of the integral. We can view Eq. (7.1) as a way of getting increased accuracy out of the trapezoidal method or as a method in and of itself. That is, we could substitute the expression for the trapezoidal rule, rearrange it so that we do not compute any function values more than once, and consider Eq. (7.1) to be a quadrature method in its own right.

But what if the error still isn't small enough? What if we want to correct the correction? We might imagine generating a sequence of approximations like

$$T_h, T_{h/2}, T_{h/4}, T_{h/8}, \dots$$

that converges to the correct value in the limit. We would probably want to use an acceleration technique like Aitken's Δ^2 process that will turn the sequence $T_h, T_{h/2}, T_{h/4}, T_{h/8}, \dots$ into a more rapidly converging sequence. But Aitken's Δ^2 process only applies to linearly convergent methods. Let's look at another technique. Recall from Sec. 5.2 that the simple Simpson's rule has the form

$$\int_a^b f(x)dx = \frac{h}{3}\left(f(a) + 4f(\frac{a+b}{2}) + f(b)\right)$$
$$-\frac{1}{90}f^{(4)}(x_0)h^5 + \kappa_6 f^{(5)}(x_0)h^6 + \kappa_7 f^{(6)}(x_0)h^7 \dots$$

for some constants $\kappa_6, \kappa_7, \dots$ (see Sec. 5.2). That is, the truncation error E has the form

(7.2) $$E = c_5 h^5 + c_6 h^6 + c_7 h^7 + \dots$$

for some constants $c_5, c_6, c_7, \dots$ depending on the values of the derivatives of f at the *known* point x_0. For a fixed application of the method, the c_i are fixed and in principle they could be found. In practice we won't need to know their values.

Most of the methods we have seen have a truncation error that can be expanded in a form similar to Eq. (7.2). Suppose we're looking at a quadrature rule of the form

$$A = Q_h + E_h$$

where A is the area, Q_h is the result of applying the numerical method with spacing h to the integrand, and

(7.3) $$E_h = c_1 h + c_2 h^2 + c_3 h^3 + c_4 h^4 + \ldots$$

is the error, where the constants c_i $(i = 1, 2, 3, \ldots)$ do not depend on h. Then

$$A = Q_{h/2} + E_{h/2}$$

where

$$E_{h/2} = c_1 \frac{h}{2} + c_2 \frac{h^2}{4} + c_3 \frac{h^3}{8} + c_4 \frac{h^4}{16} + \ldots$$

is the truncation error, which is $O(h)$ if $c_1 \neq 0$. But since the c_i are constants, we can eliminate terms in the truncation error! We'll take a weighted average of Q_h and $Q_{h/2}$ that still sums to A but with a superior error term. Let's try to eliminate the $O(h)$ term. We have

$$
\begin{aligned}
A &= Q_h + E_h \\
&= Q_{h/2} + E_{h/2}
\end{aligned}
$$

so

$$
\begin{aligned}
A &= 2A - A \\
&= 2(Q_{h/2} + E_{h/2}) - (Q_h + E_h) \\
&= 2Q_{h/2} - Q_h + 2E_{h/2} - E_h \\
&= 2Q_{h/2} - Q_h + c_1 h + c_2 \frac{h^2}{2} + c_3 \frac{h^3}{4} + c_4 \frac{h^4}{8} + \ldots \\
&\quad -(c_1 h + c_2 h^2 + c_3 h^3 + c_4 h^4 + \ldots) \\
\text{(7.4)} \quad &= 2Q_{h/2} - Q_h + \tilde{c}_2 h^2 + \tilde{c}_3 h^3 + \tilde{c}_4 h^4 + \ldots \\
&= Q_h^{(2)} + E_h^{(2)}
\end{aligned}
$$

where $Q_h^{(2)} = 2Q_{h/2} - Q_h$ is a new quadrature rule that has an error term $E_h^{(2)}$ that is $O(h^2)$ rather than $O(h)$. We have used an $O(h)$ rule, evaluated at two different spacings, to generate an $O(h^2)$ rule.

Incidentally, we never used the fact that Q_h was a quadrature rule. This technique works on any numerical method Q_h that approximates a quantity A with an error term as in Eq. (7.3).

Furthermore, the method can be iterated. Let's call the basic method $Q_h = Q_h^{(1)}$ so the $O(h^2)$ method is $Q_h^{(2)} = 2Q_{h/2}^{(1)} - Q_h^{(1)}$. From Eq. (7.4),

$$A = Q_h^{(2)} - \tilde{c}_2 h^2 + \tilde{c}_3 h^3 + \tilde{c}_4 h^4 + \dots$$

$$A = Q_{h/2}^{(2)} - \tilde{c}_2 \frac{h^2}{4} + \tilde{c}_3 \frac{h^3}{8} + \tilde{c}_4 \frac{h^4}{16} + \dots$$

so we can knock out the $O(h^2)$ term–that is, the leading term in the error–using a linear combination of $Q_{h/2}^{(2)}$ and $Q_h^{(2)}$:

$$
\begin{aligned}
A &= (4A - A)/3 \\
&= \frac{4}{3}(Q_{h/2}^{(2)} + \tilde{c}_2 \frac{h^2}{4} + \tilde{c}_3 \frac{h^3}{8} + \tilde{c}_4 \frac{h^4}{16} + \dots) - \\
&\quad \frac{1}{3}(Q_h^{(2)} + \tilde{c}_2 h^2 + \tilde{c}_3 h^3 + \tilde{c}_4 h^4 + \dots) \\
&= \frac{1}{3}(4Q_{h/2}^{(2)} - Q_h^{(2)}) + \frac{1}{3}(\tilde{c}_2 h^2 + \tilde{c}_3 \frac{h^3}{2} + \tilde{c}_4 \frac{h^4}{4} + \dots) - \\
&\quad \frac{1}{3}(\tilde{c}_2 h^2 + \tilde{c}_3 h^3 + \tilde{c}_4 h^4 + \dots) \\
&= Q_h^{(3)} + \hat{c}_3 h^3 + \hat{c}_4 h^4 + \dots
\end{aligned}
$$

(7.5)

where

$$
\begin{aligned}
Q_h^{(3)} &= (4Q_{h/2}^{(2)} - Q_h^{(2)})/3 \\
&= Q_{h/2}^{(2)} + \frac{Q_{h/2}^{(2)} - Q_h^{(2)}}{3} \\
&= Q_{h/2}^{(2)} + \frac{Q_{h/2}^{(2)} - Q_h^{(2)}}{2^2 - 1}
\end{aligned}
$$

is a new $O(h^3)$ method (or possibly higher, if it should happen that $\hat{c}_3 = 0$). We can continue this process indefinitely; in general, if $Q_h^{(1)}$ is the $O(h)$ method then $Q_h^{(i)}$ is an $O(h^i)$ method, where $Q_h^{(i)}$ is defined recursively by

(7.6)
$$Q_h^{(i)} = Q_{h/2}^{(i-1)} + \frac{Q_{h/2}^{(i-1)} - Q_h^{(i-1)}}{2i - 1}$$

($i = 2, 3, \dots$). This process is known as **Richardson extrapolation** and is a general and flexible tool for accelerating convergence. Performing it based on Q_h and $Q_{h/2}$ is common but not essential; we could have used, say, Q_h and $Q_{h/3}$ and still used the appropriate linear combination to eliminate the first term in the error expansion. We would just end up with different constants in Eq. (7.6).

EXAMPLE 1: Let's use the Richardson extrapolation technique with the left endpoint quadrature rule

$$\int_a^b f(x)dx = \sum_{i=0}^{n-1} f(a + ih)h + E_h$$

($h = 1/n$). It can be shown that E_h is of the proper form for Richardson extrapolation. We'll look for an $O(h^3)$ approximation. Let's use $h = .1$ and integrate $f(x) = 4x^3$ from 0 to 1; the true value of the integral is 1. We have

$$
\begin{aligned}
Q_h^{(1)} &= \sum_{i=0}^{n-1} f(a+ih)h \\
&\doteq .8100 \\
Q_{h/2}^{(1)} &= \sum_{i=0}^{2n-1} f(a+ih/2)\frac{h}{2} \\
&\doteq .9025 \\
Q_h^{(2)} &= Q_{h/2} + (Q_{h/2} - Q_h) \\
&\doteq .9950
\end{aligned}
$$

(note that $Q_h^{(2)}$ is a significantly improved approximation for negligible additional computation). Now we need $Q_{h/2}^{(2)}$ so we can form $Q_h^{(3)}$. We have

$$
\begin{aligned}
Q_{h/4}^{(1)} &= \sum_{i=0}^{4n-1} f(a+ih)h \\
&\doteq .9506 \\
Q_{h/2}^{(2)} &= Q_{h/4} + (Q_{h/4} - Q_{h/2}) \\
&\doteq .9988 \\
Q_h^{(3)} &= Q_{h/2}^{(2)} + \frac{Q_{h/2}^{(2)} - Q_h^{(1)}}{3} \\
&\doteq 1
\end{aligned}
$$

and in fact $Q_h^{(3)} = 1$ to the number of digits displayed. It's customary to arrange these quantities in a table

$Q_h^{(1)}$		
.8100		
$Q_{h/2}^{(1)}$	$Q_h^{(2)}$	
.9025	.9950	
$Q_{h/4}^{(1)}$	$Q_{h/2}^{(2)}$	$Q_h^{(3)}$
.9506	.9988	1.0000

sometimes called an **extrapolation table**. The errors are

$Q_h^{(1)}$		
$E \doteq .1900$		
$Q_{h/2}^{(1)}$	$Q_h^{(2)}$	
$E \doteq .0975$	$E \doteq .0050$	
$Q_{h/4}^{(1)}$	$Q_{h/2}^{(2)}$	$Q_h^{(3)}$
$E \doteq .0494$	$E \doteq .0013$	$E \doteq .0000$

and the errors for $Q_h^{(1)}$ are indeed $O(h)$ while those for $Q_h^{(2)}$ are about $O(h^2)$. The size of the error in $Q_h^{(3)}$ is better than expected. $\square$

Richardson extrapolation may be applied whenever we have a numerical method Q_h for approximating a quantity A that satisfies

$$A = Q_h + c_1 h^{m_1} + c_1 h^{m_2} + ... + c_1 h^{m_N} + O(h^{m_{N+1}})$$

where

$$m_1 < m_2 < \cdots < m_N < m_{N+1}$$

are positive integers. We choose linear combinations to knock out the terms in h^{m_1}, h^{m_2}, ..., h^{m_N} successively. It is not uncommon for a numerical method to have an error term that involves only even or only odd powers of h. Indeed, the trapezoidal rule is one such method; if f is infinitely differentiable then it has the form

$$\int_a^b f(x)dx = T_h f + c_2 h^2 + c_4 h^4 + c_6 h^6 + ...$$

where each c_i $(i = 2, 4, 6, ...)$ is in general nonzero. If f is at least $2k + 1$ times differentiable then the trapezoidal rule has the form

$$(7.7) \qquad \int_a^b f(x)dx = T_h f + c_2 h^2 + c_4 h^4 + c_6 h^6 + ... + c_{2k} h^{2k} + O(h^{2k+1})$$

(with the same c_i, $i = 2, 4, 6, ..., 2k$). Hence if f is sufficiently differentiable we may apply Richardson extrapolation to the trapezoidal rule as follows: Set $T_h^{(1)} = T_h$. Then define

$$T_h^{(2)} = \frac{4T_{h/2}^{(1)} - T_h^{(1)}}{3}$$

to eliminate the $O(h^2)$ terms (as in Eq. (7.5)). In general, the sequence of methods

$$(7.8) \qquad\qquad T_h^{(i)} = \frac{4^i T_{h/2}^{(i-1)} - T_h^{(i-1)}}{4^i - 1}$$

is the result of applying Richardson extrapolation to the trapezoidal rule, and Eq. (7.8) defines a method known as **Romberg integration** (or the **Romberg algorithm**) that was formerly widely used. The resulting extrapolation table is called the **T-table** because of its lower triangular structure.

Each $T_h^{(i)}$ $(i \geq 2)$, when viewed as method in its own right as opposed to one component of Romberg integration, is called a **Romberg formula**. Romberg formulas are not interpolatory rules (except the first few); in fact the degree of accuracy of $T_h^{(i)}$ is only $2i - 1$. They are however Riemann sum rules. (Recall that many advanced quadrature methods can be viewed as Riemann sum rules.) The Romberg algorithm is an efficient way to generate highly accurate answers from the trapezoidal rule if the integrand f is sufficiently differentiable but can be very inefficient if the differentiability assumption is violated. There is a class of algorithms known as **cautious Romberg methods** that use the known degree of accuracy of $T_h^{(i)}$ to check for this situation and take corrective action if it is not met.

Using a single step of Richardson extrapolation to improve upon an estimate is quite common. One case when we might hesitate to do it is when we are very carefully controlling the error estimates, as we do not get an error estimate for the extrapolated value.

EXAMPLE 1: Let's try Romberg integration on $f(x) = \exp(x)$ from 0 to 1. The true value is 1.7183 (to four decimal places). Although we can start with any h we often start with $h = b - a$ in hopes of achieving convergence quickly with relatively few function evaluations; if we are storing previously computed values then this is efficient because, for example, $T_{h/2}^{(1)}$ re-uses all the values used in computing $T_h^{(1)}$. We have

$$
\begin{aligned}
T_h^{(1)} &= \frac{h}{2}(f(0) + f(1)) \\
&\doteq 1.8591 \\
T_{h/2}^{(1)} &= \frac{h}{4}(f(0) + 2f(.5) + f(1)) \\
&\doteq 1.7539 \\
T_h^{(2)} &= \frac{4T_{h/2}^{(1)} - T_h^{(1)}}{4 - 1} \\
&\doteq \frac{4 \cdot 1.7539 - 1.8591}{3} \\
&\doteq 1.7189
\end{aligned}
$$

(note that the computation of $T_h^{(2)}$ and requires no new function evaluations). The table is

$$
\begin{array}{cc}
T_h^{(1)} & \\
1.8591 & \\
T_{h/2}^{(1)} & T_h^{(2)} \\
1.7539 & 1.7189
\end{array}
$$

so far. If we want to get a better approximation we need a new entry in the first column. This will let us get a slightly better estimate in the second column but a much better one in the third column. We have

$$
\begin{aligned}
T_{h/4}^{(1)} &= \frac{h}{8}(f(0) + 2f(.25) + 2f(.5) + 2f(.75) + f(1)) \\
&\doteq 1.7272 \\
T_{h/2}^{(2)} &= \frac{4T_{h/4}^{(1)} - T_{h/2}^{(1)}}{3} \\
&\doteq 1.7183 \\
T_h^{(3)} &= \frac{4^2 T_{h/2}^{(2)} - T_h^{(2)}}{4^2 - 1} \\
&\doteq \frac{8 \cdot 1.7183 - 1.7272}{7} \\
&\doteq 1.7182
\end{aligned}
$$

(note that we are still using $h = b - a$, not adjusting it for each application of the composite trapezoidal rule). The new table is

$$T_h^{(1)}$$
$$1.8591$$
$$T_{h/2}^{(1)} \qquad T_h^{(2)}$$
$$1.7539 \qquad 1.7189$$
$$T_{h/4}^{(1)} \qquad T_{h/2}^{(2)} \qquad T_h^{(3)}$$
$$1.7272 \qquad 1.7183 \qquad 1.7182$$

and $T_{h/2}^{(2)}$ is slightly more accurate ($E \doteq 3.7E - 5$) than $T_h^{(3)}$ ($E \doteq 4E - 5$). This happens because we are not yet into the limiting behavior, where the last entries in successive columns should be an order of h better estimates–at least, until we get to the point where round-off error is significant. □

The convergence criterion used is usually that the most accurate entry in the last column of the extrapolation table and the most accurate entry in the next-to-last column of the extrapolation table agree to within a given tolerance. When using Romberg integration it is possible that, during the first several stages of refinement (representing the first several columns of the extrapolation table), two entries at the bottoms of their columns will agree to within the tolerance by chance rather than because they are accurate. Because of this we typically insist that the entries in the last three columns of the table all agree to within some tolerance. If not, we generate a new column.

It's worth repeating that extrapolation is a very general technique that may be applied to many different numerical methods. Often in trying to understand what method a piece of software is using you will discover that an extrapolation/acceleration technique has been woven into a standard method in a clever and efficient way that is difficult to "reverse engineer" from the code. It's important to document such modifications in the comments so that later users (and programmers) can understand the method being used by the program.

MATLAB

We've mentioned singular integrals before. By a singular integral we mean an improper integral which exists but is improper because of a singularity of the integrand either at one or both of the endpoints or somewhere in the interior of the interval of integration. The MATLAB command isfinite detects both NaN and Inf values, whereas isinf would only detect Inf values and similarly for isnan. This can be useful within a quadrature program when we are "avoiding the singularity" because we simply recognize its presence and stay away from it. In a technical sense this makes the rule open but we imagine that we are simply adjusting the rule to avoid getting an infinite result.

Another simple method of handling singularities is "ignoring the singularity", that is, doing nothing about it. We used this technique with the function $f(x) = 3x^2 + \ln((\pi - x)^2)/\pi^4 + 1$ in the MATLAB subsection of Sec. 5.2. Let's look at another example. Consider the function $f(x) = 1/\sqrt{|x|}$ on $[-1, 1]$. If we use a rule that attempts to evaluate $f(0)$ then we'll encounter a problem. We will use Simpson's 3/8 rule $\int_a^b f(x)dx \approx \frac{3h}{8}(f(a) + 3f(a + h) + 3f(a + 2h) + f(b))$ because it avoids the singularity; Simpson's rule would land on the singularity immediately.

(Since Simpson's 3/8 rule and Simpson's rule have the same order of error it would probably be more efficient to just use Simpson's rule and fudge the endpoints.) Enter:

```
>> f=@(x) 1./sqrt(abs(x));
>> h=2/3;
>> w=(3/8)*[1 3 3 1];
>> x=[-1 -1/3 1/3 1];
>> T11=dot(h*w,f(x))
```

The correct answer is 4. We can't just subdivide by dividing h in half because we would land on the singularity. Instead we'll use Simpson's 3/8 rule on each of the three panels, shifting the nodes x as needed. Enter:

```
>> h=h/3;
>> T12_1=dot(h*w,f((1/3)*(x-2)))       %Leftmost panel.
>> T12_2=dot(h*w,f((1/3)*x))           %Middle panel.
>> T12_3=dot(h*w,f((1/3)*(x+2)))       %Rightmost panel.
>> T12=T12_1+T12_2+T12_3
```

This is an improved estimate but it still isn't very good. Richardson extrapolation isn't justified because f is not differentiable and so the error estimate $E = c_4 h^4 + c_5 h^5 + ...$ of Simpson's 3/8 rule isn't valid. Let's try it anyway. Since we are going from h to $h/3$ (not $h/2$) the term $c_4 h^4$ in E_h becomes $c_4 h^4/3^4$ in $E_{h/3}$ so we have $80A \approx 81 Q_{h/3} - Q_h$, i.e., $A \approx Q_{h/3} + (Q_{h/3} - Q_h)/80$ as the Richardson step. Enter:

```
>> T2 = T12 + (T12 - T11)/80          %Richardson extrapolation.
```

This is a very slight improvement. For this nondifferentiable function we should probably be using a lower-order method than Simpson's 3/8 rule. Still, if we continue long enough we would see convergence. Enter:

```
>> integral(f,-1,1)
```

The program checks for singularities at the end and avoids them but a method like this may still land on the singularity at zero. Let's avoid the singularity by using the fact that integral will check for singularities at the endpoints and move slightly away from them. Enter:

```
>> A1=integral(f,-1,0)+integral(f,0,1)
>> abs(A1-4)
```

Recall that the default tolerance is $1E - 6$. If we think that the method is still having difficulties we can work harder to avoid the problem point. Enter:

```
>> A2=integral(f,-1,0,1E-5)+integral(f,0,1,1E-5)
>> abs(A2-4)
```

For a singular integrand this isn't bad performance. There are special methods for singular integrals.

Problems

1.) Show that Eq. (7.8) is the result of applying Richardson extrapolation to the trapezoidal rule using Eq. (7.7).

2.) What happens if you apply Richardson extrapolation to the trapezoidal rule in the form of Eq. (7.6) (that is, as if the error term was of the form of Eq. (7.3))?

3.) a.) Use Eq. (7.7) to determine how many times f must be continuously differentiable in order for $T_h^{(i)}$ to be well-defined.

b.) Use the Romberg algorithm on $f(x) = x^{1/2}$, $x^{3/2}$, $x^{5/2}$, $x^{7/2}$, and $x^{9/2}$ on $[0, 1]$. In each case generate 4 columns in the extrapolation table. Discuss your results.

4.) a.) Use Romberg integration to approximate $\int_0^1 \exp(x)dx$ to within 10^{-6}.

b.) Use Romberg integration to approximate $\int_0^1 \sin(\pi x)dx$ to within 10^{-6}.

5.) Derive the formula for a single step of Richardson extrapolation applied to Simpson's rule.

6.) a.) Show that the Romberg formula $T_h^{(2)}$ is Simpson's rule.

b.) Write out the Romberg formula $T_h^{(3)}$ explicitly.

c.) Write out the Romberg formula $T_h^{(4)}$ explicitly.

7.) The midpoint rule also has an error expansion of the form in Eq. (7.7). Derive a method analogous to Romberg integration for the midpoint rule. (The resulting table is called the **M-table**.) Compare your method to Romberg integration.

8.) Write a MATLAB program that accepts a function, interval, and tolerance and performs Romberg integration. Test your program.

9.) a.) Write a MATLAB program that accepts a sequence and applies Richardson extrapolation to it. Test it on several appropriate sequences, corresponding to the first column of an extrapolation table, and some inappropriate sequences (not corresponding to spacings of h, $h/2$, $h/4$, ..., for example, or not having an appropriate error expansion).

b.) Compare your program to the results of using Aitken's process. Test only on linearly convergent sequences.

10.) If $A = \sum_{i=1}^{\infty} a_i$ is to be found and $B = \sum_{i=1}^{\infty} b_i$ is a known series such that $a_i/b_i \to \gamma$ where $\gamma \neq 0$, then **Kummer's transformation** $A = \gamma B + \sum_{i=1}^{\infty} a_i (1 - \gamma b_i/a_i)$ converges more rapidly to A than the original series does. Use this fact to accelerate the convergence of the **Leibniz formula** $A = 1 - 1/3 + 1/5 - 1/7 + ...$ by comparing it to an appropriate series b_i. It is a fact that $A = \pi/4$.

11.) The formula $T_i^{(1)} = \frac{1}{2}\left(T_{i-1}^{(1)} + \frac{h}{2^{k-2}} \sum_{k=1}^{2^{i-2}} f\left(a + (2k-1)\frac{h}{2^{k-1}}\right)\right)$ can be used to structure the computation of the first column of the T-table efficiently is as follows. Prove that this formula is correct by taking $h = b - a$ and defining $T_i^{(1)} = T_{h/2^{i-1}}^{(1)}$ (that is, $T_i^{(1)}$ uses 2^{i-1} panels used in the trapezoidal rule). Comment on its efficiency.

12.) Perform a numerical experiment to verify that the trapezoidal rule satisfies $\int_a^b f(x)dx = T_h f + c_2 h^2 + c_4 h^4 + c_6 h^6 + ...$ if f is infinitely differentiable.

13.) In the MATLAB subsection of Sec. 1.5 we experimented with the finite difference Newton's method approximation $s = \frac{f(x+h)-f(x)}{h}$ of the slope of the function f at x. In particular, we took $f(x) = \ln(x)$, $x = 1$, and $h = .1$, then reduced h and investigated the error in this approximation of $f'(x)$. Assuming that $f'(x) = \frac{f(x_k+h)-f(x_k)}{h} + c_1 h + c_2 h^2 + ...$, use Richardson extrapolation to produce more accurate estimates of $f'(x)$. Recall that this approximation was sensitive to round-off error.

14.) a.) Describe a method for performing adaptive Romberg integration.

b.) Write a MATLAB program that implements your method. Test your program.

15.) a.) A **modified Romberg algorithm** is one that does not cut h in half at each step but instead uses some other progression. Typically such a method adds fewer new points than the Romberg algorithm would add so as to be more efficient and avoid the round-off error that will eventually be problematic if we are using $h/2^k$ as our step size in the T-table. One such method starts with a single panel then adds one panel at each step. What would the formula corresponding to Eq. (7.8) be for this modified Romberg algorithm?

b.) One modified Romberg algorithm that is particularly efficient is **Oliver's modified Romberg algorithm**, based on the number of panels $1, 2, 3, 4, 6, 8, 12, \ldots$ where this sequence is defined by the requirement that it include all numbers of the form 2^k and $3 \cdot 2^k$. What is the significance of this choice of panels? What would the formula corresponding to Eq. (7.8) be for Oliver's modified Romberg algorithm?

16.) Prove that Eq. (7.8) converges to the value of the integral if f is infinitely differentiable.

8. Periodizing Transformations

Let's turn again to the Euler-Maclaurin summation formula, which we are interpreting as the trapezoidal rule error formula

$$T_h f = \int_a^b f(x)dx + \frac{B_2}{2}h^2(f'(b) - f'(a)) + \frac{B_4}{4!}h^4(f^{(3)}(b) - f^{(3)}(a))$$
$$+ \frac{B_6}{6!}h^6(f^{(5)}(b) - f^{(5)}(a)) + \ldots + \frac{B_{2k}}{(2k)!}h^{(2k-1)}(f^{(2k-1)}(b) - f^{(2k-1)}(a))$$
$$+ h^{2k+1}\int_a^b P_{2k+1}\left(\frac{x-a}{h}\right)f^{(2k+1)}(x)dx$$

assuming that f is $2k + 1$ times continuously differentiable on $[a, b]$. We can use this to create families of high-order methods for numerical quadrature by arranging for many of the terms in the error expansion to cancel out.

As we've noticed before, and as is evident from the Euler-Maclaurin summation formula, although the trapezoidal rule has error that is $O(h^2)$ we get even better performance if f is such that

$$f'(a) = f'(b)$$

for then the trapezoidal rule has error that is $O(h^4)$. The trapezoidal rule is significantly more accurate for functions that satisfy this condition, and the more odd-order derivatives of f that have the same values at the endpoints of the interval of integration, the higher the order of the error.

EXAMPLE 1: Consider $\int_0^1 (\frac{5}{2}x^2 - x^5)dx = 2/3$. The integrand $f(x) = \frac{5}{2}x^2 - x^5$ satisfies $f'(0) = f'(1)$ but $f'''(0) \neq f'''(1)$. The trapezoidal rule for various values of h gives

h:	1	1/2	1/4	1/8	1/16
T_h:	.7500	.6719	.6670	.6667	.6667

and at $h = 1/8$ (8 panels) the absolute error is $2.0345E - 5$, while for $h = 1/16$ (16 panels) it is $1.2716E - 6$. For comparison, Simpson's 3/8 rule based on 15 panels gives an absolute error of $1.4815E - 5$.

Consider $\int_0^1 (x^2/2 - x^3/3)dx = 1/12$. The integrand $f(x) = x^2/2 - x^3/3$ satisfies $f^{(2k+1)}(0) = f^{(2k+1)}(1)$ for all k. The trapezoidal rule for $h = 1$ gives .08333333333333, which is accurate to every decimal place displayed. □

Remember, the trapezoidal rule uses a linear interpolant to approximate the integrand, so its performance on the function $f(x) = x^2/2 - x^3/3$ is already pretty impressive. The reason for this success is that f is infinitely differentiable, with $f^{(2k+1)}(0) = f^{(2k+1)}(1)$ for all k. This means that the Euler-Maclaurin summation formula gives

$$T_h f - \int_a^b f(x)dx = h^{2k+1} \int_a^b P_{2k+1}\left(\frac{x-a}{h}\right) f^{(2k+1)}(x)dx$$

for any choice of $k \geq 1$. The order of the error is better than $O(h^N)$ for any finite N for functions of this type (sometimes called **spectral accuracy**). Hence we expect the trapezoidal rule to perform extremely well for such integrands.

The only naturally arising class of functions on $[a,b]$ that satisfy $f^{(m)}(a) = f^{(m)}(b)$ for all odd m is the class of periodic functions with period $b - a$. Hence, the trapezoidal rule is recommended if the integrand is known to be periodic and smooth on the interval of integration. In the unlikely event that we know that our function satisfies $f^{(m)}(a) = f^{(m)}(b)$ for $m = 1, 2, ..., M$ even though it isn't periodic we should also consider the trapezoidal rule. An example would be $f(x) = \sin(\pi x)$ for $0 \leq x \leq 1$.

Suppose we have an integral $\int_a^b f(x)dx$ that we wish to approximate. Might it be possible to change it to an equivalent integral

$$\int_a^b f(x)dx = \int_\alpha^\beta g(t)dt$$

for which $g(t)$ satisfies $g^{(m)}(\alpha) = g^{(m)}(\beta)$ for $m = 1, 2, ..., M$? If so, and if M is large enough, then applying the trapezoidal rule to $\int_\alpha^\beta g(t)dt$ might give good results, which translates immediately into good results for the integral of interest to us. The efficiency of such a technique would depend on the difficulty of finding and using the transformation, which is called a **periodizing transformation**.

One of the most common periodizing transformations is the **IMT transformation** (named for its developers Iri, Moriguti, and Takasawa) given by

$$\phi(t) = \exp\left(-\frac{c}{t(1-t)}\right)$$

$$\psi(x) = \frac{1}{K}\int_0^x \phi(t)dt$$

for any $c > 0$, where

$$K = \int_0^1 \phi(t)dt$$

is a normalization constant[11]. The function $\phi(t)$ is defined to have its limiting value, which is 0, as $t \to 0^+, 1^-$ (from within the interval). The function $\psi(x)$ is monotonically increasing and differentiable on $[0,1]$, and so we can make the change of variables

---

[11]There is a more general formulation of the IMT transformation that depends on four parameters.

$$\int_0^1 f(x)dx = \int_0^1 f(\psi(x))\psi'(x)dx$$

$$= \int_0^1 f(\psi(x))\frac{\phi(x)}{K}dx$$

$$= \int_0^1 g(x)dx$$

within the integrand. In this case g is at least as many times continuously differentiable as f is, and furthermore

$$g^{(i)}(0) = g^{(i)}(1) = 0$$

for $i = 0, 1, 2, ..., M$ if $f \in C^M[0,1]$. We expect that the trapezoidal rule applied to the integral with integrand $g(x)$ will give excellent results.

The larger the constant c is, the better the performance, in principle, as it controls the rate at which the endpoints values $\phi(0)$ and $\phi(1)$ approach their limiting common value of 0. We speak of the **damping power** of a periodizing transformation in this sense, and would like a rapid approach. But making it too large can make other quantities in the rule so small as to cause numerical problems.

Suppose now that $f \in C^\infty[0,1]$. Then so is g, and so applying trapezoidal rule to $\int_0^1 g(x)dx$ will be highly effective. This leads to the **IMT rule** for $\int_0^1 f(x)dx$,

$$\int_0^1 f(x)dx = \int_0^1 g(x)dx$$

$$\approx T_h g$$

$$= T_h\left(\frac{1}{K}f(\psi(x))\phi(x)\right)$$

$$= \frac{1}{K}T_h f(\psi(x))\phi(x)$$

$$(8.1) \qquad = \frac{1}{Kn}\sum_{i=1}^{n-1}\phi\left(\frac{i}{n}\right)f\left(\psi\left(\frac{i}{n}\right)\right)$$

where we have used the fact that $\phi(0) = \phi(1) = 0$. If we define new nodes and weights by

$$(8.2) \qquad x_i = \psi\left(\frac{i}{n}\right)$$

$$(8.3) \qquad w_i = \frac{1}{Kn}\phi\left(\frac{i}{n}\right)$$

then the IMT rule can be written in the familiar form

$$\int_0^1 f(x)dx \approx \sum_{i=1}^{n-1} w_i f(x_i)$$

which we think of as an $n+1$ point rule with $w_0 = w_n = 0$. Note that the nodes (Eq. (8.2)) and weights (Eq. (8.3)) may be pre-computed; they do not depend on f. Of course, K and $x_i = \psi(i/n)$ $(i = 1, ..., n-1)$ will likely need to be computed by a quadrature rule[12]. If the original integral has an interval of integration other than $[0, 1]$ then we can use a linear change-of-variables first in order to bring it into the desired format.

There is a formula similar to the Euler-Maclaurin summation formula for the midpoint rule and hence there is a related **midpoint IMT rule**, which can be written (on $[0, 1]$) as

$$\int_0^1 f(x)dx \approx \sum_{i=1}^{n-1} \widetilde{w}_i f(\widetilde{x}_i)$$

where

$$\widetilde{x}_i = \psi\left(\frac{2i-1}{2n}\right)$$

$$\widetilde{w}_i = \frac{1}{Kn}\phi\left(\frac{2i-1}{2n}\right)$$

are the nodes and weights. Both of the trapezoidal and midpoint IMT rules are useful for smooth functions on a finite interval but are not very useful for functions with few continuous derivatives.

Neither of the IMT rules given here are interpolatory and in fact they do not even integrate constants correctly, in general! We have placed a strong focus on interpolatory rules in this chapter–and quite properly so–but this is a good reminder that not all numerical quadrature rules are interpolatory in nature. Periodizing transformations represent a special-case approach for special circumstances: Well-behaved functions for which we seek especially accurate estimates, most notably. In fact, one of the reasons why they aren't seen as widely as the methods in earlier sections of this chapter is that while they do extremely well on "nice" functions, all methods do pretty well on "nice" functions, meaning that we really need a driving reason to introduce the overhead of a periodizing approach. But–these occur.

EXAMPLE 2: Let $f(x) = \exp(x)$. We'll apply the trapezoidal IMT rule with $n = 3$ and $c = 1$. We have

$$\phi(t) = \exp\left(-\frac{1}{t(1-t)}\right)$$

$$\psi(x) = K^{-1}\int_0^x \exp\left(-\frac{1}{t(1-t)}\right)dt$$

$$K = \int_0^1 \exp\left(-\frac{1}{t(1-t)}\right)dt$$

$$\doteq .0070$$

and so the weights are (from Eq. (8.3))

---

[12]There are other approaches for finding these values based on approximations of $\psi(x)$, etc.

$$w_1 = \frac{1}{3K}\phi\left(\frac{1}{3}\right)$$

$$\doteq .5268$$

$$w_2 = \frac{1}{3K}\phi\left(\frac{2}{3}\right)$$

$$\doteq .5268$$

and $w_0 = w_3 = 0$. The nodes are (from Eq. (8.2))

$$x_1 = \psi\left(\frac{1}{3}\right)$$

$$= \frac{1}{K}\int_0^{1/3}\exp\left(-\frac{1}{t(1-t)}\right)dt$$

$$\doteq 0.1257$$

$$x_2 = \psi\left(\frac{2}{3}\right)$$

$$= \frac{1}{K}\int_0^{2/3}\exp\left(-\frac{1}{t(1-t)}\right)dt$$

$$\doteq .8743$$

and so

$$\int_0^1 \exp(x)dt \approx w_1 e^{x_1} + w_2 e^{x_2}$$

$$\doteq 1.8600$$

(the actual value to four places is 1.7183). The trapezoidal rule based on 4 points, used directly on f, gives 1.7342. However, the trapezoidal rule based on a single panel, which like the IMT rule only uses two values of f, gives 1.8591.

Let's see if we can find a bigger difference by using a less friendly function. Let $f(x) = 1/(2+\sin(3x))$. The actual value of the integral to four places is .3803. The trapezoidal rule based on 2 points gives .4835 and the trapezoidal rule based on 4 points gives .3931. The IMT rule (using the already computed nodes x_1, x_2 and weights w_1, w_2) gives .4335, more accurate than the trapezoidal rule using the same number of function evaluations but less accurate than the rule using four function evaluations.

If we change c to $c = 2$ in the definition of $\phi(t)$ then the nodes become $x_1 = .0634$, $x_2 = .9366$ and the weights become $w_1 = .4241$, $w_2 = .4241$. Then the IMT rule gives .3761 for $f(x) = 1/(2 + \sin(3x))$ which is more accurate than the trapezoidal rule using four function evaluations. For $f(x) = \exp(x)$ this IMT rule gives 1.5340 which is not very accurate. $\square$

The IMT rule is usually viewed as being based on $n + 1$ points for purposes of comparison, not the $n - 1$ points at which it actually computes function values, because the fact that the extra two points are known to have null function values is built into the method. For a larger n this difference is not very meaningful, but for smaller n values we should remember this when comparing it with other methods.

There are a number of other periodizing transformations that may be used. Because of the assumption of smoothness they are poor choices for a general-purpose automatic integrator but may be useful in when circumstances warrant their use. One such transformation is the **double exponential transformation**

$$\psi(x) = \frac{1}{2}\left(1 + \tanh\left(\alpha\sinh\left(\beta\left(\frac{1}{1-x} - \frac{1}{x}\right)\right)\right)\right)$$

($\alpha > 0$, $\beta > 0$), leading to the corresponding **double exponential rule** found by transforming the integrand from $f(x)$ to $f(\psi(x))\psi'(x)$,

$$\int_0^1 f(x)dx = \int_0^1 f(\psi(x))\psi'(x)dx$$

and applying the trapezoidal rule to the transformed integral. As with the IMT rules, $\psi(x)$ is a monotonically increasing function on $[0,1]$ and all continuous derivatives of $\psi(x)$ are zero at 0 and at 1. Hence the new integrand $f(\psi(x))\psi'(x)$ is zero at $x = 0$ and $x = 1$, leading to desirable behavior when T_h is applied to $f(\psi(x))\psi'(x)$. A form of the double exponential rule is used by some packages when they have detected a singularity, in the hopes that the rapid decay of the transformed function will lead to improved accuracy. Strictly speaking, of course, a true singularity means a discontinuous function, making the analysis of the rule much harder.

The Clenshaw-Curtis method can be viewed as a form of periodizing transformation that uses a connection between the Euler-Maclaurin summation formula and the Discrete Cosine Transform (DCT) from Fourier analysis. The connection to the DCT allows for fast computation via the Fast Fourier Transform (FFT) algorithm, which is a great advantage. Hence even though we found that method by other means, viewing it in the present context provides valuable extra insights.

MATLAB

Let's extend Example 2. We'll use the IMT rule with $n = 10$ and $c = 1$ on the integrand $f(x) = \exp(x)$ of. Enter:

```
>> f=@(x) exp(x)
>> phi=@(t) exp(-1./(t.*(1-t)))
>> K=integral(phi,0,1)
>> n=10;x=[];
>> w=phi((1:n-1)/n)/(K*n)
>> sum(w)
>> for i=1:n-1,x(i)=integral(phi,0,i/n,1E-10)/K;end
>> x
```

(We could have computed the nodes x more efficiently by noting that $x_{i+1} = x_i + \int_{i/(n-1)}^{i} \phi(t)dt$, thereby reusing our previously computed values.) The nodes go from 0 to 1 as expected. The integration rule is $\sum_{i=1}^{n-1} w_i f(x_i)$ (Eq. (8.1)). Enter:

```
>> A1=exp(1)-1           %True value.
>> IMT=dot(w,f(x))       %Approx. value.
>> abs(A1-IMT)
```

We are using 9 function values and the error is on the order of 10^{-4}. This is pretty good performance.

Let's try using another function. Enter:

```
>> f2=@(x) sin(x.^2)
```

```
>> IMT2=dot(w,f2(x))
>> A2=integral(f2,0,1)
>> abs(A2-IMT2)
```
Again, the error is on the order of 10^{-4}. Let's go from $n = 10$ to $n = 20$. Enter:
```
>> n=10;x=[];
>> w=phi((1:n-1)/n)/(K*n)
>> for i=1:n-1,x(i)=integral(phi,0,i/n)/K;end
>> IMT3=dot(w,f(x))
>> IMT4=dot(w,f2(x))
>> abs(A-IMT3)
>> abs(A2-IMT4)
```
We are using 19 explicit function evaluations (21 points). Let's compare the trapezoidal rule. Enter
```
>> xx=0:.05:1;             %21 points.
>> abs(A1-trapz(xx,f(xx)))
>> abs(A2-trapz(xx,f2(xx)))
```
The IMT rule is performing slightly better than the trapezoidal rule; if we assume that the weights and nodes for the IMT rule have been precomputed then it is slightly more efficient because it doesn't explicitly use the endpoints.

Let's try a function that's already periodic. Enter:
```
>> f3=@(x) sin(2*pi*x)
>> trapz(xx,f3(xx))        %Should be zero.
>> dot(w,f3(x))
```
Both rules are doing well but the trapezoidal rule is doing better. Of course, with error around 10^{-17} we are at the level of numerical noise. Enter:
```
>> w(1),x(1)               %weight and node for i=1 from IMT rule.
```
Problems due to round-off error are more of an issue with the IMT rule than with the trapezoidal rule, which uses equally spaced nodes and equal weights (excepting the endpoints).

Problems

1.) a.) Use the trapezoidal rule to approximate $\int_{-1}^{1} \sin(\pi x)dx$, $\int_{-1}^{1} \cos(\pi x)dx$, and $\int_{0}^{1}(x^5 - 2x^3 + 1)dx$.
b.) Repeat part a. using Simpson's rule. Comment.
c.) Use the trapezoidal rule to approximate $\int_{-1}^{1} \cos(\pi x)dx$ by integrating from 0 to 1 and doubling the result. Is your answer as accurate as the corresponding result from part a.?
2.) Show that the trapezoidal rule can be viewed as the average of the left and right endpoint rules.
3.) a.) Verify experimentally that the trapezoidal rule is $O(h^2)$ if $f'(a) \neq f'(b)$ but that it is $O(h^4)$ if $f'(a) = f'(b)$ but $f^{(3)}(a) \neq f^{(3)}(b)$.
b.) What happens if $f'(a) \neq f'(b)$ but $f'(a) \approx f'(b)$?
4.) If $f'(a) \neq f'(b)$ but $f'''(a) = f'''(b)$, does the performance of the trapezoidal rule improve?
5.) a.) Use the IMT rule with $n = 3$ and $c = 1$ and the IMT rule with $n = 3$ and $c = 2$ (from Example 2) to approximate $\int_{0}^{1}(x^8 + 1)dx$ and $\int_{0}^{1} 1dx$.

b.) Find the nodes and weights for the IMT rule with $n = 3$ and $c = 3$. Use this rule to approximate $\int_0^1 (x^8 + 1)dx$.

c.) Repeat part b. with $n = 3$ and $c = 5$.

d.) Repeat part b. with $n = 4$ and $c = 3$.

e.) Repeat part b. with $n = 4$ and $c = 5$.

6.) a.) For each $k = 1, 3, 5, 7, 9$, construct a function that has $f(a) = f(b)$ and odd-order derivatives zero at the endpoints for orders 1 through k, but not $k + 2$. Compare the trapezoidal rule to an IMT rule on these functions.

b.) Compare Simpson's rule on these functions to your answers from part a. When is Simpson's rule the better choice?

7.) a.) Use the trapezoidal rule to approximate $\int_0^1 (2 + \sin(2\pi x))^{-1}dx$ and then to approximate $\int_0^1 (2 + 6\sin(2\pi x))^{-1}dx$.

b.) Repeat for $\int_0^1 (1 + \cos(\pi x)/2)^{-1}dx$ and $\int_0^1 (1 + 4\cos(\pi x)/2)^{-1}dx$..

8.) Use the double exponential rule with $\alpha > 1$, $\beta > 1$ on $\int_0^1 (2 + \sin(2\pi x))^{-1}dx$, $\int_0^1 (2 + 6\sin(2\pi x))^{-1}dx$ and $\int_0^1 (1 + \cos(\pi x)/2)^{-1}dx$.

9.) Use the IMT rule with $n = 8$ and $c = 1$ to approximate $\int_0^1 x^{1/2}dx$, $\int_0^1 x^{3/2}dx$, $\int_0^1 x^{5/2}dx$, $\int_0^1 x^{7/2}dx$, and $\int_0^1 x^4 dx$.

10.) Justify the use of the double exponential rule.

11.) Another important application of the Euler-Maclaurin summation formula is in the approximation of sums. (This is not quite as easy as it may appear; the error will eventually grow with the number of derivative terms used after first shrinking. We omit a discussion of this issue.) Use the Euler-Maclaurin summation formula to approximate $\sum_{i=1}^{200} 1/i^2$ and $\sum_{i=1}^{\infty} 1/i^2$.

12.) Write a MATLAB program that accepts a function, an interval, and a value of n and c, then performs the IMT rule with that n and c. Transform the integral on $[a, b]$ to an integral on $[0, 1]$ if necessary. Optionally return the nodes and weights of the rule.

13.) Write a MATLAB program that accepts a function, an interval ,and a value of n and c, then performs the midpoint IMT rule with that n and c. Transform the integral to one on $[0, 1]$ if necessary.

14.) Write a MATLAB program that accepts a function, an interval, a number of points, and a value of α and β, then performs the double exponential rule with that α and β. Transform the integral to one on $[0, 1]$ if necessary.

15.) a.) Show that $\psi(x) = \frac{1}{K}\int_0^x \exp(-c/t(1 - t))\,dt$ (the IMT transformation) yields an integrand that is at least as differentiable as f and has all continuous derivatives zero at the endpoints.

b.) Show that $\psi(x) = \frac{1}{2}(1 + \tanh(\alpha \sinh(\beta(1/(1 - x) - 1/x))))$ (the double exponential transformation) yields an integrand that is at least as differentiable as f and has all continuous derivatives zero at the endpoints.

c.) Show that $\psi(x) = \frac{1}{2}(1 + \tanh(\frac{\alpha}{2}(1/(1 - x) - 1/x)))$ (which is called the **tanh-transformation**; $\alpha > 0$) yields an integrand that is at least as continuously differentiable as f and has all continuous derivatives zero at the endpoints.

d.) One description of the periodizing transformation methods is that they preprocess a function in order to improve the behavior of the trapezoidal rule. Which is the better way to think of them–as methods in their own right, or as a preparatory step before using the trapezoidal rule?

16.)   a.) A standard approach for transforming $\int_0^1 f(x)dx$ to $\int_0^1 f(\psi(x))\psi'(x)dx$ uses a nonnegative weight function $w(s)$ that is not everywhere zero to define a monotonic change-of-variables $\psi(x) = \int_0^x w(s)\,ds/K$ where $K = \int_0^1 w(s)\,ds$. The **Korobov method** uses the **Korobov transform** $w(s) = (s(1-s))^m$ where $m \geq 1$ is an integer. Test the Korobov method for $m = 1, 2, 3$ on several integrals.

b.) What happens if $m = 0$?

c.) Show that the order of the Korobov method is $m+1$ if $m+1$ is even and $m+2$ if $m+1$ is odd. Why does it make sense that the order is always even?

d.) This weight function is said to be symmetric because it is symmetric about the midpoint $s = 1/2$ of the interval of integration. What advantages are there to using a symmetric weight function? Which of the other rules in this section are symmetric?

CHAPTER 5 BIBLIOGRAPHY:

Computational Integration, Arnold R. Krommer and Christoph W. Ueberhuber, SIAM 1998

A modern, comprehensive treatment of the subject.

Methods of Numerical Integration, 2nd ed., Philip J. Davis and Philip Rabinowitz, Academic Press 1984

The classic reference on the subject.

Numerical Quadrature and Solution of Ordinary Differential Equations, A. H. Stroud, Springer-Verlag New York, 1974

An older, detailed text.

Numerical Analysis, 6th ed., Richard L. Burden and J. Douglas Faires, Brooks/Cole Publishing Company 1997

A traditional introductory undergraduate level textbook.

CHAPTER 6

Differential Equations

1. Numerical Differentiation

One of the most important areas of numerical analysis is the numerical solution of ordinary and partial differential equations. We'll discuss the numerical solution of ordinary differential equations in this chapter. Studying the numerical solution of partial differential equations (PDEs) requires more knowledge about the analytical theory of these equations than we will assume. The numerical solution of ODEs is also referred to as the numerical quadrature of ODEs.

We will focus on the simple first-order ODE initial value problem $y' = f(x, y)$, $y(x_0) = y_0$ and will assume that $f(x, y)$ is a well-behaved function. There will be many similarities to, and applications of, numerical integration methods.

Before we delve into the numerical quadrature of ODEs, though, we should discuss numerical differentiation. Suppose we have a differentiable function $f(x)$ or a list of data $y_0, y_1, ..., y_n$ representing observations of a differentiable function (either experimental data or the output of a program that implements the function) and we wish to approximate the derivative of the function using only function values. The obvious technique is to start from the definition

$$f'(x) = \lim_{h \to 0} \frac{f(x+h) - f(x)}{h}$$

and use the finite difference approximation

$$f'(x) \approx \frac{f(x+h) - f(x)}{h}$$

for some small $h > 0$. This is called the **forward difference** approximation because it uses a finite difference of f values (as opposed to the infinitesimal difference $f(x+h) - f(x)$ as $h \to 0$) and steps forward from the point of interest x. Similarly, we could use the fact that

$$f'(x) \approx \frac{f(x) - f(x-h)}{h}$$

(for some small $h > 0$), which is called the **backward difference** approximation to the first derivative of f at x. We can interpret these as secant approximations. (See Fig. 1, where $x_0 = .25$ and $h = .25$. The picture equally well describes the backward difference approximation for $x_0 = .5$ and $h = .25$.) The slope of f at x is approximated by the slope of the secant line.

DOI: 10.1201/9781003042273-6

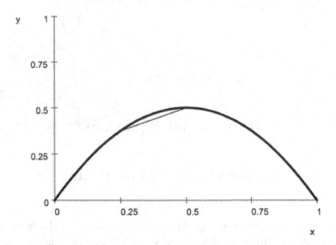

FIGURE 1. The forward difference approximation.

EXAMPLE 1: Let $f(x) = \cos(x)$. Then we may approximate $f'(0) = 0$ with finite differences using the forward difference approximation

$$f'(0) \approx \frac{f(0+h) - f(0)}{h}$$
$$= \frac{\cos(h) - 1}{h}$$

or the backward difference approximation

$$f'(0) \approx \frac{f(0) - f(0-h)}{h}$$
$$= \frac{1 - \cos(h)}{h}$$

and because

$$\frac{1 - \cos(h)}{h} = \frac{1 - (1 - h^2/2! + h^4/4! - ...)}{h}$$
$$= h/2! - h^3/4! + ...$$

this is an $O(h)$ estimate of the true value of the derivative. With $h = .1$ we have

$$f'(0) \approx \frac{f(0+.1) - f(0)}{.1}$$
$$= \frac{\cos(.1) - 1}{.1}$$
$$\doteq -.0500$$
$$f'(0) \approx \frac{f(0) - f(0 - .1)}{.1}$$
$$= \frac{1 - \cos(-.1)}{.1}$$
$$\doteq .0500$$

producing two different approximations, both or which are reasonably close to the true value $f'(0) = 0$. □

It won't happen in general that the forward and backward difference approximations are the negatives of one another. However, it *will* happen in general that these approximations are $O(h)$. For, Taylor series with remainder gives us

$$f(x_0 + h) = f(x_0) + hf'(x_0) + \frac{h^2}{2}f''(\xi)$$

for some ξ between x and x_0, assuming that f is twice continuously differentiable. If we write $x = x_0 + h$ this becomes

$$
\begin{aligned}
f(x_0 + h) &= f(x_0) + hf'(x_0) + \frac{h^2}{2}f''(\xi) \\
f(x_0 + h) - f(x_0) &= hf'(x_0) + \frac{h^2}{2}f''(\xi) \\
f'(x_0) &= \frac{f(x_0 + h) - f(x_0)}{h} - \frac{h}{2}f''(\xi)
\end{aligned}
$$

which shows that the forward difference approximation is $O(h)$. If instead we write $x = x_0 - h$ then we have

$$
\begin{aligned}
f(x_0 - h) &= f(x_0) - hf'(x_0) + \frac{h^2}{2}f''(\xi) \\
f(x_0 - h) - f(x_0) &= -hf'(x_0) + \frac{h^2}{2}f''(\xi) \\
f'(x_0) &= \frac{f(x_0) - f(x_0 - h)}{h} - \frac{h}{2}f''(\xi)
\end{aligned}
$$

so that the backward difference approximation is also $O(h)$. Consider again Fig. 1; it seems sensible to average the two estimates, or equivalently to take the expansions

$$(1.1) \qquad f(x_0 + h) = f(x_0) + hf'(x_0) + \frac{h^2}{2}f''(\xi_1)$$

$$(1.2) \qquad f(x_0 - h) = f(x_0) - hf'(x_0) + \frac{h^2}{2}f''(\xi_2)$$

and subtract the second equation from the first, giving

$$
\begin{aligned}
f(x_0 + h) - f(x_0 - h) &= \left(f(x_0) + hf'(x_0) + \frac{h^2}{2}f''(\xi_1) \right) - \\
&\quad \left(f(x_0) - hf'(x_0) + \frac{h^2}{2}f''(\xi_2) \right) \\
f(x_0 + h) - f(x_0 - h) &= 2hf'(x_0) + \left(\frac{h^2}{2}f''(\xi_1) - \frac{h^2}{2}f''(\xi_2) \right) \\
f'(x_0) &= \frac{f(x_0 + h) - f(x_0 - h)}{2h} - \frac{1}{4}h\left(f''(\xi_1) - f''(\xi_2) \right)
\end{aligned}
$$

after solving for $f'(x)$. This should be a better approximation since we expect that $f''(\xi_1) \approx f''(\xi_2)$ if h is small. In fact, if we assume additional differentiability of f in Eq. (1.1) and Eq. (1.2) we can get a better error term. For

$$f(x_0 + h) = f(x_0) + hf'(x_0) + \frac{h^2}{2}f''(x_0) + \frac{h^3}{6}f'''(\gamma_1)$$

$$f(x_0 - h) = f(x_0) - hf'(x_0) + \frac{h^2}{2}f''(x_0) - \frac{h^3}{6}f'''(\gamma_1)$$

giving

$$f(x_0 + h) - f(x_0 - h) = \left(f(x_0) + hf'(x_0) + \frac{h^2}{2}f''(x_0) + \frac{h^3}{6}f'''(\gamma_1) \right)$$

$$- \left(f(x_0) - hf'(x_0) + \frac{h^2}{2}f''(x_0) - \frac{h^3}{6}f'''(\gamma_1) \right)$$

$$f(x_0 + h) - f(x_0 - h) = 2hf'(x_0) + \left(\frac{h^3}{6}f'''(\gamma_1) + \frac{h^3}{6}f'''(\gamma_2) \right)$$

$$f'(x_0) = \frac{f(x_0 + h) - f(x_0 - h)}{2h} - \frac{1}{6}h^2 \left(\frac{f'''(\gamma_1) + f'''(\gamma_2)}{2} \right)$$

$$= \frac{f(x_0 + h) - f(x_0 - h)}{2h} - \frac{1}{6}h^2 f'''(\gamma)$$

for some $\gamma \in [\gamma_1, \gamma_2]$. This last step follows from the continuity of $f'''(x)$ and the Intermediate Value Theorem applied to the average of $f'''(\gamma_1)$ and $f'''(\gamma_2)$, that is, to the quantity

$$\frac{f'''(\gamma_1) + f'''(\gamma_2)}{2}$$

which must be achieved by $f'''(x)$ for some $x \in [\gamma_1, \gamma_2]$. We might also have suspected that we should be able to find an error term containing $f'''(\gamma)$ for another reason. After all, the previous error term involved $f''(\xi_1) - f''(\xi_2)$ which looks like something to which we could apply a version of the Mean Value Theorem in order to generate an f''' term.

This new approximation for $f'(x)$ does not use the value of f at x but only at the two points h units away from it to either side (see Fig. 2, where $x_0 = .25$ and $h = .25$). The formula

$$f'(x) \approx \frac{f(x + h) - f(x - h)}{2h}$$

is called the **centered difference** (or **central difference**) approximation to $f'(x)$, and is $O(h^2)$. The centered difference is generally superior to the other two schemes but there are times when we need other methods. For example, if x_0 is the first point, we cannot use a centered difference as there is no point to the left of x_0.

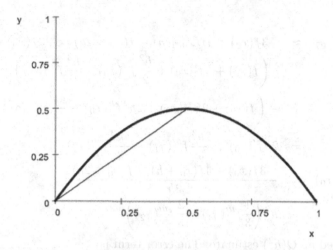

FIGURE 2. The centered difference approximation.

It's sometimes useful to have an $O(h^2)$ method for approximating f' at the end-points. Can we get an $O(h^2)$ method for approximating $f'(x_0)$ using only $f(x_0)$, $f(x_1)$, and $f(x_2)$? We could use an approach analogous to the method of undetermined coefficients, asking that low order polynomials be differentiated exactly. We could also interpolate f at x_0, x_1, and x_2 and use the derivative of the interpolant.

A third approach is to proceed as above, expanding in Taylor series about x_0 when appropriate in order to generate the approximations

$$f(x_0) = f(x_0)$$

$$f(x_0 + h) = f(x_0) + hf'(x_0) + \frac{h^2}{2}f''(x_0) + \frac{h^3}{6}f'''(\gamma_1)$$

$$f(x_0 + 2h) = f(x_0) + 2hf'(x_0) + 2h^2 f''(x_0) + \frac{4h^3}{3}f'''(\gamma_2)$$

(if we assume equal spacing) and then attempt to knock out as many unwanted terms as we can. It's not immediately clear how many terms we'll need to retain in the Taylor series, so we err on the side of including more terms when we are unsure. We'd like to form a linear combination of the form

$$f'(x_0) \approx af(x_0) + bf(x_0 + h) + cf(x_0 + 2h)$$

and so we need

$$a + b + c = 0$$
$$b/2 + 2c = 0$$

to eliminate the terms in $f(x_0)$ and $f''(x_0)$, respectively. One solution of this system is $a = -3$, $b = 4$, $c = -1$, and all other solutions are multiples of this solution. So we form

$$
\begin{aligned}
f'(x_0) &\approx -3f(x_0) + 4f(x_0 + h) - f(x_0 + 2h) = -3f(x_0) + \\
&\quad 4\left(f(x_0) + hf'(x_0) + \frac{h^2}{2}f''(x_0) + \frac{h^3}{6}f'''(\gamma_1)\right) \\
&\quad -\left(f(x_0) + 2hf'(x_0) + 2h^2f''(x_0) + \frac{4h^3}{3}f'''(\gamma_2)\right) \\
&= 2hf'(x_0) + \frac{2h^3}{3}f'''(\gamma_1) - \frac{4h^3}{3}f'''(\gamma_2)
\end{aligned}
$$

$$
(1.3) \qquad f'(x_0) = \frac{-3f(x_0) + 4f(x_0 + h) - f(x_0 + 2h)}{2h} -
$$

$$
h^2\left(\frac{1}{3}f'''(\gamma_1) - \frac{2}{3}f'''(\gamma_2)\right)
$$

which is indeed an $O(h^2)$ estimate. The error term is

$$
\begin{aligned}
E &= -h^2\left(\frac{1}{3}f'''(\gamma_1) - \frac{2}{3}f'''(\gamma_2)\right) \\
&= -h^2(\frac{1}{3}\left(f'''(\gamma_1) - f'''(\gamma_2)\right) - \frac{1}{3}f'''(\gamma_2)) \\
&\approx \frac{h^2}{3}f'''(\gamma_2)
\end{aligned}
$$

if $f'''(\gamma_1) \approx f'''(\gamma_2)$, and indeed a different analysis shows that the error may be written in the form $E = h^2 f'''(\gamma_2)/3$ for some $\gamma \in [\gamma_1, \gamma_2]$. A similar formula holds for $f(x_n)$ in terms of x_{n-2}, x_{n-1}, and x_n.

A formula for $f''(x)$ may be found by differencing the forward and backward difference formulas for $f'(x)$ over an interval of length h. The formula is

$$
\begin{aligned}
f''(x) &\approx \frac{\frac{f(x+h)-f(x)}{h} - \frac{f(x)-f(x-h)}{h}}{h} \\
&= \frac{f(x+h) - f(x) - (f(x) - f(x-h))}{h^2} \\
&= \frac{f(x+h) - 2f(x) + f(x-h))}{h^2}
\end{aligned}
$$

and it is called the centered difference approximation to $f''(x)$. This approximation is $O(h^2)$, as is easily verified by expanding the Taylor series and canceling terms. In fact, one might have wondered whether an interval of length $2h$ made more sense here than an interval of length h, but using Taylor series expansions would show that that approximation was off by a factor of 2^1.

Similar formulas may be found for higher order derivatives, and there are references listing formulas for various derivatives based on different numbers of points. (We talk of n-point formulas for approximating the derivative.) The more points used, the more attractive the error term can be made.

Numerical differentiation is inherently unstable. This is hardly a surprise when we look at, say, the forward difference approximation

---

[1]The forward and backward approximations "think" they're located at $x \pm h/2$, giving an interval of length h.

$$f'(x) \approx \frac{f(x+h) - f(x)}{h}$$

which for small h involves the subtraction of nearly equal numbers in the numerator, followed division by a small number. It's as though we were looking to design a difficult quantity to compute!

In fact, suppose that we can compute $f(x)$ to within some error $\epsilon > 0$ that does not depend on x. (This ϵ will likely will be considerably larger than the machine epsilon, particularly if $f(x)$ is the output of some simulation). If we assume that all other computations are done without error, then when we attempt to approximate $f'(x)$ we actually have

$$f'(x) = \frac{f(x+h) - f(x)}{h} - \frac{h}{2} f''(\xi)$$

($h > 0$) meaning the truncation error from terminating the Taylor series, that is, the approximation error inherent in the use of this formula, is $E_A = -hf''(\xi)/2$. We can write a formula for the round-off error in computing the forward difference as

$$E_R = \left[\frac{f(x+h) - f(x)}{h}\right] - \left[\frac{\mathrm{fl}\,(f\,(x+h)) - \mathrm{fl}\,(f\,(x)))}{h}\right]$$

since we are assuming that the only errors made are those associated with evaluating f. Because in addition $|\mathrm{fl}\,(f\,(x))| \leq f(x) + \epsilon$ by our assumption, and similarly for $\mathrm{fl}\,(f\,(x+h))$, this simplifies to

$$|E_R| \leq \frac{\epsilon + \epsilon}{h}$$
$$= \frac{2\epsilon}{h}$$

and so for the total error $E_T = E_A + E_R$ we have

$$|E_T| = |E_A + E_R|$$
$$\leq |E_A| + |E_R|$$
$$E_T \leq \frac{h}{2}|f''(\xi)| + \frac{2\epsilon}{h}$$

(1.4)
$$\leq \alpha h + \beta\frac{1}{h}$$

with $\alpha > |f''(\xi)|/2$, $\beta = 2\epsilon$. The general shape of this error bound is shown in Fig. 3. As $h \to 0$ the round-off error grows (instability); as h increases the truncation error increases.

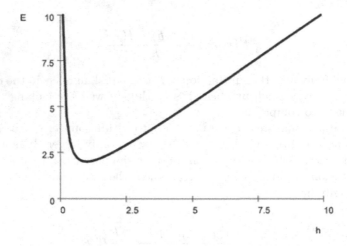

FIGURE 3. Bound on total error (fwd. difference).

There is a point at which this bound on the absolute error is a minimum. Of course, as this is an upper bound it may or may not correspond to the true minimum of the error; still, it is still instructive to find this point. Differentiating the error bound in Eq. (1.4) with respect to h and setting it to zero gives

$$\alpha - \beta \frac{1}{h^2} = 0$$
$$h^2 = \frac{\beta}{\alpha}$$
$$= \frac{2\epsilon}{|f''(\xi)|/2}$$
$$= \frac{4\epsilon}{|f''(\xi)|}$$

that is, the optimal h for the error bound is

(1.5) $h = 2\sqrt{\epsilon/M}$

where M is $\min(|f''(\xi)|)$ over $[x, x+h]$. Assuming that M is on the order of unity, this means that if the errors ϵ in computing f are on the order of machine epsilon, say $\epsilon \approx 10^{-16}$, then h should not be taken smaller than about 10^{-8}.

EXAMPLE 2: Let $f(x) = \exp(x)$. Then $f'(0) = 1$. With $h = 10^{-k}$, the forward difference approximation to $f'(0)$ is

$$f'(0) \approx \frac{\exp(10^{-k}) - 1}{10^{-k}}$$

and the absolute error drops by about an order of magnitude for each increase in k until the minimum is achieved at 10^{-8} (where the error is about $6E - 9$) then increases at the same rate until, starting at $k = 16$, the error is 1 because the numerator of the finite difference is evaluated as zero. $\square$

Reducing h too far is counterproductive. This is also the case for numerical integration, but because quadrature is stable the eventual build-up of round-off

1. NUMERICAL DIFFERENTIATION

error is slow and is not likely to have disastrous consequences. In numerical differentiation, round-off error can easily swamp our calculations. Think about it: In MATLAB® we're generally working in double precision but can only take h about as small as the machine epsilon for single precision, and even that assumes that f is being computed to full double precision accuracy. If the f values are noisy, possibly because they're being computed by some other program that has a tolerance of, say, $1E-6$, then we will have to use an even larger h.

MATLAB

The `diff` command computes the finite differences needed for the numerator of the forward difference approximation. Enter:

```
>> z=[1 2 4 10]
>> diff(z)
```

If the x values are equally spaced with spacing $h > 0$, the approximate derivatives are `diff(z)/h`. Enter:

```
>> h = .001; x = 0:h:pi;
>> y=diff(sin(x))/h;
>> plot(x,cos(x),'b',x(1:end-1),y,'r')
```

The agreement is excellent. Enter:

```
>> format long
>> max(cos(x(1:end-1))-y)
```

The worst-case error is about $5E-4$. Since h is 10^{-3} and the method is $O(h)$, this is as expected. The command `diff(y,n)` may be used to difference the differences n times, as in the derivation of the centered difference approximation for the second derivative as the difference of two other approximations. Enter:

```
>> format short
>> z
>> diff(diff(z))
>> diff(z,2)
```

These are the numerators for the centered difference approximation for the second derivative. Enter:

```
>> yy=diff(sin(x),2)/h^2;
>> plot(x,-sin(x),'b',x(1:end-2),yy,'r')
```

Once again we have very good agreement between the analytical and the computed second derivative of $f(x)$ for this value of h.

You can use `diff` to perform symbolic (ordinary or partial) differentiation. Enter:

```
>> syms x y
>> diff(x^2)
>> diff(sin(x*y),y)
```

You may also have available the command to compute Taylor series symbolically. Enter:

```
>> syms z
>> taylor(sin(z))        %Taylor series about z=0.
```

The command `taylortool` gives an interactive GUI tool for computing and plotting Taylor series.

The commands `polyder` and `polyint` differentiate and integrate, respectively, polynomials given in terms of their coefficients (in the standard MATLAB format).

See also `help poly2sym` and `help sym2poly`. Some symbolic capabilities will require the Symbolic Toolbox. A dedicated CAS package is generally preferable for symbolic operations.

Problems

1.) Use the forward, backward, and centered difference approximations for the first derivative to approximate the derivative of $\ln(x)$ at $x = 1, 1.1, 1.2, ..., 2$. Compare to the true values. Then repeat at $x = 1, 1.05, 1.1, 1.1.5, ..., 2$.

2.) Derive the centered difference approximation for the first derivative using polynomial interpolation as explained in the text, including the error term.

3.) For what order polynomials are the forward, backward, and centered difference approximations for the first derivative exact? How about the centered difference approximation for the second derivative?

4.) Use Taylor series to show that the centered difference approximation for the second derivative has error $E = -h^2 f^{(4)}(\gamma)/12$.

5.) Show that the centered difference approximation for the first derivative is the average of the forward and backward difference formulas.

6.) Find the rule corresponding to Eq. (1.5) for the centered difference approximation for the first derivative.

7.) Derive the analogue of Eq. (1.3) for the right endpoint of a set of data.

8.) Use Taylor series to derive an $O(h^4)$ five point formula for $f'(x)$ based on the values of f at the points $x - 2h$, $x - h$, x, $x + h$, $x + 2h$.

9.) Use Taylor series to derive a formula for $f'(x)$ based on the values of f at the unequally spaced points $x - h$, x, $x + 2h$. Such formulas are needed when some data is missing.

10.) a.) Construct a log-log plot illustrating the effects in Example 2. (See the help for `loglog` for the plot.)

b.) Choose five different infinitely differentiable functions and do the same. Do you always get minimum error at $h = 10^{-8}$?

c.) Choose five different differentiable functions that are not twice differentiable and do the same, approximating the derivative at a point where the function is only once differentiable. Do you always get minimum error at $h = 10^{-8}$?

11.) How could you use the left endpoint quadrature rule to approximately solve the ODE IVP $y' = f(x, y)$, $y(x_0) = y_0$? Demonstrate on $y' = \exp(2x)$, $y(0) = 1$, for $x = .1, .2, .3, ...2$. Compare your answers to the true solution. Cut the step size in half and repeat.

12.) Draw the graphs of several functions for which the forward, backward, and centered difference approximations for the first derivative will give poor results at the resolution indicated. Discuss your graphs.

13.) a.) Show that the centered difference approximation for the first derivative has an error expansion that involves only even powers of h.

b.) Can you make a similar statement regarding the forward or backward difference formulas? What about the centered difference approximation for the second derivative?

14.) Write a MATLAB program that accepts a set of x and y values and uses `trapz` to approximate the integral of the underlying function $y = f(x)$. Approximate $f''(x)$ at every interior point x using the centered difference approximation and use it to bound the absolute error $|(b - a)h^2 f''(\gamma)/12|$ in the composite trapezoidal rule.

15.) Write a MATLAB program that accepts a set of x and y values and computes the clamped spline of the data using derivatives approximated by Eq. (1.3) and the corresponding formula for the right endpoint. You may use the `spline` command.

16.) a.) Argue that the standard definition of the first derivative is equivalent to the claim that the forward and backward derivatives agree in the limit as $|h| \to 0$.

b.) Is it also true that $f'(x)$ could be defined by the limit of the centered difference approximation as $h \to 0$?

2. Euler's Method

We're now ready to discuss the quadrature of the first-order ODE IVP $y' = f(x, y), y(x_0) = y_0$. We know that analytical solutions are available in some special cases (first-order linear, separable, exact, etc.) but that in many cases no closed form solution can be found. Nonetheless $y(x)$ is a well-defined function (under appropriate assumptions on f) which we might attempt to find by approximate methods. One such method would be to look for a power series solution and use its partial sums, though this requires that $y(x)$ be analytic at x_0. We'll focus on numerical methods, where we seek to approximate the solution $y(x)$ on a grid

$$x_0, x_0 + h, x_0 + 2h, ..., x_0 + nh$$

where, for now, we are assuming constant spacing h. We'll write $x_i = x_0 + ih$. Once we have approximate values of $y(x_0), y(x_1), , y(x_2), , ..., y(x_n)$ we can plot the solution, look for where it is maximal or minimal, etc.

If $f(x, y)$ happens to be a function of x alone, that is, if the problem is of the form $y' = \phi(x), y(x_0) = y_0$, then we may simply use a numerical quadrature rule. If we wish to solve, say,

$$y' = \sin(x^2)$$
$$y(0) = 0$$

then the trapezoidal rule applied to compute

$$y(x) = y(0) + \int_0^x \sin(t^2) dt$$

with $h = .1$ gives

$$y_0 = 0$$
$$y_1 = y_0 + \frac{h}{2}(\phi(x_0) + \phi(x_1))$$
$$= \frac{.1}{2}(\phi(0) + \phi(.1))$$
$$\doteq 4.9999E - 4$$
$$y_2 = y_1 + \frac{h}{2}(\phi(x_1) + \phi(x_2))$$
$$= y_1 + \frac{.1}{2}(\phi(.1) + \phi(.2))$$
$$\doteq .0030$$

where y_i is the numerical approximation to $y(x_i)$. The actual values are $y(.1) \doteq 3.3333E - 4$ and $y(.2) \doteq .0027$.

We could use the `cumtrapz` command to generate an approximate solution in this way, or a more accurate method. But in general we will be working with the problem

$$(2.1) \qquad\qquad y' = f(x, y)$$
$$(2.2) \qquad\qquad y(x_0) = y_0$$

where $f(x, y)$ depends on both x and y. In that case we cannot use the trapezoidal method, for in attempting to evaluate

$$y_1 = y_0 + \frac{h}{2}(f(x_0, y_0) + f(x_1, y(x_1)))$$
$$(2.3) \qquad\qquad \approx y_0 + h\frac{f(x_0, y_0) + f(x_1, y_1)}{2}$$

we know f, x_0, $x_1 = x_0 + h$, and y_0, but we do not know $y(x_1)$, or even an approximation y_1 for it, as that is what we're currently trying to estimate. We could treat this equation as implicitly defining y_1, however, and use a root-finding method such as Newton's method on Eq. (2.3) in the form

$$y_1 - y_0 - \frac{h}{2}(f(x_0, y_0) + f(x_1, y_1)) = 0$$

then repeat this process to find $y_2, y_3, ..., y_n$ This is called the **trapezoidal scheme** and it is sometimes used, but there are more efficient approaches.

Consider again the ODE Eq. (2.1) with the initial condition Eq. (2.2). There *is* an integration rule that we can apply to it, namely, the left endpoint rule. This is the Riemann sum rule that uses the approximation

$$\int_a^b \phi(x)dx \approx (b - a)\phi(a)$$

that is, we approximate the area under the curve $y = \phi(x)$ by a rectangle with height equal to the value of the function at the left endpoint $x = a$. But the solution of the ODE IVP is

$$y(x) = y_0 + \int_{x_0}^x f(t, y(t))dt$$

so the left endpoint rule gives

$$y(x_1) = y_0 + \int_{x_0}^{x_1} f(t, y(t))dt$$
$$\approx y_0 + (x_1 - x_0)f(x_0, y(x_0))$$
$$y_1 = y_0 + hf(x_0, y_0)$$

(where y_1 is the approximation to $y(x_1)$). We can iterate this:

$$
\begin{aligned}
y_1 &= y_0 + hf(x_0, y_0) \\
y_2 &= y_1 + hf(x_1, y_1) \\
&\vdots \\
y_n &= y_{n-1} + hf(x_{n-1}, y_{n-1})
\end{aligned}
$$

and the resulting method is known as **Euler's method**. Since the simple left endpoint rule is $O(h^2)$ for a single step, Euler's method is $O(h^2)$ in going from y_i to y_{i+1}; since the composite left endpoint rule is $O(h)$, we expect that Euler's method will be $O(h)$ in going from y_0 to y_i ($i \geq 2$). As with numerical integration rules, it is the latter error estimate that concerns us. Euler's method is $O(h)$.

EXAMPLE 1: Let's apply Euler's method to the ODE IVP $y' = 1 + y^2$, $y(0) = 1$. The true solution is $y(x) = \tan(x + \pi/4)$. We'll use $h = .5$ and solve over $[0, 2]$. We have

$$
\begin{aligned}
y_0 &= 1 \\
y_1 &= y_0 + hf(x_0, y_0) \\
&= 1 + .5(1 + 1^2) \\
&= 2 \qquad (\approx y(.5)) \\
y_2 &= y_1 + hf(x_1, y_1) \\
&= 2 + .5(1 + 2^2) \\
&= 4.5 \qquad (\approx y(1)) \\
y_3 &= y_2 + hf(x_2, y_2) \\
&= 4.5 + .5(1 + 4.5^2) \\
&= 15.125 \qquad (\approx y(1.5)) \\
y_4 &= y_3 + hf(x_3, y_3) \\
&= 15.125 + .5(1 + 15.125^2) \\
&\doteq 130.0078 \quad (\approx y(2))
\end{aligned}
$$

from Euler's method. How good are these results? We have the following:

Euler	Actual	Error
2	3.4082	1.4082
4.5	−4.5880	9.0880
15.125	−1.1527	16.2777
130.0078	−0.3721	114.5107

and so the results are very poor for this h. It's instructive to see why, and we'll do so in the MATLAB subsection. □

We can gain some additional insight into Euler's method by deriving it in three other ways. One method is to simply expand $y(x)$, the unknown solution of the ODE IVP, in a Taylor series

$$
\begin{aligned}
y(x_0 + h) &= y(x_0) + hy'(x_0) + \frac{h^2}{2}y''(\gamma) \\
&\approx y(x_0) + hy'(x_0) \\
&= y(x_0) + hf(x_0, y(x_0)) \\
y_1 &= y_0 + hf(x_0, y_0)
\end{aligned}
$$

giving Euler's method. This derivation requires that y'' exist and clearly shows the $O(h^2)$ truncation error for a single step.

Another way to derive Euler's method is to note that the ODE IVP gives us knowledge of both one point on the solution $y(x)$, the point (x_0, y_0), and a way to compute the slope of the solution $y(x)$ there, $y'(x_0) = f(x_0, y_0)$. Hence we could use a tangent line approximation at (x_0, y_0) and linearly extrapolate to a value of y at $x_1 = x + h$, then repeat from (x_1, y_1), and so on; see Fig. 1. Since we know a point on the line and its slope $f'(x_0)$, we have that

$$
y - y_0 = f'(x_0)(x - x_0)
$$

from the point-slope formula for a line. We want to know the value of y at $x = x_0 + h$:

$$
\begin{aligned}
y_1 - y_0 &= f'(x_0)((x_0 + h) - x_0) \\
y_1 &= y_0 + hf'(x_0)
\end{aligned}
$$

which is Euler's method once again. Note that Euler's method approximates the slope of the function over $[x_i, x_{i+1}]$ using its value at the left endpoint whereas the trapezoidal scheme (Eq. (2.3)) uses an averaged value of the slope over the interval. As would be expected, the trapezoidal scheme is more accurate ($O(h^2)$).

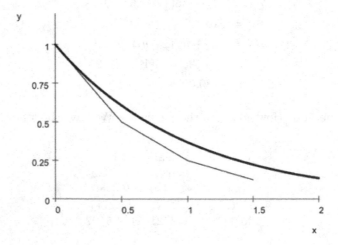

FIGURE 1. Euler's method as a tangent line approximation.

Finally, we can derive Euler's method using finite differences. We substitute the forward difference approximation for y' in Eq. (2.1), giving

$$y' = f(x, y)$$

$$\frac{y(x+h) - y(x)}{h} \approx f(x, y)$$

$$y(x+h) \approx y(x) + hf(x, y)$$

and so we define

$$y_{i+1} = y_i + hf(x_i, y_i)$$

to be the approximate method. Euler's method may be viewed as a **finite difference method**, that is, a method found by replacing exact derivatives by finite difference approximations and neglecting the truncation errors that this process introduces.

Given the relationship between Euler's method and the numerically worrisome forward difference approximation, we might wonder if growth of round-off error is an issue in Euler's method. Note first that the truncation error is worse than it might at first appear: An ODE IVP like Eq. (2.1) with Eq. (2.2) represents one solution picked from the infinite one-parameter family of solutions defined by the ODE alone. Euler's method, viewed as a tangent line approximation, starts on the true solution (neglecting any error in x_0 and y_0) but immediately steps off onto a second, "parallel" curve. The next iterate steps away onto yet another curve, which in general will be even further from the desired curve; the error builds and builds. If the curves aren't getting closer to one another as x increases, we'll have a large error eventually. This emphasizes the fact that Euler's method is $O(h)$ *at any fixed point* x^*, that is, if we consider only the error at some particular point x^* then it decreases as $O(h)$. The error grows, in general, with increasing x but decreases with decreasing step size h. The following theorem, which will be proved in Sec. 7.4, provides a bound on this truncation error effect:

THEOREM 1: Let $y(x)$ be the solution of $y' = f(x, y)$, $y(x_0) = y_0$. If $f(x, y)$ is continuously differentiable with respect to both its independent variables in some open region R containing (x_0, y_0) and $|y''(x)|$ is bounded for x within R, then there are $A > 0$, $B > 0$ for which the iterates $y_0, y_1, ..., y_n$ generated by Euler's method satisfy

$$|y(x_i) - y_i| \leq hA\left(e^{Bih} - 1\right)$$

provided the iterates and the solution $y(x)$ remain in R.

The assumptions of this theorem can be weakened and A and B can be identified in terms of y and f, but the key point is that at a particular point $x_i = x_0 + ih$,

$$|y(x_i) - y_i| \leq hA\left(e^{B(x_i - x_0)} - 1\right)$$

is a bound on the absolute error. Note that if we reduce h for additional accuracy the distance from x_i to x_0 doesn't change, though the i that identifies it does (e.g., x_{10} for a step size h has the same numerical value as x_{20} for step size $h/2$); hence the term in parentheses is unchanged. This means that the error at any fixed point x^* is $O(h)$,

$$|y(x^*) - y^*| \leq hAM$$

$(M = (e^{B(x_i - x_0)} - 1))$. But Theorem 1 also says that if instead of focusing on a particular point x^* we imagine letting $i \to \infty$ for a fixed step size h then we are looking at an exponentially growing error bound. Euler's method is $O(h)$ *at a fixed point*.

In practice the actual error is almost never as bad as this worst-case estimate might suggest and is typically considerably smaller. The curves may get closer to one another rather than farther apart, in which case the error might actually be decreasing. If every member of the one-parameter family of solutions satisfies

$$y(x) \to 0$$

as $x \to \infty$ then even when we step onto a "wrong" curve we will eventually end up in the right place (a steady state at zero).

The theorem only addresses truncation error, but round-off error is also an issue. If an error of absolute value at most ϵ is made in evaluating each y_i (including the representation error in y_0) then we will have in place of Theorem 1

$$|y(x_i) - y_i| \le \left(hA + \frac{\epsilon}{Bh} \right) \left(e^{B(x_i - x_0)} - 1 \right) + \epsilon e^{B(x_i - x_0)}$$

for the same A, B. The important thing to notice is that once again we have the error behavior

$$\alpha h + \frac{\beta}{h}$$

that we saw for the forward difference method. We have problems with the truncation error as h increases (the αh part) and problems with the round-off error as h decreases (the β / h part), and so we must be wary of making h too small.

The methods we will be discussing are for ODE IVPs. We'll see how to use them for higher-order ODE IVPs such as

$$
\begin{aligned}
ay'' + by' + cy &= f(x) \\
y(0) &= y_0 \\
y'(0) &= v_0
\end{aligned}
$$

also. The ability to solve systems of ODE IVPs numerically is extremely important in practice, and so we'll consider topic that as well.

We will not look at methods for the numerical solution of ODE BVPs (boundary value problems), that is, problems of the form

$$
\begin{aligned}
ay'' + by' + cy &= f(x) \\
y(0) &= \gamma_0 \\
y(1) &= \gamma_1
\end{aligned}
$$

which have two boundary conditions rather than two initial conditions. There are of course specialized numerical methods for ODE BVPs.

We also won't generalize to partial differential equations (PDEs), as that requires some background in the analytical understanding of such equations. However, the methods we are discussing generalize naturally to the popular and important finite difference method for PDEs.

Euler's method can occasionally be an acceptable choice for ODE IVPs on small intervals, using appropriately chosen step sizes. In such cases its simplicity may make it attractive. Much more importantly, though, the most commonly used methods for the numerical solution of ODES take its basic idea and modify it, as we'll soon see.

MATLAB

There are a number of commands for solving ODEs in MATLAB. Type help ode and hit the tab key twice to see a list (if you are using MATLAB6). Enter:
```
>> odeexamples
```
and run all the examples listed. Then select *Boundary Value Problems* from the drop-down menu and run those examples. These are ODE BVPs as opposed to ODE IVPs.

Let's look at an example. We'll use both ode23, which is a low order routine (mixing second and third order methods), and ode45, which is a higher-order routine (mixing fourth and fifth order methods[2]). Consider the ODE IVP $y' = x + y$, $y(0) = 1$. Enter:
```
>> f=@(x,y) x+y
>> ode23(f,[0 1],1)
>> figure
>> ode45(f,[0 1],1)
```
A plot of the solution is produced. The function is $f(x,y)$, the vector [0 1] gives the initial and final values of x (or t), and the last argument is the initial condition $y(0)$. Note that the higher-order routine used many more points than the lower-order routine. Enter:
```
>> close all
>> [x23,y23]=ode23(f,[0 1],1);
   >> [x45,y45]=ode45(f,[0 1],1);
   >> whos  %ode45 used many more x-values than ode23.
>> plot(x23,y23)
>> figure
>> plot(x45,y45)
```
The values of the nodes that were used, x, and the y values found at those nodes, y, are returned. The method is adaptive so the returned nodes will not typically be equally spaced.

Let's compare the computed solution with the analytical solution of this first-order linear ODE. If you have the Symbolic Toolbox, enter:
```
>> syms y(x)
>> dsolve(diff(y)==x+y)        %General solution.
>> dsolve(diff(y)==x+y,y(0)==1)  %Particular solution.
```
The true solution is $y(x) = -x - 1 + 2\exp(x)$. Make sure the second figure is the active figure, then enter:
```
>> hold on
>> plot(x45,-x45-1+2*exp(x45),'r')
```
to plot the two solutions together. The visual fit is excellent. Let's go further. Enter:
```
>> close all
```

---

[2]Hence the command is read "o-d-e four five", not "o-d-e forty-five", and similarly for ode23.

```
>> [x45,y45]=ode45(f,[0 20],1);
>> plot(x45,y45)
>> hold on
>> plot(x45,-x45-1+2*exp(x45),'r')
```
This is still pretty good to the eye, though note the vertical scale. Enter:
```
>> norm(y45-(-x45-1+2*exp(x45)))       %Absolute error.
>> ans/norm(y45)                       %Relative error.
```
The absolute error is considerable but the relative error is quite small. This error is measured over the entire sequence; let's look at it point-by-point. Enter:
```
>> max(abs((y45-(-x45-1+2*exp(x45)))))
>> max(abs((y45-(-x45-1+2*exp(x45))))./abs(y45))
```
This is the l_∞ measure of the error in the iterates (i.e., the max norm of the sequence generated by ode45). The absolute error is large, but the relative error is quite reasonable.

Look again at Example 1. What went so wrong there? We used a fairly large step size, of course ($h = .5$). Let's look at the true solution and the approximate solution. Enter:
```
>> clear all
>> f=@(y) 1+y^2
>> y0=1;h=.5;
>> y1=y0+h*f(y0)        %Approx.  solution at x1=.5.
>> y2=y1+h*f(y1)        %Approx.  solution at x2=1.
>> y3=y2+h*f(y2)        %Approx.  solution at x3=1.5.
>> y4=y3+h*f(y3)        %Approx.  solution at x4=2.
>> plot(0:.5:2,[y0 y1 y2 y3 y4],'b'),hold on
>> plot(0:.5:2,[y0 y1 y2 y3 y4],'ko')
```
Now let's overlay the true solution $y(x) = \tan(x + \pi/4)$. Enter:
```
>> xx=[0:.01:2];
>> plot(xx,tan(xx+pi/4),'r')
```
Look at the figure. Then enter:
```
>> axis([0 2 -140 140])
>> grid
```
Look carefully at this plot. The true solution is in red. Euler's method uses the slope at $x = 0$ to get an approximate value at $x = .5$. It then uses the slope at $x = .5$ to get an approximate value at $x = 1$. But, the solution has a discontinuity at $\pi/4 \doteq .79$ and Euler's method steps across this singularity without ever seeing it. From then on the y value is so far off that it's simply luck whether we end up near the true solution again or not.

We generated the point $(x_1, y_1) = (.5, 2)$ at the first step. It does not lie on the desired solution curve but it does lie on some solution curve, corresponding to the initial condition $y(.5) = 2$. If you have the Symbolic Toolbox, enter:
```
>> syms y(x)
>> dsolve(diff(y)==1+y^2,y(.5)==2)
```
to see that the general solution of the ODE $y' = 1+y^2$ is $y(x) = \tan(x+c)$. Because of the discontinuity this solution may not continue from the origin to $x = 1$ but it is valid at $x_1 = .5$. Let's solve $y_1 = \tan(x_1 + c)$ for c. Enter:
```
>> c=atan(2)-1/2
>> tan(.5+c)    %Check initial condition.
```

```
>> plot(xx,tan(xx+c),'c')
```
The cyan curve, which we should probably ignore to the left of $x = 1/2$, should go through the point $(x_1, y_1) = (.5, 2)$. Enter:
```
>> axis([.4 .6 -10 10])
```
to see this more clearly. There is a similar curve for every point generated by Euler's method, unless it should happen to land on a point of discontinuity.

Let's try to see the $O(h)$ behavior of Euler's method. We'll use the ODE IVP $y' = y$, $y(0) = 1$. The true solution is $y(x) = \exp(x)$.

Enter:
```
>> f=@(y) y
>> h=.5;y0=1;
>> y1=y0+h*f(y0)            %Approx. solution at x1=.5.
>> y2=y1+h*f(y1)            %Approx. solution at x2=1.
>> E=abs(exp(1)-y2)
```
Even with two steps over $[0, 1]$, the exponential function is growing so rapidly that linear approximations aren't very good. But we know that if we take h sufficiently small then we can get an arbitrarily good approximation of $\exp(x)$ over any $[x, x+h]$. Enter:
```
>> h=h/2,y=y0;
>> for i=1:(1/h);y=y+h*f(y);end,y   %Final point is now x4=1.
>> E=[E,abs(exp(1)-y)]
>> h=h/2,y=y0;
>> for i=1:(1/h);y=y+h*f(y);end,y   %Final point is now x8=1.
>> E=[E,abs(exp(1)-y)]
```
The vector E has three components, and they show about a 50% reduction in the error at $x = 1$. Repeat this process until E has 14 components; note the 50% reduction in error at each step. Note also that it is taking longer and longer as you go on. Enter:
```
>> 1/h
```
This is $2^{14} = 16384$; that's how many steps we are now taking to reach $x = 1$. If we reduce h to 10^{-7} then, after a considerable time, we get an error of about 10^{-7}, as we would expect from an $O(h)$ method.

What if we wanted to solve this ODE IVP using ode23? Since ode23 expects the function to accept two arguments, enter:
```
>> f=@(x,y) y
>> [x,y]=ode23(f,[0 1],1);
>> y(end)
>> abs(y(end)-exp(1))      %Error.
>> plot(x,y)
```
The value of $y(1)$ is approximated to about 4 decimal places. If f were an M-file instead, as would be likely in practice, we would simply not use one of the input arguments to the program within the program itself.

It is usually advisable to try ode45 first and then switch to another method if the performance of ode45 is unacceptable. Many users of MATLAB make exceedingly heavy use of ode45.

Problems

1.) Use Euler's method with $h = .05$ to approximate the solution of $y' = y^2$, $y(0) = 1$, over $[0, 1]$. Compare your answer with the true solution.

2.) Show that Euler's method is the result of using linear extrapolation starting from the known point (x_0, y_0) and with the known slope $y'(x_0) = f(x_0, y_0)$.

3.) Use Euler's method with $h = .4$, $h = .2$, $h = .1$, and $h = .05$ to approximate the solution of $y' = -y$, $y(0) = 1$, over $[0, 4]$. Compare your answers with the true solution.

4.) a.) Use the trapezoidal scheme with $h = .05$ to approximate the solution of $y' = x + y + 1$, $y(1) = 1$, over $[1, 3]$.

b.) Use the trapezoidal scheme with $h = .05$ to approximate the solution of $y' = \sin(xy) + 1$, $y(1) = 1$, over $[1, 2]$.

5.) Write a MATLAB program for performing Euler's method on a given function, interval $[x_0, x_n]$, initial value y_0, and step size. Return a vector of y values and an error variable signalling bad inputs (e.g., if $(x_n - x_0)/h$ is not a positive integer). Accept an optional argument that requests the output be plotted using cubic splines. Test your program.

6.) Write a MATLAB program for performing Euler's method on a given function, vector of x-values, initial value y_0, and step size. Assume the x-values will be supplied as an ordered vector of values beginning at x_0. They need not be equally spaced. Return a vector of y values and an error variable signalling bad inputs (e.g., if the x values are not strictly increasing). Accept an optional argument that requests the output be plotted using cubic splines. Test your program.

7.) a.) Use Euler's method with $h = .05$ to approximate the solution of $y' = 1 + y^2$, $y(0) = 1$, over $[0, .5]$. Compare your answer with the true solution. (See Example 1.) Plot the approximate and analytical solution together; then use a spline on the numerical solution and plot it together with the analytical solution. How accurate is the spline at x values between the nodes?

b.) Attempt to extend your numerical solution to the interval $[0, 1]$ using ode23.

8.) Write a MATLAB program for performing the trapezoidal scheme. Test your program.

9.) The constant A in Theorem 1 is equal to $K/2B$ where K is $\max(|y''(x)|)$ over the interval of interest and B is a Lipschitz constant for f with respect to y. Use Euler's method with $h = .05$ to approximate the solution of $y' = y + 1$, $y(0) = 1$, over $[0, 1]$. Compare your answer with the true solution and with the error bound from Theorem 1.

10.) Show that the trapezoidal scheme is $O(h^3)$ for a single step and $O(h^2)$ in general.

11.) Use Euler's method with $h = .4$, $h = .2$, $h = .1$, and $h = .05$ to approximate the solution of $y' = -8y$, $y(0) = 1$, and of $y' = -8y$, $y(0) = 1$ over $[0, 2]$. Compare your answers with the true solutions.

12.) a.) What method, if any, results from using the backward difference approximation in Eq. (2.1)? How about the centered difference approximation?

b.) Develop a method for solving ODE IVPs based on the right endpoint rule, and implement it in MATLAB.

13.) Euler's method may also be applied to a system of ODEs. If $y' = F(x, y)$, $y(x^{(0)}) = y^{(0)}$ where x and y are vectors then Euler's method is $y^{(i+1)} = y^{(i)} + hF(x^{(i)}, y^{(i)})$. Use this method to approximate the solution of $y_1' = y_1 + 2y_2^2$, $y_2' = 2y_1^2 + 2y_2$, with the initial conditions $y_1(0) = y_2(0) = 1$.

14.) Use Euler's method to approximate the solution of $y' = x + y^2$, $y(0) = 1$, over $[0, 2]$ for several values of h. Plot the absolute error at $x = 1.5$ as a function of h. Comment.

15.) Compare Euler's method with a very small h to ode23 and ode45 with larger values of h for a variety of problems. Discuss your results.

16.) Discuss the four different ways in which Euler's method was derived in this section. What insights were gained from each of them?

3. Improved Euler's Method

Euler's method is only $O(h)$. This is hardly surprising: It uses a value of $f(x, y)$ at one endpoint of the subinterval $[x_i, x_{i+1}]$ to estimate the slope of the line that it uses to extrapolate a value of $y(x_{i+1})$. But the slope is changing over that interval, possibly rapidly. The trapezoidal scheme

$$(3.1) \qquad y_{i+1} = y_i + h \frac{f(x_i, y_i) + f(x_{i+1}, y_{i+1})}{2}$$

uses two values to get an average slope and is $O(h^2)$. Unfortunately it's an implicit method, that is, it defines the value y_{i+1} implicitly, forcing us to find it with a root-finding method.

One simple idea turns the trapezoidal schemes into an explicit method: Why not estimate the unknown value by using Euler's method, which is explicit? This idea turns the trapezoidal scheme Eq. (3.1) into the method

$$\tilde{y}_{i+1} = y_i + hf(x_i, y_i)$$
$$y_{i+1} = y_i + h \frac{f(x_i, y_i) + f(x_{i+1}, \tilde{y}_{i+1})}{2}$$

where we say that we have predicted the value of the function at x_{i+1} using Euler's method, then corrected it using the trapezoidal scheme. This approach is called the **improved Euler's method** (or **modified Euler's method**) and may be written in a single line as

$$y_{i+1} = y_i + h \frac{f(x_i, y_i) + f(x_{i+1}, y_i + hf(x_i, y_i))}{2}$$

and it is important to note that we need only compute $f(x_i, y_i)$ once, and hence this still uses only two function evaluations per iteration. It is $O(h^2)$.

Similarly, we might attempt to apply the midpoint rule to $y' = f(x, y)$, yielding what we might refer to as the **midpoint scheme**[3]

$$(3.2) \qquad \begin{aligned} y_{i+1} &= f(x_i, y_i) \\ &= y_i + hf(x_{i+1/2}, y_{i+1/2}) \end{aligned}$$

---

[3]The names for many of these methods are not fully standardized and so may vary from source to source.

where $x_{i+1/2} = x_i + h/2$, and $y_{i+1/2}$ is an approximate value of $y(x_{i+1/2})$. Once again we estimate the problematic quantity using Euler's method:

$$\tilde{y}_{i+1/2} = y_i + \frac{h}{2}f(x_i, y_i)$$
$$y_{i+1} = y_i + hf(x_{i+1/2}, \tilde{y}_{i+1/2})$$

giving what is sometimes called the **midpoint method**

$$y_{i+1} = y_i + h\left(f(x_{i+1/2}, y_i + (h/2)\,f(x_i, y_i))\right)$$

which is explicit, and is $O\left(h^2\right)$. The midpoint scheme isn't even implicit; Eq. (3.2)) is a single equation in two unknowns $y_{i+1/2}$, y_{i+1}. But the midpoint method is easy to apply.

EXAMPLE 1: Let's use the improved Euler's method on the ODE IVP $y' = x + y^2$, $y(0) = 1$. With $h = .1$ we have

$$y_1 = y_0 + h\frac{f(x_0, y_0) + f(x_1, y_0 + hf(x_0, y_0))}{2}$$
$$= 1 + \frac{.1}{2}\left((0 + 1^2) + (.1 + \tilde{y}_1^2)\right)$$

where

$$\tilde{y}_1 = y_0 + hf(x_0, y_0)$$
$$= 1 + .1(0 + 1^2)$$
$$= 1.1$$

is the estimate from Euler's method. So

$$y_1 = 1 + .05\left(1 + (.1 + 1.1^2)\right)$$
$$= 1.1605$$

is the improved Euler's method estimate. (The actual value is 1.1165.) Then

$$y_2 = y_1 + h\frac{f(x_1, y_1) + f(x_2, y_1 + hf(x_1, y_1))}{2}$$
$$= 1.1605 + \frac{.1}{2}\left((.1 + 1.1605^2) + (.2 + \tilde{y}_2^2)\right)$$

where

$$\tilde{y}_2 = y_1 + hf(x_1, y_1)$$
$$= 1.1605 + .1(.1 + 1.1605^2)$$
$$\doteq 1.3052$$

is the Euler's method estimate starting from (x_1, y_1). (Note that this is not the value we would have gotten by applying Euler's method starting from $(x_1, \tilde{y}_1) = (.1, 1.1)$.) So

$$y_2 = 1.1605 + .05\left((.1 + 1.1605^2) + (.2 + 1.3052^2)\right)$$
$$\doteq 1.3280$$

is the improved Euler's method estimate. (The actual value is 1.2736.) Although y_1 was a better estimate than $\tilde{y}_1$, $\tilde{y}_2$ is a better estimate than y_2. A lower-order method can do better than a higher-order method by chance, but the higher-order method will win out in the limit (as h is decreased). □

Methods for solving differential equations that turn an implicit method (such as the trapezoidal scheme) into an explicit method (such as the improved Euler's method) by using an explicit method (such as Euler's method) on the right-hand side of a numerical method of the form

$$y_{i+1} = g(y_{i+1}, p_1, ..., p_n)$$

are called **predictor-corrector methods**. (Here $p_1, ..., p_n$ are any other parameters such as x_i, y_i, etc.) We say that the explicit method makes a prediction and that the implicit method corrects it, that is, improves it. In the improved Euler's method the predictor is Euler's method and the corrector is the trapezoidal scheme. There are many predictor-corrector methods.

There is no reason why we must stop after a single correction. We can use the corrected value as a new prediction and then correct it, continuing until the difference between successive corrections is acceptably small before moving on to the next point.

EXAMPLE 2: Consider the ODE IVP $y' = x - y$, $y(1) = 1$. We'll use the midpoint method $y_{i+1} = y_i + hf(x_{i+1/2}, y_i + (h/2)\,f(x_i, y_i))$ with $h = .1$. We have:

$$y_0 = 1$$
$$\tilde{y}_1 = y_0 + \frac{h}{2}f(x_0, y_0) \qquad \text{(predictor)}$$
$$= 1 + .05(1 - 1)$$
$$= 0$$
$$y_1 = y_0 + hf(x_{i+1/2}, \tilde{y}_1) \qquad \text{(corrector)}$$
$$= 1 + .1(1.05 - 0)$$
$$= 1.105$$

and the true solution is $y(x) = x - 1 + \exp(1 - x)$ so $y(1.1) \doteq 1.0048$. Let's correct again:

$$
\begin{aligned}
y_1^{(2)} &= y_0 + hf(x_{i+1/2}, y_1) && \text{(corrector)} \\
&= 1 + .1(1.05 - 1.105) \\
&\doteq 0.9945 \\
y_1^{(3)} &= y_0 + hf\left(x_{i+1/2}, y_1^{(2)}\right) && \text{(corrector)} \\
&= 1 + .1(1.05 - 0.9945) \\
&\doteq 1.0056 \\
y_1^{(4)} &= y_0 + hf\left(x_{i+1/2}, y_1^{(3)}\right) && \text{(corrector)} \\
&= 1 + .1(1.05 - 1.0056) \\
&\doteq 1.0044 \\
y_1^{(4)} &= y_0 + hf\left(x_{i+1/2}, y_1^{(3)}\right) && \text{(corrector)} \\
&= 1 + .1(1.05 - 1.0044) \\
&\doteq 1.0046 \\
y_1^{(5)} &= y_0 + hf\left(x_{i+1/2}, y_1^{(4)}\right) && \text{(corrector)} \\
&= 1 + .1(1.05 - 1.0046) \\
&\doteq 1.0045
\end{aligned}
$$

and further iterations do not change this value to the number of figures displayed, resulting in an error of about $3E - 4$. □

Using a higher-order method, using a smaller h, and repeatedly correcting are all ways to reduce the error in our results. Higher-order methods are more difficult to program and require additional differentiability; smaller step sizes take more time to reach a desired point and can lead to the growth of round-off error; and repeated correction takes more time. Additionally, repeated correction ultimately solves the predictor-corrector equation, viewed as a fixed point iteration, in the limit, but the numerical method is not the same as the ODE because of truncation error. Hence, repeated correction converges to the exact solution of a discrete approximation of the ODE (the corresponding implicit method) whereas reducing h leads to the correct solution of the ODE, in the absence of round-off error. There's no one right approach, and this is reflected in the variety of methods for solving ODEs available in MATLAB and similar packages or software libraries.

The second-order improved Euler's method uses two function evaluations to take a step of length h. Euler's method uses a single function evaluation to take a step of length h, and hence Euler's method could take two steps each of length $h/2$ for the same cost as one step of the $O\left(h^2\right)$ method. But for small h an $O(h^2)$ error will certainly be considerably smaller than the $O(h/2)$ error of Euler's method with the step size halved. We generally like to take step sizes as large as feasible, consistent with an error tolerance, to speed the computation.

Another predictor-corrector method that is in common use, and that is similar to the improved Euler's method, is **Heun's method**[4], also known as **Ralston's method**:

---

[4]Some sources use "Heun's method" for the improved Euler's method instead.

$$y_{i+1} = y_i + \frac{h}{4}\left(f(x_i, y_i) + 3f(x_{i+2/3}, y_i + \frac{2h}{3}f(x_i, y_i)) \right)$$

where $x_{i+2/3} = x_i + 2h/3$. Heun's method uses a weighted average of the slope at the left endpoint, weighted 1/4, and the slope 2/3 of the way into the interval, weighted 3/4. It is $O\left(h^2\right)$. The coefficients were chosen to minimize the truncation error, that is, the coefficient in front of h^2 in the actual $O\left(h^2\right)$ error term.

The **theta methods** are a class of schemes based on a parameter $\theta \in [0, 1]$. For any such θ, we have an $O(h)$ method of the form

$$y_{i+1} = y_i + h\left(\theta f(x_i, y_i) + (1 - \theta)f(x_{i+1}, y_{i+1})\right)$$

i.e., a weighted average of the values at the endpoints. Examples include Euler's method ($\theta = 1$) and the trapezoidal scheme ($\theta = 1/2$; this is the sole $O\left(h^2\right)$ theta method).

An important special case of the theta methods is $\theta = 0$, the **backward Euler's method** (or **implicit Euler's method**)

$$y_{i+1} = y_i + hf(x_{i+1}, y_{i+1})$$

which is an implicit method that is useful in certain circumstances, as we'll discuss later in this chapter. The backwards method can also be thought of as arising from using the backward difference approximation

$$\frac{y_{i+1} - y_i}{h} \approx f(x_{i+1}, y_{i+1})$$

in the original ODE. Euler's method is sometimes called the **forward Euler's method** when necessary to differentiate it from this method.

For now we've merely asserted the orders of these methods. We'll show in the next section how to use Taylor series to determine the order.

MATLAB

Several MATLAB commands allow for a vector of options to be passed to them, for example, `fzero`, `fminbnd`, `fminsearch`, and the ODE commands. Enter:
```
>> more on
>> help ode23
```
and read about the OPTIONS vector. Enter:
```
>> help odeset
```
and read the help information. Enter:
```
>> more off
>> f=@(x,y) x+y
>> [x,y]=ode45(f,[0 1],1);
>> options=odeset('Stats','on') %Display computational statistics.
>> [x,y]=ode23(f,[0 1],1,options);
>> options=odeset('S','off')  %We can shorten 'Stats' to just 'S'.
```
There are many other options. Many are indicated in the demos in `odeexamples`.

The MATLAB ODE solvers use both an absolute error tolerance and a relative error tolerance, choosing whichever one is met first. That is, they try to make the absolute error α_i at each node x_i at most $\max(\alpha, \rho|y_i|)$ where α is the absolute tolerance `AbsTol`, which defaults to $1E - 6$, and ρ is the relative tolerance `RelTol`, which defaults to $1E - 3$. Enter:

```
>> xx=x(end)
>> A=-xx-1+2*exp(xx);    %True solution at x=1.
>> y(end)                %Approx. solution.
>> abs(y(end)-A)         %Absolute error.
>> abs(y(end)-A)/A       %Relative error.
```
The absolute error tolerance of $1E - 6$ was not met but the relative error tolerance of $1E - 3$ was. Enter:
```
>> options=odeset('Abs',1E-8,'Rel',1E-8)    %Alpha=1E-8,rho=1E-8.
>> [x,y]=ode45(f,[0 1],1,options);
>> abs(y(end)-A)                            %Error.
>> abs(y(end)-A)/A       %Relative error.
```
Both tolerances were met. (Is this also true for ode23?) You aren't permitted to set either AbsTol or RelTol to be zero.

The ODE routines are adaptive. This means that the more rapidly the slope of the function is changing, the more points that will be placed there, analogous to the case for automatic integrators. Enter:
```
>> g=@(x,y) log(x+y)
>> [x,y]=ode45(g,[1 20],1); %No longer using the options input.
>> plot(x,y,'o-')
```
Note that the circles, representing the nodes x_i, are very tightly clustered near the left side; the slope of the log function changes rapidly for very small arguments.

The ode45 command uses a higher-order method than the ode23 command. Enter:
```
>> options=odeset('Stats','on')   %Tolerances are reset to defaults.
>> [x23,y23]=ode23(f,[0 1],1,options);
>> abs(y23(end)-A)
>> [x45,y45]=ode45(f,[0 1],1,options);
>> abs(y45(end)-A)
```
The higher-order method used significantly more function evaluations but also gave an answer that was much better than requested. This may or may not be a good thing! If the error tolerances accurately reflect the precision that is meaningful for this problem and f is expensive to compute then getting 8 decimal place accuracy slowly may be less attractive than getting 4 decimal place accuracy rapidly. Still, the ode45 routine is the one that you should try first on a problem unless you have reason to suspect that another method will be better.

We'll discuss the mathematical ideas behind ode23 and ode45, and the other ODE routines, in a later section. Let's look at some other functionalities of the routines in the ODEs suite. We might want to display the one-parameter family of solutions. Enter:
```
>> [x,y]=ode45(f,[0 1],0:.1:1,options);
>> size(y)
```
The y vector contains solutions for all 11 initial conditions in 0:.1:1. Note that the same number of function evaluations were used as before, making it much more efficient to solve all 11 of the ODE IVPs simultaneously. This is a useful feature. Enter:
```
>> plot(x,y)
```
Let's look at a more interesting ODE. Enter:
```
>> close all
```

```
>> f=@(x,y) y.*(y-1)
>> [x,y]=ode45(f,[0 .4],1:.1:2,options);
>> plot(x,y),hold on
>> [x2,y2]=ode45(f,[0 5],0:.1:1,options);
>> plot(x,y)
```

On one side of the equilibrium solution $y = 1$ the solutions diverge while on the other side they tend to zero as $x \to \infty$.

The ODE may be solved at a pre-selected set of nodes by entering them in place of a time span. For example, enter:

```
>> xx=sort(rand([1 6]))
>> [x,y]=ode45(f,xx,1);
>> x
```

The nodes x returned by ode45 are exactly the nodes xx that were given to it. However, in order to achieve the error tolerance the routine may be computing function values at additional points.

The ODE routines can handle systems as well. We'll look at systems in more detail later. As a very simple case, however, let's solve an autonomous linear 2×2 system with initial conditions $x_1(0) = .1$, $x_2(0) = .2$ solved over the time interval $[0, 1]$. Enter:

```
>> clear all
>> f=@(t,x) [1 1;1 -1]*x   %The x variable is a vector.
>> [t,x]=ode45(f,[0 1],[.1 .2]);
>> whos
```

The x vector returned by ode45 is a matrix: Its first column is the first component y_1 of the solution of $y' = Ay$, and its second column is the second component y_2.

Problems

1.) a.) Use the improved Euler's method with $h = .05$ to approximate the solution of $y' = y^2$, $y(0) = 1$, over $[0, 1]$. Compare your answers with the true solution and with your answers for Problem 1 of Sec. 6.2.

b.) Repeat using the improved Euler's method with one extra correction step.

c.) Repeat using the midpoint and Heun's methods.

2.) Write a MATLAB program for performing the improved Euler's method.

3.) Use the improved Euler's method with $h = .4$, $h = .2$, $h = .1$, and $h = .05$ to approximate the solution of $y' = -y$, $y(0) = 1$, over $[0, 4]$. Compare your answers with the true solution and with your answers for Problem 3 of Sec. 6.2.

4.) a.) Use improved Euler's method with $h = .05$ to approximate the solution of $y' = x + y + 1$, $y(1) = 1$, over $[1, 3]$. Compare your answers with your answers for Problem 4.a. of Sec. 6.2.

b.) Use the improved Euler's method with $h = .05$ to approximate the solution of $y' = \sin(xy) + 1$, $y(1) = 1$, over $[1, 2]$. Compare your answers with your answers for Problem 4.b. of Sec. 6.2 and with the results generated by ode23.

5.) a.) What is the predictor-corrector method found by using the forward Euler's method in the backward Euler's method?

b.) What is the predictor-corrector method found by using the improved Euler's method in the backward Euler's method? Compare this method to the explicit method in part a.

6.) a.) Use the improved Euler's method with $h = .05$ to approximate the solution of $y' = 1 + y^2$, $y(0) = 1$, over $[0, .5]$. Compare your answer with your answers for Problem 6 of Sec. 6.2.
b.) Repeat using Heun's method.
7.) Write a MATLAB program for performing the improved Euler's method with repeated correction. Correct until two successive iterates differ by no more than a user-supplied tolerance.
8.) Conduct an experiment to compare the improved Euler's, Heun's, and midpoint methods. Does one tend to perform better than the others?
9.) Use ode45 to plot the one-parameter family of solutions of $y' = y(y-1)(y+1)$. Make sure that your plot captures all the interesting behavior of this family.
10.) Plot the phase portrait of the system $x' = \sin(x)$, $y' = \cos(x)$ by first solving each ODE separately using ode45.
11.) a.) Use the improved Euler's method with $h = .4$, $h = .2$, $h = .1$, and $h = .05$ to approximate the solution of $y' = -8y$, $y(0) = 1$, and of $y' = -8y$, $y(0) = 1$ over $[0, 2]$. Compare your answers with your answers from Problem 11 of Sec. 6.2 and with the true solution.
b.) Repeat using the improved Euler's method with one extra correction step.
c.) Repeat using the backward Euler's method.
d.) Repeat using the **implicit midpoint rule** $y_{i+1} = y_i + hf(x_{i+1/2}, (y_i + y_{i+1}))$. Note that this method uses the midpoint idea but is not derived from the midpoint quadrature rule as the midpoint scheme and methods are. Also repeat using the explicit rule corresponding to this method with prediction by Euler's method.
12.) Use the improved Euler's, Heun's, and midpoint methods to solve $y' = -30y$, $y(0) = 1$ over $[0, 10]$. Comment.
13.) Use the Euler's, improved Euler's, Heun's, and midpoint methods to solve $y' = y(y - \pi)$, $y(0) = c$ over $[0, 10]$ for various values of $c > 0$. Note that if your initial condition is not equal to π (e.g., if you use y0=pi to approximate $y(0) = \pi$) then your answer should either tend to 0 or to ∞ and that if $y(0) > 0$ then $y(t) > 0$ for all $t > 0$.
14.) Use the improved Euler's method to approximate the solution of $y' = x + y^2$, $y(0) = 1$, over $[0, 2]$ for several values of h. Plot the absolute error at $x = 1.5$ as a function of h. Comment.
15.) a.) Show that the midpoint scheme is $O(h^2)$.
b.) Show that the midpoint method is $O(h^2)$.
16.) Create a variety of functions that are k times continuously differentiable for $k = 1, 2, 3, 4, 5$, and compare ode23 and ode45 on them.

4. Analysis of Explicit One-Step Methods

The predictor-corrector methods we have seen so far, as well as Euler's method, are of a form known as **one-step methods**, meaning methods of the form

$$y_{i+1} = g(x_i, x_{i+1}, y_i, y_{i+1})$$
$$y_{i+1} = y_i + h\Phi_f(x_i, y_i, h)$$

(the step size h_i may also be given as an argument of g, but could be computed from $h_i = x_{i+1} - x_i$ in this form). If $y_{i+1} = g(x_i, x_{i+1}, y_i)$ then the method is explicit; otherwise it is implicit. Often we have the form

$$y_{i+1} = y_i + h\Phi_f(x_i, y_i, h)$$

for explicit one-step methods. The function Φ_f is called the **increment function**. By writing Φ_f we mean to indicate that Φ may depend on the function f and possibly its derivatives.

A method is said to be explicit because it defines y_{i+1} explicitly in terms of known quantities and is said to be one-step because it computes y_{i+1} in terms of y_i. An **explicit multi-step method** is of the form

$$y_{i+1} = \sum_{j=0}^{j=k-1} \alpha_j y_{i-j} + h\Phi_f(x_i, y_i, y_{i-1}, ..., y_{i-k+1}, h)$$

($k \geq 2$), also called an **explicit k-step method**. (This formula assumes a fixed step size. For a variable step size, we must list all the relevant nodes as arguments of the increment function Φ_f.) The **implicit multi-step methods** or **implicit k-step methods** are methods of the form

$$y_{i+1} = \sum_{j=0}^{j=k-1} \alpha_j y_{i-j} + h\Phi_f(x_i, y_{i+1}, y_i, y_{i-1}, ..., y_{i-k+1}, h)$$

($k \geq 2$). The backward Euler's method is an implicit one-step method, and in fact any theta method with $\theta \in (0, 1]$ is an implicit one-step method. We could generate an implicit two-step method for the ODE IVP $y' = f(x, y)$, $y(x_0) = y_0$ by integrating $y' = f(x, y(x))$ using Simpson's rule over an interval of width $2h$:

$$
\begin{aligned}
y' &= f(x, y(x)) \\
y(x_2) &= y_0 + \int_{x_0}^{x_2} f(t, y(t))dt \\
y(x_2) &\approx y_0 + \frac{h}{3}\left(f(x_0, y(x_0)) + 4f(x_1, y(x_1)) + f(x_2, y(x_2))\right) \\
(4.1)\qquad y_2 &= y_0 + \frac{h}{3}\left(f(x_0, y_0) + 4f(x_1, y_1) + f(x_2, y_2)\right)
\end{aligned}
$$

where the last line defines the numerical method. (Here $h = (x_2 - x_0)/2$). This is an implicit two-step method; not only is it implicit in y_2 but we also need a starter method, preferably of the same order, to get a value for y_1. Then we compute

$$y_3 = y_1 + \frac{h}{3}\left(f(x_1, y_1) + 4f(x_2, y_2) + f(x_3, y_3)\right)$$

(and higher y_i) the method will still be implicit but we will not need to have additional values calculated by another method. For now let's discuss the analysis of explicit one-step methods. Consider the explicit one-step method

$$(4.2)\qquad y_{i+1} = y_i + h\Phi(x_i, y_i, h)$$

where it is understood that $\Phi = \Phi_f$. We say that it is **consistent** with the ODE IVP $y' = f(x, y)$, $y(x_0) = y_0$ if

$$\Phi(x, y, 0) = f(x, y)$$

and $\Phi(x, y, h)$ is continuous at $h = 0$. This means that $\lim_{h \to 0} \Phi(x, y, h) = \Phi(x, y, 0)$. Consistency insures that as $h \to 0$ the difference equation $y_{i+1} = y_i + h\Phi(x_i, y_i, h)$, i.e.

$$\frac{y_{i+1} - y_i}{h} = \Phi(x_i, y_i, h)$$

approximates the true equation $y' = f(x, y)$ more and more accurately. If the method isn't consistent, it doesn't correctly approximate the ODE.

We define the **local truncation error** (**LTE**) of an explicit one-step method at a point x_i to be the extent to which the true solution $y(x)$ of the ODE IVP fails to satisfy the numerical method Eq. (4.2). That is, the LTE is the difference

$$(4.3) \qquad LTE_{i+1} = y(x_{i+1}) - [y(x_i) + h\Phi(x_i, y(x_i), h)]$$

between $y(x_{i+1})$ and the value we would get for y_{i+1} *if we had used the exact value* $y_i = y(x_i)$ to make the step[5]. Note that this is a function of h. The **global error** (or **global truncation error**) is the difference

$$(4.4) \qquad GE_{i+1} = y(x_{i+1}) - y_{i+1}$$

between the true value and the computed value. The absolute value of the global error is the absolute error α_{i+1}, so this is the quantity of greatest interest to us. It will be convenient, however, to analyze it using the local truncation error.

Let $n = 1/h$. We say that a method is **convergent** if the global error tends to zero as h tends to zero, that is, if for a fixed interval $[a, b]$ ($x_0 = a$, $x_n = b$) we have

$$\lim_{h \to 0} \max_{i=1}^{n} \left(|y(x_{i+1}) - y_{i+1}| \right) = 0$$

for all valid initial conditions $y(x_0) = y_0$. What we want out of a numerical scheme for solving $y' = f(x, y)$, $y(x_0) = y_0$ is that it is convergent, that the error goes to zero sufficiently rapidly for the method to be useful, and that is well-behaved with respect to round-off error.

We say that a method is **stable** if for all sufficiently small h and all $\epsilon > 0$ there is a $K > 0$ such that the numerical solution of $y' = f(x, y)$, $y(x_0) = y_0 + \epsilon$ using Eq. (4.2) differs from the numerical solution of $y' = f(x, y)$, $y(x_0) = y_0$ using Eq. (4.2) at the final node by at most $K\epsilon$. In other words, stability means that if $\{y_i\}_{i=1}^{n}$ is the numerical solution of the unperturbed problem $y' = f(x, y)$, $y(x_0) = y_0$ and $\{w_i\}_{i=1}^{n}$ is the numerical solution of the perturbed problem $y' = f(x, y)$, $y(x_0) = y_0 + \epsilon$ then

$$|w_n - y_n| \leq K\epsilon$$

for all h less than some h_0, where K is independent of ϵ and h. We could have easily as defined stability to mean that

$$|w_i - y_i| \leq K\epsilon$$

---

[5]Some sources define $LTE_i = h\tau_i$ and call τ_i the local truncation error.

for $i = 1, ..., n$ but if it is stable for any endpoint x_n then it is stable for any point x_i as we could always consider x_i the endpoint of a subproblem.

That's a lot of definitions. Let's apply them to Euler's method $y_{i+1} = y_i + hf(x_i, y_i)$ and see what we get. Is Euler's method consistent? Yes, the increment function is $\Phi_f(x, y, h) = f(x, y)$ so trivially $\lim_{h \to 0} \Phi_f(x, y, h) = f(x, y)$. What is the local truncation error of Euler's method? From Eq. (4.3) we have

$$
\begin{aligned}
LTE_{i+1} &= y(x_{i+1}) - [y(x_i) + h\Phi_f(x_i, y(x_i), h)] \\
&= y(x_{i+1}) - y(x_i) - hf(x_i, y(x_i)) \\
&= \left[y(x_i) + hy'(x_i) + \frac{h^2}{2} y''(\xi_i) \right] - y(x_i) - hf(x_i, y(x_i)) \\
&= y(x_i) + hf(x_i, y(x_i)) + \frac{h^2}{2} y''(\xi_i) - y(x_i) - hf(x_i, y(x_i)) \\
(4.5) \qquad &= \frac{h^2}{2} y''(\xi_{i+1})
\end{aligned}
$$

and in particular the LTE at the point x_{i+1} is $O(h^2)$ (if the solution y is at least twice continuously differentiable). What about the global error? Fix an interval of integration $[a, b] = [x_0, x_n]$ and assume that y is at least twice continuously differentiable. From Eq. (4.3) we have

$$
\begin{aligned}
y(x_{i+1}) &= y(x_i) + h\Phi(x_i, y(x_i), h) + LTE_{i+1} \\
y(x_{i+1}) - y_{i+1} &= y(x_i) + h\Phi(x_i, y(x_i), h) + LTE_{i+1} - y_{i+1} \\
&= y(x_i) + h\Phi(x_i, y(x_i), h) + LTE_{i+1} - (y_i + hf(x_i, y_i)) \\
GE_{i+1} &= GE_i + h(f(x_i, y(x_i)) - f(x_i, y_i)) + LTE_{i+1} \\
&= GE_i + h(f(x_i, y(x_i)) - f(x_i, y(x_i) - GE_i)) + LTE_{i+1}
\end{aligned}
$$

using Eq. (4.4). Let $e_{i+1} = |GE_{i+1}|$. Then, by the triangle inequality,

$$
e_{i+1} \le e_i + h|f(x_i, y(x_i)) - f(x_i, y_i)| + M\frac{h^2}{2}
$$

where $M > 0$ is a bound on $|y''(x)|$ over the interval $[x_0, x_n]$. Let $L > 0$ be a Lipschitz constant for f with respect to y. Then

$$
\begin{aligned}
|f(x_i, y(x_i)) - f(x_i, y(x_i) - GE_i)| &\le L|GE_i| \\
&= Le_i
\end{aligned}
$$

so

$$
\begin{aligned}
e_{i+1} &\le e_i + hLe_i + M\frac{h^2}{2} \\
&= (1 + hL)e_i + M\frac{h^2}{2}
\end{aligned}
$$

($i = 0, 1, ..., n - 1$). It is easily verified inductively that $e_{i+1} \le (1 + hL)e_i + Mh^2/2$ implies that

(4.6) $$e_i \leq (1 + hL)^i e_0 + \left[\frac{(1 + hL)^i - 1}{hL} \right] M \frac{h^2}{2}$$

(compare the analysis of stationary iterative methods in Sec. 3.1). For the purposes of global error $e_0 = 0$, giving

$$e_i \leq \left(\frac{(1 + hL)^i - 1}{L} \right) \frac{hM}{2}$$

as our bound. Using the inequality $1 + x \leq \exp(x)$ we can write

$$\begin{aligned} (1 + hL)^i &\leq (\exp(hL))^i \\ &= \exp(ihL) \end{aligned}$$

so

(4.7) $$e_i \leq \left(\frac{\exp(ihL) - 1}{L} \right) \frac{hM}{2}$$

is a bound on the global error. (Compare Theorem 6.2.1, where e_0 is zero because the arithmetic is assumed to be exact.) Note that

$$\frac{\exp(ihL) - 1}{L} = ih + (ih)^2 L/2 + (ih)^3 L^2/6 + \dots$$

but ih is at most the width $w = x_n - x_0$ of the interval, so this value is bounded. It's tempting to say it's $O(h)$, but remember that as h decreases the value of i that refers to a specific value of x increases, so ih remains unchanged. If $ih = \lambda$ then from Eq. (4.7) we have

$$\begin{aligned} e_i &\leq (\lambda + \lambda^2 L/2 + \lambda^3 L^2/6 + \dots) \frac{hM}{2} \\ e_i &\leq O(h) \end{aligned}$$

as the series in parentheses converges (to $(\exp(\lambda L) - 1)/L$, from Eq. (4.7)). Hence the global error is $O(h)$ and so it does tend to zero as $h \to 0$. Again, the global error is the error in which we are interested: Euler's method is $O(h)$.

We have considered consistency, local truncation error, and global error. What we really need of a numerical method, though, is convergence and stability–that the global error tends to zero as $h \to 0$ so that refinement of the step size leads to improved accuracy, and that round-off errors are not amplified[6].

THEOREM 1: An explicit one-step method $y_{i+1} = y_i + h\Phi_f(x_i, y_i, h)$ with an increment function $\Phi_f(x_i, y_i, h)$ that is continuous with respect to x, y, and h for all $h < h_0$ (for some $h_0 > 0$), and that is Lipschitz with respect to y, is stable, and it is convergent if and only if it is consistent.

This is a special case of the famous **Lax Equivalence Theorem**, one of the most important theorems in all of numerical analysis. It's fundamental to the study of numerical methods for partial differential equations.

---

[6]Note that stability of the algorithm means not only protection from the effects of round-off error but also from the effects of inaccuracies in the data, such as imprecise knowledge of y_0.

It follows that Euler's method is convergent and stable. Look back at the demonstration that Euler's method is consistent, which was relatively easy, and the derivation of Eq. (4.7) followed by the demonstration that the global error tends to zero as $h \to 0$ (convergence). Showing convergence was rather more work. We generally show that an explicit one-step method is convergent and stable by showing that it is consistent and appealing to this theorem. This is why we needed so many definitions: One set for what we care about (convergence, meaning that the solution of the difference equation tends to the solution of the differential equation as $h \to 0$) and the other for what we can easily show (consistency, meaning that the difference equation tends to the differential equation as $h \to 0$).

For a convergent explicit one-step method, if the local truncation error is $O(h^{m+1})$ then the global error is $O(h^m)$. This is convenient to know, since the LTE is almost always easier to find. We say that m is the order of the method; Euler's method is of order one because of Eq.(4.5).

We emphasize that Theorem 1 applies only to explicit one-step methods. There are similar theorems for other types of methods. Typically these theorems require us to show consistency and either stability or convergence; we get the third property for free. Since consistency is usually easy to show this is still an improvement over having to show both stability and convergence.

EXAMPLE 1: Let's look at the midpoint method. Recall that the midpoint method is the explicit one-step method

$$y_{i+1} = y_i + hf(x_{i+1/2}, y_i + \frac{h}{2}f(x_i, y_i))$$

$(\Phi_f(x_i, y_i, h) = f(x_{i+1/2}, y_i + (h/2)f(x_i, y_i)))$. If f is continuous then as $h \to 0$ with x, y fixed,

$$f(x_{i+1/2}, y_i + \frac{h}{2}f(x_i, y_i)) = f(x + \frac{h}{2}, y + \frac{h}{2}f(x, y))$$

tends to $f(x, y)$, so the method is consistent. In addition $\Phi_f(x, y, h) = f(x + (h/2), y + (h/2)f(x, y))$ is continuous, and if f is continuously differentiable with respect to y then $\Phi_f(x, y, h)$ is differentiable with respect to y and hence Lipschitz with respect to y. Hence if f is continuous and continuously differentiable in y then the midpoint method is stable and convergent, by Theorem 1. □

We've defined global error, convergence, and stability for a general method, but not consistency and local truncation error. Consistency always is defined to mean that as h shrinks the discrete equation tends to the ODE; LTE is always the extent to which the solution of the ODE fails to solve the difference equation, that is, it is the quantity that remains if we replace y_k by $y(x_k)$ in the numerical method.

Does all this analysis matter? Well, this type of thing *is* the "analysis"" in "numerical analysis", and it certainly improves our confidence in a program if we know the method it is intended to implement is convergent and stable, so that if something goes wrong it must be either a programming error or a violation of assumptions such as continuity of the function[7]. Given the complicated devices and structures designed via software packages, such analysis seems worthwhile for a method that will be implemented as part of a scientific or engineering package. In addition, there is no analytical formula for the solution of most ODEs so we cannot test our programs on "hard" cases and compare the answers with what they should

---

[7]Or, extremely infrequently, a hardware problem.

be. A proof of convergence and stability is therefore important for building our confidence in the programs that perform these computations. It also guides us in choosing a step size h that gives stability but also isn't so small that our program takes overly long to complete.

MATLAB

Most of this section has been rather theoretical so we'll take an opportunity to use this MATLAB subsection to talk about object-oriented programming support in MATLAB. We have already seen that the options for a number of routines are data types called structures. Enter:

```
>> help struct
```

and read the help text. Enter:

```
>> more on
>> help datatypes
```

to see information on the different types of variables available in MATLAB as well as commands for manipulating and converting between them.

One way to think of a structure is as an array of arrays, where the element arrays may be of different sizes and contain different types of data. The element arrays are also labeled by names. For example, enter:

```
>> student(1).lastname='Ash'
>> student(1).firstname='Sara'
>> student(1).id=23571
>> student(1).exams=[81 85 91]
>> student(1).hw=[10 9 10 10 8 7 9 10 10]
>> student(2).lastname='Oak'
>> student(2).firstname='John'
>> student(2).id=45731
>> student(2).exams=[88 80 90]
>> student(2).hw=[10 9 9 9 9 10 9 10 8 10]
>> student
>> student(1)
>> student(2)
>> student.id
>> student.lastname
```

When we enter the name of the structure variable we get a description, not a value. When we ask for an indexed entry however we get all the data for that entry. We can think of each `student(i)` as a virtual index card with the data on the ith student. Each item on it (lastname, firstname, etc.) is called a field. Enter:

```
>> student(2).lastname
>> student(1).id
```

The first two lines return character (see `help char`) and numeric variables which we can operate on in the usual way. Enter:

```
>> fieldnames(student)
>> isfield(student,'id')
>> isfield(student,'finalgrade')
>> student(1).finalgrade
```

The `isfield` command checks for valid fieldnames. The `fieldnames` provides all fieldnames as a special type of array (more on this later).

We can access the contents of a field using the `getfield` command rather than the structure.fieldname form. Enter:

```
>> getfield(student,'firstname')
>> getfield(student(1),'firstname')
>> getfield(student(2),'firstname')
```

We can set field values using the `setfield` command. Enter:

```
>> setfield(student(2),'firstname','Jack')
>> student(2)                 %No change!
>> student(2)=setfield(student(2),'firstname','Jack')
>> student(2)               %Changed.
```

Fields can be removed using the `rmfield` command. Fields can be added to an existing structure; enter:

```
>> student(1).grade='B+'
>> student(1)
>> student(2)
>> student(2).grade
>> student(2).grade='B+'
>> student(2)
```

Note that adding the field to one entry adds it to all entries. Notice also that entries in a given field need not be the same size in every entry, as with the names and the homework scores.

Structures may also be created using the **struct** command. To create a structure similar to **student**, enter:

```
>> S=struct('lastname',{'Ash','Oak'},'firstname',{'Sara','John'})
>> S(1)
>> S(2)
```

Note that the fieldnames could be given as character variables (rather than the explicit character variables we used above), so this form of the command is fairly flexible within a program. Structures can also be multiply indexed, e.g., S(i,j).

Structures are the type of variable used to pass information on options to the minimization and ODE routines. Enter:

```
>> type odeset
```

and see how it forms the structure variable that it creates for the ODE routines. A structure array is convenient in this case because some of the options have numeric values and some have character values; some can even be functions.

Let's look at cell arrays. Cell arrays are similar to structures in that they may hold other arrays of varying sizes, but they can be indexed in the usual way. Cell arrays may be created using curly braces, as was done in the use of the **struct** command above. Enter:

```
>> classlist{1,1}='Ash'
>> classlist{2,1}='Sara'
>> classlist{3,1}=23571
>> classlist{1,2}='Oak'
>> classlist{2,2}='John'
>> classlist{3,2}=45731
>> classlist
>> classlist(1)
>> classlist(5)
```

```
>> classlist{2,2}
```
The latter form is the usual way to access elements of a cell array. A cell array is similar to a structure array; we sacrifice the convenience of fieldnames but gain some flexibility in indexing. Enter:
```
>> classlist{1,:}
>> classlist{1:2,1:2}
>> celldisp(classlist)
>> cellplot(classlist)
```
The `cellplot` command displays the contents of a cell array graphically. The `cell` command may be used to create a cell array of a given size, initially empty, to be filled by indexing. Cells may be indexed by more than two indices, that is, `C{2,4,6,2}` can be valid if `C` is set up appropriately. This is true of structures also.

Problems

1.) Use Theorem 1 to show that the improved Euler's method is convergent and stable. Find an expression for the LTE.

2.) Show that Eq. (4.7) agrees with Theorem 7.2.1 by taking $e_0 = 0$ and expanding $\exp(ihL) - 1)/hL$ in a Maclaurin series.

3.) Prove that Eq. (4.6) is correct.

4.) Use Theorem 1 to show that Heun's method is convergent and stable. Find an expression for the LTE.

5.) a.) Use Eq. (4.1) with $h = .05$ to approximate the solution of $y' = x + y + 1$, $y(1) = 1$, over $[1,3]$. Use Heun's method as a starter method. (This method is $O(h^3)$.) Compare your answer with the true solution.

b.) Use this method with $h = .05$ to approximate the solution of $y' = y^2$, $y(0) = 1$, over $[0,1]$. Compare your answer with the true solution.

6.) Use Theorem 1 to show that Heun's method is convergent and stable. Find an expression for the LTE.

7.) a.) The **inverse Euler's method** is the scheme $y_{i+1} = y_i + hy_i f(x_i, y_i)/(y_i - hf(x_i, y_i))$. (We choose h so as to avoid division by a small number.) Show that it is of order one.

b.) Show that the inverse Euler's method is convergent.

8.) Find an expression for the LTE of the backward Euler's method.

9.) Derive Heun's method as a Radau quadrature rule.

10.) Show that the trapezoidal scheme is convergent. Theorem 1 does not apply; use a direct approach.

11.) Show that the backward Euler's method is convergent. Theorem 1 does not apply; use a direct approach.

12.) a.) Find the order of the theta methods.

b.) Show that the theta methods are convergent. Theorem 1 does not apply; use a direct approach.

13.) Develop a numerical method for $y' = f(x,y)$, $y(x_0) = y_0$ using the centered difference approximation to the first derivative and a starter method. Test your method.

14.) Test Euler's method and the improved Euler's method on a number of functions that are not Lipschitz. Compare how the error decreases with decreasing h to what happens for Lipschitz functions.

15.) Show that if $f(x,y)$ is Lipschitz then so is the increment function for the improved Euler's method.

16.) Show that if $f(x, y)$ is Lipschitz then so is the increment function for Heun's method.

5. Taylor and Runge-Kutta Methods

We have been liberal in assuming that the ODE IVP $y' = f(x, y)$, $y(x_0) = y_0$ we wish to solve has certain desirable properties. We're going to be even more so now. If a solution $y(x)$ of the ODE IVP exists then it is differentiable, for $y'(x) = f(x, y(x))$. If it is twice differentiable then

$$\frac{d^2}{dx^2}y(x) = \frac{d}{dx}y'(x)$$
$$= \frac{d}{dx}f(x, y(x))$$
$$= \frac{\partial f}{\partial x}\frac{\partial x}{\partial x} + \frac{\partial f}{\partial y}\frac{\partial y}{\partial x}$$
$$= \frac{\partial f}{\partial x}\cdot 1 + \frac{\partial f}{\partial y}\frac{dy}{dx}$$
$$= \frac{\partial f}{\partial x} + \frac{\partial f}{\partial y}f(x, y)$$

where by $\partial f/\partial x$ is meant the partial derivative of f with respect to its first variable, holding its second variable constant, and similarly for $\partial f/\partial y$. Since f is always known, we can certainly find its partial derivatives. Recall that one way in which we derived Euler's method in Sec. 6.2 was by neglecting the final term in

$$y(x_0 + h) = y(x_0) + hy'(x_0) + \frac{h^2}{2}y''(\gamma).$$

If we now write

(5.1) $$y(x_0 + h) = y(x_0) + hy'(x_0) + \frac{h^2}{2}y''(x_0) + \frac{h^3}{6}y'''(\gamma)$$

we have the method

$$y(x_0 + h) \approx y(x_0) + hy'(x_0) + \frac{h^2}{2}y''(x_0)$$

$$y_1 = y_0 + hf(x_0, y_0) + \frac{h^2}{2}f'(x_0, y_0)$$

(5.2) $$= y_0 + hf(x_0, y_0) + \frac{h^2}{2}(f_x(x_0, y_0) + f_y(x_0, y_0)f(x_0, y_0))$$

and similarly for y_2, etc. We could, if $y(x)$ is sufficiently differentiable, continue in this way and create higher and higher-order methods by taking a longer and longer Taylor series in Eq. (5.1) and using the chain rule to find the necessary derivatives of $f(x, y(x))$. The resulting methods are called the **Taylor methods**. These are of order k if the term neglected is of the form

$$\frac{h^{k+1}}{k!} y^{(k+1)}(\gamma) = \frac{h^{k+1}}{k!} f^{(k)}(\gamma, y(\gamma))$$

$(k \geq 1)$. Euler's method is the Taylor method of order one. Of course, we know from Theorem 6.2.1 that it suffices for Euler's method that f be continuously differentiable, whereas viewing it as a Taylor method requires that it be twice continuously differentiable.

EXAMPLE 1: Let's use the Taylor method of order two on the ODE IVP $y' = x^2 y$, $y(0) = 1$. The solution is $y(x) = \exp(x^3/3)$. We have

$$
\begin{aligned}
f(x, y) &= x^2 y \\
f_x(x, y) &= 2xy \\
f_y(x, y) &= x^2
\end{aligned}
$$

so the method is

$$
\begin{aligned}
y_{i+1} &= y_i + h f(x_i, y_i) + \frac{h^2}{2}\left(f_x(x_i, y_i) + f_y(x_i, y_i) f(x_i, y_i) \right) \\
&= y_i + h x_i^2 y_i + \frac{h^2}{2}\left(2x_i y_i + x_i^2 \cdot x_i^2 y_i \right) \\
&= y_i + h x_i^2 y_i + \frac{h^2}{2} x_i y_i \left(2 + x_i^3 \right) \\
&= y_i + h x_i y_i \left(x_i + \frac{h}{2}(2 + x_i^3) \right)
\end{aligned}
$$

(from Eq. (5.2)). Hence, using $h = .05$,

$$
\begin{aligned}
y_0 &= 1 \\
y_1 &= y_0 + h x_0 y_0 \left(x_0 + \frac{h}{2}(2 + x_0^3) \right) \\
&= 1 + 0 \\
&= 1 \\
y_2 &= y_1 + h x_1 y_1 \left(x_1 + \frac{h}{2}(2 + x_1^3) \right) \\
&= 1 + (.05)(.05)(1)(.05 + .025(2 + .05^3)) \\
&\doteq 1.0003 \\
y_3 &= y_2 + h x_2 y_2 \left(x_2 + \frac{h}{2}(2 + x_2^3) \right) \\
&= 1.0003 + (.05)(.1)(1.0003)(.1 + .025(2 + .1^3)) \\
&\doteq 1.0011
\end{aligned}
$$

and the errors are

$$\begin{aligned}
\alpha_1 &= |y_1 - \exp(x_1^3/3)| \\
&\doteq 4.2E - 5 \\
\alpha_2 &= |y_2 - \exp(x_2^3/3)| \\
&\doteq 3.3E - 5 \\
\alpha_3 &= |y_3 - \exp(x_3^3/3)| \\
&\doteq 2.6E - 5
\end{aligned}$$

so our results are accurate to the number of places shown. We shouldn't be too worried about the fact that $y_0 = y_1$; the solution is changing slowly near x_0, because $y'(x_0) = 0$. $\square$

Taylor methods are rarely used because of the obvious difficulty of differentiating f automatically in software. The Taylor method of order two uses three function evaluations (f, f_x, and f_y) and has local truncation error of order three, meaning its global error is of order two. The improved Euler's method uses two function evaluations, requires no differentiations, and is also of order two, so the Taylor method is hardly competitive.

What if we want methods of order higher than two but don't want to have to differentiate f? There are many approaches, but the one we are about to describe leads to one of the most widely used classes of numerical methods for ODEs. The idea is simple: The Taylor method of order two is

$$
\begin{aligned}
y_{i+1} &= y_i + hf(x_i, y_i) + \frac{h^2}{2}\left(f_x(x_i, y_i) + f_y(x_i, y_i)f(x_i, y_i)\right) \\
(5.3) \qquad &= y_i + h\left(f(x_i, y_i) + \frac{h}{2}\left(f_x(x_i, y_i) + f_y(x_i, y_i)f(x_i, y_i)\right)\right)
\end{aligned}
$$

and the fact that it is of order two means that the right-hand side is correct to $O(h^3)$ (the LTE), if y_i is exact. What if we were to approximate the term

$$
f(x_i, y_i) + f'(x, y(x))|_{(x_i, y_i)} = f_x(x_i, y_i) + f_y(x_i, y_i)f(x_i, y_i)
$$

to $O(h^3)$? If we were to do this, using only values of f and not of its derivatives, the resulting method would also have $O(h^3)$ LTE and hence $O(h^2)$ global error. We would have a method of the same order that would use only function values. **Explicit Runge-Kutta methods** (or **ERK methods**), achieve this with the one-step formula

$$
y_i = y_{i-1} + h(k_1 g_1 + k_2 g_2 + \cdots + k_N g_N)
$$

where the g_i are defined by

$$
\begin{aligned}
g_1 &= f(x_{i-1} + c_1 h, y_{i-1}) \\
g_2 &= f(x_{i-1} + c_2 h, y_{i-1} + a_{2,1} h g_1) \\
(5.4) \qquad g_3 &= f(x_{i-1} + c_3 h, y_{i-1} + a_{3,1} h g_1 + a_{3,2} h g_2) \\
&\;\;\vdots \\
g_N &= f(x_{i-1} + c_N h, y_{i-1} + a_{N,1} h g_1 + \cdots + a_{N,N-1} h g_{N-1})
\end{aligned}
$$

and are called the **RK stages**. The $\{c_i\}_{i=1}^{N}$ are called the **RK nodes** (typically $c_1 = 0$ and often $c_N = 1$) and the $\{k_i\}_{i=1}^{N}$ are called the **RK weights**; the lower triangular matrix $A = (a_{i,j})$ is called the **RK matrix**. An ERK method of order 2 may be referred to as ERK2 or often just RK2, and so on. However, while Euler's method is the only ERK1 method, this is not so for higher orders (e.g., the improved Euler's method and Heun's method are ERK2 methods). So, this notation can be ambiguous.

The choice of the nodes, weights, and RK matrix defines an ERK method. They are typically displayed in an **RK tableau**, or **Butcher table**

$$k|c|A$$

where k is the column vector of RK weights and c is the column vector of nodes[8]. Elements of A that are not in the strict lower triangle and hence must be null are simply left empty in the tableau. For example,

$\frac{1}{6}$	0			
$\frac{1}{3}$	$\frac{1}{2}$	$\frac{1}{2}$		
$\frac{1}{3}$	$\frac{1}{2}$	0	$\frac{1}{2}$	
$\frac{1}{6}$	1	0	0	1

defines a fourth-order ERK method

$$(5.5) \qquad
\begin{aligned}
y_i &= y_{i-1} + h(\tfrac{1}{6}g_1 + \tfrac{1}{3}g_2 + \tfrac{1}{3}g_3 + \tfrac{1}{6}g_4) \\
g_1 &= f(x_{i-1}, y_{i-1}) \\
g_2 &= f(x_{i-1} + \tfrac{1}{2}h, y_{i-1} + \tfrac{1}{2}hg_1) \\
g_3 &= f(x_{i-1} + \tfrac{1}{2}h, y_{i-1} + \tfrac{1}{2}hg_2) \\
g_4 &= f(x_{i-1} + h, y_{i-1} + hg_3)
\end{aligned}$$

known as the **classical fourth-order Runge-Kutta method**. This well-known method is commonly referred to as RK4 and is what people usually mean when they refer to "the" Runge-Kutta method although, again, it is not the only ERK4 method. Indeed, **Kutta's 3/8 rule** has the Butcher table

$\frac{1}{8}$	0			
$\frac{3}{8}$	$\frac{1}{3}$	$\frac{1}{3}$		
$\frac{3}{8}$	$\frac{2}{3}$	$-\frac{1}{3}$	1	
$\frac{1}{8}$	1	1	-1	1

and is an RK4 method. In principle it has better error properties but it requires more computation, and as always we would prefer not to have negative coefficients in these rules. The A matrix for the classical method is as sparse as it can be which results in fewer arithmetic operations (though compared to function evaluations their cost is minor).

The **implicit Runge-Kutta methods** (or **IRK methods**) are similar save that the matrix A need not be lower triangular; that is, each of the g_i is given by

---

[8]An alternative tableau notation writes the weights as a row vector beneath A instead.

$$g_i = f(x_{i-1} + c_i h, y_{i-1} + a_{i,1} h g_1 + \cdots + a_{i,i-1} h g_{N-1})$$

$(i = 1, ..., N)$. We must solve a nonlinear system at each step but the previous solution provides a good initial guess. Nontrivial examples of these methods are rarely seen in practice, despite their generally superior stability properties, so from now on when we say Runge-Kutta method or RK method we will mean an ERK method.

Remember, our goal in using an RK method is to get an "approximate" Taylor method by approximating the terms involving derivatives of f to the same order as the truncation error of the Taylor method. Let's try doing this for the Taylor method of order two

$$(5.6) \qquad y_{i+1} = y_i + h \left(f(x_i, y_i) + \frac{h}{2} \left(f_x(x_i, y_i) + f_y(x_i, y_i) f(x_i, y_i) \right) \right)$$

(from Eq. (5.3)). It's not immediately clear how many stages we need; there are three separate function evaluations (f, f_x, and f_y) so perhaps we'll need the same and hence a three-stage method. On the other hand there are methods of order two that use two function evaluations–the midpoint method, improved Euler's method, Heun's method–so maybe we can get away with only two stages. Let's try a two-stage RK method

$$
\begin{aligned}
y_i &= y_{i-1} + h(k_1 g_1 + k_2 g_2) \\
g_1 &= f(x_{i-1}, y_{i-1}) \\
g_2 &= f(x_{i-1} + ch, y_{i-1} + ahg_1)
\end{aligned}
$$

where we have assumed $c_1 = 0$. We need to find the weights k_1 and k_2 that will be attached to the two estimates g_1 and g_2 of the slopes, and the c and a that determine g_2. Then

$$
\begin{aligned}
g_2(x_i, y_i) &= f(x_{i-1} + ch, y_{i-1} + ahg_1) \\
&= f(x_{i-1}, y_{i-1}) + ch f_x(x_{i-1}, y_{i-1}) + ahf(x_{i-1}, y_{i-1}) f_y(x_{i-1}, y_{i-1})
\end{aligned}
$$

up to $O(h^2)$, from the Taylor series for a function $g_2(x, y)$ of two variables. Hence the RK method is

$$
\begin{aligned}
y_i &= y_{i-1} + h(k_1 g_1 + k_2 g_2) \\
&= y_{i-1} + h(k_1 f(x_{i-1}, y_{i-1}) + \\
&\quad k_2 \left(f(x_{i-1}, y_{i-1}) + ch f_x(x_{i-1}, y_{i-1}) + ahf(x_{i-1}, y_{i-1}) f_y(x_{i-1}, y_{i-1}) \right) \\
&\quad + O(h^2)) \\
&= y_{i-1} + h(k_1 + k_2) f(x_{i-1}, y_{i-1}) + \\
&\quad k_2 h^2 \left(c f_x(x_{i-1}, y_{i-1}) + af(x_{i-1}, y_{i-1}) f_y(x_{i-1}, y_{i-1}) \right) + O(h^3)
\end{aligned}
$$

which must match the Taylor method of order two

$$y_i = y_{i-1} + h\left(f(x_{i-1}, y_{i-1}) + \frac{h}{2}\left(f_x(x_{i-1}, y_{i-1}) + f_y(x_{i-1}, y_{i-1})f(x_{i-1}, y_{i-1})\right)\right)$$

$$= y_{i-1} + hf(x_{i-1}, y_{i-1}) + \frac{h^2}{2}\left(f_x(x_{i-1}, y_{i-1}) + f_y(x_{i-1}, y_{i-1})f(x_{i-1}, y_{i-1})\right)$$

(Eq, (5.6)) to within $O(h^3)$, the same as its truncation error. But this is easy! We need only require that

(5.7)

$$k_1 + k_2 = 1$$
$$ck_2 = \frac{1}{2}$$
$$ak_2 = \frac{1}{2}$$

and then the two methods agree to within $O(h^3)$ and hence both have $O(h^2)$ global error. The system of Eq. (5.7) represents three *nonlinear* equations in four unknowns and has a meaningful solution for every $k_2 \in (0, 1]$. Usually when designing a method we are seeking to choose a convenient a and c and then accept whatever k_1 and k_2 follow.

There are several standard RK methods of order two among the infinitely many possibilities. They differ in matters of convenience and of course the specific constant appearing in the $O\left(h^2\right)$ error term. The most commonly seen ones are, in tableau form,

$$\begin{array}{c|cc} 0 & 0 & \\ 1 & \frac{1}{2} & \frac{1}{2} \end{array}$$

that is

$$y_i = y_{i-1} + hf(x_{i-1/2}, y_{i-1} + \frac{h}{2}f(x_{i-1}, y_{i-1}))$$

which we now recognize as the midpoint method; the improved Euler's method

$$\begin{array}{c|cc} \frac{1}{2} & 0 & \\ \frac{1}{2} & 1 & 1 \end{array}$$

and Heun's method

$$\begin{array}{c|cc} \frac{1}{4} & 0 & \\ \frac{3}{4} & \frac{2}{3} & \frac{2}{3} \end{array}$$

but should it be convenient for our purposes to choose an RK2 method that utilizies certain x_i values (where i may take on a non-integer value), we can now do so using Eq. (5.7). The **ode23** command, for example, uses the Bogacki-Shampine pair, an RK2 method with an RK3 method. Euler's method, of course, is the Taylor and RK method of order one.

There are other ways to derive Runge-Kutta methods. Many IRK methods are equivalent to interpolatory quadrature rules; many ERK methods are closely related to Gaussian quadrature rules. Some necessary conditions, such as that the sum of the entries in the ith row of A must equal c_i in order for an RK method to be at least $O(h)$, are known. Graph-theoretical methods are generally used for deriving high-order RK methods.

The global error of an RK method is $O(h^N)$, where N is the number of stages (and hence of function evaluations), *only* for $N \leq 4$. We need six stages to get an $O(h^5)$ method, seven stages to get an $O(h^6)$ method, and nine stages to get an $O(h^7)$ method. This partially explains the popularity of RK4 methods; fourth-order methods are the last ones where we have the order of the global error equal to the number of function evaluations. Additionally, RK methods generalize to systems in the obvious way only up to $N = 4$, after which more care must be exercised in finding the vectorized versions.

EXAMPLE 2: Let's use the classical fourth-order RK method on the ODE IVP $y' = x^2 y$, $y(0) = 1$ from Example 1. We have, with $h = .05$,

$$
\begin{aligned}
g_1 &= f(x_0, y_0) \\
&= x_0^2 y_0 \\
&= 0 \\
g_2 &= f(x_0 + \frac{1}{2}h, y_0 + \frac{1}{2}h g_1) \\
&= f(.025, 1) \\
&= (.025)^2 1 \\
&= 6.25E - 4 \\
g_3 &= f(x_0 + \frac{1}{2}h, y_0 + \frac{1}{2}h g_2) \\
&= f(.025, 1 + .025 \cdot 6.25E - 4) \\
&\doteq 6.2501E - 4 \\
g_4 &= f(x_0 + h, y_0 + h g_3) \\
&= f(.05, 1 + .025 \cdot 6.2501E - 4) \\
&\doteq .0025
\end{aligned}
$$

so

$$
\begin{aligned}
y_1 &= y_0 + h(\frac{1}{6}g_1 + \frac{1}{3}g_2 + \frac{1}{3}g_3 + \frac{1}{6}g_4) \\
&\doteq 1.00004166715495 \\
y(x_1) &\doteq 1.00004166753473 \\
\alpha_1 &= |y(x_1) - y_1| \\
&\doteq 3.8E - 10
\end{aligned}
$$

(note that $g_2 \neq g_3$). Compare this with the corresponding result for the second-order method in Example 1. □

The classical fourth-order RK method is not the fourth-order method in the MATLAB routine `ode45` and indeed is no longer in common use. But `ode45` does use a pair of fourth-order and fifth-order ERK methods in order to be able to estimate and hence control the error (by adjusting the step size h adaptively), just as `ode23` does with a pair of second-order and third-order ERK methods.

Finally, it's easy to get the impression that all numerical methods for solving ODEs are RK methods. This isn't true. The RK methods are the most widely employed methods in general but, for example, they tend to not be competitive with other methods in orbital mechanics computations.

MATLAB

Most of the methods we have seen so far generalize readily to methods for systems of ODEs. Runge-Kutta methods for systems are the same as Runge-Kutta methods for scalar equations provided the order of the RK method is at most four. Let's use Euler's method $\vec{y}_i = \vec{y}_{i-1} + hf(x_{i-1}, \vec{y}_{i-1})$ to approximate the solution of the system of ODEs $R' = (a - bF)R$, $F' = (-c + dR)F$ ($R = R(t)$, $F = F(t)$) where a, b, c, and d are positive (this is called the **predator-prey equation**). We'll take $a = 2$, $b = .05$, $c = 1$, $d = .02$, $R(0) = 100$, $F(0) = 20$, and $h = .1$. Enter:

```
>> h=.1;
>> y0=[100 20]';
>> f1=@(F,R) (2-.05*F)*R;      %R'=f1(F,R).
>> f2=@(F,R) (-1+.02*R)*F;     %F'=f2(F,R).
>> y1=y0+h*[f1(y0(2),y0(1));f2(y0(2),y0(1))]
>> y2=y1+h*[f1(y1(2),y1(1));f2(y1(2),y1(1))]
>> y3=y2+h*[f1(y2(2),y2(1));f2(y2(2),y2(1))]
>> x=[0 .1 .2 .3];
>> rabbits=[y0(1),y1(1),y2(1),y3(1)];
>> foxes=[y0(2),y1(2),y2(2),y3(2)];
>> plot(x,y),legend('Rabbits','Foxes','Location','NorthWest')
```

The predator-prey equation models the interaction of populations of prey $R(t)$, which we might imagine as rabbits, and predators $F(t)$, which we might imagine as foxes, starting from some initial population level y_0. Repeat this with the improved Euler's method.

Let's try to get a better look at this solution. First we'll write a single function for f that returns a vector. Enter:

```
>> f=@(t,x) 0*t+[(2-.05*x(2))*x(1);(-1+.02*x(1))*x(2)]
>> [xx,yy]=ode45(f,[0 30],y0);
>> plot(xx,yy),legend('Rabbits','Foxes'),grid
```

The population levels of the rabbits and foxes appear to be oscillating. This behavior is common and can be seen in actual populations. Let's look at a phase plane plot. Enter:

```
>> plot(yy(:,1),yy(:,2))
```

The plot is as expected for a periodic solution. Enter:

```
>> close all
>> options=odeset('outputfcn','odephas2')
>> [xx,yy]=ode45(f,[0 20],y0,options);
```

to have MATLAB generate the phase plane plot. Make a phase portrait for this system with six different trajectories by using `hold`.

Problems

1.) a.) Write out the RK tableau $k|c|A$ where $k = (1/4, 3/8, 3/8)^T$, $c = (0, 2/3, 2/3)^T$, and $a_{2,1} = 2/3$, $a_{3,2} = 2/3$, and $a_{i,j} = 0$ otherwise. Then write out the method in the form of Eq. (5.4). This method is called the **Nystrom scheme** and is third-order.

b.) Use the Nystrom scheme with $h = .5$ to approximate the solution of $y' = y^2$, $y(0) = 1$, over $[0, 1]$. Compare your answers at $t = .5$ and at $t = 1$ with the true solution and with your answers for Problem 1 of Sec. 6.2 and Problem 1 of Sec. 6.3, where $h = .05$ was used.

c.) Repeat part b. using the classical fourth-order RK method.

2.) a.) Continue Example 1 for three more steps. Find the absolute errors.

b.) Continue Example 2 for five more steps. Find the absolute errors and compare with the Taylor method of order two.

3.) a.) Repeat Example 2 with the improved Euler's method using $h = .025$ to get an approximation of $y(x_1)$. This uses four function evaluations, the same number as the classical fourth-order RK method in Example 2 used to get an approximation of $y(x_1)$. Which method is better?

b.) Repeat part a. using the midpoint and Heun's methods.

4.) Write a MATLAB program that accepts vectors k and c and a matrix A and prints out the tableau of the corresponding ERK or IRK method. Your program should label the tableau, giving the number of stages and stating whether the method is explicit or implicit.

5.) Write a MATLAB program that implements the classical fourth-order RK method.

6.) a.) Show that the theta methods are Runge-Kutta methods. What are the corresponding Butcher tables?

b.) Show the tables for the forward and backward Euler's method in particular.

7.) a.) Verify that all of the RK methods we have seen satisfy the necessary condition that the sum of the entries in the ith row of A does equal c_i.

b.) Show that if the sum of the entries in the ith row of A does not equal c_i then the corresponding RK method cannot be of order at least one. You may want to show that it does not integrate some specific, simple ODE $y' = f(x, y)$ to this accuracy.

8.) a.) Write out the RK tableau $k|c|A$ where $k = (1/6, 2/3, 1/6)^T$, $c = (0, 1/2, 1)^T$, and $a_{2,1} = 1/2$, $a_{3,1} = -1$, $a_{3,2} = 2$, and $a_{i,j} = 0$ otherwise. Then write out the method in the form of Eq. (5.4). This method is called the **classical third-order Runge-Kutta method** and is indeed third-order.

b.) Use the classical third-order RK method with $h = .5$ to approximate the solution of $y' = y^2$, $y(0) = 1$, over $[0, 1]$. Compare your answers with the true solution and with your answers for Problem 1 of Sec. 6.2, Problem 1 of Sec. 6.3, and Problem 1 of this section.

c.) Prove that this method is $O(h^3)$.

d.) Repeat parts a., b., and c. using **Ralston's Runge-Kutta method** which has the RK tableau $k|c|A$ where $k = (2/9, 1/3, 4/9)^T$, $c = (0, 1/2, 3/4)^T$, and $a_{2,1} = 1/2$, $a_{3,2} = 3/4$, and $a_{i,j} = 0$ otherwise. This is the RK3 method of ode23.

9.) a.) Show that if $f(x, y) = f(x)$ then the classical 4th order RK method reduces to Simpson's rule.

b.) Show that if $f(x, y) = f(x)$ then an RK method is a quadrature rule. What is the relationship between the order of the RK method and the degree of accuracy of the quadrature rule?

10.) Show that the implicit midpoint rule $y_{i+1} = y_i + hf(x_{i+1/2}, (y_i + y_{i+1}))$ is an IRK method. What is its tableau?

11.) Show that the solution of $y' = \alpha y$ ($\alpha \in \mathbb{R}$) by an RK method with N stages is $y_i = y_0 \left(\sum_{j=0}^{N} (h\alpha)^j / j! \right)^i$, $i = 0, 1, \ldots$ Comment.

12.) The IRK method with tableau $k|c|A$ where $k = (1/2, 1/2)^T$, $c = (1/2 - \sqrt{3}/6, 1/2 + \sqrt{3}/6)^T$, and $a_{1,1} = 1/4$, $a_{1,2} = 1/4 - \sqrt{3}/6$, $a_{2,1} = 1/2 + \sqrt{3}/6$, and $a_{2,2} = 1/4$ is fourth-order. Write a MATLAB program that implements this method. You will need to solve a nonlinear system at each step; what is the natural choice for an initial guess for the nonlinear system solver? Test your program.

13.) Conduct an experiment to verify that the classical fourth-order RK method is indeed of order four.

14.) Solve the predator-prey equation as given in the MATLAB subsection using the classical fourth-order Runge-Kutta method. Produce a times series plot and a phase plot.

15.) Show that the requirements of Eq. (5.7) are unchanged if we start with a system of two ODEs and expand both components of g_2 in Taylor series in order to find a RK method.

16.) a.) Write a MATLAB program that implements the Taylor method of order two. Assume that f, f_x, and f_y will all be provided by the user.

b.) Modify your program to accept f as a symbolic expression and use the `diff` command within your program to find f_x and f_y.

6. Adaptivity and Stiffness

The numerical methods we have seen so far form the basis for programs that attempt to solve ODEs automatically, but there are many more issues to consider as we look at the efficient implementation of an ODE solver in a useful and robust program. Users will typically wish to simply submit an ODE and have returned a list of nodes x and corresponding approximate function values y that can be plotted, splined, numerically integrated or differentiated, searched for maxima and minima, and so on. To make this a reality, we need ways of estimating the error so that the nodes may be chosen adaptively[9].

Most programs insist that the ODE be given as a first-order system $y' = f(x, y)$, $y(x_0) = y_0$, where y may be a vector-valued quantity. That means that a higher-order ODE IVP such as

(6.1)
$$\begin{aligned} y'' + by' + cy &= f(x) \\ y(x_0) &= y_0 \\ y'(x_0) &= v_0 \end{aligned}$$

must be converted into a first-order system by the user. (Less often, the software will do so automatically before attempting to solve the problem.) To convert Eq. (6.1) to an equivalent first-order system we let $v(x) = y'(x)$, giving

$$\begin{aligned} v' + bv + cy &= f(x) \\ y(x_0) &= y_0 \\ v(x_0) &= v_0 \end{aligned}$$

that is

---

[9]This isn't all we need, but it's what we'll focus on in this section.

$$y' = v$$
$$v' = -cy - bv + f(x)$$
$$y(x_0) = y_0$$
$$v(x_0) = v_0$$

which is indeed a first-order system in the unknown functions $y(x)$ and $v(x)$. This system may now be handled by the methods of the MATLAB subsection of Sec. 6.5. In this case the system is linear so we might also use an eigenvalue/eigenvector routine on the matrix of the system, but something like `ode45` is typically significantly more efficient.

How shall we choose the nodes adaptively? We'll look only at the scalar case $y' = f(x,y)$, $y(x_0) = y_0$. We want to use a numerical method to generate an estimate y_1 of the value of y at x_1. We'll approximate $y(x_1)$ twice, using two different methods, and use them to approximate the error. If the error is too large we'll reduce h by half and approximate $y(x_{1/2})$ then, when it's accurately approximated, we'll go from there to get an estimate for $y(x_1)$. If we have to reduce h several times to get an accurate estimate of $y(x_{1/2})$ first, so be it. We continually reduce h by one-half and take more steps if the error is too large.

If on the other hand the error is much smaller than the tolerance then we are using a step that is inefficiently small (for the given error tolerance). In such a case we would double h and take longer, and hence more efficient, step sizes. Typically we would have minimum and maximum allowable step sizes specified.

EXAMPLE 1: Let's see how this would work if we had a way to find the exact local truncation error. Suppose we are solving $y' = f(x,y)$, $y(0) = 1$ with an initial $h = .2$ and an error tolerance $\epsilon = 10^{-6}$ for the LTE at each node. We estimate $y_1 \approx y(.2)$ using two different methods and find that the LTE is about 10^{-4} so we reduce h to $h = .1$ and estimate $y(.1)$. We find that the LTE is about 10^{-5} so we reduce h again to $h = .05$ and estimate $y(.05)$. The error is estimated as 10^{-6} so we have a successful step. We then estimate $y(.1)$. The error is estimated as 10^{-6} so we have a successful step. We estimate $y(.15)$ and find that the error is estimated as 10^{-8}; we have a *very* successful step so we assume that the region $[0, .1]$ was the difficult region and try to take larger steps. We double h to $h = .1$ and estimate $y(.25)$. The error is estimated as 10^{-6} so we have a successful step and move on to estimating $y(.35)$. Note that we never did find an estimate for $y(.2)$ but that we could always use interpolation to find such an estimate. □

Although we are controlling the local truncation error, the global error can be shown to be at most $(x_n - x_0)\epsilon$ if the local truncation error is kept to at most $h\epsilon$. Keeping the LTE at most $h\epsilon$ is called **error control per unit step**, as opposed to **error control per step** which is keeping the LTE at most ϵ (as was done in Example 1). Error control per unit step is generally preferred; the smaller h is, the more steps we are taking, and hence the greater the accumulation of the global error, so when we are taking many small steps we should use a tighter restriction than we are taking fewer large steps.

We need a pair of methods that can be used to estimate the error. Ideally they would use the same nodes, or nearly the same, since controlling the error is only one goal and efficiency is the other. It's no savings to double our computational effort in the hopes that we can double our step size.

We describe such an approach as the use of **paired methods**. For RK methods they are called **embedded RK pairs**. One example is the **Runge-Kutta-Fehlberg method**[10] which, in its original form, uses a pair of fourth-order and fifth-order methods that together require only six function evaluations rather than the ten that would be necessary for unpaired methods. The fourth-order method is

$$
(6.2) \qquad
\begin{array}{c|ccccc}
\frac{25}{216} & 0 \\
0 & \frac{1}{4} & \frac{1}{4} \\
\frac{1408}{2565} & \frac{3}{8} & \frac{3}{32} & \frac{9}{32} \\
\frac{2197}{4104} & \frac{12}{13} & \frac{1932}{2197} & \frac{-7200}{2197} & \frac{7296}{2197} \\
\frac{-1}{5} & 1 & \frac{439}{216} & -8 & \frac{3680}{516} & \frac{-845}{4104}
\end{array}
$$

and the fifth-order method is

$$
(6.3) \qquad
\begin{array}{c|cccccc}
\frac{16}{135} & 0 \\
0 & \frac{1}{4} & \frac{1}{4} \\
\frac{6656}{12825} & \frac{3}{8} & \frac{3}{32} & \frac{9}{32} \\
\frac{28561}{56430} & \frac{12}{13} & \frac{1932}{2197} & \frac{-7200}{2197} & \frac{7296}{2197} \\
\frac{-9}{50} & 1 & \frac{439}{216} & -8 & \frac{3680}{516} & \frac{-845}{4104} \\
\frac{2}{55} & \frac{1}{2} & \frac{-8}{27} & 2 & \frac{-3544}{2565} & \frac{1859}{4104} & \frac{-11}{40}
\end{array}
$$

and the tableau are ordered in such a way as to highlight the redundancies. Given the number of entries that are the same in tableaus Eq. (6.2) and Eq. (6.3), we often represent an embedded RK pair by a single tableau $k4|k5|c|A$ where $k4$ is the vector of weights for the fourth-order method, $k5$ is the vector of weights for the fifth-order method, c is the vector of nodes from Eq. (6.3), and A is the RK matrix from Eq. (6.3). We use the weights $k4$ and ignore the last row of the tableau for the fourth-order method, and the weights $k5$ and all of c and A for the fifth-order method. The negative values in the A matrix are unfortunate but we will accept them.

Look again at the tableaus in Eq. (6.2) and Eq. (6.3). Except for the last row, they use the same A matrix and the same nodes and hence they use exactly the same function evaluations. The fourth-order method uses 5 function evaluations–one more than needed in principle–and the fifth-order method uses those same function evaluations plus an additional one function evaluation. Since 6 function evaluations are needed to get a fifth-order ERK method, this can't be improved upon if we desire a pair of orders 4 and 5. This is very desirable: We effectively get the fourth-order estimates for "free" given that we're computing the fifth-order ones. Of course, since the latter are intended to provide a check on the quality of the former, this is an awkward way of viewing the matter; but regardless, we are using the minimum number of function evaluations needed.

How exactly do we make use of this? The LTE of Eq. (6.2) is $O(h^5)$ since the global error is $O(h^4)$. Call the approximations generated by this method y_i as it is the main method, that is, the one whose error we are controlling with the higher-order method. Call the approximations generated by Eq. (6.3) v_i; their

---

[10]The term *Runge-Kutta-Fehlberg method* is nowadays often used to describe any pair of ERK methods that differ in order by one and use nearly the minimum number of function evaluations required for the higher-order method.

LTE is (at least) $O(h^6)$. Now RK methods are one-step methods so the LTE of the lower-order method is

$$
\begin{aligned}
LTE4_{i+1} &= y(x_{i+1}) - [y(x_i) + h\Phi(x_i, y(x_i), h)] \\
&\approx y(x_{i+1}) - [y_i + h\Phi(x_i, y_i, h)] \\
&= y(x_{i+1}) - y_{i+1}
\end{aligned}
$$

to $O(h^5)$, as $y(x_i) = y_i + O(h^5)$; and the LTE of the higher-order method is

$$
\begin{aligned}
LTE5_{i+1} &= y(x_{i+1}) - [y(x_i) + h\Phi(x_i, y(x_i), h)] \\
&\approx y(x_{i+1}) - [v_i + h\Phi(x_i, v_i, h)] \\
&= y(x_{i+1}) - v_{i+1}
\end{aligned}
$$

$O(h^6)$, as $y(x_i) = v_i + O(h^6)$. Hence

$$
\begin{aligned}
LTE4_{i+1} &\approx y(x_{i+1}) - y_{i+1} \\
&= y(x_{i+1}) - v_{i+1} + v_{i+1} - y_{i+1} \\
&\approx LTE5_{i+1} + (v_{i+1} - y_{i+1}) \\
&\approx v_{i+1} - y_{i+1}
\end{aligned}
$$

if $LTE5_{i+1}$ is small compared with $LTE4_{i+1}$. Therefore, we may estimate the local truncation error of the main method as

$$
(6.4) \qquad LTE4_{i+1} \approx v_{i+1} - y_{i+1}
$$

and compare this with our tolerance ϵ or, more commonly, with $h\epsilon$ (error control per unit step). This is the missing piece needed to use adaptivity. A shorter way of saying this is that Eq. (6.4) is justified by the fact that $y(x_{i+1}) = v_{i+1} + O(h^6)$ making it a much more accurate estimate, so we may substitute it in for $y(x_{i+1})$ in the definition of the LTE for the lower-order method.

We discussed using this for step-size control by doubling or halving the step size. A more sophisticated approach can be used. Since the LTE of the lower-order method is $O(h^5)$, if h is sufficiently small then

$$
LTE4(h)_{i+1} \approx Ch^5
$$

for some constant C. Hence the LTE for step size h/m is approximately

$$
\begin{aligned}
LTE4(h/m)_{i+1} &\approx C(h/m)^5 \\
&= \frac{LTE4(h)_{i+1}}{m^5}
\end{aligned}
$$

and so if the local truncation error is too large or too small then we may use this formula to compute a new step size h/m, $m > 0$, such that the LTE is likely to be acceptable. However, the doubling and halving approach is what is commonly employed.

Although it is the LTE of the fourth-order methods that is being controlled, most programs will report the global error, estimated as ϵ times the length of the interval over which we have integrated the ODE. Global error is usually the error we want to know about.

This approach estimates and controls the error in the fourth-order method. We do not have an error estimate for the fifth-order method's results. Unsurprisingly, in practice we have the program return the approximations from the higher-order method regardless, as was done for numerical integration; in this context this is referred to as **local extrapolation**. A purist might object that we are reporting the wrong error values for the quantities we're actually returning, but if the higher-order estimates aren't more accurate our entire premise was flawed anyway. Additionally, we may view local extrapolation as simply error correction of the fourth-order method.

The `ode45` command uses the RKF idea but actually implements a scheme called the **Dormand-Prince method**, a.k.a. the **RKDP** or **DOPRI** method, which has 7 stages but still uses only 6 function evaluations per solution estimate because the method has the **First-Same-As-Last** (**FSAL**) property, namely, that the last function evaluation used in step i becomes the first function evaluation needed for step $i + 1$. (The `ode23` command also uses an FSAL method, the **Bogacki-Shampine** (2, 3) pair. The `ode23s` command does not perform local extrapolation.) The original RKF method, by Fehlberg, attempted to minimize the error in the fourth-order method for which we have an error estimate, while the Dormand-Prince method is designed to reduce the error in the fifth-order method whose values we will actually be returning (due to the use of local extrapolation)[11]. The **Cash-Karp method** (**RKCK**) is similar.

There are many other implementation issues we could discuss–how to choose the initial step size being the most obvious open question–and hopefully you see that these programming issues require mathematical understanding, intuition, and analysis. However, there is another important topic that we must now discuss and it is related to the issue of choosing a step size. In the Problems you'll be asked to solve the linear system

$$(6.5) \qquad \begin{aligned} x' &= -10x + y \\ y' &= -y \end{aligned}$$

($x(0) = 1$, $y(0) = 1$) using Euler's method with three different step sizes $h = .1, .15, .2$ and to plot the first component x of the numerical solution along with the first component of the true solution. At $h = .1$ all seems well. For $h = .15$ the numerical method seems to "think about" oscillating before settling in to the correct long-term behavior; if all we need is the steady-state solution, not the transient, this is acceptable. But for $h = .2$ there are spurious oscillations that persist indefinitely. For sufficiently small h we get the correct behavior, as expected for a stable, convergent method. But for slightly larger h we have instability?

The complete answer is a bit more involved than that, however. Euler's method for this linear system of ODEs is a linear difference equation in two variables

$$\vec{y}_{i+1} = \vec{y}_i + hA\vec{y}_i$$

and it can be solved to give

---

[11]Erwin Fehlberg is said to have been strongly opposed to local extrapolation, explaining why his approach minimized the fourth-order method's error.

(6.6) $$\vec{y}_i = (1 - 10h)^i v_1 + (1 - h)^i v_2$$

for some vectors v_1, v_2. (Perhaps unsurprisingly, these vectors are the eigenvectors of A, just as in the analytical solution of the system of ODEs itself.) That means that if we are to have the *numerical* solution display the correct asymptotic behavior then we need

$$|1 - 10h| \leq 1$$
$$|1 - h| \leq 1$$

and the more restrictive condition is

$$|1 - 10h| \leq 1$$
$$h \leq \frac{1}{5}$$

(as $h > 0$). Indeed, $h = .2$ is a border case. We don't get convergence but instead get period two behavior asymptotically because the term

$$(1 - 10h)^i$$

alternates between 1 and -1; that is, we don't get convergence but we don't get divergence to infinity either. Things will steadily degrade as h increases past .2.

The system of Eq. (6.5) has a requirement that $h < .2$ for satisfactory answers. But what if we don't want to take such a small h? What if we need to approximate the answer for very large x and want to take large steps to save computing time? We can't. Systems such as this are referred to as **stiff** systems. There's no strict mathematical definition of a stiff system; they are systems that contain widely varying rates (as -10 and -1 here) and which therefore force us to choose a smaller than desired h based on stability considerations. Note that if the matrix of the linear system had been

(6.7) $$\begin{pmatrix} -1000 & 1 \\ 0 & -1 \end{pmatrix}$$

instead then we'd have needed $h < .002$ for stability. We'd ideally like to choose our h based only on considerations of accuracy and computational time, but a stiff system forces us to use a small h, and hence spend a lot of time solving, in order to avoid instability. And this is a *linear* system! What about a large, nonlinear system with rates varying over many orders of magnitude? These occur all the time in practice–the name itself comes from an early application involving mechanical springs of differing degrees of stiffness–and would wreak havoc with the doubling-and-halving system discussed earlier. We might have to keep halving and halving and halving the step size until we reached whatever minimum step size we had set, at which point we might still not have a sufficiently small step size. Stiff systems can be very frustrating.

What can we do? Well, we have not mentioned an important point: Stiffness is a property of systems, not numerical methods, but not every numerical method is affected by it in the way that Euler's method is. The backward Euler's method

(6.8) $$y_{i+1} = y_i + hf(x_{i+1}, y_{i+1})$$

and trapezoidal scheme, for instance, reproduce the asymptotic behavior of Eq. (6.5) for any step size. We can devise special methods for stiff systems, and use them to create special routines called **stiff solvers** for such systems. Backward differentiation and implicitness often figure into stiff solvers. (Implicit methods may be made adaptive in the same manner discussed earlier in this section.) Another method appropriate for stiff systems is

(6.9) $$y_{i+2} - \frac{4}{3}y_{i+1} + \frac{1}{3}y_i = \frac{2}{3}f(x_{i+2}, y_{i+2})$$

which again is implicit and can be viewed as the result of backward differencing[12] In fact, Eq. (6.8) and Eq. (6.9) are the first- and second-order **backward differentiation formulas** (**BDF**), a class of methods that use successively higher-order backward finite difference approximations to y' in order to generate implicit methods. The methods in Eq. (6.8) and Eq. (6.9) are sometimes designated the **BDF1** and **BDF2** methods, respectively.

So we see an important reason why implicit methods are needed: Explicit methods frequently require very small step sizes for stiff systems whereas implicit methods may allow us to take step sizes as large as desired, compatible with our accuracy needs and other considerations. Of course, we pay a price in that we must solve a nonlinear system at each step.

Adaptivity is essential for the accurate and timely solution of stiff systems. It's also useful for nonstiff systems, and every professionally written ODE solver is adaptive. Writing such programs requires teams of people with expertise in a number of areas–applied mathematics, computer science and/or engineering, and often the area of scientific inquiry to which the program is to be applied–and for this reason it is advisable whenever possible to use professionally written, time-proven code and avoid the temptation to write your own programs for scientific computation.

MATLAB

Let's look at the MATLAB stiff solvers: `ode15s`, `ode23s`, `ode23t` (for moderately stiff problems) and `ode23tb`. Enter:
>> `helpwin`
to launch the help window, and then select the `funfun` directory. Scroll down to the ODE solvers. Select each of the four solvers listed above and read the help.[13]

The `ode23t` routine uses the trapezoidal scheme paired with a third-order method for error control. The `ode23s` routine uses a variant of IRK methods due to Rosenbrock and is based on a pair of methods of order two and three. It does not use local extrapolation. The `ode23tb` routine uses the trapezoidal scheme for part of its steps and a variant of BDF2 for the remainder and is an IRK method. The `ode15s` routine uses formulas of orders one through five based on backward differentiation.

---

[12]See Eq. (6.1.1.3) and Problem 7 of that section.

[13]You'll frequently see the name Larry Shampine, one of the key figures in modern computational methods for ODEs.

To see what a difference stiffness can make, let's compare the MATLAB routines ode23 (for nonstiff problems) and ode23s (for stiff problems). Enter:
>> more on
>> help ode23s
and read the help. Let's use the stiff system suggested in the help text, known as the **Van der Pol system** $v' = w$, $w' = 1000w(1 - v^2) - v$. Enter:
>> tic;[t1,y1]=ode23s(@vdp1000,[0 3000],[2 0]); toc
>> plot(t1,y1(:,1));
Now let's try again using a nonstiff solver; enter:
>> tic;[t2,y2]=ode23(@vdp1000,[0 3000],[2 0]); toc
>> plot(t2,y2(:,1));
In all likelihood you'll want to use CTRL-C to interrupt the computation (and will not get to the plot). The nonstiff solver must take excruciatingly small steps to meet the error tolerance, but the stiff solver has no problem meeting it. This is the advantage of an implicit method over an explicit method, and perhaps finally answers the question as to why we bother to use implicit methods at all.

Let's try again with the higher-order explicit method ode45 and with the rec-ommended basic stiff solver ode15s (which uses a variable order method, from one to five, as well as a variable step size); enter:
>> tic;[t3,y3]=ode15s(@vdp1000,[0 3000],[2 0]); toc
>> tic;[t4,y4]=ode45(@vdp1000,[0 3000],[2 0]); toc
Plot the results in each case. The stiff solver is considerably more efficient–again, the explicit method may well have failed to finish in a reasonable time in this example, and so you may never get to the plot. If so, experiment until you find a tolerance large enough that the explicit method does succeed, then plot the results.

The standard advice is to try ode45 first and then, if the method is performing poorly, to try the stiff solver ode15s. Of course, if you know in advance that the system is stiff then you should go directly to a stiff solver.

We haven't talked about all of the MATLAB ODE routines. The ode113 solver, for example, is a multi-step, variable-order, non-RK method that uses order 1 through 12 (with a 13th-order method available internally for error control) in order to efficiently solve nonstiff problems with especially small error tolerances or especially expensive function evaluations.

Problems

1.) Consider Eq. (6.5). As discussed in the text, solve it with $h = .1$, $.15$, and $.2$. Plot the numerical solutions against the true solution for $0 \le t \le 10$. Discuss your results.
b.) Repeat part a. with $h = .21$, $.22$, and $.25$.
2.) Consider Eq. (6.5). Attempt to solve it with Euler's method and error control per step, keeping the error at most $.01$. (Use the true solution to find the error.) Start with $h = .25$ and double or halve h as appropriate. Discuss what happens.
b.) Repeat part a. with the improved Euler's method and error control per step.
c.) Repeat part a. with the trapezoidal scheme.
3.) Prove that Eq. (6.6) is correct. Find v_1 and v_2.
4.) Use the backward Euler's method with $h = .1$ to solve $y' = x^2 y$, $y(1) = 1$ over $[1, 3]$. Compare your answers to those generated by Euler's method, which is also first-order.

5.) Write a detailed algorithm (pseudo-code) for performing the Runge-Kutta-Fehlberg method with error control per unit step.

6.) Repeat Problem 4 using the BDF2 method Eq. (6.9), started with the improved Euler's method. Compare your answers to those generated by the improved Euler's method, which is also second-order.

7.) a.) Use the backward Euler's method with various values of h to solve Eq. (6.5). Compare your answers to those generated by Euler's method, which is also first-order.

b.) Repeat part a. using the BDF2 method, started with the improved Euler's method. Compare your answers to those generated by the improved Euler's method, which is also second-order.

8.) Write a MATLAB program for performing the Runge-Kutta-Fehlberg method with error control per step or error control per unit step at the user's discretion. Test your program.

9.) Solve $y'' + 2y' + 2y = \sin(x)$, $y(0) = 1$, $y'(0) = -1$ using the improved Euler's method.

10.) a.) Repeat Problem 1 but change the A matrix to Eq. (6.7).

b.) Repeat Problem 2 but change the A matrix to Eq. (6.7).

11.) a.) Use the Runge-Kutta-Fehlberg method to solve $y' = 1 + (x-y)^2$, $y(0) = 1$ over $[0,1]$. Use $\epsilon = 10^{-6}$ and error control per unit step. Compare your answers to the true solution.

b.) Repeat part a. over $[1.5, 3]$ using the initial condition $y(1.5) = 1$.

12.) Repeat Problem 11 using error control per step. Comment.

13.) Compare and contrast the approaches taken in the design of the various MATLAB ODE solvers and discuss which solvers are appropriate for which types of problems.

14.) a.) Derive the BDF2 method using a backward difference formula.

b.) Derive a third-order BDF formula by first deriving a method for backward differencing that is third-order (see Sec. 6.1) then using it in $y' = f(x, y)$.

15.) Show that the ERK methods in the Runge-Kutta-Fehlberg scheme have the orders claimed.

16.) Prove that an RK method that uses only 5 function evaluations has global error at most $O\left(h^4\right)$, not $O\left(h^5\right)$.

7. Multi-step Methods

We have been focusing on one-step methods. There is another important class of methods, and we have seen several examples of these methods already: Multi-step methods. These were defined in Sec. 6.4 as methods of the form

$$(7.1) \qquad y_{i+1} = \sum_{j=0}^{j=k-1} \alpha_j y_{i-j} + h\Phi_f(x_i, y_i, y_{i-1}, ..., y_{i-k+1}, f, h)$$

(explicit multi-step or explicit k-step methods), or

(7.2) $\qquad y_{i+1} = \displaystyle\sum_{j=0}^{j=k-1} \alpha_j y_{i-j} + h\Phi_f(x_i, y_{i+1}, y_i, y_{i-1}, ..., y_{i-k+1}, f, h)$

(implicit multi-step or implicit k-step methods). The BDF2 formula

$$y_{i+2} - \frac{4}{3}y_{i+1} + \frac{1}{3}y_i = \frac{2}{3}f(x_{i+2}, y_{i+2})$$
$$y_{i+2} = \frac{4}{3}y_{i+1} - \frac{1}{3}y_i + \frac{2}{3}f(x_{i+2}, y_{i+2})$$

from the previous section is an implicit multi-step method with $k = 2$ steps. All multi-step methods in common use have an increment function of the form

(7.3) $\qquad \Phi_f(x_i, y_i, y_{i-1}, ..., y_{i-k+1}, f, h) = \displaystyle\sum_{j=0}^{j=k-1} \beta_j f(x_{i-j}, y_{i-j})$

(explicit) or

(7.4) $\qquad \Phi_f(x_i, y_i, y_{i-1}, ..., y_{i-k+1}, f, h) = \displaystyle\sum_{j=0}^{j=k} \beta_j f(x_{i+1-j}, y_{i+1-j})$

(implicit), where $x_{i-j} = x_i - jh$ as usual. We will only consider multi-step methods of the form in Eq. (7.1) with Eq. (7.3) or of the form in Eq. (7.2) with Eq. (7.4) and we also insist that α_j, β_j depend only on j (not f, etc.). Any such method is called a **linear multi-step method (LMM)**.

The lure of methods with memory is easy to see: We have already found $y_0, .y_1, ..., y_i$ when we come to the task of finding y_{i+1}. Why not use some or all of the previous values to help extrapolate the next value? They represent additional information that could be used to find a trend in the solution. From this point of view, one-step methods seem restrictive. After all, it's likely that we are storing all these values anyway[14].

How shall we determine α_j and β_j ($j = 0, ..., k - 1$, plus β_k if the method is implicit)? As is so often the case, interpolation is the obvious approach, though it is not the only one. (Another natural approach, leading to the same methods, is to replace y' in $y' = f(x, y)$ by an approximation that is based on several points, such as that of Eq. (1.3).) Let's look at the explicit case first. We write

(7.5) $\qquad y(x_{i+1}) = y(x_i) + \displaystyle\int_{x_i}^{x_{i+1}} f(x, y(x))dx$

and approximate the integrand by an interpolating polynomial at the k nodes $x_{i-k+1}, ..., x_i,$

---

[14]This isn't always the case. It's not uncommon to need only the steady-state solution. But we could still store enough previous values to use an LMM.

$$f(x, y(x)) \approx \sum_{j=0}^{k-1} f(x_{i-j}, y(x_{i-j})) L_j(x)$$

(7.6)
$$\approx \sum_{j=0}^{k-1} f(x_{i-j}, y_{i-j}) L_j(x)$$

in terms of the Lagrange interpolating polynomials, where we are using the fact that $y_m \approx y(x_m)$. Not all of the nodes fall within the interval of integration in Eq. (7.5), but that's fine; the interpolant is defined over this interval of integration and its behavior is influenced by the nodes outside this interval (compare the semi-simp quadrature rule). This gives a numerical method

$$
\begin{aligned}
y_{i+1} &= y_i + \int_{x_i}^{x_{i+1}} \sum_{j=0}^{k-1} f(x_{i-j}, y_{i-j}) L_j(x) dx \\
&= y_i + \sum_{j=0}^{k-1} f(x_{i-j}, y_{i-j}) \int_{x_i}^{x_{i+1}} L_j(x) dx \\
&= y_i + h \sum_{j=0}^{k-1} \beta_j f(x_{i-j}, y_{i-j})
\end{aligned}
$$

where

(7.7)
$$\beta_j = \frac{1}{h} \int_{x_i}^{x_{i+1}} L_j(x) dx$$

($j = 0, 1, ..., k-1$) are the coefficients. This is a method of the desired form (Eq. (7.1) with increment function given by Eq. (7.3)) and is called the **Adams-Bashforth method** of order k (or the k-**step Adams-Bashforth method**, or simply **AB**k), and is indeed of order k for sufficiently smooth f. For $k = 1$ this reduces to Euler's method (a one-step method). In general, however, these are not Runge-Kutta methods. We emphasize that the Adams-Bashforth methods are not the only explicit k-step methods; selecting the form Eq. (7.3) and then finding the coefficients β_j by interpolation was a choice, not a necessity.

It's not difficult to work out the explicit forms for small values of k. For instance, the $k = 2$ Adams-Bashforth method is

$$y_{i+1} = y_i + h \left(\frac{3}{2} f(x_i, y_i) - \frac{1}{2} f(x_{i-1}, y_{i-1}) \right)$$

while for $k = 3$ we have

$$y_{i+1} = y_i + h \left(\frac{23}{12} f(x_i, y_i) - \frac{4}{3} f(x_{i-1}, y_{i-1}) + \frac{5}{12} f(x_{i-2}, y_{i-2}) \right)$$

and for $k = 4$ we have

$$y_{i+1} = y_i + h \left(\frac{55}{24} f(x_i, y_i) - \frac{59}{24} f(x_{i-1}, y_{i-1}) + \frac{37}{24} f(x_{i-2}, y_{i-2}) - \frac{9}{24} f(x_{i-3}, y_{i-3}) \right)$$

and higher-order methods may be found using Eq. (7.7). Note that despite the fact that some of the coefficients are negative, the coefficients of the function values do all sum to unity. In this sense they remain variations on the basic Euler's method idea of using

$$y_{i+1} = y_i + hf(x_i, y_i)$$

but with $f(x_i, y_i)$ replaced with a weighted average of other function values in order to get a better slope to use in the formula.

The Adams-Bashforth methods are explicit. We also say that they are open, since the formula for approximating the integral

$$\int_{x_i}^{x_{i+1}} f(x, y(x)) dx$$

does not use x_{i+1} as a node. (It does use x_i so half-open would probably be a more accurate description.) The corresponding implicit, or closed, methods are derived in the same way save that Eq. (7.6) becomes

$$f(x, y(x)) \approx \sum_{j=0}^{k} f(x_{i+1-j}, y_{i+1-j}) L_j(x).$$

We then proceed as before:

$$
\begin{aligned}
y_{i+1} &= y_i + \int_{x_i}^{x_{i+1}} \sum_{j=0}^{k} f(x_{i+1-j}, y_{i+1-j}) L_j(x) dx \\
&= y_i + \sum_{j=0}^{k} f(x_{i+1-j}, y_{i+1-j}) \int_{x_i}^{x_{i+1}} L_j(x) dx \\
&= y_i + h \sum_{j=0}^{k-1} \beta_j f(x_{i+1-j}, y_{i+1-j})
\end{aligned}
$$

where

(7.8)
$$\beta_j = \frac{1}{h} \int_{x_i}^{x_{i+1}} L_j(x) dx$$

$(j = 0, 1, ..., k)$ are the coefficients. This is a method of the desired form (Eq. (7.2) with increment function given by Eq. (7.4)) and is called the **Adams-Moulton method** of order k (or the k-step **Adams-Moulton method**, or **AMk**), and is of order k for sufficiently smooth f.

Here, the $k = 1$ case is the backward Euler's method (a one-step method) and $k = 2$ is the trapezoidal method. For $k = 3$ we have the Adams-Moulton method of order three,

$$y_{i+1} = y_i + h \left(\frac{5}{12} f(x_{i+1}, y_{i+1}) + \frac{2}{3} f(x_i, y_i) - \frac{1}{12} f(x_{i-1}, y_{i-1}) \right)$$

while for $k = 4$ we have

$$y_{i+1} = y_i + h\left(\frac{9}{24}f(x_{i+1}, y_{i+1}) + \frac{19}{24}f(x_i, y_i) - \frac{5}{24}f(x_{i-1}, y_{i-1}) + \frac{1}{24}f(x_{i-2}, y_{i-2})\right)$$

and for $k = 5$ we have $y_{i+1} = y_i + h\Phi$ where the increment function is

$$\Phi = \frac{251}{720}f(x_{i+1}, y_{i+1}) + \frac{323}{360}f(x_i, y_i) - \frac{11}{30}f(x_{i-1}, y_{i-1}) + \frac{53}{360}f(x_{i-2}, y_{i-2}) - \frac{19}{720}f(x_{i-3}, y_{i-3})$$

and higher-order methods may be found using Eq. (7.8). Adams-Bashforth methods and Adams-Moulton methods are collectively referred to as **Adams methods**.

An Adams-Moulton method of order k will generally give better results than the corresponding Adams-Bashforth method but of course at greater computational cost due to the need to solve a nonlinear equation (or, more generally, system) at each step, usually by functional iteration–essentially the predictor-corrector idea–or possibly Newton's method. More commonly one would use the Adams-Bashforth method as a predictor and the Adams-Moulton method as a corrector, saving the work of solving the nonlinear system. If we were to do this with, say, the fourth-order methods then the predictor would be $\widehat{y}_{i+1} = y_i + h\Phi$ where

$$\Phi = \frac{55}{24}f(x_i, y_i) - \frac{59}{24}f(x_{i-1}, y_{i-1}) + \frac{37}{24}f(x_{i-2}, y_{i-2}) - \frac{9}{24}f(x_{i-3}, y_{i-3})$$

and the corrector would be $y_{i+1} = y_i + h\Phi$ where

$$\Phi = \frac{251}{720}f(x_{i+1}, \widehat{y}_{i+1}) + \frac{323}{360}f(x_i, y_i) - \frac{11}{30}f(x_{i-1}, y_{i-1}) + \frac{53}{360}f(x_{i-2}, y_{i-2}) - \frac{19}{720}f(x_{i-3}, y_{i-3})$$

and we could repeatedly correct until a certain tolerance was met if desired. (Again, this is functional iteration.) This method only requires two new function evaluations per iteration. Additionally, at each iteration the use of the predictor-corrector idea requires only one additional function evaluation over using either method alone. Notice that we are generating two approximations $\widehat{y}_{i+1}$, y_{i+1} of $y(x_i)$ at each step; we might be able to base an error control scheme on these two approximations. (We use a corrector that is one order lower than the predictor to make this work.) As always in numerical analysis if we have gone to the trouble of computing something we want to get all the information we possibly can from it–this is the motivation for using multi-step methods–so it is natural to think about estimating the error from these two estimates of the solution.

The combination of an Adams-Bashforth predictor and an Adams-Moulton corrector in this way is sometimes referred to as an Adams-Bashforth-Moulton method (ABM), but you will sometimes also see this term used to refer to Adams methods in general. The stability of these methods decreases as the order k increases (note the possibilities for subtraction of nearly equal numbers in these formulas), so generalizations, often described as semi-explicit or semi-implicit, are used in practice for higher-order Adams methods.

The obvious issue that remains is that of generating the first several steps. For example, to use the Adams-Moulton method of order two

$$y_{i+1} = y_i + h \left(\frac{5}{12} f(x_{i+1}, y_{i+1}) + \frac{2}{3} f(x_i, y_i) - \frac{1}{12} f(x_{i-1}, y_{i-1}) \right)$$

given only the ODE IVP $y' = f(x, y)$, $y(x_0) = y_0$ we must first choose a step-size h. But then the method is only defined for $i = 1$ and higher,

$$y_2 = y_1 + h \left(\frac{5}{12} f(x_2, y_2) + \frac{2}{3} f(x_1, y_1) - \frac{1}{12} f(x_0, y_0) \right)$$

so we must generate y_1 by some other method. Then

$$y_3 = y_2 + h \left(\frac{5}{12} f(x_3, y_3) + \frac{2}{3} f(x_2, y_2) - \frac{1}{12} f(x_1, y_1) \right)$$

and so on will have all the values needed for their computation. For higher-order methods we'll need to generate even more starting values before we can begin using the Adams (or other multi-step) method.

How can we do this? It should seem intuitive by now that using the first-order Euler's method to start a fourth-order scheme, say, could affect the performance of the latter scheme. This is certainly so. But it can be shown that a multi-step method of order k may be started with values generated by a method of order $k - 1$ without affecting the order of the multi-step method, though starting with a method of order k gives better results (in fact, results that do not differ appreciably from the results that would have been expected had we had exact starting values).

For this reason the traditional approach has been to start multi-step methods by using an ERK scheme of the same order. However, while this is fine for a nonadaptive method, it is very inefficient if the program is to be adaptive. If we find that we have to reduce the step size at the first step of the multi-step method then we must generate new starting values using the fairly expensive ERK method. Realistically all programs should be adaptive so we must address this issue. In the context of adaptivity, the significance of higher-order methods is that they allow us to take larger step sizes; an $O(h^4)$ method with step size $h = .2$ has much smaller error than an $O(h)$ method with $h = .01$ $(.2^4 = .0016)$, assuming the asymptotic error constants are comparable. So the classical fourth-order RK method, which uses four function evaluations to cover a step of length $h = .2$, can be much more efficient *and accurate* than Euler's method with $h = .01$ which would require 20 function evaluations to cover an interval of length .2.

The key idea is to realize that we are free to change the order of methods as well as the step size as long as we continue to enforce the error control. Adams methods are usually implemented with a variable order, with the order used at any given step selected to balance the needs for accuracy and efficiency as best we can. With this in mind it is reasonable to start off an Adams-Bashforth method of order k (ABk) by using the one-step method AB1 to get y_1 from y_0, then using AB2 to get y_2 from y_0 and y_1, and so on until we reach the maximal order ABk (or drop back to a lower-order method due to error control considerations). This is what is commonly done (e.g., it's essentially the idea behind `ode113`).

There is much more that could be said about Adams methods and the other major class of linear multi-step methods, the BDF methods (which are important for stiff problems). All these methods allow the use of previously computed information in a natural way.

MATLAB

Let's try using AB3 as a predictor and AM3 as a corrector for the ODE IVP $y' = x + y$, $y(0) = 1$. We'll take a constant step size $h = .1$. We have $y_0 = 1$, and we'll use AB1 (Euler's method) to find y_1; enter:

```
>> y0=1;h=.1;f=@(x,y) x+y;
>> y1=y0+h*f(0,y0)
```

We could correct using AM2. Instead let's use AB2 to find y_2; enter:

```
>> y2=y1+h*(1.5*f(.1,y1)-.5*f(0,y0))
```

We can now use AB3 to find y_3; enter:

```
>> y3=y2+h*(23*f(.2,y2)/12-4*f(.1,y1)/3+5*f(0,y0)/12)
```

The result is 1.3858. The solution of the ODE IVP is $y(x) = 2e^x - x - 1$ and at $x_3 = .3$ this gives $y(x_3) \doteq 1.3997$ for an absolute error of about $1.4E - 2$. (Check this.) Let's correct. Enter:

```
>> y3c=y2+h*(9*f(.3,y3)/24+19*f(.2,y2)/24-5*f(.1,y1)/24+f(0,y0)/24)
```

(this is AM3, used as a corrector rather than as an implicit equation to be solved for y_{i+1}). The result is 1.3856 with an absolute error of about $1.4E - 2$ (check this); in fact, the absolute error is slightly worse for the corrected version! Let's correct again. Enter:

```
>> y3c=y2+h*(9*f(.3,y3c)/24+19*f(.2,y2)/24-5*f(.1,y1)/24+f(0,y0)/24)
```

The values are not changing much. We know this can happen because at a fixed step size h, the discrete equation is not the same as the underlying differential equation and we will converge to the solution of the former, not the latter.

Let's move on to y_4; enter:

```
>> y4=y3+h*(23*f(.3,y3)/12-4*f(.2,y2)/3+5*f(.1,y1)/12)
```

We used the corrected value because this is a good policy in general even though it didn't work out as well as we might have hoped at x_3. Now we can correct; enter:

```
>> temp=9*f(.4,y4)/24+19*f(.3,y3)/24-5*f(.2,y2)/24+f(.1,y1)/24;
>> y4c=y3c+h*temp
```

Check the absolute errors. Correct a few more times. Then continue until you reach $x_{10} = 1$. Is the correction of more value farther from the initial point? (Note that we are not controlling for error, and more accurate results would have been achieved with a smaller step size.) Compare the answer ode45 gets at $x = 1$.

Problems

1.) Show that Euler's method is an Adams-Bashforth method and that the backward Euler's method is an Adams-Moulton method.

2.) a.) Use AB1 to solve $y' = x + y$, $y(0) = 1$ over $[0, 1]$ with $h = .1$.
b.) Use AB2 to solve $y' = x + y$, $y(0) = 1$ over $[0, 1]$ with $h = .1$. Use AB1 to find y_1.
c.) Use AB3 to solve $y' = x + y$, $y(0) = 1$ over $[0, 1]$ with $h = .1$. Use AB1 to find y_1 and AB2 to find y_2.
d.) Compare your answers to the true solution. Plot the errors. Does ABk seem to be an order k method?

3.) a.) Use AM1 to solve $y' = x + y$, $y(0) = 1$ over $[0, 1]$ with $h = .1$.

b.) Use AM2 to solve $y' = x + y$, $y(0) = 1$ over $[0, 1]$ with $h = .1$. Use AM1 to find y_1.

c.) Use AM3 to solve $y' = x + y$, $y(0) = 1$ over $[0, 1]$ with $h = .1$. Use AM1 to find y_1 and AM2 to find y_2.

d.) Compare your answers to the true solution. Plot the errors. Does AMk seem to be an order k method?

4.) a.) The multi-step methods seen in this section have coefficients that are of mixed sign. Is this desirable?

b.) What is the sum of the coefficients of the methods in this section? How does this compare to what we have seen in numerical differentiation and numerical integration formulae?

5.) Write a MATLAB program that performs ABk. Start it off by using AB1 to get y_1 from y_0, then using AB2 to get y_2 from y_0 and y_1, and so on until the maximal order ABk is reached.

6.) a.) Repeat the experiment in the MATLAB subsection using $h = .5$.

b.) Repeat the experiment in the MATLAB subsection using $h = .25$.

7.) a.) Repeat the experiment in the MATLAB subsection using the ODE IVP $y' = x^2 + y$, $y(0) = 1$.

b.) Repeat part a. using $h = .05$.

c.) Repeat the experiment in the MATLAB subsection using the ODE IVP $y' = x^2 y + y^2 e^x$, $y(0) = 1$. Use one of the MATLAB ODE solvers to get accurate estimates of the true solution in order to compute the absolute error.

d.) Repeat part a. using $h = .05$.

8.) Replace y' in $y' = f(x, y)$ by the numerical differentiation scheme of Eq. (1.3). What method for solving ODEs does this give?

9.) a.) Verify that the formula given for AB1 is correct using Eq. (7.7).

b.) Verify that the formulas given for AB2, AB3, and AB4 are correct using Eq. (7.7).

10.) Verify that the formulas given for AM1 through AM4 are correct using Eq. (7.8).

11.) How could you use an Adams scheme to perform numerical integration? Discuss both ABk and AMk.

12.) a.) Find the AB5 formula.

b.) Find the AB6 formula.

13.) a.) Find the AM5 formula.

b.) Find the AM6 formula.

14.) Compare the strengths and weaknesses of all the numerical methods for ODEs that we've studied so far.

15.) Write a MATLAB program that solves ODEs by accepting a value k and the ODE, initial condition, final time, and step size, and using ABk and AMk as a predictor-corrector pair. Correct at least three times for each prediction.

16.) Show that the backward differentiation formulas are linear multi-step methods.

CHAPTER 6 BIBLIOGRAPHY:

Numerical Methods for Initial Value Problems in Ordinary Differential Equations, Simeon Ola Fatunla, Academic Press 1988

A concise but complete text.

A First Course in the Numerical Analysis of Differential Equations, Arieh Iserles, Cambridge University Press 1996

A modern introductory text with a mix of theory and application.

Numerical Solution of Ordinary Differential Equations, Lawrence F. Shampine, Chapman and Hall, 1994

A detailed modern text.

Numerical Quadrature and Solution of Ordinary Differential Equations, A. H. Stroud, Springer-Verlag New York, 1974

An older, detailed text.

Finite Difference Schemes and Partial Differential Equations, John C. Strikwerda, Wadsworth 1989

A modern introduction to finite difference methods.

Numerical Solution of Partial Differential Equations: Finite Difference Methods, 3rd ed., G.D. Smith, Oxford University Press 1985

An introduction to finite difference methods.

CHAPTER 7

Nonlinear Optimization

1. One-Dimensional Searches

Root-finding (including the very special case $Ax = b$) and optimization are the most commonly occurring problems in numerical computing. They occur both as main problems and as subproblems of other problems.

Optimization problems are those in which we wish to find the value of the argument x that makes some function $F(x)$ take on a value which is either a local minimum or a local maximum. If $F : \mathbb{R}^n \to \mathbb{R}$ then we write

$$(1.1) \qquad\qquad \min_{x \in D} F(x)$$

to denote the problem of finding some value $x^* \in D \subseteq \mathbb{R}^n$ such that F has a local minimum at x^*; we also write $x^* = \min_{x \in D} F(x)$. (Recall that x^* is a local minimum of F in D if there is a neighborhood $N(x^*)$ in D such that if $x \in N(x^*)$ then $F(x) \geq F(x^*)$; the minimum is isolated, or strict, if in addition $F(x) > F(x^*)$ for $x \neq x^*$. We assume that the maxima and minima of the functions that we consider are isolated.) Note that, similar to the language for zeroes of a function, when we speak of a minimization problem we are looking for the value of x which is the location of the minimum, not the value of the function at that minimum. We say that x^* is a local minimum of F but when we wish to emphasize that it is an x-value we say that x^* is the local minimizer.

We write $\max_{x \in D} F(x)$ to denote the problem of finding some value $x^* \in D \subseteq \mathbb{R}^n$ such that F has a local maximum at x^*; we also write $x^* = \max_{x \in D} F(x)$. We say that x^* is a local maximum of the function or, for emphasis, that it is a local maximizer.

The function F is usually called the **objective function**. We will assume throughout that it is continuous. If the set D is in fact equal to $\mathbb{R}^n$ then we say that the problem is **unconstrained**; if D is a proper subset of $\mathbb{R}^n$ we say that the problem (or solution) is **constrained**. Optimization problems occur in many circumstances: Find the combination of prices and manufacturing which gives maximum profit; find the point or configuration at which the energy is a minimum; find the best fit curve for a set of data points, that is, the curve which will minimize the error; and so on.

The case where the objective function F is a linear function $F(x) = c^T x$ for some vector c and the problem is linearly constrained is of great interest and is referred to as a **linear programming (LP) problem**. We won't discuss this type

DOI: 10.1201/9781003042273-7

483

of problem. When F is nonlinear we call it a **nonlinear optimization** or **nonlinear programming (NLP) problem**. We will consider the nonlinear optimization problem in detail, assuming as usual sufficient differentiability of F where needed.

Since the minima of F are precisely the maxima of $-F$, we focus only on the minimization problem of Eq. (1.1). If F is differentiable, we might consider differentiating it and setting that derivative to zero, then applying a root-finding method. Sometimes this can be done; for problems of interest, however, we find the same difficulties that we did with Newton's method in the previous chapter– the function may be given as a program and hence not differentiable; symbolic differentiation may not be available; the derivative may be too expensive to compute (recall that if $F : \mathbb{R}^n \to \mathbb{R}$ then $F(x)$ returns a single value but the derivative of F is its Jacobian, with n^2 entries); and so on. In addition, some zeroes of F may be extrema which are neither maxima nor minima (as in $f(x) = x^3$ at $x = 0$, or a saddle point in several dimensions), and in any event if we find a zero of F we must then determine whether it corresponds to a minimum or to a maximum of F. If we must solve the problem numerically anyway, we might as well leave it in the form given and solve it there if methods for doing so are available and equally efficient.

Let's start with the case of one-dimensional minimization. Methods for finding minima of a function are also called **search methods**, and if $f : \mathbb{R} \to \mathbb{R}$ they are called **one-dimensional searches** or **line searches**. In analogy with root-finding we might start by seeking a bisection-like method. But two points do not suffice to bracket a minimum of a function; after all, $F(-1) = 0$ and $F(1) = 0$ could equally well bracket a minimum of $F(x) = x^2 - 1$ at $x = 0$ or a maximum of $F(x) = 1 - x^2$ at $x = 0$; if $F(0) = 0$ and $F(1) = 1$ then there might be, in between, an internal minimum, an internal maximum, or neither if $F(x) = x$.

However, three points do suffice to bracket a minimum (or maximum). If $a < b < c$ and f is continuous on $[a, c]$ then we know from the Extreme Value Theorem that f achieves both its maximum and its minimum value over $[a, c]$. If $f(b) < f(a)$ and $f(b) < f(c)$ then the maximum may be achieved at one of the endpoints but the minimum cannot be and so there must be a local minimum in the interior of the interval. We have bracketed a minimum of the function. This is the basis of the **exhaustive search method for minimization** (see Sec. 1.1): Take small steps in a given direction until a bracket is found, then repeat with a finer step size over that bracket.

Suppose we generate a new point d in some manner (exhaustive search, or the midpoint of the interval if b is not already the midpoint). If $a < b < d < c$ then either $f(d) < f(b)$, in which case $[b, d, c]$ is a new, smaller bracket (recall that $f(b) < f(c)$), or $f(d) > f(b)$ in which case $[a, b, d]$ is a new, smaller bracket. (We'll neglect for now the highly unlikely possibility that $f(d) = f(b)$.) Clearly, this method will converge to a local minimum of the function lying in $[a, c]$.

In the case of root-finding, the method of bisection always gives a 50% reduction in the width of the bracket; other methods will typically converge faster but will not necessarily improve by a significant amount at every iteration. For root-finding, bisection gives the best *guaranteed* reduction at each iteration.

In the case of minimization, however, where we have a three-point bracket, it's far from clear how to choose the next point so as to give the best *guaranteed* reduction in bracket width at each step. Suppose we have a bracket $[a, b, c]$. How should we choose $d \in (a, c)$? It seems clear that d should lie in the larger of the subintervals (a, b), (b, c) since if, say, (a, b) is very small, then choosing a d in that interval may give us very little reduction in bracket width.

Let's suppose that (b, c) is the larger subinterval. Let L be the length of the current bracket,

$$
\begin{aligned}
L &= c - a \\
&= (c - b) + (b - a).
\end{aligned}
$$

Since d will lie in (b, c), the new bracket will be either $[a, b, d]$ or $[b, d, c]$. Let $\Delta = d - b$. If the new bracket is $[a, b, d]$ then its length is

$$
\begin{aligned}
L_1 &= d - a \\
&= \Delta + (b - a)
\end{aligned}
$$

while if the new bracket is $[b, d, c]$ then

$$
L_2 = c - b
$$

is its length. Since we cannot predict in advance which interval will result, the safest bet is to choose d so that

$$
L_1 = L_2.
$$

This will give a guaranteed reduction from an interval of length L to one of length $L_1 = L_2$ every time. We have:

$$
\begin{aligned}
L_1 &= L_2 \\
\Delta + (b - a) &= (c - b) \\
\Delta &= (c - b) - (b - a) \\
&= L - 2(b - a)
\end{aligned}
$$

so that we in effect divide the interval (a, c) into three subintervals of lengths $(b-a)$, Δ, and $(b - a)$, respectively. The length of the new bracket, which must include the center section, is necessarily $\Delta + (b - a)$. This gives

$$
\begin{aligned}
d &= \Delta + b \\
&= (c - b) - (b - a) + b \\
&= c + a - b
\end{aligned}
$$

and any other choice of d allows both for the possibility of a smaller bracket, and at the same time for the possibility of a larger one.

Suppose we iterate this approach. Note that d is found by taking a step of length Δ into the larger subinterval. If b was selected in the same manner as d was, as would be the case if we are iterating, then it should have come from taking a step of the same proportion from a into (a, c), the interval in which it would have been selected. That is,

$$
\frac{\Delta}{c - b} = \frac{b - a}{c - a}
$$

if previous points were selected in the same fashion as d was. If we set $\lambda = b - a$ then we can write this as

$$\frac{L - 2\lambda}{L - \lambda} = \frac{\lambda}{L}$$

or

$$\frac{1 - 2r}{1 - r} = r$$

in terms of the dimensionless quantity $r = \lambda/L$. Solving for r gives

$$
\begin{aligned}
1 - 2r &= r(1 - r) \\
r^2 - 3r + 1 &= 0 \\
r_{1,2} &= \frac{3 \pm \sqrt{5}}{2}
\end{aligned}
$$

and since r must be less than one we reject the root $(3 + \sqrt{5})/2$ and take

(1.2) $r = (3 - \sqrt{5})/2 \doteq 0.3820.$

From Eq. (1.2), we see that b was (or should have been) found by stepping approximately 38.2% of the way from a into (a, c), as d is to be found by stepping approximately 38.2% of the way from b into (b, c).

The resulting algorithm is known as **golden section search** (or **GSS**): Beginning with an initial bracket $[a, b, c]$ for a local minimum of a continuous function f, choose a new point d by moving from b about 38.2% of the way into the larger of the two subintervals (a, b) and (b, c). Compute $f(d)$ and choose a new, smaller bracket that includes d. The new bracket will always be about 61.8% of the length $c - a$ of the previous bracket. This isn't as big a reduction as we achieved with bisection, but it's still pretty good.

EXAMPLE 1: Suppose we wish to find the first positive minimum of $f(t) = t^6 - 6t^4 - 3t + 1$. (By a positive minimum we mean that the minimizer x^* is positive.) Taking the derivative and setting it to zero would give a quintic equation that would still have to be solved numerically. Since we can't avoid using a numerical method, let's use a minimization method directly on f. Plotting the function suggests that the minimum is near $x = 2$ and that $[1.5, 2, 2.5]$ should be a bracket. (Recall that we do not need a sign change as we did for bisection.) Indeed,

$$
\begin{aligned}
f(1.5) &\doteq -22.4844 \\
f(2) &= -37 \\
f(2.5) &\doteq 3.2656
\end{aligned}
$$

so this is a bracket ($f(2) < f(1.5)$ and $f(2) < f(2.5)$). The choice of $b = 2$ is not optimal however; that would be $b = 1.5 + .382(2.5 - 1.5) \doteq 1.88$, for which we have $f(1.88) \doteq -35.4402$. Since $[1.5, 1.88, 2.5]$ is a bracket, let's use it. We choose $d = 1.88 + .382(2.5 - 1.88) \doteq 2.1168$ as the next trial point. Since $f(2.1168) \doteq -35.8510$ which is less than $f(1.88)$ and $f(2.5)$ the new bracket is $[1.88, 2.1168, 2.5]$. As expected, the width $2.5 - 1.88 = .62$ is about 61.8% of the original unit-length interval.

Let's do another iteration. The right-hand subinterval $[2.1168, 2.5]$ is the larger one, so we choose the next trial point as $d = 2.1168 + .382(2.5 - 2.1168) \doteq 2.2632$. Since $f(2.2632) \doteq -28.8239$ and $f(2.1168)$ is less than $f(1.88)$ and $f(2.2632)$ the new bracket is $[1.88, 2.1168, 2.2632]$. The width is $2.2632 - 1.88 = 0.3832$ and $(.618)(.62) = 0.3832$ so this matches our expectations. The minimum lies somewhere in $[1.88, 2.2632]$. $\square$

In fact even if we had used the bracket $[1.5, 2, 2.5]$ instead and used the same rule (step 38.2% of the way into the larger subinterval each time) it would not be long before the bracket achieved the form above, with guaranteed reduction to 61.8% of the length of the previous interval each time. The method is linear with asymptotic error constant .618, comparable to bisection (though that has a slightly lower asymptotic error constant of .5). Of course, in Example 1 we would have achieved more rapid convergence by applying Newton's method to $f'(t)$, since that method is quadratically convergent. We will develop comparably fast methods for minimization as we proceed.

This method is called golden section search because of the somewhat unexpected appearance of the golden number $\Phi = (1 + \sqrt{5})/2 \doteq 1.618$ that occurs in many other areas of mathematics. In fact, recall that this was also the order of convergence of the secant method!

Golden section search is the slow-but-steady method for minimization. Since we have a three-point bracket anyway, the obvious way to speed things up is to use quadratic interpolation (often called **Powell's method** in this context). We can interpolate a quadratic to the three points, use the location of the minimum of the quadratic as our estimate of the location of the minimum of the objective function, and update the bracket. (See Fig. 1.) If the objective function is sufficiently differentiable and $f''(x^*) \neq 0$ then the minimum will be nearly parabolic on a sufficiently small scale, so this is a reasonable approach. If we drop the requirement of a bracket and simply use the last three points we have an analogue of Müller's method.

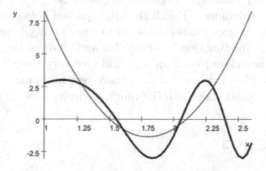

FIGURE 1. Quadratic interpolation.

Quadratic interpolation for minima, as was the case for roots, may make slow progress. Hence there is a **Brent's method for minimization** as well, which mixes quadratic interpolation and golden section search. The details are somewhat tedious but the idea is the same as in Sec. 1.6. It forms the basis for the MATLAB®

one-dimensional minimization command, `fminbnd`. Importantly, it does not assume differentiability of the objective function, just continuity (and a bracket).

MATLAB

The MATLAB one-dimensional minimization function is called `fminbnd`. It requires a function name or inline function and a bracket. For example, enter:

```
» f=@(x) (x-1)^2;
» fminbnd(f,-10,10)
```

to find the minimum at $x = 1$. Enter:

```
» fminbnd(f,-10,10,optimset('display','iter'))
```

This displays the intermediate steps and shows that a version of Brent's method is being used by `fminbnd`, switching between golden section search and quadratic (that is, parabolic) interpolation. Experiment using `fminbnd` on several other functions.

If you enter `type fminbnd` you will see a comment line that states that "f must not be evaluated too close to" certain other points and checks to see if this is so. Let's explore one reason why such a check might be used. (Another reason is simply to prevent the method from stagnating by taking very small steps.) Recall from Sec. 1.7 that, since there are only finitely many machine numbers, the distance from the machine number 1 to the next largest machine number is a positive number called the machine epsilon ϵ. Although smaller values than this can be represented and manipulated, when added to numbers as large as 1 they have no effect; enter:

```
» 1+eps==1
» 1+eps/2==1
```

Adding $\epsilon/2$ to 1 has no effect. If we are working with numbers on the order of unity or larger, asking for a bracket of a root or minimum that is smaller than $[x^* - \epsilon, x^* + \epsilon]$ isn't feasible. Many programs for root-finding, minimization, etc., incorporate a check that compares the tolerance selected by the user to one based on the machine epsilon and uses the latter tolerance if the user's choice is unrealistically small.

For minimization problems, it can be argued that $\sqrt{\epsilon}|x^*|$ should be a rough lower bound on the tolerance. (Recall that the spacing between numbers grows as the size of the numbers grow; it is only equal to ϵ near unity.) Since $f'(x^*) = 0$ at a parabolic minimum, the function is locally flat and changes very slowly there, and so values of x that are close to one another will give very close function values. For example, if $f(x) = x^3 - 3x^2 - x + 3$ then there is a minimum between $x = 1$ and $x = 3$ (plot f to see this). Let's use the rough estimate $x^* \approx 2$; enter:

```
» tol=2*eps
» x=2
    » f0=x^3-3*x^2-x+3
    » x1=x+tol
    » x1==2              %Is x1 equal to 2?
    » f1=x1^3-3*x1^2-x1+3
    » x2=x-tol
    » x2==2              %Is x2 equal to 2?
    » f2=x2^3-3*x2^2-x2+3
    » f1==f0             %Is f1 equal to f0?
    » f2==f0             %Is f2 equal to f0?
    » f1==f2             %Is f1 equal to f2?
```

Although x1 and x2 display as if they had the same value (even under format long) the equality tests show that they are different from 2; indeed, $x_2 < 2 < x_1$ as stored internally. But their function values come out precisely the same, as the tests for equality show. Think about this: The computer simply cannot see differences in function values over a range this small near the minimum because of the zero derivative. If you try this near a point where $f'(x) = 1$, say, the f-values will differ, as expected.

The value $\sqrt{\epsilon}|x^*|$ is only an estimate, but it is sensible in general to check any user-supplied tolerances for reasonableness (as well as limiting the maximum number of function evaluations). Not only is it wasteful to evaluate f in this range, but round-off error could cause an incorrect decision to be made with respect to which points are kept for a bracket, which might destroy the bracket.

Problems

1.) Perform 10 iterations of golden section search on the objective function of Example 1 starting from the initial bracket $[1.5, 2, 2.5]$ in order to estimate the location of the local minimum in that bracket. Show that the last bracket is divided into 38%/62% sections as predicted, even though the initial interval was not.

2.) Use golden section search to estimate the locations of any two minima of $f(x) = x^9 - 2x^7 + 5x^4 - 2x + 2$ to at least four decimal places.

3.) Write a detailed algorithm (pseudo-code) for performing minimization by quadratic interpolation, maintaining a bracket. Assume that a valid initial bracket will be given.

4.) a.) Perform four iterations of quadratic interpolation on the objective function of Example 1 starting from the initial bracket $[1.5, 2, 2.5]$ in order to estimate the location of the local minimum in that bracket.

b.) Perform four iterations of root-finding by quadratic interpolation on the derivative of the objective function of Example 1 starting from the initial point 1.5 in order to estimate the location of the local minimum in that bracket.

5.) a.) Use golden section search on $f^2(x)$ to approximate a zero of $f(x) = 1/x^3 - 10$ to three decimal places.

b.) Use the quadratic interpolation minimization method on $f^2(x)$ to approximate a zero of $f(x) = 1/x^3 - 10$ to three decimal places.

6.) a.) How many iterations of golden section search would be required to reduce an initial bracket of width 2 to a bracket of width not more than 10^{-4}?

b.) If the bracket of width 2 is $[1, 3]$ and $\epsilon = 10^{-16}$, what is the smallest final width we should ask for from golden section search?

c.) If the bracket of width 2 is $[1500, 1502]$ and $\epsilon = 10^{-16}$, what is the smallest final width we should ask for from golden section search?

7.) a.) Repeat Problem 5. (both parts) but use $|f(x)|$ instead of $f^2(x)$ as the objective function. Comment.

b.) Use MATLAB's fminbnd function on $f^2(x)$ and then on $|f(x)|$ to approximate a zero of $f(x) = 1/x^3 - 10$.

8.) Suppose that $f(x)$ is a smooth function and that it has a parabolic minimum x^* at which $f(x^*) = m$ for some given real value m. Would it be reasonable to attempt to find x^* by using a root-finding method on $g(x) = f(x^*) - m$?

9.) a.) Show by drawing a sketch that the points $a < b < c$ need not bracket a minimum of a continuous function if $f(b) > f(a)$ or if $f(b) > f(c)$; also show that $[a, b, c]$ *may* bracket a minimum in these cases.

b.) If $a < b < c$ and $f(b) \leq f(a)$, $f(b) \leq f(c)$, must $[a, b, c]$ bracket a minimum of the continuous function f?

10.) Write a detailed algorithm (pseudo-code) for performing minimization by a version of Brent's method (mixing quadratic interpolation and golden section search). Assume that a valid initial bracket will be given.

11.) Suppose that x^* is a parabolic minimum ($f''(x^*) \neq 0$) of a function. Expand $f(x)$ in a Taylor series about x^* and use it to justify the claim that the tolerance for a minimization method should not be smaller than about $\sqrt{\epsilon}|x^*|$.

12.) Write a MATLAB program to determine the machine epsilon (without using the eps function). Compare your results with the value from MATLAB's eps function.

13.) a.) One version of Powell's method is as follows: Suppose that the objective function f has a parabolic minimum and that x_0 is an initial guess as to its location. In addition, suppose that a step size $h > 0$ is given that is on the same scale as the distance x_0 is likely to be from the true minimum. (This is a measure of how rapidly the function changes.) Compute $f(x_0)$ and $f(x_0 + h)$; if $f(x_0) < f(x_0 + h)$ then compute $f(x_0 - h)$ else compute $f(x_0 + 2h)$. This gives a triple of points. Perform one iteration of quadratic interpolation, giving a fourth point. If the difference between the two lowest functions values is less than some tolerance, terminate; else choose the three points with the lowest function values and return to the quadratic interpolation step, *unless* choosing the point with the largest function value would allow us to find a bracket, in which case choose the bracketing triple and return to the quadratic interpolation step. Write a detailed algorithm (pseudo-code) for performing minimization by Powell's method.

b.) Use Powell's method to approximate a minimum of $f(x) = x^6 - 5.5x + 3$.

14.) Write a MATLAB program that performs golden section search on a given inline function. Include appropriate error checking. Test your program on several objective functions including -humps(x) (humps is a built-in function with two local maxima).

15.) a.) Golden section search gives the optimal guaranteed reduction in interval length for a single step. Suppose however that we planned in advance to sample the function exactly 2 times. We might ask, are two golden section steps optimal? Consider the function $f(x) = -x^4 + 6x^3 - 6.25x^2 - 8.25x + 5$ which has a single local minimum bracketed by $[0, 1, 3]$. Compare two iterations of golden section search to evaluating the function at $x = 2$ and then at $x = 1.05$ and using these values to find a smaller bracket. Would this approach give better results for all functions when the bracket is $[0, 1, 3]$?

b.) The choice of points in part a. is based on the Fibonacci recurrence relation $F_n = F_{n-1} + F_{n-2}$ with $F_0 = F_1 = 1$. The Fibonacci numbers F_i occur in the minimization method known as **Fibonacci search**, which gives the best possible reduction of bracket width for a fixed (in advance) number of iterations. Derive the method of Fibonacci search by assuming the function will be evaluated precisely N times. Hint: Think again about part a., where the initial bracket was $[0, F_3]$ which was divided into intervals of length F_1 and F_2; since $F_0 = F_1$ the last point is chosen to be very close to the next-to-last point. Develop a recursion relation for the interval lengths.

c.) It is a fact that $\lim_{n \to \infty} F_n/F_{n-1} = (1 + \sqrt{5})/2 \doteq 1.618$. Use this fact to derive the method of golden section search from the method of Fibonacci search.

d.) Why is golden section search used frequently in practice while Fibonacci search is only rarely used?

16.) Prove that the Fibonacci numbers F_i satisfy $\lim_{n\to\infty} F_n/F_{n-1} = (1+\sqrt{5})/2$. (See Problem 15.) The easiest way is to use the Binet formula from Sec. 1.7.

2. The Method of Steepest Descent

Golden section search is a linear method; the quadratic interpolation methods are superlinear ($\alpha \approx 1.3$ without brackets). If we assume that derivative information is available, we should be able to do better. One such method for a differentiable function $f : \mathbb{R} \to \mathbb{R}$ is **Davidon's method**, which takes two points a, b and fits a cubic $y(x) = \alpha(x-a)^3 + \beta(x-a)^2 + \gamma(x-a) + \delta$ to the function by requiring that the cubic not only interpolate the objective function f at $x = a$ and $x = b$ but that the derivatives of the cubic match those of f at those points also; that is, we require that

$$
\begin{aligned}
y(a) &= f(a) \\
y'(a) &= f'(a) \\
y(b) &= f(b) \\
y'(b) &= f'(b)
\end{aligned}
$$

where $f(a), f(b), f'(a), f'(b)$ are all assumed known. This results in a 4×4 linear system in which two of the unknowns, γ and δ, can be obtained trivially from the given data:

$$
\begin{aligned}
\delta &= f(a) \\
\gamma &= f'(a) \\
\alpha(b-a)^3 + \beta(b-a)^2 + \gamma(b-a) + \delta &= f(b) \\
3\alpha(b-a)^2 + 2\beta(b-a) + \gamma &= f'(b)
\end{aligned}
$$

and so $\alpha, \beta, \gamma, \delta$ are easily found; setting $y'(x) = 0$ gives a quadratic, one root of which corresponds to a minimum of y, and the other to a maximum. We can use the second derivative test to determine which is the minimum, and that is our new estimate of the minimum of the function. We keep the two most recent points. Convergence is quadratic (under appropriate assumptions on f).

Since only two points are being maintained, a safeguarded method will typically bisect the interval when the method gets bogged down. The safeguarding idea can be extended to a function $F : \mathbb{R}^n \to \mathbb{R}$.

We are going to take a different approach in this section, however. Suppose that $f : \mathbb{R} \to \mathbb{R}$. If we had no idea where a minimum of f might be, we might sample f at some point x_0 and also compute $f'(x_0)$. If $f'(x_0) < 0$ then the function is decreasing with increasing x, and so it makes sense to look for the minimum by taking our next point to the right of x_0 (see Fig. 1); if $f'(x_0) > 0$ then it makes sense to look to the left of x_0.

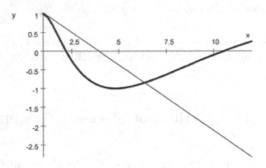

FIGURE 1. Use of the derivative.

We could use this reasoning as part of a routine to search for an initial bracket for a minimum. Of course, we do not know how far we should step in the direction in which f is decreasing, just which way to go.

It's most appropriate to follow this reasoning further in a more general setting than functions on the real line. Suppose $F : \mathbb{R}^n \to \mathbb{R}$ is a smooth function; we write F as $F(x_1, ..., x_n)$ or as $F(x)$ where it is understood that x is a vector in $\mathbb{R}^n$. The natural notion of a derivative for such a function is its gradient

$$(2.1) \qquad \nabla F(x) = \begin{pmatrix} \frac{\partial F}{\partial x_1} \\ \frac{\partial F}{\partial x_2} \\ \vdots \\ \frac{\partial F}{\partial x_n} \end{pmatrix}$$

which is a vector-valued function on $\mathbb{R}^n$. For example, if $F(x, y, z) = x^2 y^2 z^2$ then Eq. (2.1) gives

$$\nabla F(x) = \begin{pmatrix} \frac{\partial F}{\partial x_1} \\ \frac{\partial F}{\partial x_2} \\ \frac{\partial F}{\partial x_3} \end{pmatrix}$$

$$= \begin{pmatrix} 2xy^2 z^2 \\ 2x^2 yz^2 \\ 2x^2 y^2 z \end{pmatrix}$$

is its gradient. Like the derivative, the gradient is zero at a minimum (that is, it is equal to the zero vector there).

The gradient is a vector, and at any point $(x_1, ..., x_n)$ the gradient vector $\nabla F(x_1, ..., x_n)$ points in the direction in which F is increasing most rapidly, called the direction of steepest ascent; it follows that $-\nabla F(x_1, ..., x_n)$ points in the direction in which F is decreasing most rapidly, called the direction of steepest descent. If we are looking for a minimum of F (we imagine F as a bowl-shaped region in space, and we are seeking the bottom of the bowl) starting from some given point x_0, it seems sensible to look in the direction of steepest descent.

Of course, it is highly unlikely that moving from x_0 in the direction of $-\nabla F(x_0)$ will take us directly to the local minimum, if in fact there even is one. We will need an iterative method. If x_0 is the initial guess, we take

$$x_1 = x_0 - \alpha \nabla F(x_0)$$

as the next point, for some $\alpha > 0$ that represents how far we step in the direction of steepest descent. How shall we choose α? Since we want x_1 as close to the minimum as possible, we will choose that value of $\alpha > 0$ that makes

$$F(x_1) = F(x_0 - \alpha \nabla F(x_0))$$

as small as possible. Since x_0 is known, $F(x_0 - \alpha \nabla F(x_0))$ is a function of the single variable α, and so finding α involves only a (constrained) one-dimensional search no matter how large n may be. We then iterate in the form

(2.2) $$x_{k+1} = x_k - \alpha_k \nabla F(x_k)$$

where α_k is chosen so as to minimize $F(x_k - \alpha_k \nabla F(x_k))$ at each step. (Ideally this will be the first minimum of $F(x_k - \alpha_k \nabla F(x_k))$ as we step away from x_k.) This is known as **(Cauchy's) method of steepest descent**. We may require that $\|x_{k+1} - x_k\|$ (or $F(x_k) - F(x_{k+1})$) be less than some tolerance as our convergence criterion but typically we use the size of the residual $\|\nabla F(x_k)\|$, which will of course be zero at the minimum, as the principal convergence criterion.

EXAMPLE 1: Let's apply steepest descent to $F(x,y) = \sin(x^2 y^2 - 1)$. The gradient is $\nabla F(x,y) = (2xy^2 \cos(x^2 y^2 - 1), 2x^2 y \cos(x^2 y^2 - 1))^T$. Take $(x_0, y_0) = (1,1)$. The direction of steepest descent is $\nabla F(1,1) = (2\cos(0), 2\cos(0))^T = (2,2)^T$. We need to find an α_0 that satisfies

$$\begin{aligned}
\alpha_0 &= \min_{\alpha>0}\{F((x_0,y_0)^T - \alpha \nabla F(x_0,y_0))\} \\
&= \min_{\alpha>0}\{F((1,1)^T - \alpha(2,2)^T)\} \\
&= \min_{\alpha>0}\{F((1-2\alpha, 1-2\alpha)\} \\
&= \min_{\alpha>0}\{\sin((1-2\alpha)^2(1-2\alpha)^2 - 1)\} \\
&= \min_{\alpha>0}\{\sin((1-2\alpha)^4 - 1)\}
\end{aligned}$$

and the first minimum of the sine function for $\alpha > 0$ occurs at

$$\begin{aligned}
(1-2\alpha)^4 - 1 &= 3\pi/2 \\
(1-2\alpha)^4 &= 3\pi/2 + 1 \\
\alpha &= \frac{1}{2}(1 \pm \sqrt[1/4]{3\pi/2 + 1}) \\
&\doteq 1.2730, -0.2730
\end{aligned}$$

so $\alpha_0 \doteq 1.2133$. This gives

$$\begin{pmatrix} x_1 \\ y_1 \end{pmatrix} = \begin{pmatrix} x_0 \\ y_0 \end{pmatrix} - \alpha_0 \nabla F(x_0, y_0)$$

$$= \begin{pmatrix} 1 \\ 1 \end{pmatrix} - 1.2730 \begin{pmatrix} 2 \\ 2 \end{pmatrix}$$

$$= \begin{pmatrix} -1.5460 \\ -1.5460 \end{pmatrix}$$

and note that $F(x_0, y_0) = F(1,1) = 0$, $F(x_1, y_1) = F(-1.5460, -1.5460) \doteq -1$. The gradient at this point is

$$\nabla F(-1.5460, -1.5460) \doteq (7E - 16, 7E - 16)^T$$
$$\|\nabla F(-1.5460, -1.5460)\| \doteq 10E - 16$$

and clearly F cannot take on a smaller value than -1, so we have found a minimum. $\square$

Rapid convergence of the method, as above, is *not* typical. In addition, the use of a line search algorithm will generally be necessary to find α, unlike the artificial case of Example 1. The dependence of the method of steepest descent on a one-dimensional search algorithm which will be used at every iteration of steepest descent highlights the need for robust, reliable, and automated numerical methods; if we write a program to perform the method of steepest descent we certainly don't want it to stop at each iteration and prompt the user for a bracket for the line search. (This issue will be considered further in the Problems.) One approach might be to use Davidon's method with some heuristic means of choosing the second point needed to fit the cubic interpolant.

Consider again Eq. (2.2). The method of steepest descent proceeds by selecting a search direction $-\nabla F(x_k)$, stepping along it until it finds a local minimum in that direction (at which point it will be tangent to a contour of F), then repeating this procedure from that point. Although it may make many seemingly unnecessary turns and can stagnate in a region where $\|\nabla F(x)\|$ is small but nonzero, it is very robust, and typically it converges to a stationary point of F (that is, a point where $\nabla F(x) = 0$). Convergence to a stationary point that is neither a minimum nor a maximum is uncommon though the method may be very slow while it is near such a point.

We give one convergence theorem for the method of steepest descent:

Theorem 1: Let $F : \mathbb{R}^n \to \mathbb{R}$ be once continuously differentiable and suppose that $\Omega_0 = \{x \in \mathbb{R}^n : F(x) \le F(x_0)\}$ is closed and bounded. Then the method of steepest descent applied to F starting from x_0 converges to a stationary point of F.

PROOF: By construction, $\nabla F(x_k)$ $(k = 0, 1, 2, ...)$ is a monotonically decreasing (that is, nonincreasing) sequence of real numbers and hence if it is bounded from below it converges. But F is continuous and Ω_0 is closed and bounded, so F achieves its minimum (and its maximum) on Ω_0 and so $\nabla F(x_k)$ is bounded from below. Hence $\nabla F(x_k)$ converges to some value F^*, that is

$$\lim_{k \to \infty} \nabla F(x_k) = F^*.$$

But by continuity, $\nabla F(\lim_{k\to\infty} x_k) = F^*$ and so $x_k \to x^*$ for some x^*. If $\nabla F(x^*) \neq 0$ then $\nabla F(x^*)$ is a descent direction and so there is an $\alpha > 0$ such that $F(x^* - \alpha \nabla F(x^*))$, contradicting convergence of the method. $\square$

However, the method of steepest descent is in general only linearly convergent despite its use of derivative information. In the next section we will look at a more rapidly convergent method. For now we mention that the method of steepest descent is often used as a "starter" method for other algorithms. Because it converges for most initial guesses, several iterations of steepest descent may be used to improve an initial guess that will then be handed to another, faster (but less robust) algorithm. This is done not only for minimization algorithms but also for root-finding algorithms such as Newton's method for systems. If we attempt to solve

$$F(x) = 0$$

where $F : \mathbb{R}^n \to \mathbb{R}^n$, that is, where

$$F(x) = \begin{pmatrix} f_1(x_1, ..., x_n) \\ f_2(x_1, ..., x_n) \\ \vdots \\ f_n(x_1, ..., x_n) \end{pmatrix};$$

by Newton's method for systems (or some other root-finding method) with some initial guess x_0 and find that the method fails to converge, we can use the method of steepest descent on the function

$$\|F(x)\|^2 = f_1^2(x) + ... + f_n^2(x)$$

(which is non-negative and has minima precisely where F has zeroes), starting from this same x_0. After several iterations of steepest descent we return to Newton's method with the (hopefully) improved initial guess. While this approach is not guaranteed to work, it is often successful in practice–and is another example of the benefits of mixing methods that have different strengths.

EXAMPLE 2: Suppose we are trying to solve the system

$$\begin{aligned} \ln(|x|) &= 0 \\ \ln(|y|) &= 0 \end{aligned}$$

using Newton's method with the initial guess $(x_0, y_0) = (3.6, 3.6)$. For this very simple example, Newton's method for systems is just Newton's method applied separately to each component:

$$\begin{aligned} x_{k+1} &= x_k - f_1(x_k)/f_1'(x_k) \\ y_{k+1} &= y_k - f_1(y_k)/f_2'(y_k) \end{aligned}$$

which is

$$\begin{aligned} x_{k+1} &= x_k - x_k \ln(x_k) \\ y_{k+1} &= y_k - y_k \ln(y_k) \end{aligned}$$

for positive x, y, and

$$\begin{aligned} x_{k+1} &= x_k + x_k \ln(-x_k) \\ y_{k+1} &= y_k + y_k \ln(-y_k) \end{aligned}$$

for negative x, y. With the given initial conditions, $x_k \to \infty$ and $y_k \to \infty$ (check this). Let's use the method of steepest descent on

$$\begin{aligned} g(x,y) &= f_1^2(x,y) + f_2^2(x,y) \\ &= [\ln(|x|)]^2 + [\ln(|y|]^2 \end{aligned}$$

with the initial conditions $(x_0, y_0) = (3.6, 3.6)$. The gradient is

$$\nabla g(x,y) = \begin{pmatrix} 2\ln(x)/x \\ 2\ln(y)/y \end{pmatrix}$$

for x, y positive, and so $-\nabla g(3.6, 3.6) = (-.7116, -.7116)^T$ is the direction of steepest descent. We need to find an α_0 that satisfies

$$\begin{aligned} \alpha_0 &= \min_{\alpha>0}\{g((x_0, y_0)^T - \alpha\nabla g(x_0, y_0))\} \\ &= \min_{\alpha>0}\{g((3.6, 3.6)^T - \alpha(0.7116, 0.7116)^T)\} \\ &= \min_{\alpha>0}\{g(3.6 - 0.7116\alpha, 3.6 - 0.7116\alpha)^T\} \\ &= \min_{\alpha>0}\{[\ln(3.6 - 0.7116\alpha)]^2 + [\ln(3.6 - 0.7116\alpha)]^2\} \\ &= \min_{\alpha>0}\{2[\ln(3.6 - 0.7116\alpha)]^2\} \end{aligned}$$

and the minimum occurs where $\ln(3.6 - 0.7116\alpha) = 0$, that is, where

$$\begin{aligned} 3.6 - 0.7116\alpha &= 1 \\ \alpha &= (3.6 - 1)/0.7116 \\ &\doteq 3.1759 \end{aligned}$$

and so, from the method of steepest descent,

$$\begin{aligned} \begin{pmatrix} x_1 \\ y_1 \end{pmatrix} &= \begin{pmatrix} x_0 \\ y_0 \end{pmatrix} - \alpha_0 \begin{pmatrix} 2\ln(x_0)/x_0 \\ 2\ln(y_0)/y_0 \end{pmatrix} \\ &= \begin{pmatrix} 3.6 \\ 3.6 \end{pmatrix} - 3.1759 \begin{pmatrix} 0.7116 \\ 0.7116 \end{pmatrix} \\ &\doteq \begin{pmatrix} 1.3399 \\ 1.3399 \end{pmatrix} \end{aligned}$$

is the next guess. At this point we could do a few more iterations of the method of steepest descent in hopes of further improving the guess, or return to Newton's method and try this guess. Let's do the latter: Using the new initial guess $(x_0, y_0) = (1.3399, 1.3399)$ in Newton's method gives

$$x_1 = x_0 - x_0 \ln(x_0) \doteq 0.9478$$
$$y_1 = y_0 - y_0 \ln(y_0) \doteq 0.9478$$
$$x_2 = x_1 - x_1 \ln(x_1) \doteq 0.9986$$
$$y_2 = y_1 - y_1 \ln(y_1) \doteq 0.9986$$
$$x_3 = x_2 - x_2 \ln(x_2) \doteq 1.0000$$
$$y_3 = y_2 - y_2 \ln(y_2) \doteq 1.0000$$

and the method has converged to the solution $(1, 1)$. The method of steepest descent has salvaged this computation. □

For a truly uncooperative root-finding problem we might decide to use the minimization strategy only, that is, we might try to find the root by minimizing $\|F(x)\|^2$. Similarly, for certain minimization problems we might switch to root-finding on $f'(x)$ or $\nabla F(x)$. As a rule however it is advisable to solve the problem in the form in which it was posed.

Because the method of steepest descent is so well-behaved, the line search $\alpha_k = \min_{\alpha>0}(F(x_k - \alpha \nabla F(x_k)))$ used to find α_k need not be performed exactly. It may be better to generate an approximate α_k quickly and take a step to a new search region in the higher-dimensional space than to spend a great deal of time getting a very precise value of α_k. Methods that find such a step size α_k are sometimes called **inexact line search** or **practical line search** methods; a common rule of thumb is that it suffices to get within about 10% of a minimum. A similar line of reasoning is sometimes used with Newton's method for solving linear systems

$$x_{k+1} = x_k - [J(x_k)]^{-1} F(x_k)$$

in that we might solve

$$J(x_k)(x_{k+1} - x_k) = -F(x_k)$$

quickly but approximately for the step $x_{k+1} - x_k$, reasoning once again that it's better to quickly take large steps in the general direction of the solution than to slowly make large steps in the direction of the solution. (Near the minimum we may need more precise steps of course.) These methods are called **inexact Newton's methods**.

MATLAB

The MATLAB multivariate minimization method is `fminsearch`. If `f` is a function handle or the name of an appropriate m-file, `fminsearch(f,x0)` attempts to find a local minimum of `f` near `x0`. Practical problems would generally be sufficiently complicated that they would require an m-file. The algorithm method does not use derivative information and hence may be applied to fairly general functions but because of this it may be too slow for some purposes.

There are a number of MATLAB commands for handling functions of several variables. Enter:

```
» help gradient
```

This function computes numerical (approximate) gradients. Try the four-line example near the end of the help text. You'll see a contour plot of $z = xe^{(-x^2-y^2)}$ (that is, curves corresponding to those values of (x, y) on which $z = c$ for various choices of $c \in \mathbb{R}$). The arrows (supplied by the `quiver` command) point in the direction of the gradient at each point and the lengths of the arrows are proportional to $\|\nabla z(x, y)\|$. Evidently there is a local minimum on the left (at the center of the concentric contours on the left-hand side) and a local maximum on the right.

To make a simple contour plot of a function $z = f(x, y)$ of two variables, e.g., $f(x, y) = x^2 + y^2$, we must form a matrix of z values. The command `meshgrid` can be used to create a matrix of x and y values that form a grid. For example, enter:

`» [x,y]=meshgrid(-4:.1:4, -4:.1:4);`

to generate a grid of (x, y) values that cover $-4 \le x \le 4$, $-4 \le y \le 4$ with a spacing of .1 in each direction. (Compare `linspace`.) Then enter the command:

`» z=x.^2+y.^2;`

to compute the corresponding z values. Note that `z` is a matrix of the same size as `x` and `y`. Enter:

`» contour(z)`

to see the contour plot. (The scales on the x and y axes are incorrect because the z matrix does not contain information concerning the x and y values at which the function was evaluated, only what the function values z were at those points.) To approximate the gradient, enter:

`» [dx,dy]=gradient(z,.1,.1);`

which uses finite differences to approximate the components `dx` and `dy` of $\nabla z(x, y)$ (the second and third argument tell the `gradient` command what spacing was used between the z values; enter `type gradient` to see the details). Then enter:

`» hold on;quiver(dx,dy)`

to display the gradient values over the contours. The arrow at every point (x, y) points in the direction of steepest ascent from that point.

You may also use the diff command with symbolic variables to compute partial derivatives. For example:

`» syms x y`
`» diff(x^2+y^2,x),diff(x^2+y^2,y)`

gives the first component and second component of the gradient of $f(x, y) = x^2 + y^2$.

Problems

1.) a.) Use the method of steepest descent to locate a minimum of $f(x, y, z) = 2x^2 - 2x + y^2 - 4y + z^2 - 6z + 13$. (Use calculus for the line search.)

b.) Show that f has only one local minimum. What is the value of f at this minimizer?

2.) What happens if the method of steepest descent is applied to a smooth function $f : \mathbb{R} \to \mathbb{R}$?

3.) a.) Show by example that the method of steepest descent can converge to a stationary point that is neither a local minimum nor a local maximum.

b.) We say that a method is a **(general) descent method** for minimizing $F : \mathbb{R}^n \to \mathbb{R}$ if it has the form $x_{k+1} = x_k - \alpha_k s_k$ for some vector s_k and some positive constant α_k and, in addition, $F(x_{k+1}) \le F(x_{k+1})$ at each step. We say that s_k is a **descent direction** for a differentiable function F if $s_k \cdot \nabla F(x_k) < 0$. Show that the method of steepest descent is a general descent method that uses a descent direction (unless $\nabla F(x_k) = 0$ for some k).

c.) Suggest another choice of s_k that might be used.

4.) a.) Repeat Example 1 using an inexact line search (e.g., take $\alpha = 1.15$ on the first step).

b.) Repeat problem 1 using an inexact line search and the same number of steepest descent iterations as you used in problem 1. Comment on the efficiency of the two approaches.

5.) In using the method of steepest descent to improve an initial guess for performing root-finding on $F(x) = 0$ it was suggested that the function $\|F(x)\|^2 = f_1^2(x) + \ldots + f_n^2(x)$ be minimized. Why is this preferable to minimizing $\|F(x)\|$?

6.) a.) Use the method of steepest descent to find a minimum of $f(x, y) = x^2(y^4 + \sin^2(xy))$ starting from the initial guess $(100, 100)$. Use `fminbnd` (finding the brackets by trial-and-error) or any other reasonable approach for the line search.

b.) Are the minima of this function isolated?

c.) Use 2 different initial guesses (on the same scale) to find 2 different minima of f.

7.) a.) For problem 1, sketch the contours of $f(x, y)$ (use MATLAB if you wish), then locate $(x_0, y_0), (x_1, y_1), \ldots$ on the plot. Sketch the vectors that indicate the steps taken from (x_0, y_0) to (x_1, y_1) etc.

b.) The method of steepest descent $x_{k+1} = x_k - \alpha_k \nabla F(x_k)$ may be rewritten as $x_{k+1} - x_k = -\alpha_k \nabla F(x_k)$; since α_k is a scalar, this indicates that the vector $x_{k+1} - x_k$ from x_k to x_{k+1} is proportional to $-\nabla F(x_k)$ and since α_k is positive it is in the same direction as $-\nabla F(x_k)$, that is, in the direction opposite $\nabla F(x_k)$. Demonstrate this on your graph by sketching in $\nabla F(x_k)$ for several points (you may need to draw additional contours).

8.) a.) Consider a general descent method $x_{k+1} = x_k - \alpha_k s_k$ for some choice of the vectors s_k. Show that if s_k is a descent direction then there is always a choice of $\alpha_k > 0$ that will ensure that $F(x_{k+1}) \leq F(x_{k+1})$.

b.) Show by example that if s_k is not a descent direction then there need not be a $\alpha_k > 0$ that will ensure that $F(x_{k+1}) \leq F(x_{k+1})$.

9.) Write a detailed algorithm (pseudo-code) for performing the method of steepest descent. Assume that both the function and the gradient will be supplied, and that the norm of the gradient at the solution should be printed and the minimizer returned. Assume that an appropriate subroutine exists for the line search.

10.) a.) Write a detailed algorithm (pseudo-code) for performing Davidon's method.

b.) Use Davidon's method to estimate the locations of any two minima of $f(x) = x^9 - 2x^7 + 5x^4 - 2x + 2$ to at least four decimal places. Compare your results to those in problem 2 of Sec. 2.1.

11.) a.) Use the method of steepest descent to find the minimum of $z = x^2 + y^2$. Use MATLAB's `fminbnd` function for the line search.

b.) Repeat part a. with an inexact line search.

12.) a.) Construct a function of two variables that has a small but nonzero gradient in some region, and for which the method of steepest descent stagnates (makes very slow progress) in that region.

b.) Show by example that the method of steepest descent can be slow even if the function is quadratic (e.g., $f(x, y) = 2x^2 + 2xy + y^2 + 2$).

13.) a.) Using the function $f(x, y, z) = x^2 y^2 z^2 + 2x^2 + 3y^2 + 4z^2 + \cos(xy + z) - 1$, compare the method of steepest descent using line search by 15, 10, 5, and 3 iterations of golden section search, respectively. (Find the bracket by trial-and-error.) How many function evaluations are required in each case to achieve four decimal place accuracy? (Do not count any function evaluations used only to find a bracket.) Comment on the use of inexact line searches.

b.) Use MATLAB's `fminsearch` command to find a local minimum of this f.

14.) a.) The method of steepest descent requires that a one-dimensional search be performed at each iteration. We will not be able to automate the method of steepest descent with a bracketing method used for the line search unless we have an automatic means of generating the bracket; if not, the user must be prompted for a bracket at every step, which is clearly undesirable. Write a MATLAB program that accepts as input a function of a single variable and a single initial guess as to the location of a minimum and attempts to return a bracket for the minimum by some form of intelligent trial-and-error. Explain the method that you develop in detail.

b.) Modify your method and program to accept a scale parameter that indicates how rapidly the function changes as well as an initial guess; use the scale parameter to decide how far you should step away from the current point when looking for a new point.

15.) a.) A general descent method may not descend rapidly enough to be of use. The **Armijo rule** is the requirement that α satisfy $F(x_k - \alpha_k s_k) - F(x_k) < -\rho \alpha_k \nabla F(x_k) \cdot s_k$ where $\rho \in (0, 1)$ is a parameter ($\rho = 10^{-4}$ is a typical choice). We say that such an α_k gives **sufficient decrease** in the function value (as opposed to **simple decrease** $F(x_{k+1}) < F(x_{k+1})$ corresponding to the choice $\rho = 0$ or the nonincrease requirement $F(x_{k+1}) \leq F(x_{k+1})$ of a general descent method). Prove that if ∇F is continuous and $\{F(x_k 0\}_{k=0}^{k=\infty}$ is bounded from below then the method of steepest descent using the Armijo rule produces a sequence $\{x_k\}_{k=0}^{k=\infty}$ that converges to a stationary point of F.

b.) One method for finding an α_k that gives sufficient decrease is the method of **backtracking**, wherein we choose $\alpha_k = \delta^m$ for some $\delta \in (0, 1)$ and for the smallest integer $m \geq 0$ for which sufficient decrease is achieved; we start with $m = 0$ ($\alpha_k = 1$) and if it is not acceptable we try $\alpha_k = \delta$, $\alpha_k = \delta^2$, $\alpha_k = \delta^3$, and so on, until the criterion is satisfied. (Note that we do not need to find an initial bracket with this approach.) Write a MATLAB program that implements the method steepest descent with the Armijo rule for sufficient decrease, with α_k selected by backtracking. Test your program on several objective functions.

16.) a.) Suggest a method for detecting, and recovering from, stagnation near a nonextremal stationary point.

b.) Implement and test your method.

3. Newton Methods for Nonlinear Optimization

The method of steepest descent is too slow to use alone. It is used to start other methods, such as Newton's method, and to add robustness to other minimization algorithms, which switch to steepest descent when they encounter difficulties. In developing a faster method we will pay the usual price: It will require more assumptions and information–in this case, second derivatives–and it will be less robust with respect to the initial guess.

We can get a quadratically convergent method by using Newton's method to perform root-finding on $f'(x)$ (or Newton's method for systems on $\nabla F(x)$). If $f : \mathbb{R} \to \mathbb{R}$ then Newton's method applied to $f'(x)$ takes the form

$$x_{k+1} = x_k - f'(x_k)/f''(x_k)$$

assuming f is twice continuously differentiable. (Since we are dealing with f', our convergence theory would require that f be four times continuously differentiable so that f' is thrice continuously differentiable.) If we assume that x_0 is sufficiently close to a minimum x^* of f and that $f''(x^*)$ is nonzero, then we expect quadratic convergence of the method to x^*. Of course, if x_0 is not sufficiently close to a minimum then we might converge to a maximum of f instead. Note all the assumptions that we must make: That the function is several times differentiable, that we can compute f' and f'', that the minimum is parabolic ($f''(x^*) \neq 0$, and in fact $f''(x^*) > 0$ since x^* is a minimum), and that x_0 is sufficiently close to the minimum. If we are prepared to make all these assumptions, however, we *do* get a method–called **Newton's method for optimization**–that is quadratically convergent, and this is certainly desirable.

We might also derive this method as follows: If x_0 is approximately the location of a minimum of f, then near x_0 we have

$$(3.1) \qquad f(x) \approx f(x_0) + f'(x_0)(x - x_0) + \frac{1}{2}f''(x_0)(x - x_0)^2$$

that is, $f(x)$ is approximated by the quadratic model

$$q(x) = f(x_0) + f'(x_0)(x - x_0) + \frac{1}{2}f''(x_0)(x - x_0)^2.$$

The minimum of this quadratic occurs where $q'(x) = 0$. Differentiating Eq. (3.1) with respect to x and setting $f'(x)$ to zero gives

$$0 \approx 0 + f'(x_0) \cdot 1 + f''(x_0)(x - x_0)$$

(note that $f'(x_0)$, etc., are constants since we are differentiating with respect to x), leading to the method

$$x_{new} = x_0 - f'(x_0)/f''(x_0).$$

Hence if x_0 is near the location of a minimum x^* then x_{new} should be a better approximation of x^*. From this point of view Newton's method for minimization

$$x_{k+1} = x_k - f'(x_k)/f''(x_k)$$

is truly an optimization method and not merely an application of Newton's method for root-finding. This is an important point, and this latter point of view is generally preferable; however, we must remember that the method may converge to any point such that $f'(x) = 0$ (e.g., a local maximum) if x_0 is not sufficiently near a local minimum.

EXAMPLE 1: Let's use the method to find a minimum of $f(x) = (x - 10)(x - 20)(x - 50) = x^3 - 80x^2 + 1700x - 10000$. Clearly the only local minimum of this cubic lies between $x = 20$ and $x = 50$. We'll take $x_0 = 20$. We have

$$
\begin{aligned}
x_1 &= x_0 - f'(x_0)/f''(x_0) \\
&= x_0 - (3x_0^2 - 160x_0 + 1700)/(6x_0 - 160) \\
&= 12.5 \\
x_2 &= x_1 - f'(x_1)/f''(x_1) \\
&\doteq 14.4853 \\
x_3 &\doteq 14.6471 \\
x_4 &\doteq 14.6482 \\
x_5 &\doteq 14.6482
\end{aligned}
$$

and the method has converged, but to a local maximum (see Fig. 1).

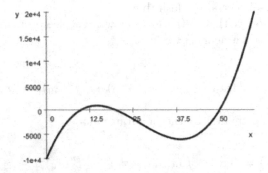

FIGURE 1. $y = (x - 10)(x - 20)(x - 50)$.

Perhaps this is not so surprising since $f''(20) = 6 \cdot 20 - 160 = -40$ so that at $x = 20$ f has the concavity that corresponds to a maximum, not to a minimum ($f''(x^*) > 0$).

Let's try again with $x_0 = 30$, where $f''(20) = 6 \cdot 30 - 160 = 20$. We have:

$$
\begin{aligned}
x_1 &= x_0 - f'(x_0)/f''(x_0) \\
&= x_0 - (3x_0^2 - 160x_0 + 1700)/(6x_0 - 160) \\
&= 50 \\
x_2 &= x_1 - f'(x_1)/f''(x_1) \\
&\doteq 41.4286 \\
x_3 &\doteq 38.9401 \\
x_4 &\doteq 38.6878 \\
x_5 &\doteq 38.6852 \\
x_6 &\doteq 38.6852
\end{aligned}
$$

and $f'(38.6852) \doteq -9.0949E - 013$, $f''(38.6852) \doteq 72.1110 > 0$ which is a good indication that we have found a local minimum. $\square$

Unlike previous methods we have seen for one-dimensional minimization, this approach requires only a single initial guess. Despite this, Newton's method is rarely employed for one-dimensional minimization; however, it and its variants are very common methods for multi-dimensional minimization.

The multi-dimensional form may be derived by applying Newton's method for systems to find a stationary point of $F : \mathbb{R}^n \to \mathbb{R}$ by solving the nonlinear system of equations $\nabla F(x) = 0$, or by using Taylor's series for a function of several variables to get the multi-dimensional analogue of Eq. (3.1). Taking the latter approach gives

$$(3.2) \qquad F(x) = F(x_0) + \nabla F(x_0)^T(x - x_0) + \frac{1}{2}(x - x_0)^T H(x_0)(x - x_0) + \dots$$

where $H(x)$ (sometimes denoted $\nabla^2 F(x)$) is the Hessian matrix

$$H(x) = \begin{bmatrix} \dfrac{\partial^2 F}{\partial x_1^2} & \dfrac{\partial^2 F}{\partial x_1 \partial x_2} & \cdots & \dfrac{\partial^2 F}{\partial x_1 \partial x_n} \\ \dfrac{\partial^2 F}{\partial x_2 \partial x_1} & \dfrac{\partial^2 F}{\partial x_2^2} & \cdots & \dfrac{\partial^2 F}{\partial x_2 \partial x_n} \\ \vdots & \vdots & \ddots & \vdots \\ \dfrac{\partial^2 F}{\partial x_n \partial x_1} & \dfrac{\partial^2 F}{\partial x_n \partial x_2} & \cdots & \dfrac{\partial^2 F}{\partial x_n^2} \end{bmatrix}$$

(or simply the **Hessian**) of $F(x)$. The Hessian matrix is symmetric if F is twice continuously differentiable.

Consider Eq. (3.2) and assume that the higher-order terms are negligible. If we take x_0 to be x^* we have

$$\begin{aligned} F(x) &= F(x^*) + \nabla F(x^*)^T(x - x^*) + \frac{1}{2}(x - x^*)^T H(x)(x - x^*) + \dots \\ &= F(x^*) + \frac{1}{2}(x - x^*)^T H(x^*)(x - x^*) + \dots \end{aligned}$$

since $\nabla F(x^*)$ is zero. Since $F(x^*)$ is the locally minimum value of F, it must be that

$$(x - x^*)^T H(x^*)(x - x^*) \geq 0$$

at least for x near x^*. (If not there would be a choice of x making this negative which, when added to $F(x^*)$, would give a smaller value of F.) If the minimum is a strict local minimum then we must have

$$(x - x^*)^T H(x^*)(x - x^*) > 0.$$

The quantity $x - x^*$ is just a vector; call it y. It appears that $H(x^*)$ is symmetric and that $y^T H(x^*)y \geq 0$ for all nonzero vectors y. Hence, the Hessian is positive semi-definite, and at a strict minimum it is positive definite. The Second Derivative Test for minima in several dimensions is that the Hessian be positive definite at the critical point. In the latter case we often say that we have a parabolic local minimum to indicate the bowl-shaped nature of the function there.

Newton's method for optimization can't simply be described as Newton's method for minimization because it is happy to land on any stationary point. To make it a minimization algorithm, we monitor the Hessian at each iterate and if it appears not to be positive definite we take some action other than a Newton step in order to get it (back) to such a region.

Using vector calculus on either Newton's method for finding roots applied to $\nabla F(x) = 0$, or on Eq. (3.2) to find its minimum (assuming $H(x^*)$ is positive definite in either case), we arrive at the formula

$$x_{k+1} = x_k - H^{-1}(x_k)\nabla F(x_k)$$

for the method. (If $H(x^*)$ is positive definite then it is necessarily nonsingular so that $H^{-1}(x)$ exists at the minimum $x = x^*$.) Of course, as always we would actually write this as

(3.3) $H(x_k)(x_{k+1} - x_k) = -\nabla F(x_k)$

and solve for the step $d_k = x_{k+1} - x_k$, then compute x_{k+1} from $x_{k+1} = x_k + d_k$.

Let's focus on the case of a sufficiently differentiable function of two variables $F : \mathbb{R}^2 \to \mathbb{R}$. For such a function $F(x, y)$ the gradient is

$$\nabla F(x,y) = \left(\begin{array}{c} \frac{\partial F}{\partial x} \\ \frac{\partial F}{\partial y} \end{array} \right)$$

and the Hessian is

$$H(x,y) = \left[\begin{array}{cc} \frac{\partial^2 F}{\partial x^2} & \frac{\partial^2 F}{\partial x \partial y} \\ \frac{\partial^2 F}{\partial y \partial x} & \frac{\partial^2 F}{\partial y^2} \end{array} \right].$$

The Hessian is positive definite wherever its eigenvalues are both positive; for a 2×2 symmetric matrix it can be shown that this occurs precisely when the determinant is positive and the $(1,1)$ entry of the matrix is also positive. Newton's method now takes the form

$$\left(\begin{array}{c} x_{k+1} \\ y_{k+1} \end{array} \right) = \left(\begin{array}{c} x_k \\ y_k \end{array} \right) - \left[\begin{array}{cc} \frac{\partial^2 F}{\partial x^2} & \frac{\partial^2 F}{\partial x \partial y} \\ \frac{\partial^2 F}{\partial y \partial x} & \frac{\partial^2 F}{\partial y^2} \end{array} \right]^{-1} \left(\begin{array}{c} \frac{\partial F}{\partial x} \\ \frac{\partial F}{\partial y} \end{array} \right)$$

where the Hessian and gradient on the right-hand side are evaluated at $(x, y) = (x_k, y_k)$.

EXAMPLE 2: Let's apply Newton's method to find a local minimum of the function $F(x, y) = xe^{(-x^2 - y^2)}$. We have

$$\partial F / \partial x = e^{-x^2 - y^2} - 2x^2 e^{-x^2 - y^2}, \partial F / \partial y = -2xye^{-x^2 - y^2}$$

so $\nabla F(x, y) = (e^{-x^2 - y^2}(1 - 2x^2), -2xye^{-x^2 - y^2})^T$. Also

$$\begin{array}{rcl} \partial^2 F / \partial x^2 & = & -6xe^{-x^2 - y^2} + 4x^3 e^{-x^2 - y^2} \\ \partial^2 F / \partial x \partial y & = & \partial^2 F / \partial y \partial x = -2ye^{-x^2 - y^2} + 4x^2 ye^{-x^2 - y^2} \\ \partial^2 F / \partial y^2 & = & -2xe^{-x^2 - y^2} + 4xy^2 e^{-x^2 - y^2} \end{array}$$

so

$$H(x,y) = \begin{bmatrix} -6xe^{-x^2-y^2} + 4x^3 e^{-x^2-y^2} & -2ye^{-x^2-y^2} + 4x^2 y e^{-x^2-y^2} \\ -2ye^{-x^2-y^2} + 4x^2 y e^{-x^2-y^2} & -2xe^{-x^2-y^2} + 4xy^2 e^{-x^2-y^2} \end{bmatrix}.$$

Hence Newton's method for this function is

$$\begin{pmatrix} x_{k+1} \\ y_{k+1} \end{pmatrix} = \begin{pmatrix} x_k \\ y_k \end{pmatrix} - H(x_k, y_k) \begin{pmatrix} e^{-x^2-y^2}(1-2x^2) \\ -2xye^{-x^2-y^2} \end{pmatrix}.$$

Let's take $(x_0, y_0) = (-1, -1)$ as the initial condition. (Note that $(x_0, y_0) = (0,0)$ would not work since $H(0,0)$ is the zero matrix and hence singular.) We have

$$\begin{pmatrix} x_1 \\ y_1 \end{pmatrix} = \begin{pmatrix} -1 \\ -1 \end{pmatrix} - \begin{bmatrix} 6e^{-2} + -4e^{-2} & 2e^{-2} + -4e^{-2} \\ 2e^{-2} + -4e^{-2} & 2e^{-2} + -4e^{-2} \end{bmatrix}^{-1} \begin{pmatrix} -e^{-2} \\ -2e^{-2} \end{pmatrix}$$

$$= \begin{pmatrix} -1 \\ -1 \end{pmatrix} - \begin{bmatrix} 2e^{-2} & -2e^{-2} \\ -2e^{-2} & -2e^{-2} \end{bmatrix}^{-1} \begin{pmatrix} -1 \\ -2 \end{pmatrix} e^{-2}$$

$$= \begin{pmatrix} -1 \\ -1 \end{pmatrix} - \frac{e^2}{2} \begin{bmatrix} 1 & -1 \\ -1 & -1 \end{bmatrix}^{-1} \begin{pmatrix} -1 \\ -2 \end{pmatrix} e^{-2}$$

$$= \begin{pmatrix} -1 \\ -1 \end{pmatrix} - \frac{1}{2} \left(\frac{1}{-1-1} \right) \begin{bmatrix} -1 & 1 \\ 1 & 1 \end{bmatrix} \begin{pmatrix} -1 \\ -2 \end{pmatrix}$$

$$= \begin{pmatrix} -1 \\ -1 \end{pmatrix} + \frac{1}{4} \begin{pmatrix} -1 \\ -3 \end{pmatrix}$$

$$= \begin{pmatrix} -1.25 \\ -1.75 \end{pmatrix}$$

$$\begin{pmatrix} x_2 \\ y_2 \end{pmatrix} \doteq \begin{pmatrix} -1.3535 \\ -2.0314 \end{pmatrix}$$

$$\begin{pmatrix} x_3 \\ y_3 \end{pmatrix} \doteq \begin{pmatrix} -1.4416 \\ -2.2629 \end{pmatrix}$$

$$\begin{pmatrix} x_3 \\ y_3 \end{pmatrix} \doteq \begin{pmatrix} -1.5206 \\ -2.4654 \end{pmatrix}$$

$$\begin{pmatrix} x_4 \\ y_4 \end{pmatrix} \doteq \begin{pmatrix} -1.5933 \\ -2.6481 \end{pmatrix}$$

$$\begin{pmatrix} x_5 \\ y_5 \end{pmatrix} \doteq \begin{pmatrix} -1.6611 \\ -2.8161 \end{pmatrix}$$

$$\begin{pmatrix} x_6 \\ y_6 \end{pmatrix} \doteq \begin{pmatrix} -1.7251 \\ -2.9727 \end{pmatrix}.$$

At this point the gradient is $(-1.0291E - 4, -2.1308E - 4)$; we seem to be making very slow progress. Let's find the minimum analytically by setting the gradient equal to zero:

$$\begin{pmatrix} e^{-x^2-y^2}(1-2x^2) \\ -2xye^{-x^2-y^2} \end{pmatrix} = \begin{pmatrix} 0 \\ 0 \end{pmatrix}$$

$$\begin{pmatrix} (1-2x^2) \\ -2xy \end{pmatrix} = \begin{pmatrix} 0 \\ 0 \end{pmatrix}$$

so that

$$\begin{aligned} x^2 &= 1/2 \\ -2xy &= 0 \end{aligned}$$

so $x = \pm 1/\sqrt{2}$ and so $y = 0$. The extrema are at $(1/\sqrt{2}, 0)$ and $(-1/\sqrt{2}, 0)$. We have $H(1/\sqrt{2}, 0) = [-1.7155\ 0; 0\ -0.8578]$ (in MATLAB notation) which is negative definite (maximum), and $H(-1/\sqrt{2}, 0) = [1.7155\ 0; 0\ 0.8578]$ which is positive definite (minimum). There is a parabolic minimum but the method seems to be slowly diverging from it.

Let's try again with a closer initial condition, say $(x_0, y_0) = (-.8, -.1)$. We have

$$\begin{pmatrix} x_1 \\ y_1 \end{pmatrix} \doteq \begin{pmatrix} -0.6961 \\ 0.0058 \end{pmatrix}$$

$$\begin{pmatrix} x_2 \\ y_2 \end{pmatrix} \doteq \begin{pmatrix} -0.7070 \\ 0.0000 \end{pmatrix}$$

$$\begin{pmatrix} x_3 \\ y_3 \end{pmatrix} \doteq \begin{pmatrix} -0.7071 \\ 0.0000 \end{pmatrix}$$

and we have clearly converged to the minimum. (In fact, x_3 agrees with $x^* = 1/\sqrt{2}$ to 7 decimal places, and $y_3 \doteq 8E - 14$ is in excellent agreement with $y^* = 0$; $\|\nabla F(x_3, y_3)\| \doteq 8E - 9$.) As promised, convergence was rapid once we were sufficiently close (which in this case appears to mean very close). $\square$

Convergence criteria typically include a consideration of the size of $\|\nabla F(x_k)\|$ and possibly the approximate relative and/or absolute errors. We also monitor the Hessian for positive definiteness. A simple approach is to test a few vectors to see if $y^T H y > 0$ appears to hold. A better approach is to use the Cholesky decomposition, stopping when the computation of this decomposition fails due to the need to take a square root of a negative number. The computation of the Cholesky decomposition can be modified to more efficiently check for positive definiteness.

If the method fails for a given x_0 then it's worth trying a few iterations of steepest descent in hopes of moving the initial guess into a better region. Remember, steepest descent *is* capable of telling a minimum from a maximum, though it may still get hung up on a saddle point, or in cases of incredibly bad luck a maximum (if it should happen to land on one, which almost certainly won't happen in practice). A hybrid method like this is very commonly employed: Start with a few iterations of steepest descent, switch to Newton's method, and switch back if the going gets rough.

Finite difference approximations of the gradient and Hessian, constant or infrequently updated Hessians, damping, and other variants are all used in conjunction with this basic idea. It is common to incorporate various heuristics into any program implementing such a method, such as periodically "restarting" the method by taking a step in a descent direction to ensure that progress toward a minimum is indeed being made.

As with root-finding, Newton's method and its variants are usually implemented with a step length parameter ϖ_k in the form

$$x_{k+1} = x_k - \varpi_k H^{-1}(x_k)\nabla F(x_k)$$

(or more often with some approximation A_k replacing $H^{-1}(x_k)$). The step length parameter will not necessarily represent damping (recall that $0 < \varpi_k \leq 1$ for the damped Newton's method). The parameter serves in part to insure that we achieve sufficient decrease in the function value in order to keep the method progressing in regions where, for example, $\|\nabla F(x_k)\|$ is small but nonzero, and may be found by a line search.

There are special situations that arise commonly and for which specialized versions of the method have been developed. One of the most important such cases is when the objective function $F(x)$ is known to be a sum of squares of other functions:

$$F(x) = \sum_{k=1}^{n} F_k^2(x)$$

(for example, if we are fitting a curve by nonlinear least squares). When F is known to have this form we should consider employing a method developed for handling this type of problem. One such method is the **Gauss–Newton method**, a modified version of Newton's method for optimization specialized for this particular sort of objective function. It attempts to switch between the method of steepest descent far from the minimum and Newton's method with an approximate version of the Hessian $H(x_x) \approx 2J^T(x_x)J(x_x)$ nearer to the minimum. The **Levenberg-Marquardt method** (also known as the **damped least squares algorithm**) further modifies this idea for a slower, but very robust, method.

MATLAB

In trying to visualize how Newton's method works it is convenient to be able to plot functions of two variables. The two main 3-D plotting commands in MATLAB are `mesh` and `surf`; `mesh` plots a simple mesh (wireframe) surface while `surf` plots a shaded surface graph. For example, enter:

```
» clc
» [x,y]=meshgrid(-2:.2:2, -2:.2:2);
» z=x.^2+y.^2;
» mesh(z)
» figure
» surf(z)
```

(The `figure` command makes a new figure window for the `surf` command; otherwise it would have replaced the mesh plot.) The `surf` command has shaded the surface between points but is otherwise similar to the mesh plot.

Edit the line that used `meshgrid` to define x and y in order to make an even finer grid, then recompute z and enter:

```
» close all
```

to close all the figures. Then repeat the mesh and surface plots in the form:
» mesh(x,y,z);figure;surf(x,y,z)
which gives the figures with correct axis labeling. See the help for the commands
xlabel, title, and text for more labeling options.

Just as there is ezplot command to go with plot, there is an ezsurf command
to go with surf (and also an ezmesh to go with mesh). Enter:
» close all
>> f=@(x,y) x.^2+y.^2
>> ezsurf(f)
>> ezsurfc(f)
>> ezsurf('r*cos(t)','r*sin(t)','t')
and see also the examples listed under help ezsurf.

Problems

1.) a.) Repeat Example 2 using the original initial guess $(x_0, y_0) = (-1, -1)$ but
use the method of steepest descent instead of Newton's method.

b.) Repeat Example 2 using the original initial guess $(x_0, y_0) = (-1, -1)$ but
perform several iterations of the method of steepest descent first to improve this
guess before using Newton's method. Your final answer should be accurate to the
last digit.

2.) a.) Use Newton's method to find a minimum of $F(x, y) = x^4 - 2xy + (x-1)^2(y - 1)^2 e^{-xy}$.

b.) Use Newton's method to find a minimum of $F(x, y) = x^4 - 2xy + (x-1)^2(y - 1)^2 e^{-xy}$ but update the Hessian matrix only every 3 iterations (that is, use $H(x_0)$
in finding x_1, x_2, and x_3, then use $H(x_3)$ to find x_4, x_4, and x_6, and so on).

3.) Explain the divergence of Newton's method seen in Example 2 for the initial
guess $(x_0, y_0) = (-1, -1)$ in terms of the shape of the objective function. (It may
be helpful to plot its contours again.)

4.) Carefully state a convergence theorem for Newton's method for minimization in
the special case of a function $f : \mathbb{R} \to \mathbb{R}$. (Use the results in Sec. 1.3 and Sec. 1.4.)
Under what conditions will the convergence be quadratic?

5.) a.) Prove that a symmetric 2×2 matrix $A = [a \ b; c \ d]$ is positive definite if
and only if $a > 0$ and $ad - bc > 0$.

b.) Show by example that a symmetric 3×3 matrix A with $a_{11} > 0$ and $\det(A) > 0$
need not be positive definite.

6.) a.) Use Newton's method to find a local minimum of $f(x, y) = 4x^2 - 12xy + 9y^2 + x^3 - 6x^2y + 12xy^2 - 8y^3$.

b.) Find the extrema analytically by solving $\nabla f(x, y) = 0$. Classify the extrema.

7.) Use Newton's method to find a local minimum of $f(x, y) = x^4 + 5x^2y^4 - 40x^2y^3 + 120x^2y^2 - 160x^2y + 80x^2 + \sin xy$.

8.) What is the most general function $f(x_1, ..., x_n)$ that has a constant Hessian
matrix?

9.) Make a surface plot of the function $F(x, y) = xe^{(-x^2-y^2)}$ of Example 2 near its
minimum that clearly shows that it is indeed a local minimum.

10.) Show that if $F : \mathbb{R}^2 \to \mathbb{R}$ then Eq. (3.2) corresponds to the usual tangent
surface (tangent plane) approximation of $F(x, y)$ at (x_0, y_0) if we neglect the term
involving the Hessian, and is a quadratic (in x and y) approximation of $F(x, y)$ at
(x_0, y_0) if we include the term involving the Hessian.

11.) Make a surface plot of the function $f(x, y) = 4x^3 - xy^2 + \ln(xy) + 1$ for (x, y) in the first quadrant. Use it to find an initial guess for Newton's method to find a local minimum of $f(x, y)$. The minimum should also lie in the first quadrant.

12.) a.) Find an initial guess such that Newton's method for the objective function of Example 2 diverges.

b.) Applying damping to Newton's method for optimization gives $x_{k+1} = x_k - \mu_k H^{-1}(x_k) \nabla F(x_k)$ for some damping sequence $\{\mu_k\}_{k=0}^{k=\infty}$, $0 < \mu_k \le 1$ and $\mu_k \to 1$ as $k \to \infty$. (Refer to Sec. 1.5.) Find a damping sequence that gives convergence for your initial guess from part a.

13.) a.) Show that if $H^{-1}(x_k)$ is positive definite then $s_k = H^{-1}(x_k) \nabla F(x_k)$ is a descent direction (unless $\nabla F(x_k) = 0$).

b.) Show that Newton's method is not always a general descent method.

c.) A **modified Newton's method for minimization** substitutes a descent direction s_k for $H^{-1}(x_k) \nabla F(x_k)$ when the latter does not appear to be a descent direction (e.g., when $H^{-1}(x_k)$ is not positive definite). One possibility is to use $x_{k+1} = x_k - \varpi_k M \nabla F(x_k)$ for some matrix M. Show that Newton's method and the method of steepest descent are both of this form for some sequence ϖ_k and matrix M.

d.) Write a detailed algorithm (pseudo-code) for a modified Newton's method that switches to the method of steepest descent when $H^{-1}(x_k)$ is not positive definite. Assume that a test for positive definiteness is available.

14.) a.) Show that the standard Second Derivative Test for a function of two variables to have a minimum, $\partial^2 f / \partial x^2 > 0$ and $(\partial^2 f / \partial x^2)(\partial^2 f / \partial y^2) - (\partial^2 f / \partial x \partial y)^2 > 0$, is equivalent to the positive definiteness of the Hessian matrix.

b.) This is a special case of the result that an $n \times n$ symmetric matrix is positive definite if and only if every upper-left $p \times p$ submatrix has a positive determinant $(p = 1, 2, ..., n)$. Prove this.

15.) a.) The quadratic form $F(x) = x^T A x$ has gradient $2Ax$ and Hessian $2A$. If A is positive definite, what happens when Newton's method is applied to $F(x)$?

b.) Show by example that the method of steepest descent can be slow for a quadratic form with A positive definite. Must it converge nonetheless?

16.) Write a MATLAB program that combines steepest descent and Newton's method for a robust and fast minimizer. Test it.

4. Multiple Random Start Methods

We have been focusing on finding local minima in the unconstrained case, that is, on the problem of finding any local minimum of $F : \mathbb{R}^n \to \mathbb{R}$, anywhere in $\mathbb{R}^n$. The constrained nonlinear optimization problem, where we add the condition that the minimum must lie in some proper subset D of $\mathbb{R}^n$, is often more difficult to solve because enforcing the requirement that the search remain within D can be challenging in several dimensions, even when D has a relatively simple geometry. In many the region D (called the **constraint region** or **feasible region**) is defined in such a way that even finding an initial guess $x_0 \in D$ (called a **feasible point**) to start off an iterative method can be difficult. Nonetheless there are algorithms for this very commonly occurring type of problem.

Let's focus instead on the **global minimization** problem, that is, the problem of finding the global minimum x^* of F. This is the (not necessarily unique) point x^* such that $F(x) \geq F(x^*)$ for all $x \in \mathbb{R}^n$. As usual, our development assumes a well-behaved parabolic global minima.

It may happen that there is only one local minimum of the function, but this need not be a global minimum. If $F(x) \to -\infty$ as $\|x\| \to \infty$, for example, then the global minimum may be undefined. (See Fig. 1.) If the function is **unimodal** (strictly increasing to either side of the minimum, like $y = x^2$, or in the case of a maximum strictly decreasing to either side of the maximum, like $y = -x^2$) then the local minimum must be the global minimum. The development of many minimization algorithms assumes that the function is unimodal in the region of interest (e.g., Davidon's cubic-fitting method in Sec. 7.2). Clearly, no special global minimization method is needed for a unimodal function.

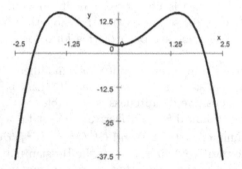

FIGURE 1. Local but not global minimum.

Any minimization algorithm might find the global minimum by chance; we want a method that does so by design. This is difficult. One approach is the **tunneling method**: Pick any initial guess x_0 and perform a local minimization starting from that point using any local minimization method. For the sake of definiteness, let's say it's the method of steepest descent. Call the location of this point w_0, and let m_0 be the value $F(w_0)$ of the objective function at this local minimum. We would like to find another, lower local minimum. If we can solve

$$(4.1) \qquad\qquad\qquad F(x) = m_0$$

(by finding a root of $F(x) - m_0 = 0$) for a point x_1 that is different from w_0, then we can perform a new local minimization starting from x_1. This gives a new point w_1 with function value $m_1 = F(w_1)$; we now solve

$$F(x) = m_1$$

for a point x_2 that is different from w_1, and start again with a new local minimization. We continue until $F(x) = m_k$ does not have a solution other than w_k.

If the global minimum is unique, this method should find it (in principle). The method tunnels from the local minimum it has found, under other local minima, to a point that we hope lies in the valley of another minimum. The method can avoid a great many irrelevant local minima for a function of several variables.

However, there are a number of difficulties with this method. Even if the local minimization algorithm converges each time and furthermore converges to a new minimum each time (as opposed to one previously located, which is possible if it is not a descent method), the root-finding algorithm will often find the current minimum, not a new point. (Deflation may be helpful here.) Additionally, if the root-finding algorithm fails to find a root, it is not clear whether it is because no other root exists, or because the method is simply not finding a new one. There is no general fix for this problem.

Let's look at a different approach. Pick an initial guess x_0 and perform a local minimization starting from that point (once again, for simplicity, let's say that the method of steepest descent is used). Call the location of this point $x_{\min}$, and let $m = F(x_{\min})$ be the value of F at this local minimum. Pick another point x_1 at random and perform a local minimization starting from that point. If the value of F at this minimizer is lower than the current lowest known value m of F, make this minimizer the new $x_{\min}$, and let m be the value $F(x_{\min})$ of F at this new, lower local minimum. Repeat with a new point x_2 chosen again at random; stop when some maximum allowable number of function values (or of initial guesses) is exceeded.

If the starting points x_0 are in fact chosen at random (uniformly over the region of interest) then this method is known as the **multiple random start** (abbreviated **MRS**) method, also called the **multistart** or **pure random search**. Any local minimizer may be employed within it.

EXAMPLE 1: Let's use the MRS method on the function $F(x) = \sin^2(100\|x\|^2) + \|x\|^2$ (which clearly has a unique global minimum at the origin) for $x \in \mathbb{R}^2$. We'll confine our search to the region $-10 \leq x, y \leq 10$; we may view this as a constrained minimization problem where x and y must both be between -10 and 10 for the solution to make sense, or as we may view the restriction $-10 \leq x, y \leq 10$ as a practical concession that must be made. The MATLAB function **rand** returns a value between 0 and 1, so:

```
» x0=20*(rand-.5),y0=20*(rand-.5)
```

gives an x_0 and a y_0 in the desired range. Suppose the values returned are $x_0 = 9.0026$, $y_0 = -5.3772$. Then MATLAB's local optimization routine **fminsearch** may be employed:

```
» F=inline('sin(norm(x)^2)^2+norm(x)^2');
» x1=fminsearch(F,[x0 y0])
```

The result is a local minimum at $(-0.1594, 0.3628)$ where $F(x_1) \doteq 0.1571$ (use **feval(F,x1)**). Choose a new x_0 and y_0 by entering:

```
» x=20*(rand(1 2])-.5)
```

that generates a random 1×2 matrix (that is, a row vector with 2 randomly selected entries) and rescales them to the desired range. Let's say the values are $x_0 = 2.1369$, $y_0 = -0.2804$. Then:

```
» x2=fminsearch(F,x)
```

searches for a minimum near the vector x. The result is a local minimum at $(2.0301, -0.2974)$ where $F(x_2) \doteq 4.2097$ (from **feval(F,x2)**). Our best current estimate of the location of the global minimum is $x_1 = (-0.1594, 0.3628)$. Let's try once more:

```
» x=20*(rand(1 2])-.5)
» x3=fminsearch(F,x)
```

» `feval(F,x3)`

giving, say, $x_0 = 7.8260$, $y_0 = 5.2419$, and a local minimum at $(0.3061, -0.0228)$ where $F(x_3) \doteq 0.0942$. This is smaller than our previous best estimate 0.1571, so $(0.3061, -0.0228)$ is our new best estimate of the location of the global minimizer. □

In the example above (x, y) pairs closer to the global minimizer had function values closer to that of the global minimum, but in general this need not be true. In the example we never found the same local minimum twice but in general we would not be at all surprised that that happened.

The n components of $x = (x_1, ..., x_n)$ are chosen uniformly over some domain that is typically of the form $l_i \leq x_i \leq u_i$ $(i = 1, ..., n)$. This may be a constraint region but could be just a limitation imposed so that we can generate the random numbers.

In fact, it's unreasonable and usually undesirable to try to use truly random numbers. We use a **pseudo-random number generator** (abbreviated **PRNG**) provided by the computing environment. The numbers are called pseudo-random[1] because they are generated by a deterministic algorithm that is designed to make the resulting sequences appear random despite not being truly random. Difficulties can arise because of the PRNG but for now we will not consider them. Methods such as this, which use (pseudo-)random numbers, are called **Monte Carlo methods** (the term **(stochastic) simulation methods** is often used interchangeably with this phrase). They arise in many areas of scientific computing and the simulation of physical processes.

What are the advantages of using a Monte Carlo approach? By choosing the initial guesses at random, we can give probabilistic estimates of the likelihood of finding the global minimum under various assumptions. Each new run of the program gives a new selection of initial guesses, so that if the method fails to find the minimum one time it has a chance the next time. (By the same token, starting the PRNG in the same state each time allows us to re-run our program with the same "random" data, which is useful for debugging purposes.) By not covering the region evenly, the method has a chance of finding small regions which would slip between the spacing of the grid, or of clustering several initial guesses in one region where there may be several closely spaced minima. If we have some specialized knowledge about our function–say, that there are only N local minima in the region of interest and each has a large parabolic valley–then we might use a different approach. But if we know (or assume) nothing about F, then we might look at a method based on pseudo-random numbers. The higher the number of dimensions, the more attractive these methods become.

The **modified multiple random start** (or **MMRS**) method is similar to the MRS method with one modification: Every time a new initial point is generated, we check to see if its function value is less than the current lowest known function value m; if it is, we proceed as before, and if not we reject it and choose another x_0 until we find one such that $F(x_0) < m$. In practice the methods give similar results.

EXAMPLE 2: Consider again the objective function $F(x) = \sin^2(100\|x\|^2) + \|x\|^2$ of Example 1, but now let us use the MMRS method instead. For the first iteration

---

[1]Although it's common to describe these numbers as "random" in casual discussions, it's important to remember that they are not.

we found $x_0 = 9.0026$, $y_0 = -5.3772$, giving a local minimum at $(-0.1594, 0.3628)$ with function value $F(x_1) \doteq 0.1571$. For the second iteration we found $x_0 = 2.1369$, $y_0 = -0.2804$; for the MMRS method we do not immediately perform a minimization but instead compute $F(2.1369, -0.2804) \doteq 4.8414$; this is greater than 0.1571, so we reject this choice. For the third iteration we found $x_0 = 7.8260$, $y_0 = 5.2419$; $F(7.8260, 5.2419) \doteq 88.9773$ and once again we reject this choice.

If this seems inefficient, note that rejecting these two initial guesses cost us only two function evaluations plus the cost of using **rand** to generate them; the use of **fminsearch** with the initial condition $x_0 = 7.8260$, $y_0 = 5.2419$ used 121 function evaluations. If function evaluations are by far the most expensive part of our computations–as is typically the case–this means that using MMRS represented a considerable savings of function evaluations that may be spent elsewhere. □

There are banks of test problems for local and global minimization problems. These are used to benchmark new software and algorithms. Some examples are the **three-hump camelback function**

$$(4.2) \qquad f(x, y) = 12x^2 - 6.3x^4 + x^6 - 6xy + 6y^2$$

which has three local minima including the global minimum at the origin (some sources use $f(x, y)/6$ for this function), and the **six-hump camelback function**

$$(4.3) \qquad f(x, y) = 4x^2 - 2.1x^4 + \frac{1}{3}x^6 + xy - 4y^2 + 4y^4$$

which has six local minima including two global minima at $(\pm 0.08984, \mp 0.71266)$. The **Levy No. 5** test problem is

$$(4.4) \qquad f(x, y) = \sum_{i=1}^{5} i \cos((i-1)x + i) \sum_{j=1}^{5} j \cos((j+1)y + j)$$

$$(4.5) \qquad + (x + 1.42513)^2 + (y + 0.80032)^2$$

and it has 760 local minima in $-10 \le x, y \le 10$, only one of which is the global minimum; dropping off the last two terms gives the **Levy No. 3** test problem

$$(4.6) \qquad f(x, y) = \sum_{i=1}^{5} i \cos((i-1)x + i) \sum_{j=1}^{5} j \cos((j+1)y + j)$$

which again has 760 local minima in $-10 \le x, y \le 10$, but for Levy No. 3 there are now 18 locations that are global minimizers.

In practice Monte Carlo methods are often an excellent choice for the global minimization of functions $F : \mathbb{R}^n \to \mathbb{R}$ with many local minima, especially if n is moderate and the global minimum has a reasonably sized parabolic valley so that some initial guess is likely to land in it sooner or later. (As was the case with root-finding, missing a small blip–a very narrow valley–is always a possibility.) A more powerful search such as Newton's method may be used with it if the function is sufficiently smooth.

We might also consider a **quasi-Monte Carlo method** based on a **quasi-random number generator** that generates numbers that are quasi-random, meaning "just random enough for our purposes"[2]. Quasi-random number generators generate deterministic sequences that cover a region with a non-uniform grid and they generate different sequences for each initial condition, but these sequences will not pass nearly as many tests of randomness as a PRNG would. For example, typically the quasi-random number generators ensure that every sub-region is sampled a minimum number of times. For a PRNG more deviation would be expected. It's also possible to have the quasi-random numbers appear in order, as though a pseudo-random sequence had been sorted. Methods based on a PRNG are more commonly employed, partly because the PRNG is generally provided as a built-in function of the language; the error estimates are typically considerably better for quasi-Monte Carlo methods, called **quasi-random searches** in the optimization context.

There are many other global minimization algorithms besides the few we have discussed here, and a great many methods for special cases. For a general function $F : \mathbb{R}^n \to \mathbb{R}$ finding a global minimizer is still a difficult problem, however.

MATLAB

Let's explore a few of the MATLAB commands that were used in the reading and Problems. The command `fminsearch` uses the Nelder-Mead method, a derivative-free method intended for not-necessarily-smooth objective functions, to perform a local optimization for a scalar-valued function of a vector. To display the number of function evaluations used, enter:

```
» options=optimset('display','iter');
» fminsearch(@(x) abs(sin(x)),1,options)
```

We can display iteration-by-iteration counts like this, if desired, by entering a line like `disp(['Searches performed= ',num2str(count)])` which uses the `disp` command to display the character (as opposed to numeric) vector found by combining (concatenating) the two character vectors `'Searches performed= '` and `num2str(count)`. The `num2str` command converts a number such as 6 to a character `'6'`. Try:

```
» disp(['Searches performed= ',num2str(6)])
» disp(['Searches performed=',num2str(6)])
» disp('Searches performed= ',num2str(6))   %Syntax error.
» disp(['Searches performed= ',6])   %Semantic error.
» disp('Searches performed= '),disp('6')
» disp('Searches performed= '),disp(6)
```

The last version above is acceptable but the formatting of the first version is preferable. See `help str2num` and `help str2double` for converting characters to numeric values.

Recall that the `rand` function returns a pseudo-random matrix of the requested size; for example, `rand([M N])` is a pseudo-random $M \times N$ matrix. The PRNG depends on a seed (an initial condition) that is always set to the same value when MATLAB is started. To reset the PRNG to its initial state,

```
» rng('default')
```

---

[2]Quasi-random numbers have low **discrepancy**, meaning that the proportion of points falling in a given region is roughly proportional to its volume.

```
>> rng('default')
>> rand(1)
>> rand(1)
>> rng('default')
>> rand(1)
>> rand(1)
```

(The use of rng replaces the now-discouraged method of using rand('state',0).) This can be useful when debugging a program. See help rand for more information, and help randn for normally distributed pseudo-random numbers.

We can pull some of these ideas into a script, that is, a list of MATLAB commands. Use the edit command to create an m-file lsearch.m and enter the following lines (only):

```
x0=rand([1 N]);
x1=fminsearch(F,x0);
f1=feval(F,x1);
disp('lsearch ended successfully')
```

Save these four lines in lsearch.m. When you run lsearch, MATLAB will use the value of N and F defined in the workspace and will perform these steps. For example, enter:

```
» clear all
» N=2;
» rng('default')
» F=@(x) sin(norm(x)^2)^2+norm(x)^2
» lsearch
» whos
```

This duplicates the first search performed in Example 1, though the specific random numbers returned may vary from version to version of MATLAB. Check that x0, x1, and f1 are returned to the workspace (the variables are not local to the script as they would be in a function). Enter:

```
» lsearch
```

and ask again for the values of x0, x1, and f1; they have been updated. With this script we can define an objective function F in the workspace and manage the overall logic of the MRS or MMRS method there, while calling the script to perform repetitive sequences of commands. As a rule, functions are to be preferred as they return values that may be used by other functions but a script like this is convenient while initially developing an algorithm. The commands pause (pause execution of the script or function and wait until a key is pressed) and echo (turn the display of script commands on and off) can be useful when using scripts.

Problems

1.) a.) Use ten additional initial guesses and the MRS method to attempt to find the global minimum of the objective function in Example 1. You may use any local optimization algorithm you wish.

b.) Which of the 10 additional initial guesses you found in part a. would have been used as the starting point for a local optimization if you had been using the MMRS method instead? Compare the MMRS estimate of the location of the global minimum to that of the MRS method.

2.) a.) Write a MATLAB program that performs the MRS method on a given function. You may use `fminsearch` or any other local optimizer. The user should supply a maximum number of allowable initial guesses.

b.) Test your program on the objective function of Example 1. How often does it find the global minimum if it is allowed $N = 20$ initial guesses? (Run your program several times and compute an average; see `help mean`.) How often does it find the global minimum if it is allowed $N = 100$ initial guesses?

3.) a.) Write a MATLAB program that performs the MMRS method on a given function. You may use `fminsearch` or any other local optimizer. The user should supply a maximum number of allowable initial guesses. In addition to returning the minimizer, the program should display the number of local searches actually performed. (One way to do this is to initialize a variable `count=0` at the beginning of the program and then increment it with `count=count+1` each time you accept an initial guess; use the command `disp(['Searches performed= ',num2str(count)])` to display it.)

b.) How many initial guesses does your program need to reliably find the global minimizer of the objective function in Example 1? How many local searches does it actually perform?

4.) Under what circumstances would you expect MMRS to outperform MRS? Under what circumstances would you expect MRS to outperform MMRS?

5.) a.) If $D \subseteq \mathbb{R}^n$, a continuous function $F : D \to \mathbb{R}$ is said to be **convex** if for every $x, y \in D$ and every $\lambda \in (0,1)$, $F(\lambda x + (1 - \lambda)y) \le \lambda F(x) + (1 - \lambda)F(y)$ (and strictly convex if the inequality is strict). The study of nonlinear optimization is seen as being split between algorithms for optimizing convex functions (the **convex programming problem**) and for nonconvex functions (the nonconvex programming problem). Show that if a function F is strictly convex then it is unimodal.

b.) Give an example of a convex function that is not unimodal. Could a convex function have two or more isolated local minima?

6.) a.) Use MATLAB to make a contour plot and a surface plot of the three-hump camelback function (Eq. (4.2)).

b.) Use the MRS method to find the global minimum of the three-hump camelback function. Use your plots in part a. to determine an appropriate range in which to generate the random points ($\alpha \le x \le \beta$, $\gamma \le y \le \delta$) and a reasonable number of initial guesses to use.

7.) a.) Use MATLAB to make a contour plot and a surface plot of the six-hump camelback function (Eq. (4.3)).

b.) Use the MMRS method to find a global minimum of the six-hump camelback function. Use your plots in part a. to determine an appropriate range in which to generate the random points ($\alpha \le x \le \beta$, $\gamma \le y \le \delta$) and a reasonable number of initial guesses to use.

8.) a.) Use the MRS method to find the global minimum of the Levy No. 5 test problem (Eq. (4.4)). Repeat with the MMRS method. Comment on the efficiencies of the two methods.

b.) Use the MRS method to find the global minimum of the Levy No. 5 test problem, limiting the method to 100 initial guesses. Do this a total of 20 times. What is the probability of finding the global minimum for this number of initial guesses?

c.) Repeat part a. using the Levy No. 3 test problem (Eq. (4.6)).

d.) Repeat part b. using the Levy No. 3 test problem.

9.) a.) Use the MRS method to find a global minimum of the **Powell function**
$F(x) = (x_1 + 10x_2)^2 + 5(x_3 - x_4)^2 + (x_2 - 2x_3)^2 + 10(x_1 - x_4)^2$.

b.) Find all global minima of the Powell function analytically.

c.) Show that the Hessian of $F(x)$ is positive semi-definite but not positive definite at the origin.

10.) Write a detailed algorithm (pseudo-code) for performing the tunneling method. Assume that an appropriate local minimization routine is available. You will need to specify a criterion for termination based on the non-existence of a root of Eq. (4.1).

11.) a.) Apply the tunneling method to the objective function of Example 1.

b.) Apply the tunneling method to the six-hump camelback function (Eq. (4.3)).

c.) Under what circumstances would you expect the tunneling method to outperform MRS and MMRS?

12.) a.) Apply the tunneling method to the Levy No. 3 problem (Eq. (4.6)).

b.) Apply the tunneling method to the Levy No. 5 problem (Eq. (4.4)).

13.) a.) Let $F : \mathbb{R}^n \to \mathbb{R}$ be a continuous function with finitely many local minima in a box $l_i \le x_i \le u_i$ $(i = 1, ..., n)$. Suppose that the local minimization routine always converges to the nearest local minimum within the box. Prove that the MRS and MMRS methods yield a sequence of function values that converge with probability one to the value of the function at its global minimum if the initial guesses are chosen uniformly at random over the box. (Note: We have assumed more than is actually needed to establish this result.)

b.) Must this result hold if pseudo-random numbers are used? Why or why not?

14.) a.) Show that if $f : \mathbb{R} \to \mathbb{R}$ then f is strictly convex exactly when the graph of f between any two points on f lies strictly below the line connecting those two points.

b.) Show that if $f : \mathbb{R} \to \mathbb{R}$ is twice continuously differentiable then f is strictly convex exactly when $f''(x) > 0$. What happens if there are points such that $f''(x) = 0$?

15.) Show that if $F : \mathbb{R}^n \to \mathbb{R}$ is twice continuously differentiable then f is strictly convex exactly when the Hessian of F is positive definite. What happens if the Hessian is positive semi-definite?

16.) a.) The **Booth function** $f(x, y) = (x + 2y - 7)^2 + (2x + y - 5)^2$ on $[-10, 10] \times [-10, 10]$ is a standard test problem for global minimization. Find its minima; which are global?

b.) The Booth function is sometimes described as a "plate function". Plot it; why is this description appropriate? Why might such a function be challenging for an optimization routine?

5. Direct Search Methods

Let's leave global optimization and return to local optimization. For functions of one variable we started off with bracketing methods that did not require derivatives (we say that they are *derivative-free methods*), namely golden section search and quadratic interpolation. Dropping the requirement of a bracket and fitting a quadratic to any three points gave another derivative-free method.

Assuming the availability of the first derivative of f gave Davidon's cubic interpolation method for functions of one variable or the method of steepest descent for functions of several variables. Assuming the second derivative exists gave Newton's method for minimization. Minimization methods based on Newton's method are far and away the ones most commonly employed in practice: They are fast and reliable (for a sufficiently close initial guess), and if managed properly–restarted periodically, preceded by several steepest descent iterations, monitored for positive definiteness of the Hessian and progress towards the minimum, and various other computational heuristics–they work quite well. If the relevant derivatives for a method are known or can be approximated accurately, it generally is worthwhile to make use of them.

But we cannot always assume that the function is differentiable or, even if it is, that we will be able to approximate the derivative(s) adequately. If $F : \mathbb{R}^n \to \mathbb{R}$ is continuous but not necessarily differentiable we say that we have a **nonsmooth optimization problem** (or **nonsmooth programming problem**). For such a problem we need a minimization algorithm that uses only function values, similar to golden section search but applicable in several dimensions. Often the function f in these cases is "noisy", meaning that we are unable to evaluate it to the precision desired (possibly because we are dealing with measured values, possibly because f is produced by a program with a too-large tolerance for our purposes). These types of problems are extremely common in industrial applications. Discontinuous objective functions also occur in practice, but we will be assuming that f is continuous in principle even if it is inaccurately known.

If $F : \mathbb{R}^n \to \mathbb{R}$ then methods that assume only continuity and use only values of the objective function F are called **direct search methods** or **function comparison methods**. One might argue that approximating derivatives with finite differences, for example, uses function values only, but these methods depend on an assumption of smoothness in order to be effective.

There is more than one way to proceed. Some methods sample the function at a number of points and use the values of F at these points to construct a simpler and smoother model of the function. Often the algorithm uses a convex function for its model. But the most common direct search methods either use some strategy for selecting new search directions and then explore in those directions, or else maintain a list of points defining a region which (it is hoped) contains the minimum. We give examples of the former type of algorithm in this section and the latter type in the next section. Theoretical developments of these types of algorithms generally assume that the function is unimodal in the region of interest, but this is not a requirement for their application in practice.

The simplest direct search method is **alternating variable search**. Suppose $F : \mathbb{R}^n \to \mathbb{R}$ is continuous and write $x = (x_1, ..., x_n)$. Choose a point $P \in \mathbb{R}^n$ and fix the values of the last $n - 1$ coordinates $(x_2, ..., x_n)$ at $(p_2, ..., p_n)$; this gives a function

$$(5.1) \qquad\qquad g_1(x_1) = F(x_1, p_2, ..., p_n)$$

of a single variable x_1, for which we have a current (initial) value p_1. We may now perform a single-variable search on $g_1(x_1)$, starting from p_1 and using any convenient line search algorithm. In effect, we are performing a one-dimensional search on $F(x_1, ..., x_n)$ parallel to the x_1 axis.

After we have obtained the result of the one-dimensional search on $g_1(x_1)$ from Eq. (5.1) we will have found a minimizer of $F(x_1, p_2, ..., p_n)$. This will be an x-value; call it

$$q_1 = \min\{g_1(x_1)\}$$

for specificity. We can now fix the values of $(x_1, x_3, ..., x_n)$ to be $(q_1, p_3, ..., p_n)$ so that we can form a new univariate function $g_2(\cdot)$ in which x_2 will be the only variable quantity, namely

(5.2) $$g_2(x_2) = F(q_1, x_2, p_3, ..., p_n)$$

and then we can perform a line search on it to find $q_2 = \min\{g_2(x_2)\}$. We repeat this process, searching only on $x_3, x_4, ..., x_n$ in succession until we have arrived at the point $Q = (q_1, q_2, ..., q_n)$.

This completes one iteration of the alternating variable search method. It's always possible, of course, that the line search algorithm fails to find an improved value from one of more x_i, whether because the search failed or the point was already optimal, and so some q_i may equal the starting x_i.

We now start over again, searching in the x_1 direction starting from q_1 with $(x_2, ..., x_n) = (q_2, ..., q_n)$ held constant and cycling through the independent variables. Each complete cycle is termed a stage.

EXAMPLE 1: Consider the function $f(x, y) = x^2 + 4xy + 4y^2$. Let's choose $P = (x_0, y_0)$ to be $(2, 2)$. From Eq. (5.1), $g_1(x) = f(x, 2) = x^2 + 8x + 16$. We wish to find the minimum of this nearest $x_0 = 2$; since g_1 is quadratic in x, the unique minimum is at $x^* = -8/2 = -4$. Call this q_1; this completes the search along x. From Eq. (5.2), $g_2(y) = f(-4, y) = 4y^2 - 16y + 16$. The minimum is at $y^* = -(-16)/8 = 2$. Call this q_2; this completes the search along y. Our new point is $Q = (-4, 2)$. Let's try another iteration. From Eq. (5.1), $g_1(x) = f(x, 2) = x^2 + 8x + 16$ so again $x^* = -8/2 = -4$. From Eq. (2), $g_2(y) = f(-4, y) = 4y^2 - 16y + 16$. The unique minimum is at $x^* = -(-16)/8 = 2$. The point $(-4, 2)$ is a fixed point of this method. Is it a solution? Let's check. We have $\nabla f(x, y) = (2x + 4y, 4x + 8y)^T$ so $\nabla f(-4, 2) = (0, 0)^T$; $(-4, 2)$ is a stationary point. The Hessian is (in MATLAB notation) $H = [2\ 4; 4\ 8]$ which is constant. It's eigenvalues are $\lambda_1 = 0$ and $\lambda_2 = 10$ so H is positive semi-definite there. (Of course, we would be suspicious of any computed quantity being *exactly* zero in a nontrivial example.) We have found a non-parabolic minimum. In fact, $f(x, y) = (x + 2y)^2$ so every point on the line $y = -\frac{1}{2}x$ is a global minimizer. □

The alternating variable search method is usually very inefficient unless the objective function exhibits a great deal of symmetry with respect to the variables $(x_1, ..., x_n)$. Suppose $n = 2$ so that $z = f(x, y)$. In Fig. 1, the line search from the initial guess $(1, .5)$ has found the minimum, that is tangent to the contour curve at $(0, 1)$. The next stage, holding $x = 0$ constant and searching along y, will find the minimum at $(0, 0)$. For fortuitously positioned contour curves such as this, the method can be very efficient.

But now suppose that the elliptical contours in the figure were rotated 45 degrees clockwise. The search from $(1, .5)$ would be tangent to some point on a level curve in the second quadrant, and the search would zig-zag to the origin from there in a slow, tedious manner.

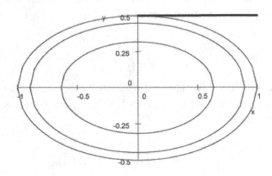

FIGURE 1. Alternating variable search (fast case).

Note that there is no guarantee that the univariate minimization steps will locate a local minimum: The current point may already be at the lowest point along that direction (even if F decreases in other directions), or the function may slope off to $-\infty$. These cases must be considered in a program realizing alternating variable search.

There are a number of possible improvements to this method, even if we continue to assume that the function is not necessarily differentiable. The motivation for the one we will consider is as follows: Experience shows that with methods of this type, if the method succeeds it is often because the method finds the bottom of a narrow contour curve valley relatively quickly and then spends most of its time traveling along the crease at the bottom of the valley toward the solution. Furthermore, points on the crease tend to fall on a more-or-less straight path. In fact this line of reasoning was originally applied to maximization problems, which are often viewed as "hill-climbing" problems: The method tends to locate a ridge, climb up to some point on it, and then spends most of its time following that ridge, and the ridge tends to be more-or-less in a straight line. Such physical and intuitive reasoning appears frequently as the motivation for a nonlinear optimization algorithm; in other cases, as in Newton's method or the method of steepest descent, calculus-based theory guides us.

If we can locate this crease (or ridge) and follow it then we should get faster convergence. Now, alternating variable search starts from a point P and produces a new point Q after we have searched along each of the n coordinate directions. But if we are in the crease–where, we expect, the method is going to be for most of its iterations–then the vector $Q - P$ from P to Q should point along the crease (roughly). Why not try to follow it? This is the basic idea behind **Rosenbrock's method** (or **Rosenbrock search**) which also handles the line searches in a modified manner. For simplicity, we will only describe the method in the case of a function defined on the plane.

Let $F : \mathbb{R}^2 \to \mathbb{R}$ be continuous. Rosenbrock's method for finding a local minimum of $F(x, y)$ given an initial guess (x_0, y_0) begins with an exploration phase; we start by searching along a direction parallel to the x-axis, that is, by doing a one-dimensional search on $F(x, y_0)$. We take a step of size $\delta_x > 0$ from x_0 in the positive x direction. If $F(x_0 + \delta_x, y_0) \leq F(x_0, y_0)$ then we say that the step was a success and replace x_0 by $x_0 + \delta_x$ and δ_x by $\alpha\delta_x$ for some $\alpha > 1$ (increase the step size); if not then we say that the step was a failure and replace δ_x by $\beta\delta_x$ for some $\beta \in (-1,0)$ (reverse the direction and decrease the step size). In either case, we now turn to y and do the same: We take a step of size $\delta_y > 0$ from y_0 in the positive y direction (using the updated value of x_0 if its step was a success). If $F(x_0, y_0 + \delta_y) \le F(x_0, y_0)$ then we say that the step was a success and replace y_0 by $y_0 + \delta_y$ and δ_y by $\alpha\delta_y$ for some $\alpha > 1$; if not then we say that the step was a failure and replace δ_y by $\beta\delta_y$ for some $\beta \in (-1,0)$.

The exploration phase is not over. We return to the x variable (which may or may not have changed) and repeat with the new δ_x, looking for a decrease from the previous value. If we had a success last time, this means an even larger step out ($\alpha > 1$); if we had a failure last time, this means a step that is both shorter and in the opposite direction ($\beta < 0$). We then do the same for y. We continue this process until we have found a success *followed by a failure* in each direction. The effect of this procedure is to find a local minimum by stepping out farther and farther along a direction of decrease until we finally take a step up; in effect, this brackets a minimum with respect to that one variable[3].

Note also that this suggests a method for finding an initial bracket for golden section search or other bracketing methods in the one-dimensional search case, which is one of the reasons we present Rosenbrock's method. This idea is used in `fzero` to find an initial bracket for root-finding. Standard values are $\alpha = 3$, $\beta = -1/2$.

After the exploration phase is completed–a success followed by a failure in each direction–we perform the search phase. We set x_1 and y_1 to be the points from which a failure was obtained. We then define the new search direction vectors p_1 and p_2 as follows: The vector p_1 is a unit vector in the direction of $(x_1, y_1)^T - (x_0, y_0)^T$ (the direction along the presumed crease, and hopefully an approximate steepest descent direction) and the vector p_2 is a unit vector orthogonal (perpendicular) to p_1. This completes the first iteration (called a stage) of the method.

For the next iteration we proceed in exactly the same way except that instead of searching in the directions of the axes we perform our first search in the direction p_1 (starting from (x_1, y_1)), then in the direction p_2, and repeat until we have a success followed by a failure in each of these directions. If the method is working as intended then we should make good progress along p_1 which will bring the search along p_2 into a better region, and we will avoid some of the see-sawing of alternating variable search. At each succeeding stage (iteration) we define a new p_1 as the vector from the old point to the new point (normalized), and a new unit vector p_2 perpendicular to it, then explore along these directions.

EXAMPLE 2: Let's apply the method of Rosenbrock search to the standard test function $f(x,y) = 100(y - x^2)^2 + (1 - x)^2$ which is known as **Rosenbrock's function**. (There is a unique local minimum at $(1,1)$ which is therefore the global minimum.) We'll start from the initial guess $(2,-1)$ and take $\alpha = 3$, $\beta = -1/2$, and $\delta_x = \delta_y = 1$. First, we compute $f(2,-1) = 100(-1 - 2^2)^2 + (1 - 2)^2 = 2501$. Let's start the exploration phase: $f(x,-1) = 100(-1 - x^2)^2 + (1 - x)^2$ and at $x = 2 + \delta_x = 3$ this is $f(3,-1) = 100(-1 - 3^2)^2 + (1 - 3)^2 = 10004$. This is greater than 2501 so it is a failure, and we set $\delta_x = -.5 \cdot 1 = -.5$. We turn to the y direction, and consider the function $f(2,y) = 100(y - 2^2)^2 + (1 - 2)^2 = 100(y - 4)^2 + 1$ at $y = -1 + \delta_y = 0$; this gives $f(2,0) = 100(0 - 4)^2 + 1 = 1601$ which is a success

---

[3]Well, not *exactly*, if the other variable is changing also.

(it is less than 2501). We use $y = 0$ as the current value of y from here on and set $\delta_y = 3 \cdot 1 = 3$.

The next round of exploration gives $f(2 + -.5, 0) = f(1.5, 0) = 506.5$ (a success since it is less than 1601; we use $x = 1.5$ from here on and set $\delta_x = 3 \cdot -.5 = -1.5$), then $f(1.5, 0 + 3) = f(1.5, 3) = 56.5$ (a success since it is less than 506.5; we use $y = 3$ from here on and set $\delta_y = 3 \cdot 3 = 9$). We have a success in each direction and are now looking for a failure in each direction.

The next round gives $f(1.5 + -1.5, 3) = f(0, 3) = 901$ (a failure since it is greater than 56.5; we set $\delta_x = -.5 \cdot -1.5 = .75$) and $f(1.5, 3 + 9) = f(1.5, 12) = 9506.5$ (a failure since it is greater than 56.5; we set $\delta_y = -.5 \cdot 9 = -4.5$). We now have a success followed by a failure in each direction. We will be searching backwards from the last success in the next stage (note that the last success has actually jumped to the other side of the minimum, so in that case this would be a good idea). The method is adjusting the scale parameters δ_x and δ_y to fit the function.

The exploration phase is over, and we must compute the new directions. The vector from $(x_0, y_0) = (2, -1)$ to $(x_1, y_1) = (1.5, 3)$ is $(1.5, 3)^T - (2, -1)^T = (-.5, 4)^T$, so $p_1 = (-.5, 4)^T / \|(-.5, 4)^T\| \doteq (-0.1240, 0.9923)^T$. A perpendicular unit vector is $p_2 = (0.9923, 0.1240)^T$. This completes a single stage of the method.

For the second stage we begin by taking a step of length $\delta_x = .75$ along p_1 starting from $(x_1, y_1) = (1.5, 3)$. (Remember that in the long run we expect the direction of p_1 to roughly stabilize and so re-using the old δ_x from the previous stage would be quite reasonable.) We compute $f((x_1, y_1) + \delta_x p_1) = f(1.4070, 3.7442) \doteq 311.5385$ (a failure since it is greater than 56.5; we set $\delta_x = -.5 \cdot .75 = -.375$) and then turn to p_2. We take a step of length $\delta_y = -4.5$ along p_2, giving $f((x_1, y_1) + \delta_y p_2) = f(-2.9654, 2.4420) \doteq 4049.63$ (a failure since it is greater than 56.5; we set $\delta_y = -.5 \cdot -4.5 = 2.25$). We continue exploring, returning to p_1 with the new δ_x. □

If $n > 2$ the method is unchanged except that after we find the direction p_1 of the crease we must define $n - 1$ directions $(p_2, ..., p_n)$ perpendicular (or, more generally, orthogonal) to it. This can be accomplished using the Gram-Schmidt process from Ch. 2. Convergence criteria may include the number of function evaluations and a comparison of the distance traveled along p_1 to the total distance traveled along the other directions in each exploratory phase.

We briefly mention the key ideas behind two similar methods. **Hooke and Jeeves (pattern) search** starts with a point P and takes a single exploratory step in each coordinate direction, retaining the value of the coordinate if the step is a success and taking a similar step with the opposite sign if it is not. After all n directions have been searched in this manner we have a new point Q. (Possibly $P = Q$ since we search at most twice in each coordinate direction.) Rather than simply moving from P to Q we assume that $Q - P$ is a good direction and take an extended step to $Q + (Q - P)$ (called the *pattern move*), then begin again.

The **Davies, Swann, and Campey method** (or **D.S.C. method**) explores each direction in a manner similar to that of Rosenbrock search until it brackets a minimum in that direction. It then uses single-variable quadratic interpolation to better locate the minimum in that direction, and hence may be thought of as combined an approximation approach along with a direct search. The rest of the algorithm is similar to Rosenbrock's method.

These methods are typically described as ridge-seeking (or valley-seeking), or razor search[4], algorithms, because they assume the existence of a "razor's edge" on which the optimum is to be found. This is surprisingly often the case in industrial applications.

MATLAB

It's convenient to be able to save variables from one MATLAB session to the next. For instance, if you worked through Example 1 by defining a function:

» f=@(x,y) x^2+4*x*y+4*y^2

then you may wish to have access to f and any other variables in another session (say, when working the Problems). Use the save command:

» save ch7sec5 f

to save f to a file named ch7sec5.mat. Enter clear all and then who to verify that f has been cleared, then enter:

» load ch7sec5
» who

to see that the load command has restored f to the workspace. You may enter:

» save temp1

to store all current variables in temp1.mat (omitting the filename causes MATLAB to use the default filename matlab.mat).

You may also use what is called the functional form of save; if you need to save a variable from within a MATLAB program, where the name of the MAT-file is supplied by the user in a character variable s, you may use the save command as follows:

» s='matfile1';
» A=[2 1;1 2];
» save(s,'A')

This saves the matrix A in the file matfile1.mat. There is a similar functional form of load, and you may specify filenames (see the help).

To remove a MAT-file from within MATLAB, you may use the operating system or the delete command. For example:

» delete matfile1.mat
» delete ch7sec5.mat
» delete('temp1.mat')

should remove the files we saved earlier. You may also use the exclamation point ! to issue a command directly to the operating system, and use its delete feature (e.g., !rm matfile.mat on a Unix system). All the usual path-related issues apply; enter:

» T=[1 3 5 7 9];
» save temp2 T
» which temp2.mat

to locate temp2.mat. Recall that the path command allows you to display and set the search path. An easier way to change the path for most needs is the addpath command; to add a directory to the path, enter (for Windows):

» addpath c:\temp

or for Unix:

---

[4]This term is also used more specifically for a 1969 algorithm by J. W. Bandler and P. A. McDonald.

» addpath /users/yourname/numerical
(use a valid path). Creating a directory for your own m-files and appending it to
the path can be useful. On a Unix machine you may use the cd command (from
MATLAB) to verify that you are in your home directory, then enter:
» !mkdir mfiles
» addpath /users/yourname/mfiles -end
to create such a directory and add it to your path. The -end makes this the last
directory on the path (and hence the last one checked).

You do not need to do add the directory to the path each time. The file
startup.m is executed each time you start MATLAB. If you wish to have a directory
such as this on your path every time you use MATLAB, make an M-file startup.m
with the line:
» addpath /users/yourname/mfiles -end
in it. You may add any other commands you would like issued on startup–for
example, a preferred default like format long and/or format compact. Save the
file in any directory on MATLAB's default path (on a Unix system you should
usually save it in your home directory). See help matlabrc for more information.
There is also a file finish.m that is executed when you exit MATLAB; see help
quit.

If you are not sure whether or not the file startup.m already exists, you may
check for it from MATLAB. Enter:
» exist('startup.m')
If you get 0, there is no such file on the path; if you get 2, as you will for:
» exist('matlabrc.m')
then there is such a file. The exist command can provide information on other
objects as well. If f is a variable (including an inline function) in the workspace,
for example, then exist f will return 1.

Problems

1.) a.) Rosenbrock's function is also called Rosenbrock's valley or the banana function. Make a contour plot of this function and use it to explain why these characterizations are accurate.
b.) Does the new search direction p_1 in Example 2 point roughly along the crease
at the base of the valley?
2.) Perform two iterations of alternating variable search on Rosenbrock's function,
starting from $(x_0, y_0) = (10, 12)$.
3.) a.) Write a MATLAB program that inputs a function of a single variable, an
initial guess, and a scale parameter (δ_x) and uses them to attempt to bracket a
minimum of the function using Rosenbrock's method in one dimension.
b.) Use Rosenbrock's method to find a minimum of $f(x) = x^9 - 2x^7 + 5x^4 - 2x + 2$.
4.) Perform two complete stages of Rosenbrock's method on the objective function
$f(x, y) = x^2 + 4xy + 4y^2$ of Example 1, starting from $(x_0, y_0) = (2, 2)$.
5.) Perform three complete stages of Rosenbrock's method on the objective function
$f(x, y) = |(\sin(xy)| + |2\cos(x)|$ starting from $(x_0, y_0) = (1, 1)$.
6.) Write a detailed algorithm (pseudo-code) for performing alternating variable
search.
7.) Write a detailed algorithm (pseudo-code) for performing Rosenbrock search on
a function of two variables.

8.) Apply two iterations of alternating variable search to the function $f(x,y) = x^2 + y^2$ starting from an arbitrary point (a,a) $(a \neq 0)$. Comment.

9.) Apply two iterations of alternating variable search to the function $F(x,y) = x^4 - 2xy + (x-1)^2(y-1)^2 e^{-xy}$ starting from the point $(5,5)$. Make a sketch indicating the steps the method takes. Does the method seem to be working?

10.) a.) Apply three iterations of Rosenbrock search to the function $f(x,y) = x^2 + y^2$ starting from the point $(2,2)$. Use $\alpha = 3$, $\beta = -1/2$, and $\delta_x = \delta_y = .5$.

b.) Repeat using Hooke and Jeeves search.

c.) Repeat using the D.S.C. method.

11.) Apply four iterations of Rosenbrock search to the function $F(x,y) = x^4 - 2xy + (x-1)^2(y-1)^2 e^{-xy}$ starting from the point $(2,2)$. Use $\alpha = 3$, $\beta = -1/2$, and $\delta_x = \delta_y = .5$.

12.) a.) Write a MATLAB program that implements alternating variable search.

b.) Test your program on the Rosenbrock function.

13.) a.) Write a MATLAB program that implements Rosenbrock search for a function of two variables.

b.) Test your program on the Rosenbrock function.

14.) It is easier to provide convergence theorems for bracketing methods, the method of steepest descent, and Newton's method than for direct search methods such as alternating variable search and Rosenbrock search. Why?

15.) Write a detailed algorithm (pseudo-code) for performing Rosenbrock search on a function of n variables. You will need to include the Gram-Schmidt orthonormalization procedure.

16.) Write a MATLAB program that implements Hooke and Jeeves search.

6. The Nelder-Mead Method

Rosenbrock's method illustrates the basic idea of the direct search methods. It's most useful when the minimizer of the multivariate objective function is known to lie in a long, narrow valley, and is widely used in one-dimensional searches as a means of obtaining an initial bracket. However, derivative-free minimization programs for general multivariate objective functions are more often based on a different approach, which we now describe.

A **simplex** in $\mathbb{R}^n$ is a set of $n+1$ distinct points. It is also referred to as a **polytope**. The term simplex or polytope is often used to describe the region in $\mathbb{R}^n$ defined by these points as well; in $\mathbb{R}^2$ a simplex is a triangle.

A **regular simplex** in $\mathbb{R}^n$ is a set of $n+1$ mutually equidistant points; in $\mathbb{R}^2$, for example, a regular simplex defines an equilateral triangle. **Simplex methods**[5] (or **polytope methods**) maintain a simplex at each iteration that hopefully encloses the location of the minimizer. In the $\mathbb{R}^2$ case this would mean maintaining and evolving a triangular region in the xy-plane which, if all goes well, will locate and then contract around the minimizer (x^*, y^*).

The prototypical polytope method is as follows: Generate a regular simplex (also called a **design** in this context) and order the points $x_1, ..., x_{n+1}$ so that $F(x_1) \leq F(x_2) \leq ... \leq F(x_{n+1})$. Find the center of mass (that is, the centroid)

[5]Not to be confused with the well-known *simplex method* for linear functions. The method we are discussing is due to Spendley, Hext, and Himsworth.

$$c = \frac{1}{n} \sum_{k=1}^{k=n} x_i$$

of the n best points of the simplex. (That is, we have omitted x_{n+1} from this average.) Our goal is to replace the worst point x_{n+1} with a new point x_{ref} that has a lower function value. Since x_{n+1} is the highest point, we reflect it about the centroid; that is. we seek x_{ref} such that

$$x_{ref} - c = c - x_{n+1}$$
$$x_{ref} = 2c - x_{n+1}$$

(see Fig. 1). We then replace x_{n+1} with x_{ref}, re-order the points according to their function values, and start again with the new highest point; it is a fact that the new set of points will also form a regular simplex. We continue evolving the simplex in this manner.

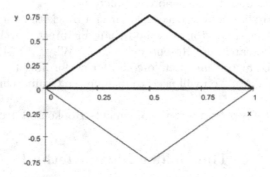

FIGURE 1. Polytope method.

EXAMPLE 1: Let's try to find a minimum of the function $f(x,y) = (x+1)^2 + 2(y+1)^2$ starting from the initial regular simplex $\{(0,0),(1/2,\sqrt{3}/2),(1,0)\}$. (Verify that each point is exactly one unit away from each of the others. This is the simplex lying on and above the x-axis in Fig. 1.) The function values at these vertices are $3, 9.2141, 6$, respectively, so $x_1 = (0,0)^T$, $x_2 = (1,0)^T$, and $x_3 = (1/2,\sqrt{3}/2)^T$. The centroid of the lowest points is $c = \frac{1}{2}((0,0)^T + (1,0)^T) = (1/2,0)^T$ (note that this is just the midpoint of the line connecting x_1 and x_2), so $x_{ref} = 2c - x_3 = 2(1/2,0)^T - (1/2,\sqrt{3}/2)^T = (1/2,-\sqrt{3}/2)^T$. The new simplex $\{(0,0),(1,0),(1/2,-\sqrt{3}/2)\}$ lies on and below the x-axis in Fig. 1. Since $f(1/2,-\sqrt{3}/2) \doteq 2.2859$ the re-ordered list of points is $x_1 = (1/2,-\sqrt{3}/2)^T$, $x_2 = (0,0)^T$, and $x_3 = (1,0)^T$.

Let's do another iteration. The new centroid is $c = \frac{1}{2}((1/2,-\sqrt{3}/2)^T + (0,0)^T) = (1/4,-\sqrt{3}/4)^T$, so $x_{ref} = 2c - x_3 = 2(1/4,-\sqrt{3}/4^T - (1,0)^T = (-1/2,-\sqrt{3}/2)$. The new simplex is $\{(1/2,-\sqrt{3}/2),(0,0),(-1/2,-\sqrt{3}/2)\}$ (again, verify that each point is the same distance from every other point). Since $f(-1/2,-\sqrt{3}/2) \doteq 0.2859$, the re-ordered list of points is $x_1 = (-1/2,-\sqrt{3})^T$, $x_2 = (1/2,-\sqrt{3}/2)^T$, and $x_3 = (0,0)^T$. The new simplex lies to the lower left of the second simplex. (Recall that the surface $z = f(x,y)$ lies above the plane.) The sole local minimum is in fact at $(-1,-1)$. $\square$

When one vertex of the simplex lies near the minimum it may be that it does not get replaced for a long time; the simplex turns about this point. For this reason the age (in number of iterations) of each point is monitored and when any point is too old the entire simplex is shrunk towards it by replacing each x_i by

$$
\begin{aligned}
x_i &= x_{old} + \frac{1}{2}(x_i - x_{old}) \\
&= \frac{1}{2}(x_{old} + x_i)
\end{aligned}
$$

where x_{old} is the oldest vertex of the design. We can use the (hyper)volume of the simplex as a convergence criterion.

Given a point in $\mathbb{R}^n$, there are simple formulas for generating a regular simplex surrounding it (or for which it is one of the vertices), so the method can be used starting with a single initial guess and an edge length for the simplex. The principal complication that may occur in using the method is that if the reflected point x_{ref} is the new worst point, then reflecting *it* will find the old x_{n+1} that was used to generate x_{ref} and the method will cycle; this is easily handled by using the next-to-worst point in such a case.

The polytope method makes no significant use of the values of F other than for deciding which point is the worst at each iteration. We can do better. There are many variations but the one that we will discuss here is the **Nelder-Mead (simplex) method**, which is the basis of MATLAB's `fminsearch` command. We begin with a set of distinct points $x_1, ..., x_{n+1}$ ordered so that $F(x_1) \leq F(x_2) \leq ... \leq F(x_{n+1})$; these points form a (not necessarily regular) simplex in $\mathbb{R}^n$. We compute the centroid c as before (once again omitting x_{n+1} from the sum) and perform a *reflection step* by finding the reflected point

$$
(6.1) \qquad x_{ref} = c + \alpha(c - x_{n+1})
$$

where $\alpha > 0$ is called the *reflection coefficient*; $\alpha = 1$ corresponds to x_{ref} in the basic polytope method, and $\alpha > 1$ corresponds to taking a larger step away from the highest point x_{n+1}. So far the only real change from the basic polytope method is that we are not requiring that the simplex be regular.

We now compute $F(x_{ref})$. If $F(x_1) \leq F(x_{ref}) \leq F(x_n)$ then x_{ref} will be neither the new best point nor the new worst point, so we replace x_{n+1} by x_{ref}, re-order the points, and begin a new iteration by computing the centroid and reflecting.

If however either $F(x_{ref}) < F(x_1)$ or $F(x_n) < F(x_{ref})$ then x_{ref} is going to be either the new best point or the new worst point. If it's going to be the new best point then the direction $c - x_{n+1}$ in which we stepped from c (Eq. (6.1)) seems to be a descent direction (particularly if the objective function is unimodal) and perhaps we should move further in that direction; if it's going to be the new worst point then the direction $c - x_{n+1}$ in which we stepped from c seems to be either the wrong direction or at least we have stepped too far along it so perhaps we should reconsider this step. Looked at another way, if x_{ref} is going to be the new worst point then perhaps the polytope is too large and we should shrink it in hopes of getting to the scale at which F is unimodal in this region.

For this reason, if $F(x_n) < F(x_{ref})$ (so that x_{ref} will be the new worst point, though it may or may not be worse than the old worst point x_{n+1}) we perform a *contraction step*. We set

$$(6.2) \qquad x_{con} = \begin{cases} c + \gamma(x_{n+1} - c) & \text{if} \quad F(x_{ref}) \geq F(x_n) \\ c + \gamma(x_{ref} - c) & \text{if} \quad F(x_{ref}) < F(x_n) \end{cases}$$

where $\gamma \in (0, 1)$ is the *contraction coefficient*. We then compute $F(x_{con})$; if it is less than both $F(x_{ref})$ and $F(x_n)$ then we accept it as the new point, re-order the points, and begin a new iteration. If not (the contraction step fails), we *shrink* the polytope toward the best point in exactly the same way as in the basic polytope algorithm:

$$\begin{aligned} x_i &= x_1 + \frac{1}{2}(x_i - x_1) \\ &= \frac{1}{2}(x_1 + x_i) \end{aligned}$$

($i = 2, ..., n+1$). We are now back to $n+1$ points, so we re-order them in the usual way and begin a new iteration.

What if x_{ref} would be the new best point? In that case stepping away from x_{n+1} has been very beneficial, and perhaps we should explore further in that direction. We perform an *expansion step* by computing a new point

$$x_{\exp} = c + \varepsilon(x_{ref} - c)$$

where $\varepsilon > 1$ is called the *expansion coefficient*. This locates a point even further along the line from x_{n+1} to c. If $F(x_{\exp}) < F(x_{ref})$ then the expansion is successful and we replace x_{n+1} by $x_{\exp}$; otherwise. if the expansion fails, we replace x_{n+1} by x_{ref}. (We use the criterion $F(x_{\exp}) < F(x_{ref})$ rather than $F(x_{\exp}) \leq F(x_{ref})$ to keep the simplex small.) We re-order the points and are ready to start again.

Typical values of the parameters are $\alpha = 1$, $\gamma = 1/2$, and $\varepsilon = 2$, but other choices are possible. We stop when $F(x_{n+1}) - F(x_1)$ is sufficiently small, or perhaps when the standard deviation of $F(x_1), F(x_2), ..., F(x_{n+1})$ is small if we think the function is "noisy"; we might also terminate the algorithm when some maximum number of function evaluations has been reached.

EXAMPLE 2: Let's go back to the function $f(x, y) = (x+1)^2 + 2(y+1)^2$ and initial (regular) simplex $\{(0, 0), (1/2, \sqrt{3}/2), (1, 0)\}$ of Example 1. Just as in Example 1, the function values are $3, 9.2141, 6$, respectively, so $x_1 = (0, 0)^T$, $x_2 = (1, 0)^T$, and $x_3 = (1/2, \sqrt{3}/2)^T$, and the centroid of the lowest points is $c = \frac{1}{2}((0, 0)^T + (1, 0)^T) = (1/2, 0)^T$. Once again $x_{ref} = 2c - x_3 = 2(1/2, 0)^T - (1/2, \sqrt{3}/2)^T = (1/2, -\sqrt{3}/2)^T$. In the polytope method we accepted this point and moved on to the next iteration; now we check $f(1/2, -\sqrt{3}/2) \doteq 2.2859$ and see that this will be the new best point, so an expansion $x_{\exp} = c + \varepsilon(x_{ref} - c) = (1/2, 0)^T + 2((1/2, -\sqrt{3}/2)^T - (1/2, 0)^T) = (1/2, -\sqrt{3})^T$ is in order. We find that $f(1/2, -\sqrt{3}) \doteq 3.3218$ so the contraction step fails and we accept x_{ref}; the new re-ordered list of points is $x_1 = (1/2, -\sqrt{3}/2)^T$, $x_2 = (0, 0)^T$, and $x_3 = (1, 0)^T$, just as it was in Example 1.

Let's do another stage. The new centroid is $c = \frac{1}{2}((1/2, -\sqrt{3}/2)^T + (0, 0)^T) = (1/4, -\sqrt{3}/4)^T$, so $x_{ref} = 2c - x_3 = 2(1/4, -\sqrt{3}/4^T - (1, 0)^T = (-1/2, -\sqrt{3}/2)$. Since $f(-1/2, -\sqrt{3}/2) \doteq 0.2859$ this will be a new best point so we try an expansion,

$x_{\exp} = c + \varepsilon(x_{ref} - c) = (1/4, -\sqrt{3}/4)^T + 2((-1/2, -\sqrt{3}/2)^T - (1/4, -\sqrt{3}/4)^T) = (-5/4, -3\sqrt{3}/4)^T$. We find that $f(-5/4, -3\sqrt{3}/4) \doteq 0.2413$; the expansion succeeds, and so $x_{\exp}$ is the new point. The re-ordered list of points is

$$x_1 = (-5/4, -3\sqrt{3}/4)^T, x_2 = (1/2, -\sqrt{3}/2)^T, x_3 = (0,0)^T$$

and they no longer form a *regular* simplex. $\square$

For any polytope method it is wise to periodically restart the method. Typically if the method fails it is because it is stagnating in some region, that is, it is making little progress, possibly as measured by a lack of sufficient decrease in the average value of the function

$$(6.3) \qquad \frac{1}{n+1} \sum_{k=0}^{k=n} F(x_i)$$

over the simplex. Choosing a slightly different polytope that covers essentially the same region (but which will define new directions because of the new points) can be a useful heuristic measure. Nelder-Mead is usually restarted by retaining the best point and defining a regular simplex that includes that best point as a vertex. The fact that the Nelder-Mead polytopes can adjust their shape to the lay of the function around them is a useful facet of the method, but a restarting provision is important in a practical application of the algorithm.

MATLAB

The average value of the function (as in Eq. (6.3)) is easily computed using the `mean` command. If we wish to use the (sample) standard deviation as the convergence criterion then we may use the `std` command; for example, enter:

```
» F=@(x,y) abs(2*x.^4-5*x.*y+1)+exp(-x.*y);
» P=rand([2 3])      %Simulate three points in R2.
» x=P(1,:)           %x-values are in row 1 of P.
» y=P(2,:)           %y-values are in row 2 of P.
» FP=feval(F,x,y)    %Evaluate F at the points P on the polytope.
» std(FP)            %Sample s.d.
```

To compute the centroid we need to order the points by their function values. With P and FP as above, enter:

```
» [FPS,IND]=sort(FP)
```

The vector FPS is the same as FP but sorted into ascending order, as desired. Recall that the vector IND is an index vector that is a permutation of the integers 1 to `length(FP)`, and may be used to sort the P vector; enter:

```
» PS=P(:,IND)
```

This has the effect of sorting the columns of P in precisely the same way that FP was sorted. Now that we have sorted P as PS, we may compute the centroid. Recall that PS is a matrix. Type `help sum`. Evidently the centroid of all but the last point may be found as follows:

```
» N=size(PS,2)-1;    %size(A,2) is the number of columns of A.
» c=sum(PS(:,1:N))
```

(Look at PS and verify that this is correct.) In fact we could also use a single command; enter:

```
» c=sum(PS(:,1:end-1))
```

(see `help end` and note also that `1:end-1` is the same as `1:(end-1)`).

Recall when using `fminsearch` that you may use `optimset` to set an `options` variable to change various defaults of the program.

Problems

1.) a.) One of the cases in Eq. (6.2) is called an *inside contraction* and the other case is called an *outside contraction*. Draw a polytope in $\mathbb{R}^2$ and its centroid (see Fig. 1). Indicate the location of the reflected point and the points and polytopes that would result from a successful inside contraction step and a successful outside contraction step. What is the reason for having the two different types of contractions?

b.) Draw a polytope in $\mathbb{R}^2$ and its centroid; indicate the location of a reflected point and the points and polytopes that would result from a successful expansion step or from a shrinking of the polytope.

c.) One variant of the Nelder-Mead method replaces the centroid c in Eq. (6.2) by the best point x_1 during a contraction step. Why might this be beneficial?

2.) Write a detailed algorithm (pseudo-code) for the basic polytope method. You need not include restarting.

3.) a.) Apply four iterations of the basic polytope method to the function $F(x, y) = |2x^4 - 5xy + 1| + e^{-xy}$ starting from the polytope $\{(0,0), (1/\sqrt{2}, 1/\sqrt{2}), (1,0)\}$.

b.) Sketch the successive polytopes on a single graph.

4.) a.) Write a detailed algorithm (pseudo-code) for the Nelder-Mead method. You need not include restarting.

5.) a.) Apply four iterations of the Nelder-Mead method to the function $F(x, y) = |2x^4 - 5xy + 1| + e^{-xy}$ starting from the polytope $\{(0,0), (1/\sqrt{2}, 1/\sqrt{2}), (1,0)\}$.

b.) Sketch the successive polytopes on a single graph.

c.) Use MATLAB's `fminsearch` command with the initial guess $(0,0)$ to find a minimum of $F(x, y) = |2x^4 - 5xy + 1| + e^{-xy}$.

6.) a.) Write a MATLAB program for performing the basic polytope method. Your method should input an initial regular simplex. Spendley, Hext, and Himsworth suggest that a point is "too old" when it has persisted for more than $1.65n + n^2/20$ iterations (when the points forming the polytope are points in $\mathbb{R}^n$). Use some measure of the size of the polytope as your convergence criterion. You need not include restarting.

b.) Test your program on the function $F(x, y) = |2x^4 - 5xy + 1| + e^{-xy}$ starting from the polytope $\{(0,0), (1/\sqrt{2}, 1/\sqrt{2}), (1,0)\}$.

7.) Given a point $x_0 \in \mathbb{R}^n$, a regular simplex containing it as a vertex can be formed by setting $x_i = x_0 + \delta_1(J - e_i) + \delta_2 e_i$ $(i = 1, ...n)$. Here J is `ones([n 1])` (in MATLAB notation), e_i is the n-vector that is equal to zero except for the ith entry, which is equal to one, and $\delta_1 = ((n+1)^{1/2} + n - 1)/n\sqrt{2}$, $\delta_2 = ((n+1)^{1/2} - 1)/n\sqrt{2}$. Write a MATLAB program that inputs an initial point (use `size` or `length` to determine n) and returns a regular simplex containing it as a vertex.

8.) Using the initial polytope $\{(15, 13), (14, -16), (-16, 17)\}$, apply four iterations of the Nelder-Mead method to Rosenbrock's function $f(x, y) = 100(y - x^2)^2 + (1 - x)^2$.

9.) a.) Write a MATLAB program for performing the Nelder-Mead method on a function of two variables. Your program should accept a function and a single initial guess and generate an initial polytope. Use $F(x_{n+1}) - F(x_1)$ for the convergence criterion. Do not include a restarting feature. Be sure to return a status variable indicating success or failure.

b.) Test your program on the function $F(x, y) = |2x^4 - 5xy + 1| + e^{-xy}$ starting from the point $(0, 0)$.

c.) Test your program on Rosenbrock's function $f(x, y) = 100(y - x^2)^2 + (1 - x)^2$ starting from the initial point $(4, 6)$.

10.) Suggest a method for detecting stagnation in the Nelder-Mead method and then restarting it. Write a detailed algorithm (pseudo-code) for implementing your method.

11.) Compare the performance of Newton's method and the Nelder-Mead method for Rosenbrock's function $f(x, y) = 100(y - x^2)^2 + (1 - x)^2$ starting from the initial points $(2, 2)$, $(10, -12)$, $(100, 100)$. You may use `fminsearch` for the Nelder-Mead method or your own version.

12.) a.) What happens if you use the basic polytope method on a function of a single variable?

b.) What happens if you use the Nelder-Mead method on a function of a single variable?

c.) Use `fminsearch` to find any two minima of $f(x) = x^9 - 2x^7 + 5x^4 - 2x + 2$ to at least four decimal places.

13.) Create a direct search method based on the idea of evolving a polytope. Explain your reasoning and then give an algorithm for performing your method.

14.) a.) One test for true convergence that is sometimes used with direct search methods based on polytopes is to restart the method from a new regular simplex around the solution found by the first run of the method; if this second use of the method converges to the same point, we declare it the solution. Write a MATLAB program that implements this by calling fminsearch (or your version of the Nelder-Mead method) and, if it appears to converge, calls it again with a new starting point. You may wish to refer to Problem 7.

b.) Test your program on Rosenbrock's function $f(x, y) = 100(y - x^2)^2 + (1 - x)^2$ starting from various initial points.

15.) What is the smallest reasonable tolerance for the size of a polytope that might be used as a convergence criterion when using double precision arithmetic?

16.) Some pseudo-random pattern searches use a box rather than a polytope because it makes it easy to sample points uniformly along any given coordinate. Suggest such an algorithm.

7. Conjugate Direction Methods

In the study of nonlinear optimization, much attention is given to the special case of a quadratic objective function

$$(7.1) \qquad F(x) = c - b^T x + \frac{1}{2} x^T A x$$

($c \in \mathbb{R}$, $b \in \mathbb{R}^n$, A an $n \times n$ matrix). One reason is obvious: A sufficiently differentiable function is well-approximated by a quadratic function near a parabolic minimum, so methods that work well on quadratic functions can be used on smooth functions. Another reason is the connection of such functions to linear systems. If the problem is unconstrained or any constraints on the problem are linear we call this a **quadratic programming (QP)** problem. We'll only consider the unconstrained case here.

Let's take $c = 0$ for convenience and suppose that A is positive definite. (Recall that such a matrix is symmetric and nonsingular.) Then if F is given by Eq. (7.1) its gradient is

$$\nabla F(x) = -b + Ax.$$

But the unique minimum of the function F occurs where $\nabla F(x) = 0$, that is, where $Ax = b$. (This is why it was convenient to choose the sign of b as we did in Eq. (7.1).) For a positive definite matrix A, a method that solves Eq. (7.1) solves the linear system $Ax = b$, and a method that solves $Ax = b$ locates the minimum of $F(x)$ in Eq. (7.1).

If the function is not truly quadratic, the A matrix we need is the Hessian $H(x)$ of F evaluated at the unknown minimizer x^*, and even then Eq. (7.1) is only an approximation of F near x^*. If the function is truly quadratic, however, the Hessian is a constant matrix and $A = H(x^*)$ will be given.

The class of methods that we are about to describe were developed for the purpose of solving $Ax = b$ when A is positive definite and large enough that Cholesky decomposition is not practical. They are also widely used for minimizing $F(x) = -b^T x + \frac{1}{2} x^T A x$.

Given a positive definite matrix A, we say that two vectors u, v are **conjugate with respect to A** (or **A-conjugate** or **A-orthogonal**) if

$$u^T A v = 0$$

which is an inner product (u, v) defined by the matrix A. This inner product in turn induces a vector norm $\|u\|_A = (u^T A u)^{1/2}$, called the **$A$-norm** or, in many physical contexts, the **energy norm**.

It is a fact that if the vectors $d_1, ..., d_n$ are nonzero and pairwise A-conjugate then they are linearly independent, and hence they form a basis for $\mathbb{R}^n$. We say that $\{d_1, ..., d_n\}$ is a set of **conjugate directions** (with respect to A).

Consider again the figure in Sec. 7.5, reproduced here as Fig. 1. (The function mapped there is a quadratic.) When the function's contours line up just so with respect to the search directions, alternating variable search–which minimizes in each coordinate direction sequentially–finds the location of the minimum *exactly* in a finite number of steps. When the contours don't line up nicely, this does not happen–even if the objective function is quadratic. *Conjugate directions are the directions that are needed to get the behavior of Fig. 1 for a general quadratic.* These directions depend on the matrix A that defines the quadratic objective function. They will allows us to determine just the right set of directions in which to search to find the location of the minimum after searching along each direction once. If we have a set of A-conjugate directions then we will find the minimum of $F(x) = -b^T x + \frac{1}{2} x^T A x$ exactly in at most n steps *starting from any initial guess* by sequential minimization along these directions. (We say that the method exhibits **quadratic termination**; obviously, this assumes exact arithmetic) The resulting algorithm is called a **conjugate direction method**.

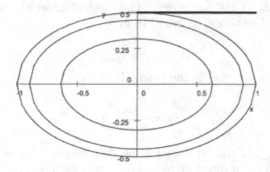

FIGURE 1. Quadratic objective function.

EXAMPLE 1: Let's consider the function $F(x) = -b^T x + \frac{1}{2} x^T A x$ with $A = [2 \ \ -1; -1 \ \ 2]$ and $b = (2,2)^T$. Note that $(1,-1)A(1,1)^T = 0$ so $d_1 = (1,1)^T$, $d_2 = (1,-1)^T$ are A-conjugate. Let's start from the initial point $(x_0, y_0) = (3,4)$. We have to find the minimum of F along d_1, that is, we must find the value α_1 that minimizes $g_1(\alpha_1) = F((x_0, y_0)^T + \alpha_1 d_1)$. (Since F is quadratic this value of α_1 is unique.) Let's find this function of α_1: $g_1(\alpha_1) = F((3,4)^T + \alpha_1(1,1)^T) = F(3+\alpha_1, 4+\alpha_1)$ and, after some simplification, we find that $g_1(\alpha_1) = \alpha_1^2 + 3\alpha_1 - 1$ and so $\alpha_1 = -3/2$ is the minimizer. The new point is $(x_1, y_1)^T = (x_0, y_0)^T + \alpha_1 d_1 = (3,4)^T + (-3/2)(1,1)^T$ which is $(3/2, 5/2)^T$.

Now we must find the value α_2 that minimizes $g_2(\alpha_2) = F((x_1, y_1)^T + \alpha_2 d_2) = F(3/2+\alpha_2, 5/2-\alpha_1)$. After some simplification, we find that $g_2(\alpha_2) = 3\alpha_2^2 - 3\alpha_2 - \frac{13}{4}$ and so $\alpha_2 = 3/6 = 1/2$ is the minimizer. The solution is at $(x_2, y_2)^T = (x_1, y_1)^T + \alpha_2 d_2 = (3/2, 5/2)^T + (1/2)(1,-1)^T$ which is $(2,2)^T$.

Let's check this: The gradient is $\nabla F(x) = Ax - b$ and at the solution $A(2,2)^T - b = (0,0)^T$ as expected. Since the Hessian A is positive definite, this is the unique minimizer. $\square$

If F is not quadratic but we have an approximation of the form of Eq. (7.1) we may apply the method to find an approximate minimizer of F. We then form an improved quadratic approximation to F at this point and repeat the process; we no longer have the property that the method terminates after finitely many iterations but this method can be very effective nonetheless. Again, if F is truly quadratic then the method exhibits quadratic termination from any initial guess.

How will we find the conjugate directions? We describe only one approach, where we assume that we may compute $\nabla F(x) = Ax - b$ at any point. We choose the first direction to be the direction of steepest descent

$$d_0 = -\nabla F(x_0)$$
$$= b - Ax_0$$

starting from the given initial guess x_0, and choose α_1 to minimize $F(x_0 + \alpha_0 d_0)$ ($\alpha_0 \geq 0$ since d_0 is a descent direction). We will then have

$$x_1 = x_0 + \alpha_0 d_0$$

as our new estimate of the minimizer. The method proceeds as follows: At each iteration, the new direction d_k is found from

$$d_k = r_k + \beta_k d_{k-1}$$

(where as usual $r_k = b - Ax_k$ is the residual), with

$$\beta_k = \frac{\|r_k\|^2}{\|r_{k-1}\|^2}.$$

The directions $\{d_k\}_{k=0}^{n-1}$ can be shown to be A-conjugate. We minimize $F(x_{k-1} + \alpha_{k-1}d_{k-1})$ along this new direction to find α_{k-1} and then set $x_k = x_{k-1} + \alpha_{k-1}d_{k-1}$. When we have gone through all n directions we will be at the minimum (possibly even sooner if we happen to land on a point with zero gradient first), excepting the effects of round-off error. This conjugate direction method is called the **conjugate gradient method** (or **CG method**). Somewhat surprisingly, perhaps, a formula can be given for the α_k:

$$(7.2) \qquad\qquad \alpha_k = \frac{\|r_k\|^2}{d_k^T A d_k}$$

($k = 1, ..., n$). Hence we need not actually perform the line search, which enhances the efficiency of this method. We have the following algorithm:

Conjugate Gradient Method:
1. Set $r_0 = b - Ax_0$, $d_0 = r_0$.
2. Begin loop ($k = 1$ to $n - 1$):
3. Set $\alpha_{k-1} = \|r_{k-1}\|^2 / d_{k-1}^T A d_{k-1}$.
4. Set $x_k = x_{k-1} + \alpha_{k-1}d_{k-1}$.
5. Set $r_k = r_{k-1} - \alpha_{k-1}Ad_{k-1}$. If convergence criterion is met, exit.
6. Set $\beta_k = \|r_k\|^2 / \|r_{k-1}\|^2$.
7. Set $d_k = r_k + \beta_k d_{k-1}$.
8. End loop.

The algorithm uses the fact that the residual may be computed as $r_k = r_{k-1} - \alpha_{k-1}Ad_{k-1}$. Since we must compute Ad_{k-1} anyway for use in computing α_k, storing that result and reusing it is more efficient than computing Ax_k in order to compute $r_k = b - Ax_k$ when A is large. An efficient implementation of this algorithm would also maintain a variable $s_k = \|r_k\|^2$ for reuse as needed.

Why do we include a convergence criterion if the method has the quadratic termination property? It's common to use this method on positive definite matrices that are extremely large. We often start out with the intention of using this as an iterative method in the spirit of Ch. 3 even though it is nominally a direct method as in Ch. 2. We may use any of our usual termination criteria. One criterion that is often used is to require that the relative residual $\|r_k\| / \|b\|$ be less than some specified tolerance. Since

$$r_k = b - Ax_k$$

is the residual error for the linear system $Ax = b$, this amounts to thinking of our problem as one of solving the linear system more than one of minimizing a quadratic objective function.

Notice that in the formulas for α_k and β_k the Hessian matrix A is only ever needed to compute a matrix-vector product (the product Ad_{k-1}); similarly for the computation of d_0 (requiring Ax_0) and of F itself (requiring Ax). The method is matrix-free. This fact can be useful when A is so large that storing it and working with it directly is impractical, whether for a minimization problem or for solving $Ax = b$.

Since it is a particular instance of the conjugate direction method, the conjugate gradient method exhibits quadratic termination. However it may be applied to other objective functions as well by forming a quadratic approximant of the objective function. Forming a quadratic approximant is not actually the way that the method is generally applied to nonquadratic functions, though. Instead, we assume that $F(x)$ and $\nabla F(x)$ may be computed at any point and perform the line search $F(x_{k-1} + \alpha_{k-1}d_{k-1})$ by a univariate minimization method (as we have done in other algorithms) rather than using Eq. (7.2), which assumes a quadratic objective function. In addition we replace the formula for β_k (which requires the Hessian A) by a formula that does not require us to find the Hessian. Two of the most popular formulas are the **Fletcher-Reeves formula**

$$\beta_k = \frac{\nabla F(x_k)^T \nabla F(x_k)}{\nabla F(x_{k-1})^T \nabla F(x_{k-1})}$$

(giving the **Fletcher-Reeves method**) and the **Polak-Ribiere formula**

$$\beta_k = \frac{\nabla F(x_k)^T (\nabla F(x_k) - \nabla F(x_{k-1}))}{\nabla F(x_{k-1})^T \nabla F(x_{k-1})}$$

(giving the **Polak-Ribiere method**). Note that in each case the denominator is $\|\nabla F(x_{k-1})\|^2$. These are used to produce conjugate directions starting from $d_0 = -\nabla F(x_0)$ using the formula

$$d_k = -\nabla F(x_k) + \beta_k d_{k-1}$$

and both of these methods can be shown to produce Hessian-conjugate directions despite not depending explicitly on the Hessian. It's usually beneficial to restart the method periodically with the current steepest descent direction.

Some of the other optimization methods we have seen actually perform better with inexact line searches so that they may make bigger steps sooner. In order to maintain the conjugacy here, however, we must perform the line searches accurately.

There are methods for generating conjugate directions that do not use the gradient (derivative-free methods). A modification of the Gram-Schmidt process may be applied but this is typically computationally expensive. Other approaches are more efficient.

Unsurprisingly, perhaps, there is a connection to Krylov subspace methods. From this it can be shown that the CG method results in monotonically decreasing absolute error $\|x - x_k\|$.

What if we wish to solve $Ax = b$ where A is *not* positive definite? The conjugate gradient method does not apply in this case, but we might consider trying to solve $Ax = b$ by applying the CG method to the normal equations

$$A^T A x = A^T b$$

since if A is nonsingular then the Gram matrix $A^T A$ is positive definite. This is referred to as **conjugate gradients on the normal equation (residual)**, or the **CGNR method**, because it has a residual-minimizing property. Of course, we pay the expected price: $\kappa(A^T A) \approx \kappa^2(A)$. A similar approach is to apply the CG method to

$$AA^T y = b$$

in order to find y, and then form $x = A^T y$; this is called **conjugate gradients on the normal equation (error)**, or the **CGNE method**, because it has an error-minimizing property (with respect to the absolute error). For both CGNR and CGNE the minimization of the indicated quantity (the residual $\|b - Ax_k\|$ and the absolute error $\|x - x_k\|$, respectively) is over all x_k in the affine space $x_0 + K_k$ where K_k is the Krylov subspace

$$K_k = \text{span}\{A^T r_0, (A^T A)A^T r_0, ..., (A^T A)^{k-1}A^T r_0\}$$

formed from successive powers of $A^T A$ applied to the vector $A^T r_0$. We might try the CGNR or CGNE method for difficult linear systems.

For both the CGNR and CGNE methods it is possible to avoid explicitly forming the cross-product matrix $A^T A$ or AA^T; we form $w = A^T A x$ in two steps as $v = Ax$, $w = A^T v$, for example. The methods are no longer truly matrix-free, however, because they are not transpose-free: We would need both a program for computing Ax and one for computing $A^T x$, which is not commonly provided when a program has been written to evaluate Ax. When A represents something physical there may be a natural way to evaluate Ax that doesn't translate easily into a corresponding trick for evaluating $A^T x$ (or for computing $x^T A$, which could then be transposed to $A^T x$). It's easy to write down $A^T x$ in an equation but it isn't always easy to compute it.

MATLAB

Let's apply the conjugate gradient method to a moderately large sparse system. We'll use a matrix that appears in applications involving partial differential equations. Enter:
```
>> full(gallery('poisson',3))
```
to view a small example. Note that the matrix is sparse, banded, and weakly diagonally dominant; clearly there is other structure in it as well. Enter:
```
>> A=gallery('poisson',50);   %Sparse matrix.
>> size(A), nnz(A)/prod(size(A))
```
to create a larger example matrix. Enter:
```
>> x=rand([2500 1])-.5;b=A*x;
>> xlu=A\b;
>> norm(x-xlu)
```
The solution found by the LU decomposition is very good! Let's try using a few iterations of conjugate gradients. Enter:
```
>> x0=ones(size(x));
>> r0=b-A*x0;d0=r0;
>> norm(r0)
>> a0=(norm(r0)^2)/(d0'*A*d0)
>> x1=x0+a0*d0;
```

```
>> r1=r0-a0*A*d0;   %Should be reusing A*d0 from above.
>> norm(r1),norm(b-A*x1)   %Check.
```
We have completed one iteration of the conjugate gradient method, and have reduced the residual. Let's check the absolute error; enter:
```
>> norm(x-x0),norm(x-x1)
```
Let's do another iteration. Enter:
```
>> b1=(norm(r1)/norm(r0))^2
>> d1=r1+b1*d0;
>> a1=(norm(r1)^2)/(d1'*A*d1)
>> x2=x1+a1*d1;
>> r2=r1-a1*A*d1;   %Should be reusing A*d0 from above.
>> norm(r2),norm(b-A*x2)   %Check.
>> b2=(norm(r2)/norm(r1))^2
>> d2=r2+b2*d1;
>> norm(r2),norm(b-A*x2)   %Check.
>> norm(x-x0),norm(x-x1),norm(x-x2)
```
An efficient implementation would of course overwrite some of these quantities, but we are just experimenting right now.

The basic MATLAB command for the CG method is pcg, a preconditioned conjugate gradient method.
```
>> xcg=pcg(A,b);     %No preconditioning.
>> norm(x-xcg),norm(x-xcg)/norm(x)
```
The quality of the solution is not very good. We could have asked the program to flag whether or not the tolerance had been met:
```
>> [xcg,status]=pcg(A,b);
>> status
```
The flag is set to 1, indicating (per the help text) that the default error tolerance has not been met. Let's try simple Jacobi preconditioning:
```
>> D=diag(diag(A));   %Still in sparse form.
>> [xcg2,status2]=pcg(A,b,1E-8,100,D);
>> norm(x-xcg2),norm(x-xcg2)/norm(x)
>> status2
```
This is improved, but we would need a more sophisticated preconditioner for this difficult matrix.

The MATLAB documentation for pcg provides an example of using the command in a matrix-free manner using a Moler matrix, named after MATLAB founder Cleve Moler. Enter:
```
>> doc pcg
```
and try the "Using pcg with Large Matrices" example given there. This is commonly done in practice. What if you needed to be able to form $A^T x$ this way too?

There is also a command bicg, implementing the **biconjugate gradients method**, that is intended for nonsymmetric matrices.

Problems

1.) a.) Verify that Eq. (7.2) gives the same results in Example 1 as were found by minimization of $g_1(\alpha_1)$ and $g_2(\alpha_2)$.

b.) Repeat Example 1 but use the conjugate directions in the opposite order ($d_1 = (1,-1)^T$, $d_2 = (1,1)^T$) and use the initial guess $(1003, 2005)$.

c.) Perform two iterations of alternating variable search on the objective function of Example 1 using the initial guess $(x_0, y_0) = (3, 4)$.

2.) a.) Use the conjugate direction method to find the minimum of the function $F(x) = -b^T x + \frac{1}{2} x^T A x$ with $A = [3 \quad -1; -1 \quad 3]$ and $b = (4, -2)^T$. Use the conjugate directions $d_1 = (1, 1)^T$, $d_2 = (1, -1)^T$ (verify that A is positive definite and the directions are A-conjugate).

b.) Repeat part a. with the conjugate gradient method.

c.) What is the solution of $Ax = b$?

3.) Write a detailed algorithm (pseudo-code) for the conjugate direction method. Assume that the conjugate directions will be supplied.

4.) Write a detailed algorithm (pseudo-code) for the conjugate gradient method. Select a reasonable convergence criterion and include a restarting provision.

5.) Write a detailed algorithm (pseudo-code) for the Fletcher-Reeves method.

6.) a.) Write a MATLAB program that implements the conjugate gradient method for a given quadratic function as in Eq. (7.1). Assume that A and b will be given and argue that the value of c is irrelevant.

b.) Test your program on the problem of Example 1.

7.) Use the conjugate gradient method to minimize $F(x) = \frac{1}{2} x^T A x - b^T x$ where $A = [5 \ 2; 2 \ 3]$ and $b = (3, 2)^T$. (Verify that A is positive definite.)

8.) Use the conjugate gradient method to minimize $F(x) = \frac{1}{2} x^T A x - b^T x$ where $A = [5 \ 2 \ 1; 2 \ 3 \ 1; 1 \ 1 \ 1]$ and $b = (-4, 2, 1)^T$. (Verify that A is positive definite.)

9.) a.) Show that if A is positive definite and diagonal then alternating variable search is in fact a conjugate direction method for A and hence exhibits quadratic termination.

b.) Show that if A is symmetric then the eigenvectors may be used to form a linearly independent set of A-conjugate vectors.

c.) Show that a set of nonzero pairwise A-conjugate vectors is always linearly independent.

d.) Show that the conjugate gradient method produces directions that are in fact conjugate.

e.) Show that the residual satisfies $r_k = r_{k-1} - \alpha_k A d_k$ in the conjugate gradient method.

f.) Show that the conjugate gradient method residuals satisfy $r_k^T r_i = 0$ for $i = 0, 1, ...k - 1$.

10.) Verify experimentally that Eq. (7.2) produces the unique values α_k that minimize $F(x_{k-1} + \alpha_{k-1} d_{k-1})$ for the conjugate gradient method. Does the formula appear to work for the more general conjugate direction method?

11.) a.) Write a MATLAB program that implements the Fletcher-Reeves method for a nonlinear function F.

b.) Test your program on the Rosenbrock function $.F(x, y) = 100(y - x^2)^2 + (1 - x)^2$.

12.) a.) Let A be a matrix for which each entry on the main diagonal is 4, each entry on the first super- and sub-diagonals is zero, each entry on the second super- and sub-diagonals is -1, and all other entries are zero. Use MATLAB to verify that this matrix is positive definite for $n = 2, ..., 10$.

b.) In order to use conjugate gradients as a matrix-free algorithm we must be able to compute Ax without forming A. Write a MATLAB program that accepts as input an order n and a vector x of length n and returns Ax for this A.

c.) Write a MATLAB program that implements the conjugate gradient method for a given quadratic function of the form in Eq. (7.1) and that calls a program such as your program from part b. to compute Ax whenever needed. Your program should only have access to A via this routine. Test it using your program from part b.

13.) Since A is assumed to be positive definite in Eq. (7.1), we could find the minimizer by solving $Ax = b$ using the Cholesky factorization of A. If A is large and sparse the conjugate gradient method is usually superior but if A is moderate in size and dense then Cholesky factorization might be superior. Show by example that the conjugate gradient method can be better for large sparse matrices and the Cholesky factorization might be better for smaller matrices.

14.) a.) The positive definite matrix A defines the inner product $(u, v) = u^T A v$ and hence the A-norm $\|u\|_A = (u^T A u)^{1/2}$. Verify that $\|\cdot\|_A$ is a valid vector norm (see Sec. 2.6).

b.) For $A = [2\ \ 1; 1\ \ 2]$, compute $\|x\|_A$ for $x = (1,1)^T$, $x = (1,-1)^T$, and $x = (-2,1)^T$. Repeat for $A = [2\ \ -1; -1\ \ 2]$.

c.) Is the standard inner product $(u, v) = u^T v$ of the form $(u, v) = u^T A v$ for some positive definite matrix A?

15.) If x^* is the minimizer of Eq. (7.1) then under reasonable conditions the conjugate gradient method without restarting produces iterates that satisfy $\|x_k - x^*\|_A \le 2\|x_0 - x^*\|_A \left[\left(\sqrt{\kappa(A)} - 1 \right) / \left(\sqrt{\kappa(A)} + 1 \right) \right]^k$ where $\|x\|_A = (x^T A x)^{1/2}$.. Comment on the speed of convergence that this suggests for the conjugate gradient method.

16.) How is the conjugate gradient method related to (an affine subspace based on) a Krylov subspace? Be specific.

8. Hansen's Method

We've focused primarily, though not exclusively, on unconstrained problems. It's quite common to be given the problem

$$\min_{x \in D} F(x)$$

where D has a complicated boundary. Often D is not given explicitly but must be found from the intersection of a list of constraints, or by a root-finding method. It can be difficult to identify the feasible region D defined by the constraints.

One general-purpose approach to constrained optimization problems is **sequential quadratic programming** (or **SQP**): The objective function is approximated by its second-order Taylor series approximation at a point, the constraints are approximated by their first-order Taylor series approximations at that point, and an appropriate algorithm is applied to find a new point, from which the process is repeated. (Often the Hessian is itself approximated in this process.) Conjugate gradient methods and their variants can play a role here. Another approach is to add to the objective function $F(x)$ a **penalty function** $P(x)$ that is small when the constraints are satisfied but large when they are violated, and then minimize $F(x) + \nu P(x)$ (for some **penalty parameter** $\nu > 0$) instead of working with $F(x)$. This is called a **penalty function method**; we solve the problem repeatedly for increasing values of the penalty parameter. When ν is sufficiently large we should be getting solutions that are close to the true minimizer.

Let's look at a very different approach. The method to be developed can be considered a constrained minimization routine if the constraints are of the appropriate form.

The IEEE standard requires that five rounding modes[6] be supported: Two different forms of round to nearest; round towards zero; round up; and round down. The standard also requires that floating point addition, subtraction, multiplication, and division satisfy the requirements

$$
\begin{aligned}
x \oplus y &= \text{round}(x + y) \\
x \ominus y &= \text{round}(x - y) \\
x \otimes y &= \text{round}(x \times y) \\
x \oslash y &= \text{round}(x/y)
\end{aligned}
$$

called the requirement of **correct rounding**. Consider addition. This means that the result of adding x and y on the computer must be the same as the result of adding x and y in infinite precision arithmetic and then applying the current rounding mode (in the current precision, e.g., double precision). Hence

$$(x \oplus y)_d \leq x + y \leq (x \oplus y)_u$$

where $(x \oplus y)_d$ is the sum computed with the *round down* (toward $-\infty$) rounding mode in effect, and $(x \oplus y)_u$ is the sum computed with the *round up* (toward ∞) rounding mode in effect. But this means that we can compute a guaranteed bound on $x + y$, and similarly for the other arithmetic operations. (If $x + y$ overflows, one of these bounds will be plus or minus infinity.) Let's repeat that: We can get guaranteed bounds in floating point arithmetic.

Unfortunately, many programming languages do not allow the user to easily change the rounding mode. However, in most cases it is possible to use special purpose software to change the rounding mode. (It can be done in C++, for example, and the compiled program could be called in MATLAB.) Assuming such software is available, we could get guaranteed error bounds on the result of an arithmetic operation by performing it twice, once with *round down* in effect and then again with *round up* in effect[7]. We would then have an interval

$$[w_l, w_u]$$

that must contain the true result $w = x + y$. For a more complicated computation such as

$$(8.1) \qquad\qquad w = x + \frac{yz - \pi}{u + v}$$

we could simply apply this technique successively, ending up with an interval that must contain the true result w. The value of π would be entered as an interval $[\pi_l, \pi_u]$ containing the true value of π.

---

[6]As of this writing, there is a sixth under consideration for inclusion.

[7]If we can switch the rounding mode at will, we could perform the operation, round it one way, then round it the other way while still stored in the FPU without re-computing it.

This technique of using directed rounding in floating point arithmetic to generate intervals that must contain the correct results is called **interval arithmetic** and numerical methods that make use of it are called **interval methods**. Interval arithmetic can be defined for analytical use also (**interval analysis**) but we will concentrate on its application to scientific computing.

In interval arithmetic we treat all variables as intervals. If c is a constant that can be represented exactly we write it as the degenerate interval $[c, c]$, called a **point interval** in this context. In order to compute an interval containing the result of an expression like Eq. (8.1) we will need to know how to do arithmetic on intervals.

EXAMPLE 1: Let $X = [1, 2]$ and $Y = [-1, 1]$. Then if $x \in X$ and $y \in Y$, $x + y$ could be as large as 3 and as small as 0 so $X + Y = [0, 3]$ contains all possible values of $x + y$. Similarly, $x - y$ could be as large as $2 - (-1) = 3$ and as small as $-1 - 2 = -3$ so $X - Y = [-3, 3]$ contains all possible values of $x - y$. □

As indicated in the example, we define the sum of two interval variables $X = [a, b]$ and $Y = [c, d]$ to be the smallest interval Z that contains every possible sum of the form $z = x + y$, $x \in X$, $y \in Y$. This means that

$$Z = [a + c, b + d]$$

with directed rounding used when we are doing this in floating point arithmetic, so that the computation of $a + c$ is rounded down and the computation of $b + d$ is rounded up. Similarly,

$$
\begin{aligned}
X - Y &= [a - d, b - c] \\
(8.2) \quad X \times Y &= [\min(ac, ad, bc, bd), \max(ac, ad, bc, bd)] \\
X/Y &= [a, b] \times [1/d, 1/c] \\
&= [\min(a/d, a/c, b/d, b/c), \max(a/d, a/c, b/d, b/c)]
\end{aligned}
$$

are defined so that the resulting interval contains every possible value of $x - y$, $x \times y$, x/y, respectively. Division may or may not be taken to be undefined if $0 \in Y$.

Interval arithmetic packages provide commands for computing intervals that contain the results of standard functions such as $\sin(X)$ and $\exp(X)$ when X is a closed interval. For some functions the resulting intervals may not be the smallest ones possible, though that is always the ideal. For example, we might develop an approximate interval version of the sine function $g(x) = \sin(x)$ on $[0, \pi/2]$ by writing

$$\gamma(X) = \left[a - \frac{1}{6}a^3, b - \frac{1}{6}b^3 + \frac{1}{120}b^5 \right]$$

(where $X = [a, b]$ is contained in $[0, \pi/2]$), using in this case the fact that the Maclaurin series for sine is an alternating series and so we know where it is an under- or over-estimate. This is a guaranteed inclusion but much wider than it need be. (Expanding around the midpoint $m(X) = (a + b)/2$ of the interval is common advice for achieving better results.) Using such extensions of standard functions to interval arguments, we can compute intervals containing quantities like

$$Y = X \sin(X) - X^2$$

and the interval Y will be guaranteed to include all possible values of $x\sin(x) - x^2$ for $x \in X$. Directed rounding extends this guarantee to the computer.

For our numerical methods to work we will need the property of **inclusion isotonicity**: An interval function $F(A, B)$ is said to be inclusion isotone if whenever $X_1 \subseteq Y_1$ and $X_2 \subseteq Y_2$, a $F(X_1, X_2) \subseteq F(Y_1, Y_2)$. The arithmetical operations and our sine function extension all satisfy this.

The traditional concerns about interval methods are fourfold: They take more computing time, the intervals will grow too large to be useful, it's a hassle to change the rounding modes, and these strict guarantees on our end result are rarely needed. These are valid concerns. However, there are times when it pays off.

We'll focus on global minimization over a box-shaped region. Consider first a twice-differentiable function of one variable $f(x)$ defined on all of $\mathbb{R}$, and some interval of interest $X \subset \mathbb{R}$. Assume we can obtain interval estimates of the value of f (say, because f can be written in terms of functions provided by an interval extension package for the language we are using). We wish to find the global minimizer of f over X.

The method we will use is called **Hansen's method**. It is based on repeated subdivision of the interval. Let ϕ be the best known upper bound on the value of f at its global minimizer (not an upper bound on f over all of X). Suppose X_1 is an interval contained within X. Hansen's method uses the following three facts:

- If $f''(x)$ is never positive on X_1 then there's no local minimum of f interior to X_1.
- If $f'(x)$ doesn't change sign on X_1 then there can be no local minimum of f interior to X_1.
- If the smallest value in X_1 is greater than the upper bound ϕ on $f(x^*)$ then X_1 can't contain the global minimum.

The first point follows because f would have the wrong concavity over X_1. The second point follows because the derivative cannot be zero in X_1. The last point follows because the lower bound on values in X_1 would exceed the upper bound on $f(x^*)$.

So we start by applying the three conditions above to X itself to see if there might be a minimizer contained in it. If it is possible that there is a minimizer in it, we split $X = [a, b]$ into two intervals $X_1 = [a, \mu]$ and $X_2 = [\mu, b]$. We check the three conditions above on X_1. If X_1 can't contain the minimum, and neither endpoint of X could be the minimum, we reject it; otherwise we subdivide $X_1 = X_{11} \cup X_{12}$ in the same way and put them on a list $\mathcal{L} = (X_{11}, X_{12})$. We repeat with X_2. After this we return to the list $\mathcal{L}$ and process the intervals on it in the same manner. An example will clarify the procedure.

EXAMPLE 2: Let $f(x) = -x^3/2 + x^2 + x/2$ on $[a, b] = [-1, 2]$. To perform Hansen's method we might let $\phi = \min(f(a), f(b)) = 1$ be the initial upper bound on $f(x^*)$. We have

$$
\begin{aligned}
f(x) &= -x^3/2 + x^2 + x/2 \\
f'(x) &= -3x^2/2 + 2x + 1/2 \\
f''(x) &= -3x + 2
\end{aligned}
$$

and we start by checking the concavity over $X = [-1, 2]$:

$$\begin{aligned} f''(X) &= -3X + 2 \\ &= -3[-1, 2] + [2, 2] \\ &= [-4, 5] \end{aligned}$$

so we can't rule out the possibility that $f''(x)$ is positive somewhere. Next we check whether there might be a zero of the derivative over X:

$$\begin{aligned} f'(X) &= -3X^2/2 + 2X + 1/2 \\ &= -\frac{3}{2}[-1, 2]^2 + 2[-1, 2] + [1/2, 1/2] \\ &= -\frac{3}{2}[0, 4] + [-2, 4] + [1/2, 1/2] \\ &= [-15/2, 9/2] \end{aligned}$$

using the fact that X^2 cannot be negative. Yes, there may be a point in X at which $f'(x)$ is zero. Lastly we check for the smallest point:

$$\begin{aligned} f(X) &= -X^3/2 + X^2 + X/2 \\ &= -[-1, 2]^3/2 + [-1, 2]^2 + [-1, 2]/2 \\ &= -\frac{1}{2}[-1, 8] + [0, 4] + [-1/2, 1] \\ &= [-9/2, 11/2] \end{aligned}$$

evaluating X^3 in the obvious way. This is an overestimate, as the true range of f over X is $[-0.0563, 1.3156]$ (to 4 places). But as $f(x)$ could be as small as $-9/2$ by this analysis we cannot throw out this interval. We subdivide it into $X_1 = [-1, 1/2]$ and $X_2 = [1/2, 2]$ and place them on a list.

Let's start with X_2. We check the concavity of f over it:

$$\begin{aligned} f''(X_2) &= -3X_2 + 2 \\ &= -3[1/2, 2] + [2, 2] \\ &= [-4, 1/2] \end{aligned}$$

and it may be of either sign. Let's check the derivative:

$$\begin{aligned} f'(X) &= -3X^2/2 + 2X + 1/2 \\ &= -\frac{3}{2}[1/2, 2]^2 + 2[1/2, 2] + [1/2, 1/2] \\ &= -\frac{3}{2}[1/4, 4] + [1, 4] + [1/2, 1/2] \\ &= [-9/2, 33/8] \end{aligned}$$

and it may change sign. The third test also doesn't eliminate it. Let's subdivide X_2 into $X_{21} = [1/2, 5/4]$ and $X_{22} = [5/4, 2]$ and add them to the list. Let's check X_{22}. The concavity check gives

$$f''(X_{22}) = -3X_{22} + 2$$
$$= -3[5/4, 2] + [2, 2]$$
$$= [-4, -7/4]$$

and we may eliminate X_{22} as the concavity cannot be positive there. The global minimizer lies in $[-1, 5/4]$. We would now continue to work with X_1 and X_{21}. $\square$

As indicated in the example, the list may be processed in any order. Getting a smaller value of ϕ as soon as possible is desirable to help keep the list small, but it's rarely clear how best to do that. Note that the method will find intervals containing all global minima if the minimizer is not unique. This is an uncommon feature among global minimization algorithms but is also rarely of great interest.

The ability of algorithms based on interval arithmetic to exclude large regions quickly is powerful: We might subdivide a large interval into two halves and immediately rule out one of them. There is also the guarantee of exactness of the results. On the other hand, the use of interval arithmetic is slower as well as less convenient since it depends on specialized software, and we are restricted to functions that can be expressed in terms of the functions provided by the interval arithmetic software package. These are significant drawbacks that limit the applicability of these methods. Frequently the objective function f is the output of another program that may itself be computing values of f by a numerical method, and an interval version of f may simply not be available.

How fast *does* the method converge? If $X = [a, b]$ then we define $w(X) = b - a$ to be the width of X. We define $\overline{g}(X)$ to be the true range of g over X, that is,

$$\overline{g}(X) = [\min(g(x), \max(g(x))]$$

where the min and max are over all $x \in X$. This may be hard to find, so we're probably using an interval extension $\gamma(X)$ with the property that $\overline{g}(X) \subseteq \gamma(X)$. Realistically, we expect that $\overline{g}(X) \subset \gamma(X)$ (a proper subset). We define the **excess width** of the interval estimate to be

$$e(X) = w(\gamma(X)) - w(g(X))$$

which measures the extent to which $\gamma(X)$ overestimates the true range of g over X. (The floating point version will be slightly wider still because of rounding.) It is a fact that for reasonable functions g that are extended to interval functions in appropriate ways, the excess width goes to zero quadratically as the width of X goes to zero; that is,

$$w(\gamma(X)) - w(\overline{g}(X)) = O(w(X)^2)$$

(as $w(X) \to 0$). Hence, the final interval returned by an interval method will not usually be too large to be useful, and we can hope to find it rapidly. In fact, answers are not uncommonly found for which the endpoints of the interval differ only in the least significant bit (**least significant bit accuracy**, or **lsba**).

Hansen's method can easily be extended to apply to a function $F : \mathbb{R}^n \to \mathbb{R}$; we subdivide boxes rather than intervals. It can be improved by applying an interval version of Newton's method to isolate regions in which a change of sign of $f'(x)$ might occur in order to rapidly identify promising intervals while lowering ϕ. It is very robust and can be used for continuous but nondifferentiable functions by simply dropping the tests involving $f''(x)$ and $f'(x)$.

In addition to the inconveniences mentioned previously, however, it can happen that the list of intervals to be processed grows so large so quickly that it can actually be a storage problem, exceeding the amount of memory available for storing of the list (if intervals are not being eliminated quickly enough). This is not an uncommon problem, and it highlights the need for a strategy to select intervals from the list for processing–do we take the largest next? The one that seems likely to give us the best improvement in ϕ? The one that has been on the list the longest? There are several strategies.

There are interval algorithms for most problems in numerical analysis. As mentioned above, there is an interval form of Newton's method for root-finding; and there are methods for numerical integration, Gaussian elimination, and so on. Interval arithmetic is also used in computational statistics.

Non-interval versions of the subdivision idea existed well before Hansen's method and are called **branch-and-bound methods**. Giving up the guarantee provided by interval arithmetic is usually not a great concern–we just ignore the slight possibility of missing something due to rounding errors. Again, how best to process the list is usually the principal issue.

MATLAB

There are third-party interval methods packages for MATLAB and other languages. Some are commercial and some are freeware. A few computing environments have the capability built-in.

The author hesitates to mention that there is an *intentionally* undocumented, unsupported command in MATLAB, feature, that allows the user to–among other things–change the rounding mode. *This is generally not recommended. It sends a signal to the FPU that affects all computations until reset. If someone else might be using your machine at the same time, read the following but do not do it.* Having said that, if you enter:

```
» s=feature('setround')
```

then, at least as of this writing, you should get the value .5, indicating the default rounding mode of round to nearest, breaking ties to even. Then we could test the various rounding modes as follows:

```
» V=ones([1 1000])*1E-20;
» T1=sum(V)
» feature('setround',-Inf)
» s=feature('setround')
» T2=sum(V)
» feature('setround',Inf)
» s=feature('setround')
» T3=sum(V)
» T2<T1      %Should be true.
» T1<T3      %Should be true.
» feature('setround',.5)    %Restore to default.
```

Of course it's possible that these values could have been equal; after all, the computation $1 + 2$ should return the same value in all rounding modes.

Your interval addition code, then, could change to round toward ∞, compute $x + y$, change to round toward $-\infty$, compute $x + y$, and then change back to the rounding mode that was current when the routine was called (available within the program by s=feature('setround'), which could be checked and stored initially). This is the safest approach. It's always best to use tested routines for computations like this, of course, even if you must write and test them yourself.

There are interval methods for many other types of problems in scientific computing, and software that implements them. If you have the opportunity to try such a package, you should!

Problems

1.) Prove that the three definitions in Eq.(8.2) correctly implement the rule that $X \boxdot Y$ should contain all values of $x \boxdot y$, $x \in X$, $y \in Y$, for $\boxdot = -, \times, /$.

2.) a.) Let $X = [1, 2]$, $Y = [1, 3]$, and $Z = [-1, 1]$. Compute $X + Y$; $X - Y$; $Z \times Z$; $X \times Y \times Z$; XZ/Z; $Y - Y$.

b.) Note that $Z \times Z$ contains negative entries. This is sensible if the two intervals might have arisen from different computations and are only coincidentally equal but makes for a poor definition of Z^2, say. Suggest a definition of Z^n for n a nonnegative integer.

3.) Are addition and multiplication commutative in interval arithmetic? Are they associative?

4.) Let $f(x) = x^3$ and let $X_n = [0, 1/2^n]$. Compute $f(X_n)$ for $n = 1, 2, 3$ and compare $w(f(X_n))$ to $w(X_n)$.

5.) Perform two iterations of Hansen's method with objective function $f(x) = 1 + 5x - x^3$ on the interval $X = [-2, 2]$; that is, check X and if it cannot be eliminated then subdivide it once and check the two resulting intervals. Plot $f(x)$ as well.

6.) Use Hansen's method to find a global minimum of the function $f(x) = x^3 - 4x^2 + x + 3$. Choose a reasonable initial interval, and don't seek great precision.

7.) Use Hansen's method to find a global minimum of the function $f(x) = x \sin(x^2)$ on $[10, 20]$. Use your knowledge of the sine function to evaluate it over an interval.

8.) Write a detailed algorithm (pseudo-code) for Hansen's method for a function of a single variable. Make sure that your algorithm addresses the possibility that the global minimum is at an endpoint of an interval. Assume that interval operations are implemented automatically. You will need to supply a list processing rule and a termination criterion; remember that the global minimizer may not be unique.

9.) a.) Prove the **subdistributive law** of interval arithmetic, $A(B+C) \subseteq AB+AC$.

b.) Prove that $A \subseteq C$, $B \subseteq D$ implies $A \boxdot B \subseteq C \boxdot D$ where $\boxdot$ is one of the four interval arithmetic operations (the **inclusion isotonicity** property).

10.) a.) The absolute value of an interval can be defined by $|X| = \max(a, b)$ where $X = [a, b]$. This also serves as the max norm $\|X\| = |X|$ of an interval. Prove that $|X| \leq w(X)$ and that $w(XY) \leq 2|X|$.

b.) Prove that $|XY| = |X||Y|$.

c.) Prove that $w(1/X) \leq |1/X|^2 w(X)$ if $0 \notin X$.

11.) Write a MATLAB program that implements Hansen's method on a polynomial in a single variable. Use the interval arithmetic formulas in this section without directed rounding if it's not easily available. Demonstrate your program on several functions.

12.) a.) Given a function $f(x)$, there are many ways to express it as a formula. The evaluation of one formula gives the same value as any other; however, this need not hold for their interval extensions. Compute $f(x) = x^2 - 1$ and $g(x) = (x-1)(x+1)$ for $X = [-1, 2]$, $X = [-1, 1]$, and $X = [-2, 0]$. Which is the better representation?

b.) One issue here is that if we define $X^2 = X \cdot X$ then $[-1, 1]^2 = [-1, 1]$. But we would expect $X^2 = [0, 1]$. How should we define X^2? How about X^3? This effect can be challenging in interval computations. Compare a naive evaluation of $X \sin(X) - X^2$ to an evaluation that is aware that each occurrence of X is the same variable. Is $X(\sin(X) - X)$ any better?

c.) Simply replacing x by X to turn an $f(x)$ into a formula defined on and returning intervals gives the **natural interval extension** of f. There are techniques for finding better extensions, that is, better interval formulae. One is the **centered form**, where we first write f as $f(x) = f(c) + (x - c)f(x - c)$ and then use the natural interval extension. Typically $c = m(X)$. What are the centered forms of $f(x)$ and of $g(x)$? What do they give for each of the three intervals in part a.? Compare the excess width of the centered form in each case to the excess widths from part a.

13.) Consider attempting to minimize the Rosenbrock function $.f(x, y) = 100(y - x^2)^2 + (1 - x)^2$ subject to the constraint $(x - 3)^2 + (y - 3)^2 \leq 1$. (The sole local minimum $(1, 1)$ is not in the constraint region.) Use the penalty function method with $P(x, y) = \nu[(x - 3)^2 + (y - 3)^2]$ if $(x - 3)^2 + (y - 3)^2 > 1$ and $P(x, y) = 0$ if $(x-3)^2 + (y-3) \leq 1$. Use $\nu = 1, 3, 5, 10, 20, 100$ and any desired local minimization routine (but note that $P(x, y)$ is not differentiable–yet another reason why we need minimization methods that can handle nondifferentiable problems).

14.) a.) The **interval Newton's method** is defined as follows: If Y is the interval of interest, set $Y_0 = Y$. Then set $N(Y_n) = m(Y_n) - f(m(Y_n))/f'(Y_n)$ and $Y_{n+1} = Y_n \cap N(Y_n)$ where $m(X)$ is the midpoint of the interval X. All zeroes of f that lie in Y are contained in Y_n at each iteration. (We allow the intervals to have infinite endpoints. Conceivably Y_{n+1} could be a pair of intervals; retain both.) Use two iterations of the interval Newton's method to locate intervals that must contain the zeroes of $f(x) = .1x - \sin(2x) - \cos(3x)$ that lie in the interval $Y = [0, 16]$.

b.) Explain why it would be useful to add this to Hansen's method.

15.) a.) Write a detailed algorithm (pseudo-code) for Hansen's method for a function of several variables.

b.) Implement your method in MATLAB. Do not use directed rounding if it is not easily available.

16.) a.) Prove that for the approximate $\sin(X)$ given in the section, $w(\gamma(X)) - w(\overline{g}(X)) = O(w(X)^2)$ as $w(X) \to 0$.

b.) How might you get an interval estimate for $f(x) = \exp(x)$?

CHAPTER 7 BIBLIOGRAPHY:

Numerical Optimization, Jorge Nocedal and Stephen J. Wright, Springer-Verlag New York, 1999
A comprehensive text.
Practical Optimization, Philip E. Gill, Walter Murray, and Margaret H. Wright, Academic Press 1981
A comprehensive reference.
Iterative Methods for Optimization, C. T. Kelley, SIAM 1999
A modern, applied reference.
Algorithms for Minimization Without Derivatives, Richard P. Brent, Dover 2002
Includes Brent's method for root-finding and for optimization.
Introduction to Global Optimization, Reiner Horst, Pãnos M. Pardalo, and Nguyen V. Thoai, Kluwer 1995
A survey of global optimization methods.
Computer Methods for the Range of Functions, H. Ratschek and J. Rokne, Ellis Horwood Ltd. 1984
An introduction to interval methods including Hansen's method.
Handbook of Test Problems in Local and Global Optimization, Christodoulos A. Floudas, Pãnos M. Pardalo, et al., Kluwer 1999
A collection of test problems.
Random Number Generation and Quasi-Monte Carlo Methods, Harald Niederreiter, SIAM 1992
An introduction to quasi-Monte Carlo methods.

Afterword

THE PURPOSE OF COMPUTING IS INSIGHT, NOT NUMBERS.
-R. W. Hamming

I consider *numerical analysis* to be the traditional side of the subject that uses functional analysis to do things like develop series representations that can be truncated to give a method (e.g., Newton's method), or uses the calculus to estimate the error in polynomial interpolation, or uses perturbation theory to perform a backwards error analysis involving a matrix norm, or uses the Fixed Point Theorem to estimate how rapidly a method converges, and so on. It's about the formal mathematical techniques used to develop and analyze methods, including round-off error analysis. It's solidly part of Applied Mathematics, though it's sometimes taught in Computer Science departments.

I consider *scientific computation* to be the side of the subject that addresses how to choose the software and method, how to piece together a larger program from smaller available programs, how to make smart use of available hardware including advanced architectures, what the effects of the particular implementation of floating point arithmetic will be, how best to code an algorithm for maximum efficiency, how to guard against underflow, overflow, unrealistically small tolerances, and such, and the art of choosing good initial guesses and good parameters, like the relaxation parameter ϖ in SOR or damping sequences in Newton's method. It's the side of the subject that "thinks computationally"[8] and makes the methods and software work in the difficult cases that occur so often. While it has interdisciplinary aspects, it's primarily part of Computer Science.

Computational Science is now widely recognized as an independent academic discipline that uses the tools of Applied Mathematics (including numerical analysis, ordinary and partial differential equations, stochastic processes,...) and Computer Science (computer architecture, programming languages, analysis of algorithms,...) along with "domain knowledge" from an area of application (say, fluid dynamics from Physics or Mechanical Engineering) to create, implement, and analyze the results of computational and simulational models in order to better understand and predict physical phenomena. It's been described as the "third way" of doing science, alongside theory and experiment. Such computational models range from those used to understand the collision of galaxies–hard to do in a traditional laboratory setting–to those used to predict the weather. Shading into Computational Engineering, numerical wind tunnels are used to design aircraft entirely within a computer. While I myself have created an undergraduate program in this area and am a vocal

---

[8]Paraphrasing W. Feller's comments about probability having two hands, one that knows about measure theory, convergence, and such, and one that "thinks probabilistically".

advocate of it, this text is primarily about the methods of numerical analysis and secondarily about the techniques of scientific computing. I recommend the text by my colleagues Holder and Eichholz for an introduction to Computational Science. Unsurprisingly, you'll find a lot of overlap.

We've seen lots of tricks and tips throughout the text. At this point, given an equation to be used in a program, you should be looking for the presence of possible subtraction of nearly equal quantities or division by small numbers; for an opportunity to add small positive numbers first with larger ones to be added later; for quantities of widely varying magnitudes; and for inefficiencies such as inverting a matrix when a solution of a linear system will suffice.

You should be looking for opportunities to reuse already computed quantities if they're at all expensive to compute the first time; for ways to rewrite equations for improved memory usage; for the possibility that a quantity might be approximated by some other, simpler method giving results about as good. You should be asking how the equation was derived and then checking to see if an earlier incarnation of the equation lends itself to more efficient or more accurate computation, possibly because another version was better-conditioned.

You should consider rewriting the equation to remove or at least separate out singularities, oscillatory terms, or other problems. A change of variable or two can often make a numerical quadrature problem much easier to solve to a desired accuracy; splitting it into the sum of two integrals may allow you to choose a better method for each part than could possibly be selected for it as a single integral. One of those methods may well be symbolic rather than numerical. You should consider other forms of preprocessing the problem, such as preconditioning for linear systems, setting tiny quantities to zero or neglecting small terms, reordering terms or rows, adding automatic differentiation, and so on.

You should look for special properties of the problem. Is the matrix symmetric? Is the function being interpolated even or odd? This will help choose the proper method and improve efficiency. In rare cases a symbolic answer may be found, or at the very least symbolic methods may help to reach the numerical answer sooner. In the infrequently occurring cases where symbolic computing helps–use it!

You should look for tolerances that have been set to unrealistically small values and other instances of overcomputing. A realistic tolerance should be set and the program should compute to that tolerance, not many orders of magnitude higher (at much greater computational expense). Interval methods should be used if a guarantee is absolutely necessary but will generally be more expensive when it isn't needed–and usually it isn't.

You should be asking if it's necessary to code a method yourself–which people often believe to be the case–or if standard software will work, which is usually the case. If possible you should use standard, tested software and simply write a calling program for it. If you must code an algorithm yourself, always work through an example of the use of the algorithm *by hand* first. If you want to develop it rapidly, or to test an idea (prototyping), a language like MATLAB is good; if you'll use it every day, recode it in Fortran, Python, or C++ or similar eventually, making use of resources like Netlib. There are some great tools out there, like GitHub and Jupyter, to help keep your work organized.

You should look for methods that are robust, either by the nature of the algorithm (e.g., Gaussian elimination with complete pivoting, or more realistically with partial pivoting) or because they mix methods in an intelligent way (Newton's method for minimization mixed with the method of steepest descent) or because they attempt to correct for errors (reorthogonalization of the modified Gram-Schmidt orthogonalization algorithm). You should consider the underlying physics of the problem in seeking a good initial guess.

You should be asking how the answers will be used and what is *really* needed from the computation. Time and time again someone will ask for the inverse of a matrix when all that is needed is the solution of a linear system; for an interpolating polynomial when all that is needed is its values at some point; for the solution of an ODE at a sequence of points when all that is needed is the limiting, steady-state value. A common complaint is that least-squares curve-fitting couldn't possibly work on *this* data because it violates the statistical assumptions underlying the method; in almost all such cases, least-squares curve-fitting will work just fine because it is so extraordinarily robust.

This is an extremely important point in practice: If you are helping someone else perform a computation it will be necessary to quiz them at length to find out what they *really* need as opposed to what they *think* they need. In fact, once this has been done, the solution will often suggest itself–the type of method to be used will be obvious. For example, if someone has a linear system to solve that has a small-to-medium sized matrix, there's little reason to consider anything other than Gaussian elimination with partial pivoting in most cases. If the matrix has special characteristics, such as being positive definite or tridiagonal, there may be a better, more specialized, approach. Once it's been determined that the solution of $Ax = b$ is what's really needed, and that the matrix is not too large and not too ill-conditioned, the method is obvious. In other areas, such as nonlinear optimization, there is a greater choice of plausible methods but even still a proper understanding of the problem will narrow the choices down to a manageable size–and then choosing to use software that is already available will likely narrow the choices much further.

Experimentation is frequently needed. If one method doesn't work, try another! Vary the initial guess. Plot the function or at least sample it pseudorandomly. Mix methods, add damping, change parameters. Can you solve for the error, approximately, and subtract it out of your solution? Try your software on a simpler problem with a known solution.

If the method does appear to work, try to check your answers. Check residuals or other notions of error. Plot the answer and check it for plausibility. Once again, try your software on a simpler problem with a known solution.

Remember, you "own" these methods and with your understanding of the ideas and theory behind them you may tweak them and add heuristics as you like. Don't ever be afraid to modify and mix methods if you need to do so.

The software is for the most part already written. How you use it is up to you. You now have the tools to use it intelligently. Don't forget to apply *all* those tools to the problem at hand!

What about Richard Hamming's famous quotation: "The purpose of computing is insight, not numbers." This is by far the most important thing that has ever been said about computing. If we compute with the hopes of generating numbers, our expectations will surely be met.

But that is not what we usually do. In most cases we want to *know* something. We want to know how the temperature of a metal plate at some point varies with its thermal conductivity when that parameter is not a constant over the plate. Often we don't want the numerical value of the temperature as much as an *understanding* of how it is affected by a change in the parameter–does it increase quickly for a small change in the parameter, or decrease slowly, or is it relatively insensitive to changes in the parameter, at least in some range? This is numerical experimentation; it's computational science, the use of computer experiments where laboratory experiments are infeasible or prohibitively costly.

Even when we do something as simple as plot a function, what we want to know is often something qualitative, not quantitative: Does it appear to be continuous? Smooth? Slowly varying? Everywhere positive? We want *insight* into the function's nature and an ability to visualize it more than its numerical value at a particular point.

Much of the underlying analysis in numerical analysis relates to perturbation theory, and stability with respect to round-off or representation errors can just as easily be considered stability with respect to other types of uncertainties. Numerical investigations of the stability of a response when the conditions are varied is of great interest in gaining insight and understanding.

When a computer model of an airplane is tested in a numerical wind-tunnel, the big issue is how well it flies, not the value of the wind shear at a certain point. Those things matter too, obviously, but the big picture issue of how well it performs is the first thing–and it gives feedback to the engineers concerning the overall design. Later, specific numerical values will be compared to known tolerances–after insight, and a physical "feel" for the design, has been gained.

When you're performing a computation, or better yet when you're *preparing* to perform a computation, you should consider what you want from that computation. Is it really just a numerical value? Or is it a more general sort of knowledge or intuition? Are you hoping to answer a crisp quantitative question, or to experiment to gain an understanding of something? Either way, be sure that the computation you are about to perform will give you what you need! Especially on a supercomputer, wasted time can be expensive.

Good luck!

Index

Printed in the United States
by Baker & Taylor Publisher Services

Printed in the United States
by Baker & Taylor Publisher Services